NEURAL AND INTEGRATIVE ANIMAL PHYSIOLOGY

Comparative Animal Physiology, Fourth Edition

NEURAL AND INTEGRATIVE ANIMAL PHYSIOLOGY

Comparative Animal Physiology, Fourth Edition

Edited by
C. LADD PROSSER
Department of Physiology and Biophysics
University of Illinois
Urbana, Illinois

WILEY-LISS

A JOHN WILEY & SONS, INC., PUBLICATION

New York • Chichester • Brisbane • Toronto • Singapore

Second Printing, December 1991

Library of Congress Cataloging-in-Publication Data

Neural and integrative animal physiology / edited by C.L. Prosser.
    p. cm.
    "Comparative animal physiology, fourth edition."
    "This book, Neural and integrative animal physiology, and its
companion volume, Environmental and metabolic animal physiology,
together comprise the fourth edition of Comparative animal
physiology" — Pref.
    Includes bibliographical references and index.
    ISBN 0-471-56071-5 : $44.95
    1. Comparative neurobiology. I. Prosser, C. Ladd (Clifford
Ladd), 1907-   . II. Comparative animal physiology.
QP356.15.N44   1991
591.1'88 — dc20   90-27164
                                                              CIP

CONTRIBUTORS

Ziad Boulos Institute of Circadian Physiology, 677 Beacon Street, Boston, Massachusetts 02115

Kenneth Davey Department of Biology, York University, 4700 Keel Street, North York, Ontario, Canada M3J 1P3

Fred Delcomyn Department of Entomology, University of Illinois at Urbana-Champaign, 422a Morrill Hall, 505 South Goodwin Avenue, Urbana, Illinois 61801

Albert S. Feng Department of Physiology and Biophysics, University of Illinois at Urbana-Champaign, 524 Burrill Hall, 407 South Goodwin Avenue, Urbana, Illinois 61801

Rhanor Gillette Department of Physiology and Biophysics, University of Illinois at Urbana-Champaign, 524 Burrill Hall, 407 South Goodwin Avenue, Urbana, Illinois 61801

Timothy H. Goldsmith Department of Biology, Kline Biology, Yale University, 219 Prospect St., New Haven, Connecticut 06511

Aubrey Gorbman Professor of Zoology, Department of Zoology NJ-15, University of Washington, Seattle, Washington 98195

William T. Greenough NPA/Beckman, University of Illinois at Urbana-Champaign, 405 N. Mathews, Urbana, Illinois 61801

Jim C. Hall University of Tennessee, Department of Zoology, M313 Walters Life Sciences Bldg., Knoxville, Tennessee 37916-0810

J. Woodland Hastings Department of Cellular and Developmental Biology, Harvard University, The Biological Laboratories, 16 Divinity Avenue, Cambridge, Massachusetts 02138

Esmail Meisami Department of Physiology and Biophysics, University of Illinois at Urbana-Champaign, 524 Burrill Hall, 407 South Goodwin, Urbana, Illinois 61801

James G. Morin Department of Biology, University of California, Los Angeles, California 90024

C. Ladd Prosser Department of Physiology and Biophysics, University of Illinois at Urbana-Champaign, 524 Burrill Hall, 407 South Goodwin Avenue, Urbana, Illinois 61801

Benjamin Rusak Department of Psychology, Dalhousie University, Halifax, Nova Scotia, Canada B3H HJ1

Andrew Spencer Department of Zoology, University of Alberta, Edmonton, Alberta, Canada T6G 2E9

PREFACE

This book, **Neural and Integrative Animal Physiology,** and its companion volume, **Environmental and Metabolic Animal Physiology,** together comprise the fourth edition of **Comparative Animal Physiology.** Previous editions of **Comparative Animal Physiology** were published in 1951, 1961, and 1973. Each book is designed to serve as a study text for upper division and graduate courses in comparative physiology; together they constitute a reference work that provides a comprehensive introduction to the literature in specific areas of comparative physiology, biochemistry, and biophysics.

In this volume the first three chapters treat membranes and transmitters, motility, and bioluminescence. The next five chapters treat sensory physiology—photoreception, mechano- and phono-reception, electro-reception, chemoreception, and internal clocks. The final two chapters are integrative—nervous and endocrine systems.

The comparative approach to neural and integrative physiology has contributed much to medicine and agriculture. Comparative physiology presents animal models for research that include animals other than the few mammals most used in medical, veterinary, and agricultural research. Examples of contributions of comparative neuroscience are establishment of both electrical and chemical synapses and elucidation of coexistence of multiple transmitters and their pharmacology.

The principal goal of all comparative physiology is to contribute to the understanding of basic biological theory. The most pervasive concept in biology is evolution. Comparative neuroscience is concerned with the origin and evolution of excitable membranes and intercellular transmitters, of neuroendocrine integration, and of ontogenetic development of integrative mechanisms. Comparative physiology deals with highly conserved processes and molecules. Molecular genetics contributes to the understanding of electrical and chemical mechanisms of integration.

Animals are subjected to numerous sensory inputs—osmotic, ionic, thermal, nutritional, photic, mechanical, and phonic—and also to social inputs. How these various inputs are received and integrated into behavior patterns is an important part of comparative animal physiology.

Movement is characteristic of all animals—motion by muscles, cilia, axoplasmic flow, and mitotic movements. Excitable membranes, neurotransmitters, and second messengers show various degrees of specialization. Genetic and molecular analysis of diverse preparations of integrative systems is leading to significant biological generalizations.

An animal does not react to a complex environment with a single organ or organ system. The parts of an organism interact with one another and in combination. The degrees of freedom of the interconnected parts are less than if separated. Out of the whole organism there emerge unique characteristics not present in any of the isolated parts. This is particularly true for animal behavior. In cognitive science, animal models, especially the comparison of behavior in different kinds of animals, may be more realistic than computer models.

This book, in keeping with the principles outlined above, is designed to guide the student to a deeper understanding of several fundamental aspects of the comparative approach to physiology:

• Comparative physiology is the functional analysis of both familiar and unfamiliar animals.
• Comparative physiology uses kinds of animals as experimental variables; it is the study of physiological diversity.
• Comparative neurophysiology uses techniques of cellular and molecular physiology to examine mechanisms of movement; ion channels and neurotransmitters; and chemical and neural integration.
• Comparative physiology of cognition examines the ways by which different neuroendocrine systems provide for animal behavior.

This book is a comprehensive survey of function systems in a diversity of animals. Each chapter has an extensive reference list. Most text statements are supported by citations to specific references. Emphasis in reference lists is on recent research papers, reviews, and classical papers. Some knowledge of cellular physiology and of classification of animals and morphology of both invertebrate and vertebrate animals is assumed.

Chapters have been written and reviewed by specialists. Many consultants have critically read all chapters. It is the hope of the authors and editor that this volume will extend the use of the comparative approach in neuro-, behavioral and endocrine science and will strengthen the position of holistic biology in general.

C. LADD PROSSER

Urbana, Illinois
September 1990

CONTENTS

Chapter 1

Excitable Membranes; Synaptic Transmission and Modulation

C. Ladd Prosser

Selective permeability of cell membranes for specific ions and communication between cells were established early in evolution. During the prebiotic period, organic molecules were synthesized in saline media, with solar and terrestrial heat energy and probably with the aid of some catalyic clays. Molecules could be aggregated into cells only after: (1) self-replicating compounds, nucleotides, could be synthesized, and (2) a bounding membrane that was selectively permeable to specific ions was formed. It is probable that membrane proteins were coded very early in cellular evolution. Plasma membranes of cells consist of a double layer of phospholipid coupled at their lipophilic ends and bound to proteins at hydrophilic ends; the double membrane is crossed by transmembrane proteins. Cell walls (peptidoglycans in some bacteria, cellulose in plants) are laid down outside the plasma membrane. In bacteria, the cell membrane contains enzymes that occur in animal cells in mitochondria and other organelles.

In early cellular evolution metabolism led to accumulation of intermediate compounds that were mainly anionic. These were balanced by accumulation of inorganic cations in establishment of charge equilibrium. Concurrently, protons were released and intracellular pH was maintained at 6.5–7.0, more acidic than the medium. One of the earliest membrane events was establishment of a proton pump, which extruded H^+ in exchange for K^+ and used energy from intracellular adenosine triphosphate (ATP). Internal pH slightly below neutrality, a net negative charge, and an accumulation of K^+ resulted. At a later time in evolution, proton transfer was coupled to an oxidative-glycolytic pathway and was accompanied by phosphorylation of adenosine diphosphate (ADP) to ATP; this type of proton pump is an important energetic pathway in mitochondria.

The plasma membrane is ~100-Å thick and an electrical gradient of 50–200 mV is established across it. An equivalent circuit of the membrane consists of longitudinal resistance (parallel but different resistances of extracellular and intracellular media), transverse resistance, and capacitance; these establish time and space constants for electrical pulses. Transmembrane resistance is ~10^3 Ω/cm^2 and capacitance is 1 μf/cm^2.

The ionic composition of cytoplasm is maintained different from the external medium (295). Potassium is the principal univalent cation within cells, sodium is more abundant in the medium. Total calcium may be higher inside, but most of it is bound so that free Ca^{2+} is lower inside than out. Several mechanisms of cellular function result in higher concentrations of K$^+$ than of Na$^+$ inside most cells. Permeability for K$^+$ in resting cells is greater than for Na$^+$; permeability constants (P) for the two ions may reverse during activity of a cell. Potassium enters against a concentration gradient but down an electrical gradient. Permeability for an ion is determined by ionic size, steric effects, and electrostatic properties. A hydrated K$^+$ ion is larger than Na$^+$. Early in cell evolution, a H$^+$-K$^+$ exchange pump originated. A Na$^+$-K$^+$ ATPase pump is present in most animal cells but not in plants; this pump is sensitive to ouabain and it may be electrogenic when unequal amounts of Na$^+$ and K$^+$ are moved in opposite directions. A significant fraction of the resting metabolism provides the ATP for Na$^+$-K$^+$ pumps in animal cells. The potassium ion is the principal cation to balance accumulated organic anions and thus to prevent osmotic swelling. The potassium ion perturbs water structure more than does hydrated Na$^+$, K$^+$ stimulates oxidative phosphorylation of proteins by mitochondria; K$^+$ promotes protein synthesis by an increase in amino acid transfer to ribosomal RNA, K$^+$ enhances activity of some enzymes—LDH, MDH, cytochrome c oxidase, and isocitrate dehydrogenase.

Membrane Potentials

The equilibrium potential for an ion is the potential given by concentrations across a boundary if the ion is free to diffuse; for example:

$$E_K = \frac{RT}{z\mathcal{F}} \ln \frac{K_i}{K_0}$$

where E_K is the diffusion or Nernst equilibrium potential for K$^+$, $z\pm$ = valence, R = universal gas constant, $\mathcal{F}$ = Faraday constant, T = absolute temperature, K_i and K_0 are concentrations of K$^+$ inside and outside the cell boundary. In living cells, fluxes across cell membranes, as measured with isotopes, yield permeability constants (P), which may be multiplied by concentrations to give the Goldman equation, which allows for steady-state permeability to Na$^+$, K$^+$, and Cl$^-$ (Table 1).

The Goldman equation assumes a constant field within a membrane and constant P values for a given steady state. If the concentration gradient of an ion across the cell membrane agrees with that calculated from the Goldman equation, the distribution of the ion is passive; if these do not agree, active transport is postulated (295).

When cells are stimulated mechanically, chemically, or electrically, the permeabilities change and electrical transients occur. The contribution of specific ions to the transients is given by the conductance (reciprocal resistance) for that ion.

TABLE 1. Permeability Constants

	P_K	P_{Na}	P_{Cl}
At rest (48)			
Frog muscle	1	0.01	1.9
Squid axon	1	0.04	0.45
At peak activity (101)			
Frog muscle	1	20.0	
Squid axon	1	12.0	

Conductance takes account of transfer number, that is, the fraction of total current carried by a given ion. The Hodgkin–Horowitz equation assumes constant conductance (not field) within a membrane and separate channels for different ions. Conductance is obtained from the slope of voltage–current curves. The chord conductance (g) in siemens (reciprocal ohms, Ω) for an ion (i) is given by $g_i = I_i/(V - V_i)$, where V_i is the equilibrium potential for i, I is current, and V is the membrane potential at any given time. The electrical response for an ion can be depolarizing or hyperpolarizing according to the relation of the Nernst potential; movement of K^+ can be depolarizing if E_K is negative to V_m or hyperpolarizing if E_K is positive to V_m. Normally, depolarizing responses result from inward movements of cations Na^+ or Ca^{2+}, and hyperpolarizing responses result from inward movement of Cl^- or outward movement of K^+.

Electrical responses may be graded and conducted passively according to the electrical constants (resistance R and capacitance C) of the membrane. Time constant τ expresses the time for a transient pulse to decline to $1/e$ and is given by the product RC. Space constant λ is the spatial spread to $1/e$ and is given by the decline in amplitude of an applied pulse with distance. Examples of graded responses are excitatory postsynaptic potentials (epsps), inhibitory postsynaptic potentials (ipsps), prepotentials (as in pacemakers), and sensory potentials. Electrical responses may also be all or none, self-propagating, as in conducted action potentials. The ions that carry depolarizing or hyperpolarizing currents may be ascertained by several methods:

1. The effect of omitting a specific ion from the medium and substituting an impermeant one.

2. Amplitude of an electrical response as a function of external ion concentration.

3. Measurement of reversal potential corresponding to the potential at which ionic current changes direction.

4. Effects of agents that specifically block certain ion channels.

5. Measurement of current under voltage clamp.

6. Measurement of current–voltage curves for different times during an action potential, commonly during maximum rate of rise and at peak.

7. Ion flux measurements by means of tracer and ion-selective electrodes.

Electrically inexcitable cells and membranes that respond only passively, that is, electronically, may act as ohmic resistors. A current–voltage plot for such a membrane is linear in a particular voltage range. Current–voltage plots are obtained by measurement of membrane current when V_m is displaced to a series of voltages. The slope of such a plot can change during activation and the membrane potential response reverses when V_m is greater than the reversal potential specified by the intersection of the $V–I$ lines for passive and active states. Electrically excitable membranes show regenerative responses and their $V–I$ curves show nonlinear regions on either depolarization or hyperpolarization (Fig. 1) relative to the equilibrium potential, arbitrarily taken as zero. The equilibrium potential is approximately -70 mV for K^+ and $+20$ mV for Na^+. When the current increases disproportionately to voltage the membrane is said to be activated, conductance increases and the response is regenerative. Rectification implies that membrane resistance for an ion in one region of membrane potential is different from that in another voltage range. Ion channels are

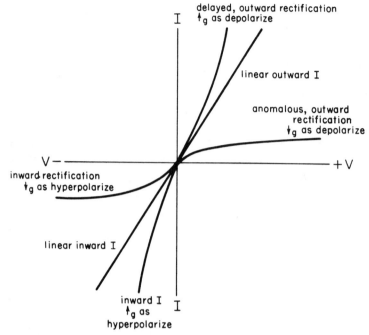

Figure 1. Diagram of different patterns of voltage–current relations in excitable membranes. Outward currents in upper right quadrant, inward currents in lower left quadrant. Deviations from linearity represent different patterns of rectification, that is, when current in one direction is not equal to that in opposite direction across a cell membrane. Examples of each pattern are given in text. Intersection on voltage scale is represented as zero but is strictly the equilibrium potential for each ion.

open at some membrane potentials, closed at others. The nonlinearity may be a disproportionate *increase* in outward current usually of potassium and is called delayed rectification. This type of rectification occurs in cardiac muscle and in many neurons (Fig. 2). In some membranes, outward rectification is a *reduction* in current on depolarization and is called anomalous rectification; an example is in striated muscle fibers. The converse patterns of nonlinearity are for inward currents that may be an increase (frog muscle) or a decrease (sympathetic ganglia) in current on hyperpolarization (Fig. 3). Echinoderm eggs show hyperpolarizing rectification (113) (Fig. 4). Many conductances have a range near 0 V where the V/I curve is relatively flat; inward rec-

tification that is turned on in this range may result in an N-shaped $V–I$ curve. The downward slope reflects a region of negative conductance; this type of V/I curve is common in rhythmic neurons and heart.

Patch clamping is clamping ionic currents by means of a small external electrode applied to an intact cell or to fragments of membrane. Many membranes under resting conditions show spontaneous unitary events, which are interpreted as opening and closing of single channels (Fig. 5a and b). For example, in marine eggs, the number of spontaneous depolarizations per second increases as the membrane is depolarized and current flow of the unitary events reverses at the equilibrium potential for

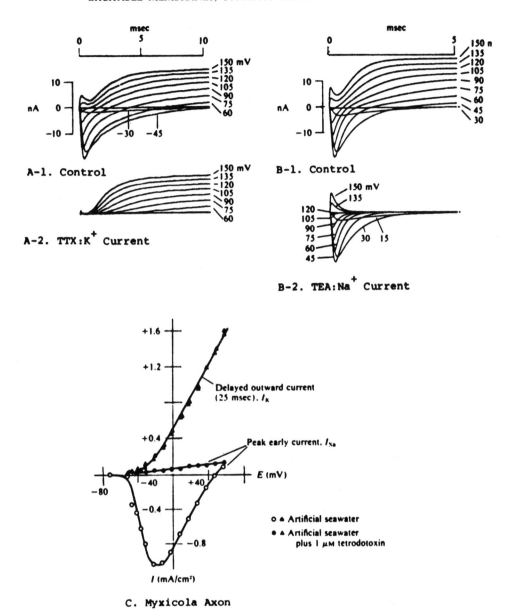

Figure 2. Voltage clamp records of currents from a single node of Ranvier of frog sciatic nerve fiber. Records show displacements of the membrane potential from a resting level of −75 mV to the levels of depolarization indicated on the individual records. (*A*-1) Time course of current flow in normal Ringer solution. (*A*-2) Records measured in presence of tetrodoloxin (TTX), which blocks the Na (inward) current. [From (123).] (*B*-1) Control and (*B*-2) after blocking K current by tetraethylammonium (TEA). [From (123).] (*C*) Voltage–current plot of excited *Myxicola* giant axon. Triangles designate delayed current (I_K), open symbols controls, solid symbols after TTX. Circles show early (I_{Na}) current, open symbols before and solid symbols after TTX. [From (29).]

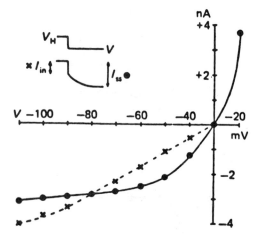

Figure 3. Voltage–current relations of neurons of bullfrog sympathetic ganglion. Records during hyperpolarizing steps from holding potential of −30 mV gives instantaneous (x) and steady state (o) current–voltage curves. Steady state is current level at end of a hyperpolarizing or depolarizing pulse. Note linearity of instantaneous and inward rectification of steady-state current during hyperpolarization. [From (3).]

Na$^+$. Patch clamping of cardiac muscle cells shows two open times of Ca^{2+} channels, fast and slow; inactivation (closing a channel) is much slower than activation (opening).

Ions as Current Carriers in Action Potentials

The comparative study of excitable cells shows that for each active ion, there are diverse channel types and that the proportion of types varies with kind of tissue.

Protons

At rest, H$^+$ is not at equilibrium with E_m; there is an outward gradient of [H$^+$] and resulting inward electrical gradient. In many plant cells, proton extrusion couples metabolic energy to the transport of solutes and permits accumulation of K$^+$ by a H$^+$–K$^+$ exchange. The pH of the medium rises and the cell is depolarized. When the luminescent flagellate Noctiluca is stimulated, an action potential of 55 mV is recorded on a resting potential of −180 mV. The action potential is due to a decrease in proton pump current (206). In bacteria, mitochondria, and chloroplasts proton transport is coupled to oxidation or photosynthetic phosphorylation. It is probable that in primitive prokaryotes, ATP-requiring pumps extruded protons and resulted in electrical and osmotic gradients (Donnan equilibrium).

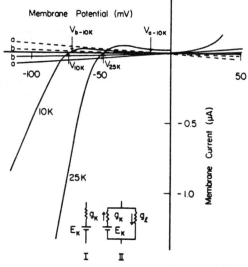

Figure 4. Steady-state current–voltage relations for the oocyte membrane of the starfish *Mediaster aequalis*. V_{10K} and V_{25K} are data obtained in 10 and 25 mM K$^+$ seawater, respectively. The broken and solid lines labeled a and b are hypothetical current–voltage relations for a conductance caused by leak around the microelectrode. The inset shows a simplified equivalent circuit for the oocyte membrane (I), and for the membrane in parallel with a leak conductance (II). [From (113).]

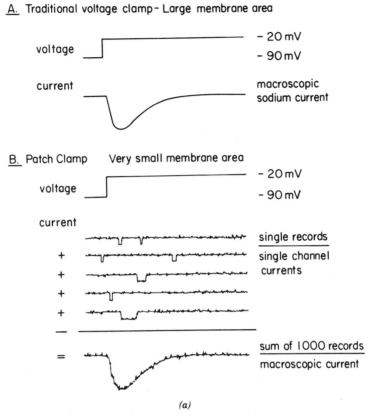

(a)

Figure 5(*a*). A, diagram of the voltage clamp analysis of Na$^+$ channels from nerve and muscle demonstrates inward currents that activate after depolarization, reach a peak within a millisecond or so, and then slowly decline toward baseline as inactivation develops. B, patch clamp records summate to give macroscopic current. [From (17).]

Calcium Currents

In animal cells Ca^{2+} action potentials are widespread and probably more primitive than Na$^+$ action potentials. Voltage-sensitive Ca^{2+} channels do not occur in prokaryotes but do occur in ciliate protozoa and in all multicellular animals.

Calcium channels originated very early in animal evolution. This is evidenced by the fact that in many kinds of excitable cells, the phase of active depolarization is by inward current carried by calcium ions (157). Transmembrane calcium currents are evoked by hormones, neurotransmitters, or mechanical stimuli. Calcium can serve as a "second messenger" inside cells. Calcium action potentials provide calcium ions for intracellular stores, for numerous enzyme reactions, and for functions such as secretion and muscle contraction. Many Ca^{2+} channels can be activated by stretch and these may be transducers for volume regulation, a primitive cellular function (54).

The best-known Ca^{2+} channels are in rat and chicken dorsal root ganglion cells and heart muscle fibers. Three different kinds of Ca^{2+} channels have been characterized and there may be other types:

1. A T channel is voltage sensitive, causes a fast or transient depolarization, has threshold positive to −40 mV, inac-

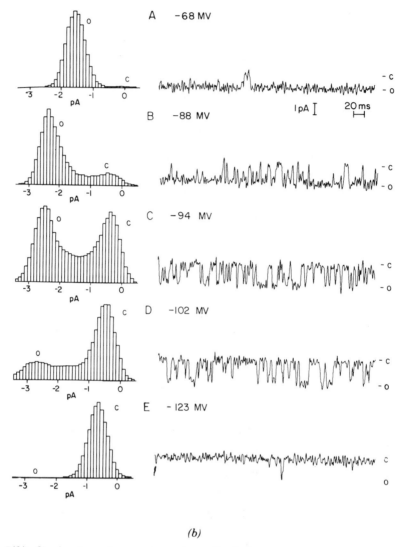

(b)

Figure 5(b). Single channel currents and amplitude histograms from a purified rabbit sodium channel reconstituted into a planar bilayer, demonstrating the voltage dependence of channel gating. Hyperpolarization (A–E) progressively shifts the channel from the open (O) to the closed state (C). This channel exhibited 50% open activity at about −95 mV. [From (17).]

tivation begins at potentials over a wide voltage range and inactivation occurs in 20–50 ms (236, 282, 283). T channels can be blocked by low concentrations of metals such as Ni^{2+}; these channels have a low affinity for dihydropyridine (DHP) compounds such as nifedipine.

2. L channels mediate longer-lasting sustained responses (Fig. 6). Threshold voltage is from −20 to 0 mV, block is by low concentrations of nifedipine and verapamil (0.2–0.5 μM). An agonist is the compound Bay 8644.

3. A third channel, novel or N, is in-

termediate between T and L. Threshold is at -10 mV and N has a high Ba^{2+} conductance. Cadmium blocks N and L and leaves 50% of T conductance (20a).

A summary of properties of the three types of Ca^{2+} channels as described for chicken dorsal root ganglion cells follows (196):

Channel	T	N	L
Whole cell activation (mV)	-70	-10	-10
Whole cell inactivation (ms)	20–50	20–50	slow (>700)

Rate			
Steady state (mV)	-100 to -60	-100 to -40	-60 to -10
Single channel conductance (ps)	8–10	13	25
DPH block	Low affinity		High affinity

Sympathetic ganglia have cells with low threshold T channels, cells with high threshold L channels that are sensitive to the DPH compound D-600, and cells with N channels that are insensitive to organic blockers. Fast and slow Ca^{2+} channels can occur together in the same cell, for example, in amphioxus muscle. In dorsal root ganglia, L-type channels are used for release of the transmitter, substance P; T channels are opened only during the after-hyperpolarization following a burst of spikes (185).

In some tissues such as frog striated muscle, the action potential in plasma membrane is normally carried by Na^+ but Ca^{2+} currents can be observed after the Na^+ current is blocked by TTX. The T tubules of vertebrate striated muscles are rich in L channels that can be blocked by dihydropyridines. In crustacean muscle, the action potentials are calcium spikes. In vertebrate cardiac muscle epinephrine increases and acetylcholine (Ach) decreases the calcium current. In mammalian intestinal smooth muscle, spikes that trigger contractions are due to Ca^{2+} and when both Ca^{2+} and Na^+ are reduced so as to maintain a constant ratio (Na^+/Ca^{2+}) there is less reduction in spike amplitude than when Ca^{2+} only is reduced (58, 290).

Calcium channels have been found in echinoderm eggs; when a starfish egg is activated by sperm or by electrical stimulation a wave of depolarization reflects an inward current of calcium ions. Voltage clamp experiments indicate one current that is activated in the range -73 to

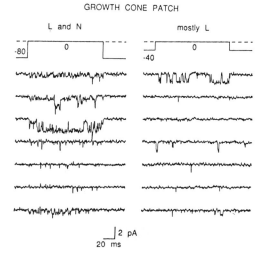

GROWTH CONE PATCH

Figure 6. L- and N-type Ca^{2+} channels distinguished by their dependence on holding potential. Current records from a cell-attached patch on a growth cone. Consecutive sweeps are in each column. From a holding potential of -80 mV (left), voltage steps to 0 mV evoke unitary currents of two amplitudes corresponding to L- and N-type Ca^{2+} channels. With a holding potential of -40 mV (right), test pulses to 0 mV evoke L-type Ca^{2+} channel openings almost exclusively. [From (164).]

-55 mM another at about -7 mV. Both currents can use Ca^{2+}, Sr^{2+}, or Ba^{2+}. The high threshold Ca^{2+} current is modulated by Na_0 (113, 114).

In ciliates such as Paramecium, the resting potential is approximately -30 mV. When the anterior end of an animal is stimulated mechanically, ciliary beat reverses and the Paramecium backs away from the stimulus. The anterior ciliary membrane gives a graded receptor potential and regenerative spike in which the depolarizing current is carried by calcium. Stimulation at the posterior end triggers a potassium current. One genetic mutant of paramecium is deficient in Ca^{2+} channels and the animal cannot reverse its swimming direction (78, 239). The Ca^{2+} channels of Paramecium resemble T channels of vertebrates (82, 289).

In several cell types—squid giant axons, some gastropod neurons, vertebrate smooth muscles—sodium has been found to enter through Ca^{2+} channels. This current is insensitive to Na^+-channel blockers, but is sensitive to Ca^{2+}-channel blockers (8, 141, 153, 135, 231).

An important location of Ca^{2+} channels is at axon terminals where Ca^{2+} triggers transmitter release. In squid giant neuron synapses, Ca^{2+} channels in presynaptic endings can be blocked by La^{2+}, Cd^{2+}, Pb^{2+}, and Ni^{2+}, not by dihydropyridines (14). The Ca^{2+} entry can be made visible by the light-emitting aequorin or a specific absorption by pigments Arsenazo III or Fura-2. Synaptic block results at all synapses when Ca^{2+} in the medium is replaced by Mg^{2+}. In squid giant fiber synapses treated with TTX and TEA to block Na^+ and K^+ conductances, the release of transmitters as measured by postsynaptic responses is proportional to Ca_0^{2+}. The Ca^{2+} entry into presynaptic terminals is shown by a aequorin flash (99). At synapses both DHP sensitive and DHP insensitive Ca^{2+} channels regulate transmitter release (165, 196). Responses of neurons of the mammalian inferior olive consist of Na^+ spikes followed by after-depolarization and after-hyperpolarization; the after-depolarization is by Ca^{2+} influx and after-hyperpolarization is a result of Ca^{2+} activated g_K (166). When hippocampal neurons are cultured, the low voltage Ca^{2+} channels develop first, high voltage channels develop later. Growth cones of developing neurons have L-type Ca^{2+} channels and Ca^- increases after stimulation (164).

Calcium currents can activate several kinases (phosphorylating enzyme) that are mediated by second messengers—cAMP, cGMP, diacylglycerol, or inositol-P_3. Calcium ions may themselves function as intracellular second messengers (196). For example, the action of epinephrine in reducing Ca^{2+} current in heart muscle is mediated by a cascade of cAMP reactions.

Active transport of Ca^{2+} by an ATPase pump occurs in membranes such as sarcoplasmic reticulum of skeletal muscle. Active uptake of calcium occurs in many freshwater animals (295). Movement of Ca^{2+} across cell membranes may also be by exchange of Na^+ for Ca^{2+}. There is evidence from several tissues, for example, heart, that in the exchange, three Na^+ ions move per Ca^{2+} ion. Exchange is mediated by a protein carrier in the membrane and since the number of charges carried by the two ions is different, the process is electrogenic. An equilibrium potential for the exchange is given by the ratios of Na^+ and Ca^{2+} concentrations inside and out. When the exchange potential is negative to E_m, Na^+ moves in and Ca^{2+} moves out; when the exchange potential is positive to E_m, Na^{2+} moves outward and Ca^{2+} moves inward. Quinidine and harmaline are nonspecific but effective blockers of the exchange process (234). The Na^+–Ca^{2+} exchange has been demonstrated in many cell types and is of general occurrence.

Sodium Currents

The electrochemical gradient for sodium is inward and the equilibrium potential for Na^+ is $+20$ to $+50$ mV. In classical studies on squid giant axons, Hodgkin and Huxley showed that the inward depolarizing current of an action potential is carried by Na^+. Spikes overshoot 0 mV and the Nernstian slope of the peak overshoot as a function of Na_0 is 58 mV. In voltage clamp measurements, the Na^+ channels can be blocked by low concentrations of tetrodotoxin (TTX from puffer fish), or by saxitoxin (STX, from red-tide flagellates), or batrachotoxin (BTX from a South American toad). DDT delays inactivation of Na^+ channels, that is, prolongs spikes.

Proteins of Na^+ channels probably evolved later than those of Ca^{2+} channels as inward current carriers although they share structures. Sodium channels occur in all multicellular animals. Most excitable cells of animals have both Na^+ and Ca^{2+} channels in different proportions. In giant axons of squid, the inward current of an action potential due to Ca^{2+} is 100 μA/cm², where the inward current due to Na^+ is 4000 μA/cm² (123). In neurons of the jellyfish, *Aglantha*, and muscle of *Amphioxus* the two currents are nearly equal. A heliozoan (protozoan) responds to mechanical stimuli with an action potential due mainly to inward Na^+ current (152).

Some membranes have Na^+ action potentials that are not TTX sensitive—some molluscan neurons and muscles of holothurians. This indicates differences in Na^+-channel proteins. The depolarization response of photoreceptors of the ventral eye of *Limulus* to a flash of light is due to inward Na^+ current (Chapter 5). This is unlike the hyperpolarizing response of a vertebrate photoreceptor in which the response is a closing of a Na^+ channel (Chapter 5).

In many Na^+ channels a permeant ion like Li^+ can replace Na^+ but Na^+ cannot be replaced by larger cations such as choline, Tris, or TEA. Threshold in axons for activation of a Na^+ channel is a small depolarization: approximately 19 mV positive to resting potential; peak depolarization approaches the Na^+ equilibrium potential. The rise time for squid giant axon spikes is 0.1–0.6 ms, inactivation time is 1–3 ms. To gate a Na^+ channel only 0.3% of the sodium current of a spike is needed (123). In patch-clamped membranes spontaneous or unit channel openings occur with currents of 20 pS (88). Neurotoxins act on different sites of channel proteins (51–53): (1) Water soluble TTX and STX act externally, probably by blocking Na^+ entry; (2) lipid soluble alkaloids such as veratridine act on hydrophobic groups; (3) scorpion venom I blocks inactivation; (4) scorpion venom II enhances activation, probably by a conformational change; and (5) some local anesthetics act from the cytoplasmic side. Sodium channels may be in one of three states—resting, activated, or inactivated; both of the latter are voltage dependent (265). From binding of toxins the number of channels has been calculated in a squid axon as 110 channels/μm² and at a Ranvier node of a myelinated fiber 2000/μm². Channels are seven times more numerous at the initial segment of a motor neuron than in the soma (123). The Na^+-channel proteins must be glycosylated in order to be functional and this process can be blocked by tunicamycin (51–53).

The Na^+ channel of mature muscle is sensitive to TTX, that of denervated and of embryonic muscle is insensitive to TTX (90). During postnatal development of rats, the mRNA for the Na^+-channel protein increases, and after denervation, the mRNA of adult-channel protein decreases; the expression of channel genes changes with state of innervation (64). The Na^+ channels are abundant at neu-

romuscular junctions, at nodes in myelinated axons, and at the axon hillock of
motor nuerons (11).

Potassium Conductance; Tissue Specific Channels

Conductance channels for potassium ions
are more diverse than for any other ion.
The K^+ channels may have originated as
H^+-K^+ exchange pumps very early in evolution. Different K^+ channels are distinguished by their biophysical properties.
Yeast membranes have a prominent spontaneous K^+ current (111). In living cells,
potassium is the most abundant diffusible
ion (295). The inward permeability constant is high and the outward gradient
has an equilibrium potential of -70 to
-90 mV. Since the resting potential is
fixed by concentrations and relative
permeabilities of K^+, Na^+, and Cl^-, the
resting membrane potential may coincide
with or be less negative than E_K; if it is
less, activation of K^+ conductance may
result in hyperpolarization. This occurs as
the depolarizing phase or outward current after Na^+ or Ca^{2+} action potentials.
If E_K is negative to E_m or if a K^+ channel
is closed by a stimulating agent, activation can result in a depolarizing response
(188).

A classification of the best known of
K^+ channels follows (155);

1. I_{K_1} is a voltage-sensitive current that
shows outward, "delayed," rectification
(Fig. 1). This current is responsible for
repolarization after a Na^+ spike and it is
activated early in Na^+ depolarization, for
example, in a squid giant axon. In squid
axons and in mammalian nodes of Ran-
vier, I_{K_1} is turned on in a few tenths of a
second and peaks in ~1.5 s. Its conductance is blocked by TEA or 4-aminopyri-
dine (4-AP). The channel is considered to
be a long pore through which ions pass
in single file.

2. I_{K_2} is a second voltage-sensitive current. This shows anomalous or inward

rectification with higher conductance in
than out (Fig. 1). Current is little affected
by depolarization positive to zero, but
current increases considerably on hyperpolarization. As with I_{K_1}, this channel is
blocked by TEA. The channel for I_{K_2} occurs in striated muscle membranes, in
marine eggs, and some neurons (278).

3. $I_{K_{Ca}}$ is in an outwardly rectifying
channel that is activated by Ca^{2+}. The
$I_{K_{Ca}}$ channels have been examined in several muscles of invertebrate animals and
in molluscan neurons. These channels account for repolarization of mammalian
smooth muscle spikes. They respond
when internal Ca^{2+} rises above a critical
level and there is kinetic evidence that at
least two Ca^{2+} are bound for a high probability that these channels are opened.
There may be several different $I_{K_{Ca}}$ channels. The barium ion may substitute for
Ca^{2+} spikes but not for activation of $I_{K_{Ca}}$.
Current in Ca^{2+}-dependent K^+ channels
is large (250 pS) low current (10–14 pS)
channels may be used after hyperpolarization (32). A type of Ca^{2+} activated K
channel (BK) occurs in lacrimal gland cell
(281).

4. I_{K_A} or A current is a fast outward
current that has been described in molluscan neurons, rat dorsal ganglion cells
and brain, Drosophila muscle and neurons, in electroplaxes, and in many muscles. The A current is activated when a
cell is depolarized following a period of
hyperpolarization, as during the repolarizing phase of an action potential (238). If
a molluscan neuron is voltage clamped at
-40 mV and step depolarized to -5 mV,
an outward K^+ current is activated in both
the delayed rectifier K^+ channel and the
K_{Ca} channel. If, however, the cell is first
held at -80 mV, a successive depolarization activates an additional outward
current in the A channel. A currents activate in the rate of -50 to -5 mV and
inactivate with hyperpolarization to -60
to -90 mV. Figure 7 shows events in a

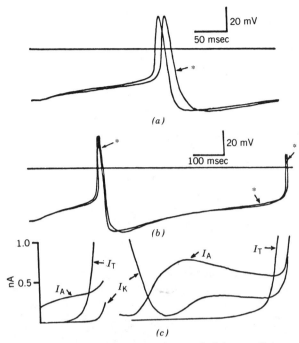

Figure 7. (*A*) Comparison of action potential recorded from cell in ganglia of *Anisodoris* with calculated (*) A.P. (*B*) Action potentials at compressed time scale for comparison of different ion curves. (*C*) Membrane currents associated with voltage behavior in (*B*). I_t total current, I_A potassium A current, I_K voltage-sensitive K^+ current. [From (61).]

rhythmic molluscan neuron or crustacean axon. At the end of the first action potential, A channels are inactivated but K^+ channels are strongly activated and the cell becomes hyperpolarized. This hyperpolarization removes inactivation of A channels and inactivates K^+ channels; depolarization then opens A channels and further depolarization is delayed so long as I_A is large. Thus the A currents space successive action potentials in the rhythmic cell (39–41, 60). A currents are sensitive to 4-AP, not TEA; they activate and inactivate faster than I_K.

In flight muscles and motor neurons of *Drosophila*, K_A channels develop prior to other K^+ channels. I_{K_A} is sensitive to TEA and under voltage clamp I_{K_A} is activated at -50 mV and is a fast response. In Shaker mutants (SH) of Drosophila, muscle and nerve action potentials are prolonged and the animal shows leg shaking when lightly etherized (238, 240). Pro-

teins of several A channels have been separated and made in Xenopus eggs after injection of appropriate mRNA. Channels of SHB1 show higher currents than channels of SHA1 (280).

5. The M channels have been studied in frog sympathetic ganglion neurons that show long-lasting depolarization following the action of Ach on muscarinic receptors (3, 4) (Fig. 8). The M channels are activated in the voltage range -60 to -10 mV. Muscarinic agonists reduce the outward steady-state current in M channels and thus depolarize. Inhibition of I_M by cholinergic agonists decreases outward rectifying current and prolongs epsp values. The M channels are described in frog stomach muscle fibers (255), decrease outward rectifying current and prolong epsp. An analogous channel is the S channel of sensory neurons in Aplysia (251). This channel is closed by the transmitter serotonin (5-HT), which broadens a

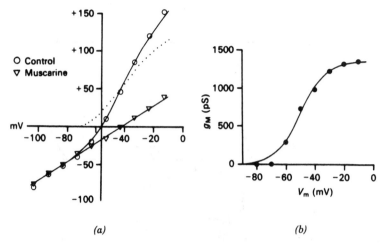

Figure 8. (*A*) Voltage–current curves for smooth muscle fiber from toad stomach to demonstrate *M* current. Muscarine caused a net inward current and reduced the slope conductance at potentials more positive than -70 mV. Values of I_m (dotted line) obtained as difference between control and muscarine *I–V* curves. (*B*) Muscarine-sensitive conductance $g_m = I_m / (V_m - E_K)$. [From (255).]

spike response. The peptide transmitter FMFRamide shortens the spike and increases the probability that S is open. The spike is broadened by TEA, which closes a channel like 5-HT. Both the M and S channels are open at rest, closed when stimulated by transmitter, and both are due to changes in K^+ conductance (23, 251).

6. In some cells, the resting potential is much less negative than E_K; an active response may be a hyperpolarization toward E_K. In the distal retina of a scallop eye, a hyperpolarizing response to light is due to an increase in g_K (96). Reverse or negative-going spikes can result from stimulation of K^+ channels in the absence of Na^+ or Ca^{2+} spikes. When the muscle of the pharynx of the roundworm *Ascaris* relaxes, the lumen is closed; the cells give negative spikes due to an increase in g_K (44). Regenerative hyperpolarizations have been reported in paramecium and in rods of the retina of Necturus. To block K^+ channels in a frog node of Ranvier TEA is effective externally at 0.4 m*M*; a *Myxicola* axon requires 24 m*M* TEA and

a squid axon is insensitive on the outside, but sensitive to TEA when applied on the inside.

Some molluscan neurons show bursts of spontaneous activity in which the intervals between spikes lengthen during a burst, and bursts are interrupted by periods of 30–50 s of quiescence. An explanation of bursting is that during a spike train Ca^{2+} enters with each spike and accumulates such that K_{Ca} channels remain open until Ca^{2+} can leave the cell thus accounting for the interburst interval (59). Some seven different types of ionic channels have been identified as responsible for spikes within bursts and the interval between bursts (2). Ventricle of a snail *Lymnaea* has stretch-activated K^+ channels with two open and three closed states as shown by patch clamping (253). Growth cones at the tips of Aplysia neurons growing in culture have two types of K^+ channels—a delayed rectifier and a Ca^{2+} activated channel; one half of the cones sampled had Na^+ action potentials, the others gave only graded depolarizations (21).

In a frog Ranvier node, three K^+ channels have been identified: $g_{K_{f1}}$ is activated in the range -80 to -30 mV, time for activation is 20 ms; this channel is blocked by 4-AP. Channel $g_{K_{f2}}$ activates in the range -40 to $+30$ mV; activation time is several hundred milliseconds; this channel is not blocked by 4-AP. Channel g_{K_s} is a slow channel requiring several seconds and is not blocked by 4-AP (76, 108).

Vertebrate heart muscle shows pacemaker potentials and a prolonged plateau that combine several currents to give the electrocardiogram. The sequence in cardiac muscle is: The principal currents are a rapid inward Na^+ current, a slower inward Ca^{2+} current, and an outward K^+ current. As I_K decreases during repolarization, a fall in membrane resistance allows background currents of Na^+ and Ca^{2+} to flow inward slowly. This brings the membrane to threshold for active I_{Na} and after more depolarization to threshold for I_{Ca}. Pacemaker cells of mammalian and frog heart under voltage clamp show an inward current I_f that is induced during hyperpolarization. Cells clamped at -40 mV and then hyperpolarized to -50 or lower show I_f. Apparently, there is a general increase in conductance and both Na^+ and K^+ contribute to I_f. Pacemaker activity in vertebrate heart is the net result of a series of conductance changes. Acetycholine increases g_K, increases K^+ efflux; epinephrine increases g_{Ca} and increases Ca^{2+} inward current with both chronotropic and inotroplic enhancement (40).

In paramecium five different currents have been described (242, 243) (1) anterior stimulation triggers an inward Ca^{2+} current in ciliary membranes that has a rise time of 5 ms. (2) Posterior stimulation evokes a rectifying outward K^+ current of rise time 20–50 ms. (3) Voltage clamping shows a hyperpolarizing anomalously rectifying K^+ current of rise time >50 ms. (4) A Ca^{2+} activated K^+ current

$I_{K_{Ca}}$ has a rise time longer than 10 s. (5) A very slow Ca^{2+}-induced Na^+ inward current $I_{Na_{Ca}}$ has rise time of several seconds and persists for many seconds or minutes. Some of these currents are made apparent in mutants with defects in one or more ion carriers (239, 242, 243). In another ciliate, *Didinium*, ciliary membranes respond to mechanical stimuli by Ca^{2+} action potentials (222).

In Drosophila, as in Paramecium, mutants occurs with defects in specific channel proteins. Two mutants with nonfunctional Na^+ channels and three for K^+ channels have been described (91, 221). Several of the defects are apparent only on warming to 35°C and the temperature effects are reversible. A mutant "nap" or no action potentials and another "para" or paralytic are paralyzed when warmed and fail to have normal sodium currents in nerve and muscle. In mutant "sh" or shaker, the potassium current I_A is deficient. A mutant "eag" shows modification in the outward rectifying I_{K^+} (91).

Structure of
Cation-Permeable Channels

The genes for voltage-sensitive cationic channels code of a family of related proteins. The mRNAs specify large protein that show considerable homology between those for K^+, Na^+, and Ca^{2+} channels (Fig. 9a). Cyclic DNAs for several channels have recently been cloned and the amino acid sequences deduced. Each channel protein consists of several segments that span the membrane and have both voltage-sensing and ion conducting regions. Voltage-sensitive K^+ channels have been observed (but not sequenced) in yeasts and other unicellular eukaryotes. Ca^{2+} channels are first seen in complex protozoa (ciliates) and Na^+ channels only in multicellular animals.

The one K^+ channel that has been cloned is the A channel that mediates rap-

Na⁺ channel

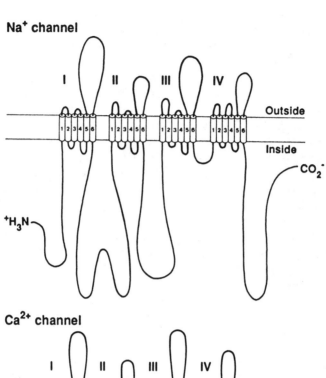

Ca²⁺ channel

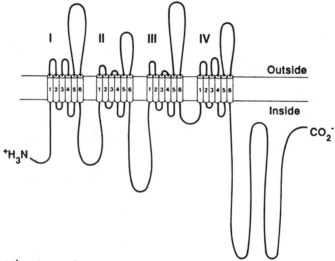

K⁺_A channel

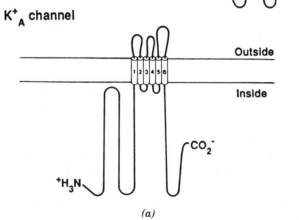

(a)

Ca²⁺ channel

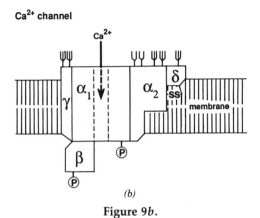

(b)

Figure 9b.

idly activating and inactivating K⁺ currents (276). The locus of the mutant Drosophila for shaker was located and the cDNA for this locus was sequenced. The sequence of this cDNA predicts a protein of six hydrophobic stretches plus an arginine-rich sequence similar to that of a Na⁺-channel protein and a DHP-binding channel protein (248). Injected into *Xenopus oocytes*, channel proteins were made with properties like those of the K_A channel. Recently, a channel gene with similar sequences has been isolated from rat brain (280). The protein deduced from the mRNA consists of 616 amino acids.

Several Ca²⁺ channels were mentioned previously. A voltage-sensitive channel that binds the Ca²⁺ blocker, DHP, was isolated from T-tubules of striated muscle. The function of this channel is given in Chapter 2. The DHP sensitive Ca²⁺ channel from T-tubules of rabbit muscle consists of three associated subunits: α (167 kD), β (54 kD), and γ (30 kD). There

are two forms of α subunits, $α_1$ and $α_2$; and $α_1$ contains the DHP-binding site. The entire complex is 416 kD. A model is shown in Figure 9b (270). The amino acid sequence of the α subunit has been deduced from a cloned DNA. This subunit α represents the ion conducting portion of the total complex. The DHP receptor has six segments very similar to the Na⁺ channel, which has eight membrane-spanning units (273). The cDNA predicts a protein of 1873 amino acids. The α subunit consists of four homologous domains, multiple sites for phosphorylation by a cAMP-dependent kinase (67). These phosphorylation sites are postulated as on the cytoplasmic side and glycosylation sites on the external side of the membrane (Fig. 9a). Glycosylation is mainly by sialic acid which results in a negative charge (Fig. 9b).

Sodium channel proteins have been isolated, and their cDNAs cloned from electric eel electroplax, from rat brain, and from different muscles (98). The Na⁺-channel protein from mammalian brain is a heterotrimeric complex of 260-, 36-, and 38-kD subunits (212, 213). The channel protein from sarcolemma of skeletal muscle has two subunits, 260 and 38 kD (98). The channel from electric eel electroplax has only the 260-kD subunit. The large subunits bind both TTS and STX. Labeling by toxins indicates that each protein spans the membrane with a central channel for Na⁺ flux (Fig. 10). The α channel protein in brain is in two forms I and II with homology 87%; between brain and electroplax, the homology is 62% (96). In

---

Figure 9. (*a*) Proposed transmembrane arrangements of the principal subunits of Na⁺, Ca²⁺, and A-current K⁺ channels (K_A^+). The protein folding models for rat brain Na⁺ channels, rabbit skeletal muscle Ca²⁺ channels, and *Drosophila* A-current K⁺ channels are presented to illustrate overall sequence similarities. (*b*) Diagram of a subunit structure of a Ca²⁺ channel as deduced from the properties of a rabbit skeletal muscle Ca²⁺ channel. Arrangement of subunits α1, α2, β, γ, and δ inferred from properties described in reference. Phosphorylation sites (P) on inner side of scheme, and glycosylation sites (ψ) on the outer force of channel, disulfides bonds S—S intramembrane. [From (52).]

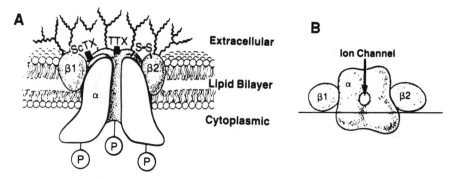

Figure 10. Model of Na⁺ channel structure. The Na⁺ channel protein from brain is illustrated (*A*) as associated with the phospholipid bilayer as inferred from biochemical and molecular biological experiments. Sites of glycosylation (S—S), phosphorylation (P), and neurotoxin (SctX, scorpion toxin; TTX, tetrodotoxin) binding are illustrated. The transmembrane pore of the sodium channel is illustrated (*B*) in *en face* view as formed in the center of the four homologous transmembrane domains of the α subunit. [From (50).]

rat brain, Na⁺ channels are of two types—I and II. The ratio I:II is 0.07 in hippocampus, 0.17 in cerebral cortex, 2.2 in spinal cord (96). Each polypeptide has four homologous repeats and each repeat consists of five hydrophobic segments and one positively charged segment. A model of Na⁺ channels consists of the six segments crossing the plasma membrane (Fig. 10). The mRNAs of Na⁺ channels of rat brain and muscle are similar in size but different immunologically (49–52). Secondary and tertiary conformation of Na⁺ channel shows four homologous domains, each with eight transmembrane segments (97, 112). The sequence of Na⁺- channel protein of Drosophila is similar to that of vertebrates (241).

Antibodies have been prepared against the Na⁺-channel proteins and immunohistochemical staining shows binding to the innervated side of the electroplax cells. Staining at endplates of striated muscle is intense and staining falls off in the perijunctional membrane (Fig. 11) (115).

Remarkable similarity in structure between the channels for K⁺, Ca²⁺, and Na⁺ is indicated (Fig. 9*a*). It has been proposed (51) that the A potassium channel is the primitive prototype and that Ca²⁺ and later Na⁺ channels evolved by duplication of the K_A gene. Each of the domains, one in K_A and four in the other two segments has six subunits; each has regions of phosphorylation and both N- and C-terminals appear to be on the cytoplasmic side.

Chloride Currents

In animal cells, the [Cl⁻] gradient is inward, Cl⁻ contributes significantly to resting potentials, buffers against depolarizing deflections, is important in regulation of pH and of cell volume. Resting membrane potential is close to the chloride equilibrium potential E_{Cl}. In plant cells, the [Cl⁻] is higher than in animals, especially in vacuolar fluid; depolarizing action potentials result from the efflux of Cl⁻ and some influx of Ca²⁺.

In twitch muscle fibers of some fishes, the resting potential is dominated by Cl⁻. In sunfish, for example, P_{Cl} is high at 20°C and substitution of an impermeant anion depolarizes the membrane. On cooling to 7°C, the resting conductance for K⁺ is re-

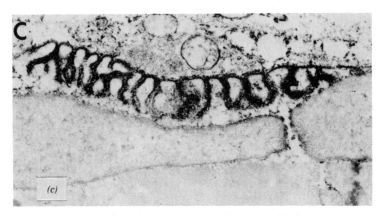

Figure 11. Electron micrograph of anterior tibial muscle sections exposed to monoclonal antibody for Na^+-channel protein. Intense reactivity was detected at endplates, where the antibody staining could be traced throughout the secondary folds of the postsynaptic membrane (C). Original magnification: $\times$ 18,000. [From (115).]

duced $<g_{Cl}$ and resting potentials are then controlled mainly by K^+ (138). Treatment of frog muscle with hypertonic glycerol-Ringer disrupts T-tubules and has little or no effect on plasma membrane. Measurements on glycerol-treated muscles show a high resting conductance for the Cl^- in plasma membrane but not in T-tubule membrane. In muscle from sting-rays, the g_{Cl} at rest is 8–10 times $>g_K$. In rat diaphragm muscle some 85% of the resting conductance is due to Cl^-. In several genetic disorders mammalian muscle shows hyperexcitability determined by a low g_{Cl}.

Hyperpolarizing responses due to Cl^- influx have been reported in cells of several types. Inhibitory postsynaptic potentials (ipsps) triggered by the transmitter γ-aminobutyric acid (GABA) in crayfish muscle are the result of the inward movement of Cl^- ions. Spinal motor neurons of mammals are hyperpolarized in response to the inhibitory transmitters glycine or GABA. The reversal potential for this inhibitory hyperpolarization can be shifted by injection into the neurons of anions with higher or lower permeabilities than Cl^-. The ipsp can be reversed if Cl^- ions are injected so that the gradient is outward (6). Spinal motor neurons with

g_{Cl} values from 20 to 70 pS have been reported both *in vivo* and in culture (123). Another type of GABA receptor causes inhibition by decreasing g_K.

In the lacrimal ducts of mouse, stimulation by ACh results in opening of Cl^- channels that can be blocked by Zn and disulfonic acid derivatives. In some plant cells chloride is transported against concentration gradients, for example, into vacuoles. Toad corneas transport Cl^- from endothelium to epithelium; 70% of the corneal short circuiting is due to Cl^- (6). Many freshwater animals absorb chloride from a dilute medium. Uptake of Cl^- may be independent of cations; in some animals, there is Cl^-/HCO_3^- exchange. An anion-sensitive adenosine triphosphatase (ATPase) with a high affinity for Cl^- has been isolated from goldfish gill (295). In some other fish, the energy for Cl^- transport comes from the movement of Na^+ down a gradient. In the rectal gland of dogfish, Cl^- is transported into the lumen by a process that moves 30 Cl^- ions/O_2 molecule consumed; a NaCl-rich secretion is formed by cotransport of Cl^- with Na^+. In various kinds of epithelia, chloride is transported by one of the following: (1) a Na^+-coupled symport, as in frog skin, (2) a Cl^-/HCO_3^- antiport or countertransport, for example, in fish gills and in mosquito larval papillae, (3) an anion-stimulated ATPase that requires ATP as an energy source, in fish gills and locust rectum (295).

Chloride-dependent channels have been reported in numerous excitable tissues; these differ widely in conductance. Patch electrode recordings of rat muscles have shown unit Cl^- channels with conductance of >400 pS (31). Membrane vesicles from the electric organ of Torpedo have been used to prepare planar bilayers, which show low conductance (9 pS).

The Cl^- in squid axons is a function of V_m; since Cl^- is close to equilibrium in most excitable cells, its contribution to membrane potentials is usually lumped to membrane leak currents and Cl^- distribution is considered as passive (126). However, chloride probably functioned as a buffer for charge differences and for pH and volume regulation in primitive cells and this anion now functions in the hyperpolarizing responses by many excitable cells.

Neurons of the nerve net of the medusa *Cyanea* depolarize when a pulse of acidic saline is applied; the response is not affected by Na^+ or K^+ but is reduced when Cl^- is replaced by aspartate. The response is an outward Cl^- current stimulated by H^+ (10).

Summary: Excitable Membranes

Establishment of a bounding membrane that separated evolving protoplasm from the surrounding medium was as essential for the origin of organisms as was establishment of self-replicating macromolecules. Selective permeability of the membrane led to retention of large organic anions and resulting inward inorganic cation gradients. Osmotic and ionic steady states were such that electrical gradients became established. Cell membranes are lipid–protein sandwiches containing carrier proteins of considerable diversity. Cell membranes have electrical resistance and capacitance and transmembrane potential due to many ion-conductance channels in parallel. Current responses to voltage changes vary with potential such that membranes have rectifier properties—current increasing or decreasing nonlinearly in depolarizing hyperpolarizing voltage–current quadrants.

Primitive cells generated protons by metabolic reactions, retained potassium, and excluded sodium; calcium became used for specific functions. Resting potentials result mainly from net passive gradients of K^+, Na^+, and Cl^-, from ion exchangers and from active ion pumps. Membranes came to respond by increases in ion permeability to mechanical deformation. Another early response was probably the alteration of ion permeability when stimulated by specific chemicals such as amino acids. Electrical responses can be graded local potentials (pacemaker, sensory, and synaptic) or they can be all-or-none action potentials. Ion channels can be opened or closed by changes in conformation and in charge distribution on proteins lining a conductance channel. Multiple channels for single ions vary in kinetics (rate of opening or closing), in dependence on transmembrane voltage, sensitivities to agonists, and antagonists of channel proteins. The genes for some channel proteins have been closed and their primary structure sequenced.

An early mechanism for establishment of ionic gradients was extrusion of protons in exchange for potassium. Proton pumps remain as electrogenic in many plant membranes and in association with electron-transport chains in mitochondria (Mitchell's chemiosmotic hypothesis).

Calcium channels occur in many excitable membranes (ciliate protozoans, muscle of many invertebrates and smooth muscle, presynaptic endings, and in intracellular membranes such as sarcoplasmic reticulum). Inward Ca^{2+} currents are depolarizing. Some Ca^{2+} channels are blocked by dihydropyridines, others by alkali metals. Three types of Ca^{2+} channels are recognized according to the voltage activation and specificity of blocking agents. Inward calcium currents trigger a variety of cascades of intracellular events leading to contraction, secretion, and other cellular events. Most of these require phosphorylation of some critical protein.

Several Na^+ channels for inward currents differ in their pharmacology and voltage dependence (266). Sodium channels are uniquely developed in animal cells. Many Na^+ channels are blocked by tetrodotoxin, some are blocked by other toxins such as saxitoxin and bufotoxin and some are not blocked by these toxins. Sodium-channel proteins from electric eel electroplax, from mammalian muscle, and rat brain have been cloned and sequenced. Brain channel protein has three subunits, from muscle it has two and electroplax one subunit. The channel protein consists of six folded segments that form the excitable membrane.

Potassium channels are more diverse than those for other ions and probably preceded Ca^{2+} and Na^+ channels in evolution. The best known K^+ channels carry outward current that repolarizes after depolarization by inward Na^+ or Ca^{2+} currents. Some responses in K^+ channels are depolarizing when the resting potential is more negative than E_K. In most cells, the permeability of K^+ is greater than to other ions and consequently K^+ determines the resting potential over a concentration range near or above normal concentrations; at lower K_0 the contribution of other ions and of the Na^+-K^+ pump are significant for resting potential. Several K channels are

1. A voltage sensitive channel responsible for repolarizing after Na^+ or Ca^+ action potentials; this channel shows outward (delayed) rectification.
2. A slow K^+ channel that shows inward (anomalous) rectification.
3. An outward K^+ channel that is activated by intracellular Ca^{2+}.
4. A so-called A channel, which is activated by depolarization from a hyperpolarized or negative voltage and inactivates during depolarization.
5. An M channel that is closed by cholinergic agonists thus resulting in depolarization and decrease in K^+ conductance.

Other K^+ channels are closed by neurotransmitters and still others function to hyperpolarize from a less negative resting potential than E_K. Mutants of specific K^+ channels are known and one genetic locus for an A channel in Drosophila has been sequenced; it consists of one membrane-spanning domain; Na^+- and Ca^{2+}-channel proteins have four domains each with hydrophobic segments.

In most animal cells, the Cl^- concentration gradient is inward. Chloride permeability varies among membrane types and according to temperature and voltage. Some hyperpolarizing responses such as inhibitory synaptic potentials can be caused by increased Cl^- conductance.

Synaptic Transmission

Very early in biotic evolution organisms acquired cellular organization. Cellular rather than multinucleate acellular organization provided appropriate volumes for protein synthesis and nuclear control. Cell organization increased the surface area per protoplasmic volume; cell size was limited by diffusion distance from membrane to interior (25). Transmission of information between cells became essential to integration. Free diffusion between cells could not occur because membranes were selective in permeability. Several modes of communication have evolved: (1) electrotonic coupling between cells by junctions across which charged ions and small molecules pass; (2) secretion of chemical agents, transmitters, which act on adjacent cells; (3) secretion of agents (modulators), which regulate liberation of transmitters and modulate the action of transmitters; (4) secretion of substances (hormones) that are transported in extracellular fluids and act on remote tissues; (5) release into air or water of pheromones—attractants or repellents. Neurons became differentiated for conduction of signals from one part of an organism to another. Transmission between neurons, from sense organs to neurons and from neurons to effectors makes use of electrical coupling or chemical transmitters. Animal behavior is determined by neural circuits that are controlled by transmission at synapses. How nervous systems originated cannot be readily deduced from knowledge of synaptic function in present-day animals. Nerve nets in primitive diploblastic animals, coelenterates, have many of the properties of complex nervous systems.

Electrotonic Transmission

In many coelenterates, specifically in hydrozoa, and in embryos of numerous animals, there is electrotonic conduction between epithelial cells. In embryos, epithelial conduction is transitory and is replaced by nervous conduction. The advantages of nervous over epithelial conduction are (1) greater transmission distance per cell; (2) faster conduction; (3)

modifiability of signals. Electrical coupling probably occurred before there were neurons and chemical transmission. Coupling between cells can be calculated by injecting current into one cell and measuring the voltage in adjacent cells; coupling can also be estimated by the passage of small molecules such as some dyes between cells (175, 176). Electrical coupling occurs between cells of many tissues, for example, mammalian liver, heart, lens, brain, and kidney. In embryonic development, electrical connections provide for diffusion of ions and metabolites between cells. Some gap junctions are voltage sensitive, others not (263).

Electrotonic junctions are seen by electron microscopy as gap junctions or nexuses in which extracellular space is small or absent. Junctional membranes contain particles that bridge between double membranes of adjacent cells. The particles can be made visible by scanning electron microscopy of membranes that have been freeze cleaved. Cleavage is along the hydrophobic layer within one member of opposing membranes. One face is on the plasma membrane side, the other on the extracellular space side. Particles on one face fit into pits in the opposing half of the opposing membrane. Five to 15% of the septal area between segments of earthworm giant axons consists of nexal membrane (38). Particles in the hemimembranes are 10–15 nm in diameter and spacings between particles are ~10 nm (25) (Fig. 12a and b). Similar patches but of slightly different dimensions occur in epithelia, secretory tissues and other cell types of many animals.

Many kinds of early embryos are functional syncytia in which the blastomeres are electrically coupled (262). Injected dyes can pass from cell to cell. Epithelial conduction has been described in amphibian embryos and adult salamanders. Conduction is slow relative to that in nerves. In most sponges there is no conducting system, but in siliceous sponges, Hexactinellida, basal membranes of epithelia are connected so that signals are conducted at 0.6 cm/s and spread from one sponge to another in a colony. Feeding currents generated by flagellated cells are stopped by conducted signals (177). Gap junctions occur in epithelia of Hydroxa not Anthozoa or Syphozoa (10, 176, 177). Body cells of *Hydra* are coupled and antibodies to rat liver gap junctions recognize junctional proteins in *Hydra* and block intracellular communication. Apical ends of body fragments of *Hydra* form heads, basal ends form foot structures, positional information is given by small molecules that pass from cell to cell (86). The large epithelial cells of dipteran salivary glands are coupled and this appears to synchronize secretion; molecules of 1200 but not of 1900 Da can pass between cells. Signals in salivary gland cells are synchronous in wild-type Drosophila, asynchronous in a mutant that is defective in electrical coupling (18). Electrotonic coupling synchronizes clusters of neurons in nerve centers; the neurosecretory bag cells of *Aplysia* are coupled and respond synchronously. In each leech ganglion, some 14 motor neurons on a side are electrically coupled. In *Aplysia*, 39 neurons of the pleural ganglion are coupled (267). In the giant axons of earthworms, septate junctions are of low resistance and current can pass from segment to segment; small molecules such as flourescein can pass through.

One limitation of electrical transmission between neurons is that it is usually excitatory, taking place only in response to depolarizing impulses. One kind of electrical inhibition occurs in the Mauthner neurons of goldfish where hyperpolarization is induced by the extracellular field generated by a coil of axons around an axon hillock. Another kind of inhibition is a long lasting or late hyperpolari-

zation after a burst of postsynaptic potentials, as in some *Aplysia* neurons.

Many synapses transmit both electrically and chemically, for example, the chicken ciliary ganglion. In the sea slug *Navanax*, 10 neurons in the buccal ganglion are electrically coupled and a chemical synapse onto one or more of these cells reduces nonjunctional resistance thereby shunting the electrical current and decreasing coupling; the chemical synapse modulates electrotonic transmission (261). In coelenterates, electrotonic conduction through an epithelium is in parallel with chemical synapses of the nerve net.

In embryos of fish and amphibians, electrotonic coupling depends on pH; acidification reduces the coupling of junctions in cardiac Purkinje cells and in smooth muscle, intercellular connections are sensitive to intracellular H^+ (263). In Purkinje fibers, spread of current and the number of functional junctions are increased by cAMP. Hypertonic solutions and detergents such as octanol can break nexal contacts (263).

Plasma membrane proteins can be removed by detergents leaving gap junction proteins; antibodies to these junction proteins have been made. Liver cells are well coupled and they provide quantities of gap junction proteins. A major protein from rat liver gap junctions has a molecular weight (MW) 27,000 D, a secondary one of 21,000 D, a dimer of 47,000 D, and a subunit of 16,000 D. Some species differences have been found (122, 208, 263). Crayfish hepatopancreas yields a junctional protein of 15,000 D. Most invertebrate junctions do not react with mammal liver protein. Lens of the mammalian eye yields a protein with MW 26,000 D (137, 237). This is the only vertebrate gap junction that does not cross-react with liver proteins. There is evidence that a gap junction nexus has a protein structure of six subunits per membrane (237) around

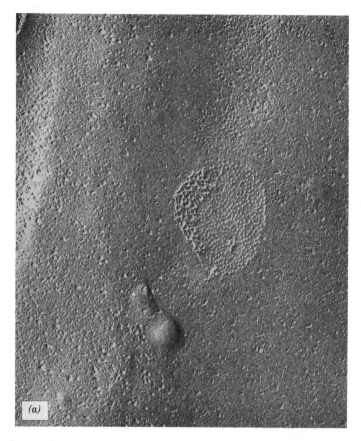

Figure 12. Freeze–fracture electron micrographs of nexus (electronic gap junctions) from retractor muscle of *Mytilus*, the membrane is split between *P* and *E* faces. In (*a*) the particles of the junctions are in the *P* face and pits in the *E* face. In (*b*) the predominant membrane face is shown as *P* face with particles and pits represent *E* face. [Personal communication from Peter Brink.]

a pore of 1–1.5 mm diameter and 15 mm long (16, 151, 224) (Fig. 13). Horizontal cells in fish retina are connected electrotonically and the gap junctional conductance is modulated by dopamine from plexiform cells (154). Electrotonic resistance of junctions in amphibian embryos is controlled by a transjunctional potential. In insect salivary glands the resistance is controlled by a transmembrane potential (resting potential). Rectifying junctions in the crayfish giant fiber system resemble those in amphibian embryos, but are faster and asymmetrical (129).

Chemical Neurotransmitters

Chemical transmission at synapses is polarized, graded, allows both excitation and inhibition, and may show considerable plasticity. The action of neurotransmitters is modulated by agents from converging axons and from nonneural cells. The origin of transmitter substances can be traced to chemical attractants and repellents of microorganisms. A survey of neurotransmitters shows that those of one class are derived directly from amino acids and that simple modifications of amino acids have resulted in a variety of

Figure 12. (*Continued*)

small-molecule transmitters. Another class of transmitter includes large neuropeptides consisting of chains of amino acids; many of these are neuromodulators, some are hormones, and a few are neurotransmitters. Neuropeptides are genically encoded and long precursor peptides are synthesized and later cleaved to smaller peptide chains that are secreted. The two transmitter types evolved in parallel. Most nervous systems use both amine-derived and polypeptide transmitters and examples of cells with both types will be given later. At many synapses, two chemical agents coexist. Specificity resides in receptors several of which react with a transmitter-modulator.

Amino Acids

Some amino acids attract, others repel bacteria (Chapter 7). Positive chemotactic responses are elicited in *Escherichia coli* by aspartate at $10^{-8}\,M$, by threonine and glutamate at $10^{-6}\,M$, and by serine with a threshold at $10^{-7}\,M$; negative chemotaxis is elicited by slightly higher concentrations of Leu, Ileu, Val, and Norleu (5). Two receptor proteins have been isolated from bacterial membranes, each ~60,000 D in size (120). A receptor for aspartate has been purified from Salmonella; it has hydrophobic sequences that span the lipid membrane layer (33). Bacterial receptors become methylated after binding to an amino acid. Leucocytes are attracted

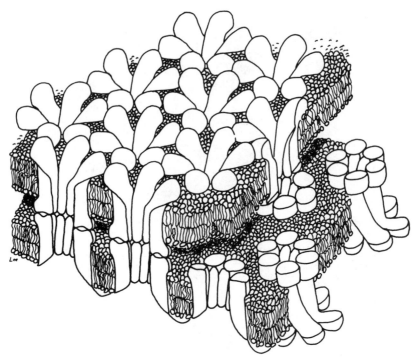

Figure 13. Drawing of the structure of the gap junctions isolated from mouse liver as deduced by X-ray diffraction. The gap junction units are made up of 12 copies of the connexin molecule arranged into two hexamers (connexons), one associated with each membrane. Each connexin molecule is divided into two domains. The transmembrane domain spans one bilayer and one half of the extracellular gap. The cytoplasmic domain is tightly associated with the lipid polar head groups on the cytoplasmic surface of the membrane. An aqueous channel extends along the center of connexons. [From (180).]

to amino acids and after stimulation their membranes become depolarized (259).

Feeding is initiated in coelenterates by amino acids; some species respond behaviorly to proline, others to tyrosine; Hydra responds to low concentrations of the tripeptide glutathione (159). A starfish *Marthasterias* responds to proline or cysteine at $10^{-7}M$ (285). In salmon, olfactory receptors are sensitive to amino acids, alanine at $10^{-7}M$. Chemoreceptors of a crab Cancer respond ·to low concentrations of glutamate, not to glutamine. Amino acids are attractants for the shrimp *Palaemonetes* (46).

At many synapses glutamate or aspartate are excitatory, glycine and the glu-tamate derivative, γ-aminobutyric acid (GABA), are inhibitory. γ-Aminobutyric acid is formed by decarboxylation of glutamate (Fig. 14). GABA is an inhibitory transmitter in many central nervous systems, for example, in ganglia of a leech (56). Glutamate is probably the excitatory transmitter at crustacean and insect neuromuscular junctions; application of glutamate yields the same response as motor nerve stimulation. γ-Aminobutyric acid is indicated as the inhibitory transmitter. Glutamate is relatively nonspecific in ion selectivity, but probably increases g_{Na}; GABA increases g_{Cl}. In the opener of the claw of a crayfish the reversal potential for a normal nerve-elicited epsp is 12.3

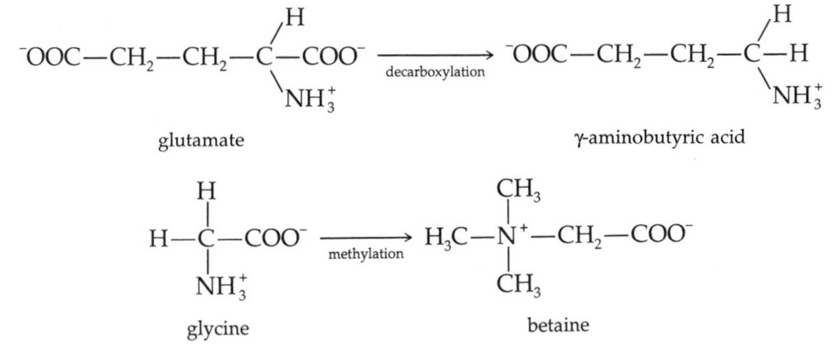

Figure 14. Pathway of conversion of glutamic acid to γ-aminobutyric acid and glycine to betaine.

mV, for the glutamate potential 19.4 mV (272). Aspartate is a probable neuromuscular transmitter in Limulus (232).

In vertebrate nervous systems glycine is inhibitory at some synapses. The action of glycine is blocked by strychnine while the action of GABA is blocked by bicuculine or picrotoxin. Glutamate receptors occur in cerebral cortex, some of them Na^+ dependent, some Na^+ independent. Several classes of glutamate receptors differ according to selectivity and ionic effects. The strychnine-binding site of the glycine receptor has homology with the nicotinic ACh receptor (105).

Glutamate and GABA are evidently transmitters in the vertebrate retina. In retina of fishes, illumination suppresses a steady-state dark current in rods and cones; the light response is a hyperpolarization and transmitter (glutamate) liberation is decreased, with the result that horizontal cells become hyperpolarized. In carp retina, aspartate is more effective than glutamate for hyperpolarizing horizontal cells. Bipolar cells of turtle retina are immunoreactive to glutamate (79). In mudpuppy retina, both receptor cells and bipolar cells are glutaminergic; "on" amacrines may liberate aspartate as transmitter (37). In primate retina, photoreceptor terminals can be stained intensely by an antibody to glutamate decarboxylase. Some rods stain for GABA and glutamic dehydrogenase, other rods do

not (37). In retinal neurons of frog, rat and guinea pig, glutamate, aspartate, and GABA are present in high concentrations (9).

Each of the excitatory amino acids (Asp or Glu) has a second carboxyl that provides a negative charge at physiological pH; the inhibitory amino acids Glyc and GABA have only one COO^- group. A GABA receptor has been extracted from brain; it has two subunits (456 and 474 amino acids). Four hydrophobic domains appear to traverse the membrane (247). There are two types of GABA receptors; $GABA_A$ receptors that act by increasing g_{Cl} and are blocked by bicuculine; $GABA_B$ receptors that act by increasing g_K and are not blocked by bicuculine (77). There is a 34% sequence identity between the A and B GABA receptors (161).

Four classes of glutamate receptors have been identified pharmacologically in brain: (1) those excited by NMDA (N-methyl-D-aspartate), (2) those receptors excited by kainic acid, (3) others excited by equisqualate, and (4) receptors for 2-amino-4-phosphonobutyric acid (187). N-Methyl-D-aspartate receptors function in long-term potentiation (LTP) in hippocampus, as well as in the epileptiform activity in the cerebral cortex (117, 187). Responses of NMDA receptors in thalamus have a long latency, while those of non-NMDA receptors have a short latency.

Methylated Quaternary Amines

Amines can be formed relatively directly from amino acids; one class of amines can become methylated by as many as three methyl groups. Methylated amines are widely distributed and their functions are not well understood.

Betaine is formed from glycine by methylation of its nitrogen atom (Fig. 14b). Betaine is a relatively inert anion and functions in volume regulation of numerous euryhaline molluscs and crustaceans under osmotic stress (295). Betaine stimulates the feeding responses of sea anemones; it is excitatory for the slowly conducting nerve net (34).

Taurine is formed from cysteine by way of cysteine sulfenic acid. It comprises one half of the free amino acids in mammalian heart. Taurine occurs in high concentration in the outer segments of vertebrate photoreceptors; illumination causes liberation of taurine. Chemoreceptors of crustaceans respond to concentrations of taurine as low as 10^{-12} M (12).

Homarine is formed by reaction of glycine and succinyl CoA with succinylglycine as an intermediate (207). Homarine can serve as a methylating agent to form the methylamines choline and betaine. Homarine is abundant in the nervous systems of crustaceans; its function is unknown.

Trimethylamines protect enzymes against the urea that accumulates in euryhaline animals such as toads that live on dry land or in brackish water. Trimethylamine oxide is a neutral solute occurring in some hyperosmotic fish. Glycylphosphocholine accumulates in proportion to plasma urea in desert amphibians.

Acetylcholine

Of all the methylated amines, the most widespread in neural transmission is acetylated choline (ACh). Choline is formed from serine by decarboxylation and methylation (Fig. 15); it is present as phosphoratidylcholine in all cell membranes. The enzyme for acetylation, choline acetyltransferase (Cho-ACTase) occurs in some plants and microorganisms as well as in animals. Porcine Cho-ACTase has been cloned and sequenced; there is a 32% identity with the corresponding enzyme from Drosophila (24). Similarly, the enzyme for hydrolysis of ACh, acetylcholine esterase (AchEs), is widely distributed and has an exceptionally high turnover rate. Acetylcholine is a relatively long, flexible molecule with a positive charge at the methylated end and a negative charge at the carboxyl end. Both specific and nonspecific choline esterases occur in many tissues; the catalytic site has both anionic and esteratic pockets. A specific acetylcholine esterase (AchEs) is blocked by physostigmine (eserine); analogs of ACh such as carbachol are not subject to hydrolysis by AchEs. Acetylcholine esterase occurs in several isoforms in various tissues and different species of mammals—3 in heart, 2 or 3 in different kinds of smooth muscles (257). Acetylcholine esterase has been found in ciliates, flatworms; it is high in activity in squid and insect ganglia. Electric organs are a rich source of AChEs. The ventral roots of the spinal cord have 300–400 times more AChEs than the dorsal roots. A nonspecific enzyme (pseudo-cholinesterase) occurs in nonnervous tissues—red blood cells, submaxillary gland, and others.

Acetylcholine esterase consists of several subunits that occur in different proportions in different animals. Mammalian AchEs has three subunits and is a large asymmetric molecule with sedimentation coefficient 16s; in chicken the s value is 20 and there are six subunits of sizes 110, 72, and 58 kD, all bound together by disulfide bridges (136). In the electroplax of

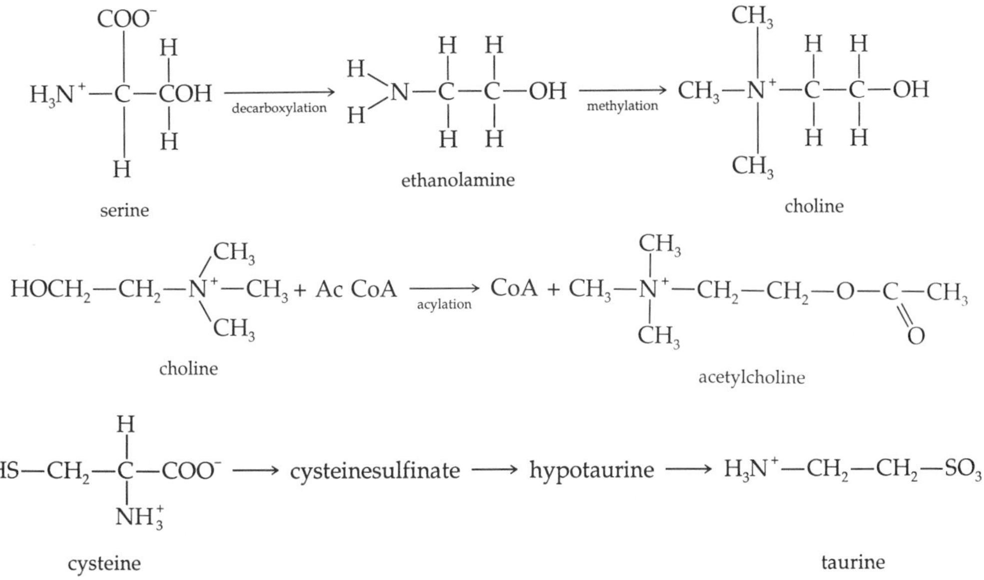

$$\text{serine} \xrightarrow{\text{decarboxylation}} \text{ethanolamine} \xrightarrow{\text{methylation}} \text{choline}$$

$$\text{choline} + \text{Ac CoA} \xrightarrow{\text{acylation}} \text{CoA} + \text{acetylcholine}$$

$$\text{cysteine} \longrightarrow \text{cysteinesulfinate} \longrightarrow \text{hypotaurine} \longrightarrow H_3N^+ - CH_2 - CH_2 - SO_3 \quad (\text{taurine})$$

Figure 15. Chemical reactions for synthesis of amino acid derivatives with quaternary nitrogen atoms. Sequences include decarboxylation, methylation, and acetylation.

the electrical eel, there are 12 subunits of similar size, 65 kD.

Acetylcholine has been reported in some bacteria and in plants such as Nitella, in roots and buds of mung beans. The ACh concentration in honeybee brain is estimated at 2.7 mM, in Octopus optic ganglion 2.1 mM, and in guinea pig cerebral cortex 0.013 mM. At cholinergic synapses ACh is packaged in clearcore vesicles, some 500 Å in diameter. Acetylcholine affects the ion permeability in many nonneural cells. Evidence that ACh is the neurotransmitter in the statocyst system of the cephalopod nervous system is given by high concentrations of ACh, ChAT, and AchEs in the nervous system and by the effects of applied ACh (42). In leeches, the cardio-excitatory neurons (HE) liberate ACh to nicotinic receptors (45).

Several types of ACh receptor have been identified, largely on the basis of blocking agents. Nicotinic receptors occur at neuromuscular junctions of vertebrates, in electric organs, and in sympathetic ganglia. The action of ACh at these synapses resembles that of nicotine and a common blocking agent is d-tubocurarine (d-Tb). In vertebrate autonomic effectors such as gastrointestinal muscle, heart and blood vessels, there are muscarinic receptors that are blocked by atropine. Two types of muscarinic receptors are designated M_1 and M_2 with specific antagonists and agonists. A third class of receptor is present in the heart of bivalve molluscs and the neurons of Aplysia and Helix and is blocked by benzoquinonium and related compounds (80).

Acetylcholine is relatively nonspecific in the ion conductance changes it induces at nicotinic junctions; postsynaptic membranes show increased conductance for Na^+, K^+, and Cl^-. In neurons of Aplysia there are three types of response to ACh: a fast depolarization due to increase in g_{Na}, a fast hyperpolarization due to increase in g_{Cl}, and slow hyperpolarization due to increase in g_K. In cells that are depolarized (D cells), the resting potential is low and similar to E_{ACh} (-33 mV); H cells are hyperpolarized and have a resting potential that is more negative (-45 to -51 mV) as is E_{ACh}. In other molluscan neurons there is slow hyperpolarization due to increase in g_K; E_{ACh} is -80 mV (135). In some molluscan hearts (Mytilus and Modiolus), the response to ACh is depolarization due to increase in g_{Na}; in hearts of other molluscs (*Crassostrea*) ACh is inhibitory by an increase in g_{Cl} and yet in others (*Mercenaria*) ACh is inhibitory by increase in g_K (102). Echinoderm muscles are very sensitive to ACh and have nicotinic receptors blocked by *d*-Tb. In vertebrate striated muscle, ACh receptors are located at endplates in subsynaptic membrane but after nerve degeneration, receptor molecules develop over the entire muscle fiber membrane. In mouse diaphragm, the abundance of receptors is $15,000/\mu m^2$ at the endplate and $30/\mu m^2$ in the extrasynaptic region (192). Leech heart muscle cells are depolarized by ACh with a reversal potential of -9 mV, blocked by curare (45). There is considerable homology between ACh receptors of Drosophila and mammalian brain (286).

Stimulation of preganglionic nerves in bullfrog or mudpuppy sympathetic ganglia initiates four synaptic potentials: (1) fast epsps that are nicotinic cholinergic; (2) slow ipsps muscarinic cholinergic of 2 s duration, (3) slow epsps, 30 s duration muscarinic cholinergic, and (4) very slow epsps (5–10 min) for which the transmitter is probably the peptide hormone LHRH (149). One class of amacrine cells in rabbit retina, stimulated by glutaminergic bipolar cells, liberates acetylcholine that excites ganglion cells (66). In heart, muscarinic receptors decrease the rate of heart beat by increasing g_K (205).

Nicotinic receptors have been isolated from electroplax, striated muscle, and from rat brain and the cDNA cloned. The sequence for α subunits of brain is similar to but different from that for muscle. The neural α subunits are 37 amino acids

longer than that from muscle (35). Chicken brain ACh receptor (Ach R) shows some differences from muscle (62, 63). The central nervous system of a locust has a AChR composed of four identical subunits (36). It has been postulated that a primitive AChR consisted of four identical subunits, that α and β lines emerged early in animal evolution, and the brain and muscle duplication occurred more recently (62).

The nicotinic receptor from the electric organ is a protein of MW 250,000 D and consists of five subunits; α(40 kD), β(50 kD), γ(60 kD), and δ(65 kD) in the ratio $2:1:1:1$. Each of the subunits is a single chain and some regions show high homology. Genes for each subunit have been cloned and sequenced (Fig. 16*a*) (211). The nicotinic receptor has multiple-binding sites for toxins. The α subunit gene is on chromosome 17, the β subunit on chromosome 2, and γ and δ on chromosome 1. There is evidence for five transmembrane segments in each subunit with a disulfide bridge near the ACh-binding site (198). Conductance through AChR channels of bovine brain is less than through channels of Torpedo electroplax probably because of difference in the δ subunit (125). The nicotinic receptor of five membrane-spanning subunits is a cylinder 80 Å in diameter and 140 Å long with an external channel 30 Å wide and at the cytoplasmic end 7 Å (Fig. 17) (147). The AChR in Torpedo brain differs in composition from that in Torpedo electroplax (198).

The widespread distribution of ACh indicates that it may have evolved very early, presumably in permeability regulation. When and for what functions specific receptors evolved can only be speculated. The proteins may have evolved for different functions and became receptors later or receptors may have coevolved with ACh as a transmitter. Coexistence of ACh with catechol-

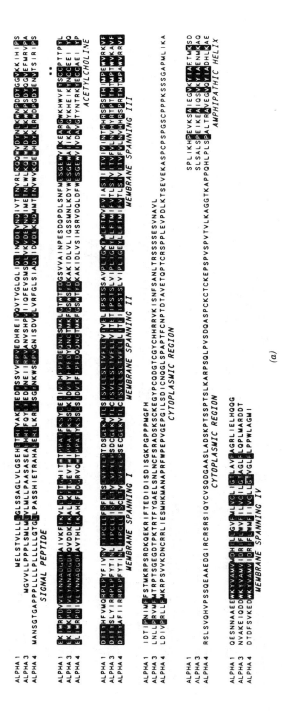

Figure 16a. Comparison of deduced amino acid sequences (standard one-letter symbols) of nicotinic ACh receptors; the mouse muscle α₁ and two neuronal α₃ and α₄ nicotinic α receptor subunits. The two asterisks indicate the cysteine residues that are thought to be close to the ACh-binding site. The molecular weights of the unglycosylated mature α₁, α₃, and α₄ subunits are 55,085, 54,723, and 67,124, respectively. [From (35).]

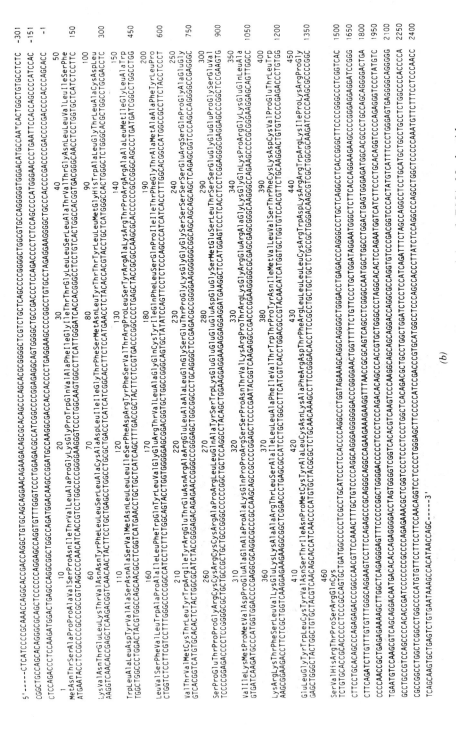

(b)

Figure 16b. Nucleotide sequence of muscarinic ACh receptor of cloned cDNA. Nucleotide residues are numbered in the 5′ to 3′ direction. The deduced amino acid sequence shown above the nucleotide sequence and the amino acid residues are numbered beginning with the initiating methionine. [From (148).]

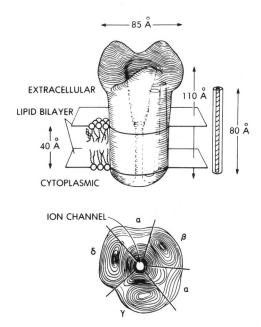

Figure 17. The three-dimensional model of AChR to show arrangement of subunits, two α, one each β, γ, and δ. Central ionophoretic channel traverses the 110 Å length of the molecule. Funnel shape visualized by EM. Receptor extends 15 Å on cytoplasmic side and 55 Å on synaptic side. [From (136).]

amines and neuropeptides is probably a recent occurrence.

Both synaptic excitation and inhibition result from cholinergic input to parasympathetic neurons in the mudpuppy Necturus. A rapid EPSP occurs at nicotinic junctions, slow facilitating IPSPs occur at muscarinic receptors (118). M_1 receptors respond to the agonist McNeill A 343, they are antagonized by pyrenzipine; M_2 receptors are antagonized by 4-Damp. Two forms of muscarinic receptors appear in chick nervous systems, an early form of 86,200 D; after synaptogenesis a 72,000 D form predominates. Muscarinic receptors are more abundant than nicotinic receptors in brain. Muscarinic receptors have been purified from pig brain; they modulate K^+ channels by combining with guanine nucleotide (G-proteins) and in-

hibiting adenylate cyclase (275a). The AChR from brain has a MW 70,000 and has the pharmacology of M_1. This AChR when cloned has a 2884 nucleotide sequence and from the coding region 460 amino acids are deduced (Fig. 16b) (148). Acetylcholine coexists with other transmitters in some neurons. Some amacrine cells of rabbit have both ACh and GABA; released ACh depends on Ca_0, release of GABA is independent of Ca_0; ACh release is by vesicle exocytosis of GABA via a carrier. A flash of light leads to the release of ACh, not of GABA (217).

Catecholamines, Other Monoamines

Monoamines function widely as neurotransmitters and are derived from essential amino acids. Tyrosine can be converted to catecholamines (CA) in either of two directions: (1) via DOPA and dopamine to norepinephrine and epinephrine, or (2) via tyramine to octopamine (Fig. 18). Tryptophan can be converted via 5-hydroxytryptophan to 5-hydroxytryptamine (5-HT or serotonin).

Neurons that use catecholamines as transmitters are called adrenergic. Vertebrate receptors for norepinephrine (NE) and epinephrine (E) are identified by pharmacological methods as two types, α and β; receptors are further divided as α_1, α_2, β_1, β_2. Commonly used α blockers are phenoxybenzamine and phentolamine, β blockers are propranolol and dichlorphentolamine. Red blood cells are stimulated by catecholamine to make membrane phospholipid; the initial step is methylation. α_1 and α_2 may be isoreceptors (250) each with MW 55 kD, they bind to the same antibodies, both are glycoproteins (197). Predominant action on platelets is by α_2, on liver by α_1; receptors. Mammalian smooth muscles have both α and β receptors; the relative amounts of receptor types differ in different muscles (Table 2). The α_1 receptor causes vasocon-

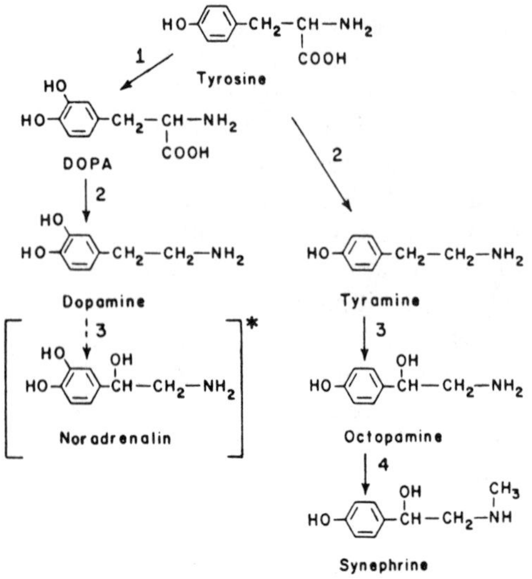

Tyrosine

DOPA

Dopamine

Tyramine

Noradrenalin

Octopamine

Synephrine

Figure 18. Pathways from tyrosine to catecholamine transmitters; dopa, dopamine, norepinephrine, and octopamine. [From (19).]

striction, stimulates the heart; it acts postsynaptically; α_2 acts presynaptically, lowers blood pressure, and causes bronchodilation. Both α and β receptors act via adenylate cyclase to increase cAMP which causes the physiological action (197, 235). The α_1 receptors are predominant in the aorta and pulmonary artery, α_2 receptors in *in ileum* and *vas deferens*. Iris dilator of rabbit shows β inhibition and α excitation.

Regions of the brain differ in their proportions of α_1 and α_2 receptors. Locus coeruleus contains neurons that are stimulated by ACh to produce NE, which modulates activity in the midbrain.

Beta receptors have MW 64,000 and show much homology with rhodopsin (71). Cyclic-DNA for a β_2 adrenergic receptor has been cloned and the active region is deduced to have 418 amino acids in seven membrane-spanning domains (Fig. 19) (140). Activation of a β receptor acts via a G-protein to activate a protein kinase (140). In brain, NE acting via β receptors increases phosphorylation of synapsin I (199).

Both dopamine (DA) and DOPA have synaptic function in vertebrate brain. Dopamine increases cAMP in postsynaptic cells of sympathetic ganglia and hyperpolarizes the cells. Dopamine causes a late inhibition in rabbit sympathetic ganglia via an increase in cAMP.

In fish retina, the cones liberate a transmitter (glutamate in some species) that excites (hyperpolarizes) horizontal cells. Horizontal cells are electrically coupled and this coupling mediates the antagonistic center-surround responses of bipolar cells (Chapter 5). Interplexiform cells synapse on horizontal cells and use DA as a transmitter that decreases lateral

TABLE 2. Classification of Adrenergic Receptors

Type	Potency	Characteristics
α-Adrenergic	NE > EPI >> ISO	Vasoconstriction, excitation of uterine contractions, contraction of nictitating membrane, pupillary dilation, inhibition of intestinal peristalsis
α_1-Adrenergic	Postsynaptic	Vasoconstriction
α_2-Adrenergic	Presynaptic	Inhibit NE release, inhibit renin release, lower blood pressure via central action
β-Adrenergic	ISO > EPI > NE	Vasodilation, inhibition of uterine contraction, myocardial stimulation
β_1-Adrenergic	ISO > EPI = NE	Fatty acid mobilization from adipose tissue, cardiac stimulation
β_2-Adrenergic	ISO > EPI >> NE	Bronchodilation, vasodepression

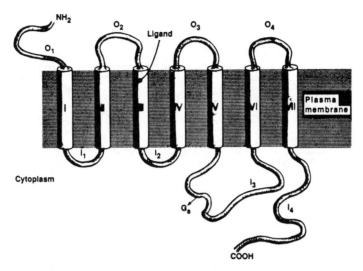

Figure 19. Schematic structure of the β-adrenergic receptor and its homologs. The receptor has a molecular size of ~46 kD and is highly hydrophobic. The hydrophobicity results from the presence of seven sequences of hydrophobic amino acids, which can be arranged in seven putative transmembrane domains. The NH_2-terminus is extracellular and the COOH-terminus intracellular. Transmembrane domain III probably participates in ligand binding, suggesting a hydrophobic environment for the ligand. The intracellular domain i_3 seems to be crucial for the interaction between the receptor and the stimulatory G protein G_s. The mechanism of signal transduction from the agonist-binding site to the domain of interaction between the β-adrenergic receptor and G_s (domain i_3) is not known. It is also possible that oligomerization of the receptor or G_s is involved, although there is no evidence to suggest this. [From (162).]

inhibition by decreasing the coupling between horizontal cells (277, 139) (Fig. 20). The net result is an increase in the responses of horizontal cells to a spot of light and a decrease in response to a surrounding annulus. The action of DA is via a cAMP-dependent protein kinase (74, 75). Dopamine is formed from L-DOPA in basal nuclei in human brain and deficiency of DA results in Parkinson's disease. Neural crest cells, which are normally cholinergic, can produce NE if cultured with nonneural brain cells; presumptive sympathetic neurons in culture are initially adrenergic but can be made to produce ACh if treated with a factor from serum (89).

Catecholamines are widely distributed in invertebrate animals. Dopamine is a transmitter in molluscan ganglia; it is present in perfusate from the pedal ganglia of bivalves. Dopamine occurs in ganglia of snails; dopamine applied to *Aplysia* ganglia stimulates many neurons. Dopamine modulates neuromuscular transmission in a prawn by decreasing release of the excitatory transmitter glutamate (195). Dopamine may be an excitatory transmitter to pacemaker neurons in the cardiac ganglion of Limulus (13).

Octopamine is a neurotransmitter or neuromodulator in many invertebrates. The nervous system of lobster *Homarus* contains large amounts of octopamine, mostly in thoracic connectives and ganglia (19, 143). Lobster supraoesophageal ganglion contains 225 µg/g and lacks NE (144, 145). Octopamine applied to the lobster abdominal ganglion is excitatory

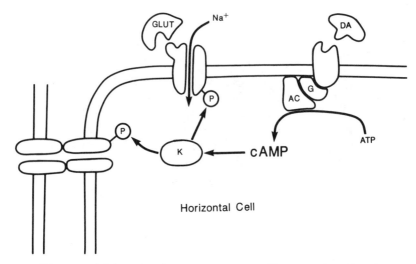

Figure 20. Diagram of fish retina showing activation of horizontal cell by glutamic acid, the transmitter from cones. Modulation by dopamine, transmitter from interplexiform neurons. Modulation uses cAMP as a second messenger. (Figure courtesy of J. Dowling.)

to extensor motor neurons, inhibitory to flexor motor neurons. In locusts the leg extensor muscle epsps are enhanced by octopamine at 10^{-10}–10^{-9} M and motor patterns are thus released (83, 84). Three classes of octopamine receptors have been identified according to their sensitivity to blocking drugs (50). Octopamine at 5×10^{-9} M is excitatory to lobster cardiac pacemakers (145). Octopamine is present in quantity in efferent fibers to photoreceptors of Limulus eye (20).

Serotonin

Serotonin, 5-hydroxytryptamine (5-HT), was discovered at the same time as a transmitter in molluscs and in the autonomic system of mammals. Serotonin is derived from tryptophan (Fig. 21). 5-Hydroxytryptamine, like catecholamines, is detected in neurons by its fluorescence. Serotonin is found in many regions of the vertebrate brain. In hippocampus CA-1 cells 5-HT has three separate actions on K^+ conductances: (1) activates Ca^{2+}-independent K^+ hyperpolarization (2) suppresses Ca^{2+}-dependent g_K that prolongs

cell discharge, (3) has slow and lasting suppression of voltage-dependent g_K that leads to depolarization (57). Several 5-HT receptors have been identified functionally in different regions of the mammalian brain. One of these, 5-HT receptor I_A, is of high affinity; it occurs in the hippocampus where it is inhibitory for adenylate cyclase. Another with lower affinity stimulates the cyclase. Others are autoreceptors, stimulants for phosphoinositide turnover. None have been purified and sequenced. Whether the 5-HT receptors of invertebrate neurons resemble those of mammals is unknown (99).

In rabbit retina, amacrine cells of one type have the autofluorescence of serotonin. 5-Hydroxytryptamine at low concentrations contracts guinea pig ileum by both direct action on the muscle and indirect action via myenteric neurons, these actions can be blocked by dibenzyline. Serotonergic synapses are blocked by lysergic acid (diethylamide); indoletropanyl blocks at concentrations as low as 10^{-14} M. Morphine blocks pain produced by serotonin.

Serotonin is also a transmitter in mol-

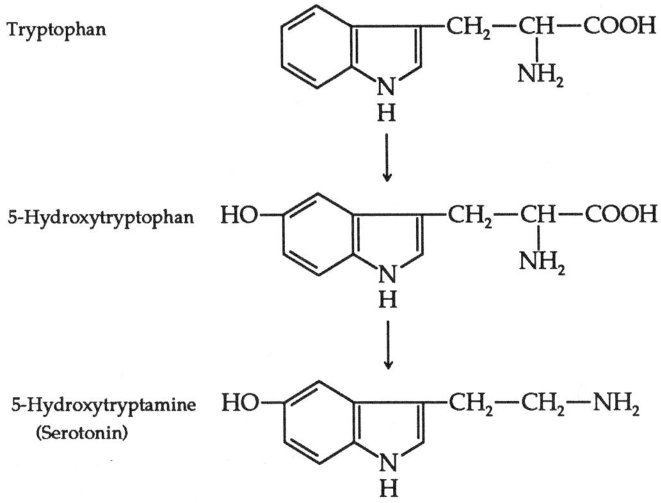

Tryptophan

5-Hydroxytryptophan   HO

5-Hydroxytryptamine   HO
(Serotonin)

Figure 21. Pathway from tryptophan to 5-HT (serotonin). [From (158a).]

luscs. Many molluscan neurons contain 5-HT, the enzymes for making it and an amine oxidase that inactivates it. 5-Hydroxytryptamine is released from clam hearts on stimulation of cardioexcitatory nerves. In a catch muscle of *Mytilus* 5-HT relaxes "catch" (Chapter 2) and amplitude of phasic contractions (202). Serotonin appears to be a transmitter of snail neurons involved in feeding behavior and may be used in conditioned behavior. Two giant neurons of Aplysia synthesize 5-HT and serotonin-containing cells are found in many molluscan ganglia. Acting on one group of neurons in the cerebral ganglion of *Aplysia,* 5-HT excites by increasing g_{Na}, in other cells 5-HT inhibits by increasing g_K, in yet other cells 5-HT decreases both g_{Na} and g_K (23, 93). The ciliated epithelium of the gill of Mytilus has dual innervation; stimulation of cerebral ganglion or perfusion with 5-HT increases ciliary activity; methysergide antagonizes this. Stimulation of nonserotonergic neurons or perfusion with dopamine or NE decreases activity of cilia (47).

Neurons in the nervous systems of the leech and earthworm show fluorescence characteristic of serotonin. Large Retzius cells, and the P cells of leech in tissue culture show a voltage-sensitive release of 5-HT. Depolarization of the Retzius cells releases 5-HT and hyperpolarizes postsynaptic membranes of P cells (70). Serotoninergic Retzius cells are multifunctional and their activity affects several muscles such that feeding is activated and mucus secretion increased. Coordinated swimming in a leech results from stimulation of 5-HT-containing interneurons that excite central pattern generator neurons (214). Serotonin may potentiate the nerve activation of a crustacean muscle.

Serotonin is the principal amine in the ganglia of lobster and fluorescence shows that octopamine and serotonin are contained in different neurons. Octopamine has the opposite actions to serotonin, stimulating flexors and inhibiting extensors (144). Injection of either of these amines results in static posturing. 5-Hydroxytryptamine causes an aggressive flexed pose with claws spread and abdomen tucked under cephalothorax; octopamine causes a submissive or extended pose with legs pointed forward and abdomen arched upward (145). Each of these two amines, also the neuropeptide proctolin, has both central and neuromuscular actions. 5-Hydroxy-tryptamine increases the release of the excitatory transmitter glutamate, increases the amplitude of epsps, and increases cAMP in muscle; it increases heart rate and excites the stomatogastric ganglion. Octopamine causes spiking in leg muscles. Serotonin increases slow excitation of motor neurons to flexor muscle and inhibition of extensor motor neurons; it inhibits flexor inhibitor motorneurons and slow excitation of extensors. Centrally these amines modulate motorneurons and lead to complex postural behavior (144). In leech 5-HT stimulates coordinated swimming.

Purines

Transmitters and modulators not derived from amino acids are the purine nucleotides adenosine, adenosine monophosphate (AMP), and ATP. Pyrimidines are not as effective as neurotransmitters. Purinergic nerve endings contain large opaque vesicles. Adenosine triphosphate was first found as an inhibitory transmitter in several autonomic effectors of vertebrates and is now known to function in the brain and in some invertebrate nervous systems.

Adenosine triphosphate and adenosine are dilators of coronary arteries, are inhibitory to smooth muscle in the gastrointestinal tract of mammals and in the lungs of amphibians and reptiles. Purinergic nerves relax the gastroesophageal sphincter and inhibit motility of the stomach. Repetitive stimulation of purinergic nerves to gut wall elicits facilitating ipsps that appear to be a result of increased g_K; frequently postinhibitory rebound follows. The purinergic neurons in myenteric plexus are activated by cholinergic neurons. Two classes of purinergic receptors are recognized: P_1 receptors are sensitive to adenosine and are blocked by methylxanthine, P_2 receptors are more sensitive to ATP and are blocked by quinidine imidazolines (43). Adenosine derivatives are excitatory to many organisms. Shrimp are attracted to low concentrations of AMP. External chemoreceptors of crustaceans are very sensitive to AMP, ATP, also to GABA, Glu, and taurine (69).

Neuropeptides

All of the preceding neurotransmitters—modulators are synthesized by one to three enzyme reactions from relatively simple compounds, mostly amino acids that preceded the evolution of nervous systems. A second class of neuroagents, the neuropeptides, evolved independently and in parallel with the amino acid-derived agents. Neuropeptides are chains of 4–45 amino acids. The same polypeptides may be synthesized by endocrines and by neural cells and a single neurosecretory cell can produce several compounds of different composition and function. The structural and functional distinctions between transmitters, modulators, and hormones in the classical sense cannot be maintained. There is disagreement as to whether long neuropeptides that are well known in vertebrates are derived phylogenetically from shorter ones that are described in invertebrate animals. There is also disagreement as to whether similarities in amino acid sequences indicate similar functions and whether they are homologies. Secretion into extracellular fluid may have preceded liberation at nerve terminals. Biopeptides function differently according to their target cells; some act on excitable membranes, some intracellularly on metabolic pathways; their action is frequently via second messengers such as cAMP. The principal function of a given neuropeptide may be different in different animals. Some neuropeptides may be of relatively

TABLE 3. Abbreviations and Symbols of Amino Acids

Amino Acid	Abbreviation	One Letter Symbol
Part A		
Alanine	Ala	A
Arginine	Arg	R
Asparagine	Asn	N
Aspartic acid	Asp	D
Cysteine	Cys	C
Glutamine	Gln	Q
Glutamic acid	Glu	E
Glycine	Gly	G
Histidine	His	H
Isoleucine	Ile	I
Leucine	Leu	L
Lysine	Lys	K
Methionine	Met	M
Phenylalanine	Phe	F
Proline	Pro	P
Serine	Ser	S
Threonine	Thr	T
Tryptophan	Trp	W
Tyrosine	Tyr	Y
Valine	Val	V

Part B

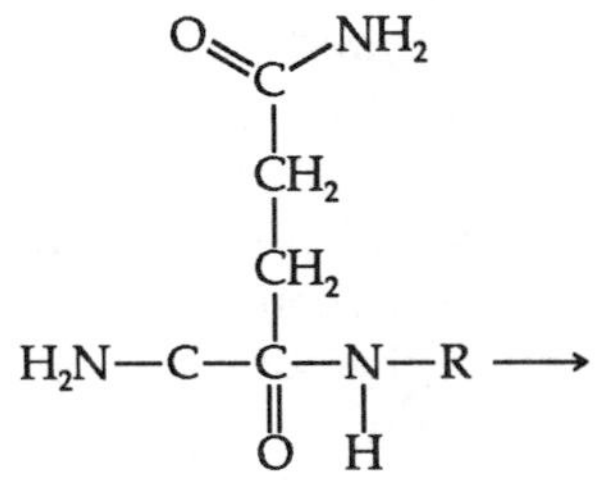

glutamine (Q)

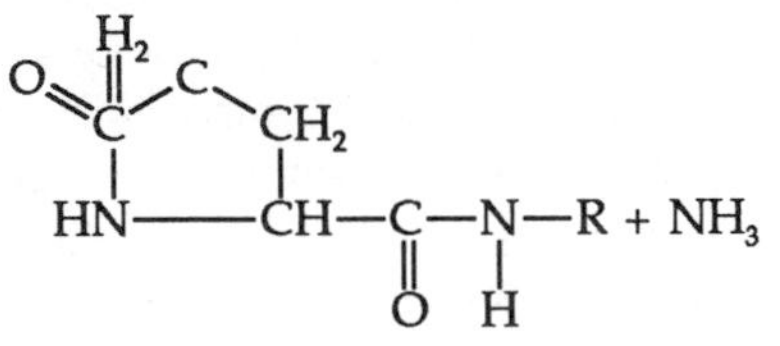

pyroglutamic acid (pE)

recent origin, some may have originally served other functions from present ones. Phylogeny based on similar amino acid sequences is valid only for a few proteins (44).

To facilitate comparison of sequences of different families of peptides, the single letter terminology for amino acids is used (Table 3A). This is particularly useful for comparing C (minus) and N (positive) terminal sequences. Glutamine (Gln;Q) presents a special problem when at the N-terminal. This amino acid, is encoded as glutamine (Q) but after processing it is cyclyzed to pyroglutamic acid (E), which is abbreviated as either <E or pE (Table 3b).

In all eukaryotes, proteins or polypeptides that are to be stored in membrane bound vesicles or are to be secreted by a cell are formed as parts of precursors that are then cleaved by proteolytic enzymes. In contrast, proteins that are retained in cells (e.g., metabolic enzymes) are not synthesized as long precursors. The precursor peptides have an N-terminal (positive) amino acid sequence (signal sequence) that is hydrophobic and enables the molecule to enter and pass through a lipid membrane. Most secreted peptides are formed with pro- and pre-segments. One precursor may yield several products that may contain multiple copies of a given polypeptide, also sequences for two or several neuropeptides as a result of differential RNA splicing. The messenger encoding the precursor for a given polypeptide may be different in different cell types—for example, brain and adrenal.

Some colocalization of the two classes of neuroagents occurs. For example, in some neurons of *Helix*, ACh coexists with a short cardiac peptide (SCP) and other neurons contain both 5-HT and SCP (170). Serotonin and Met-enkephalin occur together in cells of a gastropod ganglion (218). FMRFamide and 5-HT coexist in neurons of Aplysia (271). Both Met-Enk and CCK occur together in oxytocin-liberating terminals of pituitary (182).

The occurrence and action of bioactive

polypeptides have been investigated by several techniques (12):

1. Purification from neural tissues and comparison of the native with synthetic peptides give the most definitive identification.

2. Immunocytochemical localization by staining with antibodies against polypeptides from both vertebrate and invertebrate animals is commonly taken as evidence for the presence of a peptide but conclusions are often unreliable because of nonspecific cross-reactivity.

3. Measurement of the excitatory or inhibitory effects of polypeptides when applied to muscle, neurons, and secretory cells is descriptive of function but not of chemical structure.

4. For a few polypeptides, isolation of the gene encoding a precursor has provided convincing identification but the peptide itself must be isolated as proof that it is expressed.

5. Testing the action of synthetic peptides in which one or a few amino acids are substituted is useful, that is, structure–activity relations help define specificity.

6. Isolation of receptor molecules indicates the functions of reactive sequences; unfortunately receptors for neuropeptides are virtually unknown except for substance K (SK) (183).

No one method alone can provide proof that a given polypeptide is a transmitter or modulator. Chemical isolation combined with tests of effects are most definitive. Unfortunately, conclusions about the chemical identity of a peptide have been drawn from antibody staining and from pharmacological effects of applied compounds that are not justified.

Related neuropeptides occur in families of similar structure and sequence. The following account classifies the major families; several peptides that are incompletely identified are omitted. In general, the single letter terminology for the amino acids is used (Table 3). The negative end of a chain (C end) is often amidated, hence written as NH_2; the positive end is the N-terminal and when this is glutamine it is cyclized. The convention is to write the N terminal at the left and the C-terminal at the right.

FMRFamide Family

The first of a family of short neuropeptides to be discovered was the tetrapeptide Phe-Met-Arg-Phe-NH_2 (FMRFamide) (227). This peptide as well as peptides with similar sequences occur in molluscs and other phyla (including chordates) (Fig. 22). In addition, all molluscs contain a related but less abundant tetrapeptide FLRFamide in which Leu is substituted for Met (Fig. 22). Longer analogs have been found in the nervous systems of some pulmonate molluscs. A similar hepatapeptide QDPFLRFamide has been found in the snail Helix and related pulmonates (92).

Recordings from identified neurons in Helix show that FMRFamide increases K^+ conductance (g_K) when the peptide is applied to a ganglion; QDPFLRFamide causes a slow decrease in g_K. Comparison of the actions of different FMRFamide-related peptides indicate several types of receptors; the C-terminal sequence (with Met or Leu as the antipenultimate residue) is essential for the action of all of them. FMRFamide potentiates twitch responses of radular protractor but not retractor, whereas FLRFamide has the opposite of potentiating retractor but not protractor. The receptors of the two muscles must be different (293). FMRFamide is excitatory to the hearts of some clams (Marcrocallista) and inhibitory to the hearts of others (*Lampsilis*) (101). FMRFamide at a concentration of $10^{-9} M$ contracts the radular muscle in *Busycon* and the retractor muscles of pulmonates. QDPFLRFamide relaxes the *Busycon* muscle and acts hormonally on the heart, whereas FMRFamide is neural in its action (228). FMRFamide is antagonistic to 5-HT in the opening of S channels (one of the K^+ channels mentioned previously) in certain *Aplysia* neurons; this action is presynaptic, hence modulatory (132). The action potential of the affected sensory neuron is shortened and synaptic transmission to motor neurons is decreased.

Cnidarians have both electrical and chemical synapses in their nerve nets. An antibody for the two amino acid peptide (RFamide) stains neurons in Hydra, and also in several hydrozoans, anthozoans, syphozoans, and siphonophores (107, 110) (Fig. 23). However, the staining has broad specificity and is indicative of a tetrapeptide. pEGRFamide, extracted from anthozoans, is excitatory for ectodermal conduction in *Calliactis* (186). Cnidarian neurons show immunoreactivity to several peptides with RFcarboxy terminals (260). Another neuropeptide extracted from *Hydra* stimulates differentiation; bEGRFamide, it induces interstitial cells to become nerve cells (245).

A cDNA has been cloned for FMRFamide of *Aplysia*. A single gene encodes a precursor of 597 residues which includes 28 copies of FMRFamide and a single copy of FLRFamide (Fig. 24) (275).

In segmental ganglia of a leech, cardiac motor and regulator neurons, also neurons of a central pattern generator act by liberating a FMRFamide-like transmitter. One type of cardiac motor neuron releases both ACh and FMRFamide (150, 163).

A heptapeptide pQDFLRFamide occurs in *Helix* ganglia and YGGFMRFamide in *Octopus* nervous system. These are

FMRFamides

Mollusc prototypes	F-M-R-F-NH$_2$
	F-L-R-F-NH$_2$
Cnidarians	pE-G-R-F-NH$_2$
	pE-S-L-R-W-NH$_2$
	pE-L-L-G-G-R-F-NH$_2$
Pulmonate	S-D-P-F-L-R-F-NH$_2$
Cephalopod	A-F-M-R-F-NH$_2$
Homarus	S-D-R-N-F-L-R-F-NH$_2$
Homarus	T-N-R-N-F-L-R-F-NH$_2$
Drosophila	D-P-L-Q-D-F-M-R-F-NH$_2$
Ascaris	K-N-E-F-I-R-F-NH$_2$
Chick brain	L-P-L-R-F-NH$_2$

Figure 22. Selected members of FMRFamide family of transmitters.

equally excitatory with FMRFamide on the *Octopus* heart (181).

Two related molluscan peptides are cardioactive peptides A and B, designated SCP_A and SCP_B. The peptide SCP_A has 11 amino acids and SCP_B has 10 (Fig. 25) (200). Both are processed from the same precursor and both can occur in the same neuron of *Aplysia*. The cloned gene has 1394 nucleotides with 408 base pairs in the open reading frame, which encodes a precursor of 136 amino acids (179). The SCP peptides are released on stimulation of Aplysia neurons in culture (169). A peptide SCP_B extracted from nervous system and gut is cardioexcitatory (200).

In *Aplysia*, a cluster of 400 neurons, the bag cells, make a (5-member) neuropeptide the egg-laying hormone (ELH) that induces complex behavior leading to egg laying (246). The ELH contains 36 amino acids and has a MW 4400 D (Fig. 26). The gene produces a single copy (244). Another group of exocrine cells in the Aplysia atrial gland synthesizes two peptides (A and B) of 35 amino acids each. Since the secretions of the atrial gland cells enter the lumen of the oviduct they probably have no direct neural function

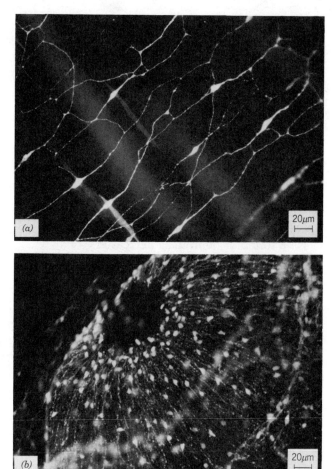

Figure 23. Immunohistochemical staining of cnidarian nerve nets with fluorescent antibody to RFamide. (*a*) Subumbrellar nerve net of medusa *Phialidium*, whole mount. (*b*) Mouth and hypostome of *Hydra*, immunoreactive neurosensory, and ganglion cells. (Figures provided by A. N. Spencer.)

(132). Bag cells are electrically coupled; stimulation of their axons causes liberation of ELH into the interstitial space of the abdominal ganglion where it is carried by the circulation to the target organs (267, 268) (Fig. 26). The gene produces a single copy (244).

In cockroaches, a neuropeptide from the ventral nerve cord stimulates contractions of hindgut muscles; this is proctolin, MW 648 D, composition RYLPT (30, 39, 264). Proctolin at 10^{-9} M is a modulator of the neurogenic heart of Limulus (287). In crustaceans (*Homarus*) proctolin coexists in the same neurons with 5-HT and modulates the nine cardiac pacemaker neurons (256). Proctolin and two

or more N-terminally extended analogs of FLRFamide are synaptic modulators in *Homarus* both centrally and peripherally (22, 219). A proctolin-like peptide has been extracted from annelids (226).

There is interaction between several transmitters and modulators in a skeletal muscle, the extensor tibiae, of a locust. One bundle of muscle fibers is rhythmic. The bulk of this muscle receives fast and slow excitor axons, which have glutamate as transmitter, one inhibitory axon probably with GABA as modulator; proctolin is present together with glutamate in the slow excitor. In addition, FMRFamide-like peptides, present in the hemolymph, modulate contractions. Octopamine de-

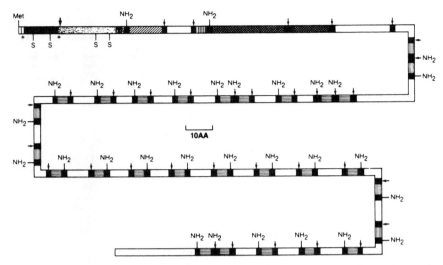

Figure 24. Schematic representation of the FMRFamide precursor protein. The predicted precursor was deduced from the combined sequence information of the FMRFamide genomic and cDNA clones. A solid region flanked by positively charged amino acids (+) indicates a region of hydrophobic residues that may serve as the signal sequence. A bold arrow following this region denotes the putative cleavage position of the signal sequence. The 28 FMRF tetrapeptides are denoted by parallel horizontal lines. The single FLRF and the carboxyl terminal LRF residues of another potential cleavage product are represented by parallel vertical lines. Predicted cleavages at single basic or dibasic residues are labeled with vertical lines or arrows, respectively. NH$_2$ above cleavage sites indicates potential amidtion sites. The cross-hatched, the slashed lines, and the stippled regions denote the CRF, α-MSH and corticotropin-like intermediate lobe peptide (CLIP) homologous sequences, respectively. [From (275).]

creases rhythmic contractions and increases tension of triggered contractions; it acts via an increase in cAMP in the muscle fibers. The modulatory neuron becomes active prior to any locomotor movement of the muscle (85). Several long neuropeptide hormones are secreted by the sinus gland of crustaceans (Chapter 10).

Peptides that react with antibodies to

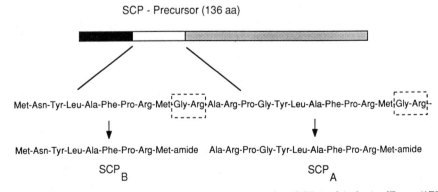

Figure 25. Schema of processing of small cardiac peptides (SCPs) of Aplysia. [From (179).]

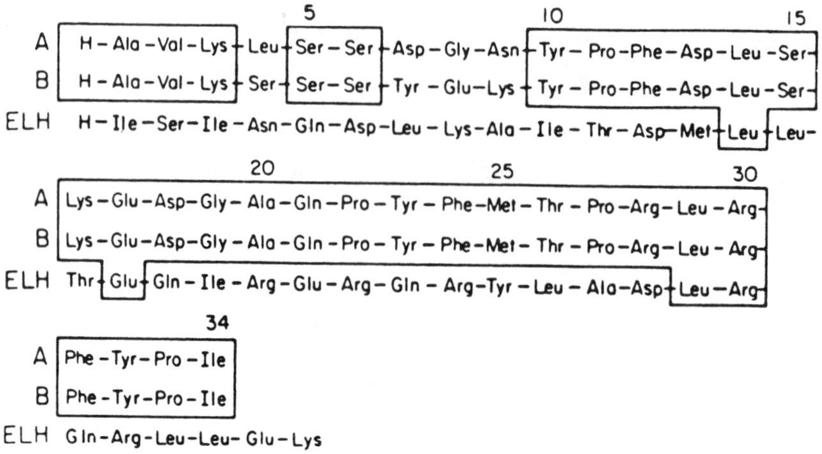

Figure 26. Comparison of amino acid sequences of ELH of bag cells and peptides A and B from atrial gland of *Aplysia*. Boxes identify homologous amino acid sequences. [From (267).]

FMRFamide have been found in vertebrate nervous systems (72, 215). It is doubtful that phylogenetic homology is indicated by the occurrence of similar amides in vertebrates and molluscs (100, 102).

Gastrin-CCK Family

These are peptides that were first discovered in mammalian small intestine, and later in the brain. Cholecystokinin (CCK) occurs in secretory mucosa and myenteric plexus neurons of the intestine. Five forms of CCK have been identified (233). An eight amino acid form constitutes 95% of the CCK, the most abundant neuropeptide in the brain (Fig. 27). Cholecystokinin stimulates secretion of pancreatic enzymes and contracts the gallbladder. In frog brain, CCK occurs in bands in the optic tectum. Gastrin resembles CCK in occurring in two forms, one with 34 and the other with 17 amino acids. The processing of preprocholecystokinin is shown in Fig. 28 (233). Peptides of activity indicative of the gastrin-CCK family have been isolated from intestinal muscle of dogfish *Squalus acanthias* (7). Leucosulfa-

kinin (LSK) of insects is a sulfated neuropeptide with similarities to CCK and gastrin. Leucosulfakinin was isolated from cockroach head; the hindgut of a cockroach responds to LSK at 10^{-10} M. It is doubtful that there is true homology between the LSKs and CCKs.

Tachykinins

Tachykinins are peptides of 10 or 11 amino acids with a common C-terminal sequence as in Fig. 29 (178, 116).

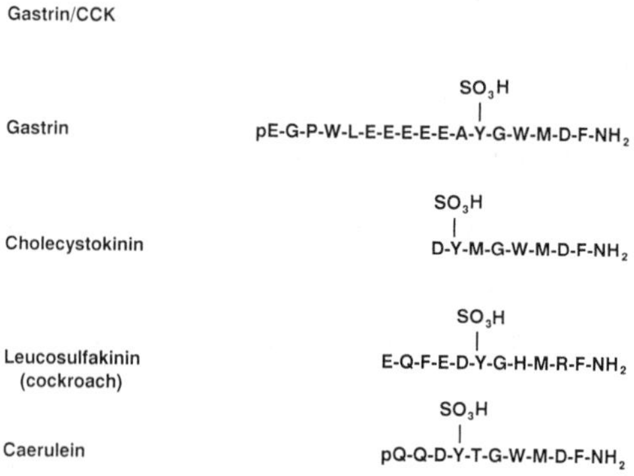

Gastrin/CCK

Gastrin

$$SO_3H$$
$$|$$
pE-G-P-W-L-E-E-E-E-A-Y-G-W-M-D-F-NH$_2$

Cholecystokinin

$$SO_3H$$
$$|$$
D-Y-M-G-W-M-D-F-NH$_2$

Leucosulfakinin (cockroach)

$$SO_3H$$
$$|$$
E-Q-F-E-D-Y-G-H-M-R-F-NH$_2$

Caerulein

$$SO_3H$$
$$|$$
pQ-Q-D-Y-T-G-W-M-D-F-NH$_2$

Figure 27. Sequences of peptides in the Gastrin-CCK family of transmitters.

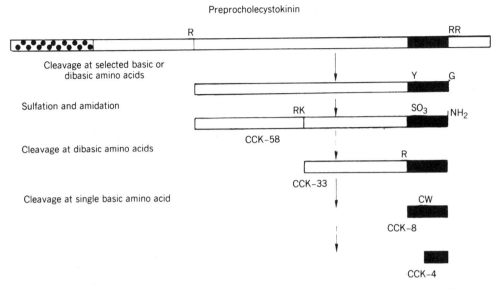

Figure 28. Processing of preprocholecystokinin, which contains the sequences of all forms of cholecystokinin. The common carboxyl terminal of these peptides is flanked in the precursor by a glycine (G) and two basic amino acids so that the amidated carboxyl terminus (NH_2) can arise from the glycine residue. This region also contains the tyrosine that is sulfated in CCK. The amino acid terminus of the various forms is produced by cleavage at a single basic amino acids for CCK-33, and between a glycine (G) and a tryptophan (W) CCK-4. [From (232).]

Substance P (SP) is an undecapeptide that has been isolated from chromaffin cells of adrenal gland, neurons of myenteric plexus, amacrine cells of retina, dorsal root ganglia, and several brain regions. Substance P elicits intestinal contractions at concentrations of $10^{-10}\,M$ and higher; it may elicit noncholinergic slow epsps in myenteric ganglion cells and the spinal cord (220). Substance P is present in rat dorsal roots at 24–130 pM/g; it depolarizes sensory interneurons in the toad spinal cord 100 times more effectively than glutamate. Substance P is synthesized in cultured chick sensory neurons and in mammals, pain sensations are mediated by SP. Substance P and another peptide, substance K, are encoded by the same gene and are formed by differential splicing of mRNA (142)

(Fig. 30). Substance P and glutamate coexist in afferent terminals of rat spinal cord. Bombesin from toadskin has resemblance to tachykinins but is not in the same family.

A peptide with sequences resembling the mammalian tachykinins occurs in the salivary glands of an octopod Eledone; the peptide is called eledoisin.

Tachykinins

Tachykinins	
Tachykinins	(-F-X-G-L-M-NH$_2$)
Substance P (SP)	-R-P-K-P-E-E-F-F-G-L-M-NH$_2$
Substance K (SK)	-H-K-T-D-S-F-V-G-L-M-NH$_2$
Eleoisin (cephalopod)	pQ-P-S-K-D-A-F-I-G-L-M-NH$_2$

Figure 29. Sequences of peptides in the tachykinin family of transmitters.

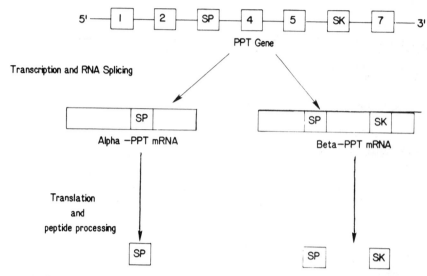

Figure 30. Processing of the preprotachykinin gene. One preprotachykinin gene can produce two mRNAs by differential RNA splicing. Following translation and peptide processing, α-prepro-tachykinin mRNA produces substance P while β-preprotachy-kinin produces substance P and substance K. [From (142).]

The receptor for substance K is the one receptor for a neuropeptide for which the cDNA has been cloned. This has similarity to the rhodopsin-type receptor (G-protein coupled). The gene consists of 2458 nucleotides and the precursor of 384 amino acids; this molecule has seven hydrophobic segments that span the cell membrane (183). Substance K is similar to rhodopsin in several sequences.

Bioopiates

Several polypeptides found in the brain have an opiate action. They may have evolved as analgesics and as stress antagonists. At least 18 opioid peptides are known (172). Some of the same substances have been identified in invertebrates. All bioopiates have homologies at the N-terminus (Fig. 31).

Endorphins are large molecules. β-Endorphin has 31 and α-endorphin has 16 amino acids. Both endorphins depress activity in the cerebral cortex and thalamus; endorphin action is antagonized by naloxone. Endorphins diminish responses to SP, that is, are analgesic; they are hypoglycemic; in low doses in rabbits they may cause hyperthermia.

Bio-opiates

α-endorphin   Y-G-G-F-L-T-S-E-K-S-Q-T-P-L-V-T

ß-endorphin   Y-G-G-F-L-T-S-E-K-S-Q-T-P-L-V-T-L-F-K-N-A-I-I-K-N-A-Y-K-K-G-E

dynorphin A   Y-G-G-F-L-R-R-I-R-P-K-L-K-W-D-N-Q

met-enkephalin   Y-G-G-F-M

leu-enkephalin   Y-G-G-F-L

Figure 31. Sequences of peptides in the bio-opiates family of transmitters.

Enkephalins are smaller bioopiate molecules than endorphins; leu-enk has four amino acids, Met-enk five in the same sequence as at the NH_2 end of β-endorphin. However, the endorphins are derived from a different precursor than the enkephalins; they are encoded by different genes. Enkephalins attenuate GABAergic inhibition in hippocampus, diminish glutamate-induced cortical activity, and reduce transmitter liberation at frog neuromuscular junctions. Enkephalins are present in myenteric neurons and at 10^{-8} M relax intestine. Enkephalins are primarily presynaptic modulators, not transmitters.

In the rat brain, bioopiates inhibit release of NE, DA, and SP, that is, they are modulators. The analgesic action of bioopiates is indicated by a rise in threshold for avoidance of foot heating in mice.

Biochemical evidence indicates that enkephalin-like neuropeptides occur in the ganglia of molluscs (pedal ganglion of Mytilus) (160). An analgesic effect of opiates is indicated for a land snail by the lengthening of latency for elevation of the foot when on a heated plate (133).

Dynorphin is an opioid tridecapeptide with the NH_2 terminus resembling Leu-enkephalin. Dynorphin is found in the pituitary, hypothalamus, and spinal cord. Dynorphin is much more potent than Leu-enk on the neurons of submucous plexus (174).

Three copies of Leu-enkephalin are formed from a precursor molecule. Bovine pre-Proenkephalin A consists of 263 amino acids and includes four copies of Met-enk, one of Leu-enk (210). Peptides derived from prodynorphin precursor may modulate pain perception, intestinal peristalsis, feeding, sleep, and motor activity. The gene for prodynorphin consists of four exons separated by introns. The prodynorphin precursor includes three copies of Leu-enkephalin, within

the sequence of neoendorphin and dynorphins A and B (55).

Several receptor types have been identified according to their affinities for opiates such as morphine and for the opioid peptides especially the enkephalins (288). Naloxone block is selective for high-affinity sites that have binding constants of <1 nM for both morphine and Leu-enkephalin; these are called μ_1 sites. A low affinity site, K_d of 10 μm for morphine is a μ_2 site. A different low affinity site, the δ site is selective for enkephalin (292); guinea pig ileum has mainly μ morphine receptors, vas deferens δ receptors (172).

A single precursor that contains segments for several bioactive polypeptides (including β-endophin) is POMC, proopiomelanocortin (Fig. 32), synthesized in the cells of hypothalamus and medulla. The cDNA for the mRNA for this large precursor protein has been cloned from the pituitary and has 1091 base pairs (121, 146, 204). The bovine POMC precursor consists of 265 amino acids; following the NH_2-terminal signal peptide is the sequence for γ-MSH (γ-melancyte stimulating hormone). Next is the ACTH region of 39 amino acid sequences; this includes α-MSH (13 amino acids) at positions 82–99 and CLIP (corticotropin-like intermediate lobe peptide) of 21 amino acids. The next long region of the precursor is γ-LTP (β-lipotropin) of 91 amino acid (aa) sequences 42–132 that contains sequences for γLPT (58 aas, sequences 42–101) and β-endorphin (30 aa, sequences 104–134). Proopiomelanocortin occurs in the anterior pituitary, the intermediate lobe of the pituitary, the hypothalamus and reproductive tract, and in each organ the precursor may be processed differently (146).

Both Met- and Leu-enkephalin, but not endorphin, are made in chromaffin cells of the adrenal medulla. The cDNA consists of 1120 base pairs (bp) with a coding window of 783 bps, coding for a precursor of 263 amino acids, MW 29,786 D. The

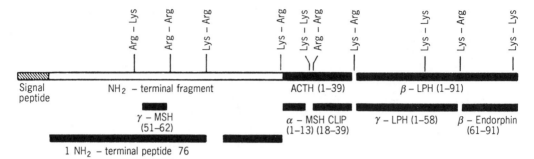

Figure 32. Structure of the ACTH/LPH precursor indicating formation of two neuro-peptides, MSH and β-endorphin. [From (238a).]

precursor contains four copies of Met-enkephalin, one copy each of Leu-enkephalin and of the analogs Met-enk-Arg-Gly-Leu and Met-enk-Arg-Phe.

Immunoreactive POMC-related peptides have been found in the earthworm, two molluscs, insects, a tunicate, and Tetrahymena but it has not been proved that the neuropeptides functions in all of these animals (103).

Hypothalamic-Anterior Pituitary Factors

Several polypeptides, that are produced in the hypothalamus, function as endocrine stimulators or as hormone releasing factors in the anterior pituitary. Corticotropin (ACTH) occurs in two forms, of 99 and of 39 amino acids; it stimulates the adrenal cortex to secretory activity. Melanocyte stimulating hormone (MSH) occurs in two forms, one of which is identical with the first 13 amino acids of ACTH. The MSH is stored in the intermediate lobe of the pituitary and as a circulating hormone, it stimulates expansion of melanophores. An assay for MSH is the darkening of bony fishes and frogs when on a black background.

Thyrotropic releasing hormone (TRH) is the shortest vertebrate peptide pGHP NH$_2$. Thyrotropin varies in length from 89 amino acids (sheep) to 209 (human).

Several releasing factors for anterior pi-

tuitary hormones have recently been suggested as neurotransmitters or neuromodulators. One is luteinizing hormone releasing hormone (LHRH), a multifunctional decapeptide. The LHRH or a LHRH-like peptide is an excitatory transmitter in sympathetic ganglia of amphibians. It may depolarize postsynaptic neurons by inactivating a slow outward K current, the M current. Probably ACh and LHRH occur in the same small preganglionic axons in frog.

Other releasing factors made in hypothalamus and acting on the anterior pituitary are thyroxin-releasing hormone (TRH) tripeptide, and the corticotropin-releasing hormone (CRH) of 41 amino acids.

Vasoactive intestinal peptide (VIP) is an octosapeptide with N-terminal histidine and C-terminal threonine in chick, and asparagine in pig, man, and rat. Vasoactive intestinal peptide is found in neurons of myenteric plexus and of ventromedial hypothalamus. It is a vasodilator and relaxant of intestine and occurs in the intestine of mammals, birds, and teleost, and elasmobranch fishes.

Neurotensin (NT) is found in hypothalamus and gastrointestinal gland cells, but not in plexus neurons. It contains 13 amino acids. Neurotensin decreases esophageal sphincter contractions and inhibits gastric secretion (158).

Somatostatin (SO) is a quadradecapeptide, formed in the hypothalamus. It regulates release of growth hormone from the pituitary. Somatostatin occurs in neurons of the cerebral cortex and in neurons of the myenteric plexus of the intestine; it inhibits secretion of glucagon and insulin from the pancreas and gastrin from the stomach glands.

Neuropeptide Y (NPY) is a 36 amino acid peptide that occurs in adrenal medulla, locus coeruleus, basal ganglia, amygdala, and hippocampus. The two isoforms that occur in brain differ in nine positions. In sympathetic ganglia NPY is colocalized with NE. Neuropeptide Y also occurs in parasympathetic ganglia (156).

Examples of the coexistence of two amino-derived transmitters in the same cell were mentioned previously. In several types of neuron, two neuropeptides coexist. Octopus neurons store a FMRFamide-like and an enkephalin-like protein, also Met-enk and CCK (181). Examples of coexistence of NE with SO and 5-HT with SP, Ach with VIP, and enk are known (173). In frog sympathetic ganglia, acetylcholine is transmitter for a fast epsp, a peptide releaser LHRH for a slow epsp (127). Nerve growth factors are described in Chapter 9, and also in ref. 249.

Miscellaneous Peptides

A multifunctional polypeptide of 198 aa residues, made in and released from anterior pituitary is prolactin, which functions in regulation of the water balance in fishes and has a role in the metamorphosis of amphibians. In mammals it stimulates corpus luteum to produce progesterone and mammary glands to secrete milk.

The posterior pituitary of all vertebrates elaborates several nonapeptides in which the amino acids form a ring (Chapter 10).

The skin of toads contains a variety of peptides (more than 10), some of which are very similar in sequences to neuropeptides. Some are toxins but the functions and evolutionary significance of the homologous peptides is unknown (82a). Toad skin contains tachykinins, angiotension, and bradykinin; also skin contains many indoles and alkylamines. Coerulein is composed of 10 amino acids; the C-terminal is the same as CCK-8. Bombesin has 14 amino acids; it occurs in brain and myenteric neurons of the intestine and is an active neuroagent in arthropods, for example, on Limulus heart.

Summary of Intercellular Transmission

Intercellular communication became essential with the evolution of integrated multicellular organisms. In some epithelia, in early embryos and in some neural nets, electrical coupling provides for intercellular transmission. Small molecules can pass through the channels of low electrical resistance. Most electrical coupling is nonpolarized but some connections are polarized and rectifying. Signals are mostly limited to depolarizations and are stereotyped.

Chemical signals may be neurotransmitters, neuromodulators, hormones, or pheromones. There are two general classes of chemical signaling agents—amines that are synthesized directly from amino acids and polypeptides that are made under the control of genic codes. We postulate that the most primitive neurotransmitters were amino acids. Some amino acids are attractants and other repellents to bacteria. Glutamate (less notably aspartate) persists as an excitor, glycine and γ-aminobutyrate (GABA) persist as inhibitors. Relatively simple methylating and decarboxylating reactions convert amino acids to amines, several of which are excitatory in very low concentrations—taurine, betaine, and homarine. The most general of methylated amines as a transmitter is ACh, which can be either excitatory or inhibitory according to the receptor. Enzymes for synthesis and degradation of ACh are widespread and ACh may have a general function of regulating membrane permeability. As a neurotransmitter, it is found in all phyla where it has been sought except cnidarians. Very high concentrations occur in the ganglia of honeybees and of squid. Esterases that hydrolyze ACh function in both neural and nonneural tissues. Several ACh receptors trigger the opening of ion channels. Nicotinic ACh receptors have been purified from electric organs, from brain, and from muscle junctions. Muscarinic receptors occur in autonomic systems and in brain; they are of two types pharmacologically and the ACh receptors of some invertebrates, specifically mollusc hearts, differ from either vertebrate type.

Several monoamines are formed from amino acids and are transmitters: catecholamine from tyrosine and serotonin from tryptophan. Dopamine and precursors of norepinephrine function as transmitters in many animals, especially molluscs, annelids, and vertebrates. Octopamine is an important agent in crustaceans. Serotonin (5-hydroxytryptamine, 5-HT) has been identified in annelids, molluscs, and vertebrates.

A few synapses use adenine or ATP and are called purinergic.

All transmitters listed in the preceding paragraphs are formed by relatively simple reactions. A different category of transmitters–modulators comprised of peptides usually consisting of 4–40 amino acids. All of these are encoded by one or more genes and are synthesized by transcription–translation cascades. Messenger RNAs for neuropeptides code for precursor molecules that are longer and more complex than the secreted substances. A tetrapeptide (Phe-Met-Arg-Phe-NH$_2$ or FMRFamide) is the prototype for short-chain peptides in cnidarians, molluscs, annelids, and arthropods. The peptide family includes besides FMRFam-ide, two short cardiac peptides (11 aa) with a precursor of 136 amino acids in Aplysia.

Several categories of neuropeptides are recognized in vertebrates; some have C-terminal sequences similar to those in some invertebrates but phylogenetic homology is doubtful. For example, cholecystokinin (CCK) is present in the gut and brain of vertebrates. A compound with similar sequences occurs in insects. Tachykinins include vertebrate substance-P and substance-K; a cephalopod compound has similar sequences. Bioopiates include endorphin, enkephalins, and dynorphin. They have a variety of modulatory effects in central nervous and autonomic effectors in both vertebrate and invertebrate animals. At least four receptors are known in mammals. Bioopiates, like other neuropeptides, are synthesized from long precursors and are cleaved before secretion. Evidence from isolation and behavioral action suggests that they occur in both invertebrates and vertebrates. A large class of vertebrate neuropeptides, that are made in hypothalamic neurons and are stored in anterior pituitary, are the hormonal releasing compounds.

Whether the presence in long neuropeptides, especially those of vertebrates, of sequences that occur in short neuropeptides, mainly in invertebrates, indicates phylogenetic homology or stochastic convergence is debatable. Sequences similar to that of FMRFamide occur in several long neuropeptides. Statistical calculations show that the probability of appearance by change of any 4-amino acid sequence in a precursor is high (228). Therefore, the frequent occurrence of similar C-terminal sequences probably indicates functional significance rather than phylogeny. The most common C-terminal sequences are hydrophobic amino acids that favor entry into the lipid membrane. Functional meaning may become apparent when receptor molecules are identified and when the tertiary structure of neuropeptides is understood; the short neuropeptides cannot show much folding. A conclusion is that short neuropeptides may be related by structural and functional similarities rather than by homology.

An unanswered question is Why are there so many different polypeptides? There is much redundancy, for example, the numerous excitatory and inhibitory compounds that act on specific muscles. An example of what appears to be the stochastic origin of a wide variety of known polypeptides is the diversity found in toad skin. Biochemical evolution contains many examples of gene duplication and rearrangement. Small differences in sequences, especially in central regions of long polypeptides may be established by genetic drift rather than by adaptive selection.

It is concluded that the two classes of transmitters–modulators evolved in parallel: those formed by direct reactions from amino acids or purines and those synthesized as are all secreted protein under genic control. All transmitters–modulators match specific receptors which probably coevolved with the transmitters.

Channel Modulators and Second Messengers

The previously discussed ion conductance changes in electrical events in cell membranes are diverse in their threshold, amplitude, electrical sign, and latency. Direct depolarizations or hyperpolarizations provide electrical sinks for graded or self-propagating impulses, nerve, or muscle action potentials. These are usually specific as to ions—Na^+ or Ca^{2+} inward and K^+ outward (voltage or Ca^{2+} activated). Numerous transmitters act on more than one receptor (209).

Neurotransmitters and hormones cause conductance changes in receptor membranes, often nonspecific as to ions. The permeability series for frog motor endplate channels is $Ca^{2+} > Rb^+ > K^+ > Na^+ > Li^+$; reversal potential is near 0 mV. The endplate channel is negligibly permeable to anions. It is calculated to be a water-filled pore 6.5 Å × 6.5 Å in size (1).

Many conductance responses act via receptor proteins to change the concentrations of second messengers in receptor cells (Fig. 33). These second messengers lead to effects such as secretion, protein synthesis, or cell contraction. Second messengers within a receptor cell may act at a distance to modulate presynaptic endings or other cells. Second messengers may also come from nonreceptor cells that modulate synaptic transmission (26).

Second messengers occur in a limited number of categories. Previous mention was made of Ca^{2+} as a second messenger. In striated muscle of vertebrates Ca^{2+} from SR-tubules combines with a coupling protein troponin-c, which sets off a series of events leading to myosin–actin interaction. In many cells, Ca^{2+} combines with calmodulin and the Ca^{2+}–calmodulin complex regulates many enzymes, mitotic movement, tubulin polymerization, and secretory activities. Calmodulin is a protein of MW 17,000, containing 148 aa residues, with four Ca^{2+}-binding sites, and with marked homology to troponin-c (189). Calmodulin occurs in all eukaryotes, and is highly conserved. There are only seven amino acid residue differences between calmodulin in the colonial coelenterate *Renilla* and the tissues of a rat (190).

A second category of second messenger is cyclic nucleotides. An activated receptor molecule reacts with a G-protein in the membrane; this activates a cyclase enzyme that converts ATP to cAMP (94). Specific G-proteins react with different re-

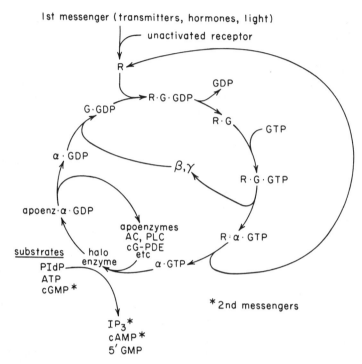

Ist messenger (transmitters, hormones, light)

Figure 33. Generalized schema of participation by G-protein and cyclic nucleotides in activation of second messengers. First messengers can be transmitters, hormones, light that activate specific receptors (R). Activated receptors may interact numerous times with G-GDP molecules catalytically to give amplification of action. Activation of enzymes gives second steps of amplification. R—activated receptor; G—generalized G-protein; GDP—guanosine diphosphate; GTP—guanosine triphosphate; α, β, and γ are subunits of G-protein; AC—adenylatecyclase; PLC—phospholipase C; cG-PDE—GMP—specific phosphodiesterase; PIdP—phosphoinositol diphosphate; IP$_3$—inositol triphosphate; ATP—adenosine triphosphate; cGMP—cyclic guanosine monophosphate. (Lloyd Barr, personal communication.)

actors; for example, a G-protein in olfactory cells signals odor transduction (130). Cyclic AMP activates a kinase that phosphorylates specific proteins at a serine or threonine. Alternatively, the G-protein activates an intermediate nucleotide cGMP, formed from guanosine triphosphate; cGMP activates a specific kinase. Cyclic AMP and cGMP are produced by adenylate or guanylate cyclases in cell membranes. Figure 33 summarizes the GTP cascade following activation of a G-protein. This generalized figure shows how GDP is regenerated in a cycle in which second messengers are formed.

Second messengers of another category come from lipids of cell membranes; these are phorbol esters. Phophatidyl-4,5-bisphosphate gives rise to (1) inositol triphosphate (IP$_3$), which with Ca^{2+} forms a protein kinase (PK) or (2) diacylglycerol (DGC), which activates C-kinase (26, 68). Each of these messengers is a modulator of transmitter or hormone action. Their action may be either obligatory or permissive. In secretory cells, IP$_3$ binds to a receptor protein on the microsomes of the endoplasmic reticulum to mobilize Ca^{2+} ions and DGC activates protein kinase C (PK-C). Inositol triphosphate in sea urchin eggs mobilizes bound Ca^{2+} and IP_4 increases the influx of Ca^{2+} across the membrane (128). The aggregation of dissociated sponge cells is enhanced by phorbol esters that activate PK-C and liberate Ca^{2+} (291). Ca^{2+}–Calmodulin, cyclic nucleotides, and phorbol esters evolved as second messengers early in eukaryote evolution. Lithium reduces responses of smooth muscle to transmitters that act via phorbol esters (191).

In mammalian heart, ACh from vagal endings binds to G-receptor proteins that activate cGTP, which opens K^+ channels. Norepinephrine acts at β receptors, which increase the cAMP that opens L-type Ca^{2+} channels. Thus ACh acts by cGMP to slow the heart by K^+ channels and NE acts via cAMP to open Ca^{2+} channels and speed the heart (119, 236).

Tracheal smooth muscles are stimulated to contract by agents such as histamine via a P-inositide cycle; IP_3 releases Ca^{2+} and DGC activates PK-C, which has a stimulatory action (Chapter 2).

The protein of sodium channels from rat brain synaptosomes consists of three subunits—α, β, and γ. The catalytic subunit of cAMP-dependent kinase phosphorylates the α subunit of the channel protein and reduces the Na^+ influx. This is an example of permissive modulation (49).

Cyclic nucleotides mediate various functions in central nervous systems. In sensory neurons of ganglia of *Aplysia*, 5-HT increases the action potential duration by prolonging closure of one of the K^+ channels, the S channel (see above). Closure of the K^+ channel is mediated by a cAMP-dependent phosphorylation of a channel protein. The 5-HT activates the adenylate cyclase that increases cAMP, which stimulates a protein kinase that phosphorylates a membrane protein so that more Ca^{2+} enters and more transmitter is released from presynaptic ending; thus 5-HT prolongs closure of a K^+

channel and increases the duration of the action potential. The putative transmitter FMRFamide opens the S channel and shortens the action potential, that is, antagonizes 5-HT (131, 252). In certain neurons of the snail *Helix*, FMRFamide suppresses inward I_{Ca} and 5-HT increases I_{Ca} an effect mimicked by injection of cGMP or by an inhibitor of cGMP diesterase (225). Short-term sensitization of gill and siphon withdrawal is by activation of a cAMP dependent kinase, long-term sensitization is by a change in the kinase molecule (103).

Protein phosphorylation consists of three primary components: a protein kinase of which there are several according to the mode of activation; a phosphatase that removes the terminal phosphate and a specific protein substrate.

Synapsin is a phosphoprotein that is probably present at all presynaptic terminals. Synapsin I has a MW 83,000; it is associated with synaptic vesicles and restricts transmitter release. Calcium entry activates a Ca^{2+}–calmodulin-dependent kinase (in some synapses a cAMP-dependent kinase) that phosphorylates synapsin at a serine site and removes its constraint, that is, phosphorylation of synapsin permits transmitter release (15, 104, 167). Synapsin I regulates transmitter release in squid giant synapses. Synapsin constitutes 1% of the proteins in mammalian brain. Postsynaptic membranes that are rich in ACh receptors have at least four different protein kinases.

Examples of second messengers are found in sense organs where transduction of a sensory stimulus to membrane responses occurs. Responses of photoreceptors in the ventral eye of Limulus are depolarizations. Illumination opens Ca^{2+} channels and intracellular Ca^{2+} increases, as shown by aequorin fluorescence. Inositol triphosphate mimics the response to light and application of IP_3 results in increased Ca^{2+} inside the cell.

Light adaptation is associated with an increase in Ca_i and a decrease in the probability of channel opening (27). In photoreceptors of a fly phytoexcitation of rhodopsin activates G-protein in a complex that generates IP_3 and depolarizes the membrane.

Vertebrate rods and cones have a dark current in the outer segment (Chapter 16) this is an inward current due to entry of Na^+ in exchange for Ca^{2+}. The inner segment shows an outward current due to a Na^+-K^+ exchange pump. Illumination results in isomerization of rhodopsin from a cis to a trans conformation and this conformational change sets off a chain of events leading to decreased permeability of the outer segment of Na^+ and resulting hyperpolarization. Illumination of rhodopsin activates a phosphodiesterase in the disk membranes, the cGMP inside the receptor is decreased, and Ca^{2+} is released to efflux (8a). In the dark, a transmitter, probably glutamate, is liberated from inner terminals of rods and cones. On illumination, liberation of transmitter is reduced and horizontal cells depolarize (Chapter 5). In fish retinae, DA liberated from interplexiform cells increases the sensitivity of horizontal cells to the transmitter and decreases coupling between horizontal cells. Dopamine acts via a cAMP cascade to make the reduction in transmitter release on illumination less effective; this cGMP is the second messenger that couples illumination to membrane hyperpolarization in rods and cones and cAMP in horizontal cells couples modulation by DA.

Summary of Intermediate Messengers

Most neurotransmitters and neurohormones bind to receptor molecules that trigger a cascade in intracellular events. Activated receptors combine with membrane proteins, for example, G-proteins or open ion channels directly. These membrane events are followed by reactions involving any of several second messengers. One that combines with Ca^{2+} is calmodulin. Others are cyclic nucleotides—cAMP and cGMP. Those of another category come from phorbol esters—inositol triphosphate (IP_3) or diacylglycerol (DGC). Most of the second messengers activate kinases, which phosphorylate proteins that have specific actions—contraction of muscles, liberation of neurotransmitters from presynaptic endings, and secretion of hormone products.

The cascade of events following the membrane actions of transmitters, modulators, and hormones has a controlling action that permits redundancy in the regulation of cellular responses. Similar relatively simple substances are used as messengers in a wide variety of animals. Presumably these substances evolved with other functions and were adapted to special chains of reactions.

References

1. Adams, D., T. Dwyer, and B. Hille. *J. Gen. Physiol.* 75:493–510, 1980. Permeability of endplate channels to cations.

2. Adams, D. et al. *Annu. Rev. Neurosci.* 3:141–167, 1980. Ionic currents in molluscan neuron.

3. Adams, P. R., D. A. Brown, and A. Constanti. *J. Physiol.* (*London*) 330:537–572, 1982; *Nature* (*London*) 296:746–749, 1982. Voltage-current analysis of neurons in bullfrog sympathetic ganglia.

4. Adams, P. R., D. A. Brown, and A. Constanti. *J. Physiol.* (*London*) 332:223–262, 1982. Membrane currents in frog sympathetic neurons.

5. Adler, J. and A. Mesclov. *J. Bacteriol.* 112:315–326, 1972; 118:560–576, 1974; *Annu. Rev. Biochem.* 44:341–356, 1975; *Nature* (*London*) 280:279–284, 1979. Amino acid attractants and repellants for bacteria.

6. Aidley, D. J. *The Physiology of Excitable Cells,* 2nd ed. Cambridge Univ. Press, New York, 1978.

7. Aldman, G. et al. *Comp. Biochem. Physiol. C* 92C:103–108, 1989. Gastrin CCK-like peptides in Squalus acanthias; concentration and actions in the gut.

8. Almers, W., and E. W. McClesky, *J. Physiol. (London)* 353:585–608, 1984. Na⁺ can pass through high threshold Ca²⁺ channels.

8a. Altman, J. *Nature (London)* 313:264–265, 1985. Ca⁺ and cAMP coupling in retinal rods.

9. Atschuler, R. A. et al. *Nature (London)* 298:657–659, 1982. Aspartate/glutamate in guinea pig photoreceptors.

10. Anderson, P. and M. McKay. *Biol. Bull (Woods Hole, Mass.)* 169:652–660, 1985. Proton-activated Cl⁻ current in colenterate neurons.

11. Angelides, K. *Nature (London)* 321:63–66, 1986. Fluorescently labeled Na channels localized and immobilized to synapses of innervated muscle fibers.

12. Archer, R. *Proc. R. Soc. London, Ser. B* 210:23–43, 1980; *Trends Neurosci.* 4:225–229, 1981. Evolution of neuropeptides.

13. Augustine, G. and R. Fetterer. *J. Exp. Biol.* 118:53–69, 1985. Neurohumoral modulation of Limulus cardiac ganglion.

14. Augustine, G., M. Charlton, and J. Smith. *J. Physiol. (London)* 367:143–181, 1985. Ca entry into presynaptic terminals of squid giant synapse.

15. Bahler, M. and P. Greengard. *Nature (London)* 326:704–707, 1987. Synapsin I bundles F-actin in phosphorylation-dependent manner.

16. Baker, T. S. et al. *J. Mol. Biol.* 184:81–98, 1985. Connexons in inter-cellular communication.

17. Barchi, R. et al., *Annu. Rev. Neurosci.* 11:455–495, 1988. Molecular structure of voltage dependent Na⁺ channel.

18. Bargiello, T. A. et al., *Nature (London)* 328:686–691, 1987. Drosophila clock gene affects intercellular junctional communication.

19. Barker, D. L. and E. A. Karvitz. *Nature (London) New Biol.* 236:61–62, 1973. Transmitters in lobster.

20. Battelle, B. A. et al. *Science* 216:1250–1252, 1982. Octopamine in Limulus eye.

20a. Bean, B. *Ann. Rev. Phys.* 51:367–384, 1989. Calcium channels.

21. Belordetti, F. et al. *J. Physiol. (London)* 374:289–313, 1986. Action potentials, macroscopic and single channel currents from growth cases of Aplysia neurons in culture.

22. Beltz, B. S. and E. A. Kravitz. *J. Exp. Biol.* 124:115–141, 1986. Adrenergic and peptidergic neuromodulation in crustacea.

23. Benson, J. and S. Levitan. *Proc. Natl. Acad. Sci. U.S.A.* 80:3522–3525, 1983. Serotonin increases an anomalously rectifying K current in Aplysia R 15.

24. Berrard, S. et al. *Proc. Natl. Acad. Sci. U.S.A.* 84:9280–9284, 1987. Cloning and sequence of porcine ChAcTase.

25. Berridge, M. J. *Sci. Am.* 253:142–152, 1985. Molecular basis of communication within the cell.

26. Berridge, M. J. *J. Exp. Biol.* 124:323–335, 1986. Regulation of ion channels by IP₃ and diacylglycerol.

27. Beyer, E. C., D. Paul, and D. Goodenough. *J. Cell Biol.* 105:2621–2629, 1987. Rat heart connexin.

28. Blanchi, D., L. Noviello, and M. Libonati. *Gen. Comp. Endocrinol.* 21:267–277, 1973. A neurohormone of cephalopods with cardroexcitation activity.

29. Binstock, L. and L. Goldman. *J. Gen. Physiol.* 54:730–764, 1969. Voltage-current curves of Ranvier node.

30. Bishop, C. A. and O'Shea. *J. Comp. Neurol.* 207:223–238, 1982. Neuropeptide prolactin: Immunocytochemical mapping of neurons in the CNS of the cockroach.

31. Blatz, A. L. et al. *Biophys J.* 43:237–241, 1983. Voltage activated chloride channels in skeletal muscle.

32. Blatz, A. L. *Nature (London)* 323:718–720, 1986. Apamin-blocked Ca-activated K channels in cultured rat muscle.

33. Bogoney, E. and D. E. Koshland. *Proc. Natl. Acad. Sci. U.S.A.* 82:4891–4895, 1985. Vectorial transmembrane receptors functional form: Aspartate receptors in chemotaxis.

34. Boothby, K. and I. MacFarlane. *J. Exp. Biol.* 125:385–389, 1986. Chemoreception of sea anemone.

35. Boulter, J. et al. *Proc. Natl. Acad. Sci. U.S.A.* 84:7763–7767, 1987. Functional expression of two neuronal nicotinic acetyl choline receptors from cDNA clones identifies a gene family; *Nature (London)* 319:368–374, 1986. Isolation of cDNA clone for neural nicotinic receptor α-subunit.

36. Breer, H., R. Kleen, and G. Henz. *J. Neurosci.* 5:3386–3392, 1985. Isolation of insect ACh receptors.

37. Brew, H. and D. Attwell. *Nature (London)* 327:707–709, 1987. Electrogenic glutamate uptake in current carrier in membrane of axolotl retinal glial cells.

38. Brink, P. et al. *J. Gen. Physiol.* 69:517–526, 1977; 72:67–86, 1978. Electrical properties of septum in giant axons of earthworms.

39. Brown, B. E. and A. N. Starratt. *J. Insect Physiol.* 21:1879–1881, 1975. Isolation of Proctolin a myotropic peptide from periplaneta americana—cockroach.

40. Brown, H. F. *Physiol. Rev.* 62:505–528, 1982. Electrophysiology of sinoatrial node.

41. Brown, J. E. and J. A. Coles. *J. Phys.* 296:373–392, 1979; *J. Gen. Physiol.* 63:337–350, 1974. Ionic dependence of Limulus ventral photoreceptors.

42. Budlemann, B. U., *Cell Tissue Res.* 227:475–483, 1982. Histochemical evidence for catecholamines as neurotransmitters in the statocyst of Octopus vulgaris.

43. Burnstock, G. *Purinergic Receptors.* Chapman & Hall, New York, 1981.

44. Byerly, L. and M. O. Masudo. *J. Physiol. (London)* 288:262–284, 1979. K-currents of negative spikes in Ascaris muscle.

45. Calabrese, R. L. and A. R. Maranto. *J. Exp. Biol.* 125:205–224, 1986. Cholinergic innervation of leech heart.

46. Carr, W. E. S. et al. *Comp. Biochem. Physiol. A* 77A:469–474, 1984. Chemo-attractants for shrimp Palaemontes.

47. Catapane, E. J. et al. *J. Exp. Biol.* 74:101–113, 1978; 83:315–323, 1979. Pharmacology of ciliated epithelium in Mytilus.

48. Catterall, W. A. *J. Biol. Chem.* 258:5117–5123, 1983; *Proc. Natl. Acad. Sci.* 82:4847–4851, 1985. Glycosylation needed for Na channels in neuroblastoma cells.

49. Catterall, W. A. *Annu. Rev. Biochem.* 55:953–985, 1986. Voltage-sensitive Na channels.

50. Catterall, W. A. *Annu. Rev. Physiol.* 50:395–406, 1988. Genetic analysis of ion channels.

51. Catterall, W. A. *Science* 223:653–661, 1984. Molecular basis of neuronal excitability.

52. Catterall, W. A. *Science* 242:50–61, 1988. Structure and function of voltage sensitive ion channels.

53. Catterall, W. A. and B. Curtis. In (B. Hille and D. M. Fambrough, eds.), Chapter 12, pp. 201–213. *Proteins of Excitable Membranes* Wiley, New York, 1987. Properties of V-sensitive calcium channels.

54. Christensen, O. *Nature (London)* 330:66–68, 1987. Ca(2+) influx through stretch activated channels.

55. Civelli, O. et al. *Proc. Natl. Acad. Sci. U.S.A.* 82:4291–4295, 1985. Sequence and expression of the rat prodynorphin gene.

56. Cline, H. T. *J. Neurosci.* 6:2848–2856, 1986 GABA as a transmitter in the leech.

57. Colino, A. and J. V. Hallwell. *Nature (London)* 328:73–77, 1987. Three separate K conductances in hippocampus.

58. Connor, C. and C. L. Prosser. *Am. J. Physiol.* 226:1212–1218, 1974. Ionic effects on longitudinal and circular muscle of intestine.

59. Connor, J. A. *Biophys. J.* 18:81–102, 1977; *Fed. Proc., Fed. Am. Soc. Exp. Biol.* 37:2139–2145, 1978; *J. Physiol. (London)* 286:41–60, 1979; in *Cellular Pacemakers* (D. O. Carpenter, ed.), Chapter 6, pp. 187–217. Academic Press, New York, 1982; *Annu. Rev. Physiol.* 47:17–28,

1985. Analysis of ionic currents in molluscan neurons.

60. Connor, J. A. *Annu. Rev. Physiol.* 47:17–28, 1985. Neural pacemakers and rhythmicity.

61. Connor, J. A. and C. F. Stevens. *J. Physiol. (London)* 213:31–53, 1971. Current changes in repetitive neurons of molluscs.

62. Conti-Troconi, B. and M. A. Raftery. *Proc. Natl. Acad. Sci. U.S.A.* 83:6646–6650, 1986. Nicotinic ACh receptor contains multiple binding sites.

63. Conti-Troconi, B. et al. *Proc. Natl. Acad. Sci. U.S.A.* 82:5208–5212, 1985. Brain and muscle ACh receptors are homologous—nicotinic.

64. Cooperman, S. et al. *Proc. Natl. Acad. Sci. U.S.A.* 84:8721–8725, 1987. Modulation of Na channel mRNA in rat skeletal muscle.

65. Cropper, E. et al. *Proc. Natl. Acad. Sci. U.S.A.* 84:5483–5486, 1987. Multiple neuropeptides in cholinergic motorneurons of Aplysia.

66. Cunningham, J. and M. Neal. *J. Physiol. (London)* 366:47–62, 1985. Effects of excitable amino acids and analogs on ACh release from amacrine cells of rabbit retina.

67. Curtis, B. and W. A. Catterall. *Proc. Natl. Acad. Sci. U.S.A.* 82:2528–2532, 1985. Phosphorylation of antagonist receptor of volt-sensitive Ca^{2+} channel by cAMP dependent protein kinase.

68. Daez, J., J. Connor et al. *Proc. Natl. Acad. Sci. U.S.A.* 86:2708–2712, 1989. Hepatocyte gap junctions permeable to IP_3 and Ca^{2+}.

69. Derby, C. D., W. Carr, and Bache. *J. Comp. Physiol., A* 155:341–349, 1984. Purinergic olfactory cells of crustaceans.

70. Dietzel, I., P. Drapeau, and J. Nichols. *J. Physiol. (London)* 372:191–205, 1986. V-dependent 5HT release at synapse in cultured leech neurons.

71. Dixon, R. et al. *Nature (London)* 321:75–79, 1986. Cloning of gene and cDNA for B-adrenergic receptor and homology in rhodopsin.

72. Dockray, G. et al. *Nature (London)* 305:328–330, 1983. A novel active pentapeptide from chicken brain identified form antibodies to FMRFamide; *J. Neurochem.* 45:152–158, 1985. FMRFamide in rat spinal cord characterized by HPLC and RIA.

73. Doolittle, R. F. *Biol. Bull. (Woods Hole, Mass.)* 172:269–283, 1987. Evolution of vertebrate proteins.

74. Dowling, J. E. *The Retina*. Harvard Univ. Press, Cambridge, MA, 1987.

75. Dowling, J. E. *Trends Neurosci.* 9:236–240, 1986. Dopamine: A retinal neuromodulator?

76. DuBois, J. M. *J. Physiol. (London)* 318:297–316, 1981. Three types of K channels in frog Ranvier node.

77. Dutar, P. and R. A. Nicoll. *Nature (London)* 322:152–158, 1986. Role of GABA receptors in CNS.

78. Eckert, R. and P. Brehm. *Annu. Rev. Biophys. Bioeng.* 8:353–383, 1979. Ionic mechanisms of excitation in Paramecium.

79. Ehinger, B., J. Dowling et al. *Proc. Natl. Acad. Sci. U.S.A.* 85:8321–8325, 1988. Bipolar cells in turtle retina are immunoreceptive to glutamate.

80. Elliot, E. J. *J. Comp. Physiol.* 129:61–66, 1979. Cholinergic responses in heart of clam.

81. Emson, P. C. and M. E. DeQuidt. *Trends in Neurosci.* 7:31–35, 1984. Neuropeptide Y: A new member of the pancreatic polypeptide family.

82. Erlich, B. E. et al. *Proc. Natl. Acad. Sci. U.S.A.* 85:5718–5722, 1988. Paramecium Ca channels.

82a. Erspamer, V. *Trends Neurosci.* 4:267–269, 1981. Peptides from toad skin.

83. Evans, P. D. *J. Physiol. (London)* 318:99–122, 1981. Octopamine in insects.

84. Evans, P. D. *J. Physiol. (London)* 366:331–341, 1985. Octopamine on locust muscle.

85. Evans, P. D. and C. N. Myers. *J. Exp. Biol.* 124:143–176, 1986. Peptidergic and aminergic modulation of insect skeletal muscle.

86. Fraser, S. E., C. R. Green, H. Bode, and N. Gilula. *Science* 237:49–55, 1987. Selective disruption of gap junctional communication interferes with patterning in Hydra.

87. Fukushima, Y. *Proc. Natl. Acad. Sci. U.S.A.* 78:1274–1277, 1981. Patch-clamping of tunicate eggs.

88. Furman, R. E. et al. *Proc. Natl. Acad. Sci. U.S.A.* 83:488–492, 1986. Na channels from rabbit T-tubular membranes.

89. Furshpan, E. J. et al. *Harvey Lect.* 76:149–191, 1981. Transmitter reptoire of sympathetic neurons in culture.

90. Gage, P. and R. Eisenberg. *J. Gen. Physiol.* 53:265–278, 1969. Properties of T-tubule membranes.

91. Ganetsky, D. and C. F. Wu. *Annu. Rev. Genet.* 20:13–44, 1986. TINS 8:222–236, 1985. Neurogenetics of membrane excitability in Drosophila.

92. Geraerts, N. P. et al. *Semin. Ser.—Soc. Exp. Biol.* 33:261–281, 1988. Bioactive peptides in molluscs.

93. Gerschenfeld, H. M. et al. *J. Physiol. (London)* 243:427–456, 1974; 274:265–278, 1978. Neurotransmitters in Aplysia.

94. Gilman, A. G. *Annu. Rev. Biochem.* 56:615–649, 1987. G-protein coupling agents.

95. Gordon, A. S. et al. *Nature (London)* 267:539–540, 1977; *Proc. Natl. Acad. Sci. U.S.A.* 74:262–267, 1977. Phosphorylation of ACh receptor in Torpedo Membranes.

96. Gordon, D., W. Catterall et al. *Proc. Natl. Acad. Sci. U.S.A.* 84:8682–8686, 1987. Tissue specific expression of RI and RII Na channel subtypes.

97. Gordon, R. D., R. L. Barchi et al. *J. Neurosci.* 8:3742–3749, 1988. Topological localization of α-segment of the eel voltage-dependent Na⁺ channel.

98. Gordon, R. D., R. L. Barchi et al. *Proc. Natl. Acad. Sci. U.S.A.* 84:308–312, 1987. Localization of C-terminal region of the V-dependent Na⁺⁺ channel from Electrophorus using antibodies raised against a synthetic peptide.

99. Green, J. P. In *Basic Neurochemistry* (G. J. Sregel et al., eds.), Chapter 12, pp. 253–271. Raven Press, New York, 1989. Histamine and seratonin.

100. Greenberg, M. J. and D. A. Price. *Mem. Calif. Acad. Sci.* 13:85–96, 1988. The phylogenetic and biomedical significance of extended neuropeptide families.

101. Greenberg, M. J. and D. A. Price. In *Peptides* (F. E. Bloom, ed.), Raven Press, New York, 1980; *Annu. Rev. Physiol.* 45:271–288, 1983. Invertebrate neuropeptides.

102. Greenberg, M. J. *Comp. Biochem. Physiol.* 33:259–294, 1970. ACh structure-activity relation.

103. Greenberg, M. J. et al. *Peptides (N.Y.)* 9:125–135, 1988. Relations between FMRFamide related peptides and other peptide families.

104. Greengard, P. et al. *Fed Proc., Fed. Am. Soc. Exp. Biol.* 33:1059–1067, 1974; *Nature (London)* 260:101–108, 1976; *Harvey Lect.* 75:277–331, 1981. Role of cyclic AMP in neurotransmission.

105. Greeningloh, G. et al. *Nature (London)* 328:215–220, 1987. The strychnine-binding subunit of glycine receptor shows homology with nicotinic acetyl choline receptor.

106. Grimmelikhuijzen, C. J. P. and D. Graff. *Proc. Natl. Acad. Sci. U.S.A.* 83:9817–9821, 1986. Peptide in sea Anemone.

107. Grimmelikhuijzen, C. J. P. et al. *Histochemistry,* 78:361–381, 1983; *Proc. Natl. Acad. Sci. U.S.A.* 83:9817–9821, 1986. FMRF amide—like peptides in Coelenterates.

108. Grissmer, S. *J. Physiol. (London)* 381:119–134, 1986. K and Na channels in frog internode.

109. Grimmelikhuijzen, C. J. P. and A. N. Spencer. *Brain Res.* (in press). Sequence of three cnidarian RFamide peptides.

110. Grimmelikhuijzen, C. J. P. *Semin. Ser.—Soc. Exp. Biol.* 33:199–218, 1988; *Cell Tissue Res.* 241:171–188, 1985.

111. Gustin, M. C. et al. *Science* 233:1195–1197, 1986. Ion channels in yeast; whole cell patch clamp.

112. Guy, H. R. and P. Seetharamulu. *Proc. Natl. Acad. Sci. U.S.A.* 83:508–512, 1986. Molecular model of the actinptemtoa; sodium channel.

113. Hagiwara, S. and L. A. Jaffe. *Annu. Rev. Biophys. Bioeng.* 8:385–416, 1979; *J. Gen. Physiol.* 67:621–638, 1976; 70:269–281, 1977; *J. Membr. Biol.* 18:61–80, 1974. Inward rectification by starfish egg membranes.

114. Hagiwara, S. et al. *J. Gen Physiol.* 50:583–601, 1967; *J. Physiol. (London)* 190:479–518, 1967; 238:109–127, 1974. Ion permeability in muscle membranes of fishes.

115. Haimovich, B., R. L. Barchi, et al. *J. Neurosci.* 7:2957–2966, 1987. Na channels differ in muscle plasma membrane and T-tubule.

116. Harmer A. J. *Trends Neurosci.* 7:57–60, 1984. Three tachykinins in mammalian brain.

117. Harris, E. W., A. H. Ganny, and W. Cotman. *Brain Res.* 323:132–137, 1984. NMDA receptors in hippocampus.

118. Hartzell, H. C., S. Kuccler, et al. *J. Phys.* 271:817–846, 1977. Synaptic excitation and inhibition by direct action of ACh on 2 types of receptors on parasympathetic neurons.

119. Hartzell, H. C. and R. Fischmeister. *Nature (London)* 323:273–275, 1986. Opposite effects of cAMP and cGMP on $I_{(Ca)}$ in heart cells.

120. Hedblom, M. L. and J. Adler. *J. Bacteriol.* 144:1048–1060, 1980. Receptor molecules for amino acids in E. coli.

121. Herbert, E. and M. Uhler. *Cell (Cambridge, Mass.)* 30:1–2, 1982. Biosynthesis of polyprotein precursors.

122. Hertzberg, E. L. et al. In *Gap Junctions*, pp. 9–28. Alan R. Liss, New York, 1988. Review on biochemical, immunological, and topological studies of gap junctions.

123. Hille, B. *Ionic Channels of Excitable Membranes.* Sinauer Associates, Sunderland, MA, 1984; *J. Gen. Physiol.* 50:1287–1301, 1967; *Nature* 210:1220–1222, 1966. Review of ion channels.

124. Huddart, H. and A. C. Oldfiele. *Comp.* *Biochem. Physiol. C* 73C:303–311, 1982. Action of biogenic amines on gut muscle of locusta.

125. Imoto, K. et al. *Nature (London)* 324:670–674, 1986. Location of S subunit determining ion transport thru ACh R channel.

126. Inoue, I. *J. Gen. Physiol.* 85:519–537, 1985. V-dependent Cl conductance in squid axon.

127. Iversen, L. L. et al. *Trends Neurosci.* 6:293–294, 1983. Neuropeptides in mammals.

128. Irvine, R. and R. Moor. *Biochem. J.* 240:917–920, 1986. Inositol derivatives in sea urchin eggs.

129. Jaslove, S. and P. Brink. *Nature (London)* 323:63–65, 1986. Rectifications at electronic motor giant synapse of crayfish.

130. Jones, D. and R. Reed. *Science* 244:790–795, 1989. An olfactory neuron specific G-protein in oderant signal transduction.

131. Kaczmarck, L. K. *J. Exp. Biol.* 124:375–392, 1986. Phorbol esters, agents of protein phosphorylation and regulation of ion channels.

132. Kaldany, R., J. Nambu, and R. Scheller, *Annu. Rev. Neurosci.* 8:431–455, 1985. Neuropeptides in identified *Aplysia* neurons.

133. Kavaliers, M. and M. Hiost. *Brain Res.* 372:370–374, 1986. FMRFamide and PLG antagonists.

134. Kavaliers, M. *Brain Res. Bull.* 21:923–931, 1988; *Brain Res.* 410:111–115, 1987. Nociception.

135. Kehoe, J. S. *J. Physiol. (London)* 225:85–172, 1972. Properties of synapses in Aplysia.

136. Kistler J. et al. *Biophys. J.* 37:371–383, 1982. Structure and function of AChR.

137. Kistler, J., D. Christie, and S. Bullivants. *Nature (London)* 331:721–723, 1988. Homologies between gap junction proteins of lens, heart and liver.

138. Klein, M. G. *J. Exp. Biol.* 144:563–579, 581–598, 1985. Ion conductances in muscle of cold and warm acclimated sunfish.

139. Knapp, A. G. and J. E. Dowling. *Nature (London)* 325:437–439, 1987. Dopamine enhances excitability and gated conductances in cultured retinal horizontal cells.

140. Kobilka, B. K. et al. *Proc. Natl. Acad. Sci. U.S.A.* 84:46–50, 1987. cDNA for β_2-adrenergic receptor protein in multiple membranes.

141. Komerth, A., H. Lux, and M. Morad. *J. Physiol. (London)* 386:603–633, 1987. Proton induced changes in Ca channels in chick dorsal root ganglia cells.

142. Krause, R. M. et al. *Proc. Natl. Acad. Sci. U.S.A.* 84:881–885, 1987. Three rat preprotachykinin mRNAs encode the neuropeptides Substance P and neurokinin A.

143. Kravitz, E. A. In *Fast and Slow Chemical Signalling in the Nervous System* (L. L. Iversen and F. Goodman, eds.), Oxford Sci. Publ. pp. 244–259. 1986. Seratonin, octopamine, and proctolin: Two amines and a peptide, and aspects of lobster behaviour.

144. Kravitz, E. A. et al. *J. Exp. Biol.* 89:159–175, 1980. Amines and peptides as neurohormones in lobsters.

145. Kravitz, E. A. et al. In *Model Neural Networks and Behavior* (A. I. Selverston, ed.), pp. 339–360. Plenum, New York, 1985. The well-modulated lobster.

146. Krieger, D. *Science* 22:975–985, 1983. Brain peptides.

147. Kubalek, E. et al. (PNT Unwih). *J. Cell Biol.* 105:9–18, 1987; *Nature (London)* 329:286–287, 1987. Nicotinic AChR; 5 membrane-spanning subunits.

148. Kudo, T. et al. *Nature (London)* 323:411–416, 1986. Sequencing and expression of cDNA encoding muscarinic ACh receptors.

149. Kuffler, S. W. *J. Exp. Biol.* 89:257–286, 1980. Synaptic responses in autonomic ganglia of amphibians.

150. Kuhlmann, J., C. L. Li and J. Calabrese, *J. Neurosci.* 5:2301–2309, 2310–2317, 1985. FMRFamide in leech.

151. Kumar, N. M. et al. *J. Cell Biol.* 103:767– 776, 1986. Gap junction polypeptides sequenced from cDNA.

152. Kung, C. and Y. Saimi. *Annu. Rev. Physiol.* 44:519–534, 1982. The physiological basis of taxes in Paramecium.

153. Lansman, J. B. et al. *J. Gen. Physiol.* 88:321–347, 1986. Na^+ can pass through high threshold Ca^{2+} channels.

154. Lasater, E. M. and J. E. Dowling. *Proc. Natl. Acad. Sci. U.S.A.* 82:3005–3029, 1985; 84:7319–7323, 1987. Dopamine decreases conductance of electrical junctions between horizontal cells.

155. Latorre, J. *Membr. Biol.* 71:1–30, 1983. Comparison of Na, K, and Cl-channels.

156. Leblanc, G. et al. *Proc. Natl. Acad. Sci. U.S.A.* 84:3511–3515, 1987. Neuropeptide Y coexistence with VIP and choline acetyltransferase.

157. Lee, K. et al. *Proc. R. Soc. London, Ser. B* 233:35–48, 1984. Ca^{2+} current plateau of cardiac AP.

158. Leeman, S. E. *J. Exp. Biol.* 89:193–200, 1980. Substance P and neurotensin.

158a. Lehninger, A. L. *Principles of Biochemistry*, p. 539, Worth, New York, 1982.

159. Lenhoff, H. M. et al. In *Coelenterate Ecology and Behavior* (G. Mackie, ed.), pp. 571–579. Plenum, New York, 1976. Chemoreception in hydra.

160. Leung, M. K. and G. B. Stefano. *Proc. Natl. Acad. Sci. U.S.A.* 81:955–958, 1984. Isolation and identification of enkephalins in pedal ganglia of Mytilus edulis.

161. Levitan, E. S. et al. *Nature (London)* 355:76–82, 1988. Structure and functional basis for GABA receptor heterogenity.

162. Levitzki, A. *Science* 241:800–806, 1988. From epinepherine to cAMP.

163. Li, C. and R. Calabrese. *J. Neurosci.* 7:595–603, 1987. FMRFamide-like peptides in leech.

164. Lipscombe, D. et al. *Proc. Natl. Acad. Sci. U.S.A.* 85:2398–2402, 1988. Calcium channels in growth cones and sympathetic neurons.

165. Llinas, R. *J. Physiol. (London)* 315:549–

567, 1981. Electrical activity of neurons in brain slices of inferior olive and thalamus.

166. Llinas, R. et al. *Proc. Natl. Acad. Sci. U.S.A.* 73:2918–2922, 1976. Presynaptic Ca current.

167. Llinas, R. et al. *Proc. Natl. Acad. Sci. U.S.A.* 82:3035–3039, 1985. Intraterminal injection of synapsin I or Ca/calm. dependent kinase II alters neurotransmitter release at squid giant synapse.

168. Lloyd, P. E., I. Kupperman, and K. Weiss. *Proc. Natl. Acad. Sci. U.S.A.* 81:2934–2937, 1984. Parallel actions of molluscan neuropeptide and serotonin in mediating arousal in Aplysia.

169. Lloyd, P. E., I. Kupperman, et al. *Proc. Natl. Acad. Sci. U.S.A.* 83:9794–9798, 1986. Release of neuropeptides during intracellular stimulation of Aplysia neurons in culture.

170. Lloyd, P. E. *Proc. Am. Soc. Exp. Biol.* 41:2948–2952, 1982. SCP(A) and SCP(B) in gastropods.

171. Loewenstein, W. R. *Fed. Proc., Fed. Am. Soc. Exp. Biol.* 32:60–64, 1973. Intracellular conduction in liver and epithelium.

172. Lord, J., A. Waterfield, J. Hughes, and H. Kosterlitz. *Nature (London)* 267:495–499, 1977. Multiple agonists and receptors of opioid peptides.

173. Lundberg, J. and T. Hokfelt. *Trends Neurosci.* 6:325–333, 1983. Coexistence, peptides and monoamines.

174. Lynch, D. R. and S. H. Snyder. *Annu. Rev. Biochem.* 55:773–799, 1986. Processing of preproenkephalins.

175. Mackie, G. O. In *Coelenterate Ecology* (G. O. Mackie, ed.), pp. 647–657. Plenum, New York, 1976. Nerve net conduction.

176. Mackie, G. O., P. Anderson, and C. L. Singla. *Biol. Bull. (Woods Hole, Mass.)* 167:120–123, 1984. Gap junctions in Cnidarians.

177. Mackie, G. O. et al. *Philos. Trans. R. Soc. London, Ser. B* 301:365–400, 401–418, 1983. Conduction and coordination in Hexactinellid sponges.

178. Maggio, J. E. *Annu. Rev. Neurosci.* 11:13–28, 1988. Substance P and tachykinins carboxyl terminal amino acids.

179. Mahon, A. C., P. E. Lloyd, et al. *Proc. Natl. Acad. Sci. U.S.A.* 82:3925–3929, 1985. Cardioactive peptides A and B of Aplysia are derived from common precursor.

180. Makowski, L. et al. *Biophys. J.* 45:205–218, 1984: Connexon Conformation and packing.

181. Martin, D. Frösch, C. Kiehling, and K. Voigt. *Neuropeptides (Edinburgh)* 2:141–150, 1981. Molluscan neuropeptide-like and enkephalin-like material coexists in octopus nerves.

182. Martin, R. et al. *Neuroscience* 8:213–227, 1983. Coexistence of Met-enk, Leu-enk, and CCK.

183. Masu, Y., K. Nakagama et al. *Nature (London)* 329:386–383, 1987. cDNA cloning of bovine substance K.

184. McCleskey, E. W., R. W. Tsien, et al. *J. Exp. Biol.* 124:177–190, 1986. Types of calcium channels in starfish eggs.

185. McClesky, E. W. and W. Walners. *Proc. Natl. Acad. Sci. U.S.A.* 82:7149–7153, 1985. Calcium channels in skeletal muscle.

186. McFarlane, L. D., D. Graff, and G. J. P. Grimmelikhuijzen. *J. Exp. Biol.* 133:157–168, 1987. Antho-RFamide excitatory effect on muscles and conducting system of Calliactis.

187. McGreer, P. G. and E. G. McGreer. In *Basic Neurochemistry* (G. J. Siegel et al., eds.), Chapter 14, pp. 287–311. Raven Press, New York, 1989. Receptors for glutamate.

188. McReynolds, J. S. and A. Gorman. *Science* 183:658–659, 1974. Ionic basis of hyperpolarizing receptor potential in scallop eye.

189. Means, A. R. and J. R. Dedman. *Nature (London)* 285:73–77, 1980. Calmodulin intracellular Ca receptors.

190. Means, A. R. et al. *Physiol. Rev.* 62:1–38, 1982. Calmodulin in eukaryotic cells.

191. Menkes, H., S. Snyder, et al. *Proc. Natl. Acad. Sci. U.S.A.* 83:5727–5730, 1986. Lithium lowers transmitter response in smooth muscle that act via phorbol esters.

192. Merlie, J. and J. Sanes. *Nature (London)* 317:66–68, 1985. ACh receptors and RNA in synaptic regions of adult muscle fibers.

193. Merzenich, M. M. and J. H. Kaas. *Trends Neurosci.* 5:434–436, 1982. Reorganization of somatosensory cortex after peripheral nerve injury.

194. Miles, K., P. Greengard et al. *Proc. Natl. Acad. Sci. U.S.A.* 84:6591–6595, 1987. Phosphorylation of nicotinic AChR in rat myotubules.

195. Miller, M. and I. Parnas. *J. Physiol. (London)* 163:363–375, 1985. Dopamine modulations of neuromuscular transmission in prawn.

196. Miller, R. J. *Science* 235:46–52, 1987. Multiple Ca channels and neuronal function.

197. Minneman, K., R. N. Pittman, and P. B. Molinoff. *Annu. Rev. Neurosci.* 4:419–461, 1981. β-adrenergic receptor subtypes: Properties, Distribution and Regulation.

198. Mishima, M. et al. *Nature (London)* 313:354–369, 1985; 318:538–543, 1985; 321:406–411, 1986. ACh receptors and subunits.

199. Mobley, P. and P. Greengard. *Proc. Natl. Acad. Sci. U.S.A.* 82:945–947, 1985. Effects of noradrenaline in axon terminals in rabbit frontal cortex.

200. Morris, H. R., P. Lloyd, et al. *Nature (London)* 300:643–645, 1982. A new cardioactive peptide SCP$_B$ from Aplysia nervous system and gut.

201. Muneoka, J. and M. Matsura. *Comp. Biochem. Physiol. C* 81C:61–70, 1985. Molluscan FMRFamide and YGGFMRFamide in Mytilus muscle.

202. Muneoka, Y. et al. *Comp. Biochem. Physiol. C* 73C:149–156, 1982. Neurotransmitter in muscle of Mytilus.

203. Nachman, R. J. et al. *Biochem. Biophys. Res. Commun.* 140:357–364, 1986; *Science* 234:71–73, 1986. LSK II; a blocked sulfonated insect neuropeptide with homology to CCK and gastrin.

204. Nakanishi, S. et al. *Nature (London)* 278:423–427, 1979. Sequenced nucleotides of cloned cDNA of bovine corticotropin-B lipotropin precursor.

205. Nathanson, N. M. *Annu. Rev. Neurosci.* 10:195–236, 1987. Molecular properties of muscarinic ACh receptor.

206. Nawata, T. and T. Sibaska. *J. Comp. Physiol.* 134:137–149, 1979. Action potentials and bioluminescence in Noetiluca.

207. Netherton, J. C. and S. Gurin. *J. Biol. Chem.* 257:11971–11975, 1982. Synthesis of amino acid derivatives.

208. Nicholson, B. et al. *Nature (London)* 329:732–733, 1987. Two components of hepatic gap junction protein.

209. Nicoll, R. A. *Science* 241:545–551, 1988. Coupling of neurotransmitter receptors to ion channels in brain.

210. Noda, M. et al. *Nature (London)* 295:202–206, 1982. Cloning and sequencing of cDNA for beef adrenal preproenkephalin.

211. Noda, M. et al. *Nature (London)* 302:251–255, 528–532, 1983. Acetylcholine receptor structure.

212. Noda, M. et al. *Nature (London)* 312:121–127, 1984. Na channel structure.

213. Noda, M. et al. *Nature (London)* 320:188–192, 1986. Na channel messenger RNAs in rat brain.

214. Nusbaum, M. P. and W. Kristan. *J. Exp. Biol.* 122:277–302, 1986. Swim inhibition in leech by 5HT containing insterneurons.

215. O'Donohue, T. et al. *Peptides* (N.Y.) 5:563–568, 569–584, 1984. FMRFamide-like immunoreactivity in rat CNS and GI tract.

216. Ohba, M., Y. Sakamoto, and T. Tomita. *J. Physiol. (London)* 267:167–180, 1977. Effects of Na, K, and Ca ions on slow waves in circular muscle of guinea pig stomach.

217. O'Malley, D. and R. Masland. *Proc.*

Natl. Acad. Sci. U.S.A. 86:3414–3418, 1988. Co-release of GABA and ACh by a retinal neuron.

218. Osborne, N. N. and G. J. Dockray. *Neurochem. Int.* 4:175–180, 1982. Coexistence of Substance P and 5-HT in an invertebrate neuron.

219. O'Shea, M., M. Schaffer. *Annu. Rev. Neurosci.* 8:171–198, 1985. Neuropeptide functions invertebrates, peptide identification and sequencing.

220. Otsuka, M. and S. Konishi. *Trends Neurosci.* 6:317–320, 1983. Substance P.

221. Papazian, D. M., L. Y. Jan et al. *Science* 237:749–753, 1987. Cloning of genomic and complimentary DNA from shaker, a putative potassium channel gene from Drosophila.

222. Pape, H. C. and H. Machemer. *J. Comp. Physiol. A* 158A:111–124, 1986. Electrical properties of membrane of Didinium.

224. Paul, D. L. *J. Cell Biol.* 103:123–134, 1986. Design of cell membrane channels.

225. Paupardin-Tsitsch, D. et al. *Nature (London)* 323:812–813, 1986. cGMP-dependent protein kinase increases Ca current in snail neurons.

226. Porchet, M. and N. Dhainaut-Courtois. *Semin. Ser.—Soc. Exp. Biol.* 33:219–234, 1988. Neuropeptides and monoamines in annelids.

227. Price, D. et al. *Zool. Sci.* 4:395–410, 1987. FMRFamides in molluscs.

228. Price, D. A. *Am. Zool.* 26:1007–1015, 1986. Evolution of molluscan cardioregulator neuropeptide.

229. Prosser, C. L. In *Comparative Animal Physiology* (C. L. Prosser, ed.), 3rd ed., vol. 2, pp. 111–132, 457–504. Saunders, Philadelphia, PA, 1973.

230. Prosser, C. L. In *Adaptational Biology* (C. L. Prosser, ed.), Chapter 2, pp. 2–65; Chapter 12, pp. 515–567; Chapter 15, pp. 640–682. Wiley, New York, 1986. Origins of life and metabolic pathways; Transmission between neurons; Bioelectric properties of cell membranes.

231. Prosser, C. L. et al. *Am. J. Physiol.* 233:C19–C24, 1977. Prolonged potentials when Na occupies Ca channels.

232. Rane, S. G. and G. A. Wyse. *Comp. Biochem. Physiol. C* 87C:121–130, 1987. Neuromuscular transmission in Limulus.

233. Rehfeld, J. J. *Neurochem.* 44:1–10, 1985. CCK characterization.

234. Requena, J. et al. *J. Gen. Physiol.* 85:789–804, 1985. Ca-Na exchange.

235. Reuter, H. *Nature (London)* 301:569–574, 1983. Ca channel modulation by transmitters, enzymes and drugs.

236. Reuter, H., S. Kokubun, and B. Prodihon. *J. Exp. Biol.* 124:191–201, 1986. Modulation of cardiac Ca channels.

237. Revel, J. P. et al. *Annu. Rev. Physiol.* 47:263–279, 1983. Chemistry of gap junctions.

238. Rogawski, M. A. *Trends Neurosci.* 8:2143–219, 1985. The A current.

238a. Rosa, P. A. and E. Herbert. *J. Exp. Biol.* 89:217, 1980.

239. Saimi, Y. et al. *Proc. Natl. Acad. Sci. U.S.A.* 80:5112–5116, 1983; *J. Exp. Biol.* 88:305–325, 1980. Ion channels in Paramecium.

240. Salkoff, L. B. and R. J. Wyman. *J. Physiol. (London)* 337:687–709, 1983. Ion currents in Drosophila flight muscles.

241. Salkoff, L. B. et al. *Science* 237:744–748, 1987. Sequence of Na^+ channels proteins in Drosophila.

242. Satow, Y. and C. Kung. *J. Exp. Biol.* 78:149–161, 1979. V-sensitive Ca channels and transient inward I of Paramecium step depolarization.

243. Satow, Y. and C. Kung. *J. Exp. Biol.* 88:293–303, 1980. Ca-induced K-outward current in Paramecium.

244. Schaefer, M. et al. *Cell (Cambridge, Mass.)* 47:457–467, 1985. Aplysia neurons express a gene encoding multiple FMRFamide neuropeotides.

245. Schaller, H. C. et al. *Proc. Natl. Acad. Sci. U.S.A.* 78:7000–7004, 1981. Morphogenetic peptides from Hydra.

246. Scheller, R. H. et al. *Science* 225:1300–

1308, 1984; *Trends Neurosci.* 6:340–345, 1983. Neuropeptides mediate behavior in Aplysia.

247. Schofield, P. R. et al. *Nature (London)* 328:221–227, 1987. Sequence and functional expression of GABA receptor shows ligand-gated receptor super-family.

248. Schwartz, T. L. et al. *Nature (London)* 331:137–142; 143–145, 1988. Multiple K-channel components produced by alternative splicing at shaker locus in Drosophila.

249. Scott, J. et al. *Nature (London)* 302:538–540, 1983. Precursors for nerve growth factors.

250. Shreeve, S. M. et al. *Proc. Natl. Acad. Sci. U.S.A.* 582:4842–4846, 1985. Alpha and alpha-2 receptors may be isoreceptors.

251. Siegelbaum, S. et al. *Nature (London)* 249:413–417, 1982. 5-HT response mediated by cAMP activating an M type K channel in Aplysia.

252. Siegelbaum, S. et al. *Nature (London)* 299:413–417, 1982. Serotonin and cAMP close K+ channels in Aplysia sensory neurons.

253. Sigurdson, W. J., E. Morris, B. Bregden, and D. Gardner. *J. Exp. Biol.* 127:191–209, 1987. Sketch activation of a K+ channel in mollusc heart cells.

254. Simon, E. and J. Hiller. In *Basic Neurochemistry* (G. J. Siegel et al., eds.), Chapter 13, pp. 271–285. Raven Press, New York, 1989. Opioid peptides and opioid receptors.

255. Sims, S. M., J. Singer, and I. Walsh. *J. Physiol. (London)* 367:503–529, 1985. M currents in toad stomach muscle.

256. Siwicki, K. B. Beltz, and E. Kravitz. *J. Neurosci.* 7:522–532, 1987. Proctolin in seratonergic, dopaminergic and cholinergic neurons in lobster immunocytochemical staining.

257. Skaw, K. A. *Comp. Biochem. Physiol. C* 83C:225–227, 1986. Mammalian ACh as molecular forms.

258. Slaughter, M. M. and R. F. Miller. *J. Neurosci.* 3:1701–1711, 1983. Excitatory amino acids—3 types of receptors.

259. Snyderman, R. and E. Goetzl. *Science* 213:830–837, 1981. Chemotaxis of leukocytes.

260. Spencer, A. N. *Can. J. Zool.* 66:639–645, 1988. Effects of Arg-Phe-amide peptides on identified motor neurons in the hydromedusa *Polyorchis penicillatus*.

261. Spira, M. E. and M. V. L. Bennett. *Brain Res.* 37:294–300, 1972; *Science* 194:1065–1067, 1976. Synaptic control of electronic coupling between neurons.

262. Spray, D. C. et al. *J. Gen. Physiol.* 77:77–93, 1981; *Proc. Natl. Acad. Sci. U.S.A.* 79:441–445, 1982. Electrical coupling between embryonic cells.

263. Spray, D. C. et al. *Am. J. Physiol.* 248:H753–764, 1985; *Proc. Natl. Acad. Sci. U.S.A.* 583:5494–5497, 1986; *Annu. Rev. Physiol.* 47:281–303, 1985. Physiology and pharmacology of gap junctions.

264. Starrett, A. M. and B. E. Brown. *Life Sci.* 17:1253–1256, 1975. Structure of the pentapeptides Proctolin, a proposed neurotransmitter in insects.

265. Stevens, C. F. *Fed. Proc., Fed. Am. Soc. Exp. Biol.* 34:1364–1370, 1975; Stevens, C. F. and R. Tsien. *Nature (London)* 297:501–504, 1982. Gating of sodium and calcium channels.

266. Stevens, C. F. *Nature (London)* 322:210–211, 1986. Editorial; in *Proteins of Excitable Membranes* (B. Hille and D. Fambrough, eds.), Chapter 6, pp. 99–108. Wiley, New York, 1987. Na channel structure and function.

267. Strumwasser, F. In *Brain Peptides* (D. Krieger, ed.), Chapter 7, pp. 183–215. Wiley, New York, 1983. Peptidergic neurons and neuroactive peptides in molluscs.

268. Stuart, D. and F. Strumwasser. *J. Neurophysiol.* 43:488–498, 1980. Egg laying hormones in Aplysia.

269. Suzuki, M., C. L. Prosser, and W. DeVos. *Am. J. Physiol.* 250:G28–G34,

1986. Waxing and waning of slow waves in intestinal muscle.

270. Takahashi, M., W. Catterall, et al. *Proc. Natl. Acad. Sci. U.S.A.* 84:5478–5482, 1987. Subunit structure of DHP sensitive Calcium channels from skeletal muscle.

271. Takayanagi, H. and N. Takeda. *Comp. Biochem. Physiol. A* 91A:613–620, 1988. Coexistence of FMRFamide Met-enkephalin and serotonin in neurons of Achatina and Aplysia.

272. Takeuchi, A. and K. Auodera. *Nature (London) New Biol.* 242:124–126, 1973; *Comp. Biochem. Physiol. C* 72C:237–239, 1982. Amino acid transmitters in crayfish.

273. Tanabe, T. et al. *Nature (London)* 328:313–318, 1987. Primary structure of receptor for Ca channel from skeletal muscle.

274. Taussig, R. and R. H. Scheller. *DNA* 5:453–462, 1986. FMRFamide gene encodes sequences related to mammalian brain peptides.

275. Taussig, R., J. R. Nambu, and R. H. Scheller. *Semin. Ser.—Soc. Exp. Biol.* 33:299–309, 1988. Evolution of peptide hormones: An *Aplysia* CRF-like peptide.

275a. Taylor, P. and J. Brown, in *Basic Neurochemistry* (G. Siegel, ed.), Chapter 10, pp. 203–232. Raven Press, New York, 1988. Acetylcholine receptors.

276. Tempel, B., Y. Jan, and L. Jan. *Nature (London)* 332:837–839, 1988; *Science* 237:770–775, 1987. K channel gene cloned from mouse brain.

277. Teranishi, T., K. Negishi, and S. Kato. *Nature (London)* 301:243–246, 1983. Dopamine effects on carp retina.

278. Thompson, S. H. *J. Physiol. (London)* 265:465–488, 1977. Three pharmacologically distinct K channels in molluscan neurons.

279. Thorndyke, M. C. *Regul. Pept.* 3:281–288, 1982. CCK/gastrin-like immunoreactive neurons in the cerebral ganglion of protochordate ascidians Styela clara and Ascidiella aspersa.

280. Timpe, L., L. Y. Jan et al. *Nature (London)* 331:143–145, 1988. Expression of K channels from shaker cDNA in Xenopus oocytes. *Science* 237:770–725, 1987.

281. Trautmann, A. and A. Marty. *Proc. Natl. Acad. Sci. U.S.A.* 81:611–615, 1984. Activation of Ca-dependent channels by carbamoylcholine in rat lacrimal glands.

282. Tsien, R. W. *Annu. Rev. Physiol.* 45:341–358, 1983. Calcium channels in excitable cell membranes.

283. Tsien, R. W. et al. In *Problems of Excitable Membranes* (B. Hille and D. M. Fambough, eds.), Chapter 10, pp. 167–187. Wiley, New York, 1987. Multiple Ca channels in excitable cells.

284. Vale, W. et al. In *Brain Peptides* (A. Gotto et al., eds.), pp. 71–88. Elsevier, Amsterdam, 1979. Analogs of LRF and somatostatin.

285. Valentinco, T. *J. Comp. Physiol.* 157:537–545, 1985. Chemoreception in sea star Marthasterias.

286. Venter, J. C. *Proc. Natl. Acad. Sci. U.S.A.* 581:272–276, 1984. Evolution and structure of ACh receptors.

287. Watson, W. *Brain Res.* 213:449–454, 1981. Proctolin acts on Limulus cardiac muscle to increase amplitude of contraction.

288. Weber, E. et al. *Trends Neurosci.* 6:333–336, 1983. Multiple ligands for opioid receptors.

289. Wehner, R. and E. Hildebrand. *J. Exp. Biol.* 119:321–334, 1985. Voltage dependent Ca-current in Paramecium inhibited by divalent cations.

290. Weigel, R. J. et al. *Am. J. Physiol.* 237:C247–C256, 1979. Ion currents in circular muscles of intestine.

291. Weissmann, G. et al. *Proc. Natl. Acad. Sci. U.S.A.* 83:2914–2918, 1986. Synergy between phorbol esters and Ca ionophore for aggregation of sponge cells.

292. Wolozin, B. L. and G. Pastemak. *Proc. Natl. Acad. Sci. U.S.A.* 78:6181–6185, 1981. Morphine and enkephalin binding sites in central nervous system.

293. Yanagawa, M. and M. Kobayashi. *Comp. Bioch. Phys.* 90C:73–77, 1988. Potentration of radular contractions by molluscan neuropeptides.

294. Yang, H.-Y. et al. *Proc. Natl. Acad. Sci. U.S.A.* 82:7757–7761, 1985. Two peptides cross react with antiserum against FMRFamide.

295. Kirschner, L. B. In *Environmental and Metabolic Animal Physiology* (C. Ladd Prosser, ed.). pp. 13–107. Wiley-Liss, Inc., New York, 1991.

Chapter
2 | Animal Movement

C. Ladd Prosser

Motility is a general property of living organisms; it can occur within cells or can result in movement of whole animals or parts of animals. After cells in prebiotic evolution became so large that diffusion was inadequate for distribution of solutes, protoplasmic streaming evolved; this provided transport from one region of a cell to another. Examples are cyclosis in plant cells and axoplasmic flow in neurons, also cleavage in cell division, chromosome movement in mitosis and meiosis, budding (yeast), and the extension of growth cones of neurons. Translational movement occurs in the tumbling of bacteria, in swimming of sperm by use of flagella, and of ciliated protozoans by cilia. Contraction of muscles is the best understood form of motility; it results from the interaction of two proteins—actin and myosin. Muscles that are attached to skeletal elements develop tension and shorten very little, contract isometrically. Muscles that shorten with little change in force, for example, around hollow organs, contract isotonically. Speed and type of movement are adapted to functions of moving parts and to life style, for example, adaptations for fast running or slow walking.

All biological movement is based on interactions between specific proteins. In nearly every known movement system, calcium is an activator and ATP is a source of energy. A few contractile proteins have not yet been characterized, for example, spasmonemes of ciliated protozoans, where activation may not depend on ATP but does depend on calcium (117, 120). Two major systems for cellular movement have evolved—tubulins and actin–myosins. Tubulins occur in flagella and cilia, mitotic spindles, and axoplasmic transport in neurons. Actins may move independently by polymerization–depolymerization but in general actins form intracellular filaments that move by interacting with larger molecules such as myosins. In cell division, the movement of chromosomes in spindles during anaphase is by tubulin; the cleavage of cytoplasm is by an actin–myosin contraction.

Muscles

Muscles show diversity in (1) speed of contraction; (2) alignment or nonalignment of thick and thin myofilaments; (3) fiber size, with long or short distances between cell membrane and contractile protein; (4) short-term energetics; (5) long-term chemical changes, especially of contractile proteins; (6) modes of calcium regulation; (7) motor impulse patterns; (8) trophic effects of nerves; (9) endogenous myogenicity; and (10) specializations for maintained tension or for fast oscillatory movements.

Speed of Contraction

Muscles differ 10 thousandfold in speed of contraction. Speed is measured by the rate of development of tension, or by rate of shortening, by rate of relaxation, by frequency of stimulation for mechanical fusion, by degree of facilitation on repetitive stimulation, by velocity of shortening under different loads (force–velocity curves), by tension–length relations in resting and active states, and by redevelopment of tension after release at different times during a contraction. Fast contractions permit rapid behavior—attack, escape, rapid locomotion, and emission of sound; slow contractions serve visceral functions and locomotion of hollow-bodied animals.

Table 1 gives examples of muscle speeds—fast to slow. The fastest movements are those of some flies, bees, and beetles, in which beats as frequent as

TABLE 1. Speed of Muscle Contraction and Relaxation (from CAP III, Ref. 120a)

Muscle	Contraction Time	Relaxation Time
Indirect flight and stridulation muscles of insects	2–4 ms	
Swim bladder (toadfish)	4–5 ms (20°C)	5–6 ms
Internal rectus (cat)	7.5–10 ms (36°C)	
Soleus (cat)	70–80 ms (36°C)	80 ms
Posterior lateral dorsi (chicken)	50 ms (25°C)	200 ms
Anterior lateral dorsi (chicken)	300–400 ms	0.75–1 s
Sartorius (frog)	40 ms (10°C)	50 ms
Molluscan fast adductor (*Pecten*)	45 ms (15°C)	40–100 ms
Proboscis restractor (*Golfingia*)		
Twitch	85 ms (15°C)	95 ms
Tonus		2–5 s
Spindle muscle (*Golfingia*)	1–1.5 s	6.5–9 s
Byssus retractor (*Mytilus*)		
Phasic	2 s	1–7 s
Tonic	2–3 s	5–1000 min
Holothurian lantern retractor (*Thyone*)	4 s	5–7 s
Holothurian body wall retractor (*Thyone*)	0.5–1 s	2–8 s
Mammalian intestine	2–7 s (37°C)	5–10 s
Turtle intestine	30 s	360 s
Coelenterate		
(*Aurelia*) phasic	0.5–1 s	0.6–1 s
(*Metridium*) tonic	30 s	1–2 min
Sponge (*Microciona*)	1–3 s	10–60 s

1000/s have been recorded; however, these are not full activation and relaxation cycles but rather result from mechanical resonance as described later. In many jumping insects, the power for a jump comes from both the muscle and the cuticle of leg joints; elastic energy is stored in the cuticle which, like a spring, adds to the power of a muscle contraction (49a). The fastest muscle contractions that are synchronous with activating nerve impulses are in insects. Muscles that move the wings of a whitefly beat at frequencies up to 181 Hz (141, 171). Muscles for singing by a cicada contract at 550 Hz (75), a katydid at 200 Hz (122). In a tettigonid cricket's flight and song are mediated by the same muscle; in flight the muscle beats at 25 Hz, in song at 100 Hz (74). Sound-producing muscles of some fish contract at 100–200 Hz. The speeds of the preceding muscles are remarkable in that they occur at relatively low (ambient) temperatures.

In mammals, the cricothyroid muscles of bats that produce high-frequency cries, contract as fast as 200/s. A hovering hummingbird beats its wings at 33–45 times/s, a finch at 22–25/s. The fastest muscles in most mammals are the extrinsic ocular muscles that contract in 5–6 ms at shortening speeds of 65 μm/s. Most muscles that move limbs of vertebrates and arthropods (except indirect flight muscles of small insects) have contraction times in the range of 40–100 ms, slightly faster contractions occur at high rather than at low temperatures. Some nonstriated postural muscles of invertebrates contract nearly as fast as striated muscles of fishes or reptiles (117).

Visceral muscles—gastrointestinal, urogenital, and vascular—are slow, with contraction times in seconds. Cardiac muscles are intermediate in speeds. Mollusc adductors contract as rapidly as the postural muscles of other animals and have the further capacity of maintaining tension for minutes or hours. The slowest muscles are those that mediate the swaying movements of sea anemones with contraction times of minutes. It is noteworthy that the speeds of contractions—minutes to milliseconds—are produced by similar molecular mechanisms.

No single explanation can account for the full range of speeds of movement. Speed of contraction depends on the amount of series and parallel viscoelastic elements and on the alignment or randomness of myosin and actin filaments. Speed is related to fiber diameter (distance from cell membrane to contractile proteins). Speed is related to patterns of energy mobilization—oxidative or glycolytic. Another determinant of speed is the rate of release of calcium and of rebinding of calcium to the sarcoplasmic reticulum. Speed is also determined by intrinsic properties of contractile proteins—fast or slow myosins that differ in kinetic activity of the actin-activated myosin adenosine triphosphate (ATPase) that releases energy. Speed is also related to the activation of a muscle—whether by action potentials or junction potentials or whether it shows endogenous activity.

Wide-Fibered and Narrow-Fibered Muscles

The classification of muscles as striated or nonstriated (smooth) is based on gross morphology. A functional classification relates structure to distance for the inward conduction of signals between fiber membrane and activation of contractile filaments. Classification as wide fibered and narrow fibered is given in Table 2. Wide-fibered muscles usually have transverse alignment of thick and thin filaments, that is, are cross-striated. These muscles have a fiber diameter 50–200 μm, are multinucleate, have invaginating transverse tubules that conduct excitatory signals inward from cell membranes to Ca^{2+}-releasing sarcoplasmic reticulum (SR), and thus activate actomyosin fila-

TABLE 2. Classification of Narrow-Fibered and Wide-Fibered Muscles

Narrow-Fibered Muscles	Wide-Fibered Muscles
Random thick and thin filaments or striations present transversely or diagonally	Transverse alignment of thick and thin filaments
Fiber diameters 2–10 μm in cylindrical (spindle) fibers or from top to bottom in ribbon fibers	Fiber diameters 50–2000 μm
Uninucleate	Usually multinucleate
Surface:volume ratio high	Surface:volume ratio low
No T-tubules, sparse SR; membrane caveolae in vertebrates	Membrane invaginations, large or as T-tubules; extensive SR
Direct membrane coupling to proteins	Indirect coupling to proteins
Ca^{2+} action potentials predominant, sometimes Ca^{2+} plus Na^+	Na^+ action potentials or junction potentials, rarely Ca^{2+}
Dense bodies or plaques for thin filament attachment, narrow Z-lines in the diagonally striated	Z-bands as transverse disks
Ca control by phosphorylation of myosin by Ca-dependent myosin kinase. Ca-sensitive light chain of some myosins	Tropomyosin–troponin modulation of Ca control of actin–myosin interaction. Some myosin phosphorylation of uncertain function
Activation:	
Nerve activated	All nerve activated
Myogenically rhythmic	
Electronic coupling by nexuses in myogenically rhythmic, not in neurally activated	No electrotonic coupling
Series, parallel	Parallel

ments. Narrow-fibered muscles usually do not show transverse alignment of thick and thin filaments, that is, they are called smooth muscles; these fibers measure 2–10 μm from cell surface to contractile filaments, are spindle or ribbon shaped, usually uninucleate, have no T-tubules, have sparse SR; have direct coupling between cell membrane and contractile filaments, either by influx of Ca^{2+} or by release of Ca^{2+} from intracellular stores. Wide-fibered muscles are excited by motor nerve impulses; many narrow-fibered muscles are nerve-activated, others are endogenously rhythmic, with neural modulation. Wide-fibered muscles are usually activated by Na^+ or Ca^{2+} action potentials or by endplate potentials. Nar-

row-fibered muscles have pacemaker potentials of Ca^{2+} or combined Ca^{2+}-Na^+ action potentials. In wide-fibered muscles tension is generated by fibers in parallel; in narrow-fibered muscles tension is generated by fibers in series as well as in parallel. Some narrow-fibered muscles are ribbon shaped, relatively wide, but with the distance <5 μm from top or bottom membrane to contractile filaments.

In some narrow-fibered muscles the myofilaments are arranged in a spiral, hence the fibers appear to have diagonal striations. In some of these, the central region of a fiber contains nuclei, mitochondria and other organelles, and a thin cortex contains myofilaments (Fig. 1). In some muscles and myoepithelial cells the

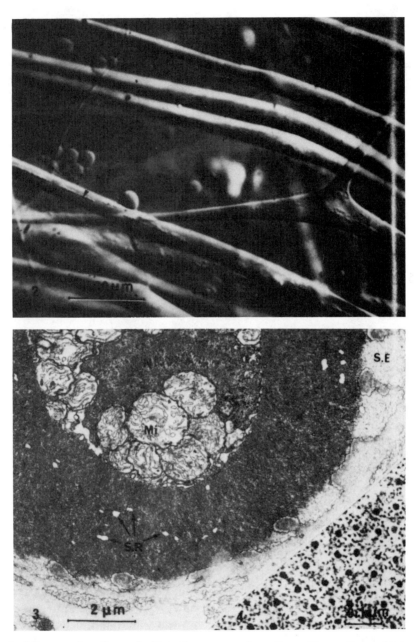

Figure 1. Micrographs of muscle fibers from ctenophore *Beroe*. Upper figure: Light micrograph of fibers, one branched, calibration 30 μm. Lower figure: EM cross-section of one fiber, central core contains nucleus (N), and mitochondria (Mi), outer region contains sarcoplasmic reticulum (SR) and contractile filaments, calibration 2 μm. Inset: Transverse section showing thick and thin myofilaments, calibration 0.1 μm. [From (622).]

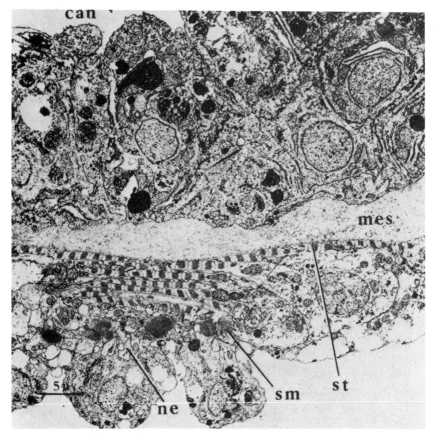

Figure 2. Section through diploblastic body wall of a medusoid coelenterate Stomoca: (can) Central canal bounded by epithelium; (mes) mesoglea between the two cell layers; (st) striated processes in circular myoepithelial cell; (sm) nonstriated process in radial direction; and (ne) nerve fiber of nerve net. [From (94a).]

contractile filaments are along one side of each fiber. Transverse striations of thick and thin filaments have evolved many times and nonstriated (smooth) muscles cannot be considered as more primitive than striated muscles.

Some animals that are considered primitive (cnidarians) have both cross-striated and nonstriated myoepithelia (Fig. 2). Myoepithelia are of common occurrence—flat fibers with wide sarcomeres (pharynx of a polychaete Scylla), flat epithelia of vertebrate embryos, and contractile cells in sweat glands (7a).

Some muscles are intermediate be-tween narrow fibered and wide fibered. A few narrow muscle fibers develop striations when contracted. Contractile fibers from the body wall of an ascidian show no striations when relaxed and fiber dimensions 380 μm long, 3–10 μm in diameter, but alternate light and dark bands appear when contracted and contracted fiber dimensions are 150 μm long, 10–15 μm in diameter (49). Fibers of frog auricle are striated, but otherwise are like narrow-fibered muscle. Myocardial cells from fish, Perca, are 133 μm long, 4.2 μm in diameter, striated, lack T-tubules, have sparse SR, and are mononucleate; they are an example of narrow-fibered striated muscle (79). In the prochordate amphioxus (*Branchiostoma*) lateral muscle fibers are 1–2 μm thick, 10–300 μm wide, and 600 μm long; thick and thin filaments are aligned, there is little sarcoplasmic reticulum (SR), spikes are sensitive to TTX (like Na^+ nerve spikes); when Na^+ currents are blocked, there are Ca^{2+} spikes (60). Isolated myocardial fibers of quail are spindle shaped, 10.2 μm in diameter, 289 mm long, with striations in 2.18-μm sarcomeres (80).

In summary, the classification of muscles as striated or smooth (nonstriated) is based on gross histology. Transverse alignments of thick and thin filaments are present in many muscles and myoepithelia and alignment may become evident in nonstriated muscles on contraction. Striations make for fast contractions. Classification of muscles as narrow fibered and wide fibered is functional. The coupling of membrane excitation to contractile filaments depends on fiber size and shape. Narrow-fibered muscles are <10 μm in diameter if cylindrical, 5 μm from the upper or lower surface to contractile filaments if ribbon shaped. Some wider fibers have a central core of organelles with myofilaments <5 μm from the cell surface.

Patterns of Innervation in Vertebrate Striated Muscles

Motor nerve impulses activate muscles to contract or modulate those that are endogenously rhythmic (Table 3). Many postural muscles of vertebrates have focal (uni- or biterminal innervation) with a single motor axon to a group of muscle fibers. A motor unit consists of a motor neuron together with the muscle fibers that it innervates. Gradation of contraction in vertebrate animals is by change in the number of motor units activated in the central nervous system and by change in the frequency of firing of single units. An all-or-none motor impulse triggers release of a transmitter (acetylcholine, ACh) that diffuses to the postsynaptic membrane where a graded endplate potential (EPP) is evoked. The EPP triggers an action potential that is conducted in the muscle membrane. Applied ACh is effective only in the region of the motor endplate. In most vertebrate uniterminal fibers, the muscle action potential is a Na^+ spike. Examples of muscles with uniterminal and unineuronal innervations are phasic white muscles. These include most fibers of frog sartorius, rat extensor digitorum, and chicken posterior latissimus dorsi.

A second category of vertebrate striated muscles consists of those with distributed (multiterminal), sometimes polyneuronal, innervation. These muscles give graded nonconducting potentials that are distributed endplate potentials. These muscles show sensitivity to ACh over the entire fiber; the fibers can be used as bioassay for ACh. The resting potentials of such fibers (in amphibians) are low −50 to −60 mV, compared with −70 to −80 mV in fast muscles. Electrical responses show various degrees of facilitation on repetitive stimulation. Examples of multiterminal innervation are most fibers of frog rectus abdominis and chicken anterior latissimus dorsi (ALD). Most muscles with unineuronal innervation are fast, lack myoglobin, and are "white," whereas most multiterminal facilitating muscles of vertebrates are slow and "red."

Striated muscles in fishes are of 2 to 5 fiber types according to the species. Three general classes are recognized (72):

1. Slow red fibers have unineuronal, multiterminal innervation, and are activated by local junction potentials. The fibers are maximally activated by stimuli at 15–20 Hz.

TABLE 3. Classification by Innervation Pattern and Energy Pathway

Cross-striated, wide fibered, long, unbranched fibers. Skeletal muscles
   Uniterminal, unineuronal; muscle action potentials
      Fast twitch, white, glycolytic, fast myosin (frog sartorius, chick PLD, rat
         EDL)
      Slow twitch, white, glycolytic, slow myosin (rat soleus, some fish muscles)
   Multiterminal, unineuronal
      White, tonic, electrical responses are graded psps
      Red, myoglobin, indirect, slow to fatigue
      Spiking or large psps, usually nonfacilitating
      Nonspiking, facilitating
   Multiterminal, polyneuronal, mostly white or pink
      Conduction by nerves (psp responses or spikes) (teleost white, neurogenically
         driven arthropod hearts)
      Multiple types of innervation, excitatory and inhibitory (crustacean postural,
         insect asynchronous)
      Oscillatory muscles (insect synchronous)
   Endogenously rhythmic or driven by nonneural pacemakers
      Cardiac muscles, fibers often branched, nexal connections
   Narrow-fibered, nerve activated, postural
      Multiunitary (Sipunculid)
      Diagonally striated, shearing (many annelids, nematodes)
      Random filaments, either dual or unitary innervation
      Large filaments, dual innervation excitatory and relaxing (paramyosin
         containing catch muscles of molluscs)
   Narrow-fibered, endogenously rhythmic, nerve modulated, random filaments,
      (vertebrate unitary smooth muscle)
   Myoepithelia, nerve-activated, nexal connections
      Cross-striated or nonstriated (coelenterates)

2. Pink fast fibers have single basket-type endplates at one end from which action potentials propagate along the muscle fiber.

3. White fibers have polyneuronal innervation, both junction potentials and conducted spikes and these fibers need stimulation at 200–300 Hz for maximum contraction.

Both types 2 and 3 are used mainly for fast bursts of swimming. Type 3 fibers are best known in recently evolved teleosts. Sarcoplasmic reticulum is more abundant in white fibers, myofibrils constitute 80–96% of volume compared with 49–60% in red fibers. In a dogfish, the peripheral fibers of myotomal muscle are red and have distributed innervation; central fibers are white and each fiber has two motor axons ending in a single endplate (Fig. 3) (11). Similar dual innervation of single endplates occurs in fast fibers of elasmobranchs, teleosts, and urodeles (12).

The sequence of precontraction events in the activation of vertebrate phasic striated muscles is as follows: (1) an all-or-none motor nerve impulse, (2) liberation of transmitter, ACh followed by binding of ACh to its receptor (properties of ACh receptors are given in Chapter 1); (3) generation of graded endplate potential, a nonspecific increase in ion conductance, hydrolysis of ACh;

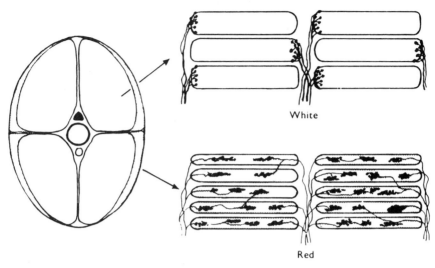

White

Red

Figure 3. Innervation of the two main types of myotonal muscle fibers of a dogfish. Focal innervation in central fibers, distributed innervation in peripheral fibers. [From (11).]

(4) conduction of all-or-none impulses by Na^+ conductance change in muscle membrane (sarcolemma); (5) conduction of signal inward via T-tubules; (6) excitation coupling to sarcoplasmic reticulum (SR); and (7) release of calcium ions. In slow (tonic) muscle, step (4) may be omitted. Calcium is stored in the terminal enlargements (cisternae) of SR. Ionic calcium is in a steady state with bound Ca^{2+} in the SR. On activation from the T-tubules, the SR release of Ca^{2+} that binds to troponin-C of the tropomyosin–troponin complex as discussed later. The SR membrane contains a high Ca^{2+} affinity Ca^{2+}-ATPase. The lumen side of the tubule has calsequestrin, a protein of high binding capacity but low affinity for Ca^{2+}. Uptake storage of Ca^{2+} requires energy from ATP.

In summary, in skeletal muscles of vertebrates with uni- or biterminal, unineuronal innervation, activation is by muscle action potentials. Vertebrate muscles with multiterminals are usually activated by many local junction potentials. Fibers activated by all-or-none action potentials and those activated by graded junction potentials have ACh as a transmitter and both have invaginating T-tubules, which excite SR that releases Ca^{2+}. Gradation of movement is determined by the number and frequency of firing of the motor units; a unit consists of one motor neuron plus all the muscle fibers it innervates.

Multiterminal, Multineuronal Muscles

Crustacean Muscles

A third type of striated muscle has multiple innervation by axons that evoke different responses of a muscle fiber (138). One fiber may receive two or more motor nerve fibers with diffuse endings. Best examples are arthropod muscles in which there may be 2–5 excitatory axons to one muscle fiber and one axon may innervate many muscle fibers. In crustacean muscle, one axon, the "fast" motor fiber, elicits electrical and mechanical responses that approach all-or-none responses and show little facilitation (Fig. 4). Another axon, the "slow" motor fiber, elicits graded highly facilitating responses. In some muscles, a given axon may evoke a fast response at one end of the muscle and a slow response at the other end. One mo-

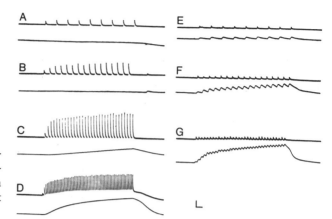

Figure 4. Electrical (upper) and mechanical (lower) records form muscle fibers of a king crab when stimulated by slow (A–D) and fast (E–G) motor axons. [From (90c).]

tor axon may innervate several muscles. In addition to the excitatory axons, there is usually one inhibitory axon and inhibition may be presynaptic or postsynaptic. Where inhibitor and excitor axons can be identified, the synaptic vesicles of excitatory axons are spherical, those of inhibitory ones are small and oval or irregular. Postsynaptic inhibition hyperpolarizes the muscle membrane; it increases conductance of small ions, principally Cl^- and is depolarizing if the membrane potential is made negative to the Cl^- equilibrium potential. Presynaptic inhibition diminishes the release of excitatory transmitter and inhibitory terminals can be seen on excitatory endings.

Most crustacean muscle fibers are large—some as much as 1000–2000 μm in diameter. There are wide invaginations of the plasma membrane, which branch to form T-tubules inside the muscle fiber (Fig. 5). Ion permeabilities of the invaginations are unlike those of the plasma membrane. Barnacle muscle fibers may be 1 or 2 mm in diameter and sarcomeres average 9.2 μm in length. Fibers as large as 4 mm occur in leg muscles of the king crab.

Response amplitude varies with frequency of excitation. When stimulation is at high frequency, speed of contraction and rate of facilitation increase (Fig. 4).

Another type of control of contraction depends on the patterning of motor impulses (8). For example, the claw opener of the crab *Eupagurus* has a single excitor axon; stimulation at low frequency elicits a slowly facilitating contraction yet interpolated single shocks in a train elicit quick strong contractions. One axon can elicit either slow or fast responses according to how it is stimulated.

One excitatory transmitter, glutamate, serves for the different speeds of contraction. The difference between fast and slow response is determined by the amount of branching of a motor axon over a muscle fiber. Facilitation increases the probability of release of transmitter quanta. In the crayfish claw opener (abductor) it is calculated that at 5/s each impulse releases on the average 2.2 quanta of transmitter. (One quantum equals the amount for one miniature postsynaptic potential.) The inhibitory transmitter is γ-aminobutyric acid (GABA) for both presynaptic and postsynaptic inhibition. γ-Aminobutyric acid hyperpolarizes by increasing conductance for Cl^- and is antagonized by picrotoxin. In the crayfish opener, the equilibrium potential for the inhibitory postsynaptic potential (IPSP) is -87 mV or ~5–10 mV negative to the resting potential (124).

In some crustaceans, one claw is longer

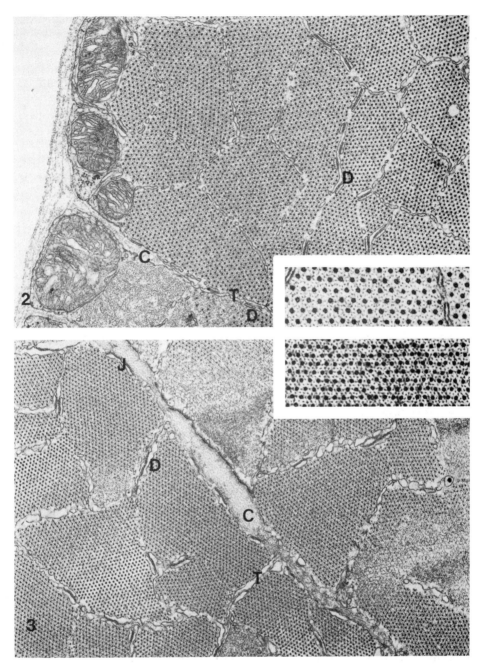

Figure 5. Portion of a long-sarcomere fiber from carpopodite flexor of crayfish. (C), sarcolemmal cleft enters fiber, gives rise to T-tubules (T). (J) peripheral junction similar to dyads (D) between SR and T-tubules. Inset: high magnification, upper part with 12 thin filaments per one thick filament, lower part 6 thin per thick. [From (38).]

than the other, crusher and cutter in lobster, snapper and pincer in a shrimp *Alpheus*. If the larger claw is removed, the remaining one reverses in form. In lobster the crusher fibers are entirely slow (sarcomeres >6 μm), while the cutter fibers are fast (sarcomeres <4 μm). A pincer is converted to a snapper by degeneration of fast fibers and hypertrophy of slow fibers, transformation is prevented if both claws are exercised, that is, if no substrate is provided for walking, or if motor neurons are destroyed. The fiber transformation depends on input to the nervous system, motor neurons that control muscle fiber differentiation (58).

In some arthropods, flexors in the legs are not antagonized by extensors but by the hydrostatic pressure developed in the hemolymph. In a scorpion claw, the closing is active, the opening is a result of fluid pressure, the flexor has two excitatory axons, one facilitating, the other nonfacilitating. Antagonistic muscles are also lacking in appendages of spiders and barnacles.

In summary, crustacean (and some insect) postural muscles receive a few, highly branched motor axons. One axon may innervate many muscle fibers. Each muscle fiber is usually innervated by two or more overlapping excitatory axons that differ in the facilitation of both electrical and mechanical responses. Glutamate is the excitatory transmitter for either fast or slow responses and the speed of response depends on the distribution of motor endings over a muscle fiber. Many multiply innervated fibers also receive an axon that inhibits both presynaptically and postsynaptically by GABA as transmitter.

Insect Muscles
Many insects, especially large slowly flying Lepidoptera and Neuroptera, resemble crustaceans in having muscles with multiple innervation by few axons. A muscle of a leg or wing may have fast

and slow excitor or fast excitor and an inhibitor. The muscles that move wings are direct flight muscles and one spike per fiber triggers a contraction and wingbeat. These muscles are synchronous.

In *Locusta*, stimulation of a slow (6 μm) axon elicits little contraction in leg muscles at frequencies below 10/s but with increased frequencies, the contractions increase to a maximum at 150/s. In the coxal depressors of a cockroach, the largest motor axons elicit fast contractions, the next largest slow contractions, and the smallest axon is inhibitory (112). In the katydid, *Neoconocephalus*, stridulation (chirping) is produced by forewing movements resulting from synchronous contractions as fast as 145–212/s with one action potential per contraction. The metathoracic posterior coxal levator of the cockroach *Periplaneta* receives four excitor and two inhibitor axons; three of the excitors are small and evoke slow responses, one is large and elicits a fast response. A common inhibitor serves both synergistic and antagonistic muscles of the leg (113). As in crustaceans, not every fiber in an insect muscle receives the same innervation (112). Glutamate is the excitatory transmitter and GABA the inhibitory one. Some of the synchronous muscles follow nerve impulses at high frequencies and have extensive sarcoplasmic reticulum relative to the amount of myofibrils.

In fast-flying insects—Diptera, Hymenoptera, and some Coleoptera—flight muscles are attached by a complex skeletal lever system. In more primitive insects and those with large wings—Orthoptera and Lepidoptera—muscle attachment is direct, that is, on a wing articulation (74). In some Orthoptera and Homoptera sound-producing organs have muscle-driven tymbae that vibrate at high frequency. These high frequency muscles contract several times to a single nerve activation; they are oscillatory or asynchronous. High frequency contractions are due to mechanical resonance in the muscle proteins.

Asynchronous muscles show pronounced stretch activation and release inactivation, phenomena seen to a lesser degree in some synchronous muscles. During extension, as in a tension–length measurement, tension leads length and, on shortening, tension is higher than on extension, hence the tension–length curve is a closed loop (Fig. 6). The tension lead over length is apparent during sinusoidal repetitive stretching (Fig. 6). The area of the loop is the difference between the work put into the muscle and the work obtained, that is, the work done by the muscle per cycle. The product of work times frequency of oscillation is the mechanical power in joules per kilogram per cycle ($P = J/kg \cdot cycle$). Power is maximum when onset of a twitch coincides with onset of the shortening half of a length cycle (76).

Stretch activation is a delayed, transient increase in tension above isometric levels following muscle stretch; release inactivation is a delayed, transient drop in tension following release from stretch. Because of stretch activation and release inactivation, more work is generated by the muscle in shortening than is required to restretch the muscle to its original length. Following shortening the muscle is inactivated and can be easily elongated, a process which, in turn, hastens reactivation. Thus with asynchronous muscle there is net mechanical work output during a contraction–relaxation cycle, whose energy comes from the hydrolysis of ATP. Asynchronous muscles are arranged in antagonistic sets or attached to elastic skeletal elements, and it is the antagonistic muscles or the stretched skeletal elements that restretch the muscle following its contraction and allow oscillatory contraction.

The frequency of contraction of asynchronous muscles varies with the mechanical load. In fast flying or singing insects with asynchronous wing muscles, the wing stroke frequency can be increased by cutting off part of the wings (unloading the muscle) or decreased by adding weight to the wings. The asynchronous muscles of some small insects (midges) have the highest contraction frequencies known, up to 1000 Hz in intact insects and up to 2000 Hz in insects with reduced wings. A blowfly flight muscle may beat at 125/s when the potentials occur at 3/s. Asynchronous muscles give ordinary isometric contractions if attached to an immovable load rather than a resonant lever. The isometric contractions of asynchronous muscles are not unusually fast. A nerve impulse or burst of impulses turns the muscle on for a rather long time, for a period during which the muscle can oscillate many times if attached to an appropriate load. A motor impulse triggers a contraction; the muscle shortens slightly, then the elastic skeletal element or antagonistic muscle stretches the activated muscle, which redevelops tension and shortens again and the cycle is repeated so long as the excitatory effect of the action potential remains (116).

Asynchronous muscles, extracted with glycerol to remove ions and soluble organic molecules, can be made to oscillate if supplied with ATP, Mg^{2+}, and Ca^{2+} in appropriate concentrations, and if attached to an appropriately resonant load. The ability to oscillate is a property of the large, insoluble proteins of the muscle. The myosin filament seems to be the sensor of stretch activation.

Fast muscles of those insects in which contractions are synchronous with motor nerve impulses have extensive sarcoplasmic reticulum (SR); in asynchronous muscles that give multiple contractions per nerve impulse, much less of the fiber volume consists of SR. Stereologic measurements from electron micrographs of

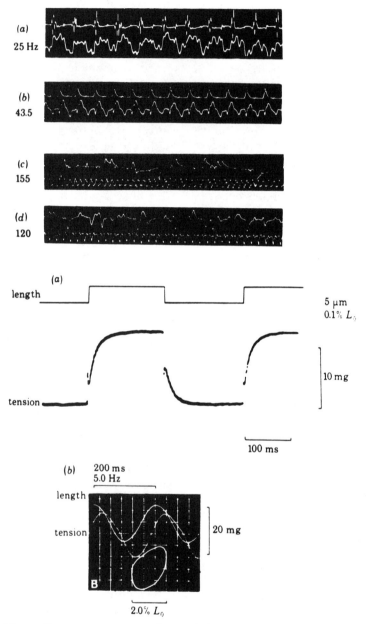

Figure 6. Upper figure: upper trace; electrical; lower trace; mechanical records from thoracic muscles of various flying insects. (*a, b*) synchronous type; asynchronous type, (*c*) fly and (*d*) wasp. Wing-beat frequency given for each type. Lower figure: tension–length responses in glycerinated fibers from a water bug. Upper portion: (*a*) quick stretch and release; (*b*) sinusoidal length modulation. Oscillograms of time course of length and tension, and a tension–length loop. Tension leads length during extension. [(116).]

katydid muscles show the following (75, 77):

Muscles	SR Thickness (μm)	Myofibrillar Area (μm²)
Synchronous	0.06–0.09	0.4–0.8
Asynchronous	0.016	1.1

Synchronous flight muscles of cicada contain, by volume: 40% mitochondria, 20% myofibrils, and 30% sarcoplasmic reticulum. Asynchronous tymbal muscles of cicada contain 58% mitochondria, 48% myofibrils, and 2% SR (75). Muscles of mesothorax and metathorax are similar in twitch dimensions in nymphal tettigonid crickets; however, in the adults, the singing mesothoracic muscles are much faster than the walking metathoracic muscles (77). The tergotrochanteral depressor of *Musca domestica* is a fast (contraction time, c.t. 2.5 ms) synchronous muscle that contains a small percentage of its volume as mitochondria (5%), a large fraction as SR (20%); the muscle receives three excitatory axons (136). This muscle is used in a jump preparatory for flight; this speeds behavioral escape.

It is probable that asynchronous muscles were derived from synchronous ones. Asynchronous activity is more economical, uses less Ca^{2+} cycling, more fibrils, and fewer spikes. Where precise control is required, as in walking, synchrony of activation and contraction are essential. Where high frequency oscillations occur, as in indirect flight muscles, there is less need for control.

In summary, insect postural muscles are of two types. (1) Those that respond synchronously with motor impulses, and (2) those that respond asynchronously and with repetitive contractions to single motor impulses so long as an active state persists. Either type of insect muscle can attain high frequencies of contraction.

Asynchronous muscles have less SR and contractile elements and oscillate several times per excitation. The oscillation is a passive property of the contractile system.

Narrow-Fibered, Nerve Activated Nonstriated Muscles

Narrow-fibered, nonstriated "smooth" muscles are of two kinds. (1) nerve activated and multiunitary and (2) endogenously active, nerve modulated, and unitary. There are few multiunitary muscles in vertebrate animals. Examples are nictitating membrane, pilomotor muscles, and sphincters. The retractor of the spiral valve of elasmobranchs is multiunitary. A few muscles have both multiunitary and unitary properties, for example vas deferens (117).

Postural narrow-fibered, nonstriated muscles that are nerve activated are used for locomotion in many invertebrate animals. These 3–5 μm in diameter fibers are activated by one or two motor axons. Contractions are relatively fast—40–60 ms. Examples are locomotor muscles in echinoderms, sipunculids, annelids, and molluscs.

Among echinoderms, the proboscis retractors of holothurians are five nonstriated muscles richly innervated by radial nerves that give off many branches. Cholinergic nerve endings are abundant, there is little or no SR and no nexal connections between fibers. In some species of sea cucumbers, the retractors are spontaneously active and in all, they can be stimulated by brief stretches (118). Blocking by *d*-tubocurarine, enhancement by physostigmine and high sensitivity to ACh indicate motor control by cholinergic neurons. Contraction of a lantern retractor of a sea urchin in response to electric shock or K depolarization depends on Ca^{2+} influx, contraction to ACh is a result

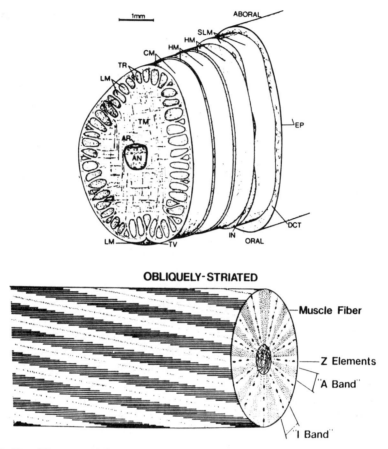

Figure 7. Top: Diagram of functional arrangement of muscles in a tentacle of a squid. Bottom: Diagram of a single obliquely striated muscle fiber with central core of organelles. [From (85).]

of release of intracellular stores of Ca^{2+} (148).

In a squid mantle, contraction of circular fibers causes ejection of water, contraction of radial fibers causes intake of water. The circular fibers show both phasic and tonic contractions, triggered by separate nerve fibers. Tentacles and arms of squid have peripheral bundles of longitudinal fibers, central radial, and circular fibers; a large nerve runs down the central core (85) (Fig. 7). In the gastropod Aplysia, the smooth muscle fibers of the gill are excited by each of three identifiable neurons in the pleural ganglion, two of them are cholinergic, the other is glu-

taminergic (17). The radular retractor of the snail *Helix* shows both fast nonfacilitating and slow facilitating action potentials. The radula protractors are excited by ACh and relaxed by serotonin. The action of transmitter FRMFamide is detailed in Chapter 1. In the snail, *Planorbis*, the radula retractor receives four motor axons that evoke EPSPs of different amplitude; there need be no spikes and muscle contractions are determined by junction potentials (152).

The proboscis retractors of the sipunculid *Phascolopsis* function over a 10-fold rest length; the fibers are narrow, 4–5 μm in diameter, and are doubly innervated.

They give fast, rapidly fatiguing, and slow facilitating electrical and mechanical responses (119, 120a). When contracted, some fibers fold like an accordion and birefringent bands appear across the muscle. Tension–length curves show two optimal lengths, one when the pleats are extended, the other presumably when thick and thin filaments are aligned.

The iris of vertebrates contains smooth muscle fibers that are innervated from the sympathetic system. In some animals—frogs and albino rats—the iris contracts in response to illumination independently of nerves. The muscle fibers contain rhodopsin, illumination of which elicits contractions (9a).

In summary, postural smooth muscles are nerve activated, that is, they are multiunitary. They perform locomotion of many invertebrate animals, especially the hollow bodied ones. Some small spindle-shaped fibers have dual innervation, phasic and tonic. Most of these nonstriated postural muscles are highly capable of shortening—as much as 10-fold.

Diagonal Striations

Diagonal striation is an adaptation for isotonic contraction of extensible muscles in which fibers shorten by both sliding and shearing of myofilaments. Diagonally striated muscles occur in the body wall of annelids, the longitudinal muscle of the roundworm *Ascaris*, and of amphioxus. Individual fibers are mostly oval or ribbon shaped and the distance from the upper or lower membrane to the contractile filaments is 3–4 μm. When diagonally striated fibers contract, the angle of filaments to the long axis of the fiber increases from 5–15° when extended up to 45° when shortened (Fig. 8). In light microscopy of the ribbon-shaped fibers of the earthworm, the myofilaments are seen running diagonally in one direction

on the upper face of the fiber and in the opposite direction on the lower face. In longitudinal sections of a polychaete muscle, electron micrographs show prominent Z lines in diagonal rows; the distance between Z lines of adjacent rows is less than in extended fibers. The pedal muscle of a whelk *Bullia* has cross-striated fibers used in crawling and burrowing and obliquely striated fibers with paramyosin that give turgidity to the foot (32).

In a muscle fiber with contractile elements in series, the generated force (F) equals the force (f) of each single element. When filaments are diagonal at an angle (θ) to the long axis, the force generated is proportional to $n \times f$ (cos θ), where n is the number of parallel elements and f is the force of each element (8a). There is evidence that the filaments of vertebrate visceral muscles show some staggering in the long axis. In many kinds of muscles shearing may add to sliding force development.

In helical or diagonally striated muscle, shearing is a passive consequence of filament sliding. When overlap is reduced, thick and thin filaments may change partners (91) (Fig. 9).

In an earthworm, ribbon-shaped fibers of the longitudinal layer are 4 μm wide; fibers of the circular layer are more oval. Longitudinal fibers show EPSPs because of cholinergic activation and IPSPs due to GABA. There is evidence that the EPSPs of some fibers are a consequence of increased Na^+ conductance, in others of Ca^{2+} conductance (70).

Possibly shearing may supplement sliding in striated as well as nonstriated muscles. Reconstructions of the T-tubule system from serial sections of vertebrate striated muscles show that the T-tubules are in helicoids rather than horizontal planes. Since the T system is aligned with striations, the striations must also have a helical arrangement (111).

Diagonally oriented myofilaments in

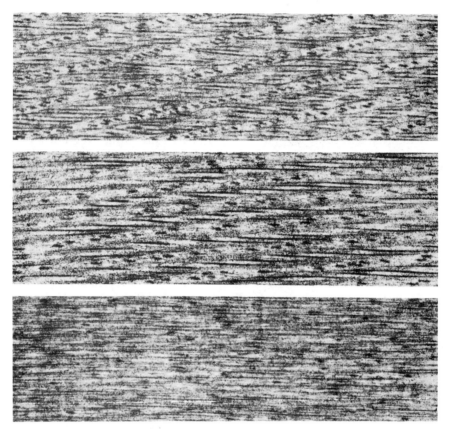

Figure 8. Longitudinal sections of diagonally straited muscle fiber of a polychaete at different lengths (L_0). Upper figure, muscle at $0.4\ L_0$; middle figure, at $1.2\ L_0$; lower figure, at $4.1\ L_0$. Separations of dark Z lines increase at longer lengths. (H. Takahashi, personal communication.)

spirally striated muscles provide for shearing when sliding between thick and thin filaments occurs. In many obliquely striated muscles, the contractile filaments are peripheral; some fibers are ribbon shaped and function for shortening or giving turgidity.

"Catch", Maintained Tension in Mollusc Muscles

Bivalve molluscs can close the shell rapidly (100 ms in Pecten) by the action of striated fibers and they can hold the shells closed against tension for long times—hours to days by nonstriated mus-

cle (109a). In the adductor muscles, the thick filaments are very large—several hundred nanometers. Muscle fibers are spindle shaped and narrow. The thick filaments have a central core of a large protein, paramyosin, and myosin is wrapped around this core (Fig. 10a). Between the thick filaments are actin filaments. Paramyosin constitutes 90% of the total myofibrillar protein. Calcium activation of A—M binding is by phosphorylation of myosin by a Ca^{2+} activated MLCK (myosin light-chain kinase) as in other molluscan muscles (82).

Electron micrographs and X-ray diffraction pictures show that paramyosin

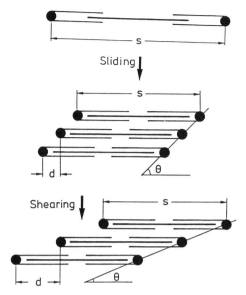

Figure 9. Diagram of contraction by sliding and shearing, showing a decrease in the angle of diagonal striations with respect to the long axis of fiber shortening.

molecules occur in planar ribbons that are wrapped into cylinders. A paramyosin filament is a three-dimensional crystal with a principal period of 14.4 nm, which repeats every five molecular units in a spiral (26) (Fig. 10b).

An adductor muscle or a byssus retractor (*Mytilus*) is activated to contract by excitatory nerve impulses with ACh as the transmitter (Fig. 11) (149a). Contraction can be induced by applied ACh, caffeine, or high K, also by Ca^{2+} at $10^{-6}M$. Once contracted, tension can be maintained at high levels—15 kg/cm^2. Energy expenditure while the muscle is in catch is extremely low. If the muscle is released in phasic contraction, tension redevelops but if released in catch, tension does not redevelop. Resistance to stretch is increased some sevenfold during catch. Relaxation requires impulses in relaxing nerve fibers; the transmitter is 5-hydroxytreyplamine (5-HT serotonin) (160a). Cutting the nerve to the muscle does not result in relaxation.

Contractions produced by calcium at $10^{-6}M$ lock into catch on removal of Ca^{2+} and relaxation can also be induced by 5-HT. Catch can be induced by added Ca^{2+} in detergent-skinned fibers and release from catch can be brought about by cAMP.

One hypothesis for catch is that when myosin filaments shorten, the paramyosin assumes a paracrystalline state. Another hypothesis is that rapid removal of calcium by muscle SR locks the muscle into catch. Two more recently presented hypotheses relate catch and release to phosphorylation events. One suggestion is that in intact fibers the phasic contraction, triggered by ACh is by a Ca^{2+}-dependent phosphorylation of myosin; in parallel may be a Ca^{2+}-calmodulin activated kinase that phosphorylates paramyosin. The paramyosin is considered to become stretch resistant as bridge cycling

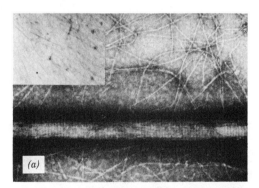

Figure 10. Electron micrographs of mulluscan catch muscle. (*a*) Longitudinal section showing single filament of paramyosin coated with myosin. Thin filaments in background insert phase contrast of filaments. (*b*) Individual filaments of paramyosin after removal of myosin sheath. Note the regular arrangement of paramyosin units. [From (149b).]

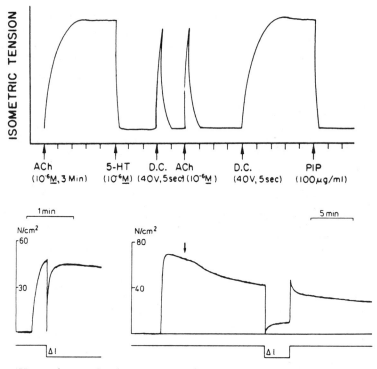

Figure 11. Upper figure: Catch properties of the worm muscle *Parapodius*. Contraction in response to ACh (3 min) with entry into catch; relaxation on application of 5-HT; phasic contractions to electrical and brief stimulation with ACh (PIP = piperazine citrate.) [From (149a).] Lower figure: First record. Active state and catch state second record, in anterior byssus retractor of bivalve *Mytilus* (ABRM). Quick release (Δl) of phasic contraction by 5% causes an immediate drop in tension followed by redevelopment of tension; quick release (Δl) of muscle in catch causes a drop in tension with no redevelopment of tension until restretched. [From (129a).]

is reduced. Serotonin causes dephosphorylation, which is postulated to decrease paramyosin stiffness. Interaction between myosin and paramyosin is proposed (1, 2, 129a). A different recent interpretation, based mainly on measurements on demembranated fibers (18) is as follows: Contraction is activated by $10^{-6}M$ Ca^{2+} and the muscle is locked into catch on removal of Ca^{2+}; release from catch, induced by cAMP, is caused by the action of cAMP-dependent kinase acting on ATP to phosphorylate the myosin heavy chain in catch. Interaction between myosin and paramyosin requires that dephosphorylation causes catch and phosphorylation

causes relaxation. The precise nature of interaction between myosin and paramyosin is unclear.

Some muscles that do not show catch contain paramyosin, for example, striated muscles of Limulus. These muscles have very long sarcomeres and it is suggested that the paramyosin provides support of long thick filaments (36). Striated muscles of Limulus have sarcomeres that can shorten from 12 to 4 μm. In thick filaments, paramyosin occurs in 10–20-nm aggregates (93). Paramyosin is not known in any vertebrate muscles; tension maintenance in these requires repetitive stimulation.

Unitary Nonstriated Muscles

Unitary narrow-fibered nonstriated muscles are endogenously rhythmic; their myogenic activity is modulated by nerves. Myogenic rhythms are rare in invertebrate animals. The heart and intestine of some annelids and all arthropods (crustaceans and insects) are driven by central or peripheral ganglia. The heart of molluscs consists of narrow-fibered muscles that beat rhythmically in isolation; they receive excitatory and inhibitory neural modulation.

Unitary narrow-fibered or smooth muscles mediate movement of vertebrate digestive tracts, blood vessels, and urogenital systems. A few vertebrate smooth muscles are spontaneously rhythmic and not innervated, for example, chick amnion. Several types of rhythmicity have been described. Most contractions are triggered by Ca^{2+} spikes; some contractions result from activation of intracellularly bound calcium. These spikes may be a second or more in duration in the stomach of anurans and elasmobranchs; the Ca^{2+} spikes are of 10–100 ms duration in mammalian stomach and intestine. In mammalian ureter there are flat-topped electrical waves much like the action potentials of the heart. The intestine of many mammals generates rhythmic "slow waves" of several seconds duration, near sinusoidal in shape; these are caused by a rhythmic Na^+-K^+ pump and they may have Ca^{2+} spikes at the peak that trigger segmental contractions. These slow waves originate in cells at the boundary between longitudinal and circular layers. Other slow waves result from a background of ACh; the ACh type of slow waves are blocked by atropine and can be elicited after block of the Na^+-K^+ pump-type waves by ouabain (31). In dog colon, slow waves of low frequency originate at the submucosal border,

higher frequency waves at the mesenteric border (129b, 141b).

Fibers of vertebrate unitary muscles are spindle-shaped, approximately 3–5-μm maximum diameter and 200 μm long. Their membranes have pinocytotic vesicles, caveolae. The net effect of the spindle shape and the caveolae is to give a surface/volume ratio many times greater than in striated fibers. The muscle fibers in most visceral organs are connected by low-resistance gap junctions (nexuses). Strips of smooth muscle cells are functional syncytia. Contractions of the uterus and of many blood vessels are hormonally activated.

Visceral muscles receive extensive innervation from intrinsic neurons and from processes of sympathetic and parasympathetic neurons (129b). Nerves do not end on endplates but liberate transmitters and modulators from swellings and varicosities along their length; transmitters diffuse for considerable distances, not every muscle fiber is contacted by a nerve fiber.

In mammals, several types of activity occur *in vivo* that are not observed *in vitro*. These may be related to the physiological state. Spikes are associated with high levels of basal motility during digestion. In the small intestine of a dog, cat, and human in a postdigestive state, bursts of large spikes occur at approximately hourly intervals; in ruminants, these migrating complexes are not related to feeding. It is postulated that the migrating complexes clear the intestine of mucus and debris.

The esophagus contains striated and smooth muscle in different proportions in different mammals. In humans and opossum, the upper quarter of the esophagus is striated, the lower three quarters is smooth; in rat, sheep, and cattle, striated fibers occur along the entire length of the esophagus; and in the dog and pig the longitudinal muscle layer is striated and the circular layer is striated except near the gastroesophageal sphincter. The esophagus is multiunitary in that contractions are not spontaneous, but are activated by nerves; the sphincter maintains tension myogenically, but relaxes reflexly. Many smooth muscles and heart muscle, when stretched, show increased excitability. Ureter and bladder are stimulated by stretch to contract. Uterine muscle shows marked differences with species in effects of hormones.

Neural control of visceral muscle differs with organ and species. Cholinergic activation may be via muscarinic (m_1 or m_2) receptors (Chapter 1), adrenergic by α or β receptors; some smooth muscles have purinergic innervation (ATP or adenosine), some have receptors for serotonin (5-HT). In mammalian intestine, neurons of the myenteric plexus produce some dozen neuropeptides that are putative transmitters or modulators. Specific transmitters may act via membrane potentials or intracellularly. There are marked differences between tissue and species.

Control mechanisms in the visceral muscles of nonmammalian vertebrates are as diverse as in mammals. Figure 12 diagrams the distribution of autonomic innervation of the stomach in representative species of vertebrate classes (15a). In fishes and amphibians, the cranial parasympathetics and anterior sympathetics run together in the vagosympathetic nerve. A frog's stomach receives excitatory cholinergic fibers of sympathetic origin and vagal preganglionic fibers that inhibit by noncatecholaminergic fibers (probably ATP). In fishes (trout) vagal fibers to the stomach are inhibitory via purinergic endings; inhibition is followed by a postinhibitory rebound contraction; the vagus also has cholinergic excitatory fibers. Reptiles resemble mammals in having vagal cholinergic excitatory fibers,

vagal preganglionic fibers that activate nonadrenergic inhibitory fibers, and sympathetic cholinergic excitatory and adrenergic inhibitory fibers.

The sympathetic system is principally excitatory cholinergic in elasmobranchs, mixed cholinergic and adrenergic in bony fishes, amphibians, reptiles and birds, and adrenergic inhibitory in mammals. It is hypothesized that the primitive vagal supply to visceral muscles was nonadrenergic inhibitory with postinhibitory rebound excitation and that the vagus may have assumed cholinergic excitatory function late in evolution (16).

In summary, visceral muscles, especially of vertebrates, contract hollow organs—digestive and excretory tubes and blood vessels. Visceral muscle fibers are small spindles, coupled electrically and the muscle may be a functional syncytium. Most visceral smooth muscles are endogenously rhythmic or are excited by mechanical stretch. Their activity is modulated by antagonistic nerves that liberate transmitters that diffuse through sheets of muscle fibers. Many visceral muscles are further regulated by circulating hormones. Contractions may be triggered by Ca^{2+}-spikes or by Ca^{2+} from intracellular stores. Electrical rhythms vary in period and cellular mechanisms.

Energy Sources of Muscles

The immediate energy for signal transduction and contraction in all muscles is supplied by adenosine triphosphate (ATP) (175). The amount of ATP in resting striated muscle is small (5×10^{-6} M/g), insufficient for more than a few twitches. Maintained contraction requires 10^{-4}–10^{-3} M ATP/g · min. Adenosine triphosphate is continually regenerated by several biochemical pathways and the relative importance of each varies with muscle and species. An intermediate store of high energy phosphate used to replenish

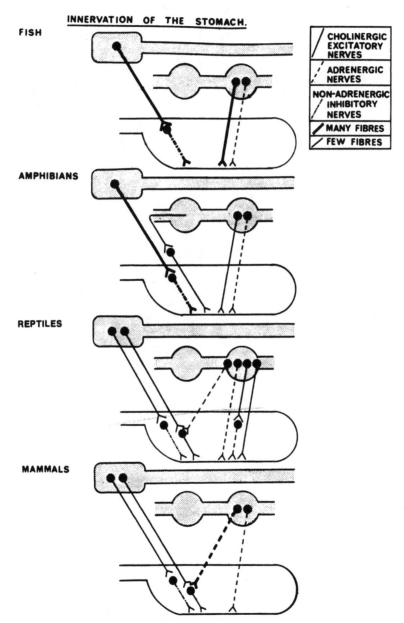

Figure 12. Diagrams of autonomic innervation of the stomach of various vertebrate animals. Vagal outflow is inhibitory in fishes and amphibians and is opposed by excitatory cholinergic sympathetic neurons. In reptiles and mammals, cholinergic fibers are in sympathetic. [From (15a).]

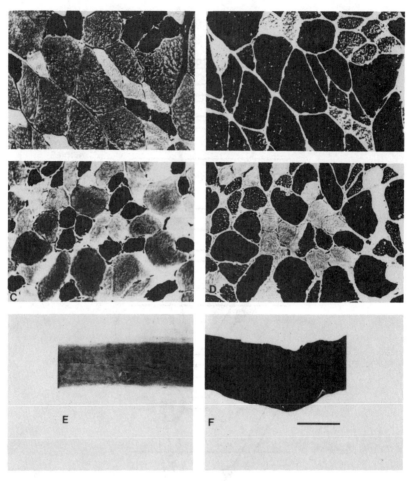

Figure 13. Myosin ATPase activity in diaphragm and psoas fibers; acid preincubation (*A, C, E,* and *F*); alkaline in (*B*) and (*D*). After acid preincubation type I fibers—dark; type IIa fibers—light; and type IIb fibers—intermediate. After alkaline preincubation type I fibers—light IIa and IIb—dark. (*A, B*) sections of diaphragm; (*C, D*) skinned fibers of diaphragm; (*E, F*) skinned psoas. (*E*) at rest length; (*F*) contracted. [From (402).]

ATP in muscle activity is a phosphagen: mostly phosphocreatine (PCr) or phosphoarginine (PArg). Muscles of all chordates contain PCr, muscles of arthropods, molluscs, coelenterates, and some lower worms have PArg. In echinoderms only PArg is found in holothurians and asteroids, PCr is found in ophiuroids, and both PCr and PArg are found in some echinoids. Hemichordates have both PCr and PArg, cephalochordates have PCr. Annelid and sipunculid phosphagens have a variety of guanidinium bases—glycocyamine, taurocyamine, and lombricine. Most annelid spermatozoa have PCr no matter what kind of phosphagen the muscle has. Phosphagen energetics is discussed in Chapter 8 of the companion to this book, *Environmental and Metabolic Animal Physiology.*

Adenosine triphosphate is formed and phosphagens are regenerated by oxidation or glycolysis according to the available O_2 and the enzyme complement present in specific muscles. Aerobic muscles generate ATP by oxidation of glycogen or fatty acids. Glycolysis produces ATP and lactate; the lactate can be oxidized by oxidative enzymes or is transported to liver where it is reconverted to glycogen (65). Use of lactate as a muscle fuel conserves glycogen. Oxidative muscles are capable of prolonged contraction, that is, are slow-twitch (tonic) muscles (SO). Obligate aerobic muscles of vertebrates contain myoglobin. Anaerobic glycolysis provides less ATP than oxidation, but phosphorylates phosphagen when O_2 is restricted. Glycolytic muscles are phasic (FG), capable of fast but rapidly fatiguing contraction. Many muscles are intermediate between the two extremes of fast-twitch oxidative–glycolytic (FOG).

Flux rates of exogenous fuels depend on ATP yields. Maximum yields are (65)

		μM ATP g_{ww}^{-1} min^{-1}
Aerobic	Fatty acid oxidation	20.4
	Glycogen oxidation	30
Anaerobic	Glycogen glycolysis	60

Metabolic Types of Phasic and Tonic Fibers of Vertebrates

Quantitative differences in metabolic patterns and in the activity of myofibrillar ATPase are indicated histochemically. A classification for mammalian muscles is (1) white fast muscles are relatively low in oxidative enzymes, high in glycolytic enzymes, and high in myosin ATPase; examples are superficial fibers of quadriceps; (2) red muscles are low in glycolytic and high in oxidative enzymes, low in myosin ATPase, slow to contract and to fatigue; examples are deep fibers of quadriceps, (3) intermediate muscles, slow oxidative–glycolytic, low in myosin ATPase, for example, the soleus. Three fiber types in rat diaphragm are distinguished by histochemical staining for myosin ATPase (Fig. 13) (51, 53).

Amphibian phasic muscles have been classified histochemically into five types; three give fast contractions, two are slow. ATPase activity is high in the three fast muscles; myosin ATPase shown by staining is intense (type 3) or pale (type 1). Fiber diameter is small in the two slow types, glycogen and lipid stores are high; SR is abundant in the fast, sparse in the slow fibers (141a). In a lizard, *Dipsosaurus*, the fibers of the iliofibularis muscle are (1) FG (fast twitch, glycolytic), (2) FOG (fast oxidative–glycolytic), and (3) SO (slow oxidative), low ATPase (54, 55). In a turtle, three fiber types are distinguished histochemically and functionally as follows (71b, 106):

Fiber Types	Maximum Tension (kN/m²)	Velocity Shortening	Alkaline ATPase	Acid ATPase	Succinate Dehydrogenase
Fast glycolytic	184	5.5	+ + +	0,sl +	+
Slow oxidative	70	1.3	0,sl +	+	+ + +
Intermediate	120	3.0	+ + +	+ + +	+ +

In fishes there are 2–5 metabolic types of fibers according to species and muscle. In slow red fibers, rich in myoglobin, 25–38% of the volume is occupied by mitochondria and the energy source is mainly oxidative. Pink fibers are intermediate glycolytic–oxidative. Pink fibers are predominant in elasmobranchs, holosteans, chondrosteans, and primitive teleosts. White fibers are rich in glycolytic enzymes and mitochondria occupy only 0.5–4% of cell volume. Glycolytic enzymes such as LDH occur at higher activity and oxidative enzymes such as succinate dehydrogenase at lower levels in white than in red fibers (72).

Rates of contraction and relaxation are slower in red than in pink and white fibers. Data for muscles of a perch Perca are (4)

Muscle	SR (vol %)	Time to Peak (ms)	$T_{1/2}$ Relax (ms)
White	5.5	14	34
Pink	4.8	29	46
Red	3.9	92	151

Myosin ATPase activity is in the series white > pink > red. Glycolytic activity (LDH) is higher in white muscle oxidative (citrate synthase) cytochrome oxidase higher in red muscles. (71a, 73).

When fish swim at low speeds, red tonic fibers provide power for steady swimming. At increasing speed, white muscle fibers are recruited at a critical speed, for example, in brook trout at 1.8 body lengths/s. Training for 21 days results in hypertrophy of red fibers and the glycolytic capacity of red fibers increases (33, 71). Carp swimming at 0.5 lengths/s use red myotonal muscle, at 1.1–1.5 lengths/s pink fibers are used, and at faster than 2–2.5 lengths/s white fibers are recruited (71, 72).

In noncyprinid fishes the metabolic patterns change with temperature acclimation (176). In striped bass, acclimated from 25 to 5°C, the percentage of oxidative red fibers in lateral muscles increased from 9 to 15, the oxidation of stored lipid increased. In goldfish acclimated to the cold, the white muscle showed no increase in the percentage of red fibers but did show an increase in myosin ATPase activity (137a) (176). In striped bass, oxidative red fibers increase from 9 to 15% on acclimations from 25 to 5°C; oxidation of lipids increases similarly (73).

Prolonged exercise in humans or rats leads to increases in oxidation of pyruvate and palmitate, increase in cytochrome c oxidase and citrate dehydrogenase in each of the three muscle fiber types. Cytochrome c doubled in amount in both superficial white and deep red fibers of the quadriceps and in the intermediate fibers of the soleus (9). Muscles of rats have subsarcolemmal (SS) and intermyofibrillar (IMF) mitochondria; after exercise the SS mitochondria increase by 30%, IMF mitochondria increase by 10% (98). In a fish after sustained swimming for 200

Activities of Enzymes from Slow (S) (red) and Fast (F) (white) Muscle Fibers of Fish[a]

Enzyme	Brook Trout		Crucian Carp		Plaice	
	F.	S.	F.	S.	F.	S.
Hexokinase	0.6	0.3	0.7	2.0	0.06	0.4
Phosphofructokinase	14.0	11.4	4.2	1.9	29.0	17.1
Lactate dehydrogenase	345	200	237	440	242	174
Citrate synthase	0.7	4.9	0.9	9.2	0.3	7.1
Cytochrome c oxidase	0.4	2.3	0.8	5.6	0.2	3.9
Mitochondrial volume (%)	9.3	31	4.6	25.5	2.0	24.6

[a] Activities in μM substrate/$g_{ww} \cdot$ min.

Enzyme Activity of Quadriceps Red and White Fibers in Sedentary Persons (s) and Runners (r) (9)

Enzyme	Type and Units	White	Red
Cytochrome c oxidase	(s) $\mu lO_2/min/g_{ww}$	167	840
	(r) $\mu lO_2/min/g_{ww}$	339	2041
Citrate synthase	(s) $\mu M/min/g_{ww}$	10.3	35.5
	(r) $\mu M/min/g_{ww}$	18.51	69.9
Carnitine palmityl-transferase	(s) $nM/min/g_{ww}$	0.11	0.72
	(r) $nM/min/g_{ww}$	0.20	1.20
Cytochrome c	(s) nM/g_{ww}	3.2	16.5
	(r) nM/g_{ww}	6.3	28.4

days, the red fibers and number of capillaries double in abundance (30).

Metabolic correlates with Speed also occur in Crustaceans

Claw closer muscles of a crab show four types of fibers according to ATPase staining: one slow oxidative, two types of fast mixed oxidative–glycolytic, and a fast glycolytic fiber (121). Type I fibers have low ATPase, slow junction potentials, and strong inhibition; type II, a mixed metabolic type have high ATPase, much facilitation of esps and contractions, and are weakly inhibited (121).

In summary, the principal phosphagens of muscles are PCr in chordates and PArg in molluscs; minor phosphagens are glycocyamine, taurocyamine, and lombricine. The striated muscles of vertebrates that are red with myoglobin are tonic and predominantly oxidative; white muscles are phasic, fatigue more rapidly, and are mainly glycolytic. Intermediate pink fibers may be oxidative–glycolytic. Activity of myosin ATPase is lower in slow oxidative than in fast glycolytic muscles.

Each muscle fiber has the genotype for different types of energy sources and myosin ATPase but individual fibers can change from one type to another in ad-

aptations to exercise, temperature, and other determinants.

Contractile Proteins

Contractile proteins include actins that are highly conserved, myosins of many forms, proteins that mediate Ca^{2+} regulation, and cytoskeletal muscle proteins that bind filaments together and to other organelles and membranes.

Actins

Actins occur in all muscles, in motile cells that are not muscle, and in the cytoplasm of many nonmotile cells, perhaps in all cells of eukaryotes and of some prokaryotes including *E. coli* (90, 109). In vertebrate striated muscles actin constitutes some 21% of the protein, in blood platelets 20–30%, in liver 1–2%, and in a soil amoeba 20–30%. Actin has a molecular weight 45,000 D, consists of 374 amino acids, and has an isoelectric point of 4.8 (41). Monomeric actins are globular (G-actin) and polymerize to fibrillar (F-actin) under appropriate ionic conditions. Adult skeletal muscle actin differs from cardiac actin in 4 amino acids (164, 165). Nonmuscle and cytoplasmic actins (brain, thymus, and kidney) are of two types α and

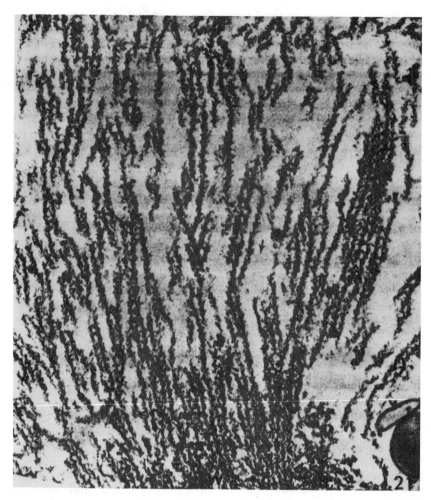

Figure 14. Thin section through part of the brush border of chick intestine. Actin filaments are attached at the terminal web and splay out at tips. Each filament has same polarity with arrowheads pointed toward the terminal web. [From (156).]

β; both differ from muscle actin by 25 amino acids (164, 165). Actin of vascular smooth muscle differs from actin in visceral smooth muscle. Blood platelet actin differs from muscle actin by 11 amino acids; fibroblast actin resembles smooth muscle actin more than it does striated muscle actin (128, 129). Fibroblasts of chick embryos contain three actins, α, β, and γ; γ resembles that in cardiac and striated muscle and γ resembles smooth muscle actin (129). In general, vertebrate actins show more specificity with tissue than with animal species. Actins of the algae *Nitella* and *Chara* react with muscle myosin to give arrowheads in the opposite direction to cytoplasmic flow. In *Nitella*, actin filaments are associated with chloroplasts that move by cyclosis. Actin of the slime mold *Physarum* resembles the cytoplasmic actin of mammals more than it does muscle actin. Actin occurs close to the cell membranes of vertebrate lenses. Actin filaments occur in the core of intestinal microvilli (Fig. 14). Actin is the major protein in the acrosome of sperm. The bacterium *Escherichia coli* has actin of molecular weight 45,000 D, which can bind to muscle myosin (101). Actin has been found in plants—tomatoes, soybeans, and Amaryllis. Soybean seedlings have a 45-kD actin that reacts with antibodies to rabbit muscle actin (27). The genes for actin constitute a family, often in multiple copies in a genome. Cardiac actin can be induced in heart ventricle by thyroxine treatment. The ciliate *Tetrahymena themophila* has much actin near the oral apparatus and a single actin gene has been identified (29). Yeast actin has 374 amino acid residues coded by a single split gene; it differs from actin of rabbit skeletal muscle by 44 amino acids and from Physarum actin by 39 amino acids (50).

Actin monomeric molecules are bilobed, each with a large and a small domain. Polymerization results in filaments, a double stranded helix with a half-turn 375 Å long. Polymerization requires a cation Mg^{2+} and ATP. The filament is polarized in that one end is barbed, the other is pointed. The polarity is such that on binding with myosin, arrowhead shaped bridges are formed and pointed in one direction (159). Approximately 10 proteins are known that bind actin filaments together and to cell membranes and organelles; in striated muscle, specific proteins, mainly actin, bind the thin filaments to Z membranes. Other proteins have a capping function, usually at the barbed end (114).

Actin filaments of muscle bind to myosin molecules in the contraction cycle and actin has a cytoskeletal function in many cells in the absence of myosin. For example, the sperm head of many invertebrate sperm has an acrosome or cap of depolymerized actin (Fig. 15). On activation, as when a sperm touches an egg, the actin is rapidly polymerized into 90-nm filaments with the barbed end pointed toward the cell body; the pointed projection penetrates the egg membrane

(157). Actin filaments extend when blood platelets explode and a blood clot retracts. Hair cells of the cochlea, (chicken and lizard) have sterocilia, 50–300/cell. Each stereocilium contains actin filaments and these are tilted at an angle according to the location of a hair cell in the cochlea (Chapter 6) (158, 159). The stereocilia bend passively when sound waves move the endolymph but they do not beat spontaneously. The brush border cells of the vertebrate intestine have microvilli, which project into the lumen. Each microvillus contains many actin filaments. Each epithelial cell of the brush border has >1200 microvilli with actin extending basally toward a terminal web in the cytoplasm. Cross-bridges occur between actin filaments at 33-nm intervals. The microvilli are rigid structures, but in the terminal web, myosin binds to the barbed ends of actin filaments and on stimulation circumferential contraction occurs in the epithelial cells (156).

Sequence analyses of actins from some 30 species of animals show that all investigated invertebrate animals have a single form of actin that resembles, but is not identical to, the cytoplasmic actin of vertebrates. The actin of prochordates resembles that of vertebrate muscle more than it resembles invertebrate actin. Cyclostomes and elasmobranchs have different actins for smooth and striated–cardiac muscle. Separation of striated from cardiac and vascular from enteric smooth muscle actins occurred in the reptiles. It appears, therefore, that vertebrate actin evolved with prochordates, not from any present-day invertebrates and that major gene duplication occurred in the reptiles (164).

In summary, actins occur in some bacteria and plants and are probably present in all animals cells. Actins are highly conserved structurally and most actins have a molecular weight approximating 45,000 D. Actins may be globular and may po-

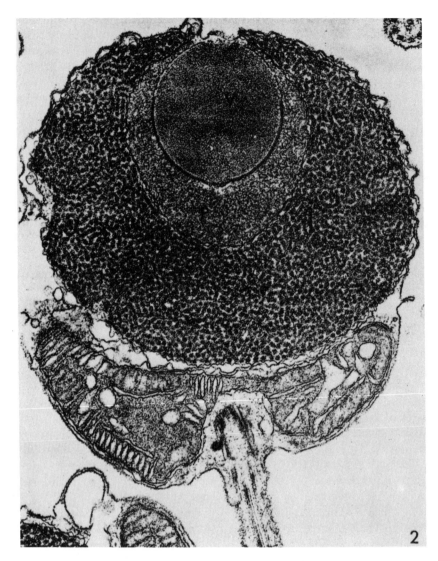

Figure 15. Electron micrograph of a section through a sperm of echinoderm *Thyone briareus*. Acrosomal vesicle (V) surrounded by periacrosomal material (P), which polymerizes to form actin filaments when activated. [From (156).]

lymerize as thin filaments that bind to large molecules such as myosin; the complex functions in contraction. Actin filaments give rigidity to stereocilia of cochlea and microvilli of intestine; actin constitutes the acrosome (head) of spermatozoa of some invertebrates and polymerization yields very long filaments.

Myosins

Myosins are numerous and more diverse and less conserved than actins; different myosins make for differences in speed of contraction. Myosins are large molecules—450,000–475,000 D. Each molecule consists of two strands twisted in a dou-

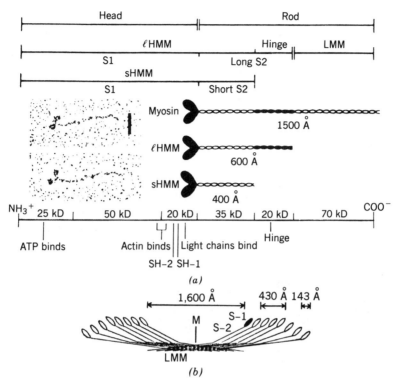

Figure 16. (*a*) Structure of myosin heavy chain. In upper figure, vertical lines indicate protease-sensitive sites. The intact molecule can be broken by chymotrypsin to form HMM and LMM. Proteolysis by trypsin produces sHMM and LMM. The two heavy chains intertwine to form a single-coiled tail and two separate heads. The hinge regions of myosin is marked by a darkened region in the tail. The bottom line shows the primary sequence of the heavy chain with locations of binding sites for ATPase and actin. It also shows the location of light chains and thiol residues. Inset: Electron microscopic frames of myosin showing two heads and tail of sHMM. [From (67).] (*b*) Diagram of myosin molecules in thick filament, S-2 hinge regions supports heads; heads oriented away from the central M region. [From (99).]

ble helix forming a rod some 140 nm long. Cleavage by a proteolytic enzyme breaks the molecule into light meromyosin (LMM, MW 140,000 D) and heavy meromyosin (HMM, MW 340,000 D). Further proteolysis cleaves the HMM to a segment (S-2) some 65 nm long and MW 60,000–100,000 D and two heads (S-1) each 21 nm long, 115,000 D (Fig. 16). In vertebrate striated muscle the S-1 heads which bind to actin and contain ATPase activity, have associated with them 4 light chains (lcs). Two of the light chains are regulatory, MW 18,000 each, and can be removed from striated muscle myosin by a thiol blocking agent (DTNB chains). The other two light chains, MW 32,000 and 15,000 D, can be removed by alkaline extraction. One light chain from the Acanthamoeba myosin has three phosphorylation sites. In rabbit muscle, the molecular weight of the alkaline-extracted light chains is 21,000 and 17,000 D and 19,000 D each for the two regulatory light chains. Removal of the alkaline light chains blocks AM ATPase (115). Smooth muscle myosin has two rather than four light chains, MW 20,000 and 17,000 D;

smooth muscle myosin has lower ATPase activity than striated muscle.

Six heavy chains have been identified in rat skeletal and cardiac muscle: fetal, neonatal, fast oxidative–glycolytic (IIa), fast glycolytic (IIb), cardiac α, and cardiac β (104). Each is encoded by a different gene. Chicken skeletal muscle light chains are encoded by a single gene. The lc gene has two initiation sites for two mRNA precursors and these are spliced to form the lc1 and lc3 proteins (107). Molecular weights of the subunits of cat muscle myosins (167) are

Subunits	Fast Twitch Myosin	Slow Twitch Myosin	Cardiac Myosin
Heavy chains	200,000	200,000	200,000
lc 1 (alk)	25,000	27,500	27,000
lc 2 (DTNB)	18,500	19,000	19,000
lc3 (alk)	16,500		

The ATPase activity of the fast striated muscle myosin is four times that of the slow muscle myosin. In chicken embryos the separation of fast and slow myosins begins at the time of myoblast fusion (81). In rabbit the genes of the two heavy chains (α and β) occur in a fetus at an α/β ratio of three, at birth a ratio of one, and in adult a ratio of four (139). Rabbit uterine muscle has 31% homology with skeletal and nematode muscle 25% with myosin from slime mold *Dictyostellum* (107a).

In the cardiac muscle of rat and rabbit three isotypes have been identified by electrophoresis, antibody reactions, and ATPase activity (105). Hearts of *Xenopus*, chicken, dog, and beef have V1 and V3, human has only V3 (171a). In rabbit heart V1 has high ATPase and V3 has low ATPase activity (22). Atrial (A) and ventricular (V) light chains differ; for example, the series for electrophoretic mobility is in mouse and rat: V lc 1 < A lc 1 < A lc 2 < V lc 2 (28).

In human cardiac myosin the β form is primarily in ventricle, the α form in auricle; the β form is the same in ventricle and in slow twitch skeletal muscle (171a).

Muscle hypertrophy brings about changes in the proportions of myosin isoforms. Cardiac hypertrophy can be caused by aortal constriction or by injection of thyroid hormones. Cardiac hypertrophy resulting from aortal constriction results in an increase in V3 light chains, in thyroid-induced hypertrophy V3 increases and V1 decreases.

In a fish (mackerel) the myosins of red and white muscle are different; red muscle myosin has two types of lcs, mg-ATPase activity of 0.18 μm $P_i/mg_{Prot}/min$ white muscle myosin has three types of light chains and high ATPase activity, 0.44 μm $P_i/mg_{Prot}/min$ (30). In general, slow red fibers from superficial muscles have myosin ATPase that is alkaline labile, and fast white muscle has myosin ATPase that is acid labile (127).

Myosin from the muscle of an ascidian has two types of light chain, each different from vertebrate myosin light chains. In the nematode *Caenorhabditis* there are three isotypes of myosin heavy chain; type A, Mw 210,000 D is in body wall and pharynx, type C, MW 206,000 D is in pharynx only, and type B, MW 203,000 D is in body wall only (94, 173).

Earthworm myosins have three light chains of MW 28, 25, and 18 kD. The 25-kD myosin occurs in most fibers of the body wall, the 28 kD form is in some body wall fibers and all the muscle of the gizzard, its ATPase is low (16a, 59).

In the soil amoeba *Acanthamoeba* there are two myosins; I is located near the plasma membrane, myosin II internally as shown by immunofluorescence (49b). Amoeba myosin consists of two heavy chains and two pairs of light chains. Acanthamoeba myosin is remarkable for lack of cysteine and for its small size— 180,000 D. The myosin of the slime mold

Physarum has two heavy chains of 225,000 D each and two pairs of lcs (84, 114, 115). Myosin of another slime mold *Dictyostelium* has two 243,000-D heavy chains and one pair each of 18,000 and 16,000 D lcs. (34). Myosin is essential for cytokinesis, for pseudopod function and for chemotaxis and phagocytosis in the unicellular stage of slime molds. In *Dictyostelium*, the HMM tail is 1800 Å long while in a mutant strain it is 500 Å. The mutant with reduced HMM shows normal movements including nuclear division but not cytokinesis (35).

Striated muscle fibers develop from uninucleate myoblasts. These fuse to form myotubes each with some 100 nuclei that unite to form myofibers with 3000–4000 nuclei/fiber. Changes in myosin occur during embryonic development (14). In rats the myosins of all striated muscles are slow at birth and some become *fast* as adults. Embryonic heavy chains differ from fast and slow adult myosins. A rat embryo has 2 of the 3 adult fast lcs at 16-days gestation. The primordia of both the fast extensor digitorum longus (EDL) and of the slow soleus (SOL) react with anti-fast myosin antibodies; at 18 days they react to both anti-fast and anti-slow antibodies. The neonatal EDL has both slow and fast myosins. Fetal myosin also shows three electrophoretic forms, one of which may be similar to adult fast myosin (52, 66, 169).

Chickens have slow anterior latissimus dorsi (ALD) and fast posterior latissimus dorsi (PLD). In 8-day embryos the half-time for contraction for both ALD and PLD is 0.50–0.53 s; at 18-days contraction time it is 0.48 s for ALD and 0.17 s for PLD. Innervation occurs at 16 days and the time of differentiation of the muscles parallels innervation (66). In adults, after denervation, the PLD atrophies while the ALD hypertrophies (42).

In mammalian phasic muscles, denervation results in atrophy. When the nerves to fast (EDL) and slow (SOL) muscles of rats are crossed, the speeds of contraction and activity of ATPase become reversed (167):

	Flexor Hallucis		Soleus	
	Normal	Cross Innervated	Normal	Cross Innervated
Contraction time (ms)	61	87	87	25
ATPase-mg$_p$/mg of myosins	2.7	0.7	0.2	2.5

One hypothesis for the effect of motor nerves on muscle myosins is that the different muscles receive different patterns of activation from the spinal cord. Electrical stimulation by implanted electrodes can lead to reversal between slow and fast myosins. A second hypothesis is that the nerves liberate a trophic factor that influences the protein synthesis. This is supported by experiments in which nerve conduction was blocked by application of TTX over long periods to the motor nerves and the reversal in muscle types (judged by width of Z bands) occurred (42). In cultured chick muscles in the absence of nerves, the transformation from embryonic to adult myosins can be induced by electrical stimulation.

Speed of contraction as correlated with metabolic pathways were discussed previously. Speed is also determined by the type of myosin. The actin-induced ATPase activity of fast myosin is four times greater than from slow myosin. The speed of shortening in contraction is correlated with ATPase activity. Values of myosin ATPase in nM P_i/min · mg of protein are for the heart muscle of rat 423, guinea pig 268, dog 139, and rabbit 94; the rate of heartbeat follows the same sequence. Comparison of cat gastrocnemius with that of sloth follows (56):

	Contraction Time (ms)	Myosin ATPase (μM P_i/s · mg of protein)
Cat	22.5	0.50
Sloth	106	0.11

In summary, myosins are diverse according to size, molecular components, and speed of movement when bound to actin. Each molecule of myosin is a double helical chain with several light chains and a total MW ~450,000 D. The synthesis of fast and slow myosins can change according to innervation and muscle activity and its stage in the life cycle. Myosins differ according to tissue—cardiac, fast striated, slow striated, and visceral smooth muscles. There are homologies between myosins of yeast, slime molds, and muscles of all animals. Genetic analyses show several myosin genes in a given animal; these are expressed at different times during development.

Myofilament Organization and Theories of Contraction

All muscles and some motile nonmuscle cells have myosin in thick filaments and actin in thin filaments. In fast muscles the thin filaments are distributed in a regular pattern, 6–8 thin/thick filament. In insect flight muscle each myosin filament is linked by bridges to six neighbors (43). In slow muscles, the arrangement of fila-

ments is irregular, and the ratio of thin per thick filament is not fixed. In striated muscles, generally wide fibered, the thick and thin filaments show transverse alignment. In smooth muscles, generally narrow fibered, the arrangement of thick and thin filaments is random.

In rat striated muscles the thick filaments are 50 nm and the thin ones are 10 nm in diameter. Myosin molecules are 10–15 nm in diameter and 160 nm long; the heads point toward the two ends of a thick filament and a zone lacking heads forms the middle M band of a sarcomere, a region where there are no cross-bridges to actin (99) (Fig. 16b). The head regions of myosin bonding to actin form a 6/2 helix of 429 nm/turn (99a).

A single striation constitutes a sarcomere and boundaries of sarcomeres are Z lines that contain several proteins, specifically α-actinin and desmin. Thin filaments are attached to Z lines. At rest length the thick filaments do not extend to Z lines. On contraction, sarcomeres shorten and thick filaments move toward the Z lines.

Energetics of Contraction

The energetics of the cross-bridge cycling in relation to actin-activated myosin ATPase has been examined by measurements on extracted proteins, by use of labeled phosphate compounds and by kinetic measurements on skinned muscle fibers. Most of the evidence is from extracts of rabbit and frog skeletal muscle. A summary is given in Figure 17a (64). The cycle begins with the myosin head S-1 attached to actin thus forming actomyosin (AM). In step one, ATP attaches to AM and the bridge between myosin (M) and actin (A) is represented as at 90° and the arm M-2 is stretched. The myosin head ML-1 then detaches from actin and ATP remains bound to myosin (Step 2). At physiological ionic strength hydrolysis

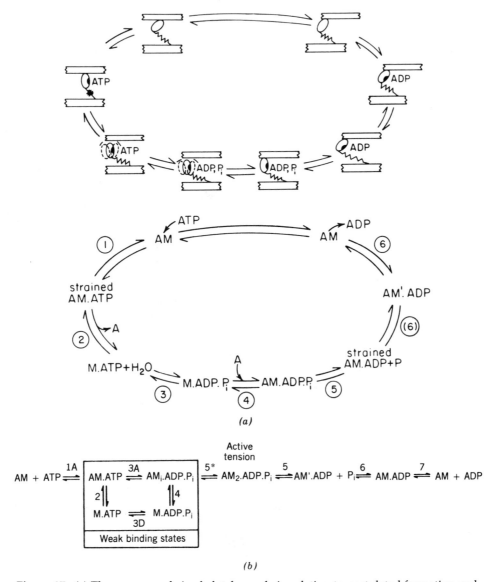

Figure 17. (*a*) The energy cycle in skeletal muscle in relation to postulated formation and functioning of cross-bridges between actin and myosin. Upper figure shows bridge attachment, sliding and detachment corresponding to numbered energy steps. Lower bars are myosin with bridge and head, upper bars represent actin, lower figure shows energy steps. [Modified from (167a); P. Best, personal communication] (*b*) Diagram of energy cycle for insect skeletal muscle. Numbers similar to those in (*a*). In insect muscle, tension development precedes the rate-limiting ATPase step and isomerization of Am·ADP·P_i is postulated as control of tension redevelopment. [From (103).]

occurs at M · ATP to yield M · ADP · P$_i$ (Step 3) and the M-1 moves to a new position. After hydrolysis, M · ADP attaches by the M-1 head to A in its new position to form AM · AD · P$_i$ (Step 4) and the bridge is represented as at 90° angle. Inorganic orthophosphate then becomes dissociated from the complex, the head angle changes to 45°, and sliding of M over A occurs (Step 5); energy is discharged. S-2 is pictured as an elastic monomer. The entire thick filament complex is bipolar so that bridges at the two ends pull actin; a myosin head contacts two actin monomers on average.

Kinetic measurements of K_m values and rate constants of the various steps indicate that the rate-limiting step is the hydrolysis of ATP to ADP (Step 3). The rate of force generation by skinned fibers of rabbit psoas correlates well with AM ATPase *in vitro*. Force redevelopment following unloaded shortening and quick stretch is a measure of the number of myosin cross-bridges bound to actin and the rate of this process is proportional to the rate of hydrolysis of ATP (13).

The development of force by the flight muscles of insects that have different resonant frequencies of wingbeat has been compared with the ATPase activity of these muscles. A different conclusion has been reached from that obtained with rabbit muscle (103). In these insect muscles, tension is stretch-activated. The time constant of tension development was measured in skinned fibers. In all species studied, the rate of hydrolysis of ATP was slower than the rate constant of tension development after a quick stretch. Tension development precedes the rate-limiting ATPase step. It is suggested that in Step 5 in Figure 17, there is isomerization of Am · ADP · P$_i$, that is, this step is in two stages, and this isomerization controls the rate of tension redevelopment. There is no significant change in fiber ATPase activity with wingbeat frequency of

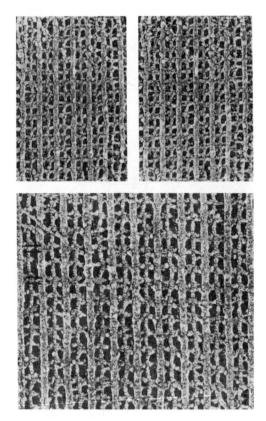

Figure 18. High-power electron micrographs of unfixed insect flight muscle to visualize thick and thin filaments and cross-bridges. [From (63a).]

different species. The work per oscillatory cycle must be smaller in insects with higher wing-beat frequencies than with low frequencies. However, both ATPase and power output rates are higher in insects in which the muscles function at higher temperature (*Apis* and *Bombus* at 39–40°C) than in those functioning at lower temperatures (*Vespa* at 29.5 and *Tipula* at 27°C).

Evidence regarding the interaction between thick and thin filaments and the role of cross-bridges in contraction is based on observations of structure in geometrically regular striated muscles—frog sartorius or insect flight muscles (Fig. 18). Tension is maximum whenever there is

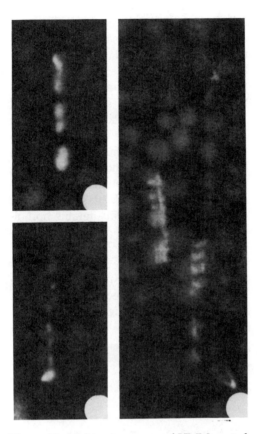

Figure 19. Light microscopy of HMM coated beads moving along chloroplast rows in Nitella. Each photograph was made by a series of 1-s exposures taken every few seconds. An aggregate consists of two or three beads that follow chloroplast rows. [From (137).]

complete overlap between thick and thin filaments. When the muscle is stretched to a length where there is no overlap, no tension is developed and when the muscle is allowed to shorten below rest length, the filaments pile up and tension is reduced. Electron micrographs provide clear evidence for sliding of thick filaments past thin ones and for changes in bridge angle during contraction. X-Ray diffraction pictures of muscles in states of relaxation, contraction, and rigor (when no ATP is available) show meridional reflection at 14.3 nm in spirals with repeats every three units or 42.9 nm. The 14.3 reflection is interpreted as coming from cross-bridges. In contraction the intensity of the 42.9-nm band decreases slightly ahead of peak tension and the 14.3 reflection decreases in intensity and broadens. This is interpreted as showing a disordering of cross-bridges as myosin heads go out to contact actin (62).

Evidence for generation of force by rotation of myosin heads rather than by a change in bridge angle comes from observations on actin arrays extracted from the alga Nitella. Myosin-coated fluorescent beads were placed on polar cables of actin filaments on the cytoplasmic face of rows of chloroplasts. There were five actin cables per row of chloroplast. In the presence of ATP, 0.7 μm diameter beads coated with skeletal muscle heavy chain myosin moved along the actin cable in the direction of rows of chloroplasts (Fig. 19) (137). Similar observations have been made on either actin or myosin filaments oriented on glass or plastic slides; particles (beads or dead bacteria) coated with the opposite protein move along the filament when ATP is supplied (86a). A myosin molecule can be cleaved into various parts and the myosin head that has ATPase activity at the point of attachment can move along an actin filament without S-2 and the molecular chain. Presumably, molecules of the head can rotate in their configuration. The S-2 hinge is, therefore, not essential for movement. However, movement is faster with an entire heavy molecule than with S-1 alone (67, 160). When smooth muscle myosin is used, it must be phosphorylated before bridges to actin can form.

Another *in vitro* motility system makes use of rotation of a very small rotor coated with F-actin and suspended in a solution containing Mg^{2+}-ATP and myosin heavy chain. Rotation of the rotor indicates that myosin pushes the actin in one direction.

Limited rotation is obtained with the S-1 fragment alone (172).

Calcium Regulation of Contractions

Calcium ions are required for contraction in all muscles. The processes by which Ca^{2+} is made available to the contractile proteins is an important part of excitation–contraction coupling. In vertebrate striated muscles Ca^{2+} is stored in particles in the SR, particularly in cisternae in regions of the SR near the T-tubules (Fig. 20a); in some muscles (particularly smooth muscles and crustacean striated muscle) calcium enters via plasmalemmal Ca^{2+} channels.

The sequence of coupling between T-tubules and SR is complicated and not fully understood. The current conducted inward in the T-tubules of striated muscle is carried mainly by Na^+, K^+, to a small amount by L-type Ca^{2+} channels. Between the T-tubules and sarcoplasmic reticulum are bridging structures called "feet" (Fig. 21) (10b, 42a, 44–47). The feet have a tetrad structure (44, 165a). The T-membrane voltage sensing protein can bind to a blocking agent dihydropyridine (DHP) (118a). This protein is a large molecule (170 kD); its cDNA has been cloned and the protein has considerable homology with Na^+-channel protein (151). Apparently, charge (not ionic current) is transferred from T to SR along this V-sensitive protein (126). Dysgenic mice have no or little DHP receptor (151, 165a). Two of the tetrad particles of feet are as-

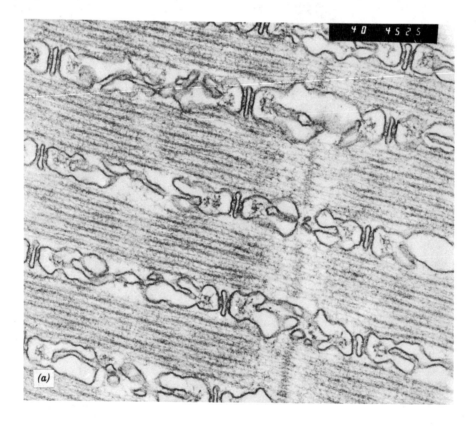

Figure 20a. Longitudinal election micrograph of a fast-twitch muscle fiber to show five rows of sarcoplasmic reticulum (SR), T-tubules (T), and feet between T and SR. (Courtesy of Franzini-Armstrong).

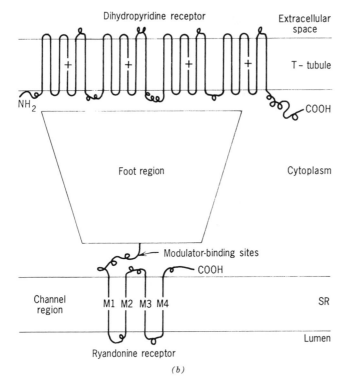

Dihydropyridine receptor

Extracellular space

T – tubule

NH$_2$

COOH

Foot region

Cytoplasm

Modulator-binding sites

COOH

Channel region

M1 M2 M3 M4

SR

Lumen

Ryandonine receptor

(b)

Figure 20*b*. Diagram of coupling proteins between T and SR. DHP. sensitive receptor on T side ryanodine sensitive and Ca^{2+} releasing protein on SR side. Large "foot" protein is part of ryanodine sensitive portion. [From (150a).]

sociated with the *t*-membrane DHP receptor. The membrane of SR and the feet also contains a ryanodine-sensitive receptor, which forms protein Ca^{2+}-activated channels (66a). The ryanodine receptor of the SR contains 5037 amino acids; it is a tetramer of 450 kD units (Fig. 20) (150a). The voltage sensitive, DHP binding, charge-carrying protein is coupled in some way to the gating of the ryanodine sensitive Ca^{2+} channels; activation via the T-tubule results in Ca^{2+} release from the SR (106a). Extraocular muscles of billfish lack contractile elements but have a rich SR tubule system that generates heat that maintains the brain temperature (109, 176).

The SR membrane contains particles of high-affinity Ca^{2+}-ATPase, which pumps Ca^{2+} back into the SR. Inside the SR there is calsequestrin, a protein of high binding capacity, but low affinity for Ca^{2+}; this provides a store of calcium in resting muscle. Release of Ca^{2+} from SR can be demonstrated if the Ca^{2+} requiring luminescent protein aequorin or a Ca^{2+}-sensitive dye is injected into a muscle fiber. In the presene of very small amounts of Ca^{2+} a flash of light is emitted from aequorin or a specific wavelength of light is absorbed by a dye. Amplitude of the flash or dye absorption is proportional to the Ca^{2+} concentration in the cytoplasm and rises more rapidly in fast than slow muscles.

Several different mechanisms have evolved that regulate the access of calcium to the myosin–actin complex. In vertebrate striated muscle the thin filaments contain, in addition to actin, the proteins tropomyosin (Tm) and troponin (Tn). A thin filament has two strands of F-actin in a helix with Tm wound round the outside (Fig. 22), seven actins per Tm. Inter-

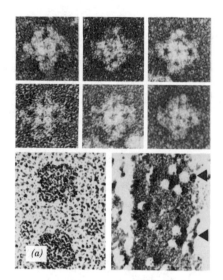

Figure 21. Ultrastructure of end feet showing both T-tubule and SR components. (*a*) The fourfold symmetry of the SR calcium release channel is evident in both negative stained and rotary shadowed images of the isolated muscle. The SR calcium release channel is composed of four 450-kD subunits each of which has a hydrophilic domain that spans between the T-tubule and SR and a hydrophobic domain that penetrates the SR bilayer. (*b*) Freeze–fracture of glutaraldehyde fixed toadfish swimbladder muscle reveals groups of four particles on the junction T membrane that correspond in their distribution with the subunits of coupling across the T–SR junction. (Courtesy of Barbara Block and Harold Erickson.)

spersed at 40-nm intervals is Tn. Tropomyosin constitutes some 9% of muscle protein and has an MW 66,000 D (39). There are three Tn subunits: TnI of MW 20,900 D binds to actin in the presence of Tm; TnC of MW 17,800 D binds to calcium, $4Ca^{2+}$/molecule of TnC; TnT of MW 30,500 D a basic protein links the other troponins to Tm and increases the Ca^{2+} sensitivity of the complex (24, 63). The thin filament proteins occur in a ratio of 7 actin:1 Tm:1TnT:1TnI:1TnC. The calcium bound to TnC comes from the SR on activation. Figure 22*b* shows the relative positions of Tm—actin and S-1 when a bridge is formed. Troponin I is different in composition in striated, fast, slow, and

cardiac muscle with more similarities in amino acid sequences in the region of binding to actin and blocking ATPase, more differences in the terminal that binds to TnC (170).

In the relaxed state, myosin and actin are not in contact and interaction between them is prevented by the attachment of Tm and TnI to actin. When Ca^{2+} ions are released from the SR, Ca^{2+} is bound to TnC and this starts a chain of events. The restraint of Tm and TnI on actin is removed such that a myosin head bridges to actin and as myosin ATPase becomes activated by the actin, energy is liberated and sliding occurs. The sequence of control is (1) Ca^{2+} release from SR; (2) reac-

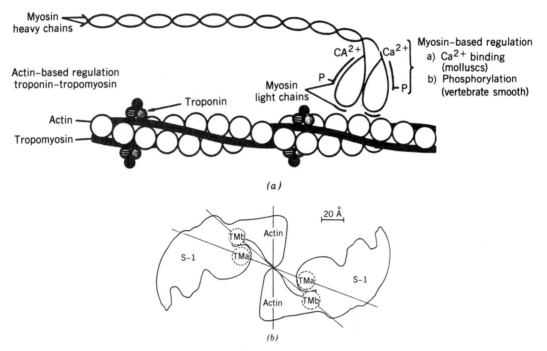

Figure 22. (*a*) Diagram of major regulatory systems. Thin filament of actin-based regulation by troponins and tropomyosins in vertebrate striated muscle; myosin-based regulation by Ca^{2+} binding in molluscs; phosphorylation of myosin light chain in vertebrate smooth muscle. [From (3).] (*b*) Summary of a model to show the relative positions of Tm, actin, and S-1 of the myosin bridge. [From (155a).]

tion of Ca^{2+} with TnC; (3) reactions of TnC with Tm, TnT, and TnI; (4) change in position of Tm and TnI relative to actin; (5) binding of myosin S-1 to actin; and (6) activation of S-1 ATPase. Tension increases rapidly at Ca^{2+} concentrations of $10^{-5.8}$–10^{-6} M, relaxation occurs at 10^{-7} M and at rest the concentration of free calcium ions is 10^{-8} M or lower. Speed of contraction is related to the activity of myosin ATPase and of relaxation to the activity of Ca-ATPase that sequesters Ca^{2+} into SR. Tropomyosin from yeast has a MW 33,000 D; it cross-reacts immunologically with brain Tm. Resemblance of yeast and mammalian Tm and myosin indicates extreme conservatism (93b).

Several alternative modes of Ca^{2+} regulation have evolved. In molluscan mus-

cle (adductor of *Pecten*) troponins are absent and regulation is by the direct combination of Ca^{2+} with a regulatory lc of myosin. Tropomyosin is present but it is not essential for Ca^{2+} activation. Calcium sequestration is mainly in plasma membrane, with small amounts in mitochondria and the sparse SR. Calcium spikes bring calcium into the fibers. A Ca^{2+} binding regulatory light chain has a MW 18,000 D and is distinct from a sulfhydryl-containing chain (SH lc) that is one half as abundant as the regulatory lc (20). Antibodies for the molluscan light chains do not cross-react with vertebrate muscle, platelet, or *Physarum* myosin lcs (166). Myosin regulation by a Ca^{2+}-sensitive lc has been found in muscles of some other invertebrates, for example, holothurians molluscs in general, bra-

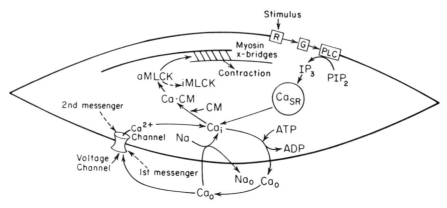

Figure 23. Generalized scheme for regulation of tension control in smooth muscle. First messengers [stimuli such as hormones, electrical stimuli, and light (to iris)] bind either to G-protein linked receptor molecule as indicate at the top of the figure, or directly to ion channels as indicated at the bottom of the figure. In the former case, a cascade of events involves release of IP$_3$ and induces release of Ca^{2+} from SR. In the latter case, the increased conductances to Ca^{2+} also leads to a rise of intracellular Ca^{2+} concentration (Ca$_i$). As the signal transduction and transmission proceed, the activation of energy transducing and contractile machinery follows. The Ca^{2+} ion binds to calmodulin (CM) and the Ca–CM complex activates MLCK. As myosin is phosphorylated the actin–myosin cross-bridges are formed (L. Barr, personal communication)

chiopods, echinoderms, and earthworms (92). Each myosin molecule binds 2Ca^{2+} ions, probably one per head. Removal of the 17–20 kD subunit from myosin by ethylenediamenetetraacetic acid (EDTA) eliminates the Ca^{2+} dependence and myosins no longer binds with actin (61).

Muscles of insects, crustaceans, annelids, and Ascaris have double control—a Ca^{2+}-sensitive myosin lc and a thin filament troponin. The troponin subunits of invertebrate muscles differ from those of vertebrate striated muscle in that they bind less calcium (123). Earthworm muscle has both TnI and TnC (37). Ascaris troponin consists of three subunits, probably TnT, TnI, and TnC; Ascaris tropomyosin is larger than rabbit Tm (36a, 86).

A third method of calcium control occurs in vertebrate smooth muscle and has been examined in chicken gizzard and "skinned" fibers of mammalian uterus (83, 95, 96). Vertebrate smooth muscles lack troponin. A different protein, a

myosin light chain kinase (MLCK) is activated when Ca^{2+} is released into the cytosol. A light chain of myosin becomes phosphorylated via MLCK and the myosin can then interact with actin (Fig. 23). After ATP hydrolysis and energy transfer associated with movement a second enzyme, a phosphatase dephosphorylates the myosin and relaxation occurs. Phosphorylation of the kinase (MLCK) occurs at a Ca^{2+} concentration of 10^{-6}–10^{-7} M and dephosphorylation via the phosphatase when the Ca^{2+} concentration drops below 10^{-7} M (3). Calcium activation by phosphorylation of myosin occurs in some other motile cells, for example, the myosin of *Acanthamoeba* (27a, 89, 90). Phosphorylation of striated muscle myosin can also occur but this is not essential for Ca^{2+} regulation (27a).

Thin filaments of smooth muscles— gizzard, stomach, and aorta—contain actin and a tropomyosin. Recently, a third protein of MW 120–140 kD has been found. This is called caldesmon; it is a strong inhibitor of AM ATPase and the inhibition is reversed in the presence of Ca^{2+} (96). Caldesmon may function in smooth muscle similarly to Tm of striated muscle (96).

Substances important for the Ca^{2+} activation of the smooth muscle myosin kinse are calmodulin (CM), inositol triphosphate (InsP$_3$), and cyclic AMP (cAMP) (Fig. 23). Calmodulin is a protein of wide occurrence which has four Ca^{2+}-binding sites; MW of CM 17,000 D. The Ca^{2+}–CM is more effective than Ca^{2+} alone in activating MLCK. Calmodulin resembles troponin-C but is a smaller molecule. Calmodulin functions in smooth muscle by buffering free Ca^{2+} concentration (97). Calmodulin also functions as a second messenger in mediating the action of some neurotransmitters. The Ca^{2+}–CM complex activates a number of kinases in addition to MLCK.

Another second messenger is cAMP. Hormones and transmitters such as norepinephrine activate adenylate cyclase, which increases the amount of cAMP in smooth muscle. A protein kinase is activated by cAMP which can phosphorylate MLCK. This weakens the binding of Ca^{2+}–CM to myosin kinase and thus modulates the action of calcium on myosin.

In some smooth muscles, IP$_3$ causes release of Ca^{2+} from SR. Figure 23 is based on evidence for the activation of smooth muscle.

Mammalian cardiac muscle has a 22-kD protein and phospholambin in the SR membrane. A cAMP-activated protein kinase phosphorylates phospholambin. Calmodulin can also independently mediate phosphorylation of phospholambin. Phosphorylated phospholambin accelerates Ca^{2+} uptake via Ca^{2+}-ATPase into the SR membrane (19, 50).

In summary, the concentration of Ca^{2+} is precisely controlled in all muscles and several methods of control are used. In vertebrate striated muscle electrical signals in invaginating T-tubules by charge transfer activate sarcoplasmic reticulum. These muscles have the proteins tropomyosin plus three troponins in the thin filaments in association with actin. The four regulatory protein molecules react in a cascade to release Ca^{2+} from sarcoplasmic reticulum. At rest, myosin and actin do not interact and activation by Ca^{2+} of the tropomyosin–troponin cascade removes the brake on the myosin–actin interaction. In many muscles of invertebrate animals, especially molluscs, Ca^{2+} regulation is by a specific myosin light chain. Muscles of some invertebrates have both a troponin chain and a Ca^{2+}-sensitive myosin light chain. In vertebrate smooth muscle, a myosin lc becomes phosphorylated by a kinase that is sensitive to a Ca^{2+}–CM complex.

Cytoskeletal Proteins

In addition to myosins, actin, tropomyosin, and troponins, muscles contain a variety of proteins that have structural properties as intermediate filaments, 10 nm in diameter (23). The Z lines of striated muscle contain several proteins, notably α-actinin, which may function in attachment of thin filaments. In smooth muscle the "dense bodies" may serve the same function as Z bands. Striated muscle C-protein is 2% of myofibrillar protein and it may hold myosin molecules together in thick filaments. Titin is a large protein that may constitute 10% of myofibrillar protein; antibodies to titin stain at Z lines and at the junctions of A to I bands. Desmin may link Z and M bands to muscle membrane; desmin filaments are abundant in smooth muscle fibers. Spectrins may function in relation to the shape of numerous cells—epithelial cells, fibroblasts, and red blood cells; actin may bind to spectrin (25). Bundling and capping proteins for actin filaments include profilin, gelsolin, villin, α-actinin, and fragmin.

Five categories of intermediate filaments are characteristic of specific cell types: glial acidic proteins (GFAP), neurofilament proteins (NFAP), vimentin in cells derived from mesenchyme, epithelia keratins, and muscle desmin (48).

Amoeboid Movement

Amoeboid movement is specialized protoplasmic streaming. It may be directed locomotion in rhizopod protozoa, in the amoeboid plasmodial stage of myxomycetes, and in wandering blood cells such as leukocytes. Amoeboid movement consists of extension, flexion, and retraction of feeding processes in Foraminifera, Heliozoa, Radiolaria, and reticulo-endothelial cells. Locomotory amoeboid movement requires attachment to a substrate; nonlocomotor amoeboid movement occurs in free pseudopods. Pseudopods may be lobopods—broad to cylindrical, round at the tip; pseudopods may be filopods, slender with pointed tips; reticulopods—threadlike, branching, and anastomosing, or pseudopods may be axopods with a central axial rod. Amoebae differ greatly in cell form, in rates of streaming and locomotion (Fig. 24). A given amoeboid cell may take on different forms under various condition; *Amoeba proteus* in distilled water becomes radiate or stellate, in dilute salt solutions it is monopodal, and it is stellate just prior to fission. Amoebae have a central fluid endoplasm (plasmasol) and an outer viscous ectoplasm (plasmagel). The greatest variability of pseudopods occurs in Formaninfera and Radiolaria in which the filamentous pseudopods may branch and form networks. Protoplasmic flow in filamentous pseudopods may be bidirectional, outward on one side and inward on the opposite side.

A typical amoeba consists of an outer membrane, the plasmalemma that has adhesive properties and is hydrophobic; the plasmalemma consists of an outer filamentous coat and inner membrane and is 35% lipid, 26% protein, and 16% polysaccharide. Beneath the plasmalemma is a hyaline layer, which is thin in the region of attachment and widens to a hyaline cap at the anterior tip of an advancing pseudopod. Under the hyaline layer is the ectoplasm, or the cylinder of plasmagel that is of high viscosity and may extend as a thin gel sheet beneath the hyaline cap. The endoplasmic core of the amoeba is the plasmasol of low viscosity. The nucleus is normally in the plasmasol. Attachment to the substrate is facilitated by traces of salts, particularly Ca^{2+} in the medium. Ectoplasm is continually being converted into endoplasm at the posterior end or at some fixed region and the endoplasm is converted to ectoplasm anteriorly or in an extending pseudopod. Freely advancing amoebae move at ~1 μm/s. Migrating plasmodia of slime molds may advance at 5–6 cm/h. Individual granules in an advancing slime mold (*Physarum*) have been seen to move in opposite directions, that is, in forward and backward moving channels and the flow generated at a given point reverses irregularly. One hypothesis to account for amoeboid movement is that the plasmagel is contractile; it is thicker in posterior regions, and it may exert pressure on the plasasol to force it forward. A second hypothesis is that contractile force is generated by microfilaments at the advancing tip and that the endoplasm is pulled forward in a fountainlike pattern. This is supported by experiments in which a large amoeba (*Chaos carolinensis*) was drawn into a small quartz tube under oil; when the tube was broken into sections each containing part of the amoeba, streaming occurred in the sections, first with axial endoplasm flowing forward and ectoplasm flowing backward, then

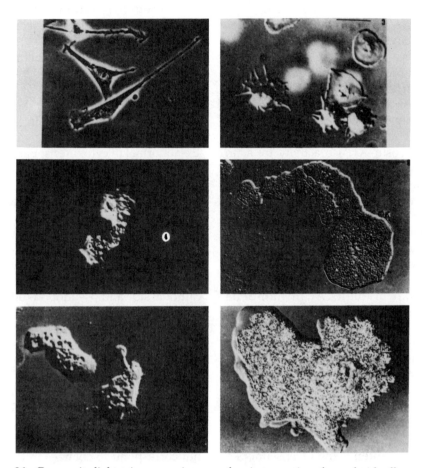

Figure 24. Composite light microscope images of various species of amoeboid cells. upper left, *Amoeba proteus*, 100 μm; upper right microplasmodium of *P. polycephalum*, 100 μm; middle left, human platelets, 5 μm; middle right, amoeboid stage of *D. discoideum*, 10 μm; lower left, *Acanthamoeba*, 10 μm; lower right rat embryo cell, 10 μm. [From (154).]

with several channels making U-turns at one end. This hypothesis postulates that frontal contraction rather than hydraulic pressure from the rear is responsible for pseudopod formation. Microdissection procedures indicate contractility and high viscoelasticity of the ectoplasm. When amoebae are subjected to high hydrostatic pressure, the viscosity falls, the ectoplasm solates, and locomotion stops (153–155).

Proteins that resemble myosins and ac-

tins have been extracted from several kinds of amoeboid cells. Slime molds (*Dictyostelium* and *Physarum*) are large multinucleate structures when mature that can advance by amoeboid movement over a substratum. Reproductive fruiting bodies are formed that liberate uninucleate amoebae; these aggregate by chemotaxis toward cAMP liberated from individual amoeboid cells. A myosin-like protein has been extracted from *Dictyostelium* that consists of two heavy chains,

210 kD each, and two classes of light chains of 16 and 18 kD. The myosin has ATPase activity, activated by Ca^{2+} and inhibited by Mg^{2+}. The myxomycete myosin can interact with muscle actin or with actin extracted from *Dictyostelium*; the ATPase activity is increased ninefold when complexed with actin; this is to be compared with the 20-fold increase of muscle myosin ATPase activity by actin (21). A protein with properties resembling muscle actin comprises 5–15% of the total protein in *Dictyostelium*. This protein has a MW 43 kD and it can form thin filaments, 6–8 nm in diameter oriented parallel to the long axis of a plasmodium.

In amoebae, as in other nonmuscle motile cells, there are 100–200 times more actin than myosin. Acanthamoeba myosin has MW 180,000 D with subunits 140,000, 16,000, and 14,000 D (114, 115). Its ATPase requires Ca^{2+} for activation. Myosin in *Dictyostelium* amoebae is most abundant in the posterior cortex. *Acanthamoeba* myosin has a globular head similar to that in vertebrate myosin but it lacks a filamentous tail.

It is postulated that gelation such as occurs at the anterior end of a pseudopod is equivalent to contraction, that the transition of ectoplasm to endoplasm at the posterior end is a dissociation of actin from myosin, a relaxed state (5, 6). Naked cytoplasm of a giant amoeba, *Chaos*, can be made to contract by addition of Ca^{2+} at concentrations $>10^{-7}$ M. When cells are bathed in a solution low in calcium, the plasmalemma becomes fragile and when torn by a fine needle, the cytoplasm flows out. When this cytoplasm is stretched, filaments appear. Addition of Ca^{2+} + ATP causes pseudopod formation and activation of ATPase. The plasmalemma is not necessary for pseudopod formation. When Mg^{2+} + ATP are added in the relaxed state, bundles of actin filaments may form, that is, the actin is transformed from less filamentous to a structured filamentous state. When myosin is present, gelation (contraction) occurs and ATPase is active.

Alterations in myosin structure of *Dictyostelium* have been obtained by isolating deficient mutants (87) and by recombination in a plasmid such that the carboxyl terminal of the heavy myosin is lacking. Amoebae with a defective myosin gene are abnormal in cytokinesis; parallel thick filaments of the cleavage furrow are missing. The cells can form pseudopods and show some amoeboid movement, but cannot aggregate.

Transmembrane potentials of -40 to -70 mV have been recorded in amoebae and slime molds (100). In an electric field, *Amoeba proteus* shows solation on the cathodal side and pseudopods advance in that direction. Oscillations of a few seconds occur in protoplasmic flow in slime mold plasmodia. When the plasmodium is kept in an hourglass form, hydrostatic pressure applied on either side can stop amoeboid flow in that direction. The pressure needed to prevent flow fluctuates periodically and the balancing pressure is 5–19 in. of H_2O (78). The oscillations do not correspond to changes in transmembrane potential.

Cytokinesis or cell cleavage requires a myosin–actin interaction. Dividing cells have actin filaments in a ring around the perimeter (21). In yeast, myosin-ATPase is localized in 10-nm filaments in a region of budding. Movement that does not require myosin but is produced by actin occurs in some sperm. In starfish sperm, the acrosome consist largely of actin that is coiled in a bundle of filaments and as a sperm approaches an egg, the actin depolymerizes to form a 50-mm long process within a few seconds (Fig. 15) (157).

Tubulin-Based Movements

Tubulins evolved in eukaryotes in parallel with actin–myosin complexes. Tubulins bring about movement of cilia and flagella, they function in mitotic and meiotic movement of chromosomes; tubulins implement axoplasmic transport in neurons; tubulin is the most abundant protein in brain tissue and actin is the third most abundant. Tubulins are present in outer segments of rods and cones and in many nonmotile cells.

Cilia

Cilia are organelles that occur on the surface of some cells; they propel fluids by lashing movements. A cilium arises from an intracellular granule or basal body, which serves as the template from which the 9-plus-2 pattern of microtubules arises. Flagella are long cilia that occur in small numbers on some motile cells. Cilia are the means of locomotion in ciliated protozoans and embryos; flagella are the means of movement of flagellate protozoa and spermatozoa. Cilia create currents for feeding in sea anemones and corals, in filter-feeding molluscs and prochordates; cilia generate respiratory currents on gills (174). Cilia have a cleansing function in some organs—in digestive, urogenital, and respiratory tracts. Ciliary movement may be pendular, waving back and forth, and flexing at the base; movement may be flexural, bending first at the tip, and progressing toward the base; movement may be undulatory, wavelike, especially in flagella in which a wave passes from base to tip or in reverse direction. Undulatory movements may be helical rather than planar. Ciliated surfaces may show metachronism, an orderly succession of initiation of ciliary beat in spatial sequence. This is evident in ciliated protozoans and in epithelia of gills and mouth. Cilia of some kinds show frequent reversal of beat direction, initiated by external or internal stimuli. The effective stroke is faster than the recovery stroke (125).

Figure 25. Diagram of cross-section of ciliary axoneme. Nine pairs of peripheral microtubules, two central microtubules, spokes between central ring, and peripheral microtubule pairs. Stippled doublets are active during effective stroke. [From (165b).]

Cilia in general show uniformity of structure. The diameter of a cilium is 0.2–0.25 μm; cilia are 10–20 μm long, flagella from 20 μm to several millimeter in length. A cilium comprises a bundle of longitudinal microtubules, the axoneme, enclosed within a membrane that is continuous with the cell membrane (Fig. 25). The axoneme consists of nine outer pairs of microtubules, and a central pair surrounded by a sheath. The two paired microtubules, A and B, are connected by a protein nexin. Spokes connect microtubules of the outer ring to the central tubular sheath and spokes are arranged at a 40° pitch (131). Projecting at 254-nm intervals from each A tubule and directed toward the B microtubule of the adjacent pair are arms consisting of a protein dynein, which has ATPase activity. Each of the microtubules consists of tubulin monomers, each with a MW 55,000 D and 4 nm in diameter (7). In cross-section, 13 of the monomers occur in each A microtubule and 11 occur in a C-shaped pattern in the B microtubule. There are two kinds of monomer, α and β, of essentially the same composition in both A and B microtubules. The monomers of an microtubule are staggered circumferentially, those in a B microtubule are aligned obliquely (7) (Fig. 26).

The components of a cilium can be separated by different chemical treatments. The microtubule monomers that have been separated can polymerize *in vitro* to form polymers similar to microtubules. Polymerization is facilitated by MAPs, microtubule associated proteins. The microtubules are solubilized by heating and the B microtubule is solubilized by less heat than is required to solubilize the A microtubule. The tubulin of the central pair differs in solublity from the peripheral pairs [Fig. 27; (144)]. Dynein is solubilized by dialysis against EDTA. Nexin is left after solubilizing tubulin at low pH. Nexin linkages occur at 960-Å intervals along the length of peripheral microtubles. Nexin, MW 165,000 D, can be separated from tubulin by extraction at low pH. Similar but slightly modified methods are used for extraction of ciliary components from sea urchin sperm, *Tetrahymena*, and scallop gill cilia (143–145).

Cilia can be removed from a cell that bears them by treatment with high salt concentrations or with detergents. The isolated cilia continue to beat if supplied with ATP. Ciliary beat is powered by energy from hydrolysis of ATP by dynein; it is calculated that two molecules of ATP are hydrolyzed per dynein per beat cycle (15). Electron micrographs show that each A microtubule extends beyond its B counterpart at the tip of the cilium. In the effective stroke, the microtubule pair on one side extends farther toward the tip than the opposing pair. For example, pair No. 6 extends farther than pair No. 1 (Fig. 28). Movement is between doublets rather

A   **B**

Figure 26. Subunit arrangements in outer doublets of a cilium. Heterodimers of A-tubule have α–β pattern; (light–dark–light); monomers in B-tubule have α to α/β to β (light–light, dark–dark) pattern. [From (145).]

than between A and B microtubules. On return stroke, the positions of doublets are reversed in position (125). The dynein bridges attach and detach during a stroke cycle. Also there is a cycle of attachment and detachment between radial spokes and central sheath.

In the cilia of *Paramecium*, the central pair of microtubules are oriented at 90° to the direction of the power stroke and the central pair rotates during a ciliary beat. In a mutant of Paramecium, pawn, the central pair fails to rotate and swimming is impaired, hence it is proposed that the

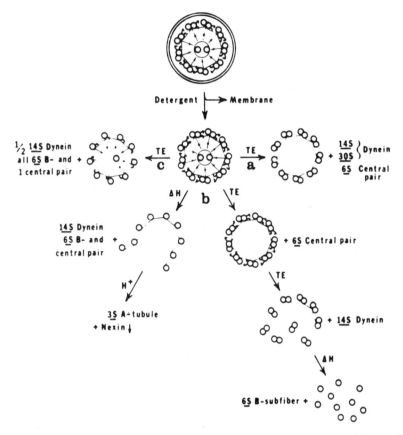

Figure 27. Flow diagram for fractionation of ciliary membrane; central pair, dynein; TE = dialysis against tris-HC1 and EDTA, ΔH centrifugation. [From (144).]

central tubules regulate the peripheral ones (110).

The α and β tubulins occur in multiple isoforms, each coded by a different gene. *Tetrahymena thermophila* has 5α and 2β genes, 3α and 1β for ciliary tubulin, and the others for cytoplasmic tubulin (149). *Chlamydomonas* has 2α and 2β tubulin genes. Adult rat brain has 7α and 14β; rat embryonic brain has 3β genes.

Ciliary beat is subject to regulation by mechanical stimuli, by changes in ionic concentrations and by neurohormones. In *Paramecium*, the cilia propel the animal forward in a medium containing 0.1 μM Ca^{2+} (108). When a *Paramecium* touches an object at its anterior end, a wave of depolarization at 40 mV/s of inward Ca^{2+}

current is initiated and ciliary beat reverses so that the *Paramecium* backs away. Caudal stimulation initiates a hyperpolarization because of increased K^+ conductance (40, 132).

Abfrontal cilia of *Mytilus* gill are activated to beat by 20 μM/1 of Ca^{2+}; lateral cilia are inhibited by high concentrations of Ca^{2+}, and are activated by <1 μM Ca^{2+} (146, 147). Lateral gill cilia are activated by serotonin (5-HT) which in *Mytilus* gill activates adenylate cyclase; cAMP can activate ciliary beat in gills that have been in Ca^{2+} arrest. Lateral cilia, permeabilized by Triton-X, show a cAMP-dependent phosphorylation of three low molecular weight proteins distinct from, but probably associated with dynein. Excitation of

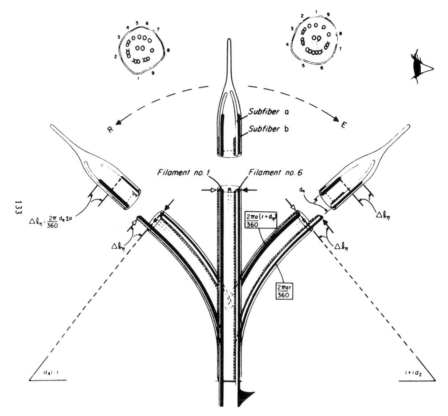

Figure 28. Sliding microtubule of ciliary beat. Doublets 1 and 6 have subfiber (a) extending beyond subfiber (b) when in a neutral position. Bending produces an apparent length difference (Δ) between subfibers so that the positions in subfibers 1 and 6 are different at the end of a movement. [From (130).]

abfrontal cilia by Ca^{2+} appears to be by binding to calmodulian of another class of protein from those phosphorylated by cAMP dependent kinase (105a, 146, 147). Mechanical stimulation of abfrontal ciliated epithelium evokes regenerative Ca^{2+} dependent potentials that propagate through many cells. Mechanical stimulation of the lateral ciliated epithelium inhibits ciliary activity.

Axoplasmic Flow

Axoplasmic transport moves materials such as neurotransmitters and metabolites that are synthesized in neuron somata out to axon terminals. Both fast and slow transport, also anterograde and retrograde transport have been identified. Fast movement is at 100–400 mm/day, slow at 0.2–2 mm/day. Anterograde transport was discovered by the accumulation of organelles in axon terminals or on the proximal side of constriction or cold block. Transport filaments contain tubulin and are distinct from neurofilaments that are cytoskeletal or intermediate filaments. Intracellular membrane components move rapidly while cytoskeletal and soluble cytoplasmic components move slowly.

When axoplasm is extruded from a squid giant axon onto a glass surface, movement of organelles such as vesicles, can be observed. Adenosine triphosphate must be supplied as an energy source. The transport filaments are 24 nm in diameter and consist of single microtubules (Fig. 29) that react with antibody to α-tubulin (133–135). Figure 29a shows the structure of transport filaments composed of chains of tubulin monomers. Figure 29b shows the movement of a vesicle (organelle) along one side of a filament. A protein fraction called kinesin, that can be destroyed by heat, is essential for the transport of an organelle. Kinesin prepared from squid axopolasm has a MW 30 kD and consists of subunits of 110 and 60–70 kD. Transport is in the direction of the positive end of a microtubule,

that is, toward the end where tubulin polymerization occurs preferentially (110A). The positive end of a microtubule is peripheral, the negative end is toward the cell body. Kinesin has ATPase activity like dynein but its catalyzed motion is of opposite polarity (on the microtubule) (161–163). A gene for kinesin identified in *Drosophila* encodes the protein that binds both ATP and microtubules (132a). *Drosophila* kinesin of 115-kD cross-reacts with kinesin of squid and sea urchin (132a). Kinesin is pictured as the motor for propelling particles along a microtubule.

Mechanisms of fast and slow axonal flow are different. Microtubules extracted from brain plus ATP undergo a slow gel

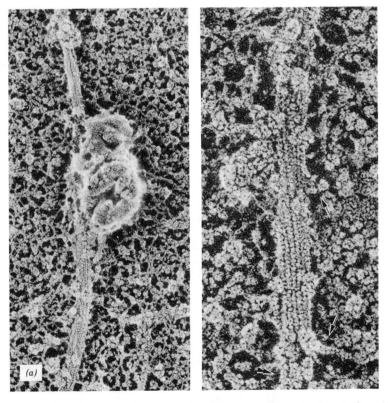

Figure 29a. Electron micrographs of transport filaments from axoplasm of squid giant axon. The figure on the left is at lower magnification than the figure on the right. Note attached vesicles (organelles) on the left. Also, note the filamentous components of α and β tubulin monomers in protofilaments.

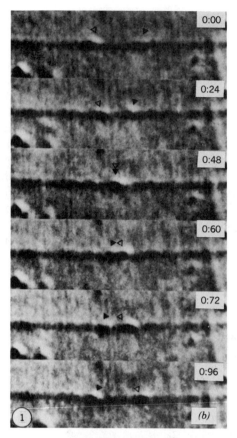

Figure 29b. Successive video micrographs showing bidirectional movement of vesicular organelles on a single transport filament. The two arrows (solid and open) point to vesicles that move in opposite directions along the same filament, pass, and continue moving in original directions. [From (135).]

contraction. Associated particulates consisting of tubulin, neurofilaments, and spectrin move along the microtubles at 1 μm/min (168).

Chromosome Movement in Mitosis and Meiosis

Tubulin is abundant in dividing cells; details of mitotic movement have been examined in cell cultures and in the large eggs of echinoderms. Astral rays differ in composition from equatorial rays. Three hypothesis for movement of chromosomes in anaphase are (1) pulling by filaments from the poles, (2) pushing by filaments in the equator, and (3) movement by independent changes in the length of filaments that attach at the kinetochores of chromosomes (57). In addition to the movements of chromosomes, the poles move apart during anaphase. The detergent axol blocks spindle formations in mitosis but in meiosis spindles form but do not migrate (90a).

Several microtubule associated protein (MAPs) have been isolated from brain and flagella. One function of MAPs is to promote microtubule assembly and stability (93a). In isolated axonemes of a flagellate an MAP promotes gliding if present with kinesis. Flow toward distal (positive) end is promoted by MAP, toward proximal (negative) end by kinesin.

The spindle becomes birefringement at metaphase and birefringence is maximal during anaphase; this birefringence has been interpreted as a result of the polymerization of tubulin. Injection of calcium into a spindle decreases birefringence, cooling or tretment with colchicine also reduces birefringence of a spindle. One hypothesis of anaphase movement is that polymerization of tubulin pushes chromosomes apart (68, 69). Another postulate is that the cytoplasmic matrix contains elastic filaments that are attached at kinetechores of chromosomes and are stretched when chromosomes move toward the equator in late prophase; cessation of stretch allows the matrix filaments to pull apart (68, 69).

Recent evidence has been obtained by antibody labeling of filaments that are attached at kinetochores and photobleaching by a 1-μm laser beam. Depolymerization of tubulin occurs during anaphase at the region of attachment at a kinetochore. Tubulin is incorporated into filaments as the chromosomes assemble in metaphase and is lost during anaphase movement (102). In addition, a translocator or motor protein similar to kinesin may be involved in the chromosome movement.

In summary, tubulins have evolved as proteins of motility in parallel with actin–myosin complexes. Tubulins are abundant in cilia and flagella, in nervous systems, and in mitotic spindles. Movements produced by tubulins, like those by actomyosins, use energy from ATP. Hydrolysis of ATP in cilia is by dynein. Self-aggregating microtubules are composed of subunits of α or β tubulin; microtubles differ in proportion and orientation of α and β tubulin subunits. In ciliary beat, sliding occurs not between members of microtubule pairs but between adjacent pairs. In axoplasmic transport, organelles are moved along microtubules in neurons under the action of motor proteins (ATPases) such as kinesin. In mitosis chromosomes move by assembly or loss of tubulin subunits attached to kinetochores. In tubulin-mediated movements in general, polymerization and depolymerization of subunit monomers require specific activating proteins.

Cytokinesis or cleavage is by a myosin–actin interaction and contraction is a cytoplasmic ring (21).

Summary

Movement, intracellular and organismic, is characteristic of life. In eukaryotes, all movement is brought about by interaction of specific proteins energized by ATP. Two main classes of proteins of movement have evolved in parallel—the actin–myosins and the tubulins. The adaptive diversity of muscles is not equalled by any other physiological system.

Actin–myosin-based contractions occur in all animals—in myocytes of sponges, myoepithelia of coelenterates, and numerous kinds of muscles in other animals. In contractile cells myosin occurs in thick filaments, actin in overlapping thin filaments. In muscle fibers with distances >5–10 μm from cell membrane to contractile filaments, membrane invaginations conduct electrical signals from surface to filaments; in most such muscles there is a transverse alignment of thick and thin filaments, seen as striations. In smaller muscle fibers (cylindrical, spindle, or ribbon shaped) there is direct coupling between membrane and filaments; these muscles are nonstriated or diagonally striated fibers. In all muscles, tension is developed by a sliding between thick and thin filaments. In diagonally striated muscles, sliding is supplemented by change in angle (shearing). Contractions that develop tension with little change in length are termed isometric; contractions in which muscles shorten considerably and develop little tension are termed isotonic.

Speeds of contraction cover a range of 10^4 from the slowest muscles of sea anemones, slow movements of visceral organs, intermediate speeds of postural muscles, to fast extraocular and sonic muscles, and the fastest flight muscles of insects. Many adaptations are associated with the speed of contraction. In general, cross-striated muscles are attached to skeletal elements and are faster than nonstriated. However, nonstriated muscles have locomotor functions in hollow bodied animals. Nerve activated muscles are usually faster than endogenously rhythmic muscles.

Patterns of innervation are markedly diverse. Some wide-fibered striated muscles are innervated by a single motor axon. This releases a transmitter, usually acetylcholine, that initiates a chain of events from endplate potential to conducted muscle action potential to an electrical signal in the invaginating T-tubule. In other kinds of vertebrate striated fibers, branched motor neurons have multiple endings; the electrical response is a graded junction-type potential; the muscle fiber is sensitive to transmitter (acetylcholine) over its entire surface. In other wide striated fibers, especially of crustaceans and insects, innervation is polyneuronal and multiterminal. "Slow" motor axons evoke graded facilitating responses, "fast" axons elicit rapid nonfacilitating responses. One transmitter (glutamate) can elicit either a fast or slow response according to how the nerve endings are distributed. In addition, crustacean and insect muscles with multiple innervation receive inhibitory axons that liberate γ-aminobutyric acid (GABA) as transmitter. Some insect muscles are synchronous and give one contraction per activation. Some narrow-fibered nonstriated muscles are nerve-activated postural. Other smooth muscles are endogenously rhythmic and are subject to nerve modulation. Some muscles of molluscs and of a few kinds of worms show "catch" or maintenance of tension; excitation is by a cholinergic nerve, relaxation by a serotonergic nerve.

Energy for contraction comes from ATP, which is recharged from phosphagens—phosphocreatine in chordates, phosphoarginine in most invertebrates, and other phosphate compounds in some kinds of invertebrates. An ultimate energy source is oxidation or glycolysis. Vertebrate oxidative muscle fibers are slow, tonic, usually have myoglobin, low activity of myosin ATPase, and high activity of oxidative enzymes. Fast phasic muscles are white, have high ATPase activity, and high activity of glycolytic enzymes. Some muscles are intermediate in metabolic types, they are oxidative–glycolytic. The myosin of phasic muscles differs in amino acid composition from tonic muscle myosin; which form of myosin is synthesized is determined by the innervation.

Actins are highly conserved proteins,

MW ~45,000 D, globular or fibrous. Actins occur in all muscle, some epithelia and in many noncontractile cells. Actin occurs also in some plant cells and bacteria. Actins can bind to large contractile or structural molecules; in some cells actin filaments give rigidity but not movement. Myosins are large, some of MW 450,000 D, and differ in their primary structure in various kinds of muscle and amoeboid cells. A myosin molecule consists of a spiral tail, a head of S_2 and S_1 chains and 2 or 4 light chains (lcs). One type of lc has ATPase action. A myosin head binds to actin, shifts its angle of attachment and detaches, then reattaches at another position and then sliding occurs; an alternative hypothesis is rotation rather than angle change.

All contractions require regulation by calcium ions. In vertebrate striated muscle Ca^{2+} liberated from sarcoplasmic reticulum after excitation by the T-tubule starts a chain reaction; actin interaction with myosin is blocked by regulatory proteins in resting muscles. Calcium from the sarcoplasmic reticulum (SR) reacts with a troponin-C, which reacts with troponin-T and then troponin-I and tropomyosin, a reaction series that releases the brake on actin–myosin interaction. In molluscan muscle, a light chain of myosin is sensitive to Ca^{2+} hence, regulation is by a myosin lc. In vertebrate smooth muscle, Ca^{2+} activates a kinase that phosphorylates myosin, a step necessary for interaction with actin. Many invertebrate muscles have a dual control—by a Ca^{2+}-sensitive lc and by a troponin sequence.

Amoeboid movement occurs not only in free-living and parasitic rhizopods, but in white blood cells, fibroblasts, some chromatophores, and phagocytic cells. Myosin and actin have been identified in several of these cells, especially in myxomycetes. Amoeboid myosin and actin show much sequence homology with myosins and actins of vertebrates. Exactly how the streaming in amoeboid cells is brought about is not known.

Tubulin-based movement is best known in cilia (flagella) and in axolasmic transport in neurons; tubulin participates in mitotic movement of chromosomes. Cilia have the characteristic structure of a ring of nine pairs of microtubules (doublets) plus two single central microtubules. The microtubules are composed of α and β tubulin. Spokes connect the central sheath to the connected peripheral doublets and from one of each of the peripheral microtubules extend arms of dynein, an ATPase. Active bending involves sliding of one microtubule beyond the other. Ciliary beat is neurally regulated (Mytilus gill), or controlled by membrane potentials (*Paramecium*).

Tubulin is one of the most abundant proteins in brain and microtubules are the basis of axoplasmic transport of vesicles and of protein aggregates. Axoplasmic flow is energized by a protein, kinesin, which has an ATPase activity. The kinesin for fast flow is probably different from that for slow flow.

Chromosome movement in anaphase is probably by shortening of microtubules connected at the kinetochores. Cleavage of the cytoplasm is by a myosin–actin system.

REFERENCES

1. Achazi, R. K. In *Basic Biology of Muscles* (B. Twarog, R. Levine, and M. Dewey, eds.), pp. 291–308. Raven Press, New York, 1982. Catch muscle long fibers.

2. Achazi, R. K., *Pfluegers Arch.* 379:197–201, Filament structure, phosphorylation in catch muscles.

3. Adelstein, R. S. and E. Eisenberg. *Annu. Rev. Biochem.* 49:921–956, 1980. Regulation of actin-myosin-ATP interaction.

4. Akster, H. A. H. Granzier and *J. Comp. Physiol. B* 155B:685–691, 1985. Muscle fiber types of perch.

5. Allen, R. D., J. W. Cooledge, and P. U. Hall. *Nature (London)* 187:896–899, 1960. Streaming in cytoplasm dissociated from the giant amoeba *Chaos chaos;* Allen, R. D. et al. *J. Cell Biol* 100:1736–1752, 1985. Microtubules from squid axoplasm move along glass cover slip.

6. Allen, R. D., P. W. Francis, and H. Nakajima. *Proc. Natl. Acad. Sci. U.S.A.* 54:1153–1161, 1965. Cyclic birefringence changes in pseudopods of *Chaos cardioenses.*

6a. Amos, L. *J. Cell Sci.* 87:105–111, 1987. Kinesin from pig brain.

7. Amos, L. and A. Klug. *J. Cell Sci.* 14:523–549, 1974. Arrangement of subunits in flagella microtubules.

7a. Anderson, M. In *Basic Biology of Muscles* (B. Twarog, R. Levine, and M. Dewey, eds.), pp. 309–322. Raven Press, New York, 1982. Striated myoepithelial cells.

8. Atwood, H. L. and S. A. Luenicka. *Am. Zool.* 27:977–989, 1987. Role of activity in determining properties of neuromuscular system in Crustacea.

8a. Bagby, R. In *Biochemistry of Smooth Muscle* (N. L. Stephens, ed.), Vol. 1, Chapter 1, pp. 1–84. CRC Press, Boca Raton, FL, 1983. Structure of vertebrate smooth muscle.

9. Baldwin, K. M. G. Klinkerfuss, R. Terjung, P. Molé and J. Holloszy et al. *Am. J. Physiol.* 222:373–378, 1972. Enzymes of red, white, and intermediate muscles in relation to exercise.

9a. Barr, L. *CRC Crit. Rev. Neurobiol.* 4:325–366, 1989. Coupling of photic stimulation to contraction of iris smooth muscle.

10. Berridge, M. J. and R. F. Irvine. *Biochem. J.* 220:345–360, 1984. Phosphoinositides.

10a. Block, B. and C. Franzini-Armstrong. *J. Cell Biol.* 107:1099–1112, 1988. Structure of membrane systems in heater cells.

10b. Block, B., T. Imagawa, K. Campbell, and C. Franzini-Armstrong. *J. Cell Biol.* 107:2587–2600, 1988. Structural evidence for direct interaction between the molecular components of the transverse tubule/sarcoplasmic reticulum junction in skeletal muscle.

11. Bone, Q. *J. Cell Sci.* 10:657–665, 1972. Dogfish neuromuscular junctions.

12. Bone, Q. *J. Mar. Biol. Assoc. U. K.* 46:321–349, 1966. Functional types of muscle fibers in fish; *Fish Physiol.* 7:361–424, 1978.

13. Brenner, B. and E. Eisenberg. *Proc. Natl. Acad. Sci. U.S.A.* 83:3542–3546, 1986. Rate of force generation correlates with actomyosins ATPase activity.

14. Brevet, A. et al. *Science* 193:1152–1154, 1976. Myosin synthesis in cultured skeletal muscle cells.

15. Brokaw, C. J. In *Molecules and Cell Movement* (S. Inoué and R. E. Stephens, eds.), pp. 165–178. Raven Press, New York, 1975. Cross-bridge behavior in sliding filament model for flagella.

15a. Burnstock, G. *Pharmacol. Rev.* 21:247–324, 1969. Evolution of autonomic innervation of visceral and cardiovascular muscles.

16. Burnstock, G. *Pharmacol. Rev.* 24:509–581, 1972. Purinergic nerves.

16a. Carlhoff, D. and J. D'Haese. *J. Comp. Physiol.* 157:589–597, 1987. Slow type muscle cells of earthworm gizzard with calcium regulated myosin isoform.

17. Carew, T. et al. *Neurophysiology* 37:1020–1040, 1974. Neuromuscular transmission to gill muscle, Aplysia.

18. Castellani, L. and C. Cohen. *Science* 235:334–337, 1987. Myosin phosphorylation and catch state of mollusc muscle.

18a. Caswell, A. and N. Brandt. *Trends Biochem. Sci.* 14:161–165, 1989. T-tubule to SR coupling.

19. Chamberlin, B. K., D. Levitsky, and S. Heischer. *J. Biol. Chem.* 258:6602–6609, 1983. Canine cardiac SR characterization.

20. Chantler, P. D. and A. G. Szent-Györgyi. *J. Mol. Biol.* 138:473–492, 1982. Scallop muscle light chains.

21. Clark, M. and J. A. Spudich. *Annu. Rev. Biochem.* 46:797–822, 1977. Actomyosin contraction in dividing cells.

22. Clark, W. A. et al. *J. Biol. Chem.* 257:5449–5454, 1982. Cardiac isomyosins.

23. Clarkson, TW, PR Sager, and T. L. M. Syverson, eds. *The Cytoskeleton: A Target for Toxic Agents.* Plenum, New York, 1986.

24. Cohen, C., G. Phillips, et al. In *Biological Recognition and Assembly*, pp. 209–231. Alan R. Liss, New York, 1980.

25. Cohen, C. M. et al. *Nature (London)* 279:163–165, 1979. Actin and spectrin in red blood cell cytoskeleton.

26. Cohen, C. M. *Proc. Natl. Acad. Sci.* 79:3176–3178, 1982. Arrangement of paramyosin molecules in catch muscles.

27. Condeelis, J. S. *Exp. Cell Res.* 88:435–439, 1974; *J. Cell Biol.* 74:901–927, 1977; *Nature, (London)* 292:161–163, 1981. Actin in plants and slime molds.

27a. Coté, G. P. et al. *J. Biol. Chem.* 259:12781–12787, 1984. Regulation of Acanthamoeba myosin by phosphorylation.

28. Cummins, P. and G. Russel. *Comp. Biochem. Physiol. B* 84B:343–348, 1986. Myosin light chains in mammal hearts.

29. Cupples, C. G. and R. E. Pearlman. *Proc. Natl. Acad. Sci. U.S.A.* 83:5160–5164, 1986. Actin gene of Tetrahymena thermophila.

30. Dabrowska, R. and A. Szpacenko. *Comp. Biochem. Physiol. B* 56B:139–142, 1977. Composition of muscle proteins of carp and rabbit muscle.

31. Dahms, V., C. L. Prosser, and N. Suzuki. *J. Physiol. (London)* 392:51–69, 1987. Two types of 'slow waves' in intestinal smooth muscle of cat.

32. Da Silva, F. W. and A. M. Hodgkin. *Comp. Biochem. Physiol. A* 87A:143–149, 1987. Structure and function of the pedal muscle of the whelk Bullia.

33. Davie, P. S. R. Wellsand V. Tetens et al. *J. Exp. Zool.* 237:159–171, 1986. Effects of sustained swimming on red muscle mass, LDH, aerobic capacity of white muscle.

34. DeLozanne, A. et al. *Proc. Natl. Acad.*

Sci. U.S.A. 82:6807–6810, 1985. Characterization of nonmuscle myosin heavy chain cDNA.

35. DeLozanne, A. and J. Spudich. *Science* 236:1086–1091, 1987. Disruption of Dictyostelium myosin heavy chain by homologous recombination.

36. Dewey, M. M., D. Colflesh, P. Brink, S. Fan, et al. In *Basic Biology of Muscles* (B Twarog, R. Levine, and M. Dewey, eds.), pp. 53–72. Raven Press, New York, 1982. Structural, functional, and chemical changes in Limulus striated muscle in sarcomere shortening and tension development.

36a. D'Haese, J. and S. Ditgens. *J. Muscle Res. Cell Motil.* 5:196–1984. Earthworm troponin, tropomyosin, and myosin linked Ca regulation.

37. Ditgens, A. and J. D'Haese. In *Contractile Proteins in Muscle and Non-muscle Systems* (E. Alla, N. Arena, and M. Russo, eds.), pp. 137–158. Praeger, New York, 1985. Body wall muscle of earthworm.

38. Eastwood, A. B., C. Franzini-Armstrong, and C. Peracchis. *J. Muscle Res. Cell Motil.* 3:273–294, 1982. Structure of membranes in crayfish muscle: Comparison of phasic and tonic fibers.

39. Ebashi, S., M. Endo, and I. Ohtsuki. *Q. Rev. Biophys.* 2:351–384, 1969. Control of muscle contraction.

40. Eckert, R. and P. Brehm. *Annu. Rev. Biophys. Bioeng.* 8:353–384, 1979. Ionic mechanisms of excitation in Paramecium.

40a. Eddinger, T. J. and R. L. Moss. *Am. J. Physiol.* 253:C210–C218, 1987. Mechanical properties of skinned single fibers of identified types from rat diaphragm.

41. Elzinga, M. *J. Biol. Chem.* 250:5915–5920, 1975. Complete sequence of 344 amino acid residues of actin.

42. Feng, T. P. Personal communication.

42a. Ferguson, D. C., H. Schwartz, and C. Franzini-Armstrong. *J. Cell Biol.* 99:1735–1742, 1984. Subunit structure of feet in triads.

43. Freundlich, A. and J. M. Squire. *J. Mol. Biol.* 169:439–454, 1983. Three-dimen-sional structure of insect flight muscle m-band.

44. Franzini-Armstrong, C. *J. Cell Biol.* 47:488–499, 1970. Studies of the triad I. Structure of the junction in frog twitch fibers.

45. Franzini-Armstrong, C. *J. Cell Biol.* 56:120–128, 1973. Studies of the triad IV. Structure of the junction in frog slow fibers.

45a. Franzini-Armstrong, C. *Fed. Proc., Fed. Am. Soc. Exp. Biol.* 34:1382–1389, 1975. Transmission at the triad in frog and fish.

46. Franzini-Armstrong, C. and G. Nanzi. *J. Muscle Res. Cell Motil.* 4:233–252, 1983. Junctional feet and particles in the triad of a fast-twitch muscle fiber.

47. Franzini-Armstrong, C. *Tissue Cell* 16(4):647–664, 1984. Freeze-fracture of frog slow tonic fibers. Structure of surface and internal membranes.

48. Fuchs, E. and I. Hanukaglu. *Cell (Cambridge, Mass.)* 32:332–334, 1983. Unraveling the structure of intermediate filaments.

49. Furuhashi, K. and K. Konishi. *Comp. Biochem. Physiol. A* 88A:625–635, 1987. Contractile protein in fibers from ascidian body wall muscle.

49a. Gabriel, J. M. *J. Exp. Biol.* 118:313–340, 1985. Development of locust jumping mechanism.

49b. Gadasi, H. and E. D. Korn. *Nature (London)* 286:452–456, 1980. Intracellular localization of myosin isoenzymes in Acanthamoeba.

50. Gallowitz, D. and I. Sares. *Proc. Natl. Acad. Sci U.S.A.* 77:2546–2550, 1980. Actin genes in yeast.

51. Gauthier, G. F. *J. Cell Biol.* 82:381–400, 1979. Identification of muscle fiber types by immunocytochemistry.

52. Gauthier, G. F. et al. *J. Cell Biol.* 92:471–484, 1982. Development of fast and slow myosins.

53. Gauthier, G. F. *Am. Zool.* 27:1033–1042, 1987. Vertebrate muscle fiber types and neuronal regulation of myosin expression.

54. Gleeson, T. T., A. Bennett, and R. Putnam. *J. Exp. Zool.* 214:293–302, 1980. Properties of skeletal muscle fibers in Dipsosaurus.

55. Gleeson, T. T., C. J. M. Nicol, and I. A. Johnston. *Cell Tissue Res.* 237:253–258, 1983. Capillarization, mitochondrial densities and O_2 diffusion distances and innervation of red and white muscle of lizard Dipsosaurus.

56. Goffart, M. *Form and Function in Sloths.* Pergamon, New York, 1971.

57. Gorbsky, G. J., P. Sammack, and G. G. Borisy. *J. Cell Biol.* 104:9–18, 1987. Chromosomes move poleward in anaphase along stationary microtubules.

58. Govind, C. K. *Am. Zool.* 27:1079–1098, 1987. Muscle fiber type transformations in clawed crustaceans.

59. Haese, J. D. and D. Carlhoff. *J. Comp. Physiol. B* 157B:171–179, 1987. Localization and histochemistry of myosin isoforms in earthworm body wall.

60. Hagiwara, S. and Y. Kidokora. *J. Physiol. (London)* 219:217–232, 1971.

61. Hardwicke, P. and A. G. Szent-Györgyi. *Nature (London)* 301:478–482, 1983. Light-chain movement and regulation in scallop myosin.

62. Haselgrove, J. C. and C. D. Rodger. *J. Muscle Res. Cell Motil.* 1:371–390, 1980. Review of X-ray diffraction pattern of muscle.

62a. Hernandez-Nicaise, M. L. et al. *J. Gen. Physiol.* 75:79–105, 1980. Structure of large muscle fibers in ctenophores.

63. Herzberg, O. and M. James. *Nature (London)* 313:653–659, 1985. Structure of regulatory muscle protein troponin.

63a. Heuser, J. H. *J. Mol. Biol.* 169:123–154, 183. Ultrastructure of striated muscle.

64. Hibberd, M. G., and D. Trentham. *Annu. Rev. Biophys. Biophys. Chem.* 15:119–161, 1986. Figure 17a.

65. Hochachka, P.-J. Exp. Biol. 115:149–164, 1985. Fuels and pathways for muscular work.

66. Hoh, J. F. *Biochemistry* 14:742–747, 1975. Neural regulation of mammalian fast and slow muscle myosins.

66a. Hymel, L. et al. *Proc. Natl. Acad. Sci. U.S.A.* 85:441–445, 1988. Ryanodine receptor of muscle SR forms Ca-activated channels.

67. Hynes, T. R. and J. Spudich. *Cell (Cambridge, Mass.)* 48:953–963. 1987. Movement of myosin fragments *in vitro*.

68. Inoue, S. and H. Ritter. In *Molecules and Cell Movement* (S. Inoué and R. E. Stephens, eds.), pp. 3–29. Raven Press, New York, 1975. dynamics of mitotic spindle organization and function.

69. Inoue, S. *J. Cell Biol.* 91:131s–147s, 1986. Cell division and mitotic spindle.

70. Ito, L. et al. *J. Exp. Biol.* 50:107–118, 1969. Excitatory junction potentials in earthworm muscles.

71. Johnston, I. A. and B. Totta *Comp. Biochem. Physiol. B* 49B:367–373, 1974. Johnston, I. A. W. Davison *J. Comp. Physiol.* 114:203–216, 1977, Energy metabolism of fish muscle. *J. Physiol.; J. Exp. Biol.* Structural, physiological, and biochemical properties of red, white, and mixed muscles of fishes.

71a. Johnston, I. A. and T. W. Moon. *J. Exp. Biol.* 87:177–194, 1980. *Cell Tissue Rls.* 219:93–109 1981; ibid, 237:253–258, 1984. Enzymes of fast and slow fish muscles; effects of exercise.

71b. Johnston, I. A. and W. Salmonski. *J. Exp. Biol.* 111:171–177, 1984. Johnston I. A. and B. Sidell. *J. Exp. Biol.* 119:239–249, 1985. Mechanical properties of red and white muscles of fishes.

72. Johnston, I. A. *Symp. Zool. Soc. London* 48:71–113, 1981. Structure and function of fish muscles.

73. Jones, P. L. and B. D. Sidell. *J. Exp. Zool.* 219:163–171, 1982. Metabolic carbon sources in fiber types of striped bass muscles.

74. Josephson, R. *J. Exp. Biol.* 108;77–96, 1984. Contraction dynamics of cricket muscles.

75. Josephson, R. In *Insect Thermoregulation* (B. Heinrich, ed.), pp. 20–44. Wiley, New York, 1981; *J. Exp. Biol.* 80:69–81, 1979; 91:219–237, 1981; 99:109–125, 1982, Body temperature and effects of

temperature on muscle contraction in insects.

76. Josephson, R. *J. Exp. Biol.* 114:493–512, 1985; 117:357–368, 1985; 118:185–208, 1985. Insect muscle power.

77. Josephson, R. and D. Young. *Am. Zool.* 27:991–1000, 1987. Fiber ultrastructure and contraction kinetics in insect fast muscles.

78. Kamiya, N. *Annu. Rep. Soc., Fac. Sci. Osaka Univ.* 8:13–41, 1960. Protoplasmic streaming in slime molds.

79. Karttumen, P. and P. Tirri. *Comp. Biochem. Physiol. A* 84A:181–188, 1986. Single myocardial cells from perch.

80. Karttumen, P. and P. Tirri. *Comp. Biochem. Physiol. A* 88A:161–166, 1987. Single myocardial cells of quail Coturnix.

81. Keller, L. R. and C. P. Emerson. *Proc. Natl. Acad. Sci. U.S.A.* 77:1020–1024, 1980. Synthesis of myosin by muscle cultures.

82. Kendrick Jones, J. et al. In *Basic Biology of Muscles* (B. Twarog, R. Levine, and M. Dewey, eds.), pp. 255–272. Raven Press, New York, 1982. Role of myosin light chains in Ca regulation of molluscuan muscles.

83. Kerrick, G. L. et al. *J. Gen. Physiol.* 77:177–190, 1981. Ca-regulatory mechanisms in nonstriated muscles.

84. Kessler, D. et al. *Cell Motil.* 1:63–71, 1981. Physarum myosin.

85. Kier, W. M. *J. Morophol.* 172:179–192, 1982. Muscles of squid arms and tentacles.

86. Kimmura, K. et al. *Comp. Biochem. Physiol. B* 88B:399–407, 1989. Troponin from nematode Ascaris body wall muscle.

86a. Kishino, A. and T. Yanagida. *Nature (London)* 334:74–76, 1988. Force measurement of single actin filaments.

87. Knecht, D. and W. Loomis. *Science* 239:1081–1085, 1987. Antisense RNA inactivation of heavy chain.

88. Korn, E. D. *Proc. Natl Acad. Sci. U.S.A.* 75:588–599, 1978. Biochemistry of actomyosin-dependent cell motility.

89. Korn, E. D. et al. In *Contractile Proteins in Muscle and Non-Muscle Cell Systems* (E. Alia, N. Arena, and M. Russo, eds.), pp. 487–494. Praeger, New York, 1985. Regulation of actin-activated ATPase.

90. Korn, E. D. *Physiol. Rev.* 62:672–737, 1982. Actin.

90a. Kuriyama, K. *J. Cell Sci.* 84:153–164, 1986. Effects of Taxol on meiosis.

90b. Lai, F., G. Meissner, et al. *Nature (London)* 331:315–319, 1988. Purification of Ca^{2+} release channel.

90c. Lang, F. A., A. Sutterlin, and C. L. Prosser. *Comp. Biochem. Physiol.* 32:615–628, 1970. Facilitation in contraction.

91. Lanzavecchia, G. et al. *J. Muscle Res. Cell Motil.* 6:569–584, 1985. Helical muscles of leeches—superelongation.

92. Lehman, W. and B. Bullard. *J. Gen. Physiol.* 62:553–563, 1974; 66:1–30, 1975. Calcium sensitive light chains of myosins.

93. Levine, R., R. Kensler, M. Stewart, and J. Haselgrove. In *Basic Biology of Muscles* (B. Twarog, R. Levine, and M. Dewey, eds.), pp. 37–52. Raven Press, New York, 1982. Molecular organization of Limulus thick filaments.

93a. Lewis, S. A. et al. *Science* 242:936–939, 1988. Microtubule-associated protein MAP_2 shares microtubule binding with tau protein.

93b. Liu, H. and A. Brebscher. *Proc. Natl. Acad. Sci. U.S.A.* 86:90–93, 1989. Purified tropomyosin from yeast

94. Mackenzie, J. M. et al. *Cell (Cambridge, Mass.)* 15:413–419, 1978. Localization of 2 myosins in the same muscle cell of Caenorhabditis.

94a. Mackie, G. O. and C. L. Singla. *J. Neurobiol.* 6:339–356, 1975. Muscles-striated and smooth-in coelenterates.

95. Marston, S. B. and W. Lehman. *Biochem. J.* 231:517–522, 1985. Caldesmon is Ca regulator of smooth muscle thin filaments.

96. Marston, S. B. *Prog. Biophys. Mol. Biol.* 41:1–41, 1982. Regulation of smooth muscle contractile proteins.

97. Marston, S. B. and E. W. Taylor. *J. Mol. Biol.* 139:573–600, 1980. Myosin and SR ATPase of 4 types of vertebrate muscles.

98. Martin, T. P. *Comp. Biochem. Physiol. B* 88B:273–276, 1987. Adaptations of muscle mitochondria to exercise.

99. McLachlan, A. D. and J. Karn. *Nature (London)* 299:227, 1982. Periodic charge distributions in the myosin rod amino acid sequence match cross-bridge spacing in muscle.

100. Miller, D. M., J. D. Anderson, and B. C. Abbott. *Comp. Biochem. Physiol.* 27:633–646, 1968. Potentials and ionic exchange in slime mold plasmodia.

101. Minkoff, L. and R. Damidian. *J. Bacteriol.* 125:353–365, 1976. Actin from E. coli.

102. Mitchison, T., L. Evans, E. Schultze, and M. Kirschner. *Cell (Cambridge, Mass.)* 45:515–527, 1986. Sites of microtubule assembly and disassembly in mitotic spindle.

103. Molloy, J. E. et al. *Nature (London)* 328:449–451, 1987. Kinetics of flight muscles from insects with different wingbeat frequencies.

104. Morkin, E., I. Flink, and L. Goldmanis. *Prog. Cardiovasc. Dis.* 25:435–464, 1983. Biochemical and physiological effects of thyroid hormone on cardiac performance.

105. Morkin, E. *Annu. Rev. Biochem.* 56:702–714, 1987. Molecular Genetics of Myosin.

105a. Moss, A. G. and S. L. Tamm. *Proc. Natl. Acad. Sci. U.S.A.* 84:6476–6480, 1987. Ca^{2+} regenerative potentials controls ciliary reversal.

106. Mutangi, G. and I. A. Johnston. *J. Exp. Biol.* 128:84–105, 1987. Effects of temperature and pH on contraction of skinned muscle from Pseudemys scripta.

106a. Nabauer, M. et al. *Science* 244:800–803, 1989. Regulation of Ca release gated by Ca conductance, not by gating charge in cardiac muscle.

107. Nabeshima, Y. et al. *Nature (London)* 308:333–338, 1984. Transcription splicing for mRNA's for two myosin light chains.

107a. Nagai, R. D Larson and M. Periasami *Proc. Natl. Acad. Sci. U.S.A.* 85:1047–1051, 1988. Sequences derived from cDNA of five different myosins.

108. Naitoh, Y. and R. Eckert. *J. Exp. Biol.* 59:63–65, 1973; 56:667–681, 1972. Sensory mechanisms in Paramecium.

109. Nakamura, K. and S. Watanabe. *J. Biochem. (Tokyo)* 83:1459–1470, 1978. Myosin and actin from E. coli. Makeshima, H., S. Nishimura, and S. Numa. *Nature* 339:439–445, 1989. Primary structure of muscle ryanodine receptor.

109a. Nunzi, M. G. and C. Franzini-Armstrong. *J. Ultrastruct. Res.* 76:134–148, 1981. Structure of adductor muscles of scallop.

110. Omoto, C. K. and C. Kung. *Nature (London)* 279:532–534, 1979. Pair of central tubules rotates during ciliary beat in Paramecium.

110a. Paschal, B. M. and R. B. Vallee. *Nature (London)* 330:181–183, 1988. Retrograde transport by microtubule-associated protein MAP·IC.

111. Peachey, L. D. *J. Exp. Biol.* 115:91–98, 1985. Excitation—contraction coupling; the link between the surface and interior of a muscle cell; L. D. Peachey and R. Eisenberg. *Biophys. J.* 22:145–154, 1977; *Hand. Physiol., Sect. 10*, Ch 2, pp. 23–72, 1983.

112. Pearson, K. G. and J. F. Iles. *J. Exp. Biol.* 50:445–471, 1969. Neuromuscular inhibition in insects.

113. Pearson, K. G. and J. F. Iles. *J. Exp. Biol.* 54:215–232, 1971. Neuromuscular transmission in cockroach.

114. Pollard, T. D. and J. A. Cooper. *Annu. Rev. Biochem.* 55:987–1035, 1986. Actin and actin binding proteins. A critical evaluation of mechanisms and functions.

115. Pollard, T. D. and P. R. Weiking. *CRC Crit. Rev.* 2:1–65, 1974. Review of actin and myosin in cell movement.

116. Pringle, J. *Proc. R. Soc. London, Ser. B*

201:107–130, 1978. Stretch activation of muscle.

117. Prosser, C. L. In *Basic Biology of Muscles* (B. Twarog, R. Levine, and M. Dewey, eds.), pp. 381–397. Raven Press, New York, 1982. Diversity of narrow-fibered and wide-fibered muscles.

118. Prosser, C. L. and G. O. Mackie. *J. Comp. Physiol.* 136:103–112, 1980. Contractions of Holothurian muscles.

119. Prosser, C. L., H. J. Curtis, and D. Travis. *J. Cell. Comp. Physiol.* 38:299–319, 1951; Prosser, C. L. and C. E. Melton. *ibid.* 44:255–275, 1954; Prosser, C. L. and N. Sperelakis. *ibid.* 54:129–133, 1959. Nervous conduction in smooth muscle of Phascolosoma (Golfingia).

120. Prosser, C. L. *Handb. Physiol. Sect. 2: Cardiovasc. Syst.* 3:635–670, 1980. Evolution and diversity of nonstriated muscles.

120a. Prosser, C. L. In *Comparative Animal Physiology* (C. L. Prosser, ed.), 3rd ed., Vol. 2, pp. 719–788. Saunders, Philadelphia, PA, 1973. Muscles.

121. Rathmayer, W. and L. Maier. *Am. Zool.* 27:1067–1077, 1987. Fiber types in crabs.

122. Ready, N. E. *J. Exp. Zool.* 238:43–54, 1986. Development of fast singing muscles in katydid.

123. Regenstein, J. M. *Comp. Biochem. Physiol. B* 56B:239–244, 1977. Lobster striated muscle myosin.

124. Reuben, J. P., P. W. Brandt, H. Garcia, and H. Grundfest. *Am. Zool.* 7:623–645, 1967. Excitation-contraction coupling in crayfish.

125. Rikmenspoel, R. and W. Rudd. *Biophys. J.* 13:955–992, 1973. Contractile mechanism in cilia.

126. Rios, E. and G. Brum. *Nature (London)* 325:717–720, 1987. Involvement of dihydropyridine receptors in excitation-contraction coupling in skeletal muscle; Rios, E. *Trends Physiol. Sci.* 3:223, 1988.

127. Rowlerson, A. et al. *J. Muscle Res. Cell. Motil.* 6:601–640, 1985. Myosins in lateral muscle of fish.

128. Rubenstein, P. and J. Spudich. *Proc.*

Natl. Acad. Sci. U.S.A. 74:120–123, 1987. Actin microheterogeneity in chick embryo fibroblasts.

129. Rubenstein, P. A. and J. Spudich. *Proc. Natl. Acad. Sci. U.S.A.* 74:120–123, 1977. Actins in chick fibroblasts and adult smooth muscle.

129a. Ruegg, J. C. *Calcium in Muscle Activation: A Comparative Approach.* Springer-Verlag, Berlin, 1986; *Physiol. Rev.*

129b. Sanders, K., and T. Smith. *J. Physiol.* 377:297–313, 1988. Neural control of slow waves in colon muscle.

130. Satir, P. In *Cilia and Flagella* (M. A. Sleigh, ed.), pp. 131–142. Academic Press, London 1974. The present status of the sliding microtubule model of ciliary motion.

131. Satir, P. In *Molecules and Cell Movement* (S. Inoué and R. E. Stephens, eds.), pp. 143–149. Raven Press, New York, 1975. Ciliary and flagellar movement.

132. Satow, T., A. D. Murphy, and C. Kung. *J. Exp. Biol.* 103:256, 1983. Touch receptor of Paramecium.

132a. Saxton, W. M. et al. *Proc. Natl. Acad. Sci. U.S.A.* 185:1109–1113, 1864–1888, 1988. Gene for heavy chain kinesin in Drosophilia.

133. Schnapp, B. J., R. D. Vale, M. P. Sheetz, and T. S. Reese. *Ann. N. Y. Acad. Sci.* 466:909–918, 1986. Dynamic aspects of microtubule biology.

134. Schnapp, B. J. and T. S. Reese. *Trends Neurosci.* 9:155–162, 1986. Mechanisms of rapid axonal transport.

135. Schnapp, B. J. et al. *Cell, (Cambridge, Mass.)* 40:455–492, 1985. Microtubules from squid axoplasm support bidirectional movement organelles.

136. Schouest, L. P., M. Anderson, and T. A. Miller. *J. Exp. Zool.* 239:146–158, 1986. Structure and physiology of tergotrochanteral depressor in Musca.

137. Sheetz, M. P. and J. A. Spudich. *Nature (London)* 303:31–35, 1983. Movement of myosin coated fluorescent beads.

137a. Sidell, B. D. In *Cellular Acclimation to Environmental Change.* (A. Cossins and P.

Sheteline, eds.), pp. 103–120. Cambridge 1983. Cellular acclimation by quantitative changes in enzymes.

138. Silverman, H., W. Costello, and D. Mykles. *Am. Zool.* 27:1011–1019, 1987. Fiber type correlates of physiological and biochemical properties of crustacean muscle.

139. Sinha, A. M. et al. *Proc. Natl. Acad. Sci. U.S.A.* 79:5847–5851, 1982. Cloning of mRNA sequences for cardiac alpha and beta myosin heavy chains.

140. Small, J. V. *J. Cell Sci.* 24:327–349, 1977. Structure of smooth muscle cell.

141. Smith, D. S. *Nature (London)* 303:539–540, 1983. Fastest synchronous muscle contraction in whitefly Trialeurodes.

141a. Smith, R. S. and W. K. Ovalle, *J. Anat.* 116:1–24, 1973. Five types of amphibian muscle fibers.

141b. Smith, T. K., J. Reed, and K. Sanders. *Am. J. Physiol.* 252:C215–C224, C290–C299, 1987. Rhythmic activity in dog colon.

142. Spudich, J. A. et al. In *Contractile Proteins in Muscle and Non-Muscle Cell Systems* (E. Alia, N. Arena, and M. Russo, eds.), pp. 171–180. Praeger, New York, 1985. Movement of myosin molecules in vitro: A quantitative essay.

143. Stephens, R. E. *Annu. Rev. Biochem.* 65:369–379, 1975. SDS-polyacrylamide electrophoresis methods for cilia, flagella.

144. Stephens, R. E. In *Cilia and Flagella* (M. A. Sleigh, ed.), pp. 39–78. Academic Press, London, 1974. Enzymatic and structural proteins of the axoneme.

145. Stephens, R. E. In *Molecules and Cell Movement* (S. Inoué and R. E. Stephens, eds.), pp. 181–204. Raven Press, New York, NY, 1983. Structural chemistry of axoneme. Tubule dimers in outer fibers.

146. Stommel, E. W. *J. Comp. Physiol.* 155:445–456, 1984. Ca regenerative potentials in Mytilus gill abfrontal ciliated epithelial cells.

147. Stommel, E. W. and R. Stephens. *J. Comp. Physiol. A* 157A:441–449, 1985.

Lateral cilia of Mytilus gill arrested by high intracellular Ca.

148. Sugi, H. et al. *Comp. Biochem. Physiol. A* 81A:397–401, 1985. Mechanical activity of lantern retraction of sea urchin.

149. Suprenant, K. A. et al. *Proc. Natl. Acad. Sci. U.S.A.* 82:6908–6912, 1985. Multiple tubulins in cilia and cytoplasm of Tetrahymena.

149a. Swanson, K. *Z. Vergl. Phys.* 74:403–410, 1971. Mechanics of catch muscles.

149b. Szent-Györgyi, A. G. et al. *J. Mol. Biol.* 56:239–258, 1971. Paramyosin and filaments of molluscan "catch" muscles. II. Native filaments: Isolation and characterization.

150. Tada, M. and A. M. Katz. *Annu. Rev. Physiol.* 44:401–423, 1982. Phosphorylation of SR and sarcolemma.

150a. Takashima, H., S. Nishimura and S. Numa. *Nature* 339:439–445, 1989. Primary structure of cyanodine receptor in striated muscle.

151. Tanabe, T. et al. *Nature (London)* 328:313–318, 1987; 336:134–139, 1988. Primary structure of receptor for calcium channel blockers from skeletal muscle.

152. Tattersall, J. E. and R. Brace. *J. Comp. Physiol. A* 160A:115–125, 1987. The physiology of radular retractor muscles of Planorbis.

153. Taylor, D. L., J. S. Condeelis, P. L. Moore, and R. D. Allen. *J. Cell Biol.* 59:378–394, 1973. Chemical control of motility in isolated cytoplasm of amoeba.

154. Taylor, D. L. and J. S. Condeelis. *Int. Rev. Cytol.* 56:57–144, 1979. Cytoplasm structure and contractility in amoeboid cells.

155. Taylor, D. L., J. A. Rhodes, and S. A. Hammond. *J. Cell Biol.* 70:123–143, 1976. Amoeboid movement.

155a. Taylor, K. A. and L. A. Amos. *J. Mol. Biol.* 147:297–324, 1981. Model for geometry of myosin binding to thin filaments.

156. Tilney, L. G. In *Molecules and Cell Move-*

ment (S. Inoué and R. E. Stephens, eds.), pp. 339–388. Raven Press, New York, 1975. Role of actin in nonmuscle cell motility.

157. Tilney, L. G. *J. Cell Biol.* 64:289–310, 1975. Actin filaments in the acrosmal reaction of Limulus sperm motion generated by alterations in packing of the filaments.

158. Tilney, L. G. and J. C. Saunders. *J. Cell Biol.* 96:807–821, 822–834, 1983; *Dev. Biol.* 116:100–118, 1986. Number of hair cells per chick cochlea.

159. Tilney, M., L. G. Tilney, and D. De-Rosier. *Hear. Res.* 25:141–151, 1987. Hair cell bundle lengths and orientation in chick cochlea.

160. Toyoshima, Y. Y., J. Spudich, et al. *Nature (London)* 328:536–539, 1987. Movement of Myosin fragments.

160a. Twarog, B. *J. Physiol.* 192:847–868, 1967. Relaxation mechanisms of catch muscles.

161. Vale, R. D., J. Scholey, and M. Sheetz. *Trends Biochem. Sci.* 11:464–468, 1986; *Cell (Cambridge, Mass.)* 42:39–50, 1985. Kinesin purified from axons, liver, sea urchin eggs, Chlamydomonas.

162. Vale, R. D. et al. *Cell (Cambridge, Mass.)* 40:559–569, 1985. Organelle and microtubule translocation promoted by soluble factors from squid axon.

163. Vale, R. D. et al. *Cell (Cambridge, Mass.)* 40:449–454, 1985. Fast movement anterograde and retrograde.

164. Vanderkerckhove, J. and K. Weber. *J. Mol. Biol.* 179:391–413, 1984. Evolution of actins.

165. Vanderkerckhove, J. and K. Weber. *Proc. Natl. Acad. Sci. U.S.A.* 75:1106–1110, 1978; *J. Mol. Biol.* 126:783–802, 1978.

165a. Wagenknecht, T. R. Grassucci, J. Frank, A. Suger, M. Inuri and S. Fleischer *Nature (London)* 338:167–170, 1989. Structure of Ca-channel foot of SR membrane.

165b. Wais-Steider, J. and P. Satir. *J. Supramol. Struct.* 11:339–347, 1979.

166. Wallman, T. and A. G. Szent-Györgyi. *Biochemistry* 20:1188–1197, 1981; *J. Mol. Biol.* 156:141–173, 1972. Ca regulation by specific light chain of myosin in scallop muscle.

167. Weeds, A. G. D. R. Trentham, C. J. Dean and A. J. Butler. *Nature (London)* 247:135–139, 1974. Myosins from cross-innervated fast and slow muscles of cat controlled by nervous system.

167a. Weeds A. and P. Wagner. In *Biochemical Evolution.* ed. (H. Gutfreund, ed.) ch. 8, pp. 262–300 Cambridge Univ. Press, New York, 1981. Control mechanisms in muscle contraction.

168. Weisenberg, R. C. et al. *Science* 238:1119–1122, 1987. Microtubule gelation-contraction relation to slow axonal transport.

169. Whalen, R. G. et al. *Proc. Natl. Acad. Sci. U.S.A.* 73:2018–2022, 1976; 76:5197–5201, 1979; *Nature (London)* 286:731–733, 1980; 292:805–806, 1981. Development of myosin isozymes in rats; embryos and neonates.

170. Wilkinson, J. M. *Eur. J. Biochem.* 103:179–188, 1980; *Nature (London)* 271:31–35, 1978. Slow fiber type troponin-c in mammals and birds.

171. Wootton, R. and D. Newman. *Nature (London)* 280:402–403, 1979. Whiteflies have highest contraction frequencies of non-fibrillar flight muscles.

171a. Yamauchi-Takihara, K. et al. *Proc. Natl. Acad. Sci. U.S.A.* 86:3504–3508, 1989. Human cardiac myosin heavy chain genes.

172. Yano, M. et al. *Nature (London)* 299:557–559, 1982. An actomyosin motor.

173. Zengel, J. M. and H. F. Epstein. *Proc. Natl. Sci. U.S.A.* 77:852–856, 1980. Nematode muscles.

174. Burggen, W. and J. Roberts. In *Environmental and Metabolic Animal Physiology* (C. Ladd Prosser, ed.). pp. 358–434. Wiley-Liss, Inc., New York, 1991.

175. Hochachka, P. W. In *Environmental and Metabolic Animal Physiology* (C. Ladd Prosser, ed.). pp. 325–352. Wiley-Liss, Inc., New York, 1991.

176. Prosser, C. L. and J. E. Heath. In *Environmental and Metabolic Animal Physiology* (C. Ladd Prosser, ed.). pp. 109–165. Wiley-Liss, Inc., New York, 1991.

3 | *Bioluminescence*

J. Woodland Hastings and James G. Morin

Bioluminescence is the emission of visible light by organisms, an unusual but sometimes spectacular phenomenon. Where it does occur, luminescence is generally functionally important, and highly effective. For example, it may deter predators, help to obtain prey, provide communicative signals, or serve other specific purposes (36, 103, 173). However, in some cases it may be viewed as a physiologically nonessential feature, analogous in this sense to pigmentation or chromatophores: nonluminous mutants, as such, are generally viable in the lab, like albinos. Moreover, some closely related species (including sympatrics), or even different strains of the same species, are both luminous and nonluminous. Indeed, some organisms even possess light organs but are nonluminous (222).

In absolute numbers of taxa, luminous species are quite rare (Table 1). For unknown reasons, luminescence is far more prevalent in the marine environment than on land or in fresh water. In the sea its abundance appears to be correlated with dim light rather than darkness or depth

per se. While rare in terms of total numbers of species, luminescence is phylogenetically diverse; luminous species are found in ~13 phyla, which include bacteria, unicellular protists, and fungi as well as animals ranging from jellyfish and brittle stars to scale worms, fireflies, squids, and fishes (126). Diversity carries over to all aspects of bioluminescence, including biochemistry, physiology, control, and functions. This array indicates that luminescence evolved independently many different times (107).

The functional importance of bioluminescence and its selection in evolution are based largely on its being detected by another organism. Unlike incandescence, emission is not related to the temperature of the excited molecule, so the energy is not thermal in origin. Unlike fluorescence and phosphorescence, emission is not dependent on prior absorption of light; bioluminescence is a catalyzed chemiluminescence, an exergonic enzymatic reaction in which chemical energy is converted to light energy.

Since the enzymes and substrates in

TABLE 1. Approximate Number and Proportion of Established Luminescent Taxa[a]

Taxon	Number of Luminescent Genera[b]	Number of Genera in Taxon	Percentage of Genera Luminescent	Representative Genera
Total (all organisms)	666	/> 100000	= ~0.67	
Bacteria	4	>246	~1.6	*Photobacterium, Vibrio*
Fungi	9	>802	~0.1	*Panus, Armillaria, Pleurotus*
Protista				
Dinoflagellata	11	>176	~6.3	*Gonyaulax, Noctiluca, Pyrocystis*
Radiolaria	9	>283	~3.2	*Thalassicola, Collozoum*
Animalia				
Cnidaria	65	1104	5.9	
Hydrozoa	37	381	9.7	*Obelia, Aequorea, Hippopodius*
Scyphozoa	4	70	5.7	*Pelagia, Atolla, Periphylla*
Anthozoa	24	647	3.7	
Octocorallia	21	220	9.5	*Renilla, Pennatula, Thourella*
Hexacorallia	3	427	0.7	*Parazoanthus, Epizoanthus*
Ctenophora	15	31	48	*Mnemiopsis, Beroe, Bollinopsis*
Nemertea	1	176	0.6	*Emplectonema*
Annelida	40	1300	3.1	
Polychaeta	26	1000	0.3	*Harmothoe, Odontosyllis, Chaetopterus, Polycirrus*
Oligochaeta	14	195	7.2	*Eisenia, Diplocardia*
Mollusca	75	3945	1.9	
Gastropoda	7	3083	0.2	
Opisthobranchia	3	327	0.9	*Phyllirrhoe, Kaloplocamus*
Pulmonata	4	1177	0.3	*Latia, Planaxis, Quantula*
Bivalvia	2	573	0.4	*Pholas, Rocellaria*
Cephalopoda	66	139	47	
Coleoidea	66	138	48	
Octopoda	3	38	7.9	*Japetella, Eledonella*
Sepioidea	6 + 5B	19	58	*Spirula, Euprymna, Heteroteuthis*
Teuthoidea	51	80	64	
Myopsida	4B	9	44	*Loligo, Doryteuthis*
Oegopsida	47	71	66	*Ommastrephes, Watasenia, Cranchia, Histioteuthis*
Vampyromorpha	1	1	100	*Vampyroteuthis*
Arthropoda	187	84600	0.22	
Chelicerata	1	6000	0.02	*Collosendeis*
Crustacea	57	5400	1.1	
Copepoda	17	1200	1.4	*Oncaea, Metridia, Pleuromamma*
Ostracoda	3	757	0.4	*Vargula, Cypridina, Conchoecia*
Malacostraca	37	3200	1.2	
Mysidacea	1	120	0.8	*Gnathophausia*
Amphipoda	8	840	1.0	*Scina, Parapronoe, Thoriella*
Euphausiacea	10	11	91	*Euphausia, Meganyctiphanes*
Decapoda	18	1200	1.5	*Sergestes, Oplophorus, Acanthephyra*
Uniramia	129	73600	0.18	
Diplopoda	3	1632	0.18	*Motyxia, Spirobolellus*
Chilopoda	5	303	1.7	*Orphaneus, Geophilus*
Insecta	121	71500	0.17	

TABLE 1. (*Continued*)

Taxon	Number of Luminescent Genera[b]	Number of Genera in Taxon	Percentage of Genera Luminescent	Representative Genera
Collembola	1	394	0.25	*Onychiurus*
Coleoptera	117	24800	0.5	*Photinus, Photuris, Pteroptyx*
Diptera	3	6033	0.05	*Arachnocampa, Orfelia*
Echinodermata	47	1126	4.2	
Crinoidea	3	164	1.8	*Annacrinus, Thallassometra*
Holothuroidea	16	154	10	*Galatheathuria, Scotoplanes, Kolga*
Asteroidea	12	300	4.0	*Plutonaster, Hymenaster, Brisinga*
Ophiuroidea	16	275	5.8	*Ophioscolex, Ophiosila, Amphipholis*
Hemichordata	3	16	*19*	*Ptychodera, Balanoglossus*
Chordata	200	8838	2.3	
Urochordata	3 + 3B	182	3.3	*Oikopleura, Cyclosalpa, Pyrosoma*
Vertebrata	194	8653	2.2	
Chondrichthyes	5	151	3.3	*Isistius, Etmopterus, Spinax*
Osteichthyes	189	3881	4.9	
Teleostei (43)[c]	189	3867	4.9	
Anguilliformes (1)	1 + 1B	147	1.4	*Saccopharynx, Lumiconger*
Clupeiformes (1)	1	68	1.5	*Coilia*
Salmoniformes (3)	18 + 3B	90	23	*Opisthoproctus, Searsia, Photostylus*
Stomiiformes (9)	53	53	100	*Argyropelecus, Malacosteus, Stomias*
Myctophiformes (2)	33	35	94	*Diaphus, Lampanyctus, Stenobrachius*
Aulopiformes (4)	5 + 1B	40	15	*Chlorophthalmus, Benthabella*
Batrachoidiformes (1)	1	19	5.3	*Porichthys*
Lophiiformes (10)	31B(2)[d]	64	48	*Oneirodes, Linophryne, Melanocetus*
Gadiformes (3)	18B	76	24	*Physiculus, Nezumia, Malacocephalus*
Beryciformes (3)	7B	39	17.9	*Photoblepharon, Anomalops, Cleidopus*
Perciformes (6)	11 + 5B	1367	1.2	*Leiognathus, Apogon, Siphamia*

[a]Numbers and examples of luminescent genera are taken from Herring (126, 132a) and more recent references; the classification scheme and the number of genera in each taxon are taken mostly from Parker (189).
[b]B = Number of genera with bacterial luminescence; all others have intrinsic luminescence.
[c]The number of families with at least some luminescent species is given in parentheses.
[d]Two genera possess both intrinsic and bacterial luminescence.

diverse bioluminescent systems differ, the terms luciferase (enzyme) and luciferin (substrate) must be prefixed by the relevant taxon: for example, bacterial luciferase or firefly luciferin. Common features of all systems are a requirement for molecular oxygen and the involvement of a reactive intermediate peroxy compound; indeed, all luciferases may be classed as oxygenases (106), in which a product molecule (P) is formed in an electronically excited state [P]*. Its return to ground state results in emission of a photon in the visible range (~400–650 nm):

$$\text{luciferin} + O_2 \xrightarrow{\text{luciferase}}$$

$$\text{peroxyluciferin} \longrightarrow [P]^* \longrightarrow P + h\nu$$

Luminescence in the blue (460 nm) corresponds to an energy of ~62 kcal/ein-

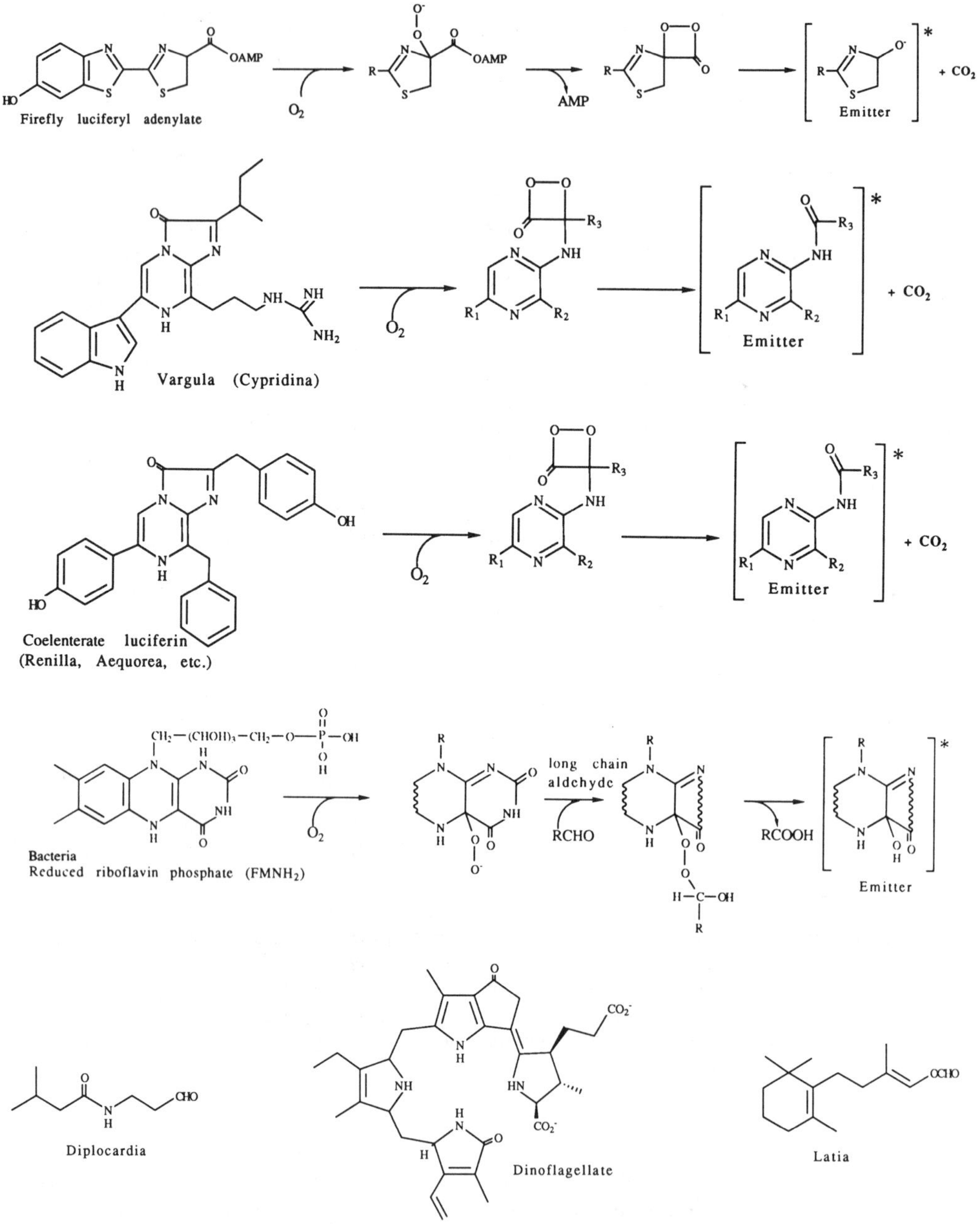

Figure 1. Structures of different luciferins, oxygen containing intermediates, and postulated emitters.

TABLE 2. Chemistry and Control of Bioluminescence in Different Organisms

Luminous Organisms	Luciferins, Other Factors	Luciferases (M_r)	Emission max (nm)	Control of Luminescence
Bacteria (*Photobacterium, Vibrio*)	$FMNH_2$; RCHO	80,000	495–500	Biosynthesis, induction, mechanical shutters
Dinoflagellates (*Gonyaulax, Pyrocystis*)	Tetrapyrrole, H^+	420,000	475	Action potential, pH change
Cnidarians (*Aequorea, Renilla*)	Coelenterazine, Ca^{2+} (imidazopyrazine nucleus)	21,000	460–490	Action potential, Ca^{2+}
Annelids (*Diplocardia*)	N-isovaleryl-3-amino propanal, H_2O_2	300,000	500	Not known
Molluscs (*Latia*)	Enol formate form of an aldehyde, or an aromatic or terpene aldehyde	170,000	500	Not known
Crustacea (*Vargula, Cypridina*)	Imidazopyrazine nucleus	60,000	465	Mixing in seawater
Insects (*Photinus, Photuris*)	(Benzo)thiazole ATP, Mg^{2+}	100,000	560	Action potential, oxygen

stein (= 260 kJ/einstein), almost an order of magnitude greater than the energy available from hydrolysis of ATP (= 7 kcal/einstein), and this amount of energy must be made available in a single step (105).

Structures for seven of the ~30 identified luciferins are known (Fig. 1). The generalized formulation above does not fully describe the mechanism in many systems. For example (Table 2), firefly luminescence requires ATP, coelenterates involve Ca^{2+}, and dinoflagellate luminescence is triggered by a pH change. Several systems involve a second protein and a secondary (or alternate) emitter.

Where *in vivo* luminescence is rapid, the control mechanisms are either mechanical (i.e., secondary control via shutters, chromatophores, or external mixing of reagents) or chemical (e.g., oxygen, calcium, or H^+). All rapid intracellular systems are ultimately triggered by changes in membrane potentials. Such triggering may occur via membrane depolarization of a single cell or may involve more complex steps, for example, neural control of photocytes, photophores, or special organs.

Bioluminescent systems are also diverse structurally. In some cells the light emission is confined to subcellular organelles, such as the scintillons in dinoflagellates and the photosomes in scale-worms (119). In the hydroids, there are discrete photocytes broadly distributed in different parts of the animal but triggered in concert by action potentials conducted through epithelial cells (74, 172). Some animals possess specialized light organs or photophores, some of which are relatively simple, such as pouchlike chambers in which luminous bacteria are cultured (some fishes and squids), while others are complex structures complete with reflectors and lenses.

From the wide and sporadic phylogenetic distribution of luminous species, and the very different structures of the various luciferases, and the genes encoding them, it is postulated that light-emitting capability has evolved independently 30 or more times (107). At the same time there are several examples of cross-phyletic similarities, some of which may be nutritionally based (72, 86, 98, 222).

Diversity, Distribution, and Abundance

While organisms in many taxa produce light, only a small fraction of the genera and species in higher taxa do so (Table 1). Although widespread in the invertebrates, luminescence is *not* known among the higher plants and the vertebrates, except for the fishes; neither is it known among several invertebrate phyla. There are only a few luminescent protists and fungi.

Some phyla may have a high proportion of luminous species (but none with 100%) and others only a few. Luminescence is especially well represented or widely studied among bacteria, dinoflagellates, cnidarians, ctenophores, annelids, insects, crustaceans, squids, ophiuroids, larvaceans, and fishes (Table 1). Commonly, a given genus contains only luminous members, but in some there may be both luminous and nonluminous species. Although luminescence is spectacular in the insects where it occurs (e.g., fireflies), luminescence is otherwise rare in the terrestrial environment (< 0.2% of all genera). A few terrestrial species of millipedes, centipedes, earthworms, snails, and bacteria are luminous. Only one truly freshwater species, the pulmonate limpet (mollusc) *Latia* from New Zealand streams, is known to emit light. Herring (132a) gives a detailed listing of known luminescent genera up to about 1987.

In some habitats the number of luminous taxa and/or individuals is exceedingly high. Luminous bacteria occur as a pure culture within light organs at densities of up to 10^{11}/mL; dinoflagellate densities during red tides in coastal waters may reach 10^7/L. In benthic taxa, such as anthozoans, polychaetes, ostracodes, and ophiuroids, numbers may exceed $100/m^2$ (173). Such high densities can exert a significant influence on the local communities in which they exist (44, 88, 93, 112, 174, 208), and luminescence clearly represents an important, and probably crucial, component in the ecology and physiology of these organisms.

Contrary to statements sometimes made in popular publications, luminescence cannot be viewed as especially associated with organisms that live in total darkness. Rather, it is more the province of dimly lit environments. On a global scale luminescence is most prevalent in the sea at mid-ocean depths (200–1200 m), where daytime illumination levels range between $\sim 10^{-1}$–10^{-12} μW/cm^2 (33, 239). At these depths luminescence occurs in over 95% of the individuals and 75% of the species in fishes (15); 86% of the individuals and 79% of the species in shrimp (124), and similar majorities in squids. The midwater luminous fish *Cyclothone* is considered to be the most abundant vertebrate on earth. Above and below mid-ocean depths, luminescence decreases significantly to < 10% of the species and individuals in surface (and neritic) waters and probably < 25% toward pelagic abyssal depths and the benthos at all depths. Among shallow marine species (< 200 m) that occur both near (demersal) and on the bottom (benthic) only ~1–3% appear to be luminescent (173). There are no known luminescent species either in deep fresh water bodies (e.g., Lake Baikal, USSR) or the darkness of terrestrial caves. There is a luminescent dipteran larva (*Arachnocampa*) that lives near the mouths of caves in New Zealand; it also occurs in culverts, forests, and undercut banks of streams where there is substantial daytime illumination.

Biochemistry and Molecular Biology

Oxygen

With one well-documented exception, and one that turns out to prove the rule, all bioluminescent organisms require molecular oxygen to produce light. Biochemically, all known luciferases are oxygenases that utilize molecular oxygen to oxidize a substrate (a luciferin), with formation of a product molecule in an electronically excited state (106). All reactions of the different luciferins with oxygen involve formation of intermediate peroxides, which in some cases cyclize before breaking down to give light emission.

Luciferins, Intermediate Peroxides, and Products

Structures for luciferins (Fig. 1) and other factors (Table 2) are known for seven different systems. Firefly luciferin, a benzothiazole, reacts enzymatically with ATP to form the luciferyl–adenylate prior to its reaction with oxygen (65, 164). The peroxy intermediate then cyclizes to form a four membered ring peroxide (214), which breaks down to yield CO_2 and an excited carbonyl, still enzyme bound, from which light is emitted with a high quantum yield ($\sim$0.9) (205).

The luciferin of cypridinid ostracode crustaceans is structurally very different from firefly (as is the luciferase), but the reaction appears also to involve a ring peroxide, oxidative decarboxylation, an excited carbonyl, and the incorporation of oxygen into products. In this case no prior (activating) reaction with ATP is involved.

In some coelenterate systems luciferin occurs with a bound sulfate, whose removal constitutes an activating step (59). This luciferin, sometimes referred to as coelenterazine, is like cypridinid luciferin

in that it is a tripeptide and contains the imidazopyrazine skeleton (Fig. 1). It is also notable for its widespread phyletic distribution and involvement in many different types of bioluminescent systems (215). The reaction mechanism in these different luminescent systems may be the same as postulated for cypridinids, involving a ring peroxide.

The coelenterazine system constitutes the apparent exception to the requirement for molecular oxygen. Living animals such as ctenophores can luminesce in the complete absence of dissolved oxygen, and so can the isolated "photoproteins," such as aequorin [molecular weight (MW) $\sim$ 20,000] from the jellyfish *Aequorea* (61). Isolated and purified in the absence of calcium, aequorin emits light simply upon the addition of calcium (212). The reason oxygen is not required is that the oxygen has already reacted with the protein (enzyme)-bound coelenterazine to form a stable peroxide with the oxygen attached to the C-2 carbon atom (180). Aequorin can be regenerated *in vitro* by reacting spent photoprotein (= *Aequorea* luciferase or apoaequorin) with coelenterazine (= *Aequorea* luciferin), and oxygen (211):

$$\text{apoaequorin} + \text{coelenterazine}$$
$$+ \, O_2 \longrightarrow \text{aequorin} \xrightarrow{Ca^{2+}}$$
$$[\text{aequorin}]^* \longrightarrow$$
$$h\nu + \text{coelenteramide} + \text{apoaequorin}$$

Two synthetic analogs of coelenterazine have been prepared and incorporated into apoaequorin, affording semisynthetic aequorins with different kinetic and spectral properties (216).

In the bacterial system an analogous (but linear) peroxide intermediate has been isolated and characterized (116). Formally, bacterial luciferase (Lase) is an external flavin monoxygenase, or mixed function oxidase: The electrons for the reduction of FMN are provided by an external reductant via FMN oxidoreductase (118).

$$\text{NAD(P)H} + \text{H}^+ + \text{FMN} \xrightarrow{\text{FMN reductase}}$$
$$\text{NAD(P)}^+ + \text{FMNH}_2$$
$$\text{FMNH}_2 + \text{RCHO} + \text{O}_2 \xrightarrow{\text{Lase}}$$
$$\text{[Lase-FHOH]}^* + \text{RCOOH} \longrightarrow$$
$$h\nu + \text{FMN} + \text{H}_2\text{O}$$

Two substrates are involved, the reduced flavin (FMNH_2) and a long chain aliphatic aldehyde (RCHO), which is oxidized to the acid (RCOOH), but the luciferase-bound flavin peroxide intermediate can be formed in the absence of the aldehyde and is quite stable:

$$\text{FMNH}_2 + \text{O}_2 \xrightarrow{\text{Lase}}$$
$$\text{Lase—FMNH—OOH}$$

Bioluminescence can then be obtained from the luciferase-flavin-peroxide intermediate simply by aldehyde addition, even in the absence of oxygen, this being analogous to the calcium triggering of the aequorin peroxide (111):

$$\text{Lase—FMNH—OOH} + \text{RCHO} \longrightarrow$$
$$\text{Lase—FMNH—OO—CHOH—R} \longrightarrow$$
$$\text{RCOOH} + \text{[Lase—FMNH—OH]}^*$$
$$\text{[Lase—FMNH—OH]}^* \longrightarrow$$
$$h\nu + \text{Lase} + \text{FMN} + \text{H}_2\text{O}$$

The reaction with aldehyde leads to formation of an excited state of the 4a-hydroxy flavin and light emission which, in the pure system, peaks at ~490 nm, well to the blue of and more energetic than the fluorescence emission of FMN itself (~530 nm) (142).

In the earthworm *Diplocardia*, the luciferin is an aldehyde (225), while in dinoflagellates it is a tetrapyrrole, sometimes with a special binding protein (71, 73, 84, 179a, 181a), but there is no information about intermediates or products in either. Dinoflagellate luciferin is very similar to that of the light emitter in krill (181). In the freshwater snail *Latia*, the product has been identified (213) and one can speculate that oxygen adds across the double bond to form a ring peroxide intermediate. There remain many other systems—two dozen or more—for which biochemical information is incomplete or lacking (107, 108). In some cases *in vitro* reactions have been demonstrated and shown not to cross-react with the known systems, indicating that they are biochemically distinct.

Luciferases: Proteins and Genes

Each bioluminescent system utilizes a different specific luciferase. Enzymes are classically viewed as catalysts, speeding up reaction rates. Luciferases have additionally been credited with enhancing the quantum yield of bioluminescent reactions; where determined, yields in the range of 10 to 90% have been reported (106). High quantum yields have also been reported recently from chemiluminescent reactions. Thus, luciferases can no longer be viewed as being *unique* in facilitating high quantum yield reactions.

Bacterial luciferase is autoinducible in some species and may comprise 5% or more of the soluble protein in fully induced cells (118, 245). It has a MW ~80 kD and is composed of two homologous (20) but nonidentical subunits (α and β, both required for activity) that may be separated under denaturing conditions and reconstituted to give full activity. Luciferases from all known species of luminous bacteria, marine and terrestrial, are homologous, and many cross-hybrids of the subunits exhibit activity (136, 194).

Other than flavin, which binds only weakly in the oxidized form (19), the protein is associated with no prosthetic groups, cofactors, or metals.

The genes for several of the bacterial luciferases, which have been cloned and expressed in *Escherichia coli* by several groups (26, 32, 58, 64, 94, 95), appear generally similar but not identical. From the gene structures, the complete amino acid sequences of the luciferase subunits are known, confirming their homology (138). The bacterial lux operon also includes genes involved in the synthesis of other factors (e.g., aldehyde) and in the regulation (autoinduction) of transcription (79–81, 170). Its use as an indicator of gene expression has been developed by several groups (82, 200).

Firefly luciferase is also a dimer (164) (MW 100 kD), with apparently identical subunits, but only one active site per dimer. Other than ATP and Mg^{2+}, no cofactors or prosthetic groups are involved. The genes for this luciferase have been cloned (68) and expressed in bacteria, animal cells, and plants. In these cases the substrate (firefly luciferin) must be added exogenously in order to visualize expression.

In aequorin, the substrate peroxy intermediate is bound and light emission is triggered by calcium. With three putative Ca^{2+}-binding sites (corresponding to binding data), aequorin is structurally related to calmodulin (49). Five different cDNAs for aequorin have been cloned and sequenced (186a, 192) and shown to code for three different isoforms, accounting for the microheterogeneity of native aequorin (31). One cDNA has been expressed in *E. coli* and the product (196 amino acids) shown to have the same molecular weight and isoelectric point as undegraded native aequorin. The coelenterate system is of particular interest because of two additional proteins involved in the luminescence of some species: a substrate-binding protein and an accessory (green fluorescent) protein with a chromophore serving as the emitter (227).

The dinoflagellate system also involves participation of a second protein, the luciferin-binding protein (MW, 70 kD), which sequesters the substrate and then releases it (upon a pH change) for rapid oxidation by luciferase with concomitant flashing (85, 108). The luciferase itself is unusually large; the single polypeptide chain is ~140 kD (70); a 35-kD proteolytic fragment is active but with a different pH activity profile (141). Cyclic DNAs for these two proteins have also been cloned and expressed in *E. coli* (179, 179a).

Reaction Mechanisms of Chemiluminescent and Bioluminescent Reactions

What kind of chemical process is it that releases enough energy, and evidently in one step, to populate an excited state? Over the past two decades, two different reaction types for chemiluminescence in the liquid phase have had prominence: electron transfer and the cleavage of ring peroxides (115, 237).

Electron transfer is evidently simpler as a model for the population of the excited state: One electron moves from a donor radical anion (An^-) to an acceptor radical cation (Ct^+), leaving one in an excited state:

$$An^- + Ct^+ \longrightarrow$$
$$An^* + Ct \longrightarrow An + h\nu$$

This is a very fast process, involving no other changes, thereby maximizing the energy available for electronic excitation. Such radical ions may be generated electrochemically; upon charge annihilation either of the two species, An or Ct, may be formed in the excited state. Until recently such a mechanism had not been proposed for bioluminescence, possibly because of the ubiquity and importance of oxygen, which does not figure directly in this mechanism, and the reluctance to think in terms of a single electron, where oxygen involves two.

A very different type of reaction, involving the cleavage of ring peroxides, was considered for some time as a possible mechanism for chemiexcitation. Upon heating, such peroxides give relatively high yields of electronically excited products:

$$O=\overset{\displaystyle O-O}{\overset{|}{C}}-\overset{|}{\underset{\displaystyle R_2}{C}}-R_1 \longrightarrow CO_2 + R_1-\overset{\displaystyle O^*}{\overset{\|}{C}}-R_2 \longrightarrow$$

$$R_1-\overset{\displaystyle O}{\overset{\|}{C}}-R_2 + h\nu$$

Two of the bioluminescent systems, firefly and cyprinids, were clear candidates for such a mechanism (Fig. 1), and it was soon shown for both that one O_2 was taken up and one CO_2 plus another carbonyl were products (136a, 218). Considering the weakness of the peroxide O—O bond, the ring strain, and the energy gained by the formation of two carbonyl double bonds, the breakdown of such compounds should generate roughly 100 kcal (415 kJ)/einstein, ample to account for one product molecule in an excited state. Nevertheless, the theoretical explanation for the mechanism of excited state production was not clear. For example, in many model compounds the excited states appeared to be overwhelmingly in triplet states, rather than in the singlet state, as in bioluminescent systems.

A proposal that seems to bridge the gap between the two above-cited mechanisms incorporates both. It is referred to as chemically initiated electron exchange luminescence (CIEEL); in this mechanism, peroxide breakdown involves electron transfer with chemiexcitation (47, 162, 199, 202, 237). It is initiated by an electron transfer from a donor species (D) to an acceptor (A). In this model the acceptor contains a weak oxygen–oxygen bond which, after electron transfer (A^-), cleaves spontaneously to form products C and B^-. The latter should be a stronger reductant than the initial species A^-, so the electron is transferred back to D^+, with energy conservation in the concomitant formation of a singlet excited state D^*:

$$A + D \longrightarrow A^- + D^+ \qquad D^* \longrightarrow h\nu + D$$
$$\downarrow \qquad\qquad$$
$$C + B^- \qquad B$$

Systems modeling this mechanism have been reported in which the luminescent breakdown of a peroxide (A) results in emission from a sensitizer (D), which behaves kinetically as a catalyst, as it should. The emission is correlated with its one-electron oxidation potential.

In a number of bioluminescent systems different colors of light may be emitted in different organs and/or in different species (126, 177, 227, 237a). In some cases (e.g., elaterid beetles) this has been shown to be due to differences in the luciferases, and may relate to the properties of the luciferin-binding site (206, 207). In others it has been shown to involve a secondary emitter, and energy transfer from a primary excited state has been postulated in some cases (227). It is possible that some cases may instead involve secondary chromophores or sensitizers that function as alternative electron donors in mechanisms modeled by CIEEL, as suggested for bacterial bioluminescence (78a, 109).

Analytical Applications

The use of firefly luciferase for rapid detection of very small amounts of ATP is well known. Other luciferases have been used analogously for the specific detection of many other substances, and such assays are in wide use in research, industry, and clinical labs (140, 163, 190a, 201). Bioluminescent systems are also being developed for luminescence immune assays in place of radioimmunoassays (226).

The principles and power of the methodology are straightforward. Any substance required for a luciferase reaction, such as O_2, ATP, aldehyde, or Ca^{2+}, may be detected directly. The diversity of factors required in different systems (Table 2) means that there are many such possibilities. Coupling other enzymic reactions to luminescent systems extends considerably the range of analytical applications. With bacterial luciferase, for example, any reaction that produces or utilizes NAD(H), NADP(H), or aldehyde, either directly or indirectly, can be coupled to the light emitting reaction (230).

The power of the luciferase-coupling method lies not only in its inherent specificity. Because the product of the reaction is light, it can be measured instantaneously and with high sensitivity. The lower limit of detection is governed by photon detection techniques; 10^4 photons/s may be readily measured (140, 217, 223). The corresponding number of molecules depends on quantum yield, but even if this is low, the sensitivity can evidently still be very high.

Luminous bacteria were first used analytically by Molisch to demonstrate the production of oxygen in photosynthesis. Because of their unusually high affinity for O_2, luminous bacteria continue to be used for O_2 detection in special applications (48), and an O_2 electrode incorporating luminous bacteria has been perfected (152). Many other analytical applications of both intact luminous bacteria and bacterial luciferase have been reported (118). As suggested above, firefly luciferase assays for ATP are used in many different applications, including detection of biomass. The purified photoprotein aequorin has been widely used for intracellular Ca^{2+} detection (31). The protein is relatively small (~20 kD), nontoxic, and may be readily injected into cells in quantities adequate to detect Ca^{2+}. Luminescence is dependent on (and specific for) free Ca^{2+} over the range of 3×10^{-7}–10^{-4} M. As mentioned earlier, methods for measuring gene expression by linking a luciferase gene to the system in question have been developed and used (80, 82, 200).

Control: Structural and Functional Organization

Bioluminescent systems are represented not only by a rich diversity of biochemical reactions, but also by a great array of structural features and a marvelous repertoire of specialized physiological mechanisms, all serving a variety of functions. The most recent and thorough general review of luminescence can be found in Herring (126). Aspects of the physiology and functional morphology of bioluminescence have been reviewed with respect to (a) control (3, 8, 45); (b) habitat [coastal (173); oceanic (239); deep sea (131)]; and (c) taxon [bacteria, dinoflagellates (109); radiolaria (128); cnidaria (172); polynoids (22); molluscs (133, 220); crustacea (132); fireflies (44); echinoderms (122); invertebrates (128); fishes (129, 135)].

Spectral Characteristics

One important feature of light emission that has functional importance and ecological significance is its spectral composition or color. With the recent development of sensitive and rapid instrumentation for recording emission spectra (130, 143, 231), a considerable number of organisms in many diverse groups have been examined (170 of the ~666 luminescent genera). The data available are summarized in Figure 2, where it can be seen that bioluminescence emission spectra have peaks ranging from the deep violet (~400 nm) to the far red (~620 nm), with one or more examples of emission falling in almost every 10-nm bin between. The great majority of pelagic and deep-sea species have blue emissions, with maxima centered in the 450–490 nm range, while those of the coastal species fall mostly in the 490–520 nm (green) region. In contrast, most terrestrial and freshwater species emit in the green and yellow, with a large number clustered around 550–580 nm, and a smaller number centered at 510–540 nm.

In terms of functional or adaptive significance, there are few definitive explanations concerning the ecological correlates of color. Both vision and color of luminescence coincide generally with the quality of the ambient light or the light penetrating the particular habitat (130, 143, 184), as illustrated by the deep sea and coastal species. However, probably related to more evolved and specialized functions, exceptions exist where organisms possess complex behavior, well-developed vision, and utilize light signals for intraspecific communication. In certain fireflies, an adaptive significance of the intraspecific differences in emission spectra has been inferred from information concerning time of evening emergence (143, 209). In the marine environment similar inferences have been made in relation to the color of luminescence used for counterillumination (242) and for a case involving red emission and visual pigments absorbing in the red (67, 187).

The great majority of the emission spectra are simple and unimodal, char-

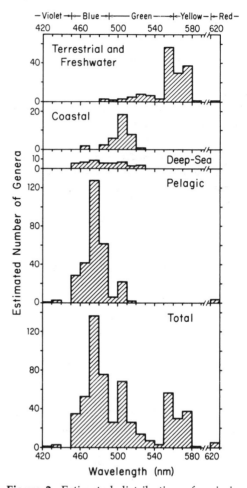

Figure 2. Estimated distribution of emission spectra maxima for all bioluminescent genera that are terrestrial and fresh water, coastal (benthic or demersal genera at depths <200 m), deep-sea (benthic or demersal genera at depths >200 m), or pelagic (genera that are always planktonic or nektonic, either oceanic or neritic). The bottom panel shows all genera combined. Habitats for the various genera were determined from a variety of sources, primarily ref. 189. In order to weigh the distributions according to genera per taxon, the known emission maxima, taken primarily from refs. 130 and 231, were assumed to be indicative for the whole taxon (as given in Table 1) and were multiplied by the number of genera in the taxon. Taxa where no reliable spectra are available were omitted.

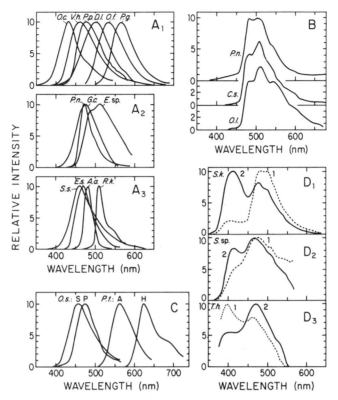

Figure 3. Bioluminescent emission spectra. (A) Spectra with single peaks; (B) spatially distinct spectra from two parts of the animal; (C) complex spectra with two or more peaks; (D) temporally distinct spectra from the same animal where curve 1 (dotted line) = first scan, curve 2 (solid line) = second scan. (A_1): D.l. = *Diplocardia longa* (oligochaete) (225); O.c. = *Oncaea conifera* (copepod) (130); O.f. = *Omphalia flavida* (fungus) (60); P.g. = *Photinus greeni* (lampyrid insect) (E. A. Widder and J. F. Case, personal communication); P.p. = *Photobacterium phosphoreum* (bacterium) (130); V.h. = *Vargula hilgendorfi* (ostracode) (231). (A_2): E.sp. = *Eunoe* sp. (polynoid annelid) (130); G.c. = *Gonichthys coccoi* (myctophid fish) (130); P.n. = *Pyrocystis noctiluca* (dinoflagellate) (231). (A_3): A.a. = *Argyropelecus affinis* (sternoptychid fish) (231); E.s. = *Euphausia superba* (euphausiid) (130); R.k. = *Renilla koellikeri* (pennatulacean cnidarian) (231); S.s. = *Selenoteuthis scintillans* (cephalopod) (130). (B): O.s. = *Oplophorus spinosus* (caridean decapod) (130), S = secretion, P = photophore; P.t. = *Phrixothrix tiemanni* (phengodid insect) (167), A = abdominal organ, H = head organ. (C): C.s. = *Cranchia scabra* (cephalopod) (130); O.l. = *Ophiopholis* cf. *longispina* (ophiuroid) (231); P.n. = *Porichthys notatus* (batrachodid fish) (231). (D): S.k. = *Searsia koefoedi* (searsiid fish) (234); S.sp. = *Sergestes* sp. (penaeid decapod) (130); T.h. = *Thourella hilgendorfi* (gorgonian cnidarian) (231).

acteristic of the fluorescent emission spectra of simple aromatic molecules (Fig. 3A). However, some—both from living organisms and *in vitro* enzyme systems—are more complex. For example, the fish *Porichthys*, cranchiid squids, ophiuroids and polynoid polychaetes, show complex spectra with two either equal or subequal peaks from the same source (Fig. 3B). In some organisms different areas of the animal or separate photophores actually emit different colors (Fig. 3C). For ex-

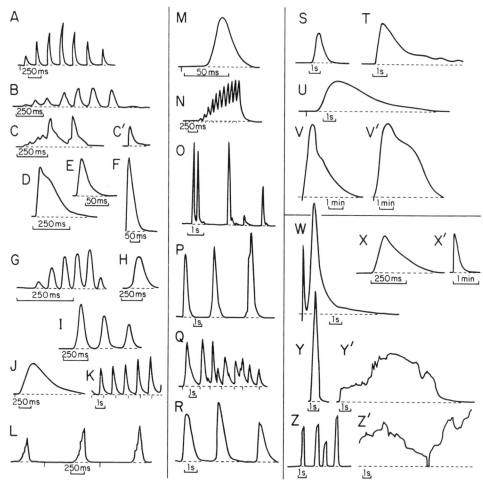

Figure 4. Representative luminescent kinetics (redrawn from various sources) or rapid flashes (A–R), slow glows (S–V), and variable pulses from rapid to slow in the same species (W–Z). A mark beneath the trace of some indicates the point of stimulation. (A) Hydrozoan *Obelia geniculata*, 1-ms stimulus (175); (B) ophiuroid *Ophiopsila californica*, 10-V, 100-ms stimulus train (34); (C) larvacean *Oikopleura dioica*, mechanical stimulation (88); (D) dinoflagellate *Pyrocystis fusiformis*, mechanical stimulation (232); (E) dinoflagellate *Noctuluca miliaris*, 1-ms stimulus (77); (F) dinoflagellate *Gonyaulax polyedra* (113); (G) firefly *Photinus evanescens*, spontaneous, flying male (156); (H) firefly *Photinus marginellus*, spontaneous (38); (I) firefly *Photuris versicolor*, spontaneous (50); (J) anthozoan *Renilla kollikeri* siphonozooid, 12-V, 0.1-ms stimulation (13); (K) *R. kollikeri* siphonozooid, 10-V, 5-ms stimulation (172); (L) anthozoan *Acanthoptilum gracile* (198); (M) myctophid *Lampanyctus ritteri*, spontaneous (21); (N) myctophid *Ceratoscopelus townsendi* (= *maderensis*) caudal organ, spontaneous (21); (O) melanostomiatid *Echiostoma* sp. suborbital organ (135); (P) malanostomiatid *Grammatostomias* sp. postorbital organ (135); (Q) myctophid *Myctophum* sp. serial photophore, (135); (R) malacosteid *Malacosteus* sp. postorbital organ (135); (S) pyrosome *Pyrosoma atlanticum* single zooid (157); (T) batrachoidid *Porichthys myriaster* isolated gular photophore, 40-V, 400-ms stimulation (16); (U) polychaete *Chaetopterus variopedatus* posterior of body, ventral nerve cord stimulation (10-V, 6-ms) (5); (V) and (V') euphausiid *Meganyctiphanes norvegica*, whole pinioned animal, photoflash stimulation

ample, in the South American phengodid beetle larva, *Phrixothrix*, the paired organs on each segment emit green light, whereas a pair on the first segment emit red. The deep-sea shrimp *Oplophorus* has photophores that emit a different color of blue than its secreted luminescence. In some stomiatoid fishes there are both red and blue emitting photophores (135) and certain squid may also have photophores emitting different colors (133, 242). In the sea pen *Umbellula* and the zoanthid anemone *Parazoanthus* different parts of the same colony emit different colors (231).

There are several possible chemical explanations for such dual emissions *in vivo*. A product of the reaction may be formed in the excited state, but its actual emission may be different depending on environmental conditions (e.g., solvent polarity or pH). In the firefly reaction, for example, the emission is yellow–green at pH 7.6 but red at pH 5, with a double peak at intermediate values (166). Another mechanism for dual emissions is the presence of a second emitter molecule as an adjunct to a given luciferin–luciferase system in some cases. This is known in the coelenterates (176, 227) and bacteria (90, 150, 195). The excitation of a second chromophore might occur by Förster-type energy transfer from the primary excited state (227) or it might involve direct population of the second emitter (see the section on biochemistry and molecular biology); whatever the mechanism, emission could come simultaneously from both emitters.

A third way in which different colors could be produced would involve the use of optical filters. Although this method results in attenuation, it appears especially important in several fish groups (66, 67, 97, 187). Finally the introduction of a second luciferin–luciferase system would also be a means for obtaining a different color. Although there is no confirmed example of this, there is one group of organisms where it could occur, for they appear to carry two different luminescent systems. The ceratioid angler fish *Linophryne* has a bacterial light organ in the esca (dorsal "lure") and an intrinsic light organ in the barbel (ventral "lure") (99). The emission spectra have not been reported.

In other organisms, such as the gorgonian *Thourella*, searsiid fishes and sergestid shrimps (Fig. 3D), there are complex spectra in which the relative magnitudes of the two peaks may differ in relation to time of continued emission (130, 231). This has not been studied in detail, but the variation could be due to changes in the relative proportions of two different excited states during continued emission, possibly because of progressive changes in conditions, for example, pH.

Kinetics and Intensity

Intensity modulation in time represents another important means for functional specialization of light emission. Grouping emissions into three classes, flashes (< 2 s), glows (> 2 s), and continuous emission (Fig. 4), Morin (173) proposed that in many organisms, particularly those lacking complex behavior and well-developed vision, flashes are repellent (startling, frightening, and/or aposematic),

---

(138a); (W) copepod *Pleuromamma xiphias*, 40-V, 5-ms stimulation (147); (X) (brief) and (X') (strong mechanical stimulation) to polychaete *Hesperonoe complanata*, intact scale (121); (Y) and (Y') oegopsid *Pterygioteuthis microlampas* (244); (Z) (daytime) and (Z') (nighttime) anomalopid *Photoblepharon palpebratus* spontaneous [except for 1 wink, the variation in (Z') is the result of the fishes swimming] (178).

whereas glows or continuous emission are attractive to photoreceptive organisms. For example, in the dinoflagellate *Gonyaulax polyedra*, which has a circadian rhythm in flashing frequency and responsiveness to stimulation, cells are less subject to predation at the time when flashing is greatest (41, 83). On the other hand, the lure (esca) of the angler fish, where luminous bacteria presumably emitting continuous light are cultured, surely functions to attract would-be predators, who are promptly converted into prey.

Flashing is under direct (dinoflagellate and coelenterate) or indirect (firefly and euphausiids) control by either the nervous system or an excitable membrane conducting system of some sort. Intracellular flashing must be under the control of effector molecules and/or membrane gating devices relevant to the particular system. It probably involves movements of specific ions or the release of cofactors involved in the light emitting reaction.

Accurate recording of the absolute intensity of luminescence emission is fraught with difficulties and the data must be interpreted with extreme care. Although many papers on bioluminescence deal only with relative units, there are methods for recording absolute intensities (140, 217, 223). When absolute levels are recorded from point sources they may be expressed either as energy (usually as $\mu W/cm^2$ at 1 m) or as quanta (photons [= quanta]/$cm^2 \cdot s$). With the establishment of low level liquid light standards (114, 149) absolute emissions may be expressed in quanta emitted in all directions, either total or per unit time. For conversion between energy and quanta at 480 nm: 1 $\mu W/cm^2 = {\sim}2.4 \times 10^{12}$ quanta/cm^2/s.

Maximum recorded luminescent intensities are ${\sim}10^{-1}$ $\mu W/cm^2$ (2×10^{11} quanta/$cm^2 \cdot s$) (239, 242). Most luminescent emissions appear to be in the range of ${\sim}10^{-3}$–10^{-9} $\mu W/cm^2$ [${\sim}10^9$–10^3 quanta/ cm $\cdot$ s] per luminescent pulse (173). This range embraces the ambient light intensities at the sea surface between 15 and 40 min post-sunset (at low latitudes). Thus at night on land or shallow seas, or at all times below a depth of ${\sim}200$–300 m in the sea, luminescent pulses can be observed above the background light (55, 239). Many organisms are known to have visual capabilities that can easily detect the dimmest luminescent pulses of 10^{-9} $\mu W/cm^2$ or less; a lanternfish emitting an intensity of 5×10^{-8} $\mu W/cm^2$ should be seen by a conspecific with a visual threshold of 10^{-10} $\mu W/cm^2$ at distances up to ${\sim}20$ m (184). Even single dinoflagellate flashes (${\sim}10^9$ quanta/0.1 s) or the flash of a single hydroid photocyte (${\sim}10^8$ quanta/ cell) are visible at 1 m to the dark adapted human eye. The luminescence of an individual bacterium (${\sim}10^4$ quanta/s, 4 pi equal to ${\sim}10^{-13}$ $\mu W/cm^2 \cdot$ cell) is well below the visual threshold of most organisms at 1 m. However, the luminescence of a colony is easily visible and luminescence of some fish light organs containing bacteria is among the brightest known (4×10^{-4} $\mu W/cm^2$) (184).

Intensity often varies in a stimulated flash sequence, most often showing strong facilitation (as in the scaleworm mentioned above), so that the first few flashes of a series increase rapidly in intensity (followed by a plateau or decline). Some organisms, however, show a maximum initial intensity followed by a decline. For stimulated luminescence, intensity can vary, even within one individual, by several orders of magnitude, usually in direct proportion to the intensity of stimulation.

Units of Functional Organization

The variety of different structural and functional arrangements of luminescent systems include both unicellular and multicellular systems. The latter range from simple photocytes located among non-luminescent cells in multicellular organisms to more complex light organs or photophores with complex dioptric accessories surrounding a well innervated and vascularized photogenic region. There are also secondary mechanically controlled systems either for gating and guiding continuous light emission or for secreting the luminescence outside the organism.

Single Cell Emitters

BACTERIA. Bacteria figure prominently in light emission by animals, notably as symbionts in specialized light organs in some groups, especially cephalopods and fishes. They also occur as intestinal symbionts: In some marine fishes the intestinal flora may be an essentially pure culture of a particular luminous bacterium (183, 196); also, fecal pellets from certain fish are known to be brightly luminescent (14, 193). Inevitably, of course, bacteria also occur as planktonic forms, free in seawater.

In some tunicates (the pyrosomes), light emission from photocytes is associated with bacteria-like intracellular organelles believed to have had an endosymbiotic origin (151, 157, 183). But in this case the emission, unlike that of typical bacteria, is not continuous, but under control by the host, by an unknown mechanism.

Bacterial light emission involves the electron-transport system. Although continuous light emission is consistent with steady electron flow, it was reported that in individual bacteria luminescence occurs in bursts or flashes every few seconds or minutes (28, 29). If this were so, it would have implications for electron transport and its control. But Haas (96) failed to confirm this; he believes that both luminescence and electron flow are truly continuous at the level of the single cell.

DINOFLAGELLATES. Luminescence occurs as rapid (< 0.1 s) flashes (Fig. 4D–F) following stimulation (78), and also spontaneously (139). Several species have been examined and, while there are clear differences in some aspects, luminescence is emitted in all cases from many subcellular organelles referred to as microsources (77) or scintillons (85).

In *Gonyaulax polyedra*, the organelles (Fig. 5A) are distributed cortically and may be visualized in the light microscope by their fluorescence, which is due to luciferin, and also by bioluminescence at the same site (78, 137) (Fig. 5B). Ultrastructural visualization and description of the organelles has been achieved by immunocytochemical staining with antibodies raised against luciferase (185, 186). Scintillons are cytoplasmic dense bodies that occur as vesicular outpocketings into the low pH vacuole, hanging as drops attached by narrow necks (Fig. 5C). This continuity with the vacuolar membrane explains flash excitation by a vacuolar membrane action potential. The dinoflagellate bioluminescent reaction is triggered by a pH change from 8 to 6 within the scintillon. This change is postulated to result from the action potential, which causes entry of H^+ from the vacuole via voltage gated proton channels, and to be restored to its original pH by a H^+-transporting ATPase (109). In the course of cell disruption at pH 8 organelles may be transformed into closed intact vesicles, presumably because the narrow necks break easily and reseal during extraction. These isolated scintillons may be induced to emit flashes mimicking the *in vivo* flash simply by rapidly shifting the pH from 8 to 6. This procedure does not destroy the integrity of the scintillons: they will emit second and third flashes after recharging with luciferin at pH 8 (85).

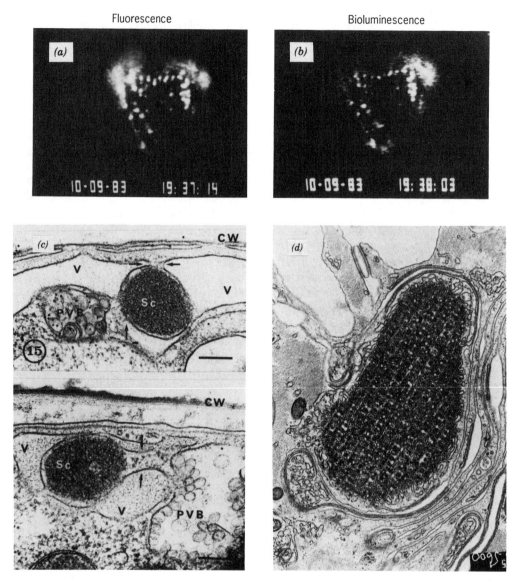

Figure 5. Endogenous fluorescence (*a*) and bioluminescence (*b*) of the scintillons of *Gonyaulax polyedra in vivo*, showing their colocalization. (*c*) Immunogold staining of scintillons in two sections after fast-freeze fixation and freeze substitution in OsO₄–acetone: The labeling was done by applying first the antibody raised in rabbits directed against luciferase followed by a second antibody, goat antirabbit with 10-nm gold particles attached (GAR-G$_{10}$ from Janssen Pharmaceutica). The flashing organelles (scintillons:Sc) are specifically labeled. They protrude into the vacuole (V) attached by narrow necks (arrows). CW, cell wall; PVB, polyvesicular body (bar = 0.2 μm). (*d*) Photosome (P) in a luminous cell of the scale-worm *Harmothoe lunulata*, after fast-freeze fixation and freeze substitutions in OsO₄–acetone during repetitive stimulation. The organelle is enclosed in a pouch wrapped by several digitations, separated by narrow extracellular spaces (E), and surrounded by arrays of intermediate reticulum (I), which connect terminal saccules forming dyade junctions with the plasma membrane (arrows). This coupling arrangement is characteristic of the photosomes during bioluminescent emission (bar = 1 μm).

RADIOLARIANS. The luminescence of the radiolarian *Thalassicolla* occurs as repetitive, 1-s duration flashes in response to stimulation. As in the coelenterates, oxygen is apparently not required for reaction (102), but calcium ions are (128).

Simple Photocyte Systems

Simple photocytes may be clustered or widely scattered; they are usually under neural or neuroid control. They are most often located in or close to the ectoderm, but occur in the endoderm in most cnidarins and ctenophores, and even *within* the spines in ophiuroids. Accessory structural features, except tissue transparency, are lacking. With few exceptions, organisms with simple photocytes have no or poorly developed photoreceptors. Single excitatory (usually mechanical) stimuli generate one or more pulses (Fig. 4*A*,*B*,*J*–*L*), which travel away from the point of stimulation, giving rise to simple, bright and conspicuous flashes (173).

Photocytes are the sole emitting system among cnidarians, ctenophores, echinoderms, and nemerteans. They also are present in some annelid groups such as the polynoids, terebellids, and syllids; gastropod molluscs; hemichordates; tunicates (larvaceans and salps—but not pyrosomes); some crustaceans such as amphipods, copepods, and halocyprid ostracodes; millipedes; pycnogonid chelicerates, and some fishes. Photocytes often lie in superficial tissues and are easily sloughed off, thus appearing to be secretory in nature.

The cnidarians and ctenophores show perhaps the simplest photocyte organization, lying singly or in clusters in the endoderm (7, 172). Light emission is triggered by calcium. In the hydroid *Obelia* the inward calcium current occurs in depolarized support cells, *not* the photocytes, and gap junctions with these neighboring epithelial cells act as channels for calcium flux to the photocytes (74). Many cnidarians have an accessory protein, the green fluorescent protein (GFP), associated with their calcium activated photoprotein complex. Light is usually green in the presence of the GFP (λ_{max} = ~510 nm) and blue (λ_{max} = ~460 nm) without it. In some hydrozoans the photoprotein, accessory GFP and membrane excitability vary during ontogeny (87). The egg contains the photoprotein but is inexcitable. In the early embryo (through the planula) all cells are excitable and produce blue light, but this declines during development. Spontaneous flashes occur during metamorphosis. Not until after metamorphosis do distinct photocytes differentiate; they contain both of the proteins and produce green light. Photocytes appear to be controlled by epithelial conduction in hydropolyps (175) and siphonophores (24), and by a colonial nerve net in anthozoans (13, 198). The light may be emitted as one to many flashes per stimulus (172) (Fig. 4*A*, *J*–*L*). The putative neurotransmitter involved with luminescent control in the anthozoan *Renilla* is adrenaline or a related catecholamine (6, 8).

In the echinoderm *Ophiopsila* the photocytes are dentritic-like cells that ramify *within* the spines and certain plates of the arms and send luminescent projections *to* the radial nerve (35, 122). Luminescence is under the control of the radial nerve, usually as multiple pulses per stimulus (Fig. 4*B*), and involves calcium (34). The emitter is reported to be a complex photoprotein (210).

Photocytes in polynoid polychaetes (scale worms) are located in the ventral epithelium of the scales (elytra) and are structurally very complex (22, 30, 190). There are distinct clusters of electrically coupled luminescent cells, at least some of which are innervated (121). These photocytes produce two types of calcium-associated action potentials in response to nerve stimulation. Luminescence is associated with each Ca^{2+} spike. Mild stimulation produces a single flash, while strong stimulation induces scale autotomization and repetitive flashes, often followed by a prolonged glow that may last for 10–60 s (Fig. 4W). Intracellularly, the photocytes contain 30–40 complex organelles, termed photosomes (23), which are paracrystalline bodies composed of geometric arrays of serpentine-like microtubules (20 nm in diameter) and saccules, both of which are connected to the endoplasmic reticulum (Fig. 5D). The photosomes are the site of luminescence by a process reminiscent of that in striated muscle. Upon repetitive stimulation, there is an intracellular recruitment of additional organelles involving ultrastructural changes with increased intensity of flashing (171).

In another group of polychaetes, the luminescent syllid *Odontosyllis*, both intracellular and extracellular luminescence are apparently produced (235). Little is known about the structure and mechanisms of light production. Males produce intracellular flashes and females extracellular (and intracellular) glows, just post-twilight a few days after full moon (221).

The luminescence in larvacean tunicates appears to be generated as rapid flashes (Fig. 4C) from subcellular organelles or portions of cells that are specialized components of the secreted mucous houses, which are repeatedly and periodically produced (88, 89). The exact source and nature of these peculiar structures is unknown.

Slightly more complex luminescent systems occur in many molluscs, annelids, crustacea, insects, and fishes as simple light organs. These show a distinct cluster of photogenic cells, usually surrounded by pigmented cells but lacking other accessory structures. Their innervation and vascularization may be quite complex.

Light Organs

Light organs are complexly organized and controlled light emitting systems (Fig. 6) found primarily in animals with well-developed vision and complex behavior, particularly among the cephalopods, crustaceans, insects, and fishes. In the literature the terms light organ and photophore are often used interchangeably. Here however, we make a distinction and consider photophores and glandular light organs each to be a subset of the broader category of light organs.

PHOTOPHORES. Photophores are particularly well developed in some squids, insects, euphausiids, shrimps, and fishes. They are especially prevalent on the ventral and lateral surfaces of midwater species. Most often photophores are relatively small, numerous, and superficial, but occasionally they are large and deeply embedded in the body; some species of fishes and squids may contain both many small and a few (1–3) large photophores. Complex photophores (Figs. 6–8) show a wide spectrum of complexity, but most possess a central photogenic region, which is well innervated and vascularized, surrounded by inner reflector(s), a sheath of pigmented cells, and outer lens(es). This arrangement allows light to be directed and often focused outwardly, while any inward directed light is either reflected outward or absorbed by the pigments, so that no light escapes *through* the organism. Various other accessories may be found with such photophores, including muscles for focusing or rotating the lenses or collimators, filters for changing the quality, color and/or intensity of the light, shutters for occluding the light, and translucent muscles or other tissues for collimating the light like fiber optics.

Physiological aspects of photophores have been best studied in the firefly and the shallow marine toadfish *Porichthys*,

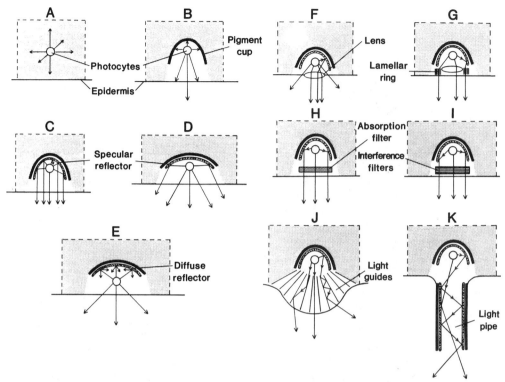

Figure 6. The effects of pigment, reflectors, and other accessory structures on light emission; arrows indicate possible ray paths. (*A*) Point source emission of group of photocytes. (*B*) Pigment cup restricts the solid angle of emission. Reflectors of different geometries provide a more efficient emission, whether they are specular (*C*) and (*D*) or diffuse (*E*). Effects of pigments, reflectors and (*F*) a lens alone, (*G*) a lens and lamellar ring, (*H*) a pigmented absorption filter, (*I*) an interference filter, (*J*) a light guide diffuser, or (*K*) a light pipe. (Modified from 131.)

and to a lesser extent in certain squids, euphausiids, shrimps, and myctophid fishes.

Fireflies. Photophores in fireflies (Fig. 7*A* and *B*) are formed by photocytes stacked in dorsoventral columns of complex rosettes radially arranged around a trachea and neuron (8, 45, 91). Light emission is under neural control but is evidently indirect. A normal flash-triggering nerve impulse travels via the ventral nerve cord to its termination in the sphincter of the tracheal end cell, which normally prevents oxygen from reaching the photogenic rosette. The nerve impulse opens the tracheole, so that a burst of oxygen triggers the biochemically poised photogenic cell to emit (38, 110, 165). While this hypothetical sequence is consistent with the structural and biochemical data, it is somewhat difficult to reconcile with the presumably slow kinetics of the diffusion of oxygen over the distance from the tracheal end cell to the photogenic cell.

Octopamine is implicated as the neuroeffector transmitter (43, 51), but cyclic nucleotides also apparently mediate the response secondarily (188). Firefly larvae emit light as a glow lasting 10–100 s. Little is known concerning the stimulus or the function of the glow. The organ is devel-

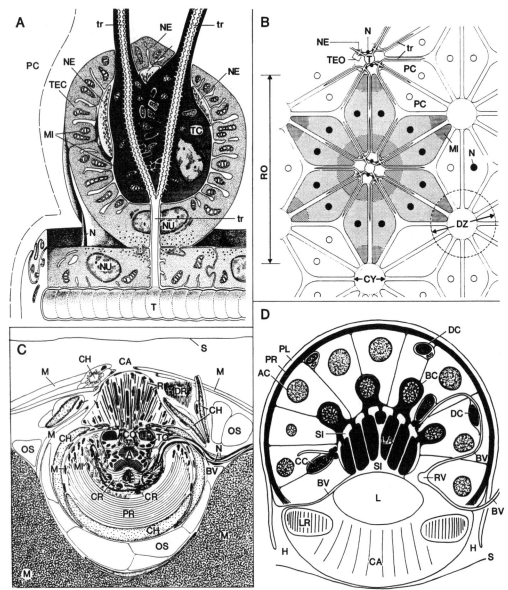

Figure 7. Complex light organs in invertebrates: Diagrammatic representation of the (*A*) tracheal end organ of a *Photuris* (firefly) (modified from 91); (*B*) rosette of photocytes of *Photuris* (modified from 100); (*C*) type C photophore in the squid *Abralia* (modified from 240); and (*D*) abdominal or thoracic photophore of a euphausiid (modified from 134). Abbreviations: A = axial stack, AC = A-cell, BC = B-cell, BV = blood vessel, CA = cap, CC = C-cell, CH = chromatophores, CO = core, CR = crystaloid cells, CY = cylinder, DC = D-cell, DR = distal reflector, DZ = differentiated zone, H = hemocoel, L = lens, LA = lantern, LR = lamellar ring, M = muscle, MI = mitochondria, N = nerve, NE = nerve ending, NU = nucleus, OS = orbital space, PC = photocytes, PL = pigmented layer, PR = proximal reflector, R = ribbon, RO = rosette, RV = ring vessel, S = surface, SI = blood sinus, T = trachea, TC = tracheal cell, TEC = tracheal end cell, TEO = tracheal end organ, TO = torus, tr = tracheole, V = vesicles.

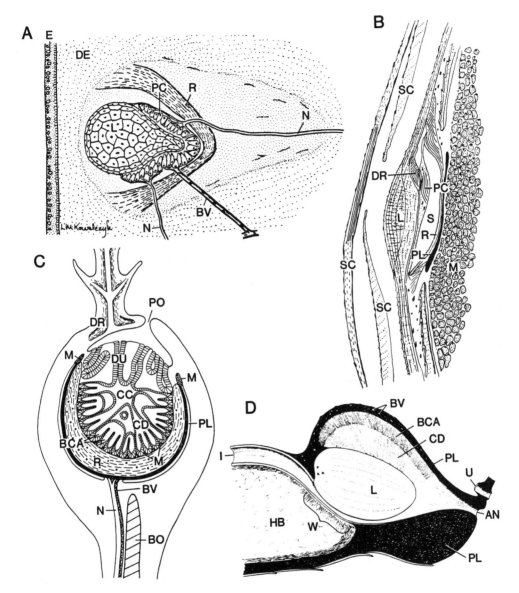

Figure 8. Diagrammatic representation of intrinsic photophores (*A* and *B*) and bacterial light organs (*C* and *D*) in fishes. (*A*) Ventral photophore of *Porichthys notatus* (midshipman) (modified from 135); (*B*) photophore of *Tarletonbeania crenularis* (myctophid) (modified from 148); (*C*) an escal bacterial light organ of *Linophyrne arborifera* (ceratioid anglerfish) (modified from 99); (*D*) a rectal light organ of *Opisthoproctus soleatus* (spook fish) (modified from 27). Abbreviations: AN = anus, BCA = bacterial culture area or tubules, BO = bone, BV = blood vessel, CC = central cavity, CD = collecting ducts, DE = dermis, DR = dorsal reflector, DU = duct, E = epidermis, HB = hyaloid body, I = intestine, L = lens, M = muscle, N = nerve, PC = photocytes, PL = pigmented layer, PO = pore, R = reflector, S = space, SC = scale, U = urogenital duct, W = window.

opmentally distinct from the adult organ, although the biochemistry appears to be the same.

Porichthys. The numerous photophores in the batrachodid fish *Porichthys* lie widely distributed just below the epidermis (Fig. 8A). The well-vascularized photocytes form a cup-shaped layer around a lens. Numerous nerve terminals occur nearby, but not on the photocytes. Luminescence that is evoked via a nerve stimulation occurs only after a considerable latency (2–3 s) and occurs as a glow (Fig. 4T) with a duration of 5–20 s (9, 16). The putative neurotransmitters in this response are epinephrine and norepinephrine (2, 4, 11, 52–54). The rare observations of luminescence from living animals indicate slow light pulses or continuous glows; flashes have not been detected (62, 144). Luciferin, which is identical to cypridinid luciferin, is apparently acquired from dietary transfer via the circulatory system (219, 222, 228).

Squids. The intricate morphology of cephalopod photophores (e.g., Fig. 7C) has been investigated in a number of species, but the mechanisms of physiological control of light emission are poorly known and have mainly been inferred from the morphology (125, 127, 133, 220, 238, 240, 241). Squid diversity is highest in mesopelagic waters; these are mostly oegopsid squids and most species possess photophores. Here, even within a single species, more than one morphological type of photophore may occur. They range from relatively simple patches to highly intricate structures. Photophores are located in a variety of sites in different cephalopods, but are especially prevalent on the ventral surfaces and around the eyes. Cells in the presumptive photogenic region in squids may be featureless or contain highly arrayed paracrystalline structures (crystalloids). Depending on the species, control of luminescence appears to be neural or hormonal, or both. Secondary control of the light via chromatophores or muscles may occur in some species. Reflectors, composed of chitin or collagen platelets, may act as specular or interference reflectors, and appear to regulate, collimate, direct and/or filter the light. Changing the spacing or orientation of the platelets allows squids to modify these characteristics. Thus midwater squids appear to be able to control the emitted intensity over several orders of magnitude, the emission spectrum, the duration (Fig. 4Y), and the angular distribution and direction of the light.

Euphausiids and Decapod Shrimps. Photophores are present in most euphausiids and several groups of decapod shrimps, mostly on their ventral surfaces (132). Euphausiids possess 10 rotatable ventral photophores (Fig. 7D). Each contains a photogenic region of several cell types with a core region, termed the lantern, composed of photocytes with quasicrystalline centers (134). This region is bathed by a blood sinus that appears to be regulated by a neurally controlled sphincter. Light emission is probably controlled by the blood supply, and perhaps a hormone (101), with the nervous system acting via the sphincter to gate the blood flow to the lantern. The photogenic region is surrounded by a multilayered reflector, probably composed of chitin, and a pigmented layer. Toward the exterior there is a lens surrounded by a lamellar ring (134). Light emission (Fig. 4U) shows a rapid rise followed by a slow decay (several seconds) or a steady glow. Serotonin (5-HT) stimulates light emission, so this molecule or an analog is probably involved in luminescence control, perhaps at the sphincter (69).

Fishes. A bewildering array of photophores occurs in many fishes, especially those from mid-ocean waters (129, 135). They range from relatively simple to highly complex and are generally arranged along the ventral and lateral surfaces and around the eyes. Many are able to produce both rapid flashes (< 100 ms) (Fig. 4L–Q) and steady glows. The myctophids are the best studied (10, 16, 21, 148). The photophores are numerous and superficial, but complex (Fig. 8B). Certain photophores can function independently or in synchrony. Both duration and intensity of the light are under direct control of the nervous system via the spinal cord. Latencies are very short. Photophores appear to show both spinal and sympathetic innervation but the neurotransmitters are unknown. The photophores do not appear to respond to serotonin or adrenaline. However, adrenaline is an effective stimulant in the hatchetfish *Argyropelecus* (17, 18). Spectral filters occur in a number of deep-sea fishes (67). Most deep-sea angler fishes (ceratioids) possess bacterial light organs, but *Linophryne* also possesses a single nonbacterial light organ on a ventral barbel (99). It is a superficial organ of mesodermal origin with a paracrystalline photogenic core bathed by a capillary matrix, which may control the light.

GLANDULAR LIGHT ORGANS. Morin (173) has defined glandular light organs as a few (1–4, occasionally more), relatively large outpocketings of the gut or body surface with complex accessory structures and which contain bacteria or intrinsic luminescent sources. We treat the bacterial light organs below under mechanically controlled systems. Nonbacterial glandular light organs are most prominent in some crustaceans and fishes.

Crustaceans. Hepatic light organs occur in a few penaeid and caridean midwater shrimps within the decapods (124, 132). These complex light organs are derived from hepatopancreas tubules and can be rotated so the light is continuously projected downward (145, 146). In the base of each tubule are the photocytes, which contain paracrystalline inclusions. A diffuse reflector and pigment layer lie internal and a blue filter external to this region (67, 145). Luminescence appears to be slow, in the form of long glows or continuous emission (229).

Fishes. One to three gut-associated glandular light organs occur among some benthic apogonids and pempherids, and some midwater evermannellids (135). The organs of the benthic species are outpocketings of the pyloric caeca, intestine and/or rectum. Chemically, the luminescent system is very similar to that of cypridinid crustaceans and thus all or part of this system may be derived from the diet (98). The nature of the control mechanism of the light is unknown.

Mechanical Control

SHUTTERS AND CHROMATOPHORES: BACTERIAL LIGHT ORGANS. Bacterial light organs occur in squids and fishes; they are often similar to photophores in their accessory structures, but instead of an endogenously luminescent photogenic source, they have a special (highly vascularized) gland, which nurtures enormous numbers of *Photobacterium leiognathi, P. phosphoreum, Vibrio fischeri,* or not yet cultured or identified bacteria (120, 125, 131, 133, 135). In addition, since these bacteria apparently produce light continually, the light organs usually have special mechanisms to occlude the light. These primarily involve muscular shutters, chromatophores or both, and sometimes reflectors, filters, and light guides.

Fishes. Bacterial light organs occur in several orders of fishes (Table 1; Figs. 8C,D,

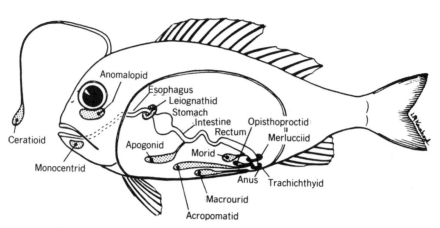

Figure 9. A diagrammatic fish indicating the different locations and openings of light organs in the fishes that harbor symbiotic luminous bacteria (112).

9), and are especially prevalent among benthic or demersal (= near bottom) species (131, 135). In most cases one or two organs are associated with the gastrointestinal tract, but three groups have paired organs located externally: suborally in the pinecone (monocentrid) fishes; subocularly in the flashlight (anomalopid) fishes; and associated with the dorsal fin (the esca) in the deep-sea angler (ceratioid) fishes (Fig. 8C). In all cases the photogenic region is composed of vascularized fish epithelial tissues that form tubules or folds (10–48 μm in diameter and 150–800 μm long) housing high concentrations of luminous bacteria. The tubules open to the gut or the exterior by one or more connecting ducts, often via an atrium. Bacterial growth, nutrition, and light output are presumably ultimately controlled via the fishes' blood supply (75, 76, 120, 182).

In most species there is a guanine or collagenous reflector backed by a pigmented layer around the culture chamber (168). The emitting face may include transparent tissues invested with chromatophores, lenses, and/or various collimating structures such as translucent muscles, bone, or translucent hyaline bodies (135, 168, 169). In bacterial light organs associated with the gut, chromatophores occur at several levels along the light emission pathway and are presumably under hormonal control to vary the external expression of the light (168). Various mechanical shutters or "windows," under neural control, may effect rapid control of the light in various species (Fig. 4Z). In flashlight fishes, light emission is controlled either by a mechanical shutter or rotation of the entire light organ or both, depending on the species. In at least one pinecone fish, light is occluded when the mouth is closed. How bacterial luminescence is controlled in the midwater angler fishes is uncertain, but may involve chromatophores and neural mechanisms. Ceratioid angler fishes and the opisthoproctids, some species of which have rectal light organs (Fig. 8D), are the only midwater fishes known to have bacterial light organs (27, 123, 135).

Squids. Bacterial light organs are known among sepiolid and myopsid squids, which are primarily benthic or demersal species, but *not* among the midwater oegopsids where photophores are the rule (125, 127, 133). In most species light organs lie against the ink sac; the light is emitted ventrally as a steady glow. In *Het-*

eroteuthis luminous material (apparently bacteria) can be squirted into the water. Where culturable, the bacterial species has proven to be *Vibrio fischeri*.

Tunicates. Among tunicates, the pelagic colonial pyrosomes possess peculiar bacterial light organs. Each zooid has two clusters of "photocytes" that contain intracellular bacteria (bacteroids). These are structurally very different from other known luminescent bacteria and luminesce in response to mechanical or photic stimulation (151, 157). Luminescent flashes have a duration of 1–30 s (Fig. 4R). The mechanism of excitation is uncertain. The photocytes are not innervated or electrically coupled to other cells, yet luminescence correlates consistently with a neural process that leads to ciliary arrest in the gill bars, with a long delay of ~600 ms. Control is hypothesized to be indirect through some mechanism involving this ciliary cessation, perhaps by somehow increasing intracellular oxygen within the photocytes.

Nematodes and Crustaceans. Some nematode, parasitic on caterpillars, carry an inoculum of a pure culture of the luminous bacterium *Xenorhabdis* and use this to infect and ultimately kill the caterpillar, which then emits a continuous light when the infection is in full bloom (191). Apparently, the hemolymph in certain crustaceans occasionally becomes infected by *Vibrio harveyi*, which includes a period of total body luminescence followed by death (103).

SECRETORY SYSTEMS. Luminescent secretions are produced by a wide variety of organisms, from many different habitats, both marine and terrestrial, and involve a number of biochemical mechanisms (12, 103, 127, 160, 224). Luminous secretions are produced as (1) surface slimes (the freshwater snail *Latia;* some earthworms,

centipedes, and hemichordates), (2) ejections into the excurrent flow streams in suspension feeding organisms (chaetopterid polychaetes, some bivalves), or (3) ejected glowing clouds from actively swimming organisms (some syllid polychaetes, copepods, ostracodes, amphipods, a mysid, shrimps, the squid *Heteroteuthis,* and searsiid and some ceratioid fishes).

The secretions are usually produced from gland cells located in or below the epidermis or from a diverticulum of the gut. Secretion is controlled by the nervous system and may be direct via exocytosis of cellular contents or indirect by action of contractile elements either within the secretory cells or in adjacent muscle cells. Many secretory systems have nipples or nozzles from which forceful ejection of the luminescent secretion evidently occurs. In some cases two different secretory systems occur in close proximity, suggesting that different components (e.g., the luciferin and luciferase) may be housed separately, then mixed externally where the luminescent reaction occurs (103). Alternatively, reactants may be "precharged" and luminesce upon expulsion when they encounter necessary factor(s) such as oxygen or a cation (e.g., Ca^{2+}). Another mechanism occurs by sloughing or ejecting particulates such as luminous bacteria (e.g., the squid *Heteroteuthis* and the ceratioid *Ceratias*) or whole photocytes (e.g., earthworms, a few cnidarians and ctenophores, and searsiid fishes) (127, 129, 133). A more complex, but comparable, mechanism occurs in some organisms (e.g., polynoid polychaetes, and ophiuroids) which, upon intense stimulation, will jetison whole body parts, which then produce repetitive flashes and/or long lasting glows (103, 173).

Depending on circumstances, some secreting species produce both rapid flashes and slow glows, or even continuous light.

In hemichordates short flashes appear to be intracellular, while long glows are extracellular (25). During mating displays male syllid polychaetes show very rapid intracellular flashing, while the females produce both rapid intracellular flashes and a glowing extracellular luminescence (103, 221). The cypridinid *Vargula* can control the duration of secreted luminescence over 2 orders of magnitude [<1 s to minutes] (57, 174). Upon initial stimulation the copepod *Pleuromamma* emits rapid 100-ms flashes within the light organs and slow (2.6 s) emissions (Fig. 4W) upon expulsion of the luminescent material from the organ (147).

Functions and Behavior

Bioluminescent systems may be far more plastic and under fewer evolutionary constraints than most other biological systems. Light emission can function in a variety of nonmutually exclusive ways (173) including: (1) for evading or deterring potential predators or aggressors, (2) for obtaining prey, and (3) for intraspecific communication.

Defense by thwarting visually orienting, would-be predators is probably the most widespread and important function of luminescence. Circumstantial and some direct evidence is accumulating that shows that contact elicited luminescence is repellent to the aggressor if it is a flash, but attractive if it is a glow (173). Evasion of a predator can be achieved by different specific uses of flashes, glows, or continuous light. These include: (1) startling or frightening, (2) temporarily blinding the aggressor (flashbulb effect), (3) alerting the aggressor to the invulnerability of the emitter (armature or unpalatability) [aposematic luminescence], (4) decoying an aggressor away from its prey, (5) attracting a predator of the aggressor and thereby removing the aggressor by a secondary predatory event (burglar alarm ef-

fect), and/or (6) preventing the aggressor from detecting the emitter (camouflage).

Luminescent flashes produced by dinoflagellates apparently deter predatory copepods and other planktonic organisms (41, 83) and those from ophiuroids deter their predators, which include a variety of decapods including crabs and lobsters (92, 93). Because dinoflagellates and ophiuroids are unpalatable to certain predators, aposematism is another likely function of luminescence in these cases. In addition, startling, flashbulb, and burglar alarm effects could also be acting. While the burglar alarm hypothesis was proposed many years ago (40), no direct evidence was available until recently when it was observed that the luminescent cloud produced by the ostracode *Vargula*, when attacked by a small nocturnal fish, attracted a larger fish that consumed the smaller fish (174).

Little experimental work has been done to demonstrate the decoy effect, but many luminescent species appear to use it, including species that produce luminescent slime or clouds when disturbed, and animals that autotomize the emitting structures upon intense stimulation (173, 239). In all these cases the response is not spontaneous, but rather occurs only after stimulation, and in most cases the light occurs as a long lasting glow at a distance from the emitter's body. In scaleworms and brittle stars, an initial mechanical disturbance may evoke bright flashes (Fig. 4X) from scales or arms, respectively (34, 121). A more life-threatening disturbance may result in the autonomy of scales or arms which, after being jettisoned, emit glows (Fig. 4X), attracting the would-be predator to it, thus effectively diverting the predator. In darkness the release of ink by an octopus or squid would serve no purpose, but a cloud of luminescence could provide a diversion and allow the animal releasing it to escape predators.

Camouflage by counter—or disruptive

illumination was suggested long ago (63) and more recently revived (56) with experimental support (104). Animals living in an aquatic medium are especially susceptible to predators that can detect their silhouettes from below against the downwelling light. Ventral luminescence can reduce, eliminate, or disrupt this silhouette. The majority of midwater squids, crustaceans, and fishes possess ventral photophores (67, 126), and bacterial light organs are primarily ventrally oriented in demersal fishes and squids (133, 135). Matching both the color, direction, and intensity of the downwelling light has been documented in midwater squids, shrimps, and fishes (46, 67, 229, 238, 242, 243).

Luminescence may be used to assist in predation in at least five, nonexclusive ways (173): (1) as a lure for attracting prey, (2) as a light for detecting prey, (3) by aggressive mimicry of a prey's communication signal, (4) by visually paralyzing prey (stun effect), and/or (5) by camouflage.

Many nocturnal animals, ranging from moths to plankters, are attracted to light at night. Not surprisingly, therefore, complex lurelike luminescent organs have evolved to attract such prey (239). These include the escae and barbels of many midwater fishes, especially ceratioids and stomiatoids. Similarly, the New Zealand dipteran glow-worm, *Arachnocampa luminosa*, attracts and entraps a variety of insects in sticky secretions (155). In flashlight fishes (anomalopids), flashing (or, more properly blinking), appears to be used defensively by individuals in an avoidance repertoire (blink and run, Fig 4Z), whereas an effectively continuous emission (Fig. 4Z'), sometimes involving schools of 50–150 fish over a coral reef (where even if blinking does occur it is asynchronous, so that the overall area is continuously illuminated) serves to attract positively phototatic or-

ganisms upon which the fish feed (by the light of their own organs) (178).

Many fishes and some crustaceans and squids have ocular light organs that may be used as flashlights to attract and/or detect prey (133, 135, 178). In fishes, ocular light organs occur in several shallow water species, including the flashlight fishes, pinecone fishes (monocentrids), pony fishes (leiognathids), and many midwater fishes, including myctophids and stomiatoids (135). Some deep-sea melastomiatid and malacosteid fishes apparently use red light from ocular photophores (along with their own red-sensitive visual pigments) to detect, with impunity, prey that possess only blue sensitive visual pigments (67, 187, 233).

Obtaining prey by mimicking the luminescent patterns of organisms that communicate using luminescence (aggressive mimicry) is documented in fireflies, where a female *Photuris* mimics the female *Photinus* sexual signal and thereby attracts, kills, and eats male *Photinus* of that species (153, 154, 156). The complex lures in the deep-sea angler and stomiatoid fishes may function in a similar manner.

The use of light to immobilize prey briefly (stun effect) might aid in feeding, but proof of this function is lacking (173). Camouflaging by counterillumination or disruptive illumination might allow a predator to approach its prey undetected, just as a prey using these techniques could avoid its predators.

Intraspecific communication by light signals, principally for reproductive purposes, is widespread among luminescent species with good vision and complex behavior. These include squids, crustaceans, insects, fishes, and some annelids. Luminescence in these cases can be either as flashes or glows. Sexual dimorphism in light organ morphology, a strong indicator of this function, occurs in all of these groups (127, 133, 135, 239). Intraspecific communication could also be used for spacing (locating conspecifics, territoriality, or aggregating), agonistic interactions or more complex behaviors (173).

Courtship has been well studied in fireflies, where species-specific signals are recognized by the time interval between a male query flash and the female's response. Other important components of the signals involve the detailed waveform of the flash (rise, decay times; single or multipeaked emission) (Fig. 4G–I) and/or the number and time interval between flashes (37, 39, 44, 155). Synchrony of many males perched on one tree can be entrained within a narrow rhythm (37).

In a comparable situation over shallow Caribbean reefs just after twilight, male ostracodes in the genus *Vargula* produce spectacular, species-specific trains of secreted luminescent pulses, attracting females (57, 174). In many parts of the world, males of the syllid polychaete *Odontosyllis* produce intracellular flashes and females eject glowing secretions during courtship (103, 127). These displays exhibit a precise lunar periodicity (221). In flashlight fishes, distinct male–female blinking signaling patterns suggest a role in reproduction in this group (178). Little is known about luminescent communication in midwater animals. But, given the pervasiveness of highly complex light organs in nearly all organisms in this habitat, it is likely that luminescence plays crucial roles in the intraspecific behaviors of these animals.

Origins and Evolution

Partly, perhaps, because the terms luciferin (substrate) and luciferase (enzyme) are employed with all luminescent organisms, it is often assumed that the different systems are homologous. They are not. This contrasts with many biological systems, where there are evident cross-phyletic relationships at the gene level, such that one can infer that systems in phylogenetically distant groups have a single origin and an evolutionary relatedness. In the case of bioluminescence, most of the different luciferins and luciferases appear unrelated, and the relevant genes that have been studied appear not to be homologous. It is thus inferred that many of the different extant light emitting systems, as well as others that may have suffered extinction, had independent origins (107).

There are, however, phylogenetically separated luminescent systems where similar luciferins or other functional molecules do occur. In some cases it may be that there is a genetic relationship; in others the luciferin may be obtained nutritionally (72, 86, 98, 222) and thus be analogous physiologically to a vitamin. On the other hand, coelenterazine and cypridinid luciferin are molecules with the same imidazopyrazine nucleus and similar reaction mechanisms, but nutritional explanations do not appear to apply (161). Both this and the occurrence of a tetrapyrrole in the unrelated dinoflagellates and euphausiids (72, 73) could be the result of convergent evolution.

An explanation alternative to that of the independent origin of similar luminescent systems is that of lateral transfer of DNA between distantly related organisms (42, 159, 236). Gene transfer by symbiosis is lateral transfer par excellence (158), and such a mechanism occurs in many bioluminescent systems with symbiotic luminous bacteria. If there do exist other general lateral gene transfer mechanisms, DNA encoding for bioluminescence would seem an ideal one with which recipient organisms could experiment, at low risk, with an interesting payoff. That, instead, luminescence apparently originated independently many times argues against such a mechanism, but does not preclude the possibility that both (or still other) mechanisms may have been involved. It is also interesting that luminescence apparently has evolved independently within closely related taxa. Annelids, arthropods (especially crustacea), molluscs, and fishes each possess more than one biochemically different bioluminescent system.

Is there some common feature of bioluminescent systems that could relate to and help account for the multiple independent origins? As elaborated above, luciferases are oxygenases. Many "prototype" weakly chemiluminescent oxidative reactions (*very* low quantum yield), which do occur in cells (1), could have provided the basis for the origin of luciferases either de novo, or by modification of already existing oxygenases (203, 204). The luciferases surely evolved on the basis of selection for light emission of the *high* quantum yield, and for being seen by another organism. Vision also appears to have evolved many times (197). From these predictions it may be inferred that luciferases evolved late in evolutionary time, after the development of vision and the divergence of the major phyla and their subgroups. This could account for the large number of different bioluminescent systems as well as their widespread and discontinuous phylogenetic distribution.

References

1. Allen, R. C. *Methods Enzymol.* 133:449–492, 1986. Phagocyte chemiluminescence.

2. Anctil, M. *J. Morphol.* 151:363–395, 1977. Photophore development in *Porichthys*.

3. Anctil, M. *Photochem. Photobiol.* 30:777–780, 1979. Review of physiological control of luminescence.

4. Anctil, M. *Rev. Can. Biol.* 38:81–96, 1979. Inhibitors and luminescence in *Porichthys*.

5. Anctil, M. *Comp. Biochem. Physiol. C* 68C:187–194, 1981. Drugs and control of luminescence in *Chaetopterus*.

6. Anctil, M. *J. Exp. Zool.* 223:11–24, 1982. Pharmacology of *Renilla* luminescence.

7. Anctil, M. *Cell Tissue Res.* 242:333–340, 1985. Ultrastructure of ctenophore luminescent cells.

8. Anctil, M. In *Nervous Systems in Invertebrates* (M. A. Ali, ed.), pp. 573–602. Plenum, New York, 1987. Neural control of luminescence.

9. Anctil, M. and J. F. Case. *Cell Tissue Res.* 166:365–388, 1976. *Porichthys* photophores and catecholamine effects.

10. Anctil, M. and J. F. Case. *Am. J. Anat.* 149:1–22, 1977. Lanternfish light organ structure.

11. Anctil, M., S. Brunel, and L. Descarries. *Cell Tissue Res.* 219:557–566, 1981. Catecholamines and 5HT in *Porichthys* photophores.

12. Anderson, J. M. *Photochem. Photobiol.* 31:179–181, 1980. Biochemistry of centipede luminescence.

13. Anderson, P. A. V. and J. F. Case. *Biol. Bull. (Woods Hole, Mass.)* 149:80–95, 1975. Luminescence and neurobiology in *Renilla*.

14. Andrews, C., D. Karl, L. Small, and S. Fowler. *Nature (London)* 307:539–541, 1984. Luminescence of oceanic fecal pellets.

15. Badcock, J. R. and N. R. Merrett. *Prog. Oceanogr.* 7:3–58, 1976. Vertical distribution of midwater fishes.

16. Baguet, F. *Prog. Neurobiol.* 5:97–125, 1975. Control of fish photophores.

17. Baguet, F. and G. Maréchal. *Comp. Biochem. Physiol. C* 60C:137–143, 1978. Effect of drugs on hatchetfish photophores.

18. Baguet, F., B. Christophe, and G. Maréchal. *Comp. Biochem. Physiol. A* 67A:375–381, 1980. Control of hatchetfish photophores.

19. Baldwin, T. O., M. Z. Nicoli, J. E. Becvar, and J. W. Hastings. *J. Biol. Chem.* 250:2763–2768, 1975. FMN binding with bacterial luciferase.

20. Baldwin, T. O., M. Ziegler, and D. Powers. *Proc. Natl. Acad. Sci. U.S.A.* 76: 4887–4889, 1979. Bacterial luciferase amino acid sequence and subunit homologies.

21. Barnes, A. T. and J. F. Case. *J. Exp. Mar. Biol. Ecol.* 15:203–221, 1974. Lanternfish flashing.

22. Bassot, J.-M. In *Recent Advances in Biological Membrane Studies* (L. Packer, ed.), pp. 259–284. Plenum, New York, 1985. Role of endoplasmic reticulum in scale-worm luminescence.

23. Bassot, J.-M. and A. Bilbaut. *Biol. Cell.* 28:155–162, 163–168, 1977. Scale worm luminescence and photosomes.

24. Bassot, J.-M., A. Bilbaut, G. O. Mackie, L. M. Passano, and M. Pavans de Ceccatty. *Biol. Bull. (Woods Hole, Mass.)* 155:473–479. 1978. Siphonophore luminescence and neurobiology.

25. Baxter, C. H. and P. E. Pickens. *J. Exp. Biol.* 41:1–14, 1964. Hemichordate nerve net and luminescence.

26. Belas, R. R. Mileham, D. Cohn, M. Hilman, M. Simon, and M. Silverman. *Science* 218:791–793, 1982. Cloning of *lux* genes.

27. Bertelsen, E. and O. Munk. *Dana Rep.* 62:1–16, 1964. Opisthoproctid light organs.

28. Berzhanskaya, L. Y., I. I. Gitelson, A. M. Fish, R. I. Chumakova, and V. Y. Shapiro. *Biofizika* 18:285–292, 1973. Rhythmic bacterial luminescence.

29. Berzhanskaya, L. Y., I. I. Gitelson, A. M. Fish, R. I. Chumakova, and V. Y. Shapiro. *Dokl. Akad. Nauk SSSR Ser. Biol.* 222:1220–1222, 1975. Pulsed bacterial luminescence.

30. Bilbaut, A. *J. Cell Sci.* 41:341–368, 1980. Scale worm luminescence and cell junctions.

31. Blinks, R. R., W. Wier, P. Hess, and F. Prendergast. *Prog. Biophys. Mol. Biol.* 40:1–114, 1982. Aequorin and Ca^{++} measurement in cells.

32. Boylan, M., A. F. Graham, and E. A. Meighen. *J. Bacteriol.* 163:1186–1190, 1985. Fatty acid reductase genes in bacteria.

33. Bradner, H. et al. *Deep-Sea Res.* 34:1831–1840, 1987. Luminescence at 4km depths.

34. Brehm, P. H. *J. Exp. Biol.* 71:213–227, 1977. Ophiuroid luminescence and neurobiology.

35. Brehm, P. H. and J. G. Morin. *Biol. Bull. (Woods Hole, Mass.)* 152:12–25, 1977. Luminescence structure in ophiuroids.

36. Buck, J. B. In *Bioluminescence in Action* (P. J. Herring, ed.), pp. 419–460. Academic Press, New York, 1978. Functions and evolution of luminescence.

37. Buck, J. B. *Q. Rev. Biol.* 63:265–289, 1988. Synchonous rhythmic flashing of fireflies.

38. Buck, J. B. and J. F. Case. *Biol. Bull. (Woods Hole, Mass.)* 121:234–256, 1961. Control of firefly flashing.

39. Buck, J. B. and J. F. Case. *Biol. Bull. (Woods Hole, Mass.)* 170:176–197, 1986. Courtship signals in fireflies.

40. Burkenroad, M. D. *J. Mar. Res.* 5:161–164, 1943. Luminescence burglar alarm hypothesis.

41. Buskey, E. J., L. Mills, and E. Swift. *Limnol. Oceanogr.* 28:575–579, 1983. Dinoflagellate luminescence effects on copepod behavior.

42. Bussinger, M., S. Rusconi, and M. L. Birnstiel. *EMBO J.* 1:27–33, 1982. An unusual evolutionary behavior of a sea urchin histone gene cluster.

43. Carlson, A. D. *J. Exp. Biol.* 92:165–172, 1981. Neural control of firefly luminescence.

44. Case, J. F. In *Insect Communication* (T. Lewis, ed.), pp. 195–222. Academic Press, Orlando, FL, 1984. Firefly behavior and vision.

45. Case, J. F. and L. G. Strauss. In *Bioluminescence in Action* (P. J. Herring, ed.), pp. 331–366. Academic Press, New York, 1978. Neural control of luminescence.

46. Case, J. F., J. Warner, A. T. Barnes, and M. Lowenstine. *Nature (London)*

265:179–181, 1977. Lanternfish counter-illumination.

47. Catalani, L. and T. Wilson. *J. Am. Chem. Soc.* 111:2633–2639, 1989. Electron transfer and chemiluminescence.

48. Chance, B. and R. Ohnishi. *Methods Enzymol.* 57:223–226, 1978. Luminous bacteria as an oxygen indicator.

49. Charbonneau, H., K. Walsh, R. McCann, F. Prendergast, M. Cormier, and T. Vanaman. *Biochemistry* 24:6762–6771, 1985. Amino acid sequence in aequorin.

50. Christensen, T. A. and A. D. Carlson. *J. Exp. Biol.* 93:133–147, 1981. Neural control of firefly luminescence.

51. Christensen, T. A., T. G. Sherman, R. E. McCaman, and A. D. Carlson. *Neuroscience* 9:183–189, 1983. Octopamine in firefly photomotor neurons.

52. Christophe, B. and F. Baguet. *J. Exp. Biol.* 104:183–192, 1983. Adrenergic stimulation of *Porichthys* photophores.

53. Christophe, B. and F. Baguet. *Comp. Biochem. Physiol.* C 81C:359–365, 1985. Adrenergic control of *Porichthys* photophores.

54. Christophe, B. and F. Baguet. *Ann. Soc. R. Zool. Belg.* 115:197–201, 1985. Seasonal variation in *Porichthys* luminescence.

55. Clarke, G. L. and E. J. Denton. In *The Sea* (M. N. Hill, ed.), Vol. 1, pp. 456–468. Wiley, New York, 1962. Light in the sea.

56. Clarke, W. D. *Nature* (*London*) 198:1244–1246, 1963. Luminescent functions in mesopelagic organisms.

57. Cohen, A. C. and J. G. Morin. *J. Crustacean Biol.* 9:297–340, 1989. Ostracode luminescence.

58. Cohn, D., A. Mileham, M. Simon, K. Nealson, and S. Rausch. *J. Biol. Chem.* 260:6139–6146, 1985. Bacterial *lux* A nucleotide sequence.

59. Cormier, M. J., K. Hori, and J. Anderson. *Biochim. Biophys. Acta* 346:137–164, 1974. Coelenterate bioluminescence.

60. Cormier, M. J. and J. R. Totter. In *Bioluminescence in Progress* (F. H. Johnson and Y. Haneda, eds.), pp. 225–231. Princeton Univ. Press, Princeton, NJ, 1966. Fungi luminescence.

61. Cormier, M. J., D. Prasher, M. Longiaru, and R. McCann. *Photochem. Photobiol.* 49:509–512, 1980. Biochemistry of the Ca^{++} activated photoprotein, aequorin.

62. Crane, J. M. *Copeia* 1965:239–241, 1965. Bioluminescent courtship in *Porichthys*.

63. Dahlgren, U. *J. Franklin Inst.* 181:525–556, 1916. Animal luminescence.

64. DeLong, E., D. Steinhauer, A. Israel, and K. H. Nealson. *Gene* 54:203–210, 1987. *Lux* genes from *P. leiognathi*.

65. DeLuca, M. and W. D. McElroy, eds. *Methods Enzymol.* 133:649, 1986. Bioluminescence and chemiluminescence.

66. Denton, E. J. and P. J. Herring. *J. Physiol.* (*London*) 284:42P, 1978. Blue photophore filters in mesopelagic animals.

67. Denton, E. J., P. J. Herring, E. A. Widder, M. F. Latz, and J. F. Case. *Proc. R. Soc. London, Ser. B* 225:63–97, 1985. Vision and photophore filters in oceanic animals.

68. De Wet, J. R., K. Wood, D. Helsinki, and M. DeLuca. *Proc. Natl. Acad. Sci. U.S.A.* 82:7870–7875, 1985. Cloning and expression of firefly luciferase gene.

69. Doyle, J. D. and R. H. Kay. *J. Mar. Biol. Assoc. U.K.* 47:555–563, 1967. Euphausiid luminescence.

70. Dunlap, J. and J. W. Hastings. *J. Biol. Chem.* 256:10509–10518, 1981. *Gonyaulax* luciferase.

71. Dunlap, J. and J. W. Hastings. *Biochemistry* 20:983–989, 1981. Biochemistry of dinoflagellate luciferin.

72. Dunlap, J., O. Shimomura, and J. W. Hastings. *Proc. Natl. Acad. Sci. U.S.A.* 77:1394–1397, 1980. Cross phyletic reactions between dinoflagellates and euphausiids.

73. Dunlap, J., J. W. Hastings, and O. Shimomura. *FEBS Lett.* 135:273–276, 1981. Dinoflagellate luciferin is related to chlorophyll.

74. Dunlap, K., K. Takeda, and P. H.

Brehm. *Nature (London)* 325:60–62, 1987. Calcium triggered luminescence via gap junctions in *Obelia* photocytes.

75. Dunlap, P. V. *Biol. Bull. (Woods Hole, Mass.)* 167:410–425, 1984. Physiology and morphology of luminescent bacterial symbionts in ponyfish.

76. Dunlap, P. V. *Arch. Microbiol.* 141:44–50, 1985. Symbiotic luminescent bacteria in ponyfish and osmotic effects.

77. Eckert, R. In *Bioluminescence in Progress* (F. H. Johnson and Y. Haneda, eds.), pp. 269–300. Princeton Univ. Press, Princeton, NJ, 1966. *Noctiluca* luminescence and excitation.

78. Eckert, R. and G. Reynolds. *J. Gen. Physiol.* 50:1429–1458, 1967. Subcellular origin of luminescence in *Noctiluca*.

78a. Eckstein, J. W., K.-W. Cho, P. Colepicolo, S. Ghisla, J. W. Hastings and T. Wilson. *Proc. Natl. Acad. Sci. U.S.A.* 84:4154–4158, 1990. Energy transfer cannot account for yellow emission in bacteria.

79. Engebrecht, J. and M. Silverman. *Proc. Natl. Acad. Sci. U.S.A.* 81:4154, 1984. Identity and functions of *lux* genes.

80. Engebrecht, J. and M. Silverman. *Methods Enzymol.* 133:83–98. 1986. Cloning techniques.

81. Engebrecht, J., K. H. Nealson, and M. Silverman. *Cell (Cambridge, Mass.)* 32:773–781, 1983. Analysis of bacterial *lux* gene.

82. Engebrecht, J., M. Simon, and M. Silverman. *Science* 227:1345–1347, 1985. Measuring gene expression with light.

83. Esaias, W. E. and H. C. Curl. *Limnol. Oceanogr.* 17:901–906, 1972. Effects of dinoflagellate luminescence on copepod predators.

84. Fogel, M. and J. W. Hastings. *Arch. Biochem. Biophys.* 142:310–321, 1971. Substrate binding protein in dinoflagellate luminescence.

85. Fogel, M. and J. W. Hastings. *Proc. Natl. Acad. Sci. U.S.A.* 69:690–693, 1972. Dinoflagellate scintillon control.

86. Frank, T. M., E. A. Widder, M. I. Latz, and J. F. Case. *J. Exp. Biol.* 109:385–389, 1984. Effect of diet on luminescence in a mysid.

87. Freeman, G. and E. B. Ridgeway. *Wilhelm Roux's Arch. Dev. Biol.* 196:30–50, 1987. Hydrozoan luminescence during development.

88. Galt, C. P. *Science* 200:70–72, 1978. Origin of larvacean luminescence.

89. Galt, C. P., M. S. Grober, and P. F. Sykes. *Biol. Bull. (Woods Hole, Mass.)* 168:125–134, 1985. Larvacean luminescence in taxonomy and structure.

90. Gast, H. and J. Lee. *Proc. Natl. Acad. Sci. U.S.A.* 75:833–837, 1978. Blue shift in bacterial light emission.

91. Ghiradella, H. *J. Morphol.* 153:187–204, 1977. Ultrastructure of firefly light organs.

92. Grober, M. S. *Anim. Behav.* 36:493–501, 1988. Anti-predatory aposematic luminescent signals in a brittle star.

93. Grober, M. S. *J. Exp. Mar. Biol. Ecol.* 45:157–168, 1988. Receiver responses to brittle-star luminescence.

94. Gupta, S. C., C. P. Reese, and J. W. Hastings. *Arch. Microbiol.* 143:325–329, 1986. Luciferase genes cloned from bacteria.

95. Gupta, S. C., D. O'Brien, and J. W. Hastings. *Biochem. Biophys. Res. Commun.* 127:1007–1011, 1985. Expression of the cloned subunits of bacterial luciferase.

96. Haas, E. *Biophys. J.* 31:301–312, 1980. Bioluminescence of single bacterial cells.

97. Haneda, Y. In *Bioluminescence in Progress* (F. H. Johnson and Y. Haneda, eds.), pp. 547–555. Princeton Univ. Press, Princeton, NJ, 1966. Filters and bacterial light organ in pine-conefish.

98. Haneda, Y., F. H. Johnson, and O. Shimomura. In *Bioluminescence in Progress* (F. H. Johnson and Y. Haneda, eds.), pp. 533–545. Princeton Univ. Press, Princeton, NJ, 1966. Origin of luminescence in pempherid and apogonid fishes.

99. Hansen, K. and P. J. Herring. *J. Zool.* 182:103–124, 1977. Two different luminescent systems in deep-sea angler fish.

100. Hanson, F. E., J. Miller, and G. T. Reynolds. *Biol. Bull. (Woods Hole, Mass.)* 137:447–464, 1969. Firefly light organ morphology.

101. Harvey, B. J. *Can. J. Zool.* 55:884–889, 1977. Structure of and effects of circulation on euphausiid photophores.

102. Harvey, E. N. *Biol. Bull. (Woods Hole, Mass.)* 51:89–97, 1926. Oxygen and luminescence.

103. Harvey, E. N. *Bioluminescence.* Academic Press, New York, 1952.

104. Hastings, J. W. *Science* 173:1016–1017, 1971. Ventral luminescence to camouflage the silhouette.

105. Hastings, J. W. *Energy Transform. Biol. Sys., Ciba Found. Symp.* 31 (New Ser.):125–146, 1975. Chemical to light energy conversion.

106. Hastings, J. W. In *Oxygenases and Oxygen Metabolism* (M. Nozaki et al., eds.), pp. 225–237. Academic Press, New York, 1982. O_2 and emitting species in bioluminescence.

107. Hastings, J. W. *J. Mol. Evol.* 19:309–321, 1983. Diversity, chemistry and evolution of bioluminescence.

108. Hastings, J. W. *Bull. Mar. Sci.* 33:818–828, 1983. Chemistry and control of marine luminescence.

109. Hastings, J. W. In *Light Emission by Plants and Bacteria* (Govindjee et al., eds.), pp. 363–398. Academic Press, Orlando, Florida, 1986. Bioluminescence in bacteria and dinoflagellates.

110. Hastings, J. W. and J. B. Buck. *Biol. Bull. (Woods Hole, Mass.)* 111:101–113, 1956. Oxygen and the firefly pseudoflash.

111. Hastings, J. W. and Q. H. Gibson. *J. Biol. Chem.* 238:2537–2554, 1963. Intermediates in the bioluminescent oxidation of $FMNH_2$.

112. Hastings, J. W. and K. H. Nealson. In *The Procaryotes* (M. P. Starr et al., eds.), pp. 1332–1345. Springer-Verlag, Berlin, 1981. The symbiotic luminous bacteria.

113. Hastings, J. W. and B. M. Sweeney. *J. Cell. Comp. Physiol.* 49:209–225, 1957. Dinoflagellate luminescent reactions.

114. Hastings, J. W. and G. Weber. *J. Opt. Soc. Am.* 53:1410–1415, 1963. A liquid photon standard.

115. Hastings, J. W. and T. Wilson. *Photochem. Photobiol.* 23:461–473, 1976. Bioluminescence and chemiluminescence.

116. Hastings, J. W., C. Balny, C. Le Peuch, and P. Douzou. *Proc. Natl. Acad. Sci. U.S.A.* 70:3468–3472, 1973. An oxygenated luciferase-flavin intermediate.

117. Hastings, J. W., A. Eberhard, T. O. Baldwin, M. Z. Nicoli, T. W. Cline, and K. H. Nealson. In *Chemiluminescence and Bioluminescence* (M. J. Cormier et al., eds.), pp. 369–380. Plenum, New York, 1973. Bacterial bioluminescence.

118. Hastings, J. W., C. J. Potrikus, S. Gupta, M. Kurfurst, and J. C. Makemson. *Adv. Microb. Physiol.* 26:235–291, 1985. Biochemistry and physiology of luminescent bacteria.

119. Hastings, J. W., J.-M. Bassot, and M.-T. Nicolas. *Ann. N.Y. Acad. Sci.* 503:180–186, 1987. Photosomes and scintillons.

120. Hastings, J. W., J. C. Makemson, and P. V. Dunlap. *Symbiosis* 4:3–24, 1987. Regulation of growth and luminescence in exosymbionts.

121. Herrera, A. A. *J. Comp. Physiol.* 129:67–78, 1979. Luminescence and neurobiology in scaleworms.

122. Herring, P. J. *J. Zool.* 172:401–418, 1974. Echinoderm bioluminescence.

123. Herring, P. J. *Eur. Mar. Biol. Symp. [Proc.], 9th*, pp. 563–572, 1975. Bacterial light organ in midwater argentinoid fishes.

124. Herring, P. J. *J. Mar. Biol. Assoc. U.K.* 56:1029–1047, 1976. Decapod crustacean luminescence.

125. Herring, P. J. *Symp. Zool. Soc. London* 38:127–159, 1977. Luminescence in cephalopods and fishes.

126. Herring, P. J., ed. *Bioluminescence in Action.* Academic Press, New York, 1978.

127. Herring, P. J. In *Bioluminescence in Action* (P. J. Herring, ed.), pp. 199–240. Academic Press, New York, 1978. Invertebrate bioluminescence.

128. Herring, P. J. *Mar. Biol.* 53:213–216, 1979. Radiolarian bioluminescence.

129. Herring, P. J. *Oceanogr. Mar. Biol.* 20:415–470, 1982. Luminescence in fishes.

130. Herring, P. J. *Proc. R. Soc. London, Ser. B.* 220:183–217, 1983. Luminescent spectra.

131. Herring, P. J. In *Physiological Adaptations in Marine Animals* (M. S. Laverack, ed.), pp. 323–350. Society of Experimental Biology, Cambridge, England, 1985. Review of deep-sea luminescence.

132. Herring, P. J. *J. Crustacean Biol.* 5:557–573, 1985. Crustacean luminescence.

132a. Herring, P. J. *J. Biolum. Chemilumin.* 1:147–163, 1987. Bioluminescent taxa.

133. Herring, P. J. In *The Mollusca* (E. R. Trueman, ed.), Vol. 11, pp. 449–489. Academic Press, San Diego, CA, 1988. Mollusc luminescence.

134. Herring, P. J. and N. A. Locket. *J. Zool.* 186:431–462, 1978. Euphausiid photophores.

135. Herring, P. J. and J. G. Morin. In *Bioluminescence in Action* (P. J. Herring, ed.), pp. 273–329. Academic Press, New York, 1978. Bioluminescence in fishes.

136. Holzman, T. F. and T. O. Baldwin. *Proc. Natl. Acad. Sci. U.S.A.* 77:6363–6367, 1980. Conserved structure in bacterial luciferase.

136a. Hopkins, T. A., H. H. Seliger, E. H. White, and M. W. Cass. *J. Am. Chem. Soc.* 89:7148–7150, 1967. Firefly reaction mechanism.

137. Johnson, C. H., S. Inoué, A. Flint, and J. W. Hastings. *J. Cell Biol.* 100:1435–1446, 1985. Luminescent particles in dinoflagellates.

138. Johnson, T. C., R. Thompson, and T. Baldwin. *J. Biol. Chem.* 261:4805–4811, 1986. Nucleotide sequence of *lux* B gene.

138a. Kay, R. H. *Proc. R. Soc. London, Ser. B.* 162:365–386, 1965. Euphausiid luminescence.

139. Krasnow, R., J. Dunlap, W. Taylor, J. W. Hastings, W. Vetterling, and V. D. Gooch. *J. Comp. Physiol.* 138:19–26, 1980. Circadian bioluminescence of *Gonyaulax polyedra*.

140. Kricka, L. J., and G. Thorpe. *Methods Enzymol.* 133:404–420, 1986. Photographic detection of luminescence.

141. Krieger, N., D. Njus, and J. W. Hastings. *Biochemistry* 13:2871–2877, 1974. Dinoflagellate luciferase activity.

142. Kurfürst, M., S. Ghisla, and J. W. Hastings. *Proc. Natl. Acad. Sci. U.S.A.* 81:2990–2994, 1984. The primary emitter in the bacterial luciferase reaction.

143. Lall, A. B., H. H. Seliger, W. H. Biggley, and J. E. Lloyd. *Science* 210:560–562, 1980. Ecology of colors in firefly luminescence.

144. Lane, E. D. *Contrib. Mar. Sci.* 12:1–53, 1967. Biology of and bioluminescence in *Porichthys*.

145. Latz, M. I. and J. F. Case. *Am. Zool.* 20:851, 1980. Photophore structure and function of midwater shrimp.

146. Latz, M. I. and J. F. Case. *J. Exp. Biol.* 98:83–104, 1982. Luminescence and body orientation in a midwater shrimp.

147. Latz, M. I., T. M. Frank, M. R. Bowlby, E. A. Widder, and J. F. Case. *Biol. Bull.* (*Woods Hole, Mass.*) 173:489–503, 1987. Kinetics of copepod luminescence.

148. Lawry, J. V. *J. Anat.* 114:55–63, 1973. Structure of lanternfish photophores.

149. Lee, J., A. Wesley, J. Ferguson, and H. Seliger. In *Bioluminescence in Progress* (F. Johnson and Y. Haneda, eds.), pp. 35–43. Princeton Univ. Press, Princeton, NJ, 1966. Luminol as an emission standard.

150. Leisman, G. and K. H. Nealson. In *Flavins and Flavoproteins* (V. Massey and C. Williams, eds.), pp. 383–386. Elsevier, Amsterdam, 1982. Yellow fluorescent proteins in luminescent bacteria.

151. Leisman, G., D. Cohn, and K. H. Nealson. *Science* 208:1271–1273, 1980. Bacterial luminescence in marine animals.

152. Lloyd, D., K. James, J. W. Williams, and N. Williams. *Anal. Biochem.* 116:17–21, 1981. An O_2 probe using luminous bacteria.

153. Lloyd, J. E. *Science* 187:452–453, 1975. Aggressive mimicry in fireflies.

154. Lloyd, J. E. In *How Animals Communicate*

(T. A. Sebeok, ed.), pp. 164–183. Indiana Univ. Press, Bloomington, 1977. Bioluminescence and communication.

155. Lloyd, J. E. In *Bioluminescence in Action* (P. J. Herring, ed.), pp. 241–272. Academic Press, New York, 1978. Insect bioluminescence.

156. Lloyd, J. E. *Science* 210:669–671, 1980. Firefly signal mimicry.

157. Mackie, G. O. and Q. Bone. *Proc. R. Soc. London, Ser. B.* 202:483–495, 1978. Pyrosome luminescence.

158. Margulis, L. *Symbiosis and Cell Evolution.* Freeman, San Francisco, CA, 1981.

159. Martin, J. P. and I. Fridovich. *J. Biol. Chem.* 256:6080–6089, 1981. Gene transfer from the ponyfish to its bioluminescent bacterial symbiont.

160. Martin, N. and M. Anctil. *Biol. Bull. (Woods Hole, Mass.)* 166:583–593, 1984. Control of *Chaetopterus* luminescence.

161. McCapra, F. and R. Hart. *Nature (London)* 286:660–661, 1980. The origins of marine bioluminescence.

162. McCapra, F. and K. Perring. In *Chemiluminescence and Bioluminescence* (J. Burr, ed.), pp. 259–320. Dekker, New York, 1985. Mechanism of chemiluminescence.

163. McElroy, W. D. and M. DeLuca, eds. *Chemiluminescence and Bioluminescence.* Academic Press, New York, 1981.

164. McElroy, W. D. and M. DeLuca. In *Comprehensive Insect Physiology* (G. Kerkut and L. Gilbert, eds.), Vol. 4, pp. 553–565. Pergamon, Oxford, 1985. Biochemistry of firefly luminescence.

165. McElroy, W. D. and J. W. Hastings. In *Physiological Triggers* (C. L. Prosser, ed.), pp. 80–84. Ronald Press, New York, 1956. Control of firefly flashing.

166. McElroy, W. D. and H. H. Seliger. In *Bioluminescence in Progress* (F. H. Johnson and Y. Haneda, eds.), pp. 427–458. Princeton Univ. Press, Princeton, NJ, 1966. Firefly luminescence.

167. McElroy, W. D., H. H. Seliger, and M. DeLuca. In *The Physiology of Insecta* (M. Rockstein, ed.), Vol. 2, pp. 411–460. Academic Press, New York, 1974. Insect bioluminescence.

168. McFall-Ngai, M. *J. Exp. Zool.* 227:23–33, 1983. Swim bladder of ponyfish as a luminescence reflector.

169. McFall-Ngai, M. and P. Dunlap. *Mar. Biol. (Berlin)* 73:227–237, 1983. Functions of luminescence in a ponyfish.

170. Meighen, E. A. *Annu. Rev. Microbiol.* 42:151–176, 1988. *Lux* operons of bioluminescent bacteria.

171. Miron, M. J., L. La Rivière, J.-M. Bassot, and M. Anctil. *Cell Tissue* 249:547–550, 1987. Monoamine-containing cells in scaleworm luminescent elytra.

172. Morin, J. G. In *Coelenterate Biology: Reviews and New Perspectives* (L. Muscatine and H. M. Lenhoff, eds.), pp. 397–438. Academic Press, New York, 1974. Coelenterate bioluminescence.

173. Morin, J. G. *Bull. Mar. Sci.* 33:787–817, 1983. Review of coastal bioluminescence.

174. Morin, J. G. *Fla. Entomol.* 69:105–121, 1986. Functions of luminescence in ostracodes.

175. Morin, J. G. and I. M. Cooke. *J. Exp. Biol.* 54:707–721, 1971. Hydropolyp luminescence and neurobiology.

176. Morin, J. G. and J. W. Hastings. *J. Cell. Physiol.* 77:305–311, 1971. Biochemistry of coelenterate bioluminescence.

177. Morin, J. G. and J. W. Hastings. *J. Cell. Physiol.* 77:313–318, 1971. Energy transfer in bioluminescence.

178. Morin, J. G., A. Harrington, K. Nealson, N. Krieger, T. O. Baldwin, and J. W. Hastings. *Science* 190:74–76, 1975. Functions of luminescence in flashlight fishes.

179. Morse, D., P. M. Milos, E. Roux, and J. W. Hastings. *Proc. Natl. Acad. Sci. U.S.A.* 86:172–176, 1989. Regulation of a substrate binding protein.

179a. Morse, D., A. M. Pappenheimer, and J. W. Hastings. *J. Biol. Chem.* 264:11822–11826, 1989. A luciferin binding protein in *Gonyaulax*.

180. Musicki, B., Y. Kishi, and O. Shimo-

mura. *J. Chem. Soc. Chem. Comm.* 1566–1568, 1986. Structure of aequorin functional region.

181. Nakamura, H., B. Musicki, Y. Kishi, and O. Shimomura. *J. Am. Chem. Soc.* 110:2683–2685, 1988. Structure of the light emitter in krill.

181a. Nakamura, H., Y. Kishi, O. Shimomura, D. Morse and J. W. Hastings. *J. Am. Chem. Soc.* 111:7607–7611, 1989. Structure of dinoflagellate luciferin.

182. Nealson, K. H. *Trends in Biochem. Sci.* 4:105–110, 1979. Luminous bacterial symbionts in fishes.

183. Nealson, K. H. and J. W. Hastings. *Microbiol. Rev.* 43:496–518, 1979. Ecology and control of bacterial bioluminescence.

184. Nicol, J. A. C. In *Bioluminescence in Action* (P. J. Herring, ed.), pp. 367–398. Academic Press, New York, 1978. Luminescence and vision.

185. Nicolas, M.-T., B. M. Sweeney, and J. W. Hastings. *J. Cell Sci.* 87:189–196, 1987. Luminescent organelles in three dinoflagellates.

186. Nicolas, M.-T., C. H. Johnson, J.-M. Bassot, and J. W. Hastings. *J. Cell Biol.* 105:723–735, 1987. Characterization of bioluminescent organelles in *Gonyaulax*.

186a. Noguchi, M., E. Tsuji, and Y. Sakaki. *Methods Enzymol.* 133:298–306, 1986. Cloning of apoaequorin cDNA.

187. O'Day, W. T. and H. R. Fernandez. *Vision Res.* 14:545–550, 1974. Red light and vision in a deepsea fish.

188. Oertel, D. and J. F. Case. *J. Exp. Biol.* 65:213–227, 1976. Neurobiology of larval firefly photocytes.

189. Parker, S. P., ed. *Synopsis and Classification of Living Organisms.* McGraw-Hill, New York, 1982.

190. Pavans de Ceccatty, M., J.-M. Bassot, J.-M. Bilbaut, and M.-T. Nicolas. *Biol. Cell.* 28:57–64, 1977. Structure of scaleworm photocytes.

190a. Pazzagli, X., E. Cadenas, L. J. Kricka, A. Roda and P. E. Stanley (eds.). *Bioluminescence and Chemiluminescence*, 1989, Wiley, New York.

191. Poinar, G. O., G. Thomas, M. Haygood, and K. H. Nealson. *Soil Biol. Biochem.* 12:5–10, 1980. Luminous bacteria symbiotic with a soil nematode.

192. Prasher, D. C., R. McCann, M. Longiary, and M. J. Cormier. *Biochemistry* 26:1326–1332, 1987. Sequence comparisons of aequorin cDNAs.

193. Raymond, J. A. and A. L. De Vries. *Limnol. Oceanogr.* 21:599–602, 1976. Luminescent fish fecal pellets.

194. Ruby, E. G. and J. W. Hastings. *Biochemistry* 19:4989–4993, 1980. Hybrid bacterial luciferases.

195. Ruby, E. G. and K. H. Nealson. *Science* 196:432–434, 1977. A yellow emitting bacterium.

196. Ruby, E. G. and J. G. Morin. *Appl. Environ. Microbiol.* 38:406–411, 1979. Luminous enteric bacteria in fishes.

197. Salvini-Plawen, L. V. and E. Mayr. *Evol. Biol.* 10:207–263, 1977. Evolution of vision.

198. Satterlie, R. A., P. A. V. Anderson, and J. F. Case. *Mar. Behav. Physiol.* 7:25–46, 1980. Anthozoan bioluminescence and coordination.

199. Schaap, A. P., T. Chen, R. Handley, R. DeSilva, and B. Giri. *Tetrahedron Lett.* 28:1155–1158, 1987. Synthesis of chemiluminescent compounds.

200. Schauer, A., M. Ranes, R. Santamaria, J. Guijarro, E. Lawlor, C. Mendez, K. Chater, and R. Losick. *Science* 240:768–772, 1988. Gene expression in the filamentous bacterium *Streptomyces coelicolor*.

201. Schölmerich, J., R. Andreesen, A. Kapp, M. Ernst, and W. G. Woods, eds. *Bioluminescence and Chemiluminescence.* Wiley, London, 1987.

202. Schuster, G. B. *Act. Chem. Res.* 12:366–373, 1979. Electron exchange chemiluminescence.

203. Seliger, H. H. *Photochem. Photobiol.* 21:355–361, 1975. Evolution of luminescence.

204. Seliger, H. H. *Photochem. Photobiol.* 45:291–297, 1987. Evolution of bacterial bioluminescence.

205. Seliger, H. H. and W. D. McElroy. *Arch. Biochem. Biophys.* 88:136–141, 1960. Quantum yield and spectrum of firefly luminescence.

206. Seliger, H. H. and W. D. McElroy. *Proc. Natl. Acad. Sci. U.S.A.* 52:75–81, 1964. Colors of firefly luminescence; luciferase effects.

207. Seliger, H. H. and W. D. McElroy. *Light: Physical and Biological Action.* Academic Press, New York, 1965.

208. Seliger, H. H., J. Carpenter, M. Loftus, and W. D. McElroy. *Limnol. Oceanogr.* 15:234–245, 1970. High concentrations of dinoflagellates.

209. Seliger, H. H., S. Lace, J. Lloyd, and W. Biggley. *Photochem. Photobiol.* 36:673–680, 1982. Firefly color vision and behavior.

210. Shimomura, O. *Photochem. Photobiol.* 44:671–674, 1986. Biochemistry of ophiuroid luminescence.

211. Shimomura, O. and F. H. Johnson. *Nature (London)* 256:236–238, 1975. Regeneration of aequorin.

212. Shimomura, O., F. H. Johnson, and Y. Saiga. *J. Cell. Comp. Physiol.* 59:223–240, 1962. Isolation of aequorin.

213. Shimomura, O., F. H. Johnson, and Y. Kohama. *Proc. Natl. Acad. Sci. U.S.A.* 69:2086–2089, 1972. Luminescent reactions in limpet and bacteria.

214. Shimomura, O., T. Goto, and F. H. Johnson. *Proc. Natl. Acad. Sci. U.S.A.* 74:2799–2802, 1977. Oxidation of firefly luciferin.

215. Shimomura, O., S. Inoué, F. H. Johnson, and Y. Haneda. *Comp. Biochem. Physiol. B* 65B:435–437, 1980. Widespread occurrence of coelenterazine.

216. Shimomura, O., B. Musicki, and Y. Kishi. *J. Biochem. (Tokyo)* 251:405–410, 1988. Semi-synthetic aequorin.

217. Stanley, P. *Methods Enzymol.* 133:587–603, 1986. Commercial photometers.

218. Stone, H. *Biochem. Biophys. Res. Commun.* 31:386–391, 1968. Biochemistry of ostracode luminescence.

219. Thompson, E. M., B. G. Nafpaktitis, and F. I. Tsuji. *Photochem. Photobiol.* 45:529–533, 1987. Ostracode luciferin induction and turnover in *Porichthys*.

220. Tsuji, F. I. In *The Mollusca* (P. Hochachka, ed.), Vol. 2, pp. 257–279. Academic Press, New York, 1983. Molluscan bioluminescence.

221. Tsuji, F. I. and E. Hill. *Biol. Bull. (Woods Hole, Mass.)* 165:444–449, 1983. Bioluminescence and spawning patterns in *Odontosyllis*.

222. Tsuji, F. I., A. T. Barnes, and J. F. Case. *Nature (London)* 237:515–516, 1972. Ostracode luciferin induction of luminescence in *Porichthys*.

223. Wampler, J. E. In *Bioluminescence in Action* (P. J. Herring, ed.), pp. 1–48. Academic Press, New York, 1978. Low level light measurements.

224. Wampler, J. E. In *Bioluminescence and Chemiluminescence* (M. A. DeLuca and W. D. McElroy, eds.), pp. 249–256. Academic Press, New York, 1981. Earthworm luminescence.

225. Wampler, J. E. and B. G. M. Jamieson. *Comp. Biochem. Physiol. B* 66B:343–350, 1980. Chemistry of earthworm bioluminescence.

226. Wannlund, J. and M. DeLuca. *Methods Enzymol.* 92:426–433, 1983. Luminescence immunoassays.

227. Ward, W. W. and M. J. Cormier. *J. Phys. Chem.* 80:2289–2291, 1976. Energy transfer in *Renilla* luminescence.

228. Warner, J. A. and J. F. Case. *Biol. Bull. (Woods Hole, Mass.)* 159:231–246, 1980. Distribution and induction of luminescence in *Porichthys*.

229. Warner, J. A., M. Latz, and J. F. Case. *Science* 203:1109–1110, 1979. Midwater shrimp counterillumination.

230. Watanabe, H., J. W. Hastings, K. Wulff, G. Michal, and F. Staehler. *J. Appl. Biochem.* 4:508–523, 1982. Determinations of enzymes by means of bacterial luciferase.

231. Widder, E. A., M. I. Latz, and J. F. Case. *Biol. Bull. (Woods Hole, Mass.)* 165:791–810, 1983. Luminescent spectra in marine animals.

232. Widder, E. A. and J. F. Case. *J. Comp. Physiol.* 143:43–52, 1981. Dinoflagellate flash kinetics.

233. Widder, E. A., M. I. Latz, P. J. Herring, and J. F. Case. *Science* 225:512–514, 1984. Red bioluminescence in 2 midwater fishes.

234. Widder, E. A., M. I. Latz, and P. J. Herring. *Photochem. Photobiol.* 44:97–101, 1986. Spectral shifts in fish luminescence.

235. Wilkens, L. A. and J. J. Wolken. *Mar. Behav. Physiol.* 8:55–66, 1981. Vision and luminescence in syllid worms.

236. Wilson, A. C., S. S. Carlson, and T. J. White. *Annu. Rev. Biochem.* 46:573–639, 1977. Biochemical evolution.

237. Wilson, T. In *Singlet Oxygen* (A. Frimer, ed.), Vol. 2, pp. 37–57. CRC Press, Boca Raton, FL, 1985. Mechanism of chemiluminescence.

237a. Wood, K. V., Y. A. Lam, H. H. Seliger and W. D. McElroy. *Science* 242:700–702, 1989. Different cDNAs elicit bioluminescence of different colors.

238. Young, R. E. *Symp. Zool. Soc. London* 38:161–190, 1977. Cephalopod counterillumination.

239. Young, R. E. *Bull. Mar. Sci.* 33:829–845, 1983. Review of oceanic luminescence.

240. Young, R. E. and J. M. Arnold. *Malacologia* 23:135–163, 1985. Squid photophore morphology.

241. Young, R. E. and T. M. Bennett. In *The Mollusca* (M. R. Clarke and E. R. Trueman, eds.), Vol. 12, pp. 241–251. Academic Press, San Diego, CA, 1988. Cephalopod luminescence.

242. Young, R. E. and F. Mencher. *Science* 208:1286–1288, 1980. Color change and counterillumination in midwater squid.

243. Young, R. E., E. M. Kampua, S. D. Maynard, F. M. Mencher, and C. F. E. Roper. *Deep-Sea Res., Part A* 27:671–691, 1980. Midwater counterillumination and maximum luminescence.

244. Young, R. E., R. R. Seapy, K. M. Mangold, and F. G. Hochberg. *Mar. Biol.* 69:299–308, 1982. Luminescent flashing in midwater squid.

245. Ziegler, M. and T. O. Baldwin. *Curr. Top. Bioenerg.* 12:65–113, 1981. Review of bacterial luciferase.

4 | *Photoreception and Vision*

Timothy H. Goldsmith

Most people would probably consider vision the sensory modality they value most highly, and it is probably the most important sense of higher primates. It is also the sense that has been studied most intensively. One reason for the attention given to vision is practical. An important requirement for experimental work on the senses is control of the stimuli in both space and time. Such control is easily achieved with light, particularly compared with chemical stimuli. Second, retinas are readily approachable; not only is it easy to isolate vertebrate photoreceptor organelles for biochemical study, but the neural retina is a small piece of the central nervous system doing sophisticated analyses at a site that is relatively accessible for neurophysiological recording. The inner ear, by contrast, is locked away in a case of bone.

Light and the Visible Spectrum

Vision involves a very restricted range of wavelengths in the electromagnetic spectrum (Fig. 1). We refer to this energy band as "visible light," reflecting the capacity of our own eyes and not a fundamental property of the physical world. The visible spectrum nevertheless occupies a sizeable portion of the wavelength band responsible for all photobiological responses (Fig. 2). There are two complementary reasons why photobiology exploits wavelengths from 300 to ~900 nm (172). First, this is the light that is available at the surface of the earth (Fig. 1). The terrestrial solar spectrum is attenuated at short wavelengths by the ozone in the upper atmosphere (142), and in acquatic environments the available light is narrowed even further by absorption in the near infrared (IR) and by additional absorption and scattering of short wavelengths (73,162). As a result, the light that penetrates clear water becomes blue as it is increasingly confined to a wavelength band centered at ~480 nm. As we shall see, these features of the environment have influenced the evolution of eyes.

There is a second and even more fun-

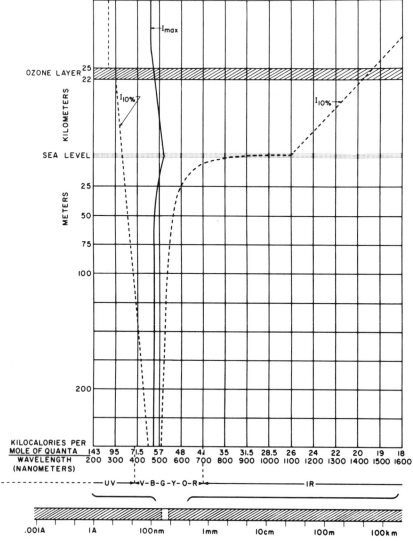

Figure 1. The solar spectrum is narrowed in the biosphere by absorption. Ozone in the upper atmosphere is principally responsible for removing the mid-ultraviolet, and water attenuates the longer wavelengths. The solid line (I_{max}) locates the wavelengths of maximum intensity; the broken lines ($I_{10\%}$) trace the wavelength boundaries within which 90% of the solar energy is concentrated at each level in the atmosphere and ocean. The letters below the wavelength scale (UV, V, B, etc.) represent ultraviolet, violet, blue, and so forth. The small portion of the electromagnetic spectrum described in this graph is illustrated on the bottom scale. [After Wald, (172). From *Life and Light*, by G. Wald. Copyright © Oct., 1959 by Scientific American, Inc. All rights reserved.]

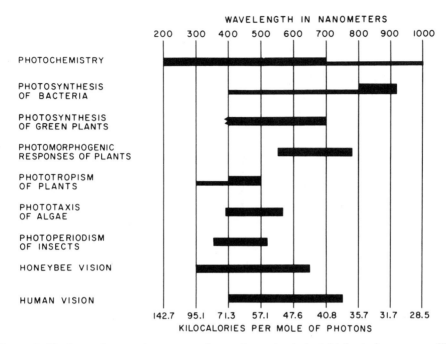

Figure 2. Regions of spectral response of some important photobiological processes. [For data or references to data see (172), (24), (138), (178), (17), (130), (60), (21), (47), (49), (104), and (170).]

damental reason why photobiological responses involve this restricted region of the electromagnetic spectrum. When a photon (or quantum) of light interacts with a molecule of matter and gives up its energy to the molecule, the light disappears and we say the photon has been *absorbed*. The energy of the photon is $E = h\nu = hc/\lambda$, where h is Planck's constant $(6.626 \times 10^{-27}$ erg $\cdot$ s or 1.584×10^{-34} cal $\cdot$ s), ν is the frequency, c is the velocity of light $(3 \times 10^{10}$ cm/s), and λ the wavelength (cm). For light of long wavelength, corresponding to the IR region of the spectrum, the energy of the absorbed photon is relatively small, and on absorption it increases vibrations and rotations between atomic groups within the molecule. In the visible and ultraviolet (UV) regions of the spectrum, however, the photons possess enough energy to change electronic energy levels and therefore have the potential for rearranging

chemical bonds. This is therefore the region of photochemistry. From a biological perspective, wavelengths longer than ~900 nm are too impoverished (<30 kcal/ mol) to create the necessary electronic excited states, whereas wavelengths in the UV are so energetically rich they can damage the molecules that absorb them. Absorption of UV light by nucleic acids is therefore potentially mutagenic, and terrestrial organisms live a fine line between exploitation of habitat and the avoidance of photodamage. (This is why dermatologists warn that excessive exposure to sunlight can lead to premature aging of the skin or the generation of epitheliomas.) The ozone screen in the upper atmosphere is itself made photochemically from oxygen, and the argument has been made that in evolution the successful colonization of the land from aquatic environments required the prior appearance of plants capable of generating oxygen in photosynthesis (173).

Vision Is One of Many Photobiological Responses

The *energy* of light may be harvested, as in photosynthesis, or the *information content* of light may be exploited, as in most of the other photobiological responses listed in Figure 2, of which vision is the most familiar example (50b). The photoperiodic responses of plants, mediated by the pigment *phytochrome*, synchronize aspects of the reproductive cycle with the most favorable time of the year (152). In animals photoperiodic responses mediated by a different pigment system regulate the time of year of reproduction or, in the case of insects, the time of year or even the time of the day of various aspects of development (9).

The terms *phototaxis* and *phototropism* refer to responses of organisms that are directed by light and serve to bring the organism into a more favorable relationship with its environment for immediate purposes such as nutrition. Phototropism is the bending of sessile organisms such as parts of plants (17, 130) or coelenterate polyps toward or away from light. Phototaxis, on the other hand, refers to the oriented movement of unrestrained organisms.

The term phototaxis has lost much of its general meaning, particularly among students of microorganisms, as more sophisticated observations have shown a variety of kinds of responses (30). *Klinokinesis* (a change in the frequency of movement of an individual organism randomly alters its *direction*), *orthokinesis* (a change in its linear *velocity*), and *phobic responses* [temporal changes in intensity cause the organism to (generally) stop moving, then to resume movement in another direction] are not oriented with re-

spect to the stimulus but can lead to *photoaccumulation* or *photodispersal* of populations of microorganisms. For example, diminishing light intensities stimulate photosynthetic bacteria to reverse their direction of swimming. The result is that they congregate in regions of high intensity. The receptor is the photosynthetic apparatus, and the adaptive importance is obvious (66).

The positive phototaxis of *Euglena*, on the other hand, is different in that swimming is *directed* toward a source of light. It involves a photoreceptor located near the base of the flagellum but which is not the colored *stigma* or "eyespot." As the animal swims, it spirals about its long axis, and if its trajectory is at an angle to the source of light, the photoreceptor is alternately shaded and exposed by the stigma (or in some motile algae, perhaps the chloroplast). Presumably by altering its direction of swimming with respect to the source of light the organism can minimize the temporal changes in illumination of the photoreceptor. Analyzed in this manner, phototaxis of *Euglena* can be accounted for as a succession of phobic responses that decrease in frequency as the animal brings its direction of movement toward the light (24, 30).

Sensitivity to light has evolved repeatedly in different groups of animals. Many photoresponses do not involve a specialized, multicellular receptor organ (91, 110, 179). For example, locomotion of *Amoeba* (97), movement of the spines of csea urchins *(Diadema)* (111), the firing of photoreceptor neurons in the sixth abdominal ganglion of the crayfish *(Procambarus)* (75), and the effect of day length on breaking the diapause of silkworm pupae (177) are instances in which the identity of the photoreceptor is not obvious by morphological criteria, although in the latter two cases it clearly resides in the central nervous system. In earthworms

(Lumbricus) light-sensitive cells are distributed over the body (135).

Photoresponses Are Mediated by Pigments

We saw at the beginning of this chapter that the wavelengths of light that are photobiologically effective are wavelengths that are capable of photochemistry. Every photobiological response therefore starts with the absorption of light by a molecule of pigment, and a consequent transformation of the pigment leads to other chemical changes in the cell. The properties of these initial reactions are therefore the key to understanding the transduction process. In many photobiological systems only a small amount of pigment is present, but there is a physiological trick that can provide information about its *absorption spectrum*. The absorption spectrum of a pigment is a molecular signature and can thus be an important clue to its chemical identity. The physiological stratagem is to measure an *action spectrum:* The relative number of photons required at each wavelength to produce some constant physiological response. Because the absorption spectrum of the receptor pigment determines the wavelength sensitivity of the photobiological process, under ideal conditions the action spectrum is an indirect measure of the absorption spectrum of the pigment. But what is meant by the caveat "under ideal conditions"? Frequently the photopigment is partially screened from the ambient light by other colored structures; where this occurs some wavelengths will have reduced effectiveness in triggering the photoresponse, and the action spectrum will not reflect the absorption spectrum of the receptor molecule. Partly for this reason, but also because action spectra usually cannot be measured with sufficiently high resolution, action spectra

alone are rarely sufficient to establish the chemical identity of the receptor pigment. Action spectra nevertheless provided the data on which Figure 2 is based, and examples of action spectra are shown in Figure 27. In studies of vision, action spectra are more frequently called *spectral sensitivity curves.*

What are the receptor pigments? *Chlorophyll* is of course the receptor pigment that harvests photons in photosynthesis (138), and the purple bacterium *(Halobacterium)* uses a membrane protein called *bacterial rhodopsin* to pump protons and exploits the resulting proton gradient as a source of energy (158). Among the pigments used in signaling, the heme protein *phytochrome* is responsible for the photomorphogenic responses of plants (152). The photopigment that mediates visual responses is called *rhodopsin,* and it will be discussed in more detail below. The identities of the receptor pigments for photoperiodic responses such as diapause of insects, the circadian rhythm of eclosion of *Drosophilia,* the photoresponses of most lower organisms, the phototropic responses of fungi and higher plants (the pigment is referred to as *cryptochrome*), as well as a variety of other responses of lower invertebrates are either unknown or uncertain. Both flavoproteins and carotenoids have been implicated as the chromophore of cryptochrome (151, 152), and both flavins and rhodopsins have been suggested as mediators of photoresponses in the chlorophyll-containing flagellates (30, 46).

Most of our knowledge of photobiological processes in animals has been obtained from studies of the visual systems of animals with large eyes. The nature of the receptor pigments, the mechanisms of excitation of the receptor cells, and the initial means by which information is processed in the production of a behavioral response are the aspects of the visual system discussed in the remaining sections of this chapter.

Photoreceptor Cells Are Diverse

Although some cells (e.g., *Amoeba*) that are responsive to light lack any apparent morphological specialization, it is more usual to possess a photoreceptor organelle formed by elaboration of the plasma membrane. These structures, which provide an increased surface area for the deposition of photopigment, fall into two general classes, depending on whether or not they form in association with the basal body of a cilium (Fig. 3). The distinction drew its original interest from the observation that the photoreceptors of the vertebrate eye are derived from cilia whereas the photoreceptors of the major invertebrate phyla are not. Early studies suggested that in general Deuterostomes (those invertebrate phyla related to the vertebrate ancestral stock) have ciliary photoreceptors, whereas Protostomes do not (34–37). As more species have been examined, so many exceptions to this generalization have appeared that the original hypothesis of two major evolutionary lines has lost its appeal. The extensive morphological diversity of photoreceptor cells and their modes of association has been interpreted as polyphyletic, with perhaps as many as 40–65 evolutionary lines (149). This position, however, emphasizes the independent evolution of photoreceptor organs at the expense of features of photoreceptor organelles that may be homologous (165). Opsins from three phyla have now been sequenced and appear to be homologous proteins (Fig. 8); it is an open question what other characters of photoreceptor cells may be monophyletic.

Ciliary Endings in Invertebrates

The paraflagellar body of the unicellular green flagellate *Euglena* is probably a photoreceptor in association with a flagellum that has not lost its motile function (36, 90). The paraflagellar body is not a specialized region of the cell membrane, but rather a swelling contained completely within the flagellum. The stigma is a cup of carotenoid, which lies outside the flagellum and is thought to be an inert shading device. The photopigment has not been identified, and the means by which excitation is coupled to oriented swimming movements is unknown.

More typically, the formation of photoreceptor organelles involves an increase in the surface area of the cell membrane, rather than development of an intracellular structure analogous to the paraflagellar body. The modification of cilia to form sensory endings is not limited to photoreceptors, for certain mechanoreceptors and olfactory endings also have ciliary origins. In general sensory cilia have lost the two central microtubules as well as the arms on the ring of nine doublets (36).

In the distal retina of the scallop *Pecten*, the photoreceptor cells bear a number of saclike projections, each consisting of a greatly flattened cilium (4). In coelenterates, echinoderms, and elsewhere (Fig. 3) the membrane of each cilium may be thrown into fingerlike folds, whereas in the photoreceptor cells of larval ascidians the ciliary membrane is invaginated to form a series of parallel disks oriented roughly parallel to the long axis of the cilium (36, 56). This latter pattern of organization is suggestive of the rods and cones of the vertebrate retina.

Rod and Cones of the Vertebrate Retina

The photoreceptor cells of the vertebrate retina are elaborate structures. An example of a *rod* cell is shown in Figure 4. The *outer segment* is derived from a cilium and consists of a stack of 500–1000 membranous disks enclosed within the plasma

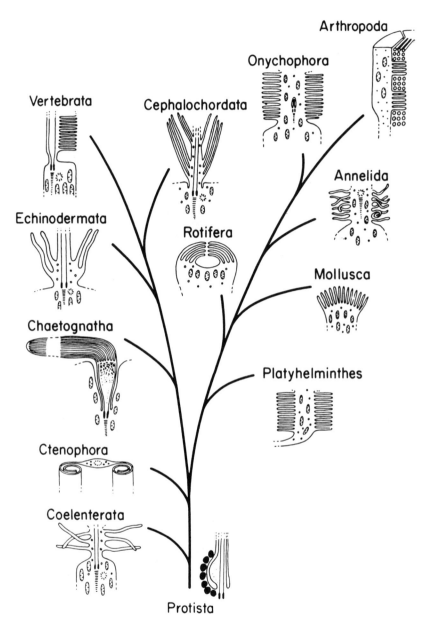

Figure 3. Schematic representation of photoreceptor organelles in selected examples of animals. [From Eakin (34).] The examples on the right branch are rhabdomeric, and those on the left are ciliary; however, there are multiple evolutionary origins and more diversity of photoreceptor structures than suggested by this illustration.

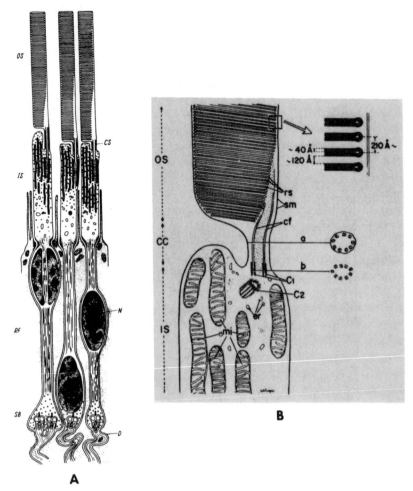

A

B

Figure 4. (*A*) Diagram of mammalian rod cells based on electron micrographs. [From (154a)]. (*B*) Enlarged diagram of the region near the junction of the inner and outer segments. [From (141)] OS, outer segment; IS, inner segment; CC, CS, connecting cilium; mi, mitochondria (in the ellipsoid); rs, rod sacs or disks; cf, ciliary filaments (microtubules); sm, surface membrane, c_1, c_2, centrioles (basal body); er, endoplasmic reticulum; N, nucleus; SB, synaptic body (rod pedicel); D, dendrite of bipolar neuron.

membrane. The outer segment is connected to the inner segment through a ciliary stalk, which shows the characteristic nine double microtubules when viewed in cross-section with an electron microscope. The disks arise as infoldings of the plasma membrane. In *cone* cells they retain continuity with the surface membrane, and their contents are therefore continuous with the extracellular space. In rods, on the other hand, most of the disks are completely pinched off from the plasma membrane (25, 143).

The inner segment and the remaining part of the cell are more conventional, containing the usual complement of organelles (25, 143). There is an aggregation of mitochondria immediately adjacent to the outer segment, forming a compact mass, which is recognized in the light mi-

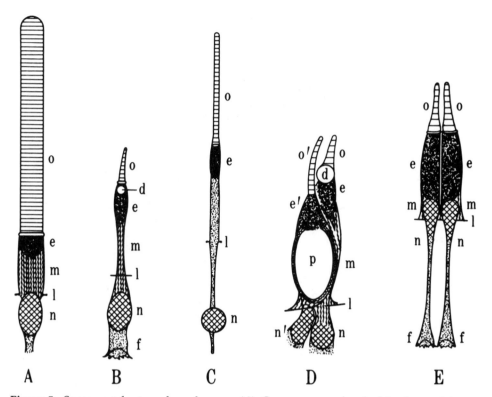

A   B   C   D   E

Figure 5. Some vertebrate rods and cones. (*A*) Common or red rod of the leopard frog, *Rana pipiens;* dark adapted, with the myoid contracted. (*B*) Cone of *R. pipiens;* dark adapted, with the myoid elongated. (*C*) Human rod from near the temporal side of the macula lutea. (*D*) Double cone of the painted turtle, *Chrysemys picta.* (*E*) Twin cone of a teleost fish, the bluegill, *Lepomis macrochirus;* light adapted, with the fused myoids contracted. d, oil droplet; e, ellipsoid (mitochondrial mass); e', ellipsoid of accessory member of pair; f, footpiece or pedicel; l, external limiting membrane of retina; m, myoid; n, nucleus; o, outer segment; o', outer segment of accessory cone; p, paraboloid (glycogen). [Redrawn from (174).]

croscope as the *ellipsoid.* The basal end of the rod cell is a presynaptic terminal with synaptic ribbons and presynaptic vesicles. Further description of the relation of the receptor cells to other neurons in the retina will be found in a following section.

Morphologically, rods are further distinguished from cones in having longer, more cylindrical outer segments, whereas cones have shorter, more nearly conical outer segments. These differences, however, are not always evident. For example, in the retinas of primates, the foveal (central) cones are slender and not at all

conical. Another difference that is present in fish, amphibians, reptiles, and birds, but not in placental mammals, is the presence in cones of an oil droplet situated at the distal end of the inner segment, just in front of the outer segment (Fig. 5). These oil droplets contain carotenoids absorbing variously in the blue, violet, and UV regions of the spectrum and selectively filtering the light before it reaches the visual pigments in the outer segments (53).

A feature characteristic of the retinas of all classes of vertebrates except mam-

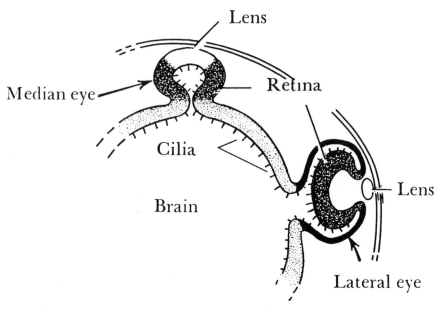

Figure 6. Diagram comparing the formation of vertebrate median and lateral eyes from the wall of the developing brain. The receptors (as well as other elements) arise from ciliated ependymal cells. The median eye develops from a simple out-pocketing of tissue, and the photoreceptor organelles face toward the lens. In the lateral eye the outpocketing of tissue collapses the rods and cones differentiate from ciliated cells facing away from the lens. [From (35).]

mals is the presence of double (amphibians, reptiles, and birds) and twin (teleost fish) cones (Fig. 5). These are pairs of cells in close apposition over their inner segments (25, 174). The two members of a double cone pair are not identical: The principal member tends to be somewhat larger with a well-developed oil droplet, whereas the accessory cone has a relatively larger inner segment containing extensive stores of glycogen (the *ellipsoid*) but lacks or has a greatly reduced oil droplet. The two members of a pair of twin cones are morphologically similar.

The parietal and pineal eyes of fish, amphibians, and reptiles have photoreceptor cells with the same fine structure as the lateral eyes (35). Because of the differences in the development of these eyes, however, the retina is not reversed in the pineal eye (Fig. 6).

Rhabdomeric Endings

When the photoreceptor membrane is not associated with basal bodies or centrioles, it characteristically takes the form of *microvilli* (34, 36, 115). Frequently, one end of the cell bears numerous microvilli, whereas the other end gives rise to a neurite of variable length. In relatively unspecialized photoreceptor cells, the microvilli project from several surfaces of the soma. In squid photoreceptors, the distal end of the cell extends as a long flat finger, which bears microvilli on two sides (184). In arthropods, the long photoreceptor cells associate in bundles, each cell projecting its band of microvilli towards the center of the bundle to form an axial rodlike structure known as the *rhabdom* (Fig. 7).

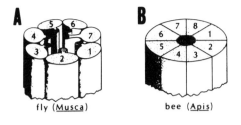

fly (Musca)   bee (Apis)

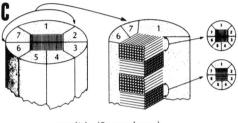

crayfish (Procambarus)

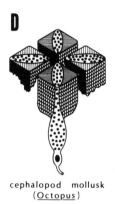

cephalopod   mollusk
(Octopus)

Figure 7. Some representative rhabdoms.
(A) A fly (Arthropoda). In Diptera and Hemiptera the rhabdomeres project into a central cavity and remain separate. Such a rhabdom is said to be "open" in contrast to the "closed" rhabdoms shown in (B) and (C). In flies there are eight retinular cells. Six of these (numbered 1–6) have rhabdomeres that run the full length of the retinula. The other two (7 and 8) have shorter rhabdomeres placed end to end. Consequently, a cross-section of the ommatidium shows only one of these rhabdomeres. In this diagram the section is from the distal half of the ommatidium and passes through cell 7, the superior central cell. Cell 8, the inferior central cell, is restricted to the basal half of the retinula and does not show

Visual Pigments Comprise a Family of Related Proteins

The photoreceptor cells of eyes are sensitive to light because they contain a visual pigment. The term *rhodopsin* is sometimes used synonymously with visual pigment, but it also has a more restricted (and original) meaning, referring to one subclass of visual pigments found in the rods of vertebrates (see below). Which meaning is intended is usually clear from the context. For historical reasons, however, cone pigments are virtually never called rhodopsins, although they are evolutionarily more closely related to rod pigments than are the rhodopsins of invertebrates (Fig. 8). *Opsin*, on the other hand, refers without ambiguity to the protein moiety of the pigment whether from rods, cones, or rhabdomeric receptors.

The significance of rhodopsin in the visual process began to be understood

here. A more proximal section would show the central rhabdomere attached to cell 8; cell 7 would be present but with apparently no rhabdomere. The inferior and superior central cells have different synaptic connections from the other six retinular cells. (B) The honeybee (Arthropoda). This fused organization is more typical of insects. Eight retinular cells each contribute a wedge of microvilli along the full length of the cell. The compact mass of microvilli from all eight cells is called a rhabdom. (C) The crayfish (Arthropoda). In decapod crustacea the microvilli project as tongues rather than continuous strips along the sides of the retinular cells. There are seven principal cells, and the tongues of microvilli interlock to form layers that can be up to 5 μm thick. An eighth cell with different synaptic connections is present distally. (D) The octopus (Mollusca). The retinas of cephalopod molluscs are not organized into ommatidia. Each photoreceptor cell has an elongate "outer segment" that consists of a central strip of cytoplasm and two lateral borders of microvilli.

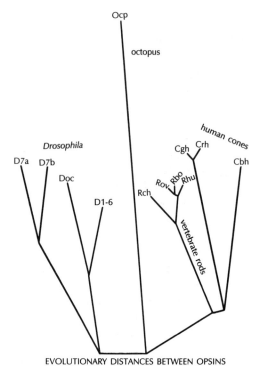

EVOLUTIONARY DISTANCES BETWEEN OPSINS

Figure 8. Evolutionary relationships of 12 opsins, calculated from a matrix of differences between corresponding amino acids; D7a (186), D7b (114), Doc (27), D1-6 (128,185), opsins of *Drosophila* [see Fig. 7A for identity of the cells; opsins D7a and D7b are found in subsets of cell No. 7; Doc is located in the ocellus (42,133) but was originally postulated to reside in cell No. 8 (27)]; Ocp (129), octopus; Rch (163), Rbo (63,120), Rov (45), and Rhu (121), rhodopsins (rod pigments) of chicken, cow, sheep, and human; Crh, Cgh, and Cbh (122), red-, green-, and blue-sensitive human cone pigments. The vertebrate and most of the invertebrate opsins differ in length by several percent and were aligned for maximal identity. The opsins of cephalopod molluscs have ~90 more amino acids on the C-terminal end, which accounts for the long branch to Ocp. Note that the opsin of human blue cones is as different from the red–green pigment pair as each is from rod pigments. The red–green cone pigments diverged from each other in Cenozoic time, as did the mammalian rod pigments; the basal evolutionary separations of vertebrate pigments probably occurred >500 million years ago.

when Boll (1876) (12) and Kühne (1877) (82) showed that the pink color of a dark-adapted retina quickly faded on exposure to light. Another important step was taken in the 1930s when Wald showed that the chromophore of rhodopsin is derived from vitamin A. In the intervening years the visual pigments of a wide variety of animals have been examined and shown to share a number of features in common.

Rhodopsins (in the broad sense of the term) are intrinsic membrane proteins (2, 45, 63) found in the disks of vertebrate outer segments and the microvilli of rhabdomeric photoreceptors. In the several cases where analyses have been done, rhodopsin accounts for ~90% of the protein of the photoreceptor membranes, which are roughly one half protein and one half lipid. Rhodopsins have molecular weights of ~38 kD (somewhat larger in cephalopods), and the primary structures from several species are now known. They fold into seven α-helical sections, each of which spans the membrane completely, and on the external surface of the membrane they bear short chains of sugar residues (Fig. 9).

The chromophore of rhodopsin is a C_{20}, fat soluble molecule known as retinaldehyde or one of two closely related derivatives (Fig. 10). In certain lower vertebrates such as freshwater fish and premetamorphic amphibia (14,81) as well as occasional invertebrates (160) the chromophore is a 3-dehydroretinal, and in the higher insects—principally Diptera and Lepidoptera—the chromophore is 3-hydroxyretinal (54, 154b, 167–169). A rod pigment (but again, usually not a cone pigment) that employs 3-dehydroretinal is traditionally referred to as *porphyropsin*, because it is more purple than the rhodopsin based on the same opsin. The insect pigments that use 3-hydroxyretinal are sometimes called xanthopsin, not because they are yellow but because the

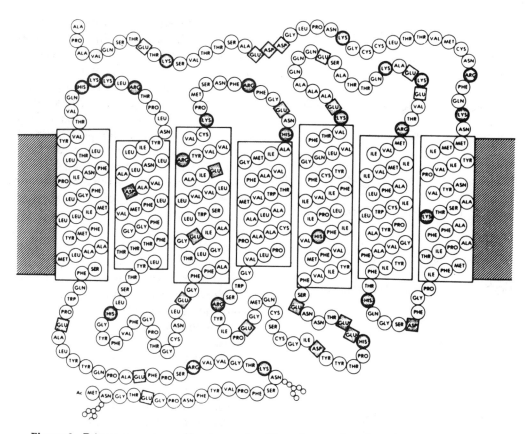

Figure 9. Primary structure of bovine opsin. All opsins are thought to have seven regions of α-helix that present nonpolar surfaces and span the membrane. The binding site of retinal is the ε-amino group of a lysine located in helix 7 (the shaded residue in the far right helix). The detailed pattern of folding of the opsin in the membrane is not known, but the seven transmembrane helices lie in a two-dimensional cluster and not the one-dimensional row depicted here. In rhodopsin the retinal therefore lies completely within the membrane and among the opsin helices. The N-terminal sequence of the opsin contains binding sites for oligosaccharides (shown with small circles at the bottom of the figure), and the C-terminal region has multiple sites that become phosphorylated following activation by light. In addition, one or more of the interhelix loops on the cytoplasmic side of the molecule are thought to be critical in the activation of G-protein; the amino acid sequence in the loop between helices 1 and 2 seems to have been especially conserved in evolution. [From (63).]

chromophore is a derivative of xanthophyll. Except for a shift to longer wavelengths in the absorption spectrum of 3-dehydroretinal, no feature of the visual process is directly attributable to either of these chemical differences in the structure of the chromophore. Where analyzed, the chromophore is attached through Schiff's base linkage to the ε-amino group of a

lysyl residue in one of the α-helices of the *opsin* (2). The chromophore therefore occupies a position buried in the membrane.

Light Isomerizes the Chromophore

Because of its conjugated system of single and double bonds, retinaldehyde (and

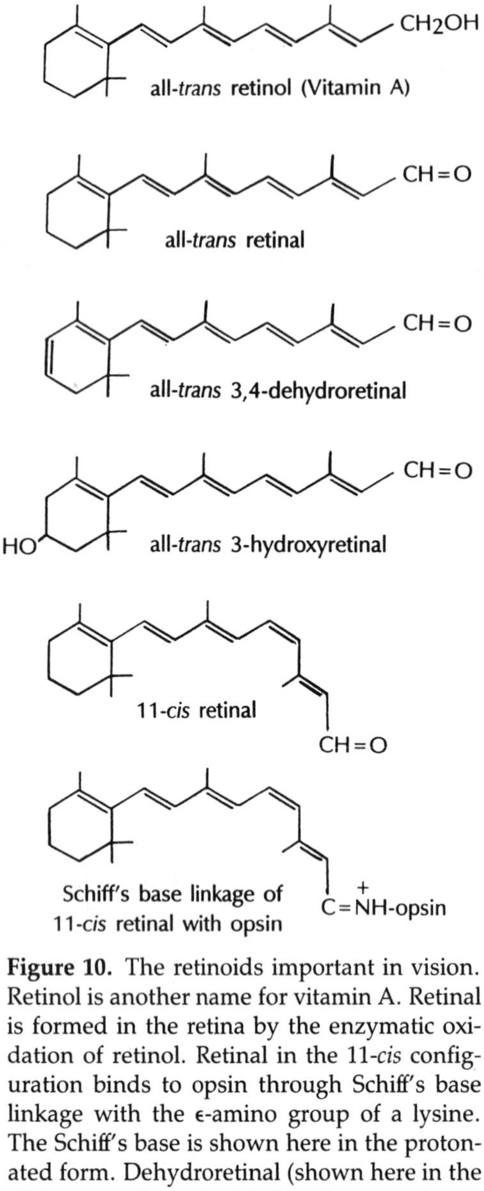

all-*trans* retinol (Vitamin A)

all-*trans* retinal

all-*trans* 3,4-dehydroretinal

all-*trans* 3-hydroxyretinal

11-*cis* retinal

Schiff's base linkage of 11-*cis* retinal with opsin

Figure 10. The retinoids important in vision. Retinol is another name for vitamin A. Retinal is formed in the retina by the enzymatic oxidation of retinol. Retinal in the 11-*cis* configuration binds to opsin through Schiff's base linkage with the ε-amino group of a lysine. The Schiff's base is shown here in the protonated form. Dehydroretinal (shown here in the all-trans configuration) is found in some lower vertebrates and a few invertebrates. Note the extra double bond in the ring, which lengthens the conjugated chain and shifts the absorption spectrum of both the chromophore and its derivative visual pigments toward the red. The recently discovered 3-hydroxyretinal, found in some insects, has an hydroxyl group in the ring at C-3.

the two derivatives 3-hydroxy- and 3-dehydroretinal) can exist in several geometrical isomers. One of these, the 11-*cis* (Fig. 10), is found in all known visual pigments. Th effect of light is to isomerize the chromophore from the 11-*cis* to the all-*trans* configuration (70). This change in shape of the chromophore forces a change in shape of the opsin, and isomerization is therefore followed by a series of conformational changes in the protein. In the case of vertebrate rhodopsin, these changes are indicated by a series of spectroscopically identifiable intermediates (Fig. 11), one of the later of which is called *metarhodopsin*. In vertebrate metarhodopsin the binding site of the chromophore is no longer protected from the aqueous environment adjacent to the membranes, and the retinaldehyde is readily hydrolyzed (98). It is this step that leads to the loss of color, or bleaching, first noted by Boll and Kühne in the last century. The absorption spectrum of bullfrog rhodopsin and its final product of bleaching are shown in Fig. 12.

Under the influence of an alcohol dehydrogenase and reduced nicotinamide adenine dinucleotide phosphate (NADPH) in the outer segments, the retinaldehyde is then reversibly converted to retinol (vitamin A) (48) (Fig. 10). The retinol is transported to cells of the pigment epithelium where it is esterified with fatty acids and sequestered. Vitamin A can be brought to the pigment epithelium from larger stores in the liver via soluble transport proteins in the blood. Other transport proteins mediate its movement between cells in the retina (15).

The Recovery Phase of the Pigment Cycle Involves a Variety of Processes

Recovery from light requires reisomerization of the chromophore from the all-*trans* to the 11-*cis* configuration. In vertebrates this isomerization takes place at

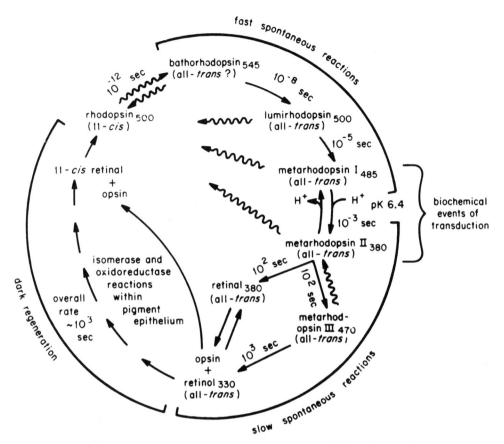

Figure 11. Summary diagram showing the intermediates of bleaching and recovery in the visual cycle of vertebrates. The wavelengths of maximal absorption are given by subscripts; the isomeric configurations of the retinoids are indicated in parentheses. Photochemical conversions are shown by wavy lines. Rate constants refer to room temperature. The complete cycle requires that the retina be in contact with the pigment epithelium. Note that the decay of metarhodopsin II can proceed either through metarhodopsin III or through a spectrally distinct equilibrium mixture of free retinal and retinal in Schiff's base linkage with various exposed amino groups, designated "retinal$_{380}$" in this figure. [From (43).]

the level of the alcohol (retinol) and in the pigment epithelium (10, 11, 16). The 11-*cis* retinol is then returned to the rods and cones. When oxidized to retinaldehyde, the chromophore combines spontaneously with opsin in the outer segment membranes (143). This process of chromophore replacement is supplemented by a much slower dark exchange (29).

In invertebrates the Schiff's base linkage between retinaldehyde and the pro-

tein is not hydrolyzed, and the metarhodopsin therefore does not bleach. The stable metarhodopsin can itself absorb light, and although this does not lead to visual excitation, it can cause a reisomerization of the chromophore and the *photoregeneration* of rhodopsin. Although in principle any rhodopsin can be photoregenerated from its metarhodopsin, in the vertebrate retina the lifetime of metarhodopsin is so short that under most con-

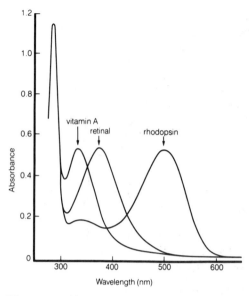

Figure 12. Absorption spectra of frog rhodopsin and the product of its bleaching in aqueous digitonin solution, pH 5.55. The α-band is mainly responsible for the spectral sensitivity of rod vision and depends on the chromophore. On bleaching it is replaced by the absorption of retinaldehyde, maximal at 385 nm. The band at 280 nm is a result of aromatic amino acid residues in the opsin and is not altered on bleaching. [From (32), based on earlier data of Wald (171).]

ditions this is not an important part of the visual cycle. In insects, by contrast, photoregeneration appears to be an important process, and in diurnal species the visual pigment spends most of its time as a photosteady-state mixture of rhodopsin and metarhodopsin (61). Most metarhodopsins have peak absorption in the blue at ~480–500 nm; consequently, in insect species with several visual pigments absorbing in different regions of the spectrum, all pigments are photoregenerated about equally (Fig. 13). In flies, however, the metarhodopsin absorbs maximally at 570 nm, which is at longer wavelengths than any of the fly rhodopsins. As a consequence it is possible to *increase* the level of rhodopsin and to deplete the eye of

metarhodopsin by exposing the animal to bright orange light, a treatment that would have the opposite effect in other species of animal.

The role of chromophore exchange, so central to the process of dark regeneration in the vertebrate retina, is much less certain in invertebrates, although examples seem to exist. In the several species that have been examined, regeneration of rhodopsin requires the synthesis of new opsin (52, 150, 156). This can be a very dramatic process; in some crabs >90% of the rhabdom is destroyed for the daylight hours and rebuilt each night (119).

De novo synthesis of opsin also plays an important role in the vertebrate retina, but it is not part of the process of dark adaptation. Rhodopsin, newly synthesized in the inner segment, is inserted into the membranes at the base of the outer segment, and old disks are shed from the distal end of the outer segment and phagocytosed by the cells of the pigment epithelium (143, 144). In some species disk shedding occurs on a diurnal cycle. These processes lead to complete replacement of rod membranes in about a month in a frog and in a week or 10 days in a mammal. The regular replacement of photoreceptor membrane therefore appears to be a universal feature of visual receptors.

Visual Pigments Vary in Their Wavelengths of Maximal Absorption

Visual pigments exhibit maximal absorption from ~340 nm in the near UV (61) to 620 nm in the red (28, 81). Three factors control this variation. First, the substitution of 3-dehydroretinal for retinal causes the absorption spectrum to red shift (14). Second, because of differences in their opsins, species living in different photic environments may have rhodopsins with somewhat different spectral properties (81, 95, 96). The rod pigments of fish pro-

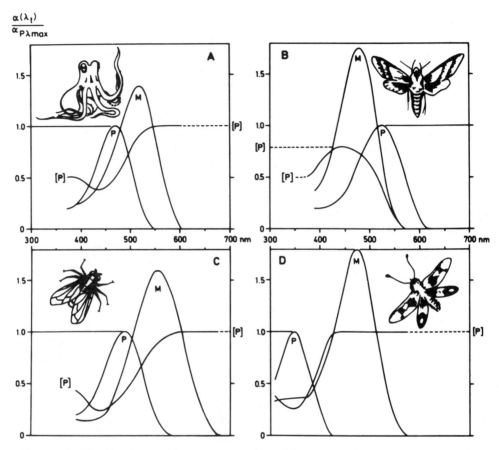

Figure 13. Rhodopsin–metarhodopsin systems of four invertebrates. P, pigment (rhodopsin); M, metarhodopsin. Because the metarhodopsins of these invertebrates are stable, actinic adapting lights produce a photosteady-state mixture of P and M. The composition of the mixture depends on the absorption spectra of the participating molecules and the wavelength of the irradiating light. The curves marked [P] show the relative proportion of rhodopsin in the photosteady state as a function of the wavelength of the adapting light. (A) the octopus *Eledone*; (B) the green receptor of the sphingid moth *Deilephila*; (C) pigment of the peripheral receptors (R1-6) of the fly *Calliphora*; and (D) the UV receptor pigment of the neuropteran *Ascalaphus*. Note that in each case the molar absorbance of the metapigment is greater than the visual pigment, but there is considerable variation in the relative spectral positions of the two forms. [From (61).]

vide the most extensive data in this category. And third, many animals have more than one opsin gene, which are expressed in different photoreceptor cells. This provides the basis for color vision.

Dehydroretinal and Porphyropsin
Dehydroretinal has one more double bond in the conjugated chain, and con-

sequently the absorption maximum is shifted from 387 to ~400 nm. When 3-dehydroretinaldehyde condenses with opsin, the pigment that forms also has its absorption maximum displaced towards the red. For example, frog rhodopsin absorbs maximally at 502 nm, whereas the corresponding porphyropsin (made with 3-dehydroretinal but with the same opsin) abosrbs maximally at ~522 nm. The shift is larger (565–620 nm) in a long wavelength cone pigment; smaller in blue-sensitive cones.

Until very recently, visual pigments based on 3-dehydroretinaldehyde were known only from lower vertebrates (14, 81), but a couple of examples are now recognized in invertebrates (160, 182). Classically, porphyropsins are said to be characteristic of freshwater fish, larval amphibians, and freshwater turtles. Both porphyropsin and rhodopsin are found in migratory fish that move between marine and freshwater environments. Which pigment predominates varies during the life cycle, but the shift from one pigment system to the other actually precedes migration to the environment for which the new pigment is intended. Similarly in amphibians, the retinas of larval frogs and salamanders contain porphyropsin, but this changes to rhodopsin during the metamorphic preparation for a more terrestrial life. The spotted newt shows a second metamorphosis from the terrestrial red eft to the fully aquatic and sexually mature adult, and this is accompanied by a shift back from rhodopsin to porphyropsin in the retina. The toads, Bufonidae, are an exception to these rules, having rhodopsin throughout their life cycle.

As the sample of fishes that have been studied has enlarged, it has become clear that about one half of the freshwater species possess rhodopsin as well as porphyropsin, and that this includes nonmigratory fish that spend their entire lives in fresh water. Moreover, a few marine forms (Labridae) have also been reported to have porphyropsin. Secondly, although there is no question that the shifts from porphyropsin to rhodopsin or vice versa that are associated with metamorphosis are genetically programmed and hormonally controlled, there is now increasing evidence that in some animals

the rhodopsin:porphyropsin ratio is under immediate environmental control. For example, in juvenile salmon a shift towards rhodopsin can be achieved by raising the ambient light intensity. In nonmigratory fish (14, 81, 117) and in the crayfish *Procambarus* (161) there are seasonal fluctuations in the proportions of these pigments, with porphyropsin maximal during the winter months. In fish, this is dependent on the total amount of light the fish receives; in crayfish it depends on temperature. Experiments on fish with individually capped eyes indicate that control is intraocular; the eyes can be influenced independently of each other and of the titer of hormones in the circulation. Microspectrophotometric measurements of single outer segments of trout rods show that rhodopsin and porphyropsin are mixed homogeneously in the outer segments of each rod cell, and that at any time all receptors have the identical mixture. Other workers find different proportions of the two visual pigments in different parts of the retina. Relatively little is known of the enzyme for interconverting retinaldehye and 3-dehydroretinaldehye (terminal ring dehydrogenase).

Some Interspecific Variations between Opsins Have Been Produced by Natural Selection

Different species can possess visual pigments with different absorption maxima because small differences in the amino acid composition of the opsin lead to proteins with slightly different abilities to interact with the chromophore. Figure 14 shows the spectral distribution of the wavelengths of maximum absorption of the visual pigments of a large number of fish. There is much interspecific variation, but the distribution of absorption maxima is not continuous. The points of maximum absorption tend to cluster around specific wavelengths, and it would be

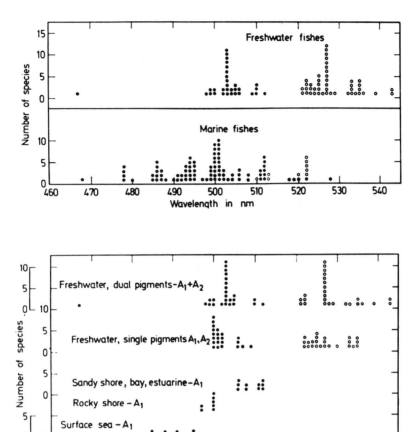

Figure 14. Distribution of visual pigments of teleost fishes. Each point represents the wavelength of maximal absorption of a pigment, obtained using partial bleaching analyses of extracts. The lower frame shows the data broken down by habitat. Data are from a number of sources. [From (28).]

most interesting to know in detail the molecular shapes of these rhodopsins and porphyropsins, as well as the different amino acid substitutions that underlie them.

An ecological generalization can also be drawn from this figure. Those species with rod pigments absorbing towards the blue tend to be found in the deeper oceanic waters, where the wavelengths of maximum transmission of sunlight are ~470–480 nm (cf. Fig. 1). On the other

hand, fish with visual pigments lying to longer wavelengths tend to come from the more turbid coastal waters or from fresh waters, where the maximum available light is in the green or yellow regions of the spectrum (81, 95, 96). The porphyropsin visual system can be viewed as an adaptation to the environment that is superimposed on interspecific variations in the opsin and that serves to shift the wavelengths of maximum sensitivity to still longer values.

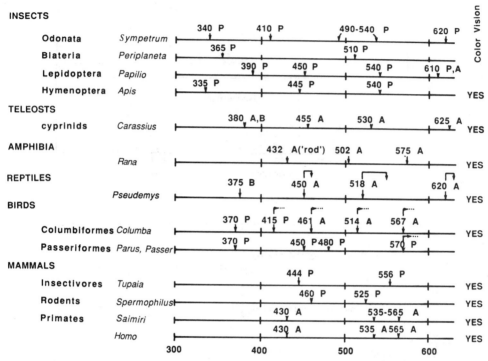

Figure 15. Some representative examples of animals in which there is good evidence for multiple visual pigments and/or different spectral classes of photoreceptors, such as required to support color vision. The presence of behavioral evidence for color vision is indicated in the last column. This compilation is not an exhaustive survey of the literature, but is intended to illustrate the wide taxonomic distribution of retinal capacity for color vision and to indicate some of the better-studied examples. The superscripts indicate experimental technique: A, absorption spectra of pigments measured directly; B, behavior; P, physiological measurement. [From (50a).]

Many Species Have More Than One Kind of Visual Pigment

Many species have two or more visual pigments segregated in different photoreceptor cells. Although the presence of multiple pigments could serve simply to broaden the band of wavelengths to which the animal is sensitive, multiple pigments are the universal basis for color vision, and capacity for color vision is widely distributed in the animal kingdom (Fig. 15). Color vision will be dealt with in more detail in a later section.

Certain of the photoreceptor cells of flies contain a sensitizing pigment, believed to be 3-hydroxyretinol, which absorbs maximally in the near UV and passes energy by radiationless transfer to the visual pigment (79). As a consequence, these cells have two-peaked spectral sensitivity functions, with maxima at ~350 and 490 nm. Other photoreceptor cells of flies have accessory carotenoids that distort the spectral sensitivity functions (62). Such intimate associations of accessory and visual pigments within the photoreceptors and at the molecular level are not known in other animals.

Visual Pigment Molecules Are Oriented in the Photoreceptor Membranes

In the outer segments of rods and cones, the chromophores lie roughly parallel to the planes of the membrane disks, at right angles to both the long axes of the rods and the direction of propagation of the incident light (93, 134) (Fig. 16). The lipids of the membrane provide a fluid environment within which the visual pigment molecules can diffuse laterally (i.e., within the plane of the membrane) and rotate about molecular axes perpendicular to the plane of the membrane (26). An extended chromophore (like retinal) absorbs polarized light most strongly when the plane of polarization is parallel to the axis of the conjugated chain. Consequently, if the conjugated system of single and double bonds were exactly perpendicular to an axial ray of unpolarized light, the probability of absorption would be increased to 1.5 times the value observed with the same number of molecules randomly oriented in solution. Because of an average departure of the absorption vectors by several degrees from coplanarity with the disks, the measured enhancement of absorption is ~1.37.

When such molecules associate in an organized array, their absorption of light depends on the angle of incidence and the plane of polarization, a phenomenon known as *dichroism*. The molecular orientation of rhodopsin is detected as a dichroic absorption of rods illuminated from the side (Fig. 16). When laterally incident light is polarized perpendicular to the rod axis (parallel to the planes of the disks), absorption is high because the conjugated system is nearly coplanar with the disks. When the plane of polarization is rotated 90° and made parallel to the rod axis, the absorption is ~22% as great, corresponding to a dichroic ratio of ~4.5. From such measurements as these, the degree of orientation of the chromophores can be estimated.

The dichroism of crustacean rhabdoms has also been measured by viewing the rhabdoms from the side and placing a measuring beam within a single band of microvilli (55) (Fig. 16). When the beam is incident along the microvillar axes, absorption is independent of the plane of polarization, as one would expect from the cylindrical symmetry of the microvilli (116). When the beam is perpendicular to the microvillar axes, absorption is dichroic, with major absorption with the electric vector parallel to the microvillar axes. If the chromophores lay in the surfaces (i.e., in the tangent planes) of the microvillar membranes, the dichroic ratio should be two. A somewhat higher result is observed, indicating some preferential alignment of the chromophores along the axes of the microvilli. Moreoover, measurements of photoinduced dichroism and local bleaching lead to the conclusion that the chromophores are much more rigidly anchored in the membrane than is the case in vertebrate rods.

The dichroic properties of the rhabdom provide the basis for detection of polarized light by arthropods; in fact, electrophysiological measurements of *polarization sensitivity* of single retinular cells (the ratio of sensitivity to lights polarized so as to give maximal and minimal responses from the same cell) indicate that microspectrophotometry underestimates by a factor of 2 or 3 the dichroic ratios of photoreceptor membranes.

The best example of the use to which polarization sensitivity is put by arthropods is the navigation system of bees and ants (84, 145, 146). Specialized ommatidia (see the section on The Evolutionary Radiation of Image-Forming Eyes, Compound Eyes) in the dorsal rim of the eye are responsible. In each ommatidium UV receptors with orthogonally oriented microvilli provide antagonistic afferent in-

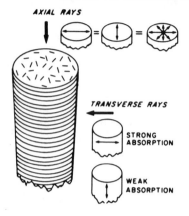

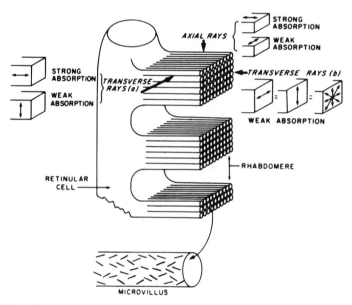

Figure 16. Vertebrate photoreceptors *in vivo* are ordinarily insensitive to the plane of polarization of light, whereas arthropod receptors exhibit polarization sensitivity. (A) In vertebrate rod and cone outer segments the chromophores lie randomly oriented, approximately in the planes of the disk membranes. Consequently, absorption of axially incident rays is independent of the plane of polarization. When illuminated from the side, however, the outer segments are dichroic: Light polarized parallel to the planes of the disks (and thus the axes of many of the chromophores) is strongly absorbed, whereas light polarized at right angles to the disks is only weakly absorbed. (B) In the rhabdomeres of decapod crustacea the absorption vectors of the pigment lie approximately in the tangent planes of the microvilli, with some tendency to align parallel to the microvillar axes. Shown here is a single retinular cell and several of its tongues of microvilli. In an isolated rhabdom these tongues are interleaved with the rhabdomeres of other receptors

put, and the orientation of the microvilli varies in a regular fashion from the front to the back of the eye. By orienting the body axis for maximum stimulation from this ommatidial array, the animal utilizes the polarization pattern of natural skylight to find the solar meridian, even when the sun is not directly visible.

Light Regulates Membrane Conductance

The process by which an environmental stimulus such as light or a specific chemical substance causes a sensory cell to respond is called *transduction*. The language of the response is the lingua franca of neurons: Changes in ion fluxes through the plasma membrane are reflected as changes in membrane voltage.

Vertebrate Rods and Cones Respond with a Decrease in Sodium Conductance and the Membrane Voltage Hyperpolarizes

In the dark, the outer segments of rods and cones have an unusually large inward leakage of cations, principally sodium. As a consequence, the membrane potential is only about -30 mV, approximately one half of the amplitude of the membrane potential of a typical neuron. The permeability of the membrane to calcium is several times larger than to sodium, with the result that $\sim$15% of the inward current is carried by Ca^{2+}. An ionic and osmotic steady state is maintained by vigorous active extrusion of sodium from the inner segment. As in neurons, Ca^{2+} is extruded by a Na^{+}-Ca^{2+} antiport, which depends on the presence of an electrochemical gradient of sodium ions. As a result of these fluxes there is a loop of current, outward across the membrane of the inner segment, parallel to the axis of the cell in the extracellular space, and inward through the membrane of the outer segment. In mammalian rods this current is large enough to turn over the intracellular cations every 3 min.

The effect of light is to decrease the conductance to sodium; consequently the photocurrent is a decrease in the dark current (Figs. 17 and 18). With the fall in sodium permeability, the cell hyperpolarizes (Fig. 19b) (20, 58, 68,153, 164).

Most Invertebrate Photoreceptors Respond with an Increase in Sodium Conductance and Depolarize

Rhabdomeric photoreceptors of invertebrates respond to light with an increase in permeability to cations, principally sodium. Consequently the membrane potential depolarizes (Fig. 19a, upper row) and exhibits a reversal potential at 0–20 mV, inside positive. The response is graded with the intensity, and after an in-

(cf. Fig. 7C). When a single band of microvilli is irradiated with a laterally incident microbeam, absorption is isotropic if the beam is propagating parallel to the microvillar axes, but is dichroic when the light arrives at right angles to the microvillar axes. This latter condition is equivalent to axial illumination of the entire rhabdom, as occurs in the living eye. Absorption is strongest when the light is polarized parallel to the microvillar axes. The measured dichroic ratio is somewhat higher than two, which means there must be some orientation of the chromophores parallel to the microvillar axes. Substantially higher dichroic ratios are inferred from microelectrode recordings of polarization sensitivity, suggesting that the optical measurement on isolated photoreceptors underestimates the actual degree of axial orientation of the chromophores. Note that unlike the pigment molecules in rod disks, those of arthropods are fixed in the membrane and do not undergo Brownian rotation.

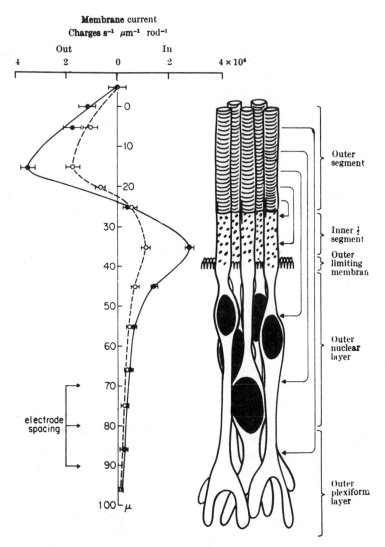

Figure 17. The photocurrent of vertebrate receptors. *Left:* Peak transmembrane photo-current, calculated from extracellular measurements of longitudinal voltage fields and resistivity between the cells. Results shown are for two slices of retina stimulated with 1×10^{11} hv cm^{-2} (*open circles*) and 4×10^{11} hv cm^{-2} (*filled circles*) of green light at 560 nm. Error bars indicate rms noise in current tracings. *Right:* Retinal rods of the rat drawn to the same scale. Arrows indicate the flow of photocurrent. [From (131).]

itial transient, the membrane potential frequently stabilizes at a value that may be maintained as long as the light remains on (7, 19, 41, 88, 107–109). The decline from the transient to the plateau phase of the response is one measure of light ad-aptation. In at least some species it is caused by a secondary increase in intra-

cellular Ca^{2+}, either by release of internal bound calcium or by influx through the light-regulated conductance (18, 94). This adaptation has the affect of expanding the range of intensity steps over which the cell can respond before it saturates; it in-creases the *dynamic range*.

Extracellular measurements along the

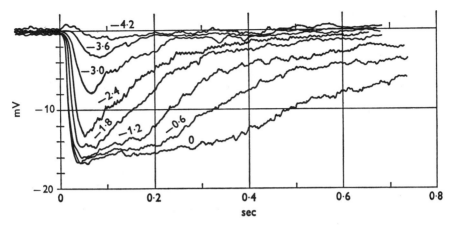

Figure 18. Graded, hyperpolarizing responses of a turtle cone to 10-ms flashes of increasing intensity, as indicated in relative logarithmic units for each trace. Brightest flash delivered 8.5×10^6 photons to the cone. Downward deflection indicates increasing negativity of the intracellular microelectrode. [From (7a).]

slender "outer segments" of squid photoreceptors made on slices of retina cut parallel to the long axes of the receptor cells demonstrate that the region of the rhabdom directly under the spot of light becomes negative to other parts of the rhabdom, which is to be expected if there is a localized current sink under the point irradiated (Fig. 20) (59). There is some evidence that the site of the conductance change is the rhabdom itself. The most direct argument is based on photoreceptor cells of the leech (88), in which the microvilli of the photoreceptive membrane project into a central "vacuole" within the cell (Fig. 21). The vacuole is really extracellular space, which is continuous through narrow channels with the fluid bathing the outside surface of the cell. A microelectrode inserted into the cytoplasm shows a positive-going (depolarizing) response when the cell is illuminated. On the other hand, a microelectrode in the vacuole (as established by dye injection) shows an increased negativity when the cell is illuminated. This is the expected result if the microvillar membranes constitute a site of inward current during illumination

and is not anticipated if the sole site of inward photocurrent is the nonmicrovillar portion of the cell body.

This process of excitation is roughly analogous to a classical excitatory chemical synapse, with light playing the role of the chemical transmitter. The receptor depolarizes in a graded fashion, analogous to the excitatory post synaptic potential, and depending on the system, either generates a spike in an afferent axon (squid) or secretes transmitter, causing a postsynaptic response in an adjacent second-order neuron (insects).

Some Invertebrate Photoreceptors Respond by Increasing Conductance and Give Hyperpolarizing Responses

Among the invertebrate photoreceptor organelles that are derived from ciliary processes are cells in the pallial eyes of the scallop *Pecten* (101). These "simple" eyes, which lie in a row along the edges of the mantle, are remarkable in possessing a double retina consisting of two distinct layers of photoreceptor cells (Fig. 24C). Each layer has its own branch of the optic nerve, and peripheral synaptic in-

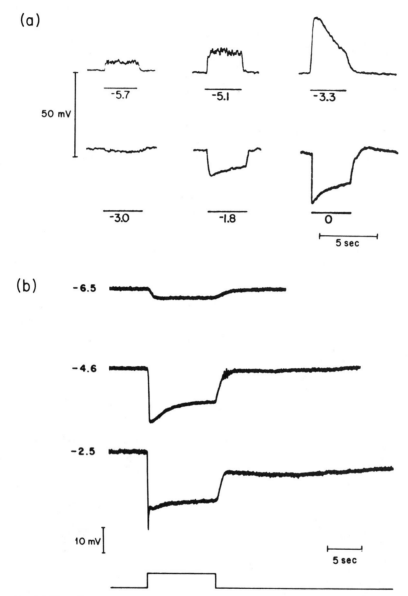

Figure 19. (*a*) Depolarizing (upper row) and hyperpolarizing responses (second row) of photoreceptor cells in the scallop *Pecten*. (*b*) Hyperpolarizing responses of photoreceptors in the lizard *Gekko*. Numbers signify log relative intensity. [From (43); (*a*) from (100). (*b*) from (80).]

teractions are believed to be absent. The proximal retina holds no surprises. The photoreceptor cells have microvillar projections, and the membrane responds to light with a conventional, depolarizing receptor potential (Fig. 19*a*, upper row).

The photoreceptor cells of the distal retina, on the other hand, have a series of lamellar membranes arising in conjunction with ciliary microtubules. Both types of photoreceptor cells contain a visual pigment with an absorption maxi-

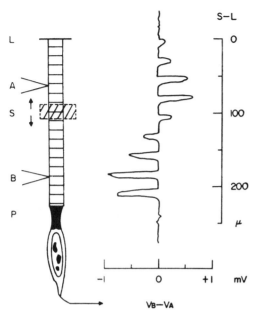

Although in the vertebrate retina the response of the receptors is hyperpolarizing, it is not much help to discuss it in terms of excitation and inhibition. Hyperpolarization of the distal photoreceptor cell of *Pecten*, however, is seemingly a case of primary inhibition. The axons of these cells discharge spikes, and the effect of light is to slow or abolish this discharge. Behaviorally, the function of molluscan photoreceptor cells showing primary inhibition is to mediate escape responses. Thus, when a shadow falls on the eye, there is a vigorous "off" discharge as the photoreceptor axons escape from inhibition, and the shells quickly close.

Figure 20. Extracellularly recorded potentials in a slice of squid retina at various depths through the layer of rhabdoms. Ordinate, distance of the movable slit-shaped stimulus (S) from the internal limiting membrane (L). A and B, positions of a pair of fine electrodes. Abscissa, potential difference between the electrode tips. P, pigmented region of the cell. Stimulus intensity equivalent to 3×10^{11} photons/cm$^2 \cdot$ s at 500 nm. Note that local irradiation produces a local current sink in the same region. [From (59).]

mum near 500 nm, but the distal cells are 2–3 log units less sensitive. Furthermore, the distal cells respond to illumination by hyperpolarizing (Fig. 19a, second row). This hyperpolarization is different in an important way from the responses of the ciliary photoreceptors of the vertebrate retina: it is based on an *increase* in membrane conductance rather than a decrease. Thus, if photons are likened to molecules of a synaptic transmitter, the hyperpolarization bears a formal resemblance to a classical inhibitory postsynaptic potential based on increased permeability to potassium or chloride ions.

Single Photon Absorptions Lead to Discrete Shots of Membrane Current

Close to threshold, photoreceptors show small transient photocurrents causing changes in membrane voltage usually less than a millivolt in amplitude (8, 157) (Fig. 19a, upper row). These reflect the gating of ionic channels in response to absorptions of single photons. As the intensity of light is increased, the frequency of occurrence of these discrete potentials increases, and with still brighter lights the responses fuse. The smooth, graded response of the membrane that is observed with moderately high intensities thus results from the addition of a large number of smaller, unitary events. Each discrete potential is associated with the movement of tens or hundreds of thousands of ions through the membrane.

The Dependence of Membrane Response on Intensity Suggests the Recruitment of Channels from a Population of Fixed Size

In both vertebrates and invertebrates the voltage response of the membrane fol-

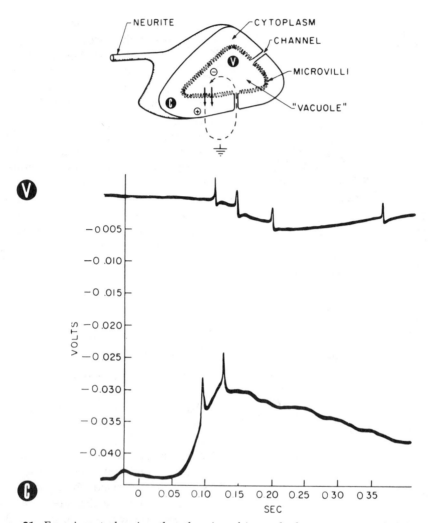

Figure 21. Experiment showing that the site of inward photocurrent in rhabdomeric endings is the microvillar membranes. In the leech (*Hirudo*) the microvilli line a "vacuole," which is in fact continuous through narrow channels with extracellular space. A microelectrode can be placed in either the vacuole or the cytoplasm, leading to records like those shown. In each case potential is plotted relative to a reference electrode (ground) in the bath. With an intracellular (cytoplasmic) recording site, the receptor potential is positive (depolarizing). Small, positive-going spikes also appear. They are attenuated because they arise in the neurite, at some distance from the soma. With the electrode in the vacuole, the receptor potential is negative. This is consistent with a current sink at the microvilli (solid arrows) and return current paths (broken arrows) as shown. It is left as an exercise for the reader to see that this result is not expected if the sole site of inward current were in the nonmicrovillar part of the soma membrane. Spikes recorded in the vacuole are positive, further evidence that their site of initiation is elsewhere. [Voltage recordings from (88).]

lows a function of the form

$$V/V_{max} = I^\alpha/(I^\alpha + \sigma)$$

where I is the intensity, σ is the intensity for a half-maximal response, and α is a coefficient characteristically close to unity. This equation describes a linear dependence of V on I at low intensities ($I \ll \sigma$), rising to a saturating value (V_{max}) at high intensity. There is a formal similarity to the Michaelis–Menten treatment of enzyme kinetics because the response to I depends on the fraction of channels that are affected by I; consequently at large I most or all of the channels are affected and little or no additional response is possible with further increases in intensity. The term V is frequently plotted as a function of log I, yielding a sigmoid function like that in Figure 37.

An Internal Messenger Links Photon Capture to Membrane Conductance

Why an Internal Transmitter Is Needed
Measurements of the absolute threshold of vision show that a dark-adapted rod cell can be activated by a single photon (67, 132). As a corollary, one must also assume that virtually every rhodopsin molecule in the outer segment is potentially capable of being the absorber (39). In rods, the vast majority of the rhodopsin molecules are located in membrane disks that have no continuity with the plasma membrane, where the ion channels controlled by light are located. Furthermore, a rod saturates when only 100 of its 10^9 (in a frog) rhodopsin molecules have been activated. Each and every rhodopsin molecule of a dark-adapted rod is therefore potentially capable of controlling sodium conductance over ~1% of the surface of the outer segment and at distances that can be several micrometers from the site of absorption. Similarly, in photoreceptor cells of the invertebrate *Limulus* there are unitary membrane currents in response to single photons that

are far greater than can be accounted for by the opening of a single channel. A single rhodopsin therefore controls several or many channels via an internal messenger.

Cyclic-GMP Controls Sodium Channels in Vertebrate Rods
The outlines of the transduction process in vertebrate rods and the identity of the internal messenger now seem to be fairly clear (22, 136, 159). Transduction involves a molecular cascade linking rhodopsin with the light-activated sodium channels. The light-regulated conductance of the plasma membrane is opened by cGMP (44), and the effect of light is to reduce the level of cGMP by activating a phosphodiesterase. The process by which this comes about is diagrammed in Figure 22.

Light isomerizes the chromophore and triggers conformational changes in the opsin [Fig. 22(1)]. An intermediate of bleaching, probably metarhodopsin II, reacts with a peripheral membrane protein known variously as G-protein or transducin [Fig. 22(2)]. This G-protein has several subunits, one of which binds GDP. When associated with light-activated rhodopsin, however, the G-protein exchanges GDP for GTP [Fig. 22(3)], whereupon the α-subunit becomes soluble [Fig. 22(4)]. The α-subunit–GTP complex has the property of interacting with an inhibitory subunit on the enzyme phosphodiesterase (PDE), perhaps removing it from the rest of the enzyme [Fig. 22(5)], but certainly removing the inhibition. Activated PDE then causes a reduction in the level of cGMP [Fig. 22(6)], which in turn causes closure of light-controlled sodium channels in the plasma membrane [Fig. 22(7)]. The steady-state level of cGMP is determined by the relative rates of formation [Fig. 22(8)] and hydrolysis [Fig. 22(6)].

The control points and recovery phases of the cycle are not completely understood. The G-protein has GTPase activity,

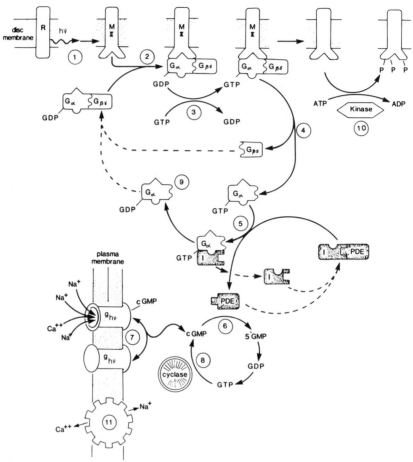

Figure 22. The phototransduction cascade and related molecular events in vertebrate rods. R, rhodopsin; M_{II}, metarhodopsin II; G_α, $G_{\beta\gamma}$, subunits of G-protein (transducin); PDE, I, phosphodiesterase and its inhibitory subunit. See the text for a more complete description that refers to the numbered processes (1–11).

which contributes to the restoration of the GDP-bound state [Fig. 22(9)]. Metarhodopsin is also phosphorylated at several sites by an ATP-dependent kinase [Fig. 22(10)], which may contribute to the quenching of PDE activity. And there is a still-mysterious 48-kD protein that binds to light-adapted membranes containing phosphorylated rhodopsin.

The system provides considerable gain. One isomerized rhodopsin molecule can catalyze several hundred G-protein activations (and thus PDE activations) at a rate of ~1/ms. The rapid lateral diffusion of metarhodopsin II in the membrane is thought to enhance the number of activations of G-protein, but as diffusion of metarhodopsin does not seem to occur in rhabdomeres, it is unlikely to be a universal feature of phototransduction. A second stage of gain occurs in the PDE reaction, so that 10^4–10^5 cGMPs are hydrolyzed as a result of a single photon absorption.

Calcium Plays a Regulatory Role
When the theoretical need for an internal messenger became clear, Ca^{2+} was an

early candidate. Elevated external Ca^{2+} was shown to cause a decrease in the dark current and a hyperpolarization of the membrane, thus mimicking the effect of light. In fact a number of experimental observations seemed consistent with a role of calcium ions as the internal transmitter. For example, light causes an accumulation of extracellular Ca^{2+} around the outer segments, as though there had been a release of stored calcium that was being pumped out of the cell. It could never be shown, however, that light caused an increase in free intracellular Ca^{2+}, as required if calcium were the internal transmitter. For more than 10 years the proponents of cGMP and Ca^{2+} were unable to resolve the issue either way until patch-clamped pieces of outer segment membrane were shown to respond directly to cGMP.

Most of the results of experiments with Ca^{2+} can now be understood, at least qualitatively (99). As in other cells, the internal concentration of Ca^{2+} is kept in the submicromolar range by both binding and the Na^+-Ca^{2+} exchange pump [Fig. 22(11)] mentioned earlier. When light causes a decrease in membrane conductance, it causes a decrease in the influx of both Na^+ and Ca^{2+}. There is therefore a transient increase in the concentration of external Ca^{2+} in the immediate vicinity of the outer segments as the steady-state balance between inward leak of Ca^{2+} and outward extrusion is temporarily perturbed. In fact, the concentration of Ca^{2+} inside the cell actually falls in the light.

The reason that elevated Ca^{2+} mimics the effect of light is less certain, but it is believed to be through a reduction in the level of cGMP. Suggested mechanisms include inhibition of the cyclase.

The Evolutionary Radiation of Image-Forming Eyes

At the morphological level, photoreceptors are frequently found in aggregates and associated with various accessory structures, which help to form images and reduce stray light (87). Large, optically sophisticated eyes capable of analyzing spatial and temporal patterns of light at a distance from the animal have evolved independently at least three times; simpler eyes have appeared even more frequently.

The easiest way to represent images of the external world is on a two-dimensional sheet of photoreceptors, analogous to a piece of film or an artist's canvas, and this is in fact what evolution has repeatedly done. In looking at the resulting biological diversity, however, it is useful to recognize that either of two different optical systems can be employed, depending on whether the retinal sheet is *concave* or *convex* (Fig. 23) (77, 175). When the photoreceptor surface is concave, an optical image can be produced with a refractile *lens* suitably placed in front of the retina. This is the optical design of the vertebrate eye as well as the large eyes of cephalopod molluscs. On the other hand, when the retinal surface is convex, it can form an image if the individual photoreceptor elements are sensitive only to a narrow cone of light incident perpendicular to the retinal surface. This is the optical principle on which the *compound eyes* of arthropods is based. Each of these optical arrangements has advantages and disadvantages, which we shall examine below, but let us first explore the evolution and biological diversity of these types of eyes in somewhat more detail.

Simple Eyes

As mentioned in the discussion of photoreceptor organelles, the most primitive forms of light sensitivity show no morphological specializations. *Simple eyes* (Fig. 24) consisting of small aggregates of photoreceptors are present in all the major animal phyla, and the paired eyes of the flatworm *Planaria* are illustrative of a

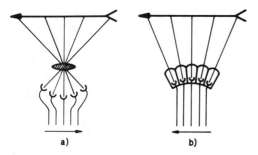

Figure 23. Two ways in which images can be formed, depending on whether the sheet of photoreceptors is concave (*a*) or convex (*b*). [From (77).]

primitive form. Each eye consists of ~200 photoreceptor cells in a cup ~45 μm wide by 25 μm deep and surrounded by dark screening pigment except for a 30-μm aperture on one side. There is no lens and no image, but the eye clearly has some directional sensitivity, for a point source subtends an angle of ~70°.

There is one clear example of a moderately large eye that forms an image through a *pinhole* (87). The lensless eye of *Nautilus* is ~1 cm in diameter with an ap-

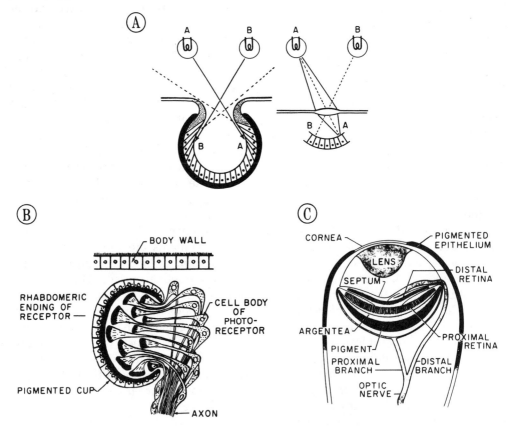

Figure 24. Simple eyes. (*A*) The evolution of directional sensitivity and the ability to form images involves (left) cups lined with receptors and surrounded with dark pigment, or (right) a lens placed above the photoreceptors, or a combination of the two. A lens permits greater light-gathering power. (*B*) Diagrammatic section through a simple eye of the flatworm, *Planaria*. The two eyes are located on the dorsal side of the animal, and the pigmented cups open laterally. The photoreceptor neurons bear rhabdomeric endings inserted into the cavity of the pigment cup. (*C*) Double retina in the "simple" eye of the scallop *Pecten*. The distal retina contains photoreceptors derived from cilia; the proximal retina contains rhabdomeric receptors. [*A* and *B* from (112); *C* from (100).]

erture that can be as small as 400 μm. Although with an aperture of 100 μm, this eye would have an optical resolution nearly as good as a lens eye, the small pupils would entail an enormous sacrifice in sensitivity. Better solutions to the problem of image formation are available and are encountered more frequently.

There are numerous examples of simple eyes with lenses, and great variation in the quality of the retinal image (87). For example, the 12 larval eyes (stemata) of the caterpillar *Isia* have few photoreceptors, which likely function in each eye as a unit; but each of the paired larval eyes of sawflies has a wide field of view (180°), while the individual photoreceptors have fields of ~5°. The ocelli of adult insects form images well behind the retina and are not involved in spatial resolution; the simple eyes of jumping spiders and wolf spiders, on the other hand, are small but highly developed visual organs. Similarly, a wide range of image quality can be found in the simple eyes of annelids.

The principal refracting surface in the simple eyes of terrestrial animals is the air–cornea interface. In aquatic animals, however, the higher refractive index of water requires the presence of spherical lenses of short focal length. In order to reduce both the focal length of the lens as well as the attendant spherical aberration, lenses of inhomogeneous refractive index have appeared in several phyletic lines. Lenses whose refractive index decreases from the center toward the periphery are found in annelids, molluscs, and fish (87).

The Vertebrate Eye

The Human Eye Exhibits a Number of
General Features of Vertebrate Eyes
The human eye, shown diagrammatically in Figure 25, is roughly spherical and is surrounded by a fibrous coat of connective tissue, the *sclera*, which is modified in front as a transparent *cornea*. The lens divides the eyeball into a front compartment filled with *aqueous humor* and a larger, more posterior compartment filled with *vitreous humor*. The *iris*, which gives the eye its color, forms a diaphragm immediately in front of the lens. The aperture of this diaphragm is called the *pupil*.

Immediately internal to the sclera is a second coat called the *choroid*, and between that and the vitreous humor is a third layer, the *retina*. The choroid is vascularized to provide the retina with nutrients, and its internal surface is faced with a sheet of deeply pigmented epithelial cells containing melanin. The retina is the light-sensitive layer and consists of receptors (rods and cones) and several kinds of neurons (Fig. 38). The outer segments of the receptors abut against the pigment epithelium, so that light that reaches the visual pigment has passed from the vitreous humor through two layers of nerve cell bodies and the inner segments of the rods and cones. The nerve cells adjacent to the vitreous humor are called ganglion cells, and their axons pass over the surface of the retina to the *optic disk* or blind spot where they collect, turn through the sclera, and form the optic nerve. By contrast, in the large and optically similar eyes of cephalopod molluscs, visual interneurons are not present in the retina, and the optic nerve consists of axons of the photoreceptors.

Close to the optic axis of the eye is a small patch of retina, the *macula lutea*, which is ~1.5 mm in diameter and is yellow because of the presence of carotenoids. At the center of the macula lutea is a depressed region of the retina ~0.3 mm in diameter called the *fovea*. The fovea contains only cones; it is the region of the retina in which visual acuity is greatest and on which objects are focused when they have the attention of the observer.

The human retina contains ~6.5 × 10^6 cones and ~110–125 × 10^6 rods (127). As

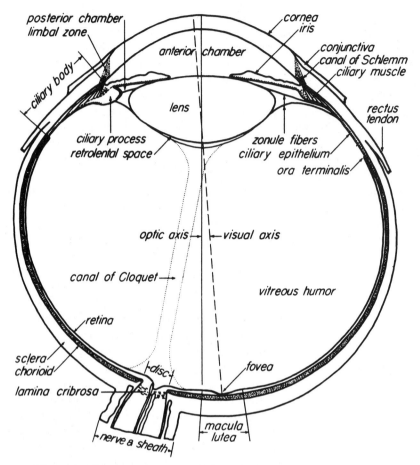

Figure 25. Horizontal section of the right human eye, ×4. [From (174).]

Figure 26 shows, the mosaic of receptors is not uniform over the surface of the retina. The density of cones is very high in the fovea, but falls to very low values in the parafoveal regions. The density of rods, on the other hand, is zero in the fovea, rises to a maximum ~20° from the fovea, and then falls to lower values towards the periphery. Note that Figure 26 shows a transect through the optic disk, where there are no receptors.

There are only ~10^6 fibers in the optic nerve; consequently, there is much convergence of both rods and cones onto single ganglion cells. Because there are ~20 times more rods than cones, the pools of rods supplying single optic nerve fibers are larger than those of cones. In fact, in the fovea there is a 1:1 relationship between cones and ganglion cell axons.

Rods and Cones Have Different Functions
In 1825, J. E. Purkinje reported that as twilight falls, red flowers, which in the strong light of day seem brighter than blue flowers, appear relatively less bright, and with advancing darkness may fade from view before blue flowers. This Purkinje shift, the sliding of maximum sensitivity of the eye to shorter wavelengths with falling levels of illumination, is one manifestation of the transference of vision from cones to rods. Rods mediate scotopic vision, vision in low levels of light.

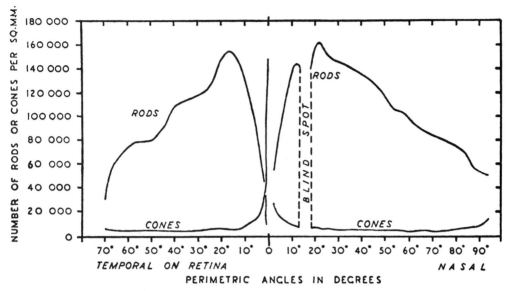

Figure 26. Distribution of rods and cones in the human retina. Cone density is highest in the fovea (0°), whereas rods are most abundant ~20° from the fovea. (Note that the distribution of rods and cones on the nasal side in and near the fovea is not plotted but is similar to the distribution on the temporal side.) [From (132).]

Rod vision is also colorless and relatively blurred. Cones mediate photopic vision when the light is brighter. Cone vision is colored and sharp (5, 132, 143).

Rod cells can be excited by the absorption of a single photon (8, 67, 132), and their intrinsic gain is more than 10 times greater than that of cones (40). Consequently, rod cells are half-saturated when they have absorbed ~30 photons, whereas cone cells require several hundred photons for half-saturation. Because a large group of rods is connected (ultimately) to a single nerve fiber, the rod pool functions as an antenna and increases the sensitivity of the ganglion cell. The absolute threshold for vision occurs when several photons are absorbed nearly simultaneously in as many different outer segments in the receptive field of a single ganglion cell, an area of the retina containing ~500 rods. Rod vision is colorless because rods contain the single visual pigment rhodopsin, and there-

fore differences in wavelength are perceived only as differences in brightness. Rod vision is blurred because the great convergence of rod cells on single nerve fibers makes the effective grain of the retina relatively coarse. And finally, rods recover sensitivity in the dark (*dark adapt*) more slowly than cones.

Cone cells take over under conditions of higher illumination. Because the cones are intrinsically less sensitive and because there is relatively less convergence of receptors on ganglion cells, the absolute sensitivity of the system is lower. But the price is paid in a good cause, for decreased convergence allows for improved visual acuity. Moreover, color vision is possible in a species in which there are several classes of cones containing visual pigments absorbing in different regions of the spectrum (see below).

The scotopic luminosity curve (the apparent brightness as a function of wavelength, measured close to threshold) is

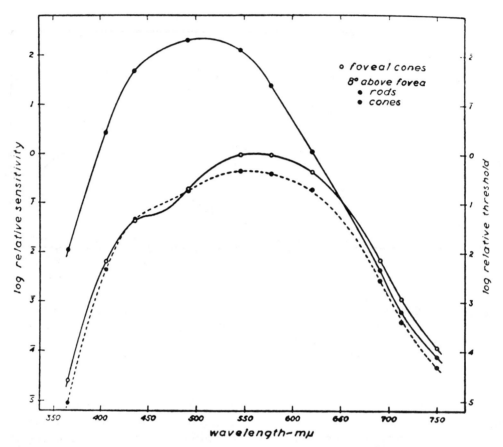

Figure 27. Spectral sensitivity (1/threshold energy) of rods and cones in the dark-adapted human eye. Rods, 8° above the fovea (filled circles, solid line); foveal cones (open circles, solid line); peripheral cones 8° above the fovea (broken line). Test field was 1° in diameter; test flash was 0.04 s. All sensitivies are expressed relative to the maximum sensitivity of the fovea. Note that cones are slightly more sensitive to deep red light than are rods. The shoulder on the foveal sensitivity curve at ≈450 nm is distortion caused by the carotenoids of the *macula lutea*. [From (170).]

determined by the spectral absorption of rhodopsin and is maximal at ~500 nm. The photopic luminosity curve (measured at higher intensities) represents a combination of the outputs of the three kinds of cones and is maximal at ~560 nm. These two spectral sensitivity curves are shown in Figure 27. Only for deep red lights is the absolute sensitivity of the cones greater than that for rods. In the blue, on the other hand, rods are ~3 log units more sensitive than cones. Figure 27 makes clear the explanation of Purkinje's observations on the relative brightness of flowers. Note also the effect of the blue-absorbing macular pigment on the sensitivity of the foveal cones.

Retinomotor Processes Help to Control Sensitivity

The vertebrate eye possesses several accessory mechanisms for adjusting the amount of light reaching the receptors (1, 174). These further enable animals to be active under varied conditions of illumination.

1. *Pigment Migration.* In many fish, anurans, and birds, and to a lesser extent in urodeles and many reptiles, the pigment granules of the pigment epithelium are capable of migrating between the outer segments of the receptors under conditions of bright illumination, and retracting towards the choroid as the ambient light level falls.

2. *Cone Contraction and Extension.* In essentially the same groups of animals that show pigment migration, this process is supplemented by movement of the cone outer segments. As the melanin advances towards the receptors, the cone outer segments retreat before it by the contraction of that part of the inner segment known as the myoid (Fig. 5B). Consequently, under photopic conditions the cones remain exposed.

3. *Rod Movement.* In many fish and anurans, the rod myoids move the rod outer segments in antiphase with the cones (Fig. 5A). Thus when the cone outer segments retract away from the pigment epithelium, the rod outer segments advance into it, and vice versa.

4. *Pupillary Adaptations.* The level of light reaching the retina can also be controlled by contraction of the pupil. With the exception of a few groups such as eels and flatfish, this process is relatively poorly developed in fish. Very good pupillary control of sensitivity is achieved in some reptiles (nocturnal snakes and crocodiles), birds, and mammals. Thus, with the exception of birds, strong pupillary responses are best developed in those groups of vertebrates that do not show migration of retinal pigment and cones.

Much Diversity Has an Ecological Basis
The human eye is a generalized vertebrate eye because it possesses both rods and cones and is not heavily specialized for entirely nocturnal or diurnal life. Species whose activity has become more restricted to the day or night hours show corresponding adaptations of their eyes (174).

Animals that are active only during the day tend to enhance their acuity at the expense of absolute sensitivity. Their eyes tend to be large (within the limits imposed by the size of the head) so that more retinal cells will fall under the image. The number of cones increases at the expense of rods, and most lizards, snakes, many birds, ground squirrels, and turtles have either all-cone retinas or nearly all-cone retinas. Birds and lizards have a region of the retina (*area centralis*) in which there are no rods and the individual cones have become very long and slender. As in the human fovea, this is a region of minimal convergence on higher-order neurons and of maximum visual acuity. The deep foveas of birds are probably the most highly developed. The retina is conspicuously thinned so that the receptors line a small depression, and the pit may act as a diverging lens and enhance visual acuity.

Animals that have forsaken the light of day for the hours of night have sacrificed acuity for sensitivity. Thus, there is an increase in the number of rods and a loss of cones. Many species have a layer of reflecting material (*tapetum lucidum*) in the choroid so that light that traverses the retina without being absorbed is reflected back into the outer segments. This reflecting layer accounts for the retinal "glow" or eye shine of animals staring back into a source of light.

Species that are active at night but that also bask in the sun need enormous control over their pupils. A round pupil that will open wide at night cannot be closed sufficiently in the daytime by means of a sphincter muscle; this accounts for the evolutionary invention of the slit pupil, which is seen to good advantage in, for example, cats and crotalid snakes.

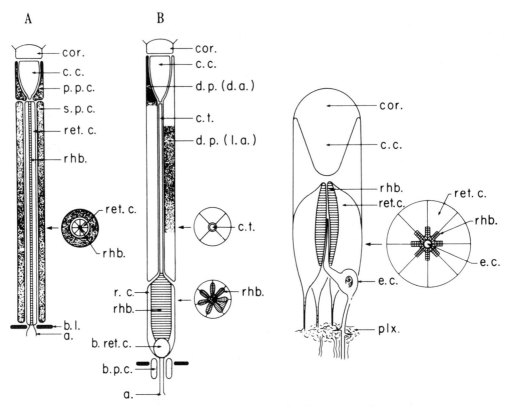

Figure 28. Types of ommatidia. (*A*) Apposition ommatidium of an insect. (*B*) Superposition ommatidium of an insect. (*C*) Ommatidium of *Limulus*. a, Axons of the retinular cells; b.l., basement lamina; b. ret. c., basal retinular cell; c.c., crystalline cone; cor., corneal lens; c.t., crystalline thread or tract; d.p. (d.a.), distal pigment in the dark-adapted position; d.p. (l.a.), distal pigment in the light-adapted position; e.c. eccentric cell; ret.c., retinular cell; rhb., rhabdom; plx., plexus of retinular and eccentric cell axons and collaterals; p.p.c., primary or iris pigment cell; b.p.c., basal pigment cell.

Compound Eyes

In the second evolutionary design (Figs. 23*b* and 28), basic units called *ommatidia*, essentially analogous to simple eyes and containing typically less than a dozen photoreceptors, associate to form a *compound eye*. Compound eyes are characteristic of arthropods. Each ommatidium has its own dioptric structure, consisting of a lens, which is a facet of the cornea and therefore extracellular, and the underlying crystalline cone, formed by four cone cells. Three types of compound eyes are recognized, the commonest names for

which are *apposition, superposition,* and *neural superposition.*

Apposition Eyes

In apposition eyes the photoreceptor organelles (*rhabdomeres*) of a single ommatidium are contiguous, forming a single optical structure (the *rhabdom*). The photoreceptor cells (*retinular cells*) and their rhabdom extend to the inner (apical) end of the crystalline cone, and the ommatidium is sheathed in screening pigment over its entire length (Fig. 28*A*). The granules of screening pigment are located in both the primary and secondary pigment

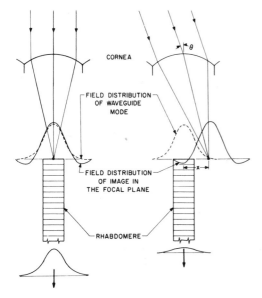

Figure 29. The amount of light trapped by a rhabdomere is proportional to the integral over the focal plane of the quantity [(field distribution of the image) times (field distribution of the waveguide mode)]. This integral is large when the spot is focused on the rhabdomere's axis (diagram on left) since the image and mode distributions have similar shapes and their peaks are in register. The integral is small when the image spot is off-axis (diagram on right) because the peaks of the two distributions are not in register and their product is consequently small. The amount of light trapped by each rhabdomere is symbolized by the distributions at the bottom of the diagram. [From (51).]

cells surrounding the crystalline cone and retinular cells, respectively, as well as in the cytoplasm of the retinular cells. In general, each ommatidium is sensitive to a narrow cone of light incident around the ommatidial axis (Fig. 29). Each facet projects a small inverted image somewhere near the apex of the crystalline cone, at the distal end of the rhabdom. (The images formed by the dioptric structures of adjacent ommatidia are thus in apposition.) In terrestrial species with rounded corneal facets, refraction occurs at the air–

corneal interface; however, this refraction is supplemented, or in aquatic species or species with flat corneal surfaces, replaced, by each dioptric element acting as a lens cylinder. A lens cylinder is a structure whose refractive index is maximal at the center and decreases towards the periphery. A corneal facet–crystalline cone acting as a lens cylinder one focal length long will form a small inverted image of a distant object at the tip of the cone. Such images can be seen in the microscope. The presence of lens cylinders in compound eyes was correctly inferred by Exner in 1891 (38a, 38b), although direct evidence based on interference microscopy is of more recent origin (83).

The refractive index of the rhabdom is higher than the surrounding cytoplasm, and so acts as a waveguide (Fig. 29). Furthermore, rays entering the corneal facet at an oblique angle to the ommatidial axis are captured by the screening pigments, thereby enhancing optical isolation of the individual ommatidia. The entire image is constructed by the CNS from the reports by the individual ommatidia.

Superposition Eyes

In superposition eyes, characteristic of nocturnal or crepuscular species or species living in environments where the light is restricted (e.g., moths and lobsters), the rhabdoms are short and separated by a wide gap from the distal ends of the crystalline cones (Figs. 28B and 30A, b). The screening pigments also exhibit pronounced migrations, filling the space between the cones and the rhabdoms during the day and withdrawing both distally and proximally at night or as a result of extensive dark adaptation. In the light-adapted state oblique rays of light are intercepted by the accessory pigments, and in effect the eyes function similarly to the apposition plan. In the dark-adapted state, however, the light that reaches any individual rhabdom has been

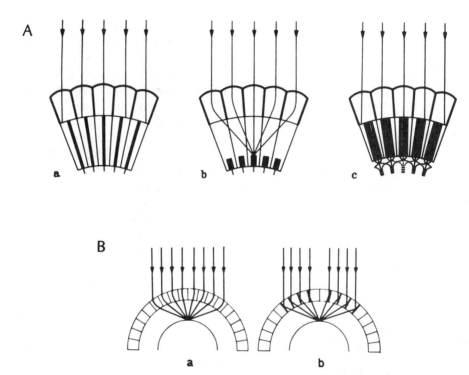

Figure 30. (*A*) Image formation by the three kinds of compound eyes: a apposition, b superposition, and c neural superposition. [From (77).] (*B*) Superposition images can be produced by either of two mechanisms: a, refraction by means of lens cylinders two focal lengths long, as in many nocturnal insects; or b, reflection by a box of mirrors at right angles to each other, as in decapod crustacea like the crayfish. In each case incoming rays are bent back through the same angle of incidence, and a cluster of facets function together to focus a distant point source at the same place on the retina. There is also a recently discovered variant called the parabolic superposition image that involves some of the features of both a and b and in which the focusing effect of the lens is compensated by parabolic reflectors in the crystalline cone (126). Superposition eyes are an adaptation to increase absolute sensitivity. [From (87).]

collected by a patch of facets (Fig. 30*A*,b) and refracted through the clear zone vacated by the migratory screening pigment. The image on the retina is therefore formed by the optical superposition of images from a number of facets (83). The effective aperture of the eye is thereby greatly enlarged with an increase in absolute sensitivity of two or three orders of magnitude.

There are two means by which superposition images are formed (Fig. 30*B*). Exner inferred that the dioptric structures of fireflies (Coleoptera) are lens cylinders two focal lengths long. Such a structure has the properties of bending an incoming ray back through its angle of incidence so that it emerges as shown in Figure 30*B*,a. A region of cornea and cones made up of lens cylinders two focal lengths long forms an erect image at some distance behind the cornea (which is why the rhabdoms in superposition eyes are located deep in the eye and not directly under the crystalline cones), and the individual dioptric elements should form small inverted images at their midpoints (i.e., at one focal length from the corneal surface).

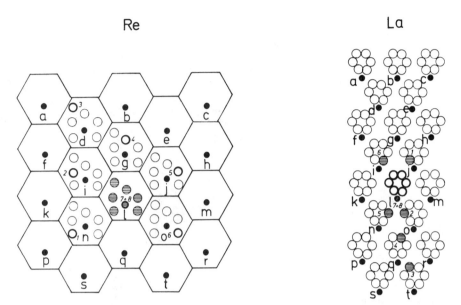

Figure 31. Neural superposition organization of the dipteran retina, Re. Under each hexagonal facet are 6 peripheral (large circles) and 2 central (small circles) photoreceptor cells. The rhabdomeres of the central cells (nos. 7 and 8) are arranged end to end and are therefore represented by a single profile in each ommatidium. The 6 peripheral retinular cells are represented in only 7 of the 20 ommatidia in the diagram. The 6 peripheral retinular cells and the pair of central cells in each ommatidium have 7 different visual fields, and the 6 peripheral cells each project to a different cartridge in the lamina ganglionaris. For example, note how the shaded profiles in the ommatidium labeled l distribute to 6 different laminar cartridges in the diagram to the right, La. Conversely, note how the 6 axons outlined with heavy lines in the laminar cartridge labeled l have come from 6 different neighbors of ommatidium l. The axons terminating in each cartridge share the same visual field. The axons of the central cells terminate deeper in the optic lobe. [From (51), but based on various sources.]

Both of these optical predictions have been confirmed in nocturnal insects with superposition eyes.

The condition for forming a superposition image is that all the light leaving the crystalline cones should be propagating on the same side of the axis of the cone as the incident light. This is achieved by an entirely different mechanism in the superposition eyes of decapod crustacea (86, 166). In these animals the refractive index of the dioptric structures is homogeneous, but the walls of the crystalline cones are lined by highly reflecting pigment. Consequently, rather than being bent back by refraction, oblique rays are reflected into the eye by mirrors (Fig. 30B,b). The presence of reflection superposition images explains the heretofore curious observation that the facets of decapods are square rather then hexagonal, as in most compound eyes. Only with an orthogonal placement of mirrors will the light emerging from the crystalline cones be in the same plane as the incident ray.

Neural Superposition Eyes
The third kind of compound eye, neural superposition (76), looks superficially similar to an apposition eye in that the rhabdomeres extend from the crystalline cone to the basement membrane. Neural

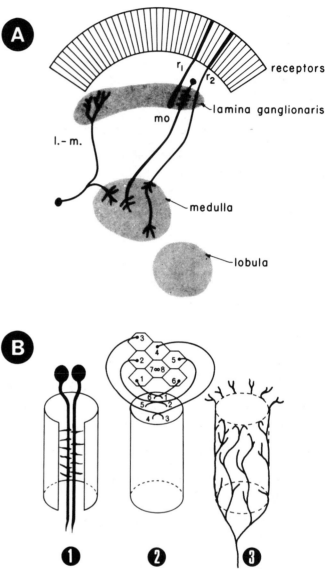

Figure 32. (*A*) Diagram of the optic lobes of an insect showing several of the more peripheral fiber types: r_1, short receptor fiber terminating in the lamina ganglionaris; r_2, long receptor fiber terminating in the medulla; mo, monopolar neuron in the lamina ganglionaris receiving input from the short retinular cell axons; l.-m., lamina–medulla linking fiber (one class of centrifugals). (*B*) Three components of Dipteran optic cartridges. The cartridges are represented by cylinders. 1, The second-order large monopolar neurons with synaptic spines. 2, Six retinular cell axons, shown here for a fly and therefore coming from six different ommatidia (cf. Fig. 31) and synapsing in the cartridge with the monopolars after a 180° reorientation. (For clarity, the fibers are not shown within the cartridge.) 3, A basket-shaped centrifugal fiber.

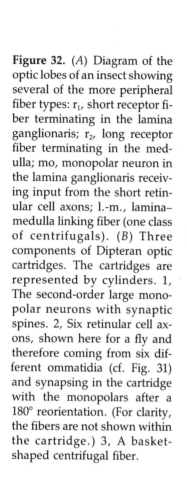

superposition eyes are found in flies. The photoreceptors differ from those of both apposition and superposition eyes, however, in that the individual rhabdomeres are not contiguous but remain optically separate structures (Fig. 7*A*) and look at slightly different points in visual space. But each rhabdomere views the same point as does another rhabdomere in each of six neighboring ommatidia, and axons from six of these seven cells converge on the same second-order neurons (Fig. 30*A*,c). The pattern of rhabdomeres with shared visual fields is diagrammed in Figures 31 and 32. Each laminar neuron thus receives convergent input from six ommatidia, but each ommatidium contributes to the responses of six laminar neurons, each of which has a different visual field.

Optical Design Influences Visual Acuity

Despite the very different optical principles that underlie the construction of simple and compound eyes, similar considerations determine their capacities to resolve images (78, 87, 155, 175). An important factor in visual acuity (angular resolution) is the *angular separation* (Φ) of the individual receptors. Resolution is proportional to $1/\Phi$. For a single-lens eye, $1/\Phi = f/s$, where f is the focal length of the lens and s is the separation between receptors (Fig. 33A). The longer the focal length, the greater the magnification of the image in the focal plane. For a compound eye of the apposition type, on the other hand, from simple geometrical considerations one can readily see that $1/\Phi = r/d_1$, where r is the radius of curvature of the eye and d_1 is the diameter of the ommatidial lens (Fig. 33B).

In either case, angular resolution can be improved by increasing the numerator (f or r) or decreasing the denominator (s or d_1). In single-lens eyes, decreasing s means making the receptors narrower and packing them closer together. But there are limits to how far this process can be carried to advantage. In resolving spatial detail, the responses of adjacent photoreceptors must be distinguished from the noise due to the random nature of the absorption of light. Or put another way, receptors must be able to detect reasonably small differences in intensity; that is, they must possess adequate *contrast sensitivity*. When only a few quanta are absorbed by a photoreceptor, the signal-to-noise ratio (S/N) is poor. From physical principles, S/N should improve in proportion to the square root of the number of quanta absorbed. The probability of photon capture (sensitivity) is proportional to the cross-sectional area presented by the photoreceptor, so making the diameters of the receptors (d_r in

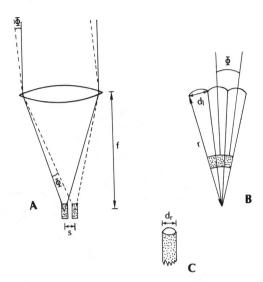

Figure 33. A small angular separation of receptors (Φ) is important for high visual acuity. (A) In a single-lens eye $\Phi = s/f$, where s is the spacing between receptors and f is the focal length. (B) In a compound eye $\Phi = d_1/r$, where d_1 is the diameter of the lenslet and r is the radius of curvature. In each case, Φ can be made smaller by enlarging the eye, but the size of the animal imposes limits. (C) Φ can also be decreased by decreasing s and d_r, the diameter of the receptor, but as described in the text, making the receptors too small decreases contrast sensitivity.

Fig. 33C) very small leads to a decrease in S/N and loss of contrast sensitivity, particularly if the animal must optimize performance in conditions of dim light. Furthermore, as we saw in Figure 29, receptors act as waveguides for light propagating along their length. For waveguides less than about a micrometer in diameter, a significant fraction of the energy (of wavelengths in the visible region of the spectrum) is in a boundary region external to the photoreceptor. Consequently, if the spacing of receptors is too close there will be optical cross-talk between near neighbors. For these reasons the outer segments of vertebrate receptors are not smaller than ~1 µm in

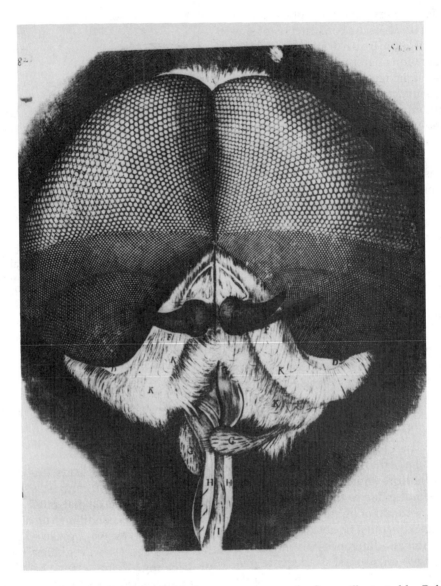

Figure 34. Frontal view of the divided compound eyes of a fly, as illustrated by Robert Hooke in 1665. Note that the dorsal ommatidia are substantially larger than those located ventrally, and the dorso-lateral part of the eye has a much larger radius of curvature than other regions. The optical implications are described in the text. As pointed out by Wehner (175), the insect in this early and famous illustration has been consistently misidentified, starting with Hooke; it is probably a tabanid.

diameter and are spaced at least 2 μm apart.

The obvious way to increase the focal length, on the other hand, is to increase the diameter of the eye, but the size of the animal imposes an upper limit on the size of eye that can be accommodated in the skull. Larger eyes also permit a larger entrance pupil, which is also necessary to preserve sensitivity, because the quantum flux on the image plane (the retina) is inversely proportional to the square of the focal length.

The size of the entrance pupil is important for another reason. The *acceptance angle* of an individual receptor is as important for visual acuity as the *angular separation* between receptors. The acceptance angle p is proportional to λ/d, where d is the diameter of the pupil. In other words, if the pupil diameter becomes too small the image is broadened by diffraction effects. In the human eye the size of the entrance pupil can adjust to give optimal acuity with the spacing of foveal cones; if the pupil stops down too far, acuity is degraded by diffraction, and if the pupil dilates under conditions of dim light, acuity is degraded by lens aberrations.

Similar considerations determine the optical performance of compound eyes. If the facet diameter d_l is made too small, the visual field of the ommatidium is broadened by diffraction. In practice, however, the need for absolute sensitivity and the preservation of S/N imposes a lower limit of facet diameter of ~10 μm, significantly larger than the diffraction limit. The obvious way to increase angular resolution is therefore to increase the radius of curvature of the eye. As with the camera eyes of vertebrates, however, there are anatomical limitations to the size of the eye that can be carried on an insect's head. This obvious restriction is partially circumvented in a very interesting way. Many insects have local regions of the eye in which the cornea is quite flat

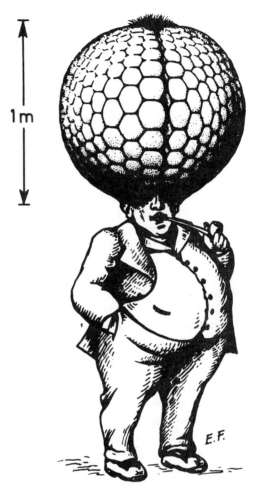

Figure 35. Kirschfeld's view of a man whose compound eye has the same resolving power as the human eye. Each facet in this drawing represents the need for 10^4 actual facets. See the text. [After (78).]

and the radius of curvature of the eye is relatively large. The ommatidial axes in such a "foveal region" are therefore separated by small angles. The facet diameters are also large, so the visual field approaches in size the (small) interommatidial angle, and visual acuity is locally high. Because of the large facets, sensitivity is kept high, and small objects that lie within the visual field of a single ommatidium are more likely to generate a detectable signal. Figure 34 is an early drawing of a tabanid fly with a so-called divided eye in which in the dorsal region both the facet diameter and the radius of curvature are relatively large.

Although the photoreceptor cells of single-lens and compound eyes may have comparable absolute sensitivities and are capable of responding to single photons, the angular resolution and visual acuity of compound eyes is about two orders of magnitude poorer. For a compound eye to have an angular resolution equivalent to that of the human eye it would have to be at least a meter in diameter, with consequences illustrated in Figure 35.

Compound eyes do not accommodate, so there is no near point within which they fail to form images. Therefore when small objects are very close to the cornea they may subtend large enough angles to be resolved. Figure 35 also suggests another positive feature of compound eyes. Because the receptors form a thin veneer on a convex surface, the visual field of the eye may encompass virtually all of the animal's surround. By contrast, in a single-lens eye the receptors are on a concave surface, and the visual field of the eye is a much smaller solid angle determined by the properties of the lens and its aperture.

Visual Systems Code Contrast at an Early Stage of Information Processing

Lateral Inhibition between Receptors in the Eye of Limulus Enhances Spatial Contrast

The ommatidia of *Limulus* are different from those of insects and crustacea. Here the second-order unit, the eccentric cell, has its cell body lying within the ommatidium, is electrically coupled to the retinular cells, and consequently is excited when the retinular cells depolarize. The axons of the eccentric cells are believed to be the only afferent fibers in the optic nerve that conduct spikes. There are no other neuron cell bodies in the region immediately proximal to the basement lamina, although there is an extensive plexus of nerve fibers made up of collateral processes of retinular cell and eccentric cell axons (Fig. 28C).

The presence of inhibitory interconnections in this plexus means that an individual ommatidium responds independently of its neighbors only if it is the sole element illuminated. Quantitative studies of this system by Hartline, Ratliff, and their colleagues (65) have elucidated many of its properties, which are similar to some of the functional characteristics of the vertebrate retina where there are more different kinds of neurons.

The steady-state interactions between two ommatidia can be described by a pair of simultaneous linear equations (140):

$$r_1 = e_1 - k_{1,2}(r_2 - r^0_{1,2})$$
$$r_2 = e_2 - k_{2,1}(r_1 - r^0_{2,1})$$

The symbols r_1 and r_2 are the responses (spikes per unit time) of the two ommatidia; e_1 and e_2 are the responses of the two ommatidia when each is illuminated alone; $k_{1,2}$ is the coefficient of inhibition of unit 2 on unit 1, and $r^0_{1,2}$ is the threshold firing rate that unit 2 must have to exert an inhibitory effect on unit 1. These equations say that the response of unit 1 is determined by the amount of excitation it receives, reduced by the inhibition it gets from the other ommatidium. This inhibition is in turn directly proportional to the firing rate of the second ommatidium, once a certain threshold value is exceeded. The inhibition between ommatidia is reciprocal but it is not symmetrical; that is, the two inhibitory coefficients and the two threshold rates are in general not equal. The magnitude of the inhibition decreases as the distance between the two

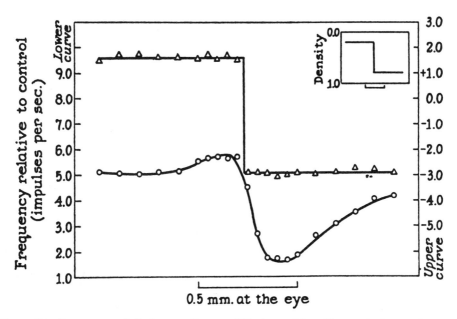

Figure 36. Frequency of discharge of ommatidia (eccentric cell axons) along a transect through the step gradient of intensity shown in the inset of the upper right. Triangles: When the ommatidia are illuminated singly, the spatial pattern of discharge follows the pattern of intensity. Circles: When the whole mosaic of receptors is illuminated by the pattern, the discharge of ommatidia follows the lower (curvilinear) graph. The enhancement and suppression of units on either side of the light–dark boundary is caused by lateral inhibition. [From (140).]

ommatidia increases. The effects of additional ommatidia are additive.

One of the important consequences of this system is to enhance spatial contrast at boundaries (139). Figure 36 (inset) shows a step gradient of light intensity imposed on the eye. The upper curve (open triangles) shows the responses of successive ommatidia across the eye when illumination is restricted to the unit being tested. The curve reflects the objective pattern of light intensity on the eye (inset). When the whole pattern is displayed at once, however, the outputs of the ommatidia follow the lower curve (open circles). On the light side of the boundary, the ommatidia are firing maximally because they are receiving little inhibition from units immediately across the dark zone. Conversely, the units just across the boundary on the darker side

have the lowest output, because being close to the cells with the greatest firing rate, they are receiving more inhibition than any other units. The net effect is to accentuate the differences in firing rate of ommatidia on the two sides of the light–dark border.

Other geometrical effects are possible, including disinhibition. Disinhibition can be produced in a row of three (nonadjacent) ommatidia when the two units at the ends of the row are so far apart that they exert negligible inhibition on each other. Excitation of one of the end ommatidia (the test ommatidium) causes it to fire. Simultaneous excitation of the middle ommatidium decreases the firing rate of the test unit. If the third ommatidium is now also excited, it slows the firing rate of the middle ommatidium,

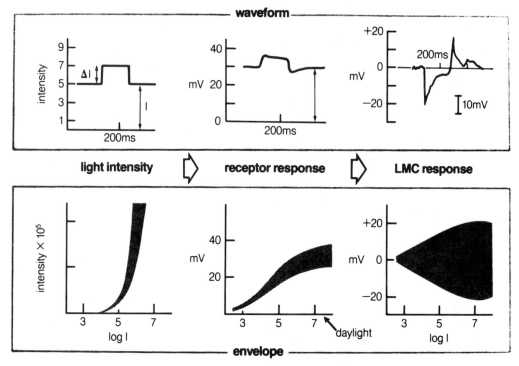

Figure 37. Contrast coding at early stages of the visual system of flies. The upper box shows the waveforms of an intensity signal of contrast $\Delta I/I = 0.4$ and of the corresponding graded voltage responses of a photoreceptor cell and a large monopolar neuron (LMG). The lower box shows the envelopes of intensity signals, receptor responses, and LMC responses, as a function of log background intensity. The envelope is the range of signal amplitudes for signals of contrast $= \pm 0.4$, which embraces ~75% of natural signals seen by flies. For the graphs of receptor and LMC potentials, 0 V on the ordinate corresponds to the resting potential of the cell. See the text for discussion. [From (89).]

causing the test ommatidium to escape from some of its previous inhibition.

The dynamic aspects of inhibition become still more complicated. The presence of self-inhibition, a form of negative feedback, produces a degree of temporal sharpening of the response and may account for some but not all of the adaptation present in eccentric cell axons. The interplay of self- and lateral inhibition, which operate with nonidentical time courses, can also produce transient effects. For example, the postinhibitory rebound of a cell released from lateral inhibition may be due to a relative lag in the development of self-inhibition (65, 137).

The Retina of the Fly Illustrates Additional Principles of Contrast Coding

The visual world consists of objects that reflect different fractions of the available incident light. The absolute amount of light varies greatly through the day, and yet eyes are able to operate effectively over a span of intensities of at least 10^4. Eyes therefore function through the detection of relative contrasts between objects, contrasts that remain constant regardless of the absolute intensity.

Figure 37 shows how this is achieved in the early stages of visual processing in the fly (89). The upper row (left) shows a step of light that increased in intensity by a factor 0.4, and the corresponding voltage responses from the photoreceptor (middle) and second-order neuron in the lamina (right). The large monopolar cell (LMC) in the lamina hyperpolarizes due to an increase in chloride conductance when the transmitter histamine is liberated from the photoreceptor terminals.

In discussing the responses of receptors earlier, we saw that voltage increases in a sigmoidal fashion with the log of the intensity. The shaded areas in the lower row in Figure 37 show how a stimulus contrast of ±0.4 (left) is translated into changes in membrane voltage in the receptors (middle) and in the second-order neuron (the LMC, right), each plotted as a function of the log of the background intensity. Consider the behavior of the photoreceptors. As the background intensity increases, the cell depolarizes to a maximum of ~30 mV. Abrupt changes of contrast of ±0.4 cause increases or decreases in membrane voltage that fall within the shaded area. Although the change in response increases at low intensities, over the range of intensities that correspond to ordinary daylight, the response to a contrast of ±0.4 is nearly independent of the background intensity.

The voltage changes in the photoreceptors are only several millivolts in amplitude, and over much of the intensity range they ride on a larger background signal. They are therefore subject to pollution by neuronal noise, particularly noise originating at the synapse. As the responses of the LMC (right) show, however, there is a severalfold amplification of the signal at the receptor–LMC synapse that helps to safeguard the signal

from this kind of contamination. Furthermore, the LMC responses are much more phasic than the receptor potentials. The basis for the antagonistic component of the response is not known, but seems to involve influences arising in the same laminar cartridge (see also Fig. 32) as well as from neighboring cartridges. The result, however, is to subtract away much of the background response averaged over a local area, leaving primarily the temporal response to contrast and extending the dynamic range of the cell.

As we shall see in the following section, these observations are likely of general significance. The receptor–LMC relationship is very similar to the receptor–bipolar unit in the vertebrate eye, where the second-order cell also codes with graded potentials and does not generate propagated spikes. Moreover, antagonistic lateral interactions, like those first described in detail in the eye of *Limulus*, are a general feature of retinal networks and are mediated by horizontal and amacrine cells in the vertebrate retina (32, 71).

The Vertebrate Retina Is an Accessible Outpost of the Central Nervous System

The retina is both part of a sense organ and part of the brain. As a thin sheet of accessible vertebrate CNS it has attracted considerable interest as a model system for correlating defined retinal stimuli with both structure and function of identified neurons (32).

Five Kinds of Neuron Share the Retina with Rods and Cones

In addition to the receptors, the retina contains five types of higher-order nerve cells. The fibers of the optic nerve are therefore at least two synapses removed from the rods and cones. The principal

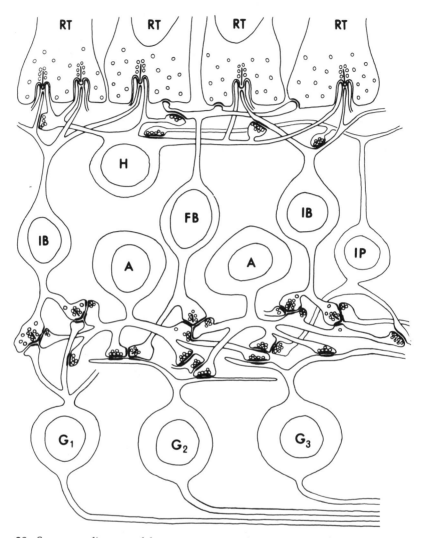

Figure 38. Summary diagram of the arrangements of synaptic contacts found in vertebrate retinas. In the outer plexiform layer, processes from invaginating bipolar (IB) and horizontal (H) cells penetrate sockets in the receptor terminals (RT) and terminate near the synaptic ribbons of the receptors. The processes of flat bipolar cells (FB) make superficial (basal) contacts onto the bases of some receptor terminals. Horizontal cells make conventional synaptic contacts onto bipolar dendrites. The processes of the interplexiform cells (IP) make conventional synapses onto horizontal cells or bipolar cells or both.

In the inner plexiform layer, bipolar terminals most commonly contact one ganglion cell (G) dendrite and one amacrine cell (A) process at the ribbon synapse (*left side of drawing*), or two amacrine cell processes (*right side of drawing*). When the latter arrangement predominates in a retina, numerous conventional synapses between amacrine cell processes (serial synapses) are observed. Amacrine cell synapses in all retinas make synapses back onto bipolar cell terminals (reciprocal synapses).

The input to ganglion cells may differ in terms of the proportion of bipolar and amacrine synapses. Ganglion cells may receive mainly bipolar cell input (G_1), a more even mixture of bipolar and amacrine cell input (G_2), or mainly or exclusively amacrine cell input (G_3). The interplexiform cells receive input from amacrine cells, and they occasionally make synapses on other amacrine cell processes. [From (32).]

cell types and the major synaptic connections are shown in Figure 38. The cell bodies lie in three distinct layers marked by aggregations of cell nuclei. The *outer nuclear layer* is closest to the choroid and consists of the cell bodies of the receptors. This layer lies just off the top of the illustration in Figure 38. The *inner nuclear layer* contains the cell bodies of four kinds of interneurons: *bipolar cells* and *interplexiform cells,* whose axes are vertical (radial), and the *horizontal* and *amacrine cells,* which extend horizontally (tangentially) through the retina. The *ganglion cell layer* is closest to the vitreous humor, and as described above, the axons of these cells collect at the optic disk and form the optic nerve.

Between the three layers of nuclei are two synaptic regions: in the *outer plexiform layer* the receptor cells synapse with bipolar and horizontal cell dendrites, and in the *inner plexiform layer* the bipolar cells connect to amacrine and ganglion cell dendrites. To summarize, the simplest path of information flow through the retina is receptor → bipolar → ganglion cell. However, at the two synaptic regions there are additional interneurons, the horizontal and amacrine cells, that mediate lateral influences, and in some retinas the simplest possible path is seldom followed. The synaptic architecture of the vertebrate retina can be pursued further in Dowling's recent book (32).

In addition to the types of neurons described above, in at least some species (birds) there is a small number of centrifugal fibers of unknown function that end on the processes of the amacrine cells in the inner plexiform layer. In addition, glial elements known as Müller cells run as columns connecting the cell layers of the retina. The Müller cells bear lateral processes that invest the neural elements. The membrane potentials of these glial cells respond to potassium ions accumulating in the intercellular spaces of the retina as a result of neuronal activity. Interestingly, these passive glial potentials contribute a significant portion of the electroretinogram (ERG), the mass electrical response of the eye that is recorded with gross electrodes.

Receptors

The receptors make characteristic synapses with a triad of postsynaptic elements inserted into the base of the rod or cone and consisting of a centrally placed bipolar terminal flanked by two horizontal cell dendrites (Fig. 38). These synapses also have a prominent presynaptic ribbon. The receptors also make more conventional synapses with a subset of bipolars. In lower vertebrates and perhaps elsewhere, receptors are also found to be electrically coupled to each other through gap junctions.

Bipolar Cells

Bipolar cells connect the receptors to the inner plexiform layer. In primates there are rod bipolars and two morphological kinds of cone bipolars, invaginating and flat, based on their synapses with the receptors. The synaptic terminals of bipolar cells in the inner plexiform layer have presynaptic ribbons, which help to identify them in electronmicrographs.

Horizontal Cells

The neurites of the horizontal cells have extensive lateral spread. One kind of horizontal cell (not found in primates) has an axon with a large terminal arborization that is greater in area than the entire dendritic arbor. As the axon does not exhibit propagated spikes (see below), the two ends of the cell presumably function with a large degree of independence of each other. Horizontal cells receive input from the receptors and mediate lateral interactions between bipolars. In at least some species (e.g., turtle) they feed back onto receptors, and in teleost fish, onto amacrine cells. Horizontal cells with the same function are electrotonically coupled through gap junctions.

Amacrine Cells

The amacrine cells receive input from bipolar cells and from each other and output onto other amacrine cells, bipolar cells, and ganglion cells. They thus mediate lateral influences in the inner plexiform layer and insert themselves to varying degrees between bipolars and ganglion cells. The latter role of amacrine cells is more prominent in lower vertebrates than in primates, with consequences for the response properties of ganglion cells (see below).

Interplexiform Cells

These cells do not stain well by the Golgi method and were discovered only recently. As their name suggests, they carry information between the two synaptic layers of the retina, mostly from amacrine to horizontal and bipolar cells.

Ganglion Cells

The axons of the ganglion cells are the fibers of the optic nerve; therefore, all information that leaves the retina must do so via ganglion cells. Ganglion cells receive input from bipolars and amacrine cells, but their receptive fields vary from a single foveal cone to several hundred rods.

The Retina Functions Almost Entirely with Slow, Graded Potentials

In a previous section we drew on examples of arthropods to illustrate how visual processing involves the detection and amplification of spatial and temporal contrasts. This same theme is found in the vertebrate retina (32, 71).

Receptors

The hyperpolarizing responses of rods and cones were discussed earlier as part of a consideration of the problem of transduction. We saw earlier in the discussion of contrast detection in the fly retina that photoreceptor signals of one or a few millivolts are subject to contamination by neural noise. The electrotonic coupling that has been observed between toad rods and between turtle cones probably serves to reduce this contamination and to enhance synaptic gain.

Cones in at least some species receive antagonistic feedback from horizontal cells. A cone stimulated with a large spot of light responds with a more phasic potential than if the spot were only a few micrometers in diameter. This is reminiscent of the transformation that takes place in responses of the dipteran LMC (see above), suggesting that the subtraction of average local background is a common feature of signal processing in visual systems.

Bipolar Cells

The area of retinal surface that contributes to the response of a single bipolar cell—its receptive field—includes a number of receptors (but not for midget bipolars in the primate fovea). The receptive fields of bipolar cells have a circular center and an annular surround with antagonistic properties (Fig. 39). The membrane response of a bipolar cell is graded with intensity, and there are no spikes. The response may be of either polarity, hyperpolarizing or depolarizing. If stimulation of the center of the receptive field causes hyperpolarization, light in the annular surround causes depolarization. Some bipolar receptive fields are color coded, as described below for ganglion cells. The receptive field surrounds are much larger (≈ 1 mm) than the dendritic trees of bipolar cells, and influences from receptors in the periphery of the receptive field are thought to be mediated by the horizontal cells. The antagonistic surround can shift the operating curve of the center, which is another manifestation of the importance of detecting contrast over a wide range of background intensities.

Horizontal Cells

Horizontal cells respond with slow graded hyperpolarizing responses at all wavelengths (L type), or with hyperpolarization to short wavelengths and depolarization to long wavelengths (C type). The L type signal luminosity; the biphasic C type signal chromatic information. The latter are of two kinds, hyperpolarizing to green and depolarizing to red, or hyperpolarizing to blue and depolarizing to yellow. The yellow–blue units sometimes show an additional hyperpolarization in the red region of the spectrum. Selective adaptation with wavelengths at one end of the spectrum can enhance the response of C units at the other end, indicating that there are different kinds of cones contributing to the response.

The receptive fields of horizontal cells are 2–5 mm on the retina, enormous compared with the 30–150 μm embrace of their dendritic arbors. These large receptive fields are due to the extensive electrical coupling between horizontal cells.

Amacrine Cells

Most amacrine cells respond with transient depolarizing responses with one or two spikes (Fig. 39). It may be, however, that the spikes are local amplifiers and are not propagated throughout the cell. Transient amacrine cells signal *on*, *off*, or *on–off*, depending on the geometrical pattern of stimulus on the retina and the properties of the particular cell. They do not have center-surround receptive fields, and are frequently responsive to moving stimuli.

Some amacrine cells have sustained responses, which may be either depolarizing or hyperpolarizing. Spikes can be triggered by depolarization. Some cells show evidence of input from more than one spectral class of cone.

Interplexiform Cells

Little is known of the response properties of these cells. They appear to be similar to amacrine cells.

Ganglion Cells

Ganglion cells depolarize and give rise to bursts or trains of spikes, which are propagated along the optic nerve. Most of the knowledge of receptive fields of ganglion cells has been obtained from extracellular recordings of spikes, however. In general, ganglion cells, like bipolars, have receptive fields ~1 mm in diameter with a circular center and an annular surround. The responses of ganglion cells can be classified as either *sustained* or *transient*, but there is in fact a variety of ways of describing ganglion cells in the literature.

In the first type the center of the receptive field may respond with a sustained discharge to light *on* and the surround to *off*, or the ganglion cell may have an *off* center and *on* surround. The size of the receptive field is not necessarily static, however. For example, the degree of inhibition from the surround can increase with light adaptation, with the result that late in the process of dark adaptation the inhibitory surround of an on-center unit may drop out.

The receptive fields of ganglion cells with sustained responses may be color coded if fed by cones (Fig. 40). In the ground squirrel, which has an all cone retina containing green-sensitive cones and blue-sensitive cones, those ganglion cells that have chromatic input may respond to green at *on* and blue at *off* or vice versa. For some ganglion cells, the green sensitive part of the receptive field is coincident with the blue sensitive part. For others, green sensitivity is confined to the center whereas blue sensitivity is distributed through the center and sur-

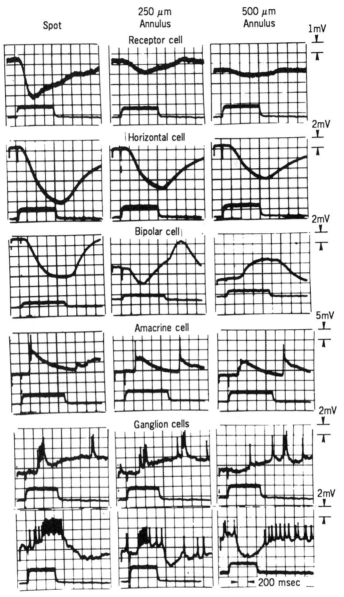

Figure 39. Recordings showing the major response types in the *Necturus* retina and the difference in response of a given cell type to a spot and to annuli of 250- and 500-μm radius. *Receptors* have relatively narrow receptive fields, so that annular stimulation evokes very little response. Small potentials recorded with annular stimulation are due to coupling and to scattered light.

The *horizontal cell* responds over a broader region of the retina, so that annular illumination with the same total energy as the spot (left column) does not reduce the response significantly (right columns).

The *bipolar cell* responds by hyperpolarization when the center of its receptive field is illuminated (left column). With central illumination maintained (right trace; note lowered base line of the recording and the elevated base line of the stimulus trace in the

round. For still others, there is a complete separation of the two populations of cones, green at the center and blue in the surround. Such units as these are apparently involved in the heightening of color contrast.

In fish, color coded ganglion cells have been described with receptive fields up to 5 mm (40–60°). The responses of these ganglion cells are somewhat more complicated than those of the ground squirrel and show double color opponentcy. The center responds to red *on*–green *off* (or red *off*–green *on*). The retinal regions containing the two types of cone are not quite coincident, with the green-sensitive cones occupying a somewhat larger area. The use of stimuli in the shape of an annulus shows that the actual surround extends much further out and is truly antagonistic to the center. That is, for a red *on*–green *off* center unit, the response to light in the surround is red *off*–green *on*. Such cells, unlike the green *on* center:blue *off* surround units of the squirrel's retina, will heighten color contrast at a spatial border.

The transiently responding ganglion cells give phasic responses to *on* and *off*, but their more characteristic feature is their sensitivity to movements of edges through the receptive field. In the retina of the frog, the adequate stimulus for one group of cells is movement of a small convex edge, a feature that makes these cells admirably suited for detecting the natural prey of frogs. Another characteristic of movement detectors is that they are relatively insensitive to the degree of contrast between the moving object and the background. In several species, cells have been found that respond to unidirectional movement of edges across the retina; movement in the opposite direction causes a suppression of any ongoing activity.

There is an interesting correlation between these types of ganglion cell and the synaptic structure of the inner plexiform layer. Species such as primates that have a large proportion of ganglion cells fed directly by bipolars have many ganglion cells that give sustained responses. On the other hand, species such as frogs, pigeons, and rabbits, in which most of the ganglion cell input comes via amacrine cells, have a higher proportion of ganglion cells that respond to movement.

records) annular illumination antagonizes the sustained polarization elicited by central illumination, and a response of opposite polarity is observed. In the middle column the annulus was so small that it stimulated the center and periphery of the field simultaneously.

The *amacrine cell* was stimulated under the same conditions as the bipolar cell, and gave transient responses at both the onset and cessation of illumination. Its receptive field was somewhat concentrically organized, giving a larger *on* response to spot illumination, and a larger *off* response to annular illumination of 500-μm radius. With an annulus of 250-μm radius, the cell responded with large responses at both on and off.

The *ganglion cell* shown in the upper row was of the transient type and gave bursts of impulses at both on and off. Its receptive-field organization was similar to the amacrine cell illustrated above. The ganglion cell shown in the lower row was of the sustained type. It gave a maintained discharge of impulses with spot illumination. With central illumination maintained, large annular illumination (right column) inhibited impulse firing for the duration of the stimulus. The smaller annulus (middle column) elicited impulses at *off*. [From (176).]

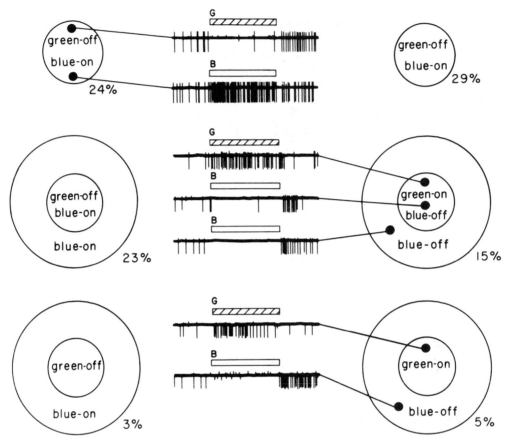

Figure 40. Summary diagram of the receptive fields of color-coded ganglion cells in the retina of the ground squirrel, illustrated with some sample spike trains. Two kinds of cone are present, maximally sensitive to green and blue light. Each ganglion cell receives input from both classes of cone, being inhibited by one and excited by the other. The ganglion cells can also be classified according to the degree of overlap between the excitatory and inhibitory regions of the receptive field. Percentages give relative frequency of recording of each type. [After (105,106).]

Eyes Adjust Sensitivity by Processes Known as Light and Dark Adaptation

We saw earlier the importance of maintaining sensitivity to contrasts over a wide range of ambient light levels. The human eye is capable of adjusting its sensitivity over a range of intensities of $\sim 10^{10}$. Pigment migration in the choroid is absent, and the response of the pupil accounts for only ~ 1.2 log units of control. Consequently, for an explanation one must look at processes intrinsic to the receptor cells and their associated neurons.

Light and Dark Adaptation Are Measured Differently but Involve the Same Underlying Changes

Two adaptation phenomena are illustrated in Figure 41. *Light adaptation* is typically described by the *increment threshold function*, the amount of light that must be

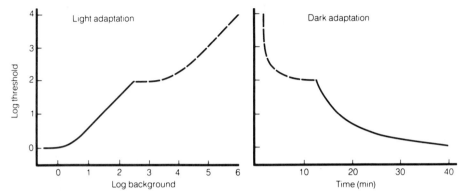

Figure 41. Light adaptation (left) is described by the increment threshold function and dark adaption by the fall in threshold with time in the dark following a period of light adaptation. These curves are for the human eye and show the contributions of both rods (solid lines) and cones (broken lines). Note that rods have lower thresholds and adapt more slowly. [From (32).]

added to an adapting background to be seen by the subject or to cause a response in a cell under study. In the human eye the curve has two limbs, the lower one due to rods and the upper one to cones. For each limb, once a threshold zone is passed the slope of the log–log plot is 1, meaning that the intensity required for threshold is directly proportional to the background intensity. This is the Weber–Fechner relationship, where $\Delta I / I$ is constant. When the background intensity is changed, the new threshold is characteristically reached in a couple of seconds.

Dark adaptation is the recovery of sensitivity when the adapting light is extinguished; it is therefore characterized by a graph of log I required for threshold plotted as a function of time. The time for dark adaptation is usually much longer than the time required to adjust to a new position on the increment threshold curve. The dark adaptation curve also exhibits rod and cone limbs.

Although light and dark adaptation have different operational definitions, they both describe changes in sensitivity. These sensitivity changes can reside either in the photoreceptors or in the associated retinal networks of neurons.

Sensitivity Changes Can Arise in the Photochemistry or Elsewhere

The recovery of rods can be conveniently studied in species or individuals that lack cones; otherwise, the first part of the recovery is not measurable because the sensitivity of the eye is determined by the photopic mechanism (147). In the rod retina of the rat, recovery of sensitivity after moderate adapting exposure is a relatively fast process that is completed in seconds or at most a very few minutes (31, 32) (Fig. 42). If the adapting exposure is more intense and a significant quantity of rhodopsin is bleached, the rapid phase of adaptation is followed by a slower phase that can be related to the amount of visual pigment in the rods.

Photochemical Adaptation
During this slower recovery, the log of the threshold is directly proportional to the fraction of the rhodopsin that remains bleached. Why this seemingly simple

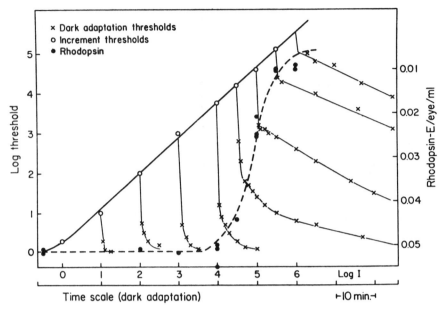

Figure 42. Visual adaptation in the rat eye as determined by sensitivity of the b wave of the electroretinogram. During light adaptation (open circles, heavy line; increment thresholds) the increase in log threshold is linearly proportional to the log of the background luminance, except at the dimmest background luminances. Dark adaptation (crosses, thin lines) is rapid unless the eye is adapted to background luminances bright enough to bleach significant quantities of rhodopsin (filled circles, dotted line) in the 5-min adaptation period. With bright background luminances, a slow component of dark adaptation is observed, the extent of which depends on the amount of rhodopsin bleached. [From (31).]

mathematical relationship should exist is not clear. To add to the mystery, the proportionality constant is 3 for human cones (69), 19 for human rods, and takes intermediate values in the photoreceptors of other vertebrates (31, 33).

In invertebrates, on the other hand, the relationship between sensitivity and pigment loss follows a simple prediction from physics: The change in threshold caused by the conversion of rhodopsin to metarhodopsin is directly proportional to the decreased probability of quantum catch. In a lobster photoreceptor it is therefore necessary to photoconvert 90% of the rhodopsin to achieve a 1 log unit elevation of threshold, whereas in a human rod a fractional bleach of 1/19 (or ~5%) produces the same result (6).

Receptor Adaptation

Photoreceptors adapt even when vanishingly small proportions of their photopigment have been isomerized (32, 57, 85). A dark-adapted rod can give saturating responses to dim flashes that activate 100 of its 10^9 rhodopsin molecules, yet receptors frequently respond to stimuli superimposed on backgrounds >3 log units above the dark-adapted threshold. Two mechanisms contribute to this capacity. First, the response decays from the initial peak, in gecko to a plateau ~0.6 the peak value. This adaptation may be caused by the decrease in internal calcium caused by shutting down some of the light-regulated conductance of the outer segment.

Second, background lights cause a change in the gain of the transduction process. This is reflected as an increase in σ in the equation $V/V_{max} = I/(I + σ)$, which describes the photoresponse. An increase in σ, the intensity for a half-maximal response, means that the curve of log V versus log I is shifted to the right on the energy axis but with no change in slope. As described above, the slope of the log–log plot is 1 until the number of channels becomes limiting and the response approaches saturation. Unit slope on a log–log plot means that the response is directly proportional to intensity; a shift to the right on the log–log plot signifies that the constant of proportionality has increased. That is, the gain has decreased and more photons are required for a given change in membrane voltage. The molecular site of the gain change presumably lies in the transduction cascade but has not been identified.

Network Adaptation
Finally, retinal neurons postsynaptic to the receptors can contribute to visual adaptation (32, 57). Such effects are evident at background intensities too low to produce measurable adaptation of the receptors and speak to the presence of amplification stages analogous to those described above in the fly retina. This site of adaptation is particularly important in the scotopic system of primates, where the rods exhibit little capacity for adjusting their gain and saturate abruptly.

Color Vision

Color Vision Requires More Than One Visual Pigment

As we have seen, the output of a photoreceptor, or more precisely the change in membrane voltage, depends on the number of photons absorbed. And the probability of absorption varies with wavelength according to the absorption spectrum of the visual pigment. Because a change in membrane voltage is triggered by isomerizations of the chromophore, a receptor has no way of distinguishing between a few incident photons at the $λ_{max}$, where the probability of absorption is high, and a larger number of photons at a wavelength far removed from the absorption maximum, where the chance of absorption is much lower, provided that both stimuli cause the same number of isomerizations. In a single receptor, therefore, wavelength and intensity cannot be distinguished from one another, a characteristic known as the *principle of univariance* (118).

In order for different wavelengths to produce distinct sensations, the visual system must be able to compare the outputs of two or more classes of receptor having different spectral sensitivities. In principle, there are several ways in which different spectral classes of receptor might be created: (a) by several visual pigments with different absorption spectra packaged in separate cells; (b) two or more visual pigments mixed in different proportions in different cells; (c) one visual pigment associated with several color filters, so situated that the cells have different spectral sensitivities.

In the real world of animals, the first of these possibilities is by far the most important; all known color vision systems, whether in vertebrates or invertebrates, are based on the presence of two or more visual pigments located in different cells. Although there are a few cases where two visual pigments are found mixed in the same cell—notably rhodopsin and porphyropsin in the same rod—it is less clear that cells normally synthesize more than one kind of opsin, and there are no demonstrated examples of color vision systems based on different mixtures of visual pigments. On the other hand, the presence of brightly colored cone oil droplets in the retinas of some vertebrates, notably many reptiles and birds, has led to speculation that these organelles provide the basis of color vision in sauropsids, but it is now clear that although oil droplets undoubtedly modify the spectral sensitivity of the underlying outer segment, there is also a multiplicity of cone pigments in these retinas (Fig. 15).

The Conceptual Foundations for Thinking about Color Vision Are Based Largely on Human Data

The first essentially correct insight into the nature of human color vision is frequently attributed to Thomas Young, writing in the first decade of the nineteenth century. Since Newton it had been recognized that white light could be decomposed by a prism into a continuum of colors, and it was common experience that most people are able to distinguish a very large number of different colors. Moreover, artists knew that all colors of paint could be generated by mixing three suitably chosen pigments. Against this background Young noted (1802) (180)

> Now as it is almost impossible to conceive each sensitive point of the retina to contain an infinite number of particles, each capable of vibrating in perfect unison with every possible undulation, it becomes necessary to suppose the number limited; for instance to the three principle colors. . . each sensitive filament of the nerve may consist of three portions, one for each principal colour.

And further (1807) (181)

> From three simple sensations, with their combinations, we obtain several primitive distinctions of colours; but the different proportions, in which they may be combined, afford a variety of tints beyond all calculation. The three simple sensations being red, green, and violet, the three binary combinations are yellow, consisting of red and green; crimson, of red and violet; and blue, of green and violet; and the seventh in order is white light, composed by all three united.

Young made two important statements. The first is that the human retina has three differentially sensitive input channels. He said ''nerves,'' but we would now say cones. The second statement is that all simple sensations of color can be produced with three variables—''primitive distinctions of colours.'' This is the *trichromacy* of human color vision. It took more than 150 years before objective measurements of the absorption of three kinds of cone outer segments validated Young's first inference. But his second statement, which is based on the first yet embodies a profound insight into how the visual system works, was verified quantitatively in the 1850s by work largely associated with the physicists Helmholz and Maxwell. To see how this was done we must consider an important principle of *psychophysics*.

Obviously stimuli that produce the same output from the receptors will be indistinguishable. We considered one example above—the effect of varying wavelength and intensity on the output of a single kind of receptor. But the idea can be generalized to arrays of receptors of different kinds; if two physically different stimuli generate identical patterns of output in the ensemble of receptors, the stimuli will produce identical sensations. But we can also turn this idea around. Stimuli that are indistinguishable must produce the same physiological condition at some stage in the nervous system. This provides us with a powerful psychophysical tool: identifying the physical properties of those stimuli that produce identical responses, as judged by the inability of an observer to detect a difference. The power of this approach stems in part from the fact that it is independent of the details of intervening neural processes; the observer functions simply as a null detector.

Consider a source of white light with a continuous (but not flat) spectrum. To characterize this light physically we must specify the energy present at each wavelength, and for a continuum of wavelengths this requires, for ultimate precision, an infinity of variables. Psychophysically, however, three variables suffice, provided that the light is sufficiently bright to stimulate the cones and is directed onto the retina within 2–3° of the central fovea. Operationally, what does this mean? Suppose that we view this light through an optical device that allows us to project it onto one half of a split field. The nineteenth century physicists found that for a test light of any spectral composition, the two halves of the field could be made to appear identical by a mixture of no more than three suitably chosen reference lights. For convenience the reference lights are frequently monochromatic, for example, narrow bands of wavelengths appearing red, green, and blue, but they need not be monochromatic, and in fact the only restriction on their spectral composition is that collectively they include wavelengths from broadly separated parts of the visible spectrum. In this specific example, if white light is projected onto the left half of the split field at an intensity I_x, and the three reference lights are projected onto the right half of the field, the observer can make the two halves of the field appear identical simply by adjusting the intensities of the three reference lights. At this point physically different lights have been made to produce identical physiological responses, and consequently an observer cannot distinguish between them. The observer has produced a *metameric match* (92).

The visual effect of this (or any other) test light can therefore be described by a *colorimetric equation* of the form

$$I_x \equiv I_r + I_g + I_b$$

which signifies that the test light of intensity I_x is matched by (is visually indistinguishable from) a mixture of three reference lights (of specified spectral composition, in this example r, g, and b for red, green and blue) with intensities I_r, I_g, and I_b. Three variables suffice to characterize the effect of the test light on the visual system—the *intensities* of the three reference lights.

In order for a colorimetric equation of this sort to be completely general, the intensity terms must be allowed to have negative as well as positive values. What is the physical meaning of this statement? A negative coefficient means that the reference light had to be added to the same side of the field as the test light in order to achieve a match. And a value of zero means that the reference light was not needed for the match. (As an extreme example of the latter case, suppose the test light were a monochromatic green of the same wavelength as the green reference light. Clearly a match would be possible without using the red and blue reference lights; their intensities would be zero.) Three variables suffice to make all matches, but some matches can be achieved with fewer than three.

Colorimetric equations behave like algebraic equations. For example, both sides of a colorimetric equation can be increased by an additional term without destroying the match (although the visual appearance of the entire field will be altered). Second, the sum of several terms can be replaced by any other term that is equivalent to it. This is because two mixtures, each of which matches a third, are themselves metameric.

Trichromacy means that we can describe any light in terms of the intensities of three reference lights needed to match it. But the choice of the three reference lights is largely arbitrary, and it is therefore not possible to calculate from colorimetric data such fundamental information as a unique set of three spectral sensitivity functions corresponding to the absorption spectra of the three kinds or cones. Furthermore, in order to standardize color matching data, colorimetric equations are not in fact expressed in terms of intensities of three reference lights. Instead, one calculates how much of the test light is "absorbed" by each of three spectral functions (called spectral tristimulus values) that are consistent with (and determined from) a large amount of averaged color-matching data. The spectral tristimulus functions are a mathematical contrivance; they are not absorption spectra of pigments or spectral sensitivity functions of cells. This detail makes the subject appear somewhat arbitrary and abstruse and should not be allowed to detract from the central ideas.

Because (to a first approximation) the relative intensities of the three reference lights in a match are not dependent on the total amount of light, for any match we can express the contributions of each of the three tristimulus functions as a fraction of the total excitation. Thus we can define three *chromaticity coordinates* C_r, C_g, and C_b, that are related to the intensities of stimulation of the three reference functions in the following fashion:

$$C_r = I_r/(I_r + I_g + I_b)$$

Note that by this definition $C_r + C_g + C_b = 1$. Because once C_r and C_g are determined, C_b is also fixed, it might appear that the number of variables required to characterize any light has been reduced from three to two. This is not so. In defining the chromaticity coordinates we observed that their values were independent of the intensity or brightness of the light. But the sensation produced by the light will change with intensity. A light can therefore be characterized by two chromaticity coordinates and a *brightness* term.

As mentioned above, color matching data do not allow us to calculate a unique set of spectra for the three cone pigments, nor do they characterize a color vision system with a single, unique chromaticity diagram. If we know the pigment spectra by other means (e.g., microspectrophotometry), however, we can describe a color vision system graphically in terms of the relative quantum catches of the three cones rather than arbitrary spectral tristimulus functions (Fig. 43). This kind of graph is called a color triangle. The loop-shaped curve is the *spectrum locus*; it is a plot of the relative quantum absorptions by the three pigments of monochromatic lights from deep violet to deep red. All real lights lie in the area bounded by the spectrum locus.

Lights that plot close to the spectrum locus appear strongly colored and can be characterized by a dominant wavelength or *hue*. Lights that lie in the center of the horseshoe, on the other hand, appear to have little color and in the limiting case are white. Consider one of the lines drawn through the points in Fig. 43 marked w. As one moves outward from the white light towards the spectrum locus, the hue intensifies, and the light becomes more *saturated*. Hue (dominant wavelength), saturation (proximity to the spectrum locus), and brightness (absolute light level) therefore provide still another way of characterizing a light in terms of three variables.

We have seen that a white light can be matched by a mixture of three reference lights. There are a number of instances in which a white light can be matched by a mixture of two monochromatic lights, in which case the two lights are said to be *complementary*. As depicted in Figure 43 for the honeybee, complementary lights

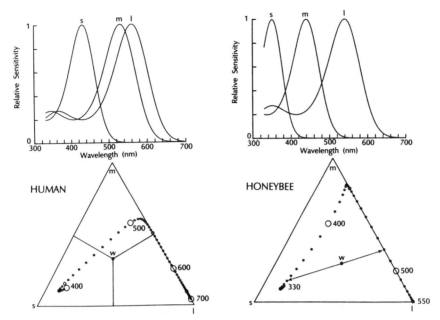

Figure 43. Above: Absorbance spectra of the color vision pigments of humans and honeybees. Each species has three pigments, absorbing at short (s), middle (m), and long (l) wavelengths. Note that short wavelengths for the honeybee extend into the near UV. Below: color triangles. Each spectral locus was calculated by normalizing the pigment spectra for equal area (rather than equal height as shown in the upper part of the figure) and computing the relative quantum catch for each pigment as a function of wavelength (148). All other lights can be represented by points in the area under the spectral locus, as, for example, the two white (w) lights that are shown. Each vertex of the triangle represents one of the three pigments (s, m, and l). As illustrated in the left triangle, for any point on or under the spectral locus, one can draw perpendiculars to the three sides of the triangle; the length of each of these perpendiculars is proportional to the number of quanta absorbed by the pigment represented at the opposite vertex. This example is a light absorbed equally by the three pigments. The concept of complementary colors is illustrated on the right; see the text for further details.

lie on opposite sides of the spectrum locus from the white in question. Thus, for the honeybee the complementary to 490 nm is located in the near UV.

Although all real lights lie in the area within the spectrum locus, a horseshoe-shaped curve defines a figure with an open side. The lights that lie in the gap between the ends of the spectrum locus are distinct colors not found in the spectrum. They are purples and magentas, formed from mixtures of red (from the long wavelength end of the visible spectrum) and violet (from the short wavelength end). In the color vision system of the honeybee, "bee-purple" is obtained by mixing UV and yellow; for bees it is a distinct color, not confused with any spectral region. For us, purple is the complementary color to green.

The rules that govern mixtures of lights (addition colors) are not quite the same as the rules that artists observe in mixing pigments (subtraction colors). Yellow and blue lights are complementary, and when mixed in the right proportions the result is white light. Together yellow and blue provide a complete spectrum. Consider what happens when blue paint is mixed with yellow paint. The blue paint is blue because it absorbs all wavelengths except those at the short wavelength end of the spectrum. A yellow pigment is yellow because it absorbs the shortest wavelengths and reflects a broad band of wavelengths at the other end of the spectrum. When the two paints are mixed, all wavelengths are absorbed except a band in the middle of the spectrum, and this reflected light produces the sensation of green.

The Molecular Bases for Color Deficiencies and Color Blindness Are Now Known

There is a simple instrument called an *anomaloscope*, which identifies the common forms of color defects (148). If a yellow light (for example, the sodium emission line at 589 nm) is projected onto one half of a split field, a normal observer will be able to match it with a mixture of red and green lights in the other one half of the field. Because the pigment of the blue-sensitive cone absorbs so little at long wavelengths, a normal observer is dichromatic in this region of the spectrum and will find a match of the following form:

$$I_{589} = I_r + I_g$$

with a unique value of I_r/I_g. There are individuals who are monochromatic in this region of the spectrum; in the anomaloscope they can match the yellow light with either the red light or the green light or a mixture of red and green with a variable ratio I_r/I_g. These individuals are red–green blind; they have a single cone pigment functioning in this long wavelength region of the spectrum, but as they have normal blue-sensitive cones, they are *dichromats*. They are of two kinds, differing in the amount of red required for a match of the yellow one half of the field.

Various lines of evidence, from measurements of the absorption spectra of the cone pigments in the living human retina by reflection densitometry (148) to molecular biology of the subject's genes all point to a simple interpretation: One subgroup of dichromats (called *deuteranopes*) has the long wavelength sensitive cone pigment, which for convenience we shall call R, and lacks G, whereas the other (*protanopes*) has G and lacks R. (The pigments R and G correspond to Crh and Cgh on the evolutionary tree in Fig. 8.) The genes for these pigments are located on the X chromosome. The two phenotypes occur in the population at a frequency of ~2%, and as expected for a sex-linked character, more commonly in males than females. Logically there should be a third type of dichromat, one missing B. These are *tritanopes*, but as the gene for B lies on an autosome, the phenotype is much rarer.

In addition to individuals who are red–green color blind, there are other people (*anomalous trichromats*) who have altered color vision. They are able to make unique matches in the anomaloscope, but the I_r/I_g ratios that they find satisfactory would not be accepted by normal observers. One of their cone pigments (R or G) is normal and the second is aberrant; consequently, they can be classified as protanomalous and deuteranomalous. Interestingly, the absorption of the aberrant pigments differ from one individual to another.

Molecular genetics provides a deeper understanding of these observations (122, 123). The genes for R and G are not only closely linked on the X chromosome, but the gene for G commonly exists in two or more copies (Fig. 44). Unequal recombination can readily generate chromosomes lacking a functional gene for one or the other of the pigments. Some of the recombinant genes, however, are functional and code for spectrally altered protein, which is the molecular basis for

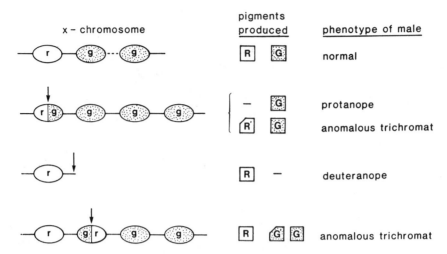

Figure 44. Molecular basis for the common (sex linked) forms of human red–green color deficiencies. As illustrated in the upper left of the diagram, the X chromosome normally contains a gene for the 565-nm cone pigment (r) and several copies of the gene for the 540-nm cone pigment (g). The gene products are the normal cone pigments, symbolized by R and G.

Because of unequal cross-over events, however, the genes can be cleaved and re-packaged in various ways, several of which are illustrated here. In the second row we see one result of a crossover that transected r. If r is thereby rendered nonfunctional, a male with this genotype lacks the cone pigment R, has no capacity for color vision in the red–green region of the spectrum, and is recognized in color vision tests as a pro-tanope. If the gene containing coding regions from both r and g retains function, however, the resulting pigment (R') characteristically has an absorption spectrum intermediate between that of R and G and the individual (male) with this chromosome is an anomalous trichromat.

In the third row the crossover has produced a chromosome containing r but not g. A male with this chromosome has no G pigment, exhibits the other form of red–green blindness, and is called a deuteranope. The bottom row shows that the molecular biology predicts that some individuals might have four functional color vision pigments (including the blue-sensitive pigment whose gene is not on the X chromosome). [Based on data of (122,123).]

a class of color defects whose bearers are known as anomalous trichromats.

Dichromats have normal visual acuity. They therefore appear to have the normal number of cones. Cells that normally would have expressed the gene for the missing pigment (e.g., G) thus appear to fill their outer segments with the other pigment (e.g., R). Interestingly, in some of the recombinants, one would expect normal R and a mixture of normal G and anomalous G' (Fig. 44), which may account for some of the variation in matches reported by anomalous trichromats.

Opponent Processes Are an Alternative and Necessary Way to Understand Color Vision

The study of color vision through a focus on trichromacy and color matching data obscures what to some are its most interesting and characteristic features. In short, there is another research tradition that goes back to another nineteenth century physiologist, Ewald Hering (72). If we consider the hues that we sense, we find that they fall into a *color circle*. Around this circle we recognize four distinct colors at the cardinal points of the compass: *red, yellow, green,* and *blue*. All intermediate colors have something of the character of adjacent pairs of these four: orange (between red and yellow), yellow–greens, blue–greens, and violet and purple (between blue and red). Some of these colors like orange have names, but in each of the four intermediate zones we see elements of two of the four fundamental colors.

Note that members of opposite pairs, *red–green* and *yellow–blue*, have a mutual exclusivity. We cannot talk about a color having red and green character simultaneously. Similarly, yellow is qualitatively different from blue. These features of how we see colors have led a number of workers to emphasize the fact that human color vision seems to be organized around three *opponent processes*, two of which are chromatic and the third achromatic. The achromatic process is white–black. Basically it is just brightness, because black (or gray) is a relative absence of stimulation. But the idea of opponentcy is that whiteness opposes or neutralizes blackness just as red neutralizes green and blue neutralizes yellow.

In what sense does red neutralize green? If we mix red light with green, the green becomes yellower until we can no longer recognize green in the mix. The light then looks yellow. Only if we continue adding red, will we begin to see red in the mix. In similar fashion yellow can neutralize blue, with the mixture appearing white at the neutral point.

There are a number of simultaneous and successive contrast phenomena that seem to be related to these opponent pro-

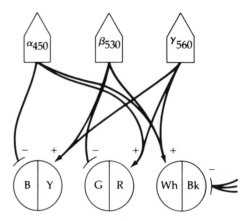

Figure 45. A model for the linkage of cone outputs to the opponent mechanisms that are described psychophysically. The latter are not intended to be any specific type of cell, but elements of opponentcy are evident in the receptive fields of retinal interneurons, particularly horizontal cells. In this scheme the B side of the Y–B opponent process is inhibited by B cones and excited by both R and G cones. The R–G process, on the other hand, is inhibited by the G cones and excited by the R cones. All three cones excite the third process, which codes luminosity. [From (72).]

cesses. For example, a neutral gray surface viewed against a green surround will appear to take on a pinkish cast. Or after viewing a bright colored object one may experience an afterimage of the complementary color.

Figure 45 shows one model of how the three cones play onto paired antagonistic neural systems. This model is based on a rich variety of psychophysical experiments, and the opponent processes do not represent any particular type of cell. The model is intended to describe how the visual system as a whole analyzes color. For further insight into the multiplicity of levels at which color vision can be studied see recent symposia (113, 183).

Other Animals Have Color Vision

Figure 15 summarizes some of the better known examples of animals with multiple visual pigments, selected to illustrate the taxonomic diversity with which color vision or wavelength-dependent behavior is associated, and because, in most of these examples of vertebrates, there is independent behavioral evidence for color vision (last column). The frequent occurrence in birds of brightly colored plumages, often associated with sexual dimorphism, provides powerful if indirect evidence that color vision is a common feature of the avian visual system, even in species that have not been tested directly. Although there is little or no quantitative behavioral data on the color vision of most invertebrates, the Hymenoptera are a dramatic exception. The color vision of bees has been studied in at least as much detail as any species other than humans (103, 104). In many other groups of arthropods a rich endowment of visual pigments indicates the presence of interesting visual capacities (102).

What conclusions can be drawn from these comparative data? First, color vision is found in the major extant vertebrate classes and color vision or wavelength-dependent processes are common among insects. Color vision is absent in the large eyes of cephalopod molluscs, and the status of crustacea is not at all clear. Further, examples of retinas with two, three, or four presumptive color vision pigments are known from both arthropods and vertebrates. Until very recently, visual pigments with maximal absorption in the near UV were thought to be the exclusive provence of arthropods, but cones with peak sensitivity at 370–380 nm have now been found in fish (13, 64, 124, 125), turtles (3), and birds (23). The best studied of the lower vertebrates therefore characteristically have four (or even five) cone pigments. Trichromacy is not the norm. A more extensive discussion of the evolution of color vision can be found in the source for which Figure 15 was prepared (50, 189).

The mammals require special comment. As reviewed extensively elsewhere, mammalian color vision is most highly developed in the primates (50, 74). Most species of rodents (many squirrels excepted), carnivores, and ungulates are either nocturnal or arrhythmic, and probably have very poor capacity for color vision. Rats and cats, both of which have been examined with some care, do not learn to make color discriminations easily and possess few cells that seem to be involved in color vision. The primitive mammalian condition as exemplified by tree shrews appears to be dichromatic, and even squirrels with many cones and diurnal habits seem to have remained dichromatic. The molecular biology of human cone pigments indicates that the trichromacy of Old World Primates is of relatively recent evolutionary origin (Fig. 8).

Summary

Vision is but one of several responses of organisms to light. Although the capacity for vision implies the presence of an image-forming eye, photoreceptors of varying degrees of anatomical complexity characterized by elaboration of plasma membrane are widely distributed among the animal phyla. Visual organs are polyphyletic structures, but at a molecular level they probably all employ opsin as the protein moiety of their photoreceptor pigment. A given species, however, can have genes for more than one opsin, including genes that diverged more than 500 million years ago. All opsins utilize as chromophore 11-*cis* vitamin A aldehyde (retinal) or one of two other closely related analogs.

Light isomerizes the 11-*cis* chromophore to the all-*trans* configuration, resulting in conformational changes in the opsin. This in turn triggers a series of molecular events that regulate membrane

conductance. In vertebrate photoreceptors there is a decrease in sodium permeability and the cell hyperpolarizes; in
invertebrates there is (characteristically,
but not always) an increase in sodium
permeability and the cell depolarizes. The
molecular cascade that controls conductance of vertebrate rods is now known in
considerable detail. The sodium channels
are caused to open by direct interaction
with intracellular cGMP; light leads to an
activation of phosphodiesterase, which
causes an increased hydrolysis of cGMP
and a decrease in sodium conductance.

Image-forming eyes are of two basic
types: lens eyes (e.g., in vertebrates and
cephalopod molluscs) that employ the
same optical principle as a camera, and
compound eyes (of arthropods) that consist of a large number of optically independent bundles of cells (ommatidia)
each of which views a small solid angle
of visual space. In each design, however,
the same basic factors determine performance. Spatial resolution requires large
eyes with many receptors with small angular sensitivity, but absolute sensitivity
and contrast resolution impose lower limits on receptor size.

Photoreceptors are exquisitely sensitive and when fully dark-adapted can respond to single photons (50b). Many
receptors also decrease their sensitivity
(gain) in the presence of background
lights. Other mechanisms of adaptation
are present deeper in the retina. Visual
systems are organized so as to detect both
temporal and spatial contrasts in the retinal image; mechanisms by which this is
accomplished are evident at the earliest
stages of afferent coding.

The presence of visual pigments with
absorption spectra maximal in different
regions of the spectrum permits an animal to have color vision. Color vision is
present in all classes of vertebrates and in
a number of arthropods. Color vision is
poorly represented among mammals, the

primates being an exception. Although
human color vision is the most extensively studied system, the trichromacy of
Old World Primates is a relatively recent
evolutionary accomplishment. Some
other animals appear to be tetrachromatic, and many other species of both
vertebrates and arthropods have receptors maximally sensitive in the near UV.

References

1. Ali, M. A. *Vision Res.* 11:1225–1288, 1971.
   Les réponses retinomotrices: Caractères
   et mécanismes.

2. Applebury, M. L. and P. A. Hargrave.
   Vision Res. 26:1881–1895, 1986. Molecular
   biology of the visual pigments.

3. Arnold, K. and C. Neumeyer. *Vision Res.*
   27:1501–1511, 1987. Wavelength discrimination in the turtle *Pseudemys scripta elegans*.

4. Barber, V. C., E. M. Evans, and M. F.
   Land. *Z. Zellforsch. Mikrosk. Anat.*
   76:295–312, 1967. The fine structure of
   the eye of the mollusc *Pecten maximus*.

5. Barlow, H. B. and J. D. Mollon, eds. *Cambridge Texts in the Physiological Sciences*,
   Vol. 3. Cambridge Univ. Press, London
   and New York, 1982.

6. Barnes, S. N. and T. H. Goldsmith, *J.
   Comp. Physiol.* 120:143–159, 1977. Dark
   adaptation, sensitivity, and rhodopsin
   level in the eye of the lobster, *Homarus*.

7a. Baumann, F. *J. Gen. Physiol.* 52:855–875,
   1968. Slow and spike potentials recorded
   from retinula cells of the honeybee drone
   in response to light.

7b. Baylor, D. A. and M. G. F. Fuortes. *J.
   Physiol.* (*London*) 207:77–92, 1970. Electrical responses of single cones in turtle
   retina.

8. Baylor, D. A., G. Mathews, and K. W.
   Yau. *J. Physiol.* (*London*) 309:591–621,
   1980. Two components of electrical dark
   noise in toad retinal rod outer segments.

9. Bennett, M. F. In *Handbook of Sensory
   Physiology* (H. Autrum, ed.), Vol. 7, Part
   6A, pp. 641–663. Springer-Verlag, Hei-

delberg, 1979. Extraocular light receptors and circadian rhythms.

10. Bernstein, P. S., W. C. Law, and R. R. Rando. *Proc. Natl. Acad. Sci. U.S.A.* 84:1849–1853, 1987. Isomerization of all-*trans*-retinoids to 11-*cis*-retinoids *in vitro*.

11. Bernstein, P. S., W. C. Law, and R. R. Rando. *J. Biol. Chem.* 262:16848–16857, 1987. Biochemical characterization of the retinoid isomerase system of the eye.

12. Boll, F. *Monatsber. Preuss. Akad. Wiss. Berlin* 12:783–788, 1876. Zur Anatomie und Physiologie der Retina.

13. Bowmaker, J. K. and Y. W. Kunz. *Vision Res.* 27:2101–2108, 1987. Ultraviolet receptors, tetrachromatic colour vision and retinal mosaics in the brown trout (*Salmo trutta*): Age-dependent changes.

14. Bridges, C. D. B. In *Handbook of Sensory Physiology* (H. J. A. Dartnall, ed.), Vol. 7, Part 1, pp. 417–480. Springer-Verlag, Berlin, 1972. The rhodopsin-porphyropsin visual system.

15. Bridges, C. D. B., R. A. Alvarez, S.-L. Fong, F. Gonzalez-Fernandez, D. M. K. Lam, and G. I. Liou. *Vision Res.* 24:1581–1594, 1984. Visual cycle in the mammalian eye: Retinoid-binding proteins and the distribution of 11-*cis* retinoids.

16. Bridges, C. D. B. and R. A. Alvarez. *Science* 236:1678–1680, 1987. The visual cycle operates via an isomerase acting on all-*trans* retinol in the pigment epithelium.

17. Briggs, W. R. *Photophysiology* 1: 223–271, 1964. Phototropism in higher plants.

18. Brown, J. E. and J. R. Blinks, *J. Gen. Physiol.* 64:643–655, 1974. Changes in intracellular free calcium concentration during illumination of invertebrate photoreceptors. Detection with aequorin.

19. Brown, J. E. and M. I Mote. *J. Gen. Physiol.* 63:337–350, 1974. Ionic dependence of reversal voltage of the light response in *Limulus* ventral photoreceptors.

20. Brown, J. E. and L. H. Pinto. *J. Physiol.* (*London*) 236:575–591, 1974. Ionic mechanism for the photoreceptor potential of the retina of *Bufo marinus*.

21. Bruce, V. G. and D. H. Minis. *Science* 163:583–585, 1969. Circadian clock action spectrum in a photoperiodic moth.

22. Chabre, M. and M. L. Applebury. *Life Sci. Res. Rep.* 34:51–66, 1986. Interactions of photoactivated rhodopsin with photoreceptor proteins: The cGMP cascade.

23. Chen, D.-M. and T. H. Goldsmith. *J. Comp. Physiol.* 159A:473–479, 1986. Four spectral classes of cones in the retinas of birds.

24. Clayton, R. K. *Photophysiology* 2:51–77, 1964. Phototaxis in microorganisms.

25. Cohen, A. I. *Biol. Rev. Cambridge Philos. Soc.* 38:427–459, 1963. Vertebrate retinal cells and their organization.

26. Cone, R. A. *Nature (London), New Biol.* 236:39–43, 1972. Rotational diffusion of rhodopsin in the visual receptor membrane.

27. Cowman, A. F., C. S. Zuker, and G. M. Rubin. *Cell (Cambridge, Mass.)* 44:705–710, 1986. An opsin gene expressed on only one photoreceptor cell type of the *Drosophilia* eye.

28. Crescitelli, F. In *Handbook of Sensory Physiology* (H. J. A. Dartnall, ed.), Vol. 7, Part 1, pp. 245–363. Springer-Verlag, Berlin, 1972. The visual cells and visual pigments of the vertebrate eye.

29. Defoe, D. M. and D. Bok. *Invest. Ophthalmol. Visual Sci.* 24:1211–1226, 1983. Rhodopsin chromophore exchanges among opsin molecules in the dark.

30. Diehn, B. In *Handbook of Sensory Physiology* (H. Autrum, ed.), Vol. 7, Part 6A, pp. 23–68. Springer-Verlag, Heidelberg, 1979. Photic responses and sensory transduction in motile protists.

31. Dowling, J. E. *J. Gen Physiol.* 46:1287–1301, 1963. Neural and photochemical mechanisms of visual adaptation in the rat.

32. Dowling, J. E. *The Retina: An Approachable Part of the Brain.* Harvard Univ. Press (Belknap), Cambridge, MA, 1987.

33. Dowling, J. E. and H. Ripps. *J. Gen. Physiol.* 56:491–520, 1970. Visual adaptation in the retina of the skate.

34. Eakin, R. M. *Evol. Biol.* 2:194–242, 1968. Evolution of photoreceptors.

35. Eakin, R. M. *Am. Sci.* 58:73–79, 1970. A third eye.

36. Eakin, R. M. In *Handbook of Sensory Physiology* (H. J. A. Dartnall, ed.), Vol. 7, Part 1, pp. 625–684. Springer-Verlag, Berlin, 1972. Structure of invertebrate photoreceptors.

37. Eakin, R. M. In *Visual Cells in Evolution* (J. Westfall, ed.), pp. 91–105. Raven Press, New York, 1982. Continuity and diversity in photoreceptors.

38a. Exner, S. *Die Physiologie der facettieren Augen von Krebsen und Insekten.* Franz Deuticke, Vienna, 1891.

38b. Exner, S. *The Physiology of the Compound Eyes of Insects and Crustaceans.* (Translated and annotated by R. C. Hardia.) Springer-Verlag Heidelberg, 1989.

39. Fain, G. L. *Life Sci. Res. Rep.* 34:67–77, 1986. Evidence for a role of messenger substances in phototransduction.

40. Fain, G. L. and J. E. Dowling, *Science* 180:1178–1181, 1973. Intracellular recordings from single rods and cones in the mudpuppy retina.

41. Fain, G. L. and J. E. Lisman. *Prog. Biophys. Mol. Biol.* 37:91–147, 1981. Membrane conductances of photoreceptors.

42. Feiler, R., W. A. Harris, K. Kirschfeld, C. Wehrahn, and C. S. Zuker. *Nature (London)* 333:737–741, 1988. Targeted misexpression of a *Drosophilia* opsin gene leads to altered visual function.

43. Fein, A. and E. Z. Szuts. *Photoreceptors: Their Role in Vision*, Cambridge Univ. Press, London and New York, 1982.

44. Fesenko, E. E., S. S. Kolesnikov, and A. L. Lyubarsky. *Nature (London)* 313:310–313, 1985. Induction by cyclic GMP of cationic conductance in plasma membrane of retinal rod outer segment.

45. Findlay, J. B. C. *Life Sci. Res. Rep.* 34:11–30, 1986. The biosynthetic, functional, and evolutionary implication of the structure of rhodopsin.

46. Foster, E. W., J. Saranak, N. Patel, G. Zarilli, M. Okabe, T. Kline, and K. Nakanishi. *Nature (London)* 311:756–759, 1984. A rhodopsin is the functional photoreceptor for phototaxis in the unicellular eukaryote *Chlamydomonas*.

47. Frank, K. D. and W. F. Zimmerman. *Science* 163:688–689, 1969. Action spectra for phase shifts of a circadian rhythm in *Drosophilia*.

48. Futterman, S. *J. Biol. Chem.* 238:1145–1150, 1963. Metabolism of the retina. III. The role of reduced triphosphopyridine nucleotide in the visual cycle.

49. Goldsmith, T. H. In *Sensory Communication* (W. A. Rosenblith, ed.), pp. 357–375. MIT Press, Cambridge, MA, 1961. Physiological basis of wavelength discrimination in eye of honeybee.

50a. Goldsmith, T. H. In *The Perception of Colour* (P. Gouras, ed.), Macmillan, London, 1990. The evolution of visual pigments and colour vision.

50b. Goldsmith, T. H. *Quart. Rev. Biol.* 65 (in press, 1990). Optimization, constraint and history in the evolution of eyes.

51. Goldsmith, T. H. and G. D. Bernard. In *The Physiology of Insecta* (M. Rockstein, ed.), 2nd ed., Vol. 2, pp. 165–272. Academic Press, New York, 1974. The visual system of insects.

52. Goldsmith, T. H. and G. D. Bernard. *Photochem. Photobiol.* 42:805–809, 1985. Visual pigments of invertebrates.

53. Goldsmith, T. H., J. S. Collins, and S. Licht. *Vision Res.* 24:1661–1671, 1984. The cone oil droplets of avian retinas.

54. Goldsmith, T. H., B. C. Marks, and G. D. Bernard. *Vision Res.* 26:1763–1769, 1986. Separation and identification of geometric isomers of 3-hydroxyretinoids and occurrence in the eyes of insects.

55. Goldsmith, T. H. and R. Wehner. *J. Gen. Physiol.* 70:453–490, 1977. Restrictions on rotational and translational diffusion of pigment in the membranes of a rhabdomeric photoreceptor.

56. Gorman, A. L. F., J. S. McReynolds, and S. N. Barnes. *Science* 172:1052–1054, 1971. Photoreceptors in primitive chordates: fine structure, hyperpolarizing receptor potentials, and evolution.

57. Green, D. G. *Vision Res.* 26:1417–1429,

1986. The search for the site of visual adaptation.

58. Hagins, W. A., R. D. Penn, and S. Yoshikami. *Biophys. J.* 10:380–412, 1970. Dark current and photocurrent in retinal rods.

59. Hagins, W. A., H. V. Zonana, and R. G. Adams. *Nature (London)* 194:844–847, 1962. Local membrane current in the outer segments of squid photoreceptors.

60. Halldal, P. *Physiol. Plant.* 11:118–153, 1958. Action spectra of phototaxis and related problems in Volvocales, *Ulva*-Gametes and Dinophyceae.

61. Hamdorf, K. In *Handbook of Sensory Physiology* (H. Autrum, ed.), Vol. 7, Part 6A, pp. 145–224. Springer-Verlag, Heidelberg, 1979. The physiology of invertebrate visual pigments.

62. Hardie, R. C. *Trends Neurosci.* 9:419–423, 1986. The photoreceptor array of the dipteran retina.

63. Hargrave, P. A., J. H. McDowell, D. R. Curtis, J. K. Wang, E. Juszcak, S. L. Fong, J. K. Mohana Roa, and P. Argos. *Biophys. Struct. Mech.* 9:235–244, 1983. The structure of bovine rhodopsin.

64. Harosi, F. I. and Y. Hashimoto. *Science* 222:1021–1023, 1983. Ultraviolet visual pigment in a vertebrate: A tetrachromatic cone system in the dace.

65. Hartline, H. K. *Science* 164:270–278, 1969. Visual receptors and retinal interaction.

66. Haupt, W. *Int. Rev. Cytol.* 19:267–299, 1966. Phototaxis in plants.

67. Hecht, S., S. Shlaer, and M. H. Pirenne. *J. Gen. Physiol.* 25:819–840, 1942. Energy, quanta, and vision.

68. Hodgkin, A. L., P. A. McNaughton, B. J. Nunn, and K.-W. Yau. *J. Physiol. (London)* 350:649–680, 1984. Effect of ions on retinal rods from *Bufo marinus*.

69. Hollins, M. and M. Alpern. *J. Gen. Physiol.* 62:430–447, 1973. Dark adaptation and visual pigment regeneration in human cones.

70. Hubbard, R. and A. Kropf. *Proc. Natl. Acad. Sci. U.S.A.* 44:130–139, 1958. The action of light on rhodopsin.

71. Hubel, D. H. *Eye, Brain, and Vision.* Scientific American Library, New York, 1988.

72. Hurvich, L. M. *Color Vision.* Sinauer Associates, Sunderland, MA, 1981.

73. Hutchinson, G. E. *Treatise on Limnology,* Vol. 1. Wiley, New York, 1957.

74. Jacobs, G. H. *Comparative Color Vision.* Academic Press, New York, 1981.

75. Kennedy, D. *J. Gen. Physiol.* 44:277–299, 1960. Neural photoreception in a lamellibranch mollusc.

76. Kirschfeld, K. *Exp. Brain Res.* 3:248–270, 1967. Die Projektion der optischen Umwelt auf das Raster der Rhabdomere im Komplexauge von *Musca*.

77. Kirschfeld, K. *Proc. Int. Sch. Fis. "Enrico Fermi"* 43:144–166, 1969. Optics of the compound eye.

78. Kirschfeld, K. In *Neural Principles in Vision* (F. Zettler and R. Weiler, eds.), pp. 354–370. Springer-Verlag, Heidelberg, 1976. The resolution of lens and compound eyes.

79. Kirschfeld, K., R. Feiler, R. Hardie, K. Vogt, and N. Franceschini. *Biophys. Struct. Mech.* 9:171–180, 1983. The sensitizing pigment in fly photoreceptors. Properties and candidates.

80. Kleinschmidt, J. and J. E. Dowling. *J. Gen. Physiol.* 66:617–648, 1975. Intracellular recordings from Gecko photoreceptors during light and dark adaptation.

81. Knowles, A. and H. J. A. Dartnall. In *The Eye* (H. Davson, ed.), 2nd ed., Vol. 2B, pp. 581–647. Academic Press, New York, 1977. Habitat, habit and visual pigment.

82. Kühne, W. *Unters. Physiol. Inst. Univ. Heidelberg* 1:15–103, 1877. Über den Sehpurpur.

83. Kunze, P. In *Handbook of Sensory Physiology* (H. Autrum, ed.), Vol. 7, Part 6A, pp. 441–502. Springer-Verlag, Heidelberg, 1979. Apposition and superposition eyes.

84. Labhart, T. *Nature (London)* 331:435–437, 1988. Polarization-opponent interneurons in the insect visual system.

85. Lamb, T. D. *Life Sci. Res. Rep.* 34:267–286, 1986. Photoreceptor adaptation—vertebrates.

86. Land, M. F. *Nature (London)* 263:764–765, 1976. Superposition images are formed by reflection in the eyes of some oceanic decapod crustacea.

87. Land, M. F. In *Handbook of Sensory Physiology* (H. Autrum, ed.), Vol. 7, Part 6B, pp. 471–592. Springer-Verlag, Heidelberg, 1981. Optics and vision in invertebrates.

88. Lasansky, A. and M. G. F. Fuortes. *J. Cell Biol.* 42:241–252, 1969. The site of origin of electrical responses in visual cells of the leech, *Hirudo medicinalis*.

89. Laughlin, S. B. *Trends Neurosci.* 10:478–483, 1987. Form and function in retinal processing.

90. Leedale, G., B. J. D. Meeuse, and E. G. Pringsheim. *Arch. Mikrobiol.* 50:68–102, 1965. Structure and physiology of *Euglena spirogyra*.

91. Lees, A. D. *Photophysiology* 4:47–137, 1968. Photoperiodism in insects.

92. Le Grand, Y. *Light, Colour and Vision*. Wiley, New York, 1957.

93. Liebman, P. A. and G. Entine. *Science* 185:457–459, 1974. Lateral diffusion of visual pigment in photoreceptor disk membranes.

94. Lisman, J. E. and J. E. Brown. *J. Gen. Physiol.* 66:489–506, 1975. Effects of intracellular injection of calcium buffers on light adaptation in *Limulus* ventral photoreceptors.

95. Lythgoe, J. N. In *Handbook of Sensory Physiology* (H. J. A. Dartnall, ed.), Vol. 7, Part 1, pp. 566–603. Springer-Verlag, Heidelberg, 1972. The adaptation of visual pigments to their photic environment.

96. Lythgoe, J. N. *The Ecology of Vision*. Oxford Univ. Press (Clarendon), London and New York, 1979.

97. Mast, S. O. and N. Stabler. *Biol. Bull. (Woods Hole, Mass.)* 73:126–133, 1937. The relation between luminous intensity, adaptation to light, and rate of locomotion in *Amoeba proteus* (Leidy).

98. Mathews, R. G., R. Hubbard, P. K. Brown, and G. Wald. *J. Gen Physiol.* 47:215–240, 1963. Tautomeric forms of metarhodopsin.

99. McNaughton, P. A., B. J. Nunn, and A. L. Hodgkin. *Life Sci. Res. Rep.* 34:79–92, 1986. Evaluation of internal transmitter candidates: Ca.

100. McReynolds, J. S. and A. L. F. Gorman. *J. Gen. Physiol.* 56:376–391, 1970. Photoreceptor potentials of opposite polarity in the eye of the scallop, *Pecten irradians*.

101. McReynolds, J. S. and A. L. F. Gorman. *J. Gen. Physiol.* 56:392–406, 1970. Membrane conductances and spectral sensitivities of *Pecten* photoreceptors.

102. Menzel, R. In *Handbook of Sensory Physiology* (H. Autrum, ed.), Vol. 6, Part 6A, pp. 503–580. Springer-Verlag, Heidelberg, 1979. Spectral sensitivity and color vision in invertebrates.

103. Menzel, R. In *Central and Peripheral Mechanisms of Colour Vision* (S. Zeki, ed.), pp. 211–233. Pergamon, Oxford, 1985. Color pathways and color vision in the honeybee.

104. Menzel, R. and M. Blakers. *J. Comp. Physiol.* 108:11–33, 1976. Colour receptors in the bee eye—morphology and spectral sensitivity.

105. Michael, C. R. *J. Neurophysiol.* 31:268–282, 1968. Receptive fields of single optic nerve fibers in a mammal with an all-cone retina. III. Opponent color fibers.

106. Michael, C. R. *Sci. Am.* 220:104–114, 1969. Retinal processing of visual images.

107. Millecchia, R., J. Bradbury, and A. Mauro. *Science* 154:1199–1201, 1966. Simple photoreceptors in Limulus polyphemus.

108. Millecchia, R. and A. Mauro. *J. Gen. Physiol.* 54:310–330, 1969a. The ventral photoreceptor cells of *Limulus*. II. The basic photoresponse.

109. Millecchia, R. and A. Mauro. *J. Gen. Physiol.* 54:331–351, 1969b. The ventral photoreceptor cells of *Limulus*. III. A voltage-clamp study.

110. Millott, N. *Science* 162:759–771, 1968. Animal photosensitivity, with special reference to eyeless forms.

111. Millott, N. and M. Yoshida. *J. Exp. Biol.*

34:394–401, 1957. The spectral sensitivity of the echinoid *Diadema antillarum* Philippi.

112. Milne, L. J. and M. Milne. *Handb. Physiol. Sect. 1;* pp. 621–645, 1959. Photosensitivity in invertebrates.

113. Mollon, J. D. and L. T. Sharpe, eds. *Colour Vision: Physiology and Psychophysics.* Academic Press, New York, 1983.

114. Montell, C. K., C. Jones, C. Zuker, and G. Rubin. *J. Neurosci.* 7:1558–1566, 1987. A second opsin gene expressed in the ultraviolet sensitive R7 photoreceptor cells of *Drosophilia melanogaster.*

115. Moody, M. F. *Biol. Rev. Cambridge Philos. Soc.* 39:43–86, 1964. Photoreceptor organelles in animals.

116. Moody, M. F. and J. R. Parriss. *J. Exp. Biol.* 39:21–30, 1962. The discrimination of polarized light by *Octopus.*

117. Munz, F. W. and W. N. McFarland. In *Handbook of Sensory Physiology* (F. Crescitelli, ed.), Vol. 7, Part 5, pp. 193–274. Springer-Verlag, Heidelberg, 1977. Evolutionary adaptations of fishes to the photic environment.

118. Naka, K.-I. and W. A. H. Rushton. *J. Physiol. (London)* 185:556–586, 1966. An attempt to analyse colour reception by electrophysiology.

119. Nässel, D. R. and T. H. Waterman. *J. Comp. Physiol.* 131:205–216, 1979. Massive diurnally mediated photoreceptor turnover in crab light and dark adaptation.

120. Nathans, J. and D. S. Hogness. *Cell (Cambridge, Mass.)* 34:807–814, 1983. Isolation, sequence analysis, and intron-exon arrangement of the gene encoding bovine rhodopsin.

121. Nathans, J. and D. S. Hogness. *Proc. Natl. Acad. Sci. U.S.A.* 81:4851–4855, 1984. Isolation and nucleotide sequence of the gene encoding human rhodopsin.

122. Nathans, J., D. Thomas, and D. S. Hogness. *Science* 232:193–202, 1986. Molecular genetics of human color vision: the genes encoding blue, green, and red pigments.

123. Nathans, J., T. P. Piantanida, R. L. Eddy,

T. B. Shows, and D. S. Hogness. *Science* 232:203–210, 1986. Molecular genetics of inherited variation in human color vision.

124. Neumeyer, C. *Naturwissenschaften* 72:165, 1985. An ultraviolet receptor as a fourth receptor type in goldfish color vision.

125. Neumeyer, C. *J. Comp. Physiol.* 158A: 203–213, 1986. Wavelength discrimination in the goldfish.

126. Nilsson, D.-E. *Nature (London)* 332:76–78, 1988. A new type of imaging optics in compound eyes.

127. Osterberg, G. *Acta Ophthalmol., Suppl. 6,* 1–103, 1935. Topography of the layer of rods and cones in the human retina.

128. O'Tousa, J. E., W. Baehr, R. L. Martin, J. Hirsch, W. L. Pak, and M. L. Applebury. *Cell (Cambridge, Mass.)* 40:839–850, 1985. The Drosophilia *ninaE* gene encodes an opsin.

129. Ovchinnikov. Yu.A., N. G. Abdulaev, A. S. Zolotarev, I. D. Artamonov, I. A. Bespalov, A. E. Dergachev, and M. Tsuda. *FEBS Lett.* 232:69–72, 1988. Octopus rhodopsin. Amino acid sequence deduced from cDNA.

130. Page R. M. *Photophysiology* 3:65–90, 1968. Phototropism in fungi.

131. Penn, R. D. and W. A. Hagins. *Nature (London)* 223:201–205, 1969. Signal transmission along retinal rods and the origin of the electroretinographic a-wave.

132. Pirenne, M. H. *Vision and the Eye,* 2nd ed. Chapman & Hall, London, 1967.

133. Pollock, J. A. and S. Benzer. *Nature (London)* 333:779–782, 1988. Transcript localization of four opsin genes in the three visual organs of *Drosophila;* RH2 is ocellus specific.

134. Poo, M. and R. A. Cone. *Nature* (London) 247:438–441, 1974. Lateral diffusion of rhodopsin in the photoreceptor membrane.

135. Prosser, C. L. *J. Exp. Biol.* 12:95–104, 1935. Impulses in the segmental nerves of the earthworm.

136. Push, E. N. and W. H. Cobbs. *Vision Res.* 26:1613–1643, 1986. Visual transduction in vertebrate rods and cones: A tale of

two transmitters, calcium and cyclic GMP.

137. Purple, R. L. and F. Dodge. In *The Functional Organization of the Compound Eye* (C. G. Bernhard, ed.), pp. 451–464. Pergamon, Oxford, 1966. Self inhibition in the eye of *Limulus*.

138. Rabinowitch, E. and Govindjee. *Photosynthesis*. Wiley, New York, 1969.

139. Ratliff, F. *Mach Bands*. Holden-Day, San Francisco, CA, 1965.

140. Ratliff, F. and H. K. Hartline. *J. Gen. Physiol.* 42:1241–1255, 1959. The responses of *Limulus* optic nerve fibers to patterns of illumination on the receptor mosaic.

141. Robertis, E. de *J. Gen. Physiol.* 43(Suppl.2):1–6, 1960. Some observations on the ultrastructure and morphogenesis of photoreceptors.

142. Robinson, N., ed. *Solar Radiation*. Elsevier, Amsterdam, 1966.

143. Rodieck, R. W. *The Vertebrate Retina: Principles of Structure and Function*. Freeman, San Francisco, CA, 1973.

144. Roof, D. J. *Life Sci. Res. Rep.* 34:287–302, 1986. Turnover of vertebrate photoreceptor membranes.

145. Rossel, S. and R. Wehner. *J. Comp. Physiol.* 155A:605–613, 1984. Celestial orientation in bees: The use of spectral cues.

146. Rossel, S. and R. Wehner. *Nature (London)* 323:128–132, 1986. Polarization vision in bees.

147. Rushton, W. A. H. *J. Physiol. (London)* 156:193–205, 1961. Rhodopsin measurement and dark-adaptation in a subject deficient in cone vision.

148. Rushton, W. A. H. *J. Physiol. (London)* 220:1P–31P, 1972. Pigments and signals in colour vision.

149. Salvini-Plawen, L. von. In *Visual Cells in Evolution* (J. Westfall, ed.), pp. 137–154. Raven Press, New York, 1982. On the polyphyletic origin of photoreceptors.

150. Schwemer, J. *J. Comp. Physiol.* 154A:535–547, 1984. Renewal of visual pigment in photoreceptors of the blowfly.

151. Senger, H., ed., *The Blue Light Syndrome*. Springer-Verlag, Heidelberg, 1980.

152. Shropshire, W., Jr. and H. Mohr, eds. *Encyclopedia of Plant Physiology, New Series*, Vol. 16A. Springer-Verlag, Berlin, 1983.

153. Sillman, A. J., H. Ito, and T. Tomita. *Vision Res.* 9:1443–1451, 1969. Studies on the mass receptor potential of the isolated frog retina. II. On the basis of the ionic mechanism.

154a. Sjöstrand, F. S. *Ergeb. Biol.* 21:128–160, 1959. The ultrastructure of the retinal receptors of the vertebrate eye.

154b. Smith, W. C. and T. H. Goldsmith. *J. Molec. Evol.* 30:72–84, 1990. Phyletic aspects of the distribution of 3-hydroxyretinal in the class Insecta.

155. Snyder, A. W. In *Handbook of Sensory Physiology* (H. Autrum, ed.), Vol. 7, Part 6A, pp. 441–502. Springer-Verlag, Heidelberg, 1979. Physics of vision in compound eyes.

156. Stein, P. J., J. O. Brammer, and S. E. Ostroy. *J. Gen. Physiol.* 74:565–582, 1979. Renewal of opsin in the photoreceptor cells of the mosquito.

157. Stieve, H. *Life Sci. Res. Rep.* 34:199–230, 1986. Bumps, the elementary excitatory responses of invertebrates.

158. Stoeckenius, W. and R. A. Bogomolni. *Annu. Rev. Biochem.* 52:587–616, 1982. Bacteriorhodopsin and related pigments of halobacteria.

159. Stryer, L. *Annu. Rev. Neurosci.* 9:87–119, 1986. Cyclic GMP cascade of vision.

160. Suzuki, T. and E. Eguchi. *Experientia* 43:1111–1113, 1987. A survey of 3-dehydroretinal as a visual pigment chromophore in various species of crayfish and other freshwater crustaceans.

161. Suzuki, T., K. Arikawa, and E. Eguchi. *Zool. Sci.* 2:455–461, 1985. The effects of light and temperature on the rhodopsin-porphyropsin visual system of the crayfish, *Procambarus clarkii*.

162. Sverdrup, H. U., M. W. Johnson, and R. H. Fleming. *The Oceans: Their Physics, Chemistry, and General Biology*. Prentice-Hall, New York, 1942.

163. Takao, M., A. Yasui, and F. Tokunaga. *Vision Res.* 28:471–480, 1988. Isolation

and sequence determination of the chicken rhodopsin gene.

164. Toyoda, J., H. Nosaki, and T. Tomita. *Vision Res.* 9:453–463, 1969. Light-induced resistance changes in single photoreceptors of *Necturus* and *Gekko*.

165. Vanfleteren, J. R. In *Visual Cells in Evolution* (J. Westfall, ed.), pp. 107–136. Raven Press, New York, 1982. A monophyletic line of evolution? Ciliary induced photoreceptor membranes.

166. Vogt, K. *Z. Naturforsch., Biosci.* 30C:691, 1975. Zur Optic des Flusskrebsauges.

167. Vogt, K. *Z. Naturforsch., Biosci.* 38C:329–333, 1983. Is the fly visual pigment a rhodopsin?

168. Vogt, K. *Z. Naturforsch. Biosci.* 39C:196–197, 1984. The chromophore of the visual pigment in some insect orders.

169. Vogt, K. and K. Kirschfeld. *Naturwissenschaften* 71:211–213, 1984. Chemical identity of the chromophores of fly visual pigment.

170. Wald, G. *Science* 101:653–658, 1945. Human vision and the spectrum.

171. Wald, G. *Am. J. Ophthalmol.* 40:18–41, 1955. The photoreceptor process in vision.

172. Wald, G. *Sci. Am.* 201:92–108, 1959. Life and light.

173. Wald, G. *Proc. Natl. Acad. Sci. U.S.A.* 52:595–611, 1964. The origins of life.

174. Walls, G. L. *The Vertebrate Eye and Its Adaptive Radiation, Bull.* 19. Cranbrook Institute of Science, Bloomfield Hills, MI, 1942.

175. Wehner, R. In *Handbook of Sensory Physiology* (H. Autrum, ed.), Vol. 7, Part 6C, pp. 287–616. Springer-Verlag, Heidelberg, 1981. Spatial vision in arthropods.

176. Werblin, F. S. and J. E. Dowling. *J. Neurophysiol.* 32:339–355, 1969. Organization of the retina of the mudpuppy, *Necturus maculosus*. II. Intracellular recording.

177. Williams, C. M. *Symp. Soc. Exp. Biol.* 23:285–300, 1969. Photoperiodism and the endocrine aspects of insect diapause.

178. Withrow, R. B., W. H. Klein, and V. Elstad. *Plant Physiol.* 32:453–462, 1957. Action spectra of photomorphogenic induction and its photoinactivation.

179. Yoshida, M. In *Handbook of Sensory Physiology* (H. Autrum, ed.), Vol. 7, Part 6A, pp. 581–640. Springer-Verlag, Heidelberg, 1979. Extraocular photoreception.

180. Young, T. *Philos. Trans. R. Soc. London* 92:12–48, 1802. On the theory of light and colours.

181. Young, T. In *Lectures on Natural Philosophy and the Mechanical Arts*, Vol. 2, pp. 613–638. Printed for Joseph Johnson, St. Paul's Church Yard by William Savage, London, 1807. On the theory of light and colours.

182. Zeiger, J. and T. H. Goldsmith. *Vision Res.* 29:519–527, 1989. Spectral characterization of porphyropsin from an invertebrate.

183. Zeki, S., ed. *Central and Peripheral Mechanisms of Colour Vision.* Pergamon, Oxford, 1985.

184. Zonana, H. V. *Johns Hopkins Hosp. Bull.* 109:185–205, 1961. Fine structure of the squid retina.

185. Zuker, C. S., A. F. Cowman, and G. M. Rubin. *Cell (Cambridge, Mass.)* 40:851–858, 1985. Isolation and structure of a rhodopsin gene from *D. melanogaster*.

186. Zuker, C. S., C. Montell, K. Jones, T. Laverty, and G. M. Rubin. *J. Neurosci.* 7:1550–1556, 1987. A rhodopsin gene expressed in photoreceptor cell R7 of the *Drosophila* eye: Homologies with other signal-transducing molecules.

Chapter

5 *Mechanoreception and Phonoreception*

Albert S. Feng and Jim C. Hall

This chapter describes how mechanical stimuli, for example, displacement and substrate-borne vibration and sound, are detected and processed by animals living in diverse environments. Mechanoreception is important for sensing contact with surfaces, position of individual extremities or appendages, velocity and acceleration of movement, orientation with respect to gravity, displacement and velocity of water and air particles (including sound and substrate vibration), and depth of water. Mechanical signals are also widely employed by aquatic and terrestrial animals to communicate with members of the same species; serving to attract mates, establish territories, signal the detection of prey or predators, and as social displays.

Mechanoreceptors

General Structure

Mechanoreceptors are differentiated epithelial cells. They may take the form of free nerve endings that transduce mechanical to electrical energy following membrane deformation, or modified free nerve endings, which have accessory structures, that serve to filter or amplify mechanical signals, or to impart directional sensitivity to the receptor.

Other mechanoreceptors consist of modified cilia, such as tactile hairs and bipolar sensory cells, and are associated with the cuticle of the invertebrate exoskeleton (Fig. 1). Here, the modified cilium, or sensory process, is a membrane limited microtubule-packed structure that extends from the basal body found in the tip of the dendritic process (220). Typically, the microtubules (350–1000 in number) are attached directly to the site of mechanical stimulation on the cuticle. Hence, the modified cilium of these mechanoreceptors has been implicated in the transduction of mechanical stimuli (347).

Some modified cilia possess specialized derivatives of microvilli called stereocilia (75). The vertebrate hair cell, which is a major constituent of the acous-

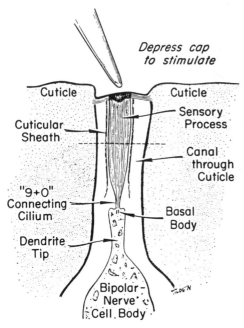

Figure 1. Diagram of campaniform sensillum of cockroach tactile hair. Mechanoreceptor consists of a cilium with many microtubules. [From (220).]

ticolateralis system, is such a mechano-receptor. In the vertebrate hair cell, the stereocilia are arranged in a hair bundle. The hair bundle is polarized with short stereocilia at one edge of the bundle grading into longer stereocilia at the opposite edge. In the vertebrate auditory and vestibular organs, the hair cell stereocilia are often accompanied by a single modification, called the kinocilium. The kinocilium is always located at the tall edge of the bundle. However, in hearing organs of mammals, that is, the cochlea, the kinocilium disappears during development and is not present in adults (144).

The stereocilia of vertebrate hair cells contain longitudinally oriented bundles of parallel microfilaments. The microfilaments contain actin (97, 306, 315, 333, 334, 358). The actin filaments are arranged in arrays and give the appearance of arrowheads, all pointing in the direction of the cuticular plate, when viewed

in longitudinal sections (Fig. 2a). The actin filaments, which are interconnected, impart a rigidity to the stereocilia such that they resist bending or displacement along their length (334).

The tip of each stereocilium is linked to the lateral wall of an adjacent, taller stereocilium by a filamentous cross-bridge, which may be involved in sensory transduction (Fig. 2b and c). Additional cross-bridges near the basal insertions of stereocilia link them to their nearest neighbors (242, 286, 309). The function of these latter cross-bridges remains unknown.

The cuticular plate, located at the apical end of the cell, is the region at which the stereocilia insert into the cell. It is a filamentous network penetrated by actin fibers comprising the stereocilia rootlets (333). When displaced, the rodlike stereocilia pivot at their point of entry into the cuticular plate. The function of the cuticular plate is not clear. It contains both actin and myosin suggesting that tension can be generated in this region through actin–myosin interactions. Indeed, electrical or acoustical stimulation, or exposure to high K^+, brings about a change in cell shape (38, 193), which can presumably modify the mechanical characteristics of the transduction apparatus.

The kinocilium of a vertebrate hair cell is cylindrical, ~0.3 μm in diameter, and in some organs can attain a length of 50 μm. Kinocilia have all the major features of motile cilia including a 9 + 2 array of microtubules, dynein arms, and a basal body (98, 129). The distal tip of the kinocilium, which is often enlarged or swollen, is typically attached to an accessory structure. For example, in the otolithic organs of the vestibular system, the kinocilium is connected to the otolithic membrane (139). These features have generated a great deal of interest in the kinocilium because true cilia or their derivatives, with similar properties, have

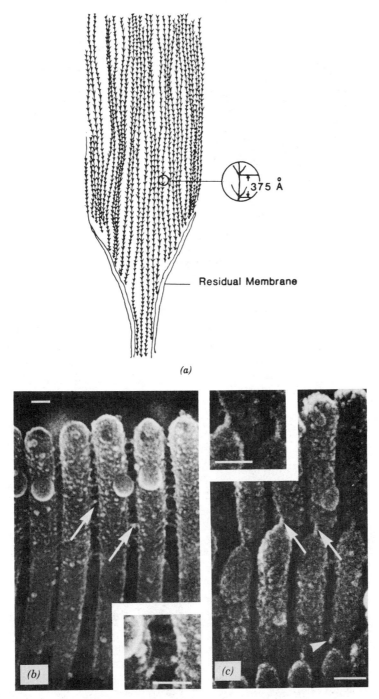

375 Å

Residual Membrane

(a)

(b)

(c)

Figure 2. (*a*) Diagram of the appearance of actin filaments in a stereocilium that has been decorated with myosin subfragments (S1). Arrowheads polarized toward cuticular plate. [From (286).] (*b* and *c*) Cross-links (arrows) on inner hair cell of guinea pig cochlea. The surface membrane is rough in the region of links. [From (233a).]

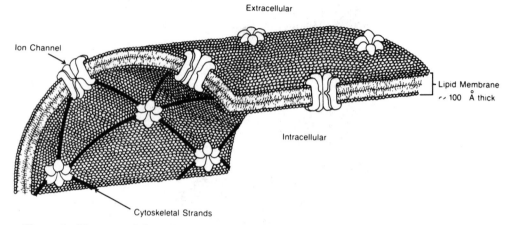

Figure 3. Diagram of the cell membrane showing stretch activated channels and cytoskeletal strands in series with the channel. [Courtesy of Alan Kegler, University of Buffalo Public Relations Office. See (276a).]

proven important for mechanosensory transduction in invertebrates. It is noteworthy that in the adult mammalian cochlea, it is the longest stereocilia that contact the overlying tectorial membrane, an accessory structure, as the kinocilium is absent (189).

Transduction

Mechanoelectrical transduction is, in general, mediated by the gating of Na^+ and/or K^+ channels to produce receptor potentials. It is believed that transduction involves the generation of tensile forces, which act upon ionic channels via cytoskeletal structures (Fig. 3) (31, 59, 60, 121, 145, 277). It is hypothesized that when there is no external force applied to transduction channels, there is a low probability of channel opening. However, when the receptor membrane is deformed, tensile forces are generated, and the probability of channel opening increases. Conductance through the channel does not change. Stimulus dependent differences in the amplitude of the receptor potential is, in general, attributable to differences in the time over which channels remain in the conducting state.

As an example of mechanoelectrical transduction consider the hair cell of the vertebrate peripheral auditory system as depicted in Figure 4. Two lines of evidence suggest that the stereocilia of individual hair cells, and not the kinocilium, are essential for transduction. First, displacement of the stereocilia results in a receptor potential: displacement towards the kinocilium depolarizes, movement in the opposite direction hyperpolarizes, the hair cell (98). Second, removal of the kinocilium does not affect the transduction process (144).

The ion channels involved in the development of the hair cell receptor potential are presumably located near the tips of the stereocilia (145, 264). According to the current working model of mechanoelectric transduction in hair cells, displacement of the stereocilia introduces tensile forces that act upon ion channels and increases the probability of channel opening. The greater the displacement, the longer the channels remain open and hence, the larger the current. Interestingly, the cross-bridges linking the tips of the stereocilia are well situated to register shearing forces between individual stereocilia and are believed to mediate the

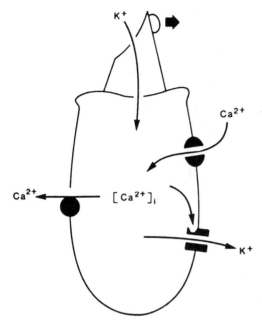

Figure 4. A model for electrical resonance in hair cells. When the hair bundle is deflected, transduction channels open and positive ions, largely K$^+$ *in vivo*, enter the cell. The depolarization evoked by this transduction current activates voltage-sensitive Ca^{2+} channels. As Ca^{2+} ions flow into the cell, they augment the depolarization. At the same time, however, the influx of Ca^{2+} raises the intracellular concentration of this ion [Ca^{2+}]$_i$, especially the local concentration just beneath the surface membrane. The high [Ca^{2+}] brings into play the Ca^{2+}-sensitive K$^+$ channels. As K$^+$ exits through these pores, it begins to repolarize the membrane, thereby diminishing the activation of Ca^{2+} channels. The fluid bathing the apical surface of a hair cell characteristically has a much higher [K$^+$] than that contacting the basolateral cellular surface. As a consequence, K$^+$ can both enter and leave the cell passively. By the time the membrane potential is somewhat more negative than its steady-state value, the intracellular [Ca^{2+}] is reduced by the sequestering of the ion within organelles and by its extrusion through ion pumps. As the Ca^{2+}-sensitive K$^+$ channels close, the cell returns to approximately its initial condition, and another cycle of the electrical resonance commences. [From (145).]

tensile forces that affect channel gating (60, 145).

Initiation of Sensory Nerve Impulses

Sensory information is conveyed to the central nervous system by means of impulses, or action potentials, in sensory nerve fibers. But, how are these action potentials generated? As discussed above, mechanoelectric transduction involves the generation of local potential changes at the level of the receptor membrane following stimulus application. We have referred to these potential changes as *receptor potentials*. Thus, receptor potentials may be defined as a change in the membrane potential of the receptor cell, or sensory nerve terminal, induced by the action of the stimulus. Receptor potentials do not always directly give rise to action potentials. Potential changes, which do result in the production of action potentials, are referred to as *generator potentials*. In some cases, for example, the stretch receptors of frogs and crayfish, the receptor potential is the same as the generator potential.

Both receptor and generator potentials share the following features. First, as mentioned above, they represent local changes in the potential across the cell membrane. Second, they are graded in nature. That is, their amplitude changes concomitant with stimulus intensity. This is in contrast to the action potential that is an all-or-none phenomenon. Third, receptor and generator potentials are conducted electrotonically. Hence, their amplitude is inversely related to the distance over which they are conducted. These features play an important role in the production of action potentials and therefore the representation of sensory information. This is best illustrated by the Pacinian corpuscle, which responds to mechanical deformation of the skin and is found in numerous vertebrate species

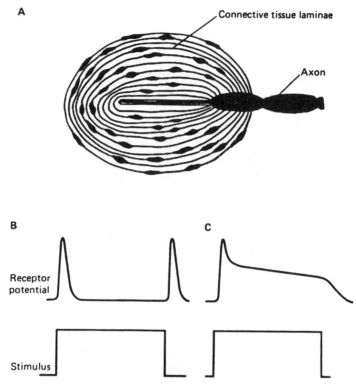

Figure 5. The Pacinian corpuscle is a rapidly adapting receptor in the skin that is sensitive to vibration. (*A*) A cross section of this receptor shows concentrically arranged layers of connective tissue surrounding the nerve terminal. (*B*) An intact Pacinian corpuscle responds with a receptor potential only on onset and offset of a mechanical stimulus. (*C*) If the connective laminae are removed, the receptor responds to the same mechanical stimulus in a slowly adapting manner. [From (155a).]

including humans. The Pacinian corpuscle comprises a number of concentrically arranged ellipsoidal lamellae around a naked nerve terminal (Fig. 5). Just before the nerve emerges from the organ, it becomes myelinated. The naked, or nonmyelinated, nerve terminal has been shown to be the primary transduction element (190) and possesses numerous independent transduction sites along its entire length. When mechanically displaced the transduction sites produce graded receptor potentials whose amplitude varies in accordance with the magnitude of the mechanical deformation. Receptor potentials are passively (electrotonically) propagated towards the first node of Ranvier,

the site of action potential generation, where they are summed to form the generator potential. When the amplitude of the generator potential exceeds the threshold of firing, action potentials are generated (24, 66). Small mechanical deformations result in correspondingly small potentials and hence the generation of fewer action potentials than stimuli of greater magnitude. The firing rate, or number of spikes produced per unit time, is also slower for small than large stimuli. Hence, the number of action potentials and the frequency at which they are produced convey information concerning stimulus intensity.

Many sensory stimuli, when applied at

a constant intensity, produce a sensation that declines with time. This phenomenon is referred to as *habituation*; its underlying basis is response adaptation, which is correlated with a decline in the amplitude of the generator potential, and hence the firing frequency of action potentials, with time. Mechanoreceptors can be classified as slowly (SA) or rapidly (RA) adapting depending on the rate at which the generator potential, and hence the firing rate of action potentials, adapt. Those that discharge action potentials throughout the entire period of the mechanical stimulus are slowly adapting, or tonic, receptors. Those that fire action potentials only at the time of stimulus application (*on*) or release (*off*) are rapidly adapting, or phasic, receptors. Tonic receptors are well suited for signaling the duration and intensity of stimulation while phasic receptors signal a change in the intensity of stimulation with time. Phasic receptors that respond to both stimulus onset as well as offset are also ideally suited for signaling stimulus duration.

In general, adaptation is due to intrinsic properties of the cell's excitable membrane. However, in some cases, like the Pacinian corpuscle, accessory structures are involved. In the intact corpuscle, the generator potential adapts rapidly and responds only to transient changes of mechanical compression. In contrast, the generator potential of a decapsulated corpuscle or naked nerve ending adapts slowly (Fig. 5) and depolarization is maintained throughout the entire period of stimulation (191, 228). Rapid adaptation in the intact corpuscle is the result of transverse slippage between the layers of connective tissue comprising the accessory structure and a corresponding decrease in force applied to the encapsulated naked nerve terminal. Thus, the accessory structure serves to filter out the slow (<20 Hz), or sustained, components of the mechanical stimulus, while allowing more transient events (20–1000 Hz), such as stimulus onset and offset, to pass; in essence, they serve as high-pass filters.

Frequency Sensitivity

Several mechanisms have evolved that shape the frequency sensitivity of mechanoreceptors. In the Pacinian corpuscle, we have already described the role of the connective tissue laminae in mediating the frequency response characteristics of the receptor.

Electrical tuning represents yet another filtering mechanism that is evident in several mechanosensory systems. For example, an electrical current applied in a stepwise fashion produces a damped oscillation of the membrane potential of hair cells in some vestibular and auditory organs of reptiles and amphibians (8, 61, 187). Each hair cell behaves as a resonator such that the amplitude of the membrane potential is greatest for a single frequency of oscillation. Interestingly, the resonance frequency of the hair cells matches the frequency of sound that best excites the cells and partially accounts for their frequency selectivity.

Electrical tuning primarily involves two species of ionic current; an early inward Ca^{2+} current and a delayed Ca^{2+} activated outward K^+ current (Fig. 4). The initial influx of Ca^{2+} during the depolarization phase causes an intracellular accumulation of Ca^{2+}, which then triggers the repolarizing outward K^+ current. For a single current step these processes are repeated several times but with decreasing magnitude until the membrane potential oscillations are completely damped.

In several reptile species, including alligators, and in chicks the frequency tuning of some of the hair cells in the auditory organ involves yet a different

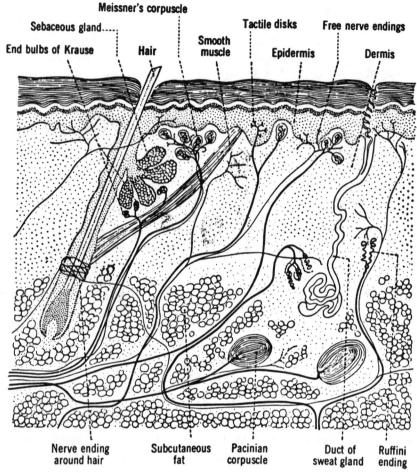

End bulbs of Krause · Hair · Sebaceous gland · Meissner's corpuscle · Smooth muscle · Tactile disks · Epidermis · Free nerve endings · Dermis

Nerve ending around hair · Subcutaneous fat · Pacinian corpuscle · Duct of sweat gland · Ruffini ending

Figure 6. Diagrammatic representation of innervation of skin with sparse hair. Heavy lines are myelinated fibers, thin lines nonmyelinated fibers. [From (111a).]

mechanism, mechanical tuning. In these, the frequency selectivity of the hair cells is determined by the resonance characteristics of their stereocilia and kinocilium (102, 334, 336, 342). The stereocilia and kinocilium of hair cells sensitive to high frequencies are short. In contrast, the stereocilia and kinocilium of hair cells sensitive to low frequencies are longer.

Finally, more than one of the above mechanisms can contribute to the frequency selectivity of a mechanoreceptor. For example, accessory structures and electrical tuning combine to bring about frequency selectivity in the ear of several

vertebrate species including, presumably, humans.

Classification of Mechanoreceptors

Mechanoreceptors can be divided into two general classes—*exteroceptors*, which are excited by mechanical stimuli originating outside the body, and *interoceptors*, which are excited by stimuli originating inside the body. Exteroceptors can be further divided into *distance receptors* (such as those involved in hearing) and *contact receptors* (such as those involved in touch). Interoceptors can be subdivided

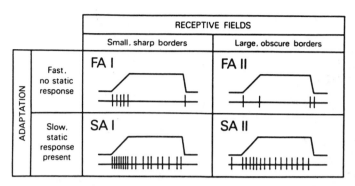

Figure 7. Types of mechano-receptive afferent units in the glabrous skin of the human hand. FA I units are sensitive to rate of skin indentation; FA II units are highly sensitive to acceleration and higher derivatives, fast adapting; SA I units have high dynamic sensitivity; SA II units show very regular sustained discharge. [From (152).]

into *equilibrium receptors*, which give information about the position and movement of the entire body, *proprioceptors*, which supply information concerning the relative position of limbs or joints, and *visceroceptors*, which oversee conditions in the rest of the body.

Vertebrate Mechanoreceptors

Exteroceptors

Cutaneous mechanoreceptors (Fig. 6), which mediate the sensation of touch are the best-known of the vertebrate exteroceptors. Mechanoreceptors of the skin can be divided into two functional categories (Fig. 7), Type I and Type II, based on differences in their receptive field size, that is, that area of the skin which when stimulated excites the receptor (152, 256). Type I receptors have small, well-defined receptive fields and are located superficially on the surface of the skin. Type II mechanoreceptors, in contrast, have large, poorly defined receptive fields.

There are two classes of Type I receptors based upon differences in function and morphology. One class, the FA I, is associated with laminated accessory structures and are fast adapting. Meissner's corpuscle, located in the papillary dermis of the skin, is an example of an FA I receptor. The second class, the SA I, does not have laminated accessory

structures and is slowly adapting. Merkel's disk, also lying in the papillary dermis, is a representative SA I receptor.

Type II mechanoreceptors may also be divided into two classes, FA II and SA II. These classes, like their Type I analogs, represent morphologically and functionally distinct classes. Like FA I receptors, FA II receptors have laminated accessory structures and adapt rapidly. SA II receptors, like their SA I analogs, are not laminated and are slowly adapting. The Pacinian corpuscle and the Ruffini ending, both found in the reticular dermis, represent FA II and SA II receptors, respectively.

Another exteroceptor of interest is represented by the vibrissae located near the mouth region of many mammals. Vibrissae are sensitive to vibration of the surrounding medium (air) as well as contact displacement. Vibrations in the range of 0.5 to 100 Hz are effective stimuli with a maximum displacement sensitivity of ~20 mm (or angular deflection of 0.1–0.2°) at 50 Hz (114, 188).

Afferents innervating a group of vibrissae derive their receptive fields from a single vibrissa (114). Each individual afferent shows directionally selective responses to vibrissa displacement. Maximum responses are evoked by displacement towards the center of the group (116, 253, 360). Because of this directional sensitivity it is not surprising to find that

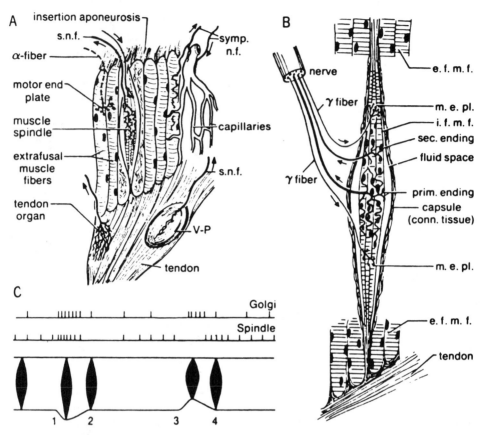

Figure 8. (A) Diagrammatic drawing of the three main receptors in skeletal muscles: muscle spindle organ, tendon organ, and Vater–Pacinian corpuscle. The spindle organ is in parallel with extrafusal muscle fibers, the tendon organ in series with fibers. (B) Simplified diagram of muscle spindle organ. (C) Response of a Golgi tendon organ and of a muscle spindle organ. [From (24a).]

nocturnal animals, such as rats and cats, show a considerable dependence on their vibrissae for orientation during locomotion in the dark.

Many submammalian vertebrates, such as frogs and snakes, show an exquisite sensitivity to substrate vibration (175). In these animals, vibrational signals are used to detect predators or prey as well as the presence of potential mates. Vibration sensitive mechanoreceptors are of varying types and can occur in the skin as well as in other body tissues.

The seismic sensitivity of the snake has been determined by measuring the evoked response of midbrain neurons to body vibration (132). Response thresholds range from 1 to 100 Å of displacement, a truly remarkable sensitivity. The cutaneous mechanoreceptors, which mediate the snake's vibration sense, are maximally sensitive to displacements occurring at a frequency of 150–200 Hz (252). Recordings from afferent fibers innervating the mechanoreceptors have demonstrated that the action potentials occur for each cycle of a sinusoidal stimulus in a 1:1 fashion for signal frequencies of up to 300 Hz. Thus, the temporal discharge pattern of the afferent may provide important information concerning the spectral content of the signal. In addition, each action potential occurs at a particular phase of the stimulus (phase-locked discharge) for stimulus frequencies of up to 600 Hz. By comparing the phase angle of the stimulus at which receptors from different loci respond, directional information concerning the source of the signal can be derived.

Interoceptors

The muscle spindle organ (MSO), found in the skeletal muscles of tetrapod vertebrates (Fig. 8), is a particularly interesting example of an interoceptor both in light of its complexity as well as its historical significance; the first records of electrical activity in sensory nerves involved stimulation of the MSO (3). The MSO is a proprioceptor. It is arranged in parallel with the muscle fibers and provides information about the static and dynamic changes of muscle length (14, 123). It is a slowly adapting mechanoreceptor, producing sustained generator potentials during steady passive stretch. Yet, noticeable adaptation can be observed following its transient "on" response (158). This adaptation is the result of a decline in muscle tension in time coupled with alterations in the local ionic environment of the nerve terminal during stretch (146).

Each muscle spindle comprises a group of small specialized muscle fibers called intrafusal fibers. These fibers are not contractile and do not contribute to the development of muscle tension; the two ends are innervated by γ-motor neurons. The intrafusal fibers regulate the excitability of the spindle afferents by deforming the sensory bulbs that contact them (159). The large skeletal muscles, which do develop significant muscle tension, are the extrafusal fibers. They are innervated by α-motor neurons situated in the ventral horn of the spinal cord.

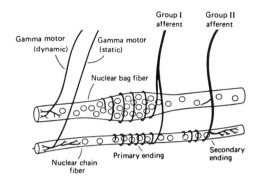

Figure 9. Nuclear bag and nuclear chain intrafusal fibers within a muscle spindle organ, each receptor has own efferent control. Group I (primary) afferents innervate both nuclear bag and nuclear chain receptor. [From (155a).]

There are two basic types of intrafusal fibers, nuclear bag and nuclear chain fibers (Fig. 9). These two fiber types differ with respect to the longitudinal arrangement of their nuclei as well as their innervation pattern. Nuclear bag fibers can be further subdivided into bag_1 (dynamic) and bag_2 (static) fibers (11, 12).

Two different types of afferent nerve endings are evident in the MSO, primary and secondary nerve endings. Primary endings are spiral shaped and innervate *both* nuclear bag and nuclear chain fibers; the ratio being dependent on the specific muscle. In contrast, the secondary endings exhibit a flower-spray shape and innervate nuclear chain fibers exclusively.

Primary and secondary endings are functionally distinct and differ with respect to the type or size of afferent nerve fibers associated with them. Primary endings, associated with Ia afferent nerve fibers, give both static and dynamic responses to passive stretch. Secondary endings, associated with Type II afferent fibers, show an almost pure static response to passive stretch of the MSO. But, primary and secondary endings are equally sensitive to mechanical stimulation. Therefore, the dynamic response of

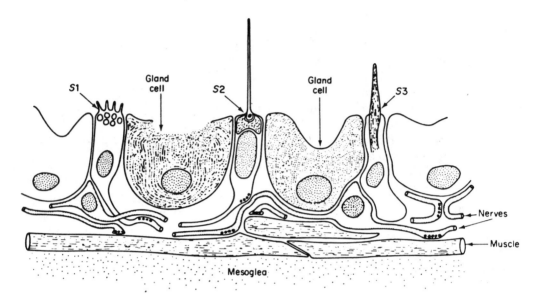

Figure 10. Tactile hairs on epithelium of "finger" of the ctenophore *Leucothea*. Three types of sensory cella; S1 bears a group of short sensory cilia; S2 bears a solitary long cilium, while S3 is in the form of a sensory peg. [From (141a).]

the MSO is attributed to the mechanical properties of bag$_1$ fibers (206, 247). Apparently, when bag$_1$ fibers are stretched continuously they rapidly return to their unstretched length because of relaxation in their equatorial region; that is, the region innervated by the primary afferent (28). Likewise, the static response of the MSO is attributed to the mechanical properties of the bag$_2$.

Another stretch receptor, the Golgi tendon organ, comprises a slender capsule that occurs in series with skeletal muscle fibers. Before terminating in musculotendinous junctions, muscle fibers enter the capsule of the Golgi tendon organ, forming a braided array of collagen fibers extending through the entire capsule. Contraction of the skeletal muscle straightens the collagen fibers, deforming the axons of Ib afferent fibers with which they are intertwined. Because of this arrangement, Golgi tendon organs are very sensitive to the development of muscle tension, but are insensitive to passive stretch.

Both the MSO and Golgi tendon organ mediate muscle stretch reflexes that are necessary for optimum motor function. Details concerning their contribution to these reflexes is beyond the scope of the chapter and has been dealt with elsewhere.

Invertebrate Mechanoreceptors

Invertebrate mechanoreceptors may be generally classified as hair cells, bipolar sensory cells, or muscle stretch receptors. Hair cells and bipolar sensory cells are modified cilia typically associated with one of a variety of accessory structures giving rise to several anatomical and functional categories of mechanoreceptors. One of these, the tactile hair (Fig. 10) resides superficially on the exoskeleton and responds to vibration. In contrast, the campaniform or lyriform sensillum lies beneath the cuticle and is primarily sensitive to mechanical stress or strain on the cuticle but can also serve as a proprioceptor. The scolophorus sense cell (Fig. 11c)

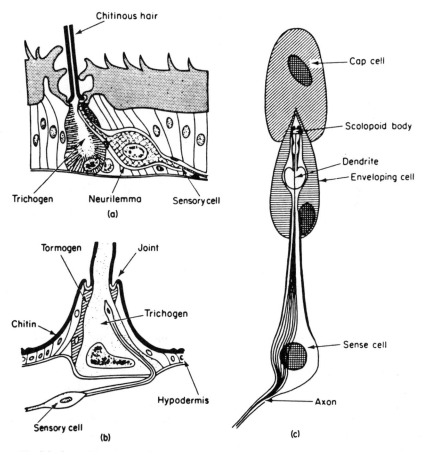

Figure 11. Mechanoreceptors of insects. (*a*) Tricobothria sensillium from the cercus of cricket; (*b*) hair sensillum from the caterpillar *Pieris*; (*c*) chordotonal sensillum. [Parts *a* and *b* are from (340a), Part *c* is from (65a).]

differs from tactile hairs and campaniform or lyriform sensillae in that it registers stretch in the tendon or joint to which it is attached. Scolophorus sense cells also respond to pressure changes in the medium surrounding the animal. These receptors are not associated with any specialized structure of the cuticular surface of the exoskeleton. As such, they belong to a group of mechanoreceptors referred to as a subcuticular, or chordotonal organs. Some chordotonal organs, such as Johnston's and tympanal organs, lie beneath the thin cuticle and serve as hearing or vibration receptors (see below). Statocysts (Fig. 12) are also com-

prised of modified cilia associated with unique accessory structures. Statocysts serve primarily to signal changes in body position relative to the direction of gravitational forces.

Tactile Hairs

Mechanoreceptive hairs and bristles are common in many aquatic and terrestrial invertebrates. They serve as touch or vibration (particle displacement) detectors that sense contact with the body surface and air or water movements. For example, the tactile hairs on the epithelium of ctenophore fingers (Fig. 10) detect slight

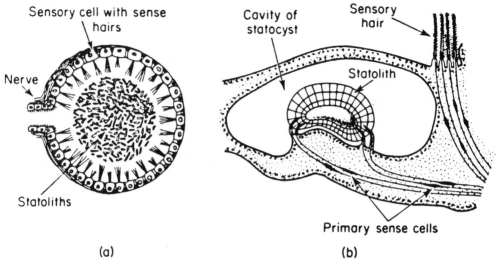

Figure 12. Statocysts: (a) *Pecten* with a free statolith; (b) statocyst of the crustacean *Leptomysis* with attached dendrites. [From (135).]

mechanical disturbances in the surrounding water and mediate prey localization and capture. The arrow-worm, *Spadella*, also utilizes vibration sensitive tactile hairs for prey capture, particularly in the dark. Like the tactile hairs of ctenophorans, those of *Spadella* are nonmotile ciliated epithelial cells that presumably evolved from motile forms. Sensory hairs on the chelae of the crayfish, *Cherax destructor*, are grouped into pits forming a compound mechanoreceptor organ. These receptors are sensitive to water vibrations with a peak-to-peak amplitude threshold of 0.2 mm over a frequency range of 150–300 Hz (326).

Tactile hairs may take several forms. For example, five types of mechanoreceptive tactile hairs may be found on the antennal flagellum of rock lobsters: spinelike receptors, feather hairs, smooth hairs, hydrodynamic receptors, and vibration receptors. All of the receptors are innervated by one or more sensory cells and respond to various components of flagellar or water movements as well as boundary contacts (338). Likewise, two morphologically distinct types of sensory

hairs may be found on the flagellum of the second antenna of the crayfish, *Astacus leptodactylus*: smooth upright hairs and feathered hairs. Smooth upright hairs are displaced by the movement of the surrounding water, which excites the underlying transduction element. A displacement as small as 3.5° for a 40-Hz stimulus can excite the receptor. The feathered hairs, in contrast, are not driven by water vibration directly, but rather are excited by the bending of the flagellum in response to water vibrations (327). The feathered hairs are more sensitive to displacement than smooth hairs, with a reported sensitivity of 0.06° at 90 Hz.

Many species of decapod crustaceans have very large second antennae, each consisting of five distinct segments, with the last segment bearing a single long flagellum (65). In the distal part of the flagellum are straight (type-2) setae, which are surrounded by curved (type-1) setae. These mechanoreceptors are analogous to the lateral line organs of fish (see below) as both are sensitive to particle displacement and vibration in the water.

The cercal hairs of orthopterans (e.g.,

grasshoppers and crickets) respond to puffs of air or to substrate vibrations of 20–800 Hz. Maximal sensitivity occurs at low (20–80 Hz) or high (200–800 Hz) frequencies (157, 254). The hairs function in predator detection and mediate, in part, reproductive behavior patterns.

Predator detection is also the role of sensory hairs found on the cercus of the cockroach, *Periplaneta* (Fig. 11*b*). Each hair contains a tricogen (supporting) cell and a sensory cell (332). Bending the hair in one direction depolarizes the membrane of the sensory cell while bending in the opposite direction hyperpolarizes it. Thus, these cells have an inherent directionality (62, 344). Directional sensitivity may be a common feature of sensory hairs as it is also a characteristic of the sensory hairs on the pedipalp of the scorpion, *Euscopius*; though, in *Locusta*, directionality of response is absent (308). In the latter case, however, neural adaptation occurs if the hair is held toward the dendrite of the sense cell, and deadaptation occurs when the hair is held in the opposite direction (140). Even though directional sensitivity is a common feature of tactile hairs in several taxa, differences in the underlying mechanisms, such as those described above, suggests that it has probably evolved independently several times.

As in some aquatic organisms (e.g., *Astacus leptodactylus*), tactile hairs found on a single structure in terrestrial invertebrates may also be characterized on the basis of differences in form and function. For example, cockroach cerci contain both threadlike and bristlelike hair sensilla. Both respond only to the phasic component of mechanical displacement and yield nonlinear responses to sinusoidal stimulation. The threadlike sensillum shows spontaneous activity, the bristlelike does not (44, 45). These properties are probably related to the mechanical characteristics of the base of the hair, which most likely account for differences

between threadlike and bristlelike sensilla as well. Similarly, the filiform hairs found on the cerci of crickets show distinct morphological and physiological differences. Long filiform hairs, which are velocity sensitive, are spontaneously active and excited by low-frequency stimulation. In contrast, short hairs, which are sensitive to acceleration, are not spontaneously active or are they sensitive to low frequency stimulation (302).

Campaniform and Slit Sensilla

Many invertebrates possess biological strain gauges in the form of campaniform organs or slit sensilla (15, 43, 184). While there are numerous morphological variations the basic plan (e.g., Fig. 1) is one of holes in the cuticle covered by a thin membrane to which the dendrite of a sensory cell is attached (15, 56). In the slit sensilla, the covering membrane is concave and averages ~0.25 μm thick dependent on species. The covering membrane of the campaniform sensilla is convex, and is composed of several thicker layers.

Campaniform and slit sensilla may be found in one of three configurations: isolated, loosely grouped, or as compound sensilla with the receptors lying side by side. The effective excitatory stimulus appears to be a pinching of the distal ciliary process (or dendrite) due to compression of the covering membrane. A displacement of the covering membrane as low as 1 nm can result in enough compressive force to excite the receptor (55).

Campaniform sensilla are found in numerous insects (e.g., honeybees, saturniid moths, ambrosia beetles, cockroaches, flys, crickets, and grasshoppers) and crustaceans (e.g., crabs and lobsters) while slit sensilla are commonly found in arachnids, most notably hunting and web spiders (15, 16).

Statocysts

Statocysts occur widely among invertebrates and have apparently evolved many times. Statocysts consist of ciliated hair cells, or bipolar sensory cells, and a statolith (Fig. 12). The statolith may be a mass of small crystals (statoconia) embedded in gelatinous material or a compact homogeneous structure. A change in the position of the statolith excites specific populations of sensory neurons by bending the cilia on the hair cells, or dendrites of bipolar neurons, thereby providing orientational information with respect to gravity (142). In *Aplysia*, the cilia are arranged irregularly in the statocyst and the organ shows no directional sensitivity (347). In contrast, the statocyst of cephalopods shows a clear directional preference; cilia are arranged in regular rows with basal feet projecting from the basal bodies in line with the direction of response (13, 41, 42).

Statocysts vary among taxa with respect to their structural and functional complexity. In the scallop, *Pecten*, the statocyst is a simple sac with hair cells and their supporting elements aligned along the inner wall. The statolith is heavy and calcareous (135). In *Leptomysis*, a crustacean, the sensory neurons are bipolar cells with dendrites anchored in a fluid-filled compartment containing the statolith (135). Changes in body position induce fluid movement in the statocyst causing displacement of the statolith or statoconia. This in turn mechanically stimulates and excites the hair cells. In crayfish and lobsters, hair cells are actually attached to the statolith instead of the chamber walls as in *Leptomysis*, though functionally, the organs are equivalent. Some species of crabs possess an elaborated fluid-filled sac that has both horizontal and vertical canals (101). The hair cells are not attached to the statolith, as in crayfish and lobsters, and are suitable for detecting angular and linear accelerations. Finally, the cephalopod statocyst is perhaps the most complex of those studied to date. It consists of a number of specialized structural elements and subserves several functions including the detection of low-frequency vibrations, angular acceleration, and gravity (156).

Chordotonal Organs

Chordotonal organs have achieved prominence because of their role in mediating the vibration and acoustic sense of many invertebrate animals, particularly the insects. Their structural and functional properties relative to this role will be discussed later in this chapter.

Proprioceptors

Muscle receptor (myochordotonal) organs, found in the dorsal muscle wall of each abdominal segments of several arthropod species, are among the most well-studied invertebrate proprioceptors. The muscle receptor organs (MROs) of the crayfish are typical examples.

Each abdominal segment of the crayfish has two pairs of MROs (Fig. 13); MRO_1, which is slowly adapting, and MRO_2, which adapts rapidly. Each receptor consists of a thin muscle embraced by the fingerlike dendritic projections of an afferent neuron. In addition, each receptor is innervated by small and large diameter fibers of central neurons which, when activated, inhibit the receptor organs.

When stretched, MROs deform the dendrites of the afferent neurons producing a receptor potential (Fig. 14(I)) (177). The amplitude of the receptor potential is graded; that is, it varies with stimulus strength. Three ionic currents are involved in the generation of the receptor potential: an inward, rapid, stretch-induced Na^+ current, a slow Cl^- outward

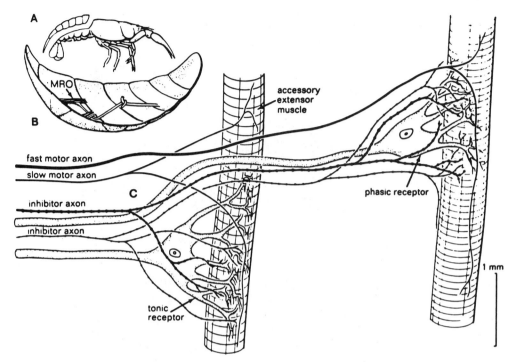

Figure 13. Diagrams of muscle receptor organ (MRO) in abdomen of decapod crustaceans. (*A* and *B*) Location of MRO. (*C*) Phasic and tonic receptors attached to their respective muscles. Motor, inhibitor, and sensory nerve fibers. [From (348a).]

current, and a K^+ outward current (34). These currents drive the membrane potential away from its resting value (-70 mV) to a reversal potential of $+3$ to $+30$ mV. Adaptation of the receptor potential during sustained stretch is attributed to an accumulation of intracellular Ca^{2+}, which activates an outward K^+ current (234).

The receptor potential propagates electrotonically from its site of initiation towards the cell soma. Nerve impulses are generated at the base of the cell's axon in response to a critical depolarization (threshold) of 10 mV for MRO_1 and 20 mV for MRO_2 (Fig. 14(I, *b*, *c*)). The receptor potential can be maintained and spike discharges continue for several hours during prolonged stretch of MRO_1, but MRO_2 ceases producing impulses after a few seconds.

The effects of applied stretch and neu-

rogenic muscle contraction on the activity of the crustacean stretch receptor are synergistic; differing from the antagonistic interactions of Type I_a and II fibers innervating the mammalian muscle receptors (119). The membrane potential of the sensory neuron is maintained at a steady-state potential of about -64 mV (78) by the inhibitory accessory fibers. If the receptor is slightly stretched and partially depolarized, stimulation of the efferent accessory fibers suppress sensory discharges (Fig. 14(II)) by preventing the generator potential from reaching firing threshold (178). In contrast, if the receptor is in a relaxed state the inhibitor actually depolarizes it. In both cases, however, inhibitor activity increases Cl^- conductance, presumably mediated by the neurotransmitter γ-aminobutyric acid (GABA), thereby effectively short-circuiting impulse initiation (149).

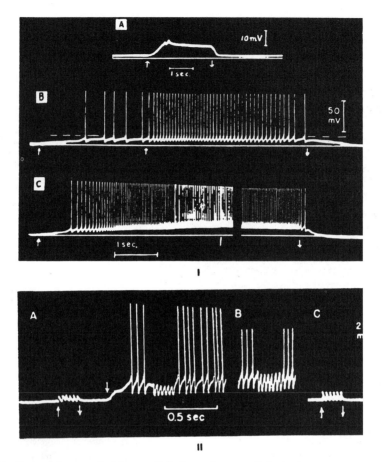

Figure 14. Potentials recorded intracellularly and extracellularly from crustacean stretch receptors. [(I) From (177) and (II) from (178).]

Other proprioceptors of several distinct morphological types may also be found in arthropods. For example, all arthropod leg joints are supplied with one or more of the following proprioceptors: muscle receptor organs, chordotonal organs (147), campaniform or slit sensilla (250, 300), multipolar joint receptors (179), tarsal hairs (30), hair plates (30, 251), and strand receptors (29).

Invertebrate Vibration Sense

Vibration is a term used to describe periodic or aperoidic particle motion in solid substrates, water, or air. Invertebrate vibration receptors have typically evolved

from mechanoreceptors that served other functions, for example, proprioception, and were preadapted for mediating the vibration sense. Biologists usually measure vibrations in terms of displacement, velocity, or acceleration. Displacement is perhaps the easiest to understand and simply refers to the magnitude of motion of the body being studied. Velocity is the time rate of change of displacement and for sinusoidal vibrations is given as

$$v = \omega A \cos \omega t \qquad (1)$$

where ω is 2π times the frequency of vibration, t is the period of the sine function and A is the half-wave amplitude of the sine function. Thus, velocity is proportional to displacement and to the frequency of vibration. Finally, acceleration is the time rate of change of velocity such that

$$a = -\omega^2 A \sin \omega t \qquad (2)$$

Therefore, acceleration is proportional to displacement and to the square of the velocity and the frequency.

Vibratory stimuli may be classified as contact vibrations, receiver is touched by sender with rhythmically varying pressure, or, as near field vibration, receiver is stimulated by the alternating flow of air or water molecules produced by rhythmic movements of the sender (201). We will concentrate on near field vibrations for which a large body of experimental data relevant to invertebrate vibration reception is available.

Generally, near field vibrations serve a communicative purpose, enabling sender and receiver to interact over a distance rather than being in direct physical contact with one another. This process is complicated by the presence of background noise, which serves to mask the biologically relevant signal. However, selection pressures have favored the evolution of signaling and receiving systems that are adapted for enhancing the detection of signals in noisy environments. For example, the frequency content of most communication signals and background noise differ. Background noise is typified by narrow bandwidth and low frequency. In contrast, signals used in vibrational communication are generally broadband and extend to higher frequencies than the noise. In addition, vibration receptors are optimally sensitive to the spectral content of biological signals while relatively insensitive to that of noise. Many communication signals, unlike noise, are also patterned in the time domain (176, 294) and thus easier to detect.

Vibrational signals may be transmitted as one or more different wave types that differ with respect to particle motion, propagation speed, and attenuation. The physical properties of the wave types suggest that some may be more ideally suited than others for communication purposes depending on the characteristics of the medium through which they are transmitted. For example, the spider, *Cupiennius salei*, lives on banana plants. Both males and females of the species communicate by means of vibrational signals transmitted through the various parts of the plant (stems, leaves, etc.). Behavioral data indicate that *bending waves* are particularly important for communication [reviewed in (16, 17)] in *Cupeinnius*. Particle motion in bending waves is in a plane perpendicular to both the direction of wave propagation and the surface of the plant, which is rhythmically bent as a whole. Bending waves are propagated slowly relative to other types of waves occurring in plants, and are dispersive, that is, depend on frequency. In the banana plant, attenuation at 75 Hz (dominant frequency of the male signal) averages only 0.3 dB/cm. Attenuation is even less at the dominant frequency of the female's signal. These results imply that *Cupiennius* uses a substrate, and wave type, well suited for its vibratory communication.

Longitudinal waves (motion perpendicular to the long axis of threads in the web, but in the plane of the web) are utilized by the orb weaving spider, *Nuctenea sclopeteria*, to detect and locate prey or mates in the web. They are transmitted via the radii of the web from the catching area to the hub. Longitudinal vibrations attenuate in a loaded web at a rate of ~2.5 dB/cm at a distance of 7 cm. Attenuation is minimal at vibration frequencies (produced by the fly, *Calliphora erythrocephala*) near 300 Hz but increases rapidly at lower and higher frequencies (204). In contrast, the web of the house spider, *Achaearanea* sp., which is distinct from that of *Nuctenia*, transmits vibratory signals from live prey with peak energy at 50 Hz (236). Spiders are sensitive to other wave types as well (e.g., *transverse*, *lateral*, and *torsional*), but the biological significance of these waves is poorly understood at this time.

The sensitivity of spiders to vibrations transmitted through solid substrates is mediated primarily by the lyriform organs, the most sensitive of which is the metatarsal lyriform organ. The metatarsal lyriform organ is a compound organ consisting of up to 21 slits, which are arranged in a similar pattern irrespective of species. The organ lies dorsally on the distal metatarsus and is oriented perpendicular to the leg. It is maximally stimulated by upward, and to a much lesser extent lateral, displacement of the tarsus.

The frequency response of spider lyriform organs follows a power function

$$y(f) = f^k \qquad (3)$$

where f = frequency and k is a constant that varies between 0.39 and 0.44 ($k = 0$ for frequency independent displacement receptors; $k = 1$ for velocity receptors). In all spiders so far investigated, the frequency response has a high-pass filtering characteristic (18, 22). In *C. salei*, and probably other species of spider as well, the metatarsal lyriform organ has a displacement threshold sensitivity of 10^{-3}–10^{-2} cm at frequencies of 10–40 Hz. Response threshold decreases sharply with increasing frequency beyond 10–40 Hz reaching 10^{-6}–10^{-7} cm at 1 kHz. Thus, the spectral energy distribution of behaviorally relevant stimuli fall within the receptor's tuning curve, while low-frequency background noise falls outside of the frequency range to which the receptor is sensitive.

The nocturnal scorpion, *Paruroctonus mesaensis*, is sensitive to substrate vibrations produced by prey burrowing through the sand (36). Compound slit sensilla and sensory hairs on the tarsi are sensitive to *Rayleigh* (motion component perpendicular to surface) and *pressure* (motion component parallel to surface) waves, respectively. Displacements as small as 0.5–2 Å can excite them. The spectral sensitivity of the receptors has not been established, though it likely parallels that of the metatarsal lyriform organ found in spiders. Interestingly, both Rayleigh and compressional waves travel at low velocities (~150 m/s) in loose sand (36). In addition, compressional waves are attenuated as they travel away from the source. Directional information concerning the signal source may therefore be derived by computing temporal differences in the activation of receptor populations on each of the legs. Moreover, distance information can be derived by comparing the degree of excitation of sensory hairs on each of the legs.

Fiddler (*Uca pugilator*) and ghost (*Ocypode ceratophthalmus*) crabs are also sensitive to vibrations transmitted through sand; the latter also being sensitive to airborne sounds. Rayleigh waves probably carry the behaviorally relevant information though this has yet to be determined. Vibration sensitivity in fiddler and ghost crabs is mediated by a specialized myochordotonal organ (Barth's organ), which is attached to a thin cuticular window in the exoskeleton of the merus of each walking leg. Barth's organ has a maximum sensitivity to substrate vibrations at 1.0–2.0 kHz and, in *Ocypode*, airborne sounds at 1.5–2.0 kHz (141).

Male fiddler and ghost crabs produce vibrational signals that mediate various aspects of their social behavior. In fiddler crabs, vibrational signals are produced by the rhythmic "waving" of the legs on one side of the body while being held off the substrate, or by tapping the substrate with the major cheliped. Barth's organ is involved in the detection of the acceleration component of these low-frequency (100–2000 Hz) signals (283). Since background noise is typically <100 Hz, Barth's organ, like the lyriform organs of spiders, is optimally suited for detecting the behaviorally relevant vibrational signals in a noisy environment. Moreover, the computation of directional information is probably similar in crabs and scorpions, since the receptors are spatially distributed in a similar manner and both utilize sand as a medium for communication.

The aquatic backswimmer, *Notonecta*, has sensitive water-surface wave detectors on the distal segments of the pro- and mesothoracic legs, which can detect water movements 1 μm in amplitude over a frequency range of 100–150 Hz (202). Attenuation of water–surface waves is more pronounced than in more solid substrates and is highly dependent on frequency. For example, surface waves are 1.67 dB/cm at 10 Hz and 8.57 dB/cm at 140 Hz measured ~3 cm from the source. Thus, *Notonecta* receptors are specifically adapted for detecting surface waves over large distances where only frequencies below ~200 Hz will prevail.

Notonecta can localize prey with an accuracy of 19–30°. Directional information is derived in a manner similar, in part, to that shown for the scorpion; that is, by comparing differences in the time and phase relationships of water waves as they stimulate receptors arranged symmetrically with respect to the longitudinal axis of the body (349). The fact that several unrelated species utilize a similar strategy for solving the problem of localization is representative of convergent evolution.

Finally, several receptors have evolved into organs that are sensitive to the near field, or displacement, component of airborne sounds. The Johnston's organ is perhaps the most well known of these receptors. Johnston's organ is a campaniform sensilla typically found on the an-

tennae of flies, mosquitoes, and bees, to name a few. Located at the pedicellus–funiculus joint, it responds phasically to lateral torsions as small as 2–5° (290). In *Calliphora*, the response is proportional to the logarithm of the rate of movement and saturates at a speed of 100°/s. It also follows movements of up to 500 Hz (46). Johnston's organ plays an important role enabling insects to orient in the wind while flying. In mosquitoes, and particularly honeybees, it may also mediate various aspects of social communication.

Hearing

Physical Properties of Sound

All sounds in an elastic medium (e.g., air) have both pressure and displacement components. Displacement is a vectorial component. Its amplitude decays rapidly, independent of sound frequency, as it propagates away from the source; that is, as a function of $1/r^2$ (where r is the distance from the point source). Displacement is the only component of sound present in an incompressible medium such as a solid substrate. The pressure component is a scalar quantity. Its amplitude decays linearly with distance (i.e., as a function of $1/r$) in a frequency dependent manner (low frequencies are propagated the farthest). There is a point in space, $r = \lambda/2c$ (where λ is the sound wavelength and c is the sound velocity), where pressure and displacement are equal in amplitude. From this point to the sound source (i.e., the near field) the displacement component predominates over the pressure component. In the far field (beyond $r = \lambda/2c$) the opposite is true. The pronounced sensitivity to sound pressure has coevolved with the emergence of specialized accessory structures (e.g., the tympanic membrane) among vertebrate and invertebrate taxa alike. The primary function of these accessory structures is the efficient transformation of sound pressure into mechanical displacement, which can then effectively stimulate the displacement-sensitive mechanoreceptors of the ear (337).

In the preceding section, several examples of invertebrate displacement receptors were given. In this section, we focus on pressure-sensitive organs in both the invertebrates and vertebrates. Also included, will be a brief discussion of vertebrate displacement receptors with an emphasis on the structural and functional organization of the lateral line system.

Hearing in Insects

Although acoustic signals play an important role in mediating a variety of invertebrate behavior patterns, it is the insects that have probably evolved the most elaborate receptors for detecting sound pressure. Far-field pressure waves may carry information concerning the identity and location of potential predators, prey, mates, or rivals for a number of insect species.

The frequency range of hearing in insects is generally very broad, in many species extending into the ultrasonic range. For instance, the ear of noctuid moths is sensitive to the ultrasonic pulses of bats, their main predators (213). Stridulating crickets, which utilize audible sounds for communicating with conspecifics, are also sensitive to the ultrasonic signals produced by bats (216, 245). In contrast, nonstridulating crickets are sensitive to infrasounds of 8–12 Hz, which they employ as communication signals over short distances (155).

Most insects produce sounds by stridulation (74, 210, 282). Several stridulatory organs have evolved independently and differ in anatomical position, though all of them employ the same basic mechanism; a "scraper," which moves over a

"file" (Fig. 15). The scraper and file located on the wings of crickets are representative examples. Other insects, for example, cicadas, produce sound by the buckling of a series of stiff ribs embedded in a flexible tymbal; the sound is radiated by means of a large air-filled abdominal cavity serving as a resonant structure (125, 257, 356). The buckling of the ribs is controlled by a powerful muscle that can contract at a rate of a few hundred hertz. Cicada sounds therefore, can have a crude tonal quality and comprise repetitive tone pulses with carrier frequencies ranging from several hundred hertz to the ultrasonic range, according to species (282, 355). Some moths produce repetitive sounds using a similar mechanism (23). For instance, the tymbal organ of the arctiid moth, *Melese laodamia*, can generate pulses with a carrier frequency of 30–90 kHz and at a repetition rate of 1200 pulses/s.

Detection of sound pressure is mediated by different types of phonoreceptors, often located in very different parts of the insect body. Some of the most sensitive vibration receptors, such as hair sensilla, can also be used to detect high intensity sound pressure (95, 115). The Johnston's organ of male mosquitoes, primarily a displacement receptor, is also sensitive to sound pressure, particularly at the frequency of the female's wing beat (210). In hawkmoths, the vibration-sensitive palp-pilifer organ, located near the mouthparts, is sensitive to ultrasounds. Airborne sounds produce vibrations of the air-filled palp cavity that are then transmitted to the distal lobe of the pilifer and excite vibration-sensitive sensillae (265, 266). The palp-pilifer organ is sensitive to frequencies ranging from 25 to 70 kHz dependent on the mechanical properties of the palp, which can serve as a frequency filter.

The specialized auditory receptor of most insects takes the form of a tympanal

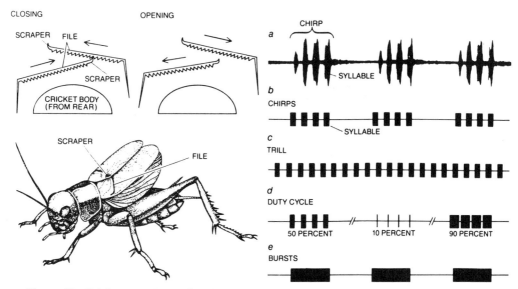

Figure 15. Cricket song is produced by the male when he rubs his wings together. The highly schematic diagrams at the upper left show the sound-making mechanism: Each wing incorporates a scraper and a file, so that every time the wings close they are set in brief vibration at a frequency of ~5 kHz (5000 cycle/s). The diagram at the lower left shows how the wings of the cricket are raised and scraped together when the sound is made. The song (*a*) is therefore made up of "syllables" of 5-kHz tone. Typically there are four syllables per chirp, and syllables come at a rate of 35/s. The temporal patterns of some computer-synthesized songs are schematized below the temporal pattern of the natural song. One synthesized song (*b*) reproduces the principal features of the natural song. Another (*c*) is a "trill": It fails to divide the song into chirps. Others (*d*) preserve the chirps and syllable rate but expand or contract the duration of each syllable, thus varying the duty cycle, or ratio of the sound to the silence within a chirp. Finally, one test pattern (*e*) is made up of "bursts" of 5-kHz tone of the correct chirp duration but not separated into syllables. [From (143).]

organ. The tympanal organs have diverse morphological and topographical variations among the various insect taxa (9). There is a general plan, however, consisting of a tympanic membrane attached to a chordotonal organ within an air-filled cavity (e.g., Fig. 16*a*); except for green lacewings where the tympanic membrane is backed by a fluid-filled cavity. The transducing elements are scolophorous sense cells, which range in number from one to several thousand.

The tympanal organ of noctuid moths (Fig. 16*b*) is located in the thorax and contains two types (A_1, A_2) of sensory cells. A_1 and A_2 are sensitive over a broad range of ultrasonic frequencies (15–100 kHz) coinciding with the frequency range employed by bats during echolocation (267). Ultrasonic sensitivity is attributed to the mechanical resonance of the scolopidium, whereas the sensitivity in the audible range is derived from the mechanical properties of the tympanal organ (1). Neither the A_1 nor A_2 cell is tuned to a particular frequency. However, there is evidence of range fractionation with regard to stimulus intensity. For example, the larger A_1 cell is ~10 times more sensitive to acoustic signals than the A_2 cell (265). Thus, when the moth is exposed to faint ultrasonic pulses, only the A_1 cell is

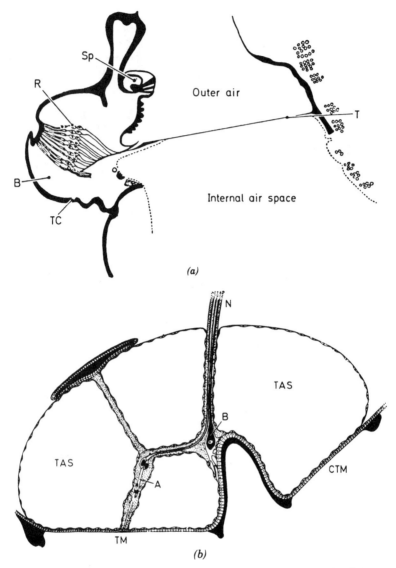

Figure 16. (*a*) Left ear of a cicada. B, hemolymph; R, receptor cells; SP, spiracle; T, tympanum; TC, tympanal capsule. Tympanum bridges across organ and separates outer air from internal. [From (65a).] (*b*) The ear of noctuid moths. TAS, tympanic air sac; TM, tympanic membrane; CTM, countertympanic membrane; A, A cells; B, B cell; N, nerve. [From (208).] (*c*) The locust ear. III, Right ear as viewed from inside, arrow indicates angle from which I is viewed. I, Right ear with receptor cells a, b, c, d. II, Left ear as seen from outside. [From (208).] (*d*) Comparative anatomy of cricket ears. (*A*) Diagram of the proximal part of a tibia showing the orientation of the tympana relative to the body. Schematic diagrams of transverse sections through the tibia at the level of the auditory organ in (*B*) the cricket, *G. bimaculatus*, (*C*) the weta *H. crassidens*, and (*D*) the bush-cricket. ATB, anterior tracheal branch; ATM, anterior tympanic membrane; BC, blood channel; PTB, posterior tracheal branch; PTM, posterior tympanal membrane; SC, sense cells; TC, tympanal cavity; TS, tympanal slit. [From (148).]

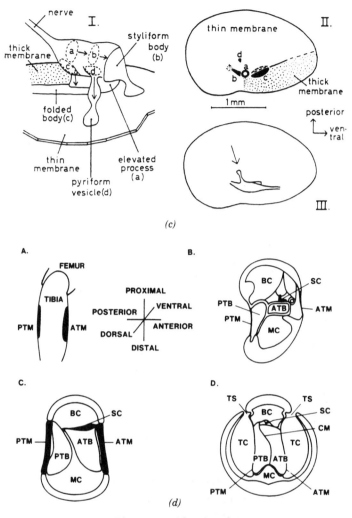

(c)

(d)

Figure 16. (*Continued*)

activated and the moth alters its flight pattern away from the signal source. In contrast, when both the A_1 and A_2 cells are activated by very loud sounds, indicating that a potential predator is nearby, the moth makes a complex and unpredictable dive toward the ground. Interestingly, the defensive mechanism of some arctiid moths does not involve alterations in flight path. Rather, arctiid moths produce ultrasonic clicks when they detect a bat's ultrasonic hunting signal (104, 105). Flying bats avoid these moths because the clicks are associated with distasteful insects (72, 105). Thus, while the evolution

of structure and function with respect to the moth ear appears to be quite conservative, the behavior patterns mediated by receptor activation have diverged considerably.

In cicadas, the ear (Fig. 16*a*) is situated at the level of the abdomen. Vibration of the tympanum is transmitted to a large number of scolopidia (as many as 1500) via connective tissue or the tympanic apodeme (69). *Cystosoma saunderii* males possess an enlarged, hollow, abdomen that acts as a resonator for amplifying the species song, which is emitted at a frequency of 800 Hz. The mechanical properties of the male's ear, however, greatly reduces its directionality. In contrast, females lack the hollow abdomen. Instead, they possess an air sac that makes the ear highly directional at low frequencies such as those contained in the male's song (96).

The water bug, *Corixa punctata*, has a pair of mesothoracic tympanal organs located below the hindwing joints. The organ contains two scolopidia which are excited by both airborne or waterborne sounds. When the animal is submerged, a large air bubble isolates the tympanic membrane from the surrounding water. Waterborne sounds are converted to airborne sounds via the air bubble, which also functions as a resonator imparting a maximal frequency sensitivity of 2 kHz to the entire organ (249). The frequency sensitivity of the receptor cells differ between the left and right side. The left cell has a minimum threshold of 38 dB SPL (sound pressure level relative to 20 micropascals) at 2.35 kHz while the right cell is maximally sensitive to 1.73 kHz with threshold near 45 dB SPL (248). Differences in spectral sensitivity are attributed to differences in the mechanical properties of the left and right tympanic membrane.

Among the various insect taxa, the auditory system of locusts and crickets have been the most thoroughly characterized. In these groups, three receptors mediate sensitivity to airborne sounds: subgenual organs, intermediate organs, and the crista acoustica (148). The crista acoustica is primarily responsible for detecting sounds used for long distance communication and as such are sensitive to frequencies ranging from 1 to 90 kHz according to species.

The tympanic membrane of locusts is located in the abdominal wall. The tympanic membranes on either side of the body are functionally coupled by a connecting, air-filled tracheal system. Thus, each tympanic membrane is subject to pressure differences between its outer

and inner face. Seventy receptor cells, arranged in four groups (Fig. 16c), are attached to four cuticular bodies on the tympanum (208). Each group is sensitive to a particular frequency range dependent on the resonant characteristics of its attachment site (208, 311). Three groups (a, b, c) are sensitive to low frequencies; 3.5–4.0, 4.0, and 5.5–6.0 kHz, respectively. The fourth group (d) is sensitive to high frequencies (5–40 kHz) with optimum sensitivity in the range of 10–20 kHz. Thus, the tympanum can be considered as showing a "place code" for frequency discrimination. While the behavioral significance of this frequency analysis is not clear, the frequency sensitivity of the receptors does match the spectral content of the species calling songs and may serve to differentiate between the songs of conspecifics and those of other, sympatric species.

In crickets, the tympanic membrane (Fig. 16d) does not participate in frequency analysis, but is required for hearing (163, 183). It vibrates in a simple mode over a broad frequency range and thus, differs from the tympanic membrane of locusts (209, 238, 359). Some species of crickets have only a single tympanum on each leg. But, most species of crickets have a small anterior and a large posterior, externally visible, tympana on each tibia. Both the outer and inner faces of the posterior tympanum are subject to changes in sound pressure; the latter is mediated by a tracheal system, which has an external opening located just caudal of the prothoracic leg. In addition, the tympana on either side of the body are coupled by a tracheal system as in the locust. The cricket ear is therefore typical of a *pressure gradient receiver*, which compares the sound pressure arriving at either side of the tympanic membrane.

Individual receptors within the crista of tettigoniids are tuned to different frequencies, with low and high frequencies represented at the proximal and distal ends of the sensilla, respectively (232, 268). The frequency selectivity of individual receptors is dependent on the region of the tympanic membrane to which they are coupled, as discussed above. In contrast, the tuning of auditory receptors in crickets is due to the mechanical and/or electrical properties of the receptors themselves and not the mechanical properties of the tympanic membrane (232).

The primary afferent nerve fibers of acoustically active insect species typically show a V-shaped tuning curve (77, 232, 357) with different fibers being tuned to different frequencies according to the receptor that they innervate (154, 259, 359). The central projections of these fibers terminate in an auditory neuropil which, at least in tettigoniids (231, 268), is tonotopically organized (i.e., the orderly arrangement of characteristic frequencies seen at the level of the receptor is maintained in the spatial distribution of afferent terminals).

The neural basis of acoustic communication in insects has been intensively investigated, particularly in crickets. Crickets produce several types of "songs" including territorial, calling, and courtship songs. The neural mechanisms mediating the recognition of the calling song have received the greatest attention because of the role of the calling song as a reproductive isolation mechanism (138).

In crickets, most primary afferent fibers arising from the ear, as well as central auditory neurons, are tuned to the dominant spectral components of the species calling song (~4–5 kHz). But, the spectral energy distribution of the calling song of sympatric cricket species often overlaps; the carrier frequency of the call can be varied by 1–2 kHz without reducing its attractiveness (331). Therefore, call recognition in crickets relies heavily upon the ability of the auditory system to discriminate between the species-specific temporal parameters of the calls (6, 74). Among the Gryllidae, the syllable repetition rate seems to be the most important temporal feature for recognition (143, 331). While the number and duration of syllables in the call are less important than repetition rate for recognition, recent studies in *Teleogryllus oceanicus*, *Gryllus bimaculatus*, and *Acheta domesticus* revealed that a hierarchy exists with regard to the effective temporal parameters eliciting positive phonotaxis. In most cases, when one cue is suboptimal, females utilize other temporal cues for selecting conspecific mates (67, 68, 313).

It is worth noting that the calling songs of some cricket species may exhibit a segregation of information; that is, different components of the song may convey different messages (Fig. 15). For example, the calling song of *T. oceanicus* consists of chirp and trill components that differ with respect to their temporal properties (19). The temporal parameters of the chirp component are attended to by females, those of the trill by males (243, 244). Thus, the chirp and trill components of the calling song may mediate reproductive as well as territorial behavior patterns, respectively.

How are the temporal features of the calling song encoded in the cricket nervous system? The duration and onset time of both chirps and syllables are encoded by auditory neurons at various levels of the cricket's auditory pathway (27, 51, 73, 246, 314, 353). Generally these neurons do not show selective responses to the conspecific calling song. However, several auditory neurons in the brain exhibit highly specific responses to artificial calls having the same syllable repetition rate as that of the species call (Fig. 17); thus, providing a neural correlate for call recognition (289).

Females must not only be able to recognize the call of a conspecific male, but must also be able to locate that calling

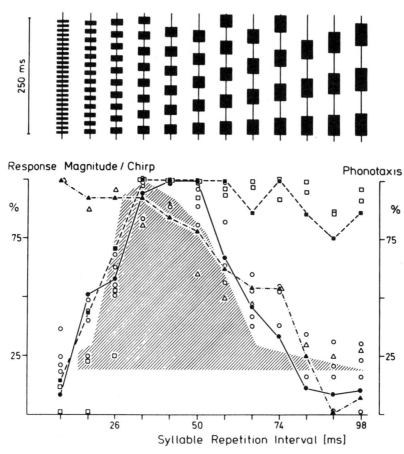

Figure 17. Relative magnitude of responses of various auditory brain neurons to chirps shown in upper figure. Each of the three curves (maximum response = 100% in each case) shows averaged (12 stimulus presentations) responses of single identified neurons with particular temporal filter properties. Data points with squares were from neuron BNV1d, triangle from BNC2b and circles from BNC2a. Open symbols (unconnected by lines) show responses of other examples of these identified neurons, to indicate degree of variability. Hatched area shows relative effectiveness of the syllable repetition intervals in eliciting behavioral phonotactic tracking [right ordinate, replotted from (331)]. Sound frequency: 5 kHz; intensity: 80 dB SPL. [From (289).]

male in space. In spite of their small size, many calling species of insects are quite capable of locating a sound source (10, 222, 343). *Gryllus bimaculatus*, in a Y-maze test choice for sidedness, showed an acuity for lateralization of 15°, for sounds presented at 4.7 kHz, the carrier frequency of the species calling song (260).

What are the mechanisms that enable insects to locate the source of sounds? As an example, consider sound localization in the ensifera. The ensiferan ear is a four-input receiver, which behaves as a push–pull pressure gradient system with a cardioid directional pattern (137, 182). Consequently, there is a substantial intensity difference between the posterior tympanal membranes on either side of the body (183, 291). This interaural intensity difference produces differences in the firing rate and onset of response between fibers of the tympanal nerves on either side of the body (269). These differences mediate directionally selective responses in the central nervous system (25, 26, 261). No insect neuron has been shown to have a circumscribed auditory receptive field as seen in some birds. But, the directional response of central interneurons is characterized by a sigmoidal curve with a maximal response for sound coming from the opposite side of the body. Thus, directional coding requires a neural comparator that can convert the bilateral differences in neural responses into unambiguous directional information (270).

Hearing in Fish

Several hypotheses concerning the evolution of labyrinthine organs in fish have been proposed. Once, the labyrinth was believed to have been derived from the lateral line organs (337). Recently, however, it has been proposed that the lateral line and auditory systems evolved independently (230). At least two lines of evidence support this notion (54, 81, 153, 223, 284). First, the lateral line and auditory systems have different and distinct functions. The lateral line system mainly detects near-field displacement gradients while the inner ear organs process far-field sounds. The second, and perhaps most significant evidence, comes from neuroanatomical studies indicating that the lateral line and auditory system have distinctly different projections into the central nervous system. This would not be expected if the auditory system was derived directly from the lateral line system.

Since the mechanoreceptive hair cells are embedded deep inside body tissue with no apparent external opening, the question arises as to how fish are able to detect sounds. In some taxa, the air-filled swimbladder serves this purpose. As far-field pressure changes, the swimbladder expands and contracts accordingly. Thus far-field pressure changes are converted into displacement components internally, which can then stimulate the hair cells through action of the otolith (81, 82).

The swimbladder of most species in which it is present, has a resonance frequency of 100 Hz to >1 kHz, matching the frequency range of hearing in most fish species (58, 81, 133, 328). But the frequency tuning of afferents innervating the auditory end organs (sacculus, lagena, or utricle) does not merely reflect the resonance frequency of the swimbladder. Each afferent is tuned to a different frequency dependent on the electrical tuning characteristics of the hair cell innervated (82). Thus, two mechanisms, the physical properties of the swimbladder and the electrical properties of the mechanoreceptors, contributes to afferent frequency sensitivity.

Four groups of fish, cyprinids, siluroids, characinids, and gymnotids, possess specialized structures, the Weberian ossicles, which transmit swimbladder displacement to the inner ear. The Weberian ossicles are analogous to the middle ear structures of mammals, which couple displacement of the tympanic membrane to inner ear structures (136, 339).

For years, investigators have been intrigued by the ability of fish to determine the direction from which far-field pressure waves originate. Time and intensity cues utilized by many animals for sound localization apparently are not utilized by fish because (1) their "interaural" distance is very small resulting in very small differences in time of arrival that are too minute to detect and (2) their bodies are relatively "transparent" to pressure waves in water; thus, interaural intensity differences are also very small. Since fish do not utilize interaural time and inten-

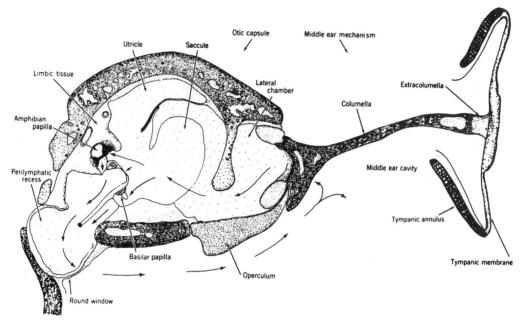

Figure 18. Schema of frog ear. [From (345).]

sity disparities to localize sounds an alternative mechanism mediating directional hearing in fish must be in operation.

The otolith organs (saccule, lagena, and utricle) comprise the auditory portion of the fish ear. Each of these organs contains a patch of sensory epithelium, or maculae, consisting of mechanoreceptive hair cells and supporting cells. The hair cells themselves are oriented in different directions across any given macula. Overlying the hair cells is a calcium carbonate otolith and otolith membrane. Hair cells are stimulated when shearing forces are applied to their cilia by motion of the otolith. This motion is produced in at least two ways; either by swimbladder motion, which is transmitted to the otolith, or by acceleration of the fishes body in a sound field. The latter results in a lag between the movement of the otolith and the otolith organs. This lag results in a shearing of hair cell cilia and hence receptor activation. The acoustic particle acceleration involved in inertial stimulation (79, 80, 153, 295) is a vector function with inherent direction as well as magnitude. Thus, directional hearing in fish appears to be coded within and between otolith organs by the relative response across an array of three-dimensionally organized receptors.

Hearing in Amphibians

As aquatic animals ventured onto land, the auditory system (Fig. 18) changed in response to selection pressures unique to the new terrestrial environment. Developmental processes in anurans (frogs and toads) exemplify the transformation from water-dwelling to land-based animals. Larval anurans, or tadpoles, are aquatic; becoming terrestrial after metamorphosis. The lateral line organs, which serve as water current detectors in early tadpoles, degenerate prior to metamorphosis (stage 26). At about the same time, the two auditory organs (amphibian and bas-

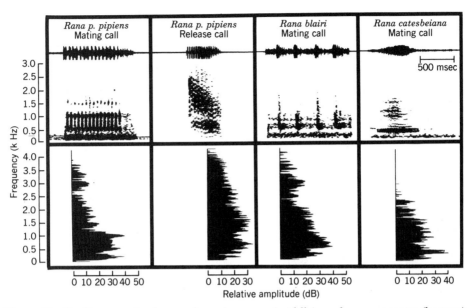

Figure 19. Oscillograms (top), sound spectrograms (middle), and power spectra (bottom) of anuran vocalization (mating and release calls) illustrating intra- and interspecific differences in spectral content and temporal pattern. Only a single pulse of the bullfrog (*Rana catesbieana*) mating call is shown. [From (128).]

ilar papillae) begin to differentiate along with the overlying tectorial membrane (91, 181). The middle ear ossicles, which are essential for converting the airborne sound pressure at the tympanic eardrum into displacement of the fluid inside the otic capsule, are formed from the first pharyngeal pouch and mesenchymal structures (136, 352). Thus, by the time of metamorphosis, structures essential for pressure-to-displacement conversion are completely formed and adult frogs are well equipped for detecting airborne sounds.

Displacement of otic capsule fluid results in a shearing force on amphibian and basilar papilla hair cell stereocilia due to movement of the overlying tectorial membrane. Typically, the two organs are responsive to different frequency ranges (89). The basilar papilla is sensitive to high frequencies (1.0–1.5 kHz to 3.0–5.0 kHz dependent on species). In contrast,

the amphibian papilla is sensitive to low (100–500 Hz) and intermediate (500–1000 Hz) frequencies, the range of which is again dependent on species. In addition, the amphibian, but not basilar, papilla is tonotopically organized with low frequencies represented rostrally, intermediate frequencies, caudally (186, 350). Interestingly, the spectral sensitivity of the anuran ear closely matches the spectral components of the species-specific mating call (89) giving rise to the concept of a *matched peripheral filter*. In fact, the majority of information concerning the structural and functional organization of the anuran auditory system stems from studies focused upon mating call recognition.

The mating calls of bullfrogs have a bimodal spectral peak (Fig. 19) with energy concentrated between 100–500 Hz (low-frequency peak) and 1000–1500 Hz (high-frequency peak). The gestalt features of

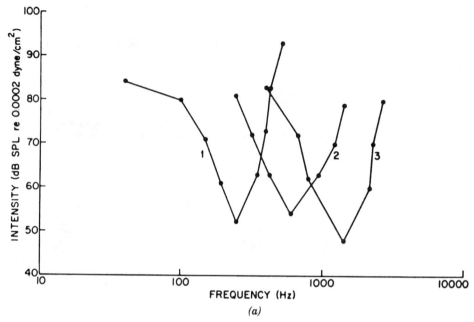

Figure 20a. Tuning curves for three populations of auditory fibers in the bullfrog. Unit 1 (low-frequency fiber) has a best excitatory frequency (BEF) of 250 Hz, unit 2 (mid-frequency fiber) has a BEF of 600 Hz, while unit 3 (high-frequency fiber) has a BEF of 1400 Hz. The response of unit 1 to a tone at its BEF could be totally inhibited by the simultaneous presentation of a second tone of 700 Hz at an intensity 10 dB greater than the excitatory tone. The responses of units 2 and 3 could not be inhibited by a second tone. [From (89).]

the bullfrog mating call essential for recognition have been identified. The simultaneous presence of low- *and* high-frequency energy is crucial for species identification as determined by the evoked calling response of males (49). Thus, mating call detection involves a logical "AND" operation, that is, the simultaneous excitation of both the low-frequency sensitive fibers from the amphibian papilla and the high-frequency sensitive fibers from the basilar papilla (103). But, the essential gestalt features for recognition vary with species. For example, some species of North American treefrogs require the simultaneous presence of intermediate- and high-frequency sound energy for mating call identification (112) while other anuran species may rely upon temporal, rather than spectral,

cues for species identification (112, 192). Thus, the mechanisms involved in call recognition are also likely to vary with species. Let us examine some of these mechanisms more closely as they occur in ranid frogs.

At the periphery, auditory nerve fibers exhibit V-shaped frequency tuning curves (Fig. 20a). For many low-frequency sensitive fibers, there is additionally a V-shaped inhibitory tuning curve at the high-frequency flank of the excitatory tuning curve. Inhibition is observed when the ear is simultaneously stimulated with two tones falling within both the excitatory and inhibitory areas (50). The phenomenon of two-tone suppression is believed to be the result of mechanical events occurring in the inner ear as it is still observed following transection of the

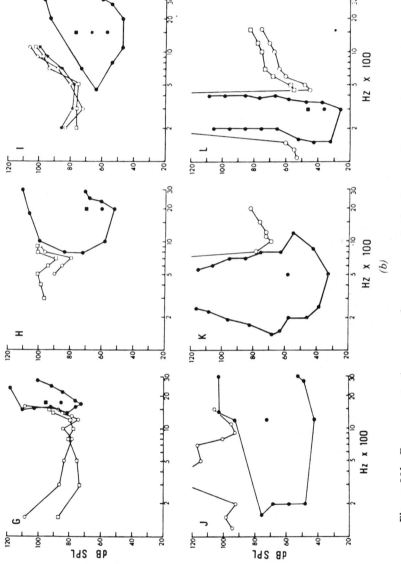

Figure 20b. Frequency tuning curves from neurons in the frogs auditory midbrain; excitatory (closed symbols), inhibitory (open symbols). [From (110).]

auditory nerve. Intermediate- and high-frequency sensitive fibers do not show two-tone suppression.

The majority of auditory neurons at the level of the caudal brainstem, that is, within the dorsal medullary nucleus and the superior olivary nucleus, are similar to each other and to the primary afferents with respect to their frequency tuning characteristics (107). Low-, intermediate-, and high-frequency sensitive neurons are represented in an orderly, or tonotopic, fashion in these structures (85, 86, 106).

Neurons in the auditory midbrain (*torus semicircularis*) display complex excitatory and inhibitory tuning curves (Fig. 20*b*), which are distinctly different from those exhibited by neurons at lower levels of the auditory pathway (110). Some torus neurons do not respond to single tones, but require two or three tones presented simultaneously for excitation. Interestingly, a few torus neurons respond *only* to the combination of low- *and* high-frequency tones characteristic of the species mating call. Such AND neurons are concentrated within the posterior nucleus of the thalamus. Thus, the posterior nucleus likely plays an essential role in the detection of the species mating call in the *frequency* domain (108).

Auditory neurons additionally encode the *temporal* characteristics of complex sounds such as the species mating call. Throughout the auditory pathway, neurons responding selectively to one or more temporal features of the call (pulse repetition rate, pulse duration, or pulse rise time) can be found (20, 126, 224, 271, 272, 340). In the thalamus, neurons coding the temporal features of complex sounds are concentrated in the central thalamic nucleus (127) and are thus distinct from those processing the spectral components (see above). Thus, within the thalamus, there are apparently two parallel subsystems for encoding complex sounds in the central auditory

pathway of ranid frogs; one subsystem is responsive to the temporal features of the sound, the other, the spectral features. Interestingly, information provided by either subsystem alone should be sufficient for call recognition. However, by combining information provided by each pathway, females should be able to discriminate, with a high degree of reliability, the calls of conspecifics from those of numerous other species calling at the same time and place. Moreover, the responses of neurons in the central thalamic nucleus to complex sounds are often the result of their sensitivity to a single temporal parameter of the sound (e.g., pulse repetition rate or pulse duration) rather than a combination of these parameters as shown in Fig. 21 (128). Here, the response selectivity of a thalamic neuron to the conspecific mating call is directly attributable to its preference for pulses of short (10–20 ms) duration. The manner in which these pulses are assembled (i.e., the pulse repetition rate) has little or no effect on the unit's response. The converse may also be true depending on the cell being investigated. Conceivably, the information provided by an assemblage of such neurons may be useful for reliably discriminating between signals with many distinct features.

In some species of frogs, the auditory system exhibits sexually dimorphic properties. For example, the Puerto Rican coqui frog produces a two-note call (224). The first, or "co," note is meaningful to the male during territorial encounters. The second, or "qui," note is of a higher frequency than the co note and serves to attract females. In males, the greatest percentage of primary afferents are tuned to frequencies matching those contained within the co note (224). In contrast, primary afferents tuned to high frequencies, that is, those of the qui note, represent the largest proportion of the afferent population (224). Thus, the auditory system of each sex is specialized for detecting that

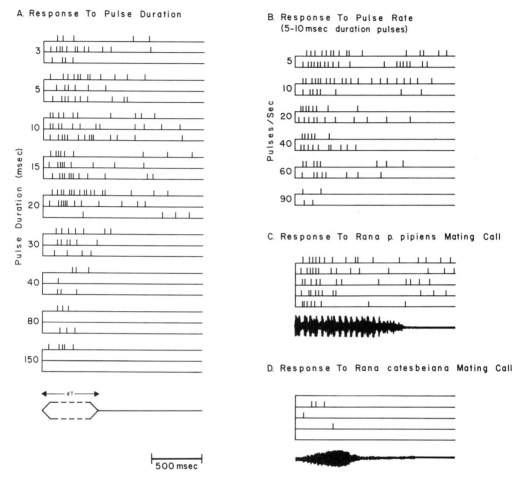

Figure 21. Neuron in the frog's central thalamic nucleus of *Rana pipiens* shows differential responses to stimulus presentations. This neuron responds vigorously to the leopard frog (C), but not to the bullfrog (D), mating call. This difference can be explained by the preference of this neuron for stimuli comprised of short duration pulses (A), which approximate those typical of the leopard frog mating call (7–10 ms). The neuron is not selective for amplitude modulation rates 5–60 pulses/s (B); encompassing the range of modulation rates typical of both calls in (C) and (D). The neuron is not selective for envelope rise and fall times (not shown). (Courtesy of Jim Hall.)

portion of the call that is behaviorally meaningful.

At first glance, the ability of frogs to accurately localize other calling frogs based upon interaural time and/or intensity differences would appear to be compromised by both the low-frequency (long wavelength) content of their communication signals and their typically

short interaural distance. Nevertheless, frogs are well adapted, both mechanically and neurally to overcome these problems [for a review, see (80)].

Gravid female frogs are capable of locating calling males in both the horizontal and vertical planes with a horizontal accuracy of 8–12° (88, 113, 237, 262). The integration of bilateral input is essential

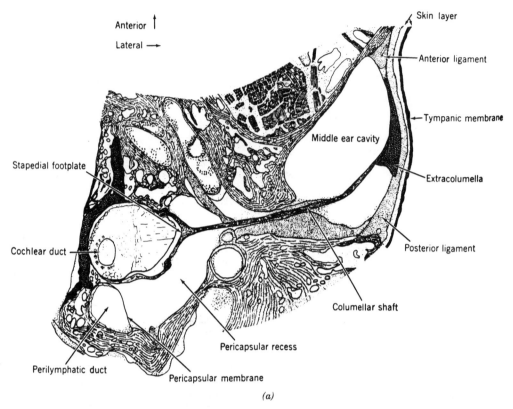

Anterior ↑

Lateral →

Skin layer

Anterior ligament

Tympanic membrane

Middle ear cavity

Stapedial footplate

Extracolumella

Cochlear duct

Posterior ligament

Columellar shaft

Pericapsular recess

Perilymphatic duct

Pericapsular membrane

(a)

Figure 22. (*a*) A frontal section through the ear region of the turtle *Chrysemys scripta*, at a level that cuts across the entire columella. (Scale 7.5×). (*b*) The auditory papilla of *Chrysemys scripta elegans* in the far ventral region of the cochlea at left side of figure. Note that the stereocilia of some hair cells are embedded in an overlying tectorial membrane, others are free standing. (Scale 25×) [From (346).]

for sound localization; as animals with the input to one ear blocked cannot localize sounds (88).

The frog's ability to localize sound is, in part, dependent on mechanical specialization of the middle ear, which amplifies the interaural intensity difference. That is, the two middle ears communicate directly through the mouth cavity forming a push–pull type of receiver (i.e., combination pressure and pressure-gradient receiver), which is inherently highly directional (4, 90, 235). Additionally, a minute change in sound intensity produces a large shift in the response latency of peripheral fibers (84). Thus, there is a pronounced time difference between the onset of responses in the auditory nerve of either side of the animals. This time difference is detected by binaurally sensitive neurons in the central auditory system (87). As a consequence, auditory neurons in the midbrain show distinct directional responses (83). For example, they may respond maximally to contralateral sounds while ipsilateral sounds evoke little, if any response (sigmoidal directional responses). Or, they may respond maximally, or minimally, to sounds originating along the midline (figure-8 directional responses). The above structural and functional properties of the anuran auditory system provide a substrate for accurate sound localization in

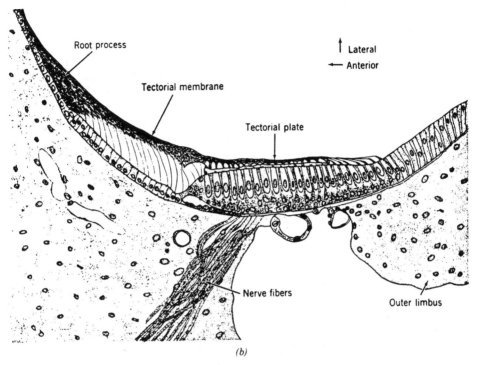

Root process

Tectorial membrane

Tectorial plate

↑ Lateral
← Anterior

Nerve fibers

Outer limbus

(b)

Figure 22. (*Continued*)

spite of problems involving signal spectrum and interaural distance.

Hearing in Reptiles

The significance of sound for communication and acoustically guided behavior in most reptiles is not well understood. One exception is the gecko, a lizard that is unique in possessing a vocal cord (111) and which produces threat or territorial calls as well as various other sounds whose functions are not clear (199). In addition, when foraging, geckos may utilize sounds produced by other animals as a means of locating prey items. For example, geckos show phonotactic responses to the calling song of male crickets. This presumably aids in their capture of female crickets that are not singing, but also orienting to the male's song (281). In general, the hearing range

of most reptiles is restricted to frequencies below 8 kHz (195, 345, 346).

The anatomy and physiology of the reptilian peripheral auditory system have been extensively investigated (196, 214, 335, 341, 346). The external ear is generally not distinct and plays a small role in signal transmission. In many lizards and crocodiles, the tympanic membrane is recessed from the body surface creating a small auditory meatus. In other lizards, snakes, and turtles, the tympanic membrane is either absent or is flush with the body surface (335).

The middle ear in reptiles, like that of amphibians, comprises two main ossicles, the columella and extracollumella (136, 346). The distal end of the extracollumella attaches to the tympanic membrane; proximally it joins the columella. The proximal end of the columella broadens to form a footplate covering the oval window of the inner ear (Fig. 22). Functionally, the middle ear serves to convert the sound-induced displacement of the tympanic membrane into an amplified displacement of the oval window which, in turn, leads to pressure changes in the fluid-filled inner ear. The impedance mismatch, which exists between such an air–water interface, is thus compensated for by the middle ear.

Fluid displacement in the inner ear, which results from oval window displacement, leads to the excitation of hair cells within the basilar papilla. Hair cells along the basilar papilla have stereocilia either embedded in an overlying tectorial membrane or free standing (Fig. 22b). Thus, two mechanisms leading to hair cell excitation exist; one involves shearing forces applied to the stereocilia during displacement of the overlying tectorial membrane, while the other is the result of stereocilia displacement induced directly by fluid movements in the inner ear. While a basilar membrane is present in almost all species of reptiles, its role in frequency tuning is not well established (61, 195, 335, 341).

The processing of sounds by the central auditory system of reptiles is not well understood. Apparently, some neurons in the central auditory system show V-shaped tuning curves, much like those seen in the eighth nerve (194, 198). In *Varanus*, a monitor lizard, asymmetrical tuning curves, similar to those seen in the auditory pathway of frogs and mammals, are also apparent (196). These tuning curves are most likely the result of neural interactions that function to enhance the frequency selectivity of central auditory neurons relative to those of the periphery.

Hearing in Birds

Birds are vociferous. Their songs are complex and have been analyzed in great detail. For many birds, sound is often the primary means of signaling social status, physiological condition and, species identity. Sound may also play a role in establishing and maintaining territories, attracting mates, as well as announcing the presence of predators.

The tympanic membrane of birds is generally more distinct than that of reptiles. It is attached to the stapedius, a small bone of the middle ear, the proximal end of which is attached to the columella. The proximal end of the columella is, in turn, connected to a footplate, which is anchored to the oval window. A round window, which serves as a pressure releasing device is also present. Interestingly, in woodpeckers, the round window is also coupled to a special bony plate of the columella. This serves to dampen mechanical energy originating from pecking (170).

The transfer function of the middle ear has been measured in a number of species of birds. It matches the behavioral audibility curve (285). Most birds hear up to 10 kHz, but are most sensitive to frequencies ranging from 1 to 4 kHz (70, 171, 278, 285). Homing pigeons and guinea fowl are also sensitive to infrasounds; responding to frequencies as low as 2–10 Hz (330, 354). Low frequency hearing in pigeons probably provides them with an additional navigation cue for homing. The functional significance of infrasonic sensitivity in guinea fowls is not known.

The basilar papilla (Fig. 23) of the bird inner ear is an elongated organ and has some features similar to, others different from, the mammalian cochlea (298). Hair cells of the basilar papilla are graded in size and differ with respect to position and innervation (53, 307, 325). The tallest hair cells are found over the entire length of the basilar papilla. They receive large afferent and small efferent nerve endings at their base. In contrast, the shortest hair cells are concentrated along the proximal two thirds of the organ. They receive small afferent and large efferent nerve

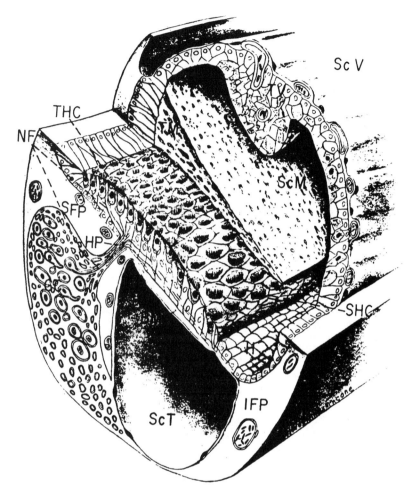

Figure 23. Schematic drawing of transverse section of pigeon's basilar membrane. Short hair cells (SHC) on free basilar membrane and tall hair cells (THC) over the superior fibrocartilaginous plate (SFP). Part of the tectorial membrane has been removed to display surface of papilla and hair bundles on sensory cells, attached to the epithelial cells on the superior wall of the cochlear duct. Hyaline cells (H), inferior fibrocartilaginous plate (IFP), scale media (ScM), scale vesibuli (ScV), peripheral processes (NF), habenula pereforata (HP), scale tympani (ScT). [From (325).]

endings. Intermediate hair cells vary with respect to location, morphology, and innervation.

Many of the hair cells in the avian inner ear are associated with a basilar membrane that vibrates in response to displacement of the oval window. The compliance, or stiffness, of the basilar membrane changes along its length, de-creasing from base to apex. The basilar membrane of birds, like that of mammals, supports a traveling wave (225). Afferent fibers innervating hair cells along the apical end of the basilar membrane are sensitive to low frequency tones, while those innervating hair cells basally, are sensitive to higher frequency tones (197). Thus, as is the case in mammals (see be-

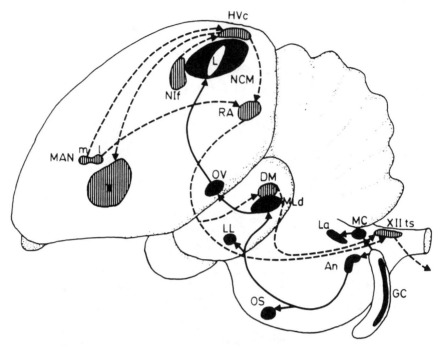

Figure 24. Major centers and connections of the two central pathways responsible for acoustic communication in birds. Solid lines: auditory pathway; dashed lines, vocal pathway. GC, cochlear ganglion (output of cochlea); MC, nucl. magnocellularis; An, nucl. angularis; La, nucl. laminaris; OS, superior olive; LL, nucl. lemnisci lateralis; MLd, nucl. mesencephalicus lateralis pars dorsalis; OV, nucl. ovoidalis; L, field L; NCM, neostriatum caudale pars mediale; NIf, nucl. interfacialis; X, area X; MAN, nucl. magnocellularis anterior neostriatalis; l, pateral part; m, medial part; HVc, hyperstriatum ventrale pars caudale; RA, nucl. robustus archistriatalis; DM, dorsomedial part of nucl. intercollicularis; XIIts, tracheosyringeal part of the nucl. hypoglossus. [From (184a).]

low), a *place map* for frequency exists along the basilar membrane.

The primary afferent fibers of birds have symmetrical, V-shaped tuning curves (279). Moreover, they are capable of responding to individual cycles of sinusoidal tones, up to 4 kHz in the case of pigeons (120, 197, 280), in a phase-locked fashion. All fibers of the eighth nerve show two-tone suppression. Unlike frogs, inhibitory response areas flank the excitatory tuning curve on either side.

Neurons of the central auditory pathway of birds (Fig. 24), particularly those of the nucleus magnocellularis (homolog of the mammalian anteroventral cochlear

nucleus), nucleus angularis (homolog of the mammalian dorsal and posteroventral cochlear nuclei), and nucleus mesencephalis lateralis pars dorsalis (MLD; homologous to the mammalian inferior colliculus) exhibit excitatory tuning curves with inhibitory flanks similar to those seen peripherally (274, 278, 288, 312). However, neural interactions at central levels of the auditory pathway result in unique neuronal response properties not shown by peripheral fibers. For example, some cells of the nucleus magnocellularis are inhibited, but not excited, by sound stimuli. Further, a small number of neurons in nucleus angularis are excited by tones presented at low intensity levels but inhibited by the same tones when their amplitude is increased. Finally, in MLD, there are two classes of cells, simple and complex. The response of simple cells to natural calls can be predicted from the relationship between the unit's excitatory response area and the call spectra. Responses of complex cells to natural calls cannot be so readily predicted. Presumably, complex cells derive their input from nucleus angularis neurons, which possess complex excitatory and inhibitory tuning curves. The result of this convergence is a population of neurons in MLD that show enhanced responses to one or more of the natural calls. No MLD neuron responds exclusively to a single call.

All neurons in the auditory thalamus have multiple excitatory and inhibitory areas (21), which contribute to their call selectivity. However, it is in the telencephalon where such selectivity has been firmly established. In field L of the caudal neostriatum of the forebrain, some neurons respond to single tones. These are tonotopically organized, much like those of the lower brainstem auditory centers (171, 221, 274) from which they derive their input. Many neurons, however, do not respond to single tones, but respond instead to particular tone combinations. Such complex tuning can account for the high degree of call specificity exhibited by these neurons. For example, in the guinea fowl, a small number of neurons in layers L1 and L3 of field L are tuned to frequencies between 1 and 2 kHz. These cells respond selectively to "iambus'like" calls, that is, natural calls with a fundamental frequency of 1 kHz and sideband energy at 300-Hz intervals flanking the harmonic (287). Sixteen percent of these neurons do not respond to single tones. Thus response selectivity may be attributable to the presence of spectral sidebands (characteristic of the call), which

fall within the excitatory regions of these neurons. Other factors, such as the rate and depth at which the calls are amplitude modulated may also determine the response selectivity of these field L neurons (180).

Birds that learn their songs often possess neurons at higher levels of the nervous system, which respond to their own song. For example, neurons in the hyperstriatum ventrale pars caudalis, a song control region in the forebrain of the white-crown sparrow, are particularly selective for the individual's own song (200, 207). In birds singing abnormal songs, because of early experiential manipulation, neurons are selective for the abnormal song! In addition to song spectrum, these neurons must also be selective for the specific temporal sequence of song elements since the presentation of individual song elements, or reverse playbacks of the song, elicit poor responses from these neurons.

In spite of their small head size, birds are able to locate a sound source with a great degree of accuracy (167, 173). Owls, which locate prey by their sounds, have the highest acuity for localization among all animals (1–2°; 169, 172). Only marsh hawks, with an angular resolution near 2° (263), come close. Other diurnal raptors typically localize sounds with an angular resolution of larger than or equal to 10°.

Behavioral studies have demonstrated that angular acuity for localization is dependent on sound frequency. For owls, localization is most precise for sounds of 4–8.5 kHz. For pigeons, however, sounds within this frequency range are not well located (151). In songbirds, such as the tit, the minimum resolvable angle for localizing broadband complex natural calls, or pure tones of 2 kHz, is 16–20°. Localization acuity falls to 45° at other frequencies (165).

The presence of acoustic transients, as well as signal spectrum, also contributes to localization acuity. The presence of transients in an acoustic signal enables accurate, binaural comparisons of time of arrival. The importance of time and frequency cues for localization have placed constraints upon the structure of signals used in different contexts. For example, when it is adaptive for a signal to be localized, such as a mating trill, the signal is usually broadbanded and has a rapid, or transient, rise time. Other signals that are best ventroloquial, such as alarm calls, are typically pure-tone like without distinct transient components (32, 203, 351).

The exceptional localization acuity of barn owls is attributed to neural specializations. Most birds have a push–pull type of pressure gradient receiver for amplifying the interaural intensity difference at the two ears (57, 167, 185, 273). But, in owls, the ears are not tightly coupled at the frequency range important for sound localization (217). Rather, they act as independent pressure receivers much as in mammals. Konishi and his colleagues (173, 174) have demonstrated that localization involves a bicoordinate process; that is, azimuthal localization primarily involves binaural processing of ongoing time differences in the microsecond range, whereas localization in the vertical plane involves the processing of binaural intensity differences of high-frequency components of the sound. The intensity difference at the two ears is a direct result of the owl's asymmetrical ears, which are sensitive to the incident angle of sound originating at different elevations (168, 229, 239).

Interaural time and intensity differences are processed independently (i.e., in parallel) within the lower brainstem; time in the nucleus magnocellularis and intensity in the nucleus angularis. These pathways converge in the midbrain resulting in neurons with circumscribed auditory receptive fields (173, 323, 324). In the auditory midbrain (Fig. 25), as well as the optic tectum (166), these spatially tuned neurons are arranged topographically according to the two-dimensional location of their receptive fields. An area of space in front of the owl, encompassing ~30°, is overly represented in the auditory system. This is correlated with the region of space within which the owl exhibits the highest degree of acuity as demonstrated behaviorally. Maps such as this differ from those in which the organization of the sensory epithelium is mapped topographically, for example, a tonotopic map, because these maps are constructed entirely from neural interactions. As such, topographic maps, like those of auditory space in the midbrain of the barn owl, are referred to as *computational maps*. Of course, not all birds have adopted the same strategy for localizing sound signals. Some utilize other mechanisms which, while not as accurate as those of the barn owl, nevertheless permit the animal to behave adaptively in its environment (48, 57).

Hearing in Mammals

The mammalian auditory system is characterized by a generalized peripheral organ (Fig. 26). Most mammals have pinnae that serve as sound collectors and as horns to focus sounds originating within the frontal sound field (47, 240, 301). The pinnae of some echo locating bats are sometimes as long as the length of the animal's body. They are highly mobile and mediate, in part, prey localization (293).

Three middle ear ossicles, the incus, maleus, and stapes, serve to transmit and amplify pressure at the tympanic membrane, converting it into displacement of the oval window. As is the case for most lower vertebrates, the middle ear functions primarily as an impedance matching device.

The auditory organ, the cochlea, is a

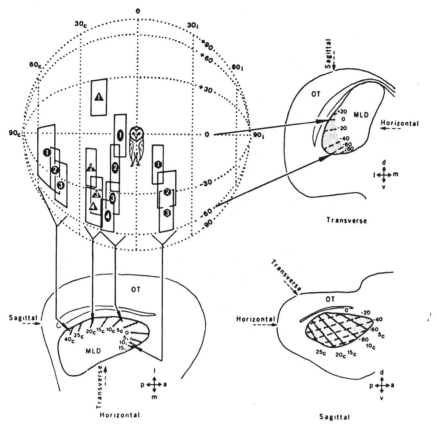

Figure 25. Neural map of auditory space, which is depicted as an imaginary globe surrounding the owl. Projected onto the globe are the best receptive field areas of 14 neurons. The best area, the zone of a maximal response within a receptive field, remains unaffected by variations in sound intensity and quality. The numbers surrounded by the same symbols (○, □, etc.) represent neurons from the same electrode penetration; the numbers themselves denote the order in which the neurons were encountered. Below and to the right of the globe are illustrations of three histological sections through the inferior colliculus (nucleus mesencephalicus lateralis dorsalis, MLD). The shaded portion is the external nucleus. Iso-azimuth contours are shown as solid lines in the horizontal and sagittal sections; iso-elevation contours are represented by dashed lines in the transverse and sagittal sections. [From (173).]

coiled structure. The length of the cochlear duct varies greatly from one species to another; 0.25 turns in the platypus, 1.5 turns in whales, 2.75 turns in humans, and almost 4 turns in guinea pigs. The cochlear duct itself, however, follows a basic design scheme (Fig. 27). There are three rows of outer hair cells and one row of inner hair cells that occupy the two sides of the bony tunnel of Corti (2). Over-lying these hair cells is a tectorial membrane that is anchored on one side. Vibration of the oval window produces a traveling wave along the length of the basilar membrane (241), displacing it vertically. This produces a shearing movement between the tectorial membrane and the stereocilia of the hair cells that are embedded in it. This shearing force effectively stimulates the hair cells (64).

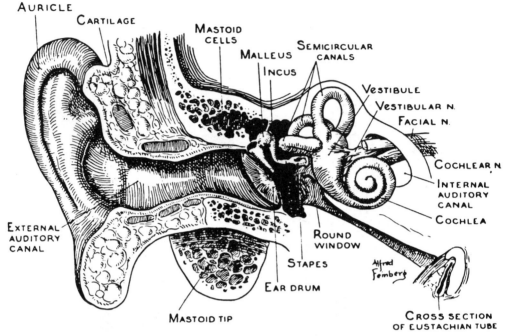

Figure 26. Cross-section drawing of outer, middle, and inner human ear. [From (2).]

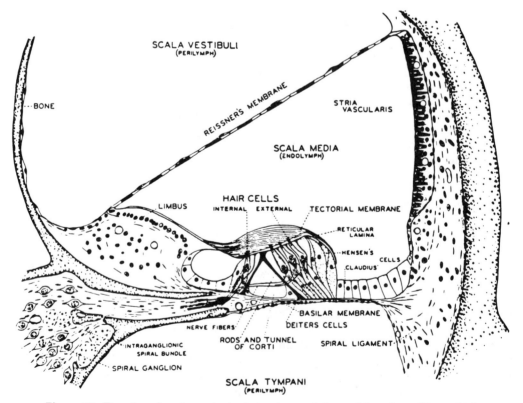

Figure 27. Drawing showing principal structures of the cochlear duct. [From (2).]

Frequency selectivity is achieved primarily through the micromechanical properties of the basilar membrane whose compliance changes with length (low-apex; high-base) and which supports a traveling wave. Therefore, different regions of the basilar membrane will be displaced maximally dependent on the stimulus frequency producing a place code for sound frequency at the cochlea.

Intracellular recordings from stimulated outer and inner hair cells show that: (1) receptor potentials comprise both alternating (ac) and direct (dc) components and (2) receptor potential amplitudes vary with stimulus frequency. Generally, the stimulus frequency that evokes the largest amplitude inner hair cell response is also the frequency that elicits the largest responses from the afferents innervating that hair cell (63, 161, 275). The receptor potentials of outer hair cells are typically larger than those of inner hair cells.

Approximately 95% of afferent fibers in the eighth nerve innervate inner hair cells; only 5% innervate outer hair cells (310). Thus, the activity of outer hair cells does not significantly contribute to information transfer from the periphery to the central auditory pathway. What then, is the function of outer hair cells? When excited, outer hair cells will contract (99) and become shorter. It is therefore plausible that outer hair cells may act as effector organs controlling the stiffness of basilar membrane (*remember*, their stereocilia are embedded in the overlying tectorial membrane) and/or the amplitude of the shearing force between the tectorial membrane and the inner hair cells (99). Outer hair cells may also sensitize inner hair cells electrically either by altering the ionic milieu within which all hair cells are bathed, or through the generation of field potentials (37).

Despite the broad frequency range over which they can communicate, and their highly developed and complex auditory systems, acoustic communication in mammals does not appear to be more sophisticated than in birds with, perhaps, the exception of some primates (e.g., humans), which are capable of producing and comprehending speech sounds. In mammals, as in several other vertebrate taxa, the analysis of communication signals is performed primarily in the central auditory pathway. This is necessary as the cochlea is basically a receiver, which performs a Fourier analysis of the signal and relays the results centrally with no further processing.

All vertebrates possess the same major auditory centers at the level of the brainstem. However, those of mammals (Fig. 28) are complex and have several subdivisions (130, 131). Frequency is represented tonotopically within each subdivision of each center. Thus, each center contains several frequency maps (5, 40, 241) giving rise to multiple, parallel, ascending pathways in the auditory system.

The auditory cortex is the pinnacle of the ascending auditory pathway. In mammals, it is elaborate and consists of distinct divisions. Frequency is mapped tonotopically in the primary auditory cortex. Other cortical auditory areas are specialized for encoding different acoustic attributes of sound signals such as intensity and location (39). Functional specialization is most evident in the auditory cortex of echolocating bats. As such, bats have served as useful models for examining the structural and functional organization of the mammalian auditory cortex and will be discussed in more detail below.

Microchiropteran bats rely primarily on sound to orient in space as they fly and hunt for prey in the dark. They emit ultrasonic signals and analyze the returning echoes to interpret the characteristics of objects in front of them (118). Two types of echo location signals, characteristic of different bat species, have been identified (Fig. 29): (1) Frequency modulated (FM) signals that are brief and con-

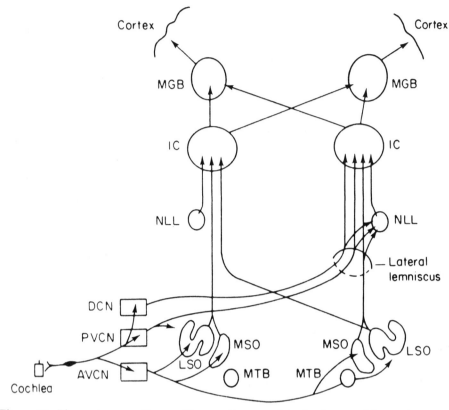

Figure 28. The main ascending auditory pathways of the brainstem. Many minor pathways are not shown. IC: inferior colliculus; MGB: medial geniculate body; NLL: nucleus of the lateral lemniscus. The branching and joining of arrows indicate divergence and convergence of pathways. [From (241).]

sist of a downward frequency sweep, and (2) long, constant frequency (CF) signals that are terminated by a brief, downward frequency modulated element (227, 293, 305).

The timing of emitted echolocation pulses is correlated with particular phases of hunting. During the searching phase, bats generally produce long duration signals suitable for detecting targets over great distances. During pursuit, the duration of the signal is shortened and the repetition rate is increased resulting in an increase in the rate at which the target is sampled. Just prior to capture, the pulses occur at the highest rate, described as a terminal buzz. Again, the increased pulse

rate leads to an increase in sampling rate, presumably reducing capture errors.

Some carnivorous bats hunt by means of passive sound localization as well as active echolocation. That is, they either produce weak echo location signals that minimize detection by potential prey, such as mice and crickets, or they do not actively produce signals at all, relying instead on sounds produced by prey (passive cues) for localization (94, 227).

Constant frequency (CF) bats can discriminate live from dead insects on the basis of insect wing beats. The beating wings produce a pronounced amplitude and frequency modulation of the CF signal (293). Different insect species have dif-

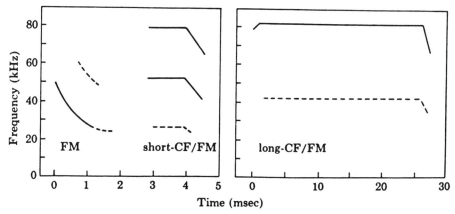

Figure 29. Representation of sound spectrograms illustrate the three commonly observed types of echo location sounds used by bats. Each type of sound contains a constant-frequency (CF) and frequency-modulated (FM) signal as basic elements. In the FM-type sound, the CF component follows the FM sweep, and its presence is controlled by the bat. There are several harmonics in a single sound; solid lines indicate strong harmonics, and dashed lines show weaker harmonics. [From (305).]

ferent wing-beat patterns. Thus, bats can discriminate one insect species from another based on differences in the modulation patterns.

While stationary, CF bats emit a narrow-band signal near CF, which is termed the reference frequency or RF. In flight, the frequency of returning echoes may be different than the RF due to the Doppler shift effect. Typically, CF bats compensate for the Doppler shift by lowering the emission frequency of the next pulse (compensation for negative Doppler shifts is rare). Doppler shift compensation results in a change in the echo frequency so that it matches the RF (226). The frequency range over which the Doppler shift occurs provides the bat with information concerning the velocity of the pursued target relative to the bat (292, 293).

Neurophysiological studies at various levels of the auditory system of CF bats have shown that it conforms to that of other mammals with two exceptions. First, tonotopic frequency representation is characterized by a disproportionate representation of neurons tuned to the bat's RF forming an *acoustic fovea* analogous to the fovea of the retina. Second, the frequency tuning of neurons comprising the acoustic fovea is exceptionally sharp (93, 297, 318). Interestingly, neurons in the inferior colliculus that are tuned to RF are particularly selective for amplitude or frequency modulated signals (296) similar to the echoes returning from flying insects (Fig. 30a).

Target discrimination in frequency modulating (FM) bats primarily depends on spectral cues (124, 304). Different targets have different frequency absorption characteristics yielding FM echoes with different spectral content. These echoes are detected by specialized neurons in the inferior colliculus that differ with respect to their sensitivity to both the frequency range and direction (high- to low-frequency sweep, or vice versa) of an FM sweep (233, 316). Generally, the response of these neurons can be predicted from their excitatory and inhibitory response areas in the frequency domain (Fig. 31).

In order to successfully track targets in space, bats require information concern-

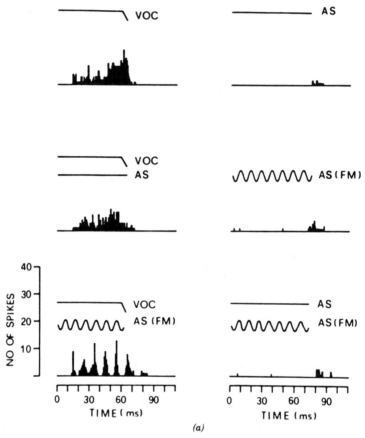

Figure 30a. Discharge patterns of a collicular neuron to (VOC) vocalization alone, (AS) to an artificial pure tone at the CF frequency, of the echolocation sound, (VOC + AS) to a combination of vocalization and the pure tone, [AS(FM)] to a sinusoidally frequency modulated tone (carrier frequency: CF frequency, modulation depth: ±500 Hz), [VOC + AS(FM)] to a combination of vocalization and the frequency modulated echo, and [AS + AS(FM)] to a combination of pure tone at CF frequency and frequency modulated echo. Intensity of artificial echoes 80 dB SPL throughout. Note especially different responses to the two acoustically identical stimulus situations in lower two graphs and patterned response only to VOC-AS(FM).

ing target range. For FM bats, this information is provided by the time delay between the emitted pulse and returning echo (303). Range discrimination capability can be predicted from the spectral properties of the bat's echolocation pulse (303). Superior range discrimination is highly correlated with broadband signals. Because CF bats typically emit narrowband signals, they generally do not perform as well as FM bats during range discrimination tasks.

Range information is systematically encoded in the brain of FM bats. A population of neurons in the intercollicularis nucleus and the auditory cortex respond only to paired sound stimuli (simulating pulse–echo pairs) if the time delay between the stimuli fall into specific time windows (Fig. 30b). These neurons are referred to as "delay tuned" neurons (92, 319, 322). In the auditory cortex of mustached bats, delay tuned neurons are systematically organized to form a

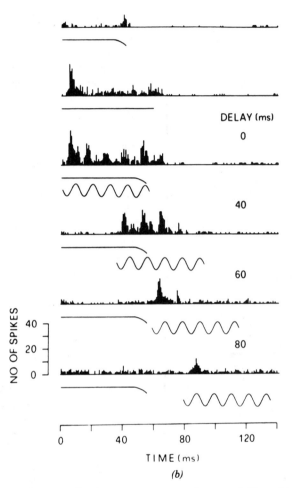

Figure 30b. Responses of a collicular neuron to vocalization followed by a sinusoidally frequency modulated artificial stimulus (carrier frequency 1 kHz higher than CF frequency, modulation depth ± 500 Hz) at different time delays to the onset of vocalization. Histograms sampled with 40 stimulus presentations. [From (296a).]

topographic map of target range (319). Like the auditory space map in the midbrain of barn owl, that of target range in the bat auditory cortex is a computational map.

In general, different information-bearing elements of returning echoes are represented by discrete, computational maps (Fig. 32) within the auditory cortex of bats (317). For example, in *Pteronotus*, which employs a CF–FM sonar signal, the target range is computed from the FM components of the sonar signal. These compu-

tations are performed in a discrete region of the brain, the FM–FM area. Velocity information is provided by the activity of neurons in a different region of the auditory cortex, the CF–CF area. Here, neurons respond only to specific combinations of the first, second, and third harmonics of the CF. Lateral to the CF–CF area is a region that is tonotopically organized. Within this region, RF frequency is mapped along the radius of the circle defining the boundary of the frequency map while signal amplitude, which is an

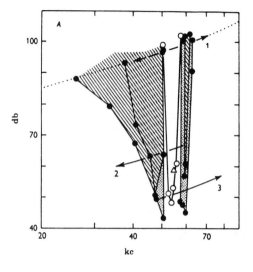

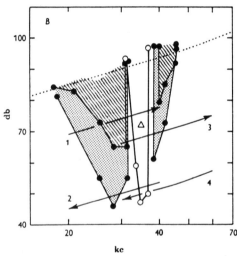

Figure 31. Audiograms of two neural units in inferior colliculus of a bat. Open circles give excitation areas; shaded or stippled regions give inhibitory areas. (*A*) A neuron that responded to FM tone pulses. (*B*) A neuron that showed lower threshold for downward sweep of FM tone pulse. Arrows indicate direction and range of frequency sweeps. [From Suga (316).]

indicator of target size, is represented along the subtended angle of the same circle. Thus, different elements of the echo are processed in parallel in the bat's auditory cortex. Parallel processing of complex sound features appears to be a general property of the vertebrate auditory system.

The vertebrate auditory pathway has a pronounced descending component. There are typically reciprocal connections between neurons at one level of the auditory pathway and those of successively higher levels (131). The olivocochlear efferent system illustrates the functional significance of descending projections.

There are two anatomical subdivisions of the olivocochlear system: (1) a lateral division, with cell bodies lying at the lateral aspect of the superior olivary complex that project ipsilaterally to the inner hair cells of the cochlea, and (2) a medial division, with cell bodies lying at the medial aspect of the superior olivary complex that innervate outer hair cells of the cochlea bilaterally (122, 164). At low sound intensities, electrical stimulation of the crossed olivocochlear bundle decreases the amplitude of the auditory evoked receptor potentials recorded from inner hair cells (35), the sound evoked responses of afferent fibers (348), and the sharpness of afferent tuning as indicated by an elevation of the tip of the excitatory tuning curve (160). These influences are apparently mediated by changes in the activity of outer hair cells that are innervated by the crossed olivocochlear pathway. Following stimulation of the crossed olivocochlear bundle, the outer hair cells are hyperpolarized (100). Because of their contractile properties, changes in the membrane potential of outer hair cells can lead to an alteration of the micromechanical properties of the cochlea (discussed above) which, in turn, alters the activity of inner hair cells and their associated afferents (276).

Efferent activation also plays a role in mediating the middle ear reflex. When animals are actively vocalizing, or when they hear intense sounds, the middle ear muscles contract. As a result, transmission through the middle ear is reduced

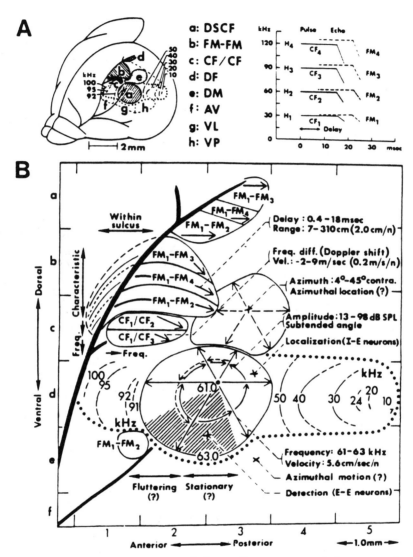

Figure 32. Functional organization of the auditory cortex of the mustached bat. (*A*) Dorsolateral view of the left cerebral hemisphere. The areas within the dashed lines comprise the auditory cortex. DSCF, FM–FM, CF/CF, DF, and DM areas (a, b, c, d, and e, respectively) are specialized for the systematic representation of biosonar information. The branches of the median cerebral artery are shown by the branching lines. The longest branch is on the sulcus. (*B*) Graphic summary of the functional organization of the auditory cortex. The tonotopic representation of the primary auditory cortex and the functional organization of the DSCF, FM–FM, CF/CF, and DM areas are indicated by lines and arrows. [From (317).]

and the ear becomes less sensitive to stimulation (218). In echo locating bats, this mechanism is particularly important for processing weak echoes, which return immediately following emission of loud echo location pulses although bats also possess neural mechanisms within the central auditory pathway itself, which attenuates responses to self-vocalized sounds (320, 321).

Most mammals have an acute ability to localize sounds. Cats, opossoms, macaques, elephant seals, and dolphins can resolve azimuth angles of less than 5° (33, 52, 134, 255, 258, 329). As in other vertebrates, sound localization involves binaural comparison of intensity and time cues (76, 117, 215). It is believed that the neural processing of binaural time and intensity differences occur in parallel; specifically, in the medial and lateral superior olivary nuclei, respectively (205). But, single unit recordings in the inferior colliculus (where the output of the two olivary pathways converge) and auditory cortex have not revealed neurons with small, well-defined receptive fields of auditory space like those exhibited by cells in the barn owl's auditory midbrain. Instead, collicular neurons respond to sounds originating in either one half of the contralateral sound field, or the entire field itself (5, 109, 211, 219, 299). Furthermore, the receptive fields correspond to the position of the contralateral pinna such that the location of the fields shift when the pinna moves. Therefore, mammals, in contrast to barn owls, do not appear to possess a very precise sensory map of auditory space.

In the deep layers of the superior colliculus, multimodal sensory neurons do show restricted auditory receptive fields with respect to space and are maximally excited by sounds originating from specific azimuthal locations. Interestingly, the spatial receptive fields of these neurons are topographically mapped (162,

212). These collicular neurons are believed to be responsible for aligning the eyes, ears, and head while orienting toward a sound source. In primates, receptive fields of deep superior collicular neurons shift in accordance with a change in eye position. Therefore the auditory and visual maps are always in register (150). This alignment is thought to be essential for acquiring a common motor reference for the generation of saccadic eye movement.

Summary

Mechanical stimuli play an important role in the lives of most animals. They provide information mediating a diversity of activities; from limb movement to communication. Mechanical stimuli are detected by specialized epithelial cells or mechanoreceptors. Mechanoreceptors are mainly tactile or proprioceptive; equilibrium receptors are stimulated by gravity and by acceleration and deceleration; sound and vibration receptors are stimulated by regular, usually sinusoidal, vibrations of various frequencies.

Mechanoreceptors are sensitive to surface deformation. This sensitivity is generally mediated by stretch activated ion channels located within the cell membrane; or at the tips of stereocilia as in vertebrate auditory hair cells. Sensitive mechanoreceptors show receptor potentials that are graded in amplitude according to the amount of displacement, indicating increased permeability to ions, particularly sodium and/or potassium. When a generator potential (i.e., summated receptor potential) reaches a critical size, an impulse is triggered, as shown in Pacinian corpuscles, muscle spindles, and crustacean stretch receptors.

Mechanical components of end-organs, including accessory structures, may provide directional sensitivity, matching of mechanical impedance of stimulus to tissue, regulation of sensitivity, and speed of sensory adaptation; they may thereby control frequency of response. The latter may also be under the influence of cellular mechanisms such as electrical tuning, typical of some hair cells in the auditory and vestibular organs of amphibians and reptiles.

Orientation in space and posture at rest or in locomotion depend on equilibrium receptors, proprioceptors, and visual input. These are highly integrated for reflex control of body position; there is some redundancy of sensory information, and much central compensation for deviation from normally balanced inputs.

Receptors for sound have evolved along different lines in arthropods and vertebrates. Crustaceans may detect sound by statocysts, and insects by resonant tympanal organs. Specialized sensing elements for equilibrium and acceleration (labyrinth), as well as sound (cochlea) have evolved in vertebrates.

Different parameters of sound stimulation are important to various animals. In orthopteran insects, central neurons generally do not code for frequency per se but for pattern of sound pulses. In the vertebrates, individual neurons have optimal input frequencies, and these overlap to provide for appropriate frequency discrimination. Sound is a basis for species and mate recognition in many orthopterans, frogs, and birds, for predator detection by many animals, and for prey identification as in bats. Many different mechanisms have evolved, which enable these animals to locate behaviorally relevant sounds.

References

1. Adams, W. B. *J. Exp. Biol.* 57:297–304, 1972. Mechanical tuning of the acoustic receptor of *Prodenia eridania* (cramer) (Noctuidae).
2. Ades, H. W. and H. Engström. In *Hand-book of Sensory Physiology* (A. D. Keidel and W. D. Neff, eds.), Vol. 5, Part 1, pp. 125–158. Springer-Verlag, Berlin, 1974. Anatomy of the inner ear.
3. Adrian, E. D. *The Basis of Sensation: The Action of the Sense Organs.* Christophers, London, 1928.
4. Aertsen, A. M. H. J., M. S. M. G. Vlaming, J. J. Eggermont, and P. I. M. Johannesma. *Hear. Res.* 21:17–40, 1986. Directional hearing in the grassfrog (*Rana temporaria* L.). II. Acoustics and modelling of the auditory periphery.
5. Aitkin, L. M. *The Auditory Midbrain: Structure and Function in the Central Auditory Pathway.* Humana Press, Clifton, NJ, 1985.
6. Alexander, R. D. Acoustical communication in Arthropods. *Annu. Rev. Entomol.* 12:495–526, 1967.
7. Art, J. J., A. C. Crawford, R. Fettiplace, and P. A. Fuchs. *Proc. R. Soc. London, Ser. B* 216:377–384, 1982. Efferent regulation of hair cells in the turtle cochlea.
8. Ashmore, J. F. *Nature (London)* 304:536–538, 1983. Frequency tuning in a vestibular organ of the frog.
9. Autrum, H. *Z. Vergl. Physiol.* 28:326–352, 1940. Über Lautäußerungen und Schallwahrnehmung bei Arthropoden II. Das Richtungshören von Locusta und Versuch einer Hörtheorie für Tympanalorgane vom Locustidentyp.
10. Bailey, W. J. and P. Thompson. *J. Exp. Biol.* 67:61–75, 1977. Acoustic orientation in the cricket *Teleogryllus oceanicus* (Le Guillou).
11. Bakker, G. J. and F. J. R. Richmond. *J. Neurophysiol.* 45:973–986, 1981. Two types of muscle spindles in cat neck muscles: A histochemical study of intrafusal fiber composition.
12. Banks, R. W., D. W. Harker, and M. J. Stacey. *J. Anat.* 123:783–796, 1977. A study of mammalian intrafusal muscle fibers using a combined histochemical and ultrastructural technique.
13. Barber, V. C. *Symp. Zool. Soc. London* 23:37–62, 1968. The structure of mollusc

statocysts, with particular reference to cephalopods.

14. Barker, D. In *Handbook of Sensory Physiology* (C. C. Hunt, ed.), Vol. 3, Part 2, pp. 1–190. Springer-Verlag, New York, 1974. The morphology of muscle receptors.

15. Barth, F. G. In *Sense Organs* (M. S. Laverack and D. Cosens, eds.), pp. 112–141. Blackie, Glasgow, 1981. Strain detection in the arthropod exoskeleton.

16. Barth, F. G. In *Neurobiology of Arachnids* (F. G. Barth, ed.), pp. 162–188. Springer-Verlag, Berlin, 1985. Slit sensilla and the measurement of cuticular strains.

17. Barth, F. G. In *Neurobiology of Arachnids* (F. G. Barth, ed.) pp. 203–229. Springer-Verlag, Berlin, 1985. Neuroethology of the spider vibration sense.

18. Barth, F. G. and Geethalbali. *J. Comp. Physiol. A* 148A:175–185, 1982. Spider vibration receptors: Threshold curves of individual slits in the metatarsal lyriform organ.

19. Bentley, D. R. and R. R. Hoy. *Anim. Behav.* 20:478-492, 1972. Genetic control of the neuronal network generating cricket (*Teleogryllus oceanicus*) song patterns.

20. Bibikov, N. G. and O. N. Gorodetskaya. *Neirofiziolagiya* 12:264–271, 1980. Single unit responses in the auditory center of the frog mesencephalon to amplitude-modulated tones.

21. Biederman-Thorson, M. *Brain Res.* 24:247–265, 1970. Auditory responses of units in the ovoid nucleus and cerebrum (field L) of the ring dove.

22. Bleckmann, H. and F. G. Barth. *Behav. Ecol. Sociobiol.* 14:303–312, 1986. Sensory ecology of a semiquatic spider (*Dolomedes* triton). II. The release of predatory behavior by water surface waves.

23. Blest, A. D., T. S. Collett, and J. D. Pye. *Proc. R. Soc. London, Ser. B* 158:196–207, 1963. The generation of ultrasonic signals by a new world arctiid moth.

24. Bolanowski, S. J., Jr. and J. J. Zwislocki. *J. Neurophysiol.* 51:812–830, 1984. Intensity and frequency characteristics of Pacinian corpuscles. II. Receptor potentials.

24a. Bordal, A. *Neurological Anatomy in Relation to Clinical Medicine*, 3rd ed., p. 157. Oxford Univ. Press, London and New York, 1981.

25. Boyan, G. S. *J. Comp. Physiol. A* 130A:137–150, 1979. Directional responses to sound in the central nervous system of the cricket *Teleogryllus commodus* (Orthoptera: Gryllidae). I. Ascending interneurons.

26. Boyan, G. S. *J. Comp. Physiol. A* 130A:151–159, 1979. Directional responses to sound in the central nervous system of the cricket *Teleogryllus commodus* (Orthoptera: Gryllidae). II. Descending interneurons.

27. Boyan, G. S. *J. Insect Physiol.* 30:27–41, 1984. Neural mechanisms of auditory information processing by identified interneurones in orthoptera.

28. Boyd, I. A. In *Muscle Receptors and Movement* (A. Taylor and A. Prochazka, eds.), pp. 17–32. Macmillan, London, 1981. The action of three types of intrafusal fibre in isolated cat muscle spindles on the dynamic and length sensitivities of primary and secondary sensory endings.

29. Bräunig, P. *J. Exp. Biol.* 116:331–341, 1985. Strand receptors associated with femoral chordotonal organs of locust legs.

30. Bräunig, P. and R. Hustert. *J. Comp. Physiol. A* 157A:73–82, 1985. Actions and interactions of proprioceptors of the locust hindleg coxo-trochanteral joint. I. Afferent responses in relation to joint position and movement.

31. Brehm, P., R. Kullberg, and F. Moody-Corbett. *J. Physiol. (London)* 350:631–648, 1984. Properties of nonjunctional acetylcholine receptor channels on innervated muscle of *Xenopus laevis*.

32. Brown, C. H. *Z. Tierpsychol.* 59:338–350, 1982. Ventroloquail and locatable vocalizations in birds.

33. Brown, C. H., M. D. Beecher, D. B. Moody, and W. C. Stebbins. *J. Acoust. Soc. Am.* 68:127–132, 1980. Localization of noise bands by old world monkeys.

34. Brown, H. M., D. Ottoson, and B. Rydqvist. *J. Physiol. (London)* 284:155–179, 1978. Crayfish stretch receptor: An investigation with voltage-clamp and ion-sensitive electrodes.

35. Brown, M. C., A. L. Nuttall, and R. I. Masta. *Science* 222:69–72, 1983. Intracellular recordings from cochlear inner hair cells: Effects of stimulation of the crossed olivocochlear efferents.

36. Brownell, P. and R. D. Farley. *J. Comp. Physiol. A* 131A:23–30, 1979. Detection of vibrations in sand by tarsal sense organs of the nocturnal scorpion, *Paruroctonus mesaensis*.

37. Brownell, W. E. *Hear. Res.* 6:335–360, 1982. Cochlea transduction: An integrative model and review.

38. Brownell, W. E., C. R. Boder, D. Bertrand, and Y. D. Ribaupierre. *Science* 227:194–196, 1985. Evoked mechanical response of isolated outer hair cells.

39. Brügge, J. F. In *Cortical Sensory Organization* (C. N. Woolsey, ed.), Vol. 3, pp. 59–70. Humana Press, Clifton, NJ, 1982. Auditory cortical areas in primates.

40. Brügge, J. F. and C. D. Geisler. *Annu. Rev. Neurosci.* 1:363–394, 1978. Auditory mechanisms of the lower auditory brainstem.

41. Budelmann, B.-U. *Brain Res.* 160:261–270, 1979. Hair cell polarization in the gravity receptor systems of the statocysts of the cephalopods *Sepia officinalis* and *Loligo vulgaris*.

42. Budelmann, B.-U., V. C. Barber, and S. West. *Brain Res.* 56:25–41, 1973. Scanning electron microscopical studies of the arrangements and numbers of hair cells in the statocysts of *Octopus vulgaris*, *Sepia officinalis* and *Loligo vulgaris*.

43. Bullock, T. H. and G. A. Horridge. *Structure and Function in the Nervous Systems of Invertebrates*. Freeman, San Francisco, CA, 1965.

44. Buno, W., Jr., L. Monti-Bloch, and L. Crispino. *J. Neurobiol.* 12:101–121, 1981. Dynamic properties of cockroach cercal "bristlelike" hair sensilla.

45. Buno, W., Jr., L. Monti-Bloch, A. Mateos, and P. Handler. *J. Neurobiol.* 12:123–141, 1981. Dynamic properties of cockroach cercal "threadlike" hair sensilla.

46. Burkhardt, D. *J. Insect Physiol.* 4:138–145, 1960. Responses to mechanical stimulation in antenna of *Calliphora*.

47. Calford, M. B. and J. D. Pettigrew. *Hear. Res.* 14:13–19, 1984. Frequency dependent of directional amplification at the cat's pinna.

48. Calford, M. B., L. Z. Wise, and J. D. Pettigrew. *J. Comp. Physiol. A* 157A:149–160, 1985. Coding of sound location and frequency in the auditory midbrain of diurnal birds of prey, families Accipitridae and Falconidae.

49. Capranica, R. R. The evoked vocal response of the bullfrog: A study of communication by sound. *MIT Res. Monogr.* No. 33, 1965.

50. Capranica, R. R. In *Frog Neurobiology* (R. Llinas and W. Precht, eds.), pp. 551–576. Springer-Verlag, Berlin, 1976. Morphology and physiology of the auditory system.

51. Casaday, G. B. and R. R. Hoy. *J. Comp. Physiol. A* 121A:1–13, 1977. Auditory interneurons in the cricket *Teleogryllus oceanicus*: Physiological and anatomical properties.

52. Casseday, J. H. and W. D. Neff. *J. Acoust. Soc. Am.* 54:365–372, 1973. Localization of pure tones.

53. Chandler, J. P. *J. Comp. Neurol.* 222:506–522, 1984. Light and electron microscopical studies of the basilar papilla in the duck, *Anas platyrhynchos*. I. The hatchling.

54. Chapman, C. J. and A. D. Hawkins. *J. Comp. Physiol. A* 85A:147–167, 1973. A field study of hearing in the cod, *Gadus morhua*.

55. Chapman, K. M., R. B. Duckrow, and D. T. Moran. *Nature (London)* 244:453–454, 1973. Form and role of deformation

in excitation of an insect mechanoreceptor.

56. Clarac, F. In *Structure and Function of Proprioceptors in the Invertebrates* (P. J. Mill, ed.), pp. 299–321. Chapman & Hall, London, 1976. Crustacean cuticular stress detectors.

57. Coles, R. B., D. B. Lewis, K. G. Hill, M. E. Hutchings, and D. M. Gower. *J. Exp. Biol.* 86:153–170, 1980. Directional hearing in the Japanese quail *Coturnix coturnix japonica*). II. Cochlear physiology; Coles, R. B. *Proc. Int. Ornithol. Congr., 17th, 1978,* pp. 714–717, 1980. Functional organization of auditory centres in the midbrain of birds.

58. Coombs, S. In *Hearing and Sound Communication in Fishes* (W. N. Tavolga, A. N. Popper, and R. R. Fay, eds.), pp. 174–178. Springer-Verlag, New York, 1981. Interspecific differences in hearing capabilities for select teleost species.

59. Cooper, K. E., J. M. Tang, J. L. Rae, and R. S. Eisenberg. *Jour. Membr. Biol.* 93:259–269, 1986. A cation channel in frog lens epithelia responsive to pressure and calcium.

60. Corey, D. P. and A. J. Hudspeth. *J. Neurosci.* 3:962–976, 1983. Kinetics of the receptor current in bullfrog saccular hair cells.

61. Crawford, A. C. and R. Fettiplace. *J. Physiol. (London)* 312:377–412, 1981. An electrical tuning mechanism in turtle cochlear hair cells.

62. Dagan, D. and J. M. Camhi. *J. Comp. Physiol.* 133:103–110, 1979. Responses to wind recorded from the cercal nerve of the cockroach *Periplanta americana.* II. Directional selectivity of the sensory neurons innervating single columns of filiform hairs.

63. Dallos, P., J. Santos-Sacchi, and Å. Flock. *Science* 218:582–584, 1982. Intracellular recordings from cochlear outer hair cells.

64. Davis, H. *Laryngoscope* 68:359–382, 1958. Transmission and transduction in the cochlea.

65. Denton, E. J. and J. Gray. *Proc. R. Soc. London, B Ser.* 226:249–261, 1985. Lateral-line-like antennae of certain of the *Penaeidea* (Crustacea, Decapoda, Natantia).

65a. Dethier, V. G. *The Physiology of Insect Senses.* Wiley, New York, 1963.

66. Diamond, J., J. A. B. Gray, and M. Sato. *J. Physiol. (London)* 133:54–67, 1956. Site of initiation of impulses in pacinian corpuscle.

67. Doherty, J. A. *J. Comp. Physiol. A* 156A:787–801, 1985. Trade-off phenomena in calling song recognition and phonotaxis in the cricket, *Gryllus bimaculatus* (Orthoptera, Gryllidae).

68. Doolan, J. M. and G. S. Pollack. *J. Comp. Physiol. A* 157A:223–233, 1985. Phonotactic specificity of the cricket *Teleogryllus oceanicus*: Intensity-dependent selectivity for temporal parameters of the stimulus.

69. Doolan, J. M. and D. Young. *Philos. Trans. R. Soc. London, Ser. B* 291:525–540, 1981. The organization of the auditory organ of the bladder cicada, *Cystosoma saunderii*.

70. Dooling, R. J. In *Comparatice Aspects of Hearing in Vertebrates* (A. N. Popper and R. R. Fay, eds.), pp. 261–288. Springer, New York, 1980. Behavior and psychophysics of hearing in birds.

72. Dunning, D. C. *Z. Tierpsychol.* 25:129–138, 1968. Warning sounds of moths.

73. Elepfandt, A. and A. V. Popov. *J. Insect Physiol.* 25:429–441, 1979. Auditory interneurones in the mesothoracic ganglion of crickets.

74. Elsner, N. In *Bioacoustics: A Comparative Approach* (B. Lewis, ed.), pp. 69–92. Academic Press, London, 1983. Insect stridulation and its neurophysiological basis.

75. Engström, H. and B. Engström. *Hear. Res.* 1:49–66, 1978. Structure of the hairs on cochlear sensory cells.

76. Erulkar, S. D. *Physiol. Rev.* 52:237–260, 1972. Comparative aspects of spatial localization of sound.

77. Esch, H., F. Huber, and D. W. Wohlers. *J. Comp. Physiol. A* 137A:27–38, 1980. Primary auditory neurones in crickets: Physiology and central projections.

78. Eyzaguirre, C. and S. W. Kuffler. *J. Gen. Physiol.* 39:87–119, 1955. Processes of excitation in the dendrites and in the soma of single isolated sensory nerve cells of the lobster and crayfish.

79. Fay, R. R. *Science* 225:951–954, 1984. The goldfish ear codes the axis of acoustic particle motion in three dimensions.

80. Fay, R. R. and A. S. Feng. In *Directional Hearing* (G. Gourevitch and W. A. Yost, eds.), pp. 179–213. Springer-Verlag, 1987. Mechanisms for directional hearing among non-mammalian vertebrates.

81. Fay, R. R. and A. N. Popper. In *Comparative Studies of Hearing in Vertebrates* (A. N. Popper and R. R. Fay, eds.), pp. 3–42. Springer-Verlag, New York, 1980. Structure and function in teleost auditory system.

82. Fay, R. R. and A. N. Popper. In *Hearing and Other Senses: Presentation in Honor of E.G. Wever* (R. R. Fay and G. Gourevitch, eds.), pp. 123–148. Groton, CT, Amphora Press, 1983. Hearing in fishes: Comparative anatomy of the ear and the neural coding of auditory information.

83. Feng, A. S. *J. Comp. Physiol. A* 144A:419–428, 1981. Directional response characteristics of single neurons in the torus semicircularis of the leopard frog (*Rana pipiens*).

84. Feng, A. S. *Hear. Res.* 6:241–246, 1982. Quantitative analysis of intensity-rate and intensity-latency functions in peripheral auditory nerve fibers of northern leopard frogs (*Rana p. pipiens*).

85. Feng, A. S. *Brain Res.* 367:183–191, 1986. Afferent and efferent innervation patterns of the cochlear nucleus (dorsal medullary nucleus) of the leopard frog.

86. Feng, A. S. *Brain Res.* 364:167–171, 1986. Afferent and efferent innervation patterns of the superior olivary nucleus of the leopard frog.

87. Feng, A. S. and Capranica, R. R. *J. Neurophysiol.* 41:43–54, 1978. Sound localization in anurans. II. Binaural interaction in superior olivary nucleus of the green treefrog (*Hyla cinerea*).

88. Feng, A. S., H. C. Gerhardt, and R. R. Capranica. *J. Comp. Physiol. A* 107A:241–252, 1976. Sound localization behavior of the green treefrog (*Hyla cinerea*) and the barking treefrog (*H. gratiosa*).

89. Feng, A. S., P. M. Narins, and R. R. Capranica. *J. Comp. Physiol. A* 100A:221–229, 1975. Three populations of primary auditory fibers in the bullfrog (*Rana catesbeiana*): Their peripheral origins and frequency sensitivities.

90. Feng, A. S. and W. P. Shofner. *Hear. Res.* 5:201–216, 1981. Peripheral basis of sound localization in anurans. Acoustic properties of the frog's ear.

91. Feng, A. S. and W. P. Shofner. In *Neural and Behavioral Development* (R. Aslin, ed.), pp. 39–72. Ablex Publ. Corp., Norwood, NJ, 1986. Development of the anuran auditory system.

92. Feng, A. S., J. A. Simmons, and S. A. Kick. *Science* 202:645–647, 1972. Echo detection and target-ranging neurons in the auditory system of the bat *Eptesicus fuscus*.

93. Feng, A. S. and M. Vater. *J. Comp. Neurol.* 235:529–553, 1985. Functional organization of the cochlear nucleus of rufous horseshoe bats (*Rhinolophus rouxi*): Frequencies and internal connections are arranged in slabs.

94. Fenton, M. B. *Q. Rev. Biol.* 59:33–53, 1984. Echolocation: Implications for ecology and evolution of bats.

95. Fletcher, N. H. *J. Comp. Physiol. A* 127A:185–189, 1978. Acoustical response of hair receptors in insects.

96. Fletcher, N. H. and K. G. Hill. *J. Exp. Biol.* 72:43–55, 1978. Acoustics of sound production and of hearing in the bladder cicada *Cystosoma saunderii* (Westwood).

97. Flock, Å., H. C. Cheung, B. Flock, and

G. Utter. *J. Neurocytol.* 10:133–147, 1981. Three sets of actin filaments in sensory cells of the inner ear. Identification and functional orientation determined by gel electrophoresis, immunofluorescence and electron microscopy.

98. Flock, Å. *Cold Spring Harbor Symp. Quant. Biol.* 30:133–146, 1965. Transducing mechanisms in lateral line canal organ receptor.

99. Flock, Å. In *Hearing-Physiological Bases and Psychophysics* (R. Klinke and R. Hartmann, eds.), pp. 2–9. Springer-Verlag, Berlin, 1983. Hair cells, receptor with a motor capacity.

100. Flock, Å and I. J. Russell. *Nature (London)* 243:89–91, 1973. Efferent nerve fibers: Postsynaptic action on hair cells.

101. Fraser, P. J. and D. C. Sandeman. *J. Comp. Physiol.* 96:205–221, 1975. Effects of angular and linear accelerations on semicircular canal interneurons of the crab *Scylla serrata*.

102. Frishkopf, L. S. and D. J. DeRosier. *Hear. Res.* 12:393–404, 1983. Mechanical tuning of freestanding stereociliary bundles and frequency analysis in the alligator lizard cochlea.

103. Frishkopf, L. S., R. R. Capranica, and M. H. Goldstein, Jr. *Proc. IEEE* 56:969–980, 1968. Neural coding in the bullfrog's auditory system—A teleological approach.

104. Fullard, J. F. and M. B. Fenton. *Can. J. Zool.* 55:1213–1224, 1977. Acoustic and behavioral analyses of the sounds produced by some species of Nearctic Arctiidae (Lepidoptera).

105. Fullard, J. F., M. B. Fenton, and J. A. Simmons. *Can. J. Zool.* 57:647–649, 1979. Jamming bat echolocation: the clicks of arctiid moths.

106. Fuzessery, Z. M. and A. S. Feng. *J. Comp. Physiol. A* 143A:339–347, 1981. Frequency representation in the dorsal medullary nucleus of the leopard frog *Rana p. pipiens*.

107. Fuzessery, Z. M. and A. S. Feng. *J. Comp. Physiol. A* 150A:107–119, 1983. Frequency selectivity in the anuran medulla: Excitatory and inhibitory tuning properties of single neurons in the dorsal medullary and suprior olivary nuclei.

108. Fuzessery, Z. M. and A. S. Feng. *J. Comp. Physiol. A* 150A:333–344, 1983. Mating call selectivity in the thalamus and midbrain of the leopard frog (*Rana p. pipiens*): Single and multiunit analyses.

109. Fuzessery, F. M. and G. D. Pollak. *J. Neurophysiol.* 54:757–781, 1985. Determinants of sound location selectivity in bat inferior colliculus: A combined dichotic and free-field stimulation study.

110. Fuzessery, Z. M. and A. S. Feng. *J. Comp. Physiol. A* 146A:471–484, 1982. Frequency selectivity in the auditory midbrain: Single unit responses to single and multiple tone stimulation.

111. Gans, C. and P. F. A. Maderson. *Am. Zool.* 13:1195–1203, 1973. Sound production mechanisms in recent reptiles: Review and comment.

111a. Gardner, D. *Fundamentals of Neurology,* 5th ed., p. 137. Saunders, Philadelphia, PA, 1968.

112. Gerhardt, H. C. *Am. Zool.* 22:581–595, 1982. Sound pattern recognition in some North American tree frogs (Anura: Hylidae): Implications for mate choice.

113. Gerhardt, H. C. and J. Rheinlaender. *Science* 217:663–664, 1982. Localization of an elevated sound source by the green treefrog.

114. Gibson, J. M. and W. I. Welker. *Somatosensory Res.* 1:51–67, 1983. Quantitative studies of stimulus coding in first-order vibrissa afferents of rats. 1. Receptive field properties and threshold distributions.

115. Gödde, J. *Biophys. Struct. Mech.* 10:275–296, 1984. Ultrasound elicits tonic responses and diminishes the phasic responses to adequate stimuli in threadhair mechanoreceptors of *Acheta domesticus*.

116. Göttschaldt, K.-M., A. Iggo, and D. W. Young. *J. Physiol. (London)* 235:287–315,

1973. Functional characteristics of mechanoreceptors in sinus hair follicles of the cat.

117. Gourevitch, G. In *Comparative Studies of Hearing in Vertebrates* (A. N. Popper and R. R. Fay, eds.), pp. 357–373. Springer, New York, 1980. Directional hearing in terrestrial mammals.

118. Griffin, D. R. *Listening in the Dark*. Dover, New York, 1958.

119. Grillner, S. *Physiol. Rev.* 55:247–304, 1975. Locomotion in vertebrates: Central mechanisms and reflex interaction.

120. Gross, N. B. and D. J. Anderson. *Brain Res.* 101:209–222, 1976. Single unit responses recorded from the first order neuron of the pigeon auditory system.

121. Guharay, F. and F. Sachs. *J. Physiol. (London)* 352:615–623, 1984. Stretch-activated single ion channel currents in tissue-cultured embryonic chick skeletal muscle.

122. Guinan, J. J., Jr., W. B. Warr, and B. E. Norris. *J. Comp. Neurol.* 221:358–370, 1983. Differential olivocochlear projections from lateral versus medial zones of the superior olivary complex.

123. Guthe, K. F. In *Biology of the Reptilia* (C. Gans, ed.), Vol. 11, pp. 266–355. Academic Press, New York, 1981. Reptilian muscle: Fine structure and physiological parameters.

124. Habersetzer, J. and B. Vogler. *J. Comp. Physiol. A* 152A:275–282, 1983. Discrimination of surface-structured targets by the echolocating bat *Myotis myotis* during flight.

125. Hagiwara, S. *Physiol. Comp. Oecol.* 4:142–153, 1956. Neuro-muscular mechanism of sound production in the ciccada.

126. Hall, J. and A. S. Feng. *Neurosci. Lett.* 63:215–220, 1986. Neural analysis of temporally patterned sounds in the frog's thalamus: Processing of pulse duration and pulse repetition rate.

127. Hall, J. and A. S. Feng. *J. Comp. Neurol.* 258:407–419, 1987. Evidence for parallel processing in the frog's auditory thalamus.

128. Hall, J. and A. S. Feng. *Hear. Res.* 36:261–276, 1988. Influence of envelope rise time on neural responses in the auditory system of anurans.

129. Hamilton, D. W. *Anat. Rec.* 164:253–258, 1969. The cilium on mammalian vestibular hair cells.

130. Harrison, J. M. and M. E. Howe. In *Handbook of Sensory Physiology* (A. D. Keidel and W. D. Neff, eds.), Vol. 5, Part 1, pp. 283–336. Springer-Verlag, Berlin, 1974. Anatomy of the afferent auditory nervous system of mammals.

131. Harrison, J. M. and M. E. Howe. In *Handbook of Sensory Physiology* (A. D. Keidel and W. D. Neff, eds.), Vol. 5, Part 1, pp. 363–388. Springer-Verlag, Berlin, 1974. Anatomy of the descending auditory system.

132. Hartline, P. H. *J. Exp. Biol.* 54:349–371, 1971. Physiological basis for detection of sound and vibration in snakes.

133. Hawkins, A. D. and A. A. Myrberg, Jr. In *Bioacoustics: A Comparative Approach* (D. B. Lewis, ed.), pp. 347–406. Academic Press, London, 1983. Hearing and sound communication under water.

134. Heffner, R. S. and H. E. Heffner. *J. Comp. Physiol. Psychol.* 96:926–944, 1982. Hearing in the elephant (*Elephas maximus*): Absolute sensitivity, frequency discrimination and sound localization.

135. Heidermanns, J. *Grundzüge der Tierphysiologie*. Stuttgart: Fischer, Stuttgart, 1957.

136. Henson, O. W., Jr. In *Handbook of Sensory Physiology* (A. D. Keidel and W. D. Neff, eds.), Vol. 5, Part 1, pp. 39–110. Springer-Verlag, Berlin, 1974. Comparative anatomy of the middle ear.

137. Hill, K. G. and G. S. Boyan. *J. Comp. Physiol. A* 121A:79–97, 1977. Sensitivity to frequency and direction of sound in the auditory system of crickets (Gryllidae).

138. Hill, K. G., J. J. Loftus-Hills, and D. F. Gartside. *Aust. J. Zool.* 20:153–163, 1972. Pre-mating isolation between the Australian field crickets *Teleogryllus commo-*

dus and *T. oceanicus* (Orthoptera: Gryllidae).

139. Hillman, D. E. *Brain Res.* 13:407–412, 1969. New ultrastructural findings regarding a vestibular ciliary apparatus and its possible functional significance; Hillman, D. E. and E. R. Lewis. *Science* 174:416–419, 1971. Morphological basis for mechanical linkage in otolithic receptor transduction in the frog.

140. Hoffman, C. *Z. Vergl. Physiol.* 54:290–352, 1967. Structure and function of mechanoreceptors of scorpion.

141. Horch, K. W. *Z. Vergl. Physiol.* 73:1–21, 1971. Hearing and vibration sense in *Ocypode*.

141a. Horridge, G. A. *Proc. R. Soc. London, Ser. B* 162:333–350, 1965. Nonmotile sensory cilia and neuromuscular junctions in a ctenophore effector system.

142. Horridge, G. A. In *Gravity and the Organism* (S. A. Gordon and M. J. Cohen, eds.), pp. 203–221. Univ. of Chicago Press, Chicago, IL, 1971. Primitive examples of gravity receptors and their evolution.

143. Huber, F. and J. Thorson. *Sci. Am.* 253:60–68, 1985. Cricket auditory communication.

144. Hudspeth, A. J. *Annu. Rev. Neurosci.* 6:187–215, 1983. Mechanoelectrical transduction by hair cells in the acustico-lateralis sensory system.

145. Hudspeth, A. J. *Science* 230:745–752, 1985. The cellular basis of hearing: The biophysics of hair cells.

146. Husmark, I. and D. Ottoson. *J. Physiol. (London)* 212:577–592, 1971. Adaptation of muscle spindles.

147. Hustert, R. *J. Comp. Physiol.* 147:389–399, 1982. The proprioceptive function of a complex chordotonal organ associated with the mesothoracic coxa in locusts.

148. Hutchings, M. and B. Lewis. In *Bioacoustics: A Comparative Approach* (B. Lewis, ed.), pp. 181–205. Academic Press, London, 1983. Insect sound and vibration receptors.

149. Iwasaki, S. and E. Florey. *J. Gen. Physiol.* 53:666–682, 1969. Inhibitory potentials in stretch receptors of crayfish.

150. Jay, M. F. and D. L. Sparks. *Nature (London)* 309:345–347, 1984. Auditory receptive fields in primate superior colliculus shift with changes in eye position.

151. Jenkins, W. M. and B. M. Masterton. *J. Comp. Physiol. Psychol.* 93:403–413, 1979. Sound localization in pigeon (*Columbia livia*).

152. Johansson, R. S. and A. B. Vallbo. *Trends Neurosci.* 6:27–32, 1983. Tactile sensory coding in the globrous skin of the human hand.

153. Kalmijn, A. In *Sensory Biology of Aquatic Animals* (J. Atema, R. R. Fay, A. N. Popper, and W. N. Tavolga, eds.), pp. 83–130. Springer-Verlag, Berlin and New York, 1987. Hydrodynamic and acoustic field detection.

154. Kalmring, K., B. Lewis, and A. Eichendorf. *J. Comp. Physiol.* 127:109–121, 1978. The physiological characteristics of the primary sensory neurons of the complex tibial organ of *Decticus verrucivorus* L. (Orthoptera: Tettigoniidae).

155. Kämper, G. and M. Dambach. *J. Insect Physiol.* 12:925–929, 1985. Low-frequency airborne vibrations generated by crickets during singing and aggression.

155a. Kandell, E. R. and J. H. Schwartz, eds., *Principles of Neural Science*, 2nd ed., p. 293. Elsevier, New York, 1985.

156. Katsuki, Y. *Physiol. Rev.* 45:380–423, 1965. Comparative neurophysiology of hearing.

157. Katsuki, Y. and N. Suga. *J. Exp. Biol.* 37:279–290, 1960. Neural mechanisms of hearing in insects.

158. Katz, B. *J. Physiol. (London)* 111:261–282, 1950. Depolarization of sensory terminals and the initiation of impulses in the muscle spindle.

159. Katz, B. *Philos. Trans. R. Soc. London* 243:221–240, 1961. The termination of the afferent nerve fibre in the muscle spindle of the frog.

160. Kiang, N. Y. S., E. C. Moxon, and R. A. Levine. In *Sensorineural Hearing Loss*

(G. E. W. Wolstenholme and J. Knight, eds.), Ciba Found. Symp., pp. 241–268. Churchill, London, 1970. Auditory-nerve activity in cats with normal and abnormal cochleae.

161. Kiang, N. Y. S., T. Watanabe, E. C. Thomas, and L. F. Clark. *Discharge Patterns of Single Fibers in the Cat's Auditory Nerve*, Res. Monogr. No. 35. MIT Press, Cambridge, MA, 1965.

162. King, A. J. and A. R. Palmer. *J. Physiol. (London)* 342:361–381, 1983. Cells responsive to free-field auditory stimuli in guinea pig superior colliculus: Distribution and response proper ties.

163. Kleindienst, H., D. W. Wohlers, and O. N. Larsen. *J. Comp. Physiol. A* 151A:397–400, 1983. Tympanal membrane motion is necessary for hearing in crickets.

164. Klinke, R. and N. Galley. *Physiol. Rev.* 54:316–357, 1974. Efferent innervation of vestibular and auditory receptors.

165. Klump, G. M., W. Windt, and E. Curio. *J. Comp. Physiol. A* 158A:383–390, 1986. The great tit's (Parus major) auditory resolution in azimuth.

166. Knudsen, E. I. *J. Neurosci.* 2:1177–1194, 1982. Auditory and visual maps of space in the optic tectum of the owl.

167. Knudsen, E. I. In *Bioacoustics: A Comparative Approach* (D. B. Lewis, ed.), pp. 311–346. Academic Press, London, 1983. Space coding in the vertebrate auditory system.

168. Knudsen, E. I. and M. Konishi. *J. Comp. Physiol. A* 133A:13–21, 1979. Mechanisms of sound localization in the barn owl (*Tyto alba*).

169. Knudsen, E. I., G. G. Blasdel, and M. Konishi. *J. Comp. Physiol. A* 133A:1–11, 1979. Sound localization by the barn owl (*Tyto alba*) measured with the search coil technique.

170. Kohllöffel, L. U. E. *Hear. Res.* 13:73–76, 1984. Notes on the comparative mechanics of hearing. I. A shock-proof ear.

171. Konishi, M. *Z. Vergl. Physiol.* 66:257–272, 1970. Comparative neurophysiological studies of hearing and vocalization in songbirds.

172. Konishi, M. *Am. Sci.* 61:414–424, 1973. How the owl tracks its prey.

173. Konishi, M. *Trends Neurosci.* 9:163–168, 1986. Centrally synthesized maps of sensory space.

174. Konishi, M., W. E. Sullivan, and T. Takahashi. *J. Acoust. Soc. Am.* 78:360–364, 1985. The owl's cochlear nuclei process different sound localization cues.

175. Koyama, H., E. R. Lewis, E. L. Leverenz, and R. A. Baird. *Brain Res.* 250:168–172, 1982. Acute seismic sensitivity in the bullfrog ear.

176. Kraft, B. *Symp. Zool. Soc. London* 42:59–67, 1978. The recording of vibratory signals performed by spiders during courtship.

177. Kuffler, S. W. *Exp. Cell. Res., Suppl.* 5:493–519, 1958. Synaptic inhibitory mechanisms, properties of dendrites and problems of excitation in isolated sensory nerve cells.

178. Kuffler, S. W. and C. Eyzagúirre. *J. Gen. Physiol.* 39:155–184, 1955. Synaptic inhibition in an isolated nerve cell.

179. Kuster, J. E. and A. S. French. *J. Comp. Physiol. A* 150A:207–215, 1983. Sensory transduction in a locust multipolar joint receptor: Their dynamic behaviour under a variety of stimulus conditions.

180. Langner, G., D. Bonke, and H. Scheich. *Exp. Brain Res.* 43:11–24, 1981. Neuronal discrimination of natural and synthetic vowels in field L of trained mynah birds.

181. Larsell, O. *J. Comp. Neurol.* 60:473–527, 1934. The differentiation of the peripheral and central acoustic apparatus in the frog.

182. Larsen, O. N. and A. Michelsen. *J. Comp. Physiol. A.* 123A:217–227, 1978. Biophysics of the ensiferan ear. III. The cricket ear as a four-input system.

183. Larsen, O. N., A. Surlykke, and A. Michelsen. *Naturwissenschaften* 71:538–539, 1984. Directionality of the cricket ear: A property of the tympanal membrane.

184. Laverack, M. S. In *Structure and Function of Proprioceptors in the Invertebrates* (P. J.

Mill, ed.), pp. 1–63. Chapman & Hall, London, 1976. External proprioceptors.

184a. Leppelsack, H.-J. *Acta Oto-Laryngol.* 429:57–60, 1986. Critical periods in bird song learning.

185. Lewis, D. B. In *Bioacoustics: A Comparative Approach* (D. B. Lewis, ed.), pp. 233–260. Academic Press, London, 1983. Directional cues for sound localization.

186. Lewis, E. R., E. L. Leverenz, and H. Koyama. *J. Comp. Physiol. A* 145A:437–445, 1982. The tonotopic organization of the bullfrog amphibian papilla, an auditory organ lacking a basilar membrane.

187. Lewis, R. S. and A. J. Hudspeth. *Nature (London)* 304:538–541, 1983. Voltage- and ion-dependent conductances in solitary vertebrate hair cells.

188. Lindblom, V. and D. N. Tapper. *Exp. Neurol.* 17:1–15, 1967. Terminal properties of a vibrotactile sensor.

189. Lindeman, H. H., H. W. Ades, G. Bredberg, and H. Engström. *Acta Oto-Laryngol.* 72:229–242, 1971. The sensory hairs and the tectorial membrane in the development of the cat's organ of Corti. A scanning electron microscope study.

190. Loewenstein, W. R. *Ann. N.Y. Acad. Sci.* 81:367–387, 1959. The generation of electric activity in a nerve ending.

191. Loewenstein, W. R. and M. Mendelson. *J. Physiol. (London)* 177:377–397, 1965. Components of receptor adaptation in a pacinian corpuscle.

192. Loftus-Hills, J. J. and M. J. Littlejohn. *Copeia* 1971:154–156, 1971. Pulse repetition rate as the basis for mating call discrimination by two sympatric species of Hyla.

193. Luther, P. W., H. B. Peng, and J. J.-C. Lin. *Nature (London)* 303:61–64, 1983. Changes in cell shape and actin distribution induced by constant electric field.

194. Manley, G. A. *Z. Vergl. Physiol.* 66:251–256, 1970. Frequency sensitivity of auditory neurons in the Caiman cochlear nucleus.

195. Manley, G. A. *Evolution (Lawrence, Kans.)* 26:608–621, 1973. A review of some current concepts of the functional evolution of the ear in terrestrial vertebrates.

196. Manley, G. A. *Prog. Sens. Physiol.* 2:49–134, 1981. A review of the auditory physiology of the reptiles.

197. Manley, G. A., O. Gleich, H.-J. Leppelsack, and H. Oeckinghaus. *J. Comp. Physiol. A* 157A:161–181, 1985. Activity patterns of cochlear ganglion neurones in the starling.

198. Manley, J. A. *Z. Vergl. Physiol.* 71:255–261, 1971. Single unit studies in the midbrain auditory area of *Caiman*.

199. Marcellini, D. *Am. Zool.* 17:251–260, 1977. Acoustic and visual display behavior of gekkonid lizards.

200. Margoliash, D. *J. Neurosci.* 3:1039–1057, 1983. Acoustic parameters underlying the responses of song-specific neurons in the white-crowned sparrow.

201. Markl, H. In *Neuroethology and Behavioral Physiology.* (F. Huber and H. Markl, eds.), pp. 332–353. Springer-Verlag, Berlin, 1983. Vibrational communication.

202. Markl, H. and K. Wiese. *Z. Vergl. Physiol.* 62:413–420, 1969. Wave detection by water beetles.

203. Marler, P. *Nature (London)* 176:6–8, 1955. Characteristics of some animal calls.

204. Masters, M. W. *Behav. Ecol. Sociobiol.* 15:207–215, 1984. Vibration in the orb-web of *Nuctenea sclopetaria* (Araneidae). I. Transmission through the web.

205. Masterton, B. M. and T. J. Imig. *Annu. Rev. Physiol.* 46:275–287, 1984. Neural mechanisms for sound localization.

206. Matthews, P. B. C. *Annu. Rev. Neurosci.* 5:189–218, 1982. Where does Sherrington's "muscular sense" originate? Muscles, joints, corollary discharges?

207. McCasland, J. S. and M. Konishi. *Proc. Natl. Acad. Sci. U.S.A.* 78:7815–7819, 1981. Interaction between auditory and motor activities in an avian song control nucleus.

208. Michelsen, A. *Z. Vergl. Physiol.* 71:49–128, 1971. The physiology of the locust ear; In *Handbook of Sensory Physiology* (W.

D. Keidel and W. D. Neff, eds.), Vol. 5, Part 1, pp. 339–422. Springer-Verlag, Berlin, 1974. Hearing in invertebrates.

209. Michelsen, A. and O. N. Larsen. *J. Comp. Physiol. A* 123A:193–203, 1978. Biophysics of the ensiferan ear. I. Tympanic vibrations in bushcrickets (Tettigoniidae) studied with laser vibrometry.

210. Michelsen, A. and H. Nocke. *Adv. Insect Physiol.* 10:247–296, 1974. Biophysical aspects of sound communication in insects.

211. Middlebrook, J. C. and J. D. Pettigrew. *J. Neurosci.* 1:107–120, 1981. Functional classes of neurons in primary auditory cortex of the cat distinguished by sensitivity to sound location.

212. Middlebrook, J. C. and E. I. Knudsen. *J. Neurosci.* 4:2621–2634, 1984. A neural code for auditory space in the cat's superior colliculus.

213. Miller, L. A. In *Neuroethology and Behavioral Physiology* (F. Huber and H. Markl, eds.), pp. 251–266. Springer-Verlag, Berlin, 1983. How insects detect and avoid bats.

214. Miller, M. R. *J. Comp. Neurol.* 232:1–24, 1985. Quantitative studies of auditory hair cells and nerves in lizards.

215. Mills, A. W. In *Foundations of Modern Auditory Theory* (J. V. Tobias, ed.), Vol. 3, pp. 303–348. Academic Press, New York, 1972. Auditory localization.

216. Moiseff, A. and R. Hoy. *J. Comp. Physiol. A* 152:155–167, 1983. Sensitivity to ultrasound in an identified auditory interneuron in the cricket: A possible neural link to phonotactic behavior.

217. Moiseff, A. and M. Konishi. *J. Comp. Physiol. A* 114A:299–304, 1981. The owl's interaural pathway is not involved in sound localization.

218. Moller, A. R. In *Handbook of Sensory Physiology* (A. D. Keidel and W. D. Neff, eds.), Vol. 5, Part 1, pp. 491–518. Springer-Verlag, Berlin, 1974. Function of the middle ear.

219. Moore, D. R., M. N. Semple, P. D. Addison, and L. M. Aitkin. *Hear. Res.* 13:159–174, 1984. Properties of spatial receptive fields in the central nucleus of the cat inferior colliculus. I. Responses to tones of low intensity.

220. Moran, D. T. and F. G. Varela. *Proc. Natl. Acad. Sci. U.S.A.* 68(4):757–760, 1971. Microtubules and sensory transduction.

221. Müller, S. C. and H. Scheich. *J. Comp. Physiol. A* 156A:1–12, 1985. Functional organization of the avian auditory field L: A comparative 2DG study.

222. Murphey, R. K. and M. D. Zaretsky. *J. Exp. Biol.* 56:335–352, 1972. Orientation to calling song by female crickets, *Scapsipedus marginatus* (Gryllidae).

223. Myrberg, A. A., Jr. and J. Spires. *J. Comp. Physiol. A* 140A:135–144, 1980. Hearing in damselfishes: An analysis of signal detection among closely related species.

224. Narins, P. M. and R. R. Capranica. *Brain, Behav. Evol.* 17:48–66, 1980. Neural adaptation for processing the two-note call of the Puerto Rican treefrog, *Eleutherodactylus coqui*.

225. Necker, R. In *Physiology and Behaviour of the Pigeon* (M. Abs, ed.), pp. 193–219. Academic Press, London, 1983. Hearing.

226. Neuweiler, G. In *Comparative Physiology of Sensory Systems* (R. D. Keynes and S. H. P. Maddrell, eds.), pp. 115–141. Cambridge Univ. Press, London and New York, 1984. Auditory basis of echolocation in bats.

227. Neuweiler, G. *Naturwissenschaften* 71:446–455, 1984. Foraging, echolocation and audition in bats.

228. Nishi, K. and M. Sato. *J. Physiol. (London)* 184:376–386, 1966. Blocking of the impulse and depression of the receptor potential by TTX in nonmyelinated nerve terminals in pacinian corpuscles.

229. Norberg, R. A. *Philos. Trans. R. Soc. London, Ser. B* 280:375–408, 1977. Occurrence and independent evolution of bilateral ear assymetry in owls and implications on owl taxonomy.

230. Northcutt, R. G. In *Comparative Studies of Hearing in Vertebrates* (A. N. Popper and R. R. Fay, eds.), pp. 79–118. Sprin-

ger-Verlag, New York, 1980. Central auditory pathways in anamniotic vertebrates.

231. Oldfield, B. P. *J. Comp. Physiol. A* 151A:389–395, 1983. Central projections of primary auditory fibres in Tettigoniidae (Orthoptera: Ensifera).

232. Oldfield, B. P., H. U. Kleindienst, and F. Huber. *J. Comp. Physiol. A* 159A:457–464, 1986. Physiology and tonotopic organization of auditory receptors in the cricket *Gryllus bimaculatus* DeGeer.

233. O'Neill, W. E. *J. Comp. Physiol. A* 157A:797–815, 1985. Responses to pure tones and linear FM components of the CF-FM biosonar signal by single units in the inferior colliculus of the mustache bat.

233a. Osborne, M. P., S. D. Comis and J. D. Pickles. *Cell Tissue Res.* 237:43–48, 1985. Morphology and cross-linkage of stereocilia.

234. Ottoson, D. and C. Swerup. *Brain Res.* 244:337–341, 1982. Studies on the role of calcium in adaptation of the crustacean stretch receptor. Effects of intracellular injection of calcium, EGTA and TEA.

235. Palmer, A. R. and A. C. Pinder. *J. Theor. Biol.* 110:205–215, 1984. The directionality of the frog ear described by a mechanical model.

236. Parry, D. A. *J. Exp. Biol.* 43:185–192, 1965. The signal generated by an insect in a spider's web.

237. Passmore, N. I., R. R. Capranica, S. R. Telford, and P. J. Bishop. *J. Comp. Physiol. A* 154A:189–197, 1984. Phonotaxis in the painted reed frog (*Hyperolius marmoratus*). The localization of elevated sound sources.

238. Paton, J. A., R. R. Capranica, R. R. Dragsten, and W. W. Webb. *J. Comp. Physiol. A* 119A:221–240, 1977. Physical basis for auditory frequency analysis in field crickets (Gryllidae).

239. Payne, R. S. *J. Exp. Biol.* 54:535–573, 1971. Acoustic location of prey by barn owls (*Tyto alba*).

240. Phillips, D. P., M. B. Calford, J. D. Pet-
tigrew, L. M. Aitkin, and M. N. Semple. *Hear. Res.* 8:13–28, 1982. Directionality of sound pressure transformation at the cat's pinna.

241. Pickles, J. O. *An Introduction to the Physiology of Hearing.* 2nd ed. Academic Press, London, 1982.

242. Pickles, J. O., S. D. Comis, and M. P. Osborne. *Hear. Res.* 15:103–112, 1984. Cross-links between stereocilia in the guinea pig organ of Corti, and their possible relation to sensory transduction.

243. Pollack, G. S. *J. Comp. Physiol. A* 146A:217–221, 1982. Sexual differences in cricket calling song recognition.

244. Pollack, G. S. and R. R. Hoy. *J. Comp. Physiol. A* 144A:367–373, 1981. Phonotaxis to individual rhythmic components of a complex cricket calling song.

245. Pollack, G. S., F. Huber, and T. Weber. *J. Comp. Physiol. A* 154A:13–26, 1984. Frequency and temporal pattern-dependent phonotaxis of crickets (*Teleogryllus oceanicus*) during tethered flight and compensated walking.

246. Popov, A. V. and A. M. Markovich. *J. Comp. Physiol. A* 146A:351–359, 1982. Auditory interneurons in the prothoracic ganglion of the cricket, *Gryllus bimaculatus.*

247. Poppele, R. E. and D. C. Quick. *J. Neurosci.* 7:1881–1885, 1985. Effect of intrafusal muscle mechanics on mammalian muscle spindle sensitivity.

248. Prager, J. and O. N. Larsen. *Naturwissenschaften* 68:579–580, 1981. Asymmetrical hearing in the water bug *Corixa punctata* observed with laser vibrometry.

249. Prager, J. and R. Streng. *J. Comp. Physiol. A* 148A:323–335, 1982. The resonance properties of the physical gill of *Corixa punctata* and their significance in sound reception.

250. Pringle, J. W. S. *J. Exp. Biol.* 15:101–113, 1938. Proprioception in insects. I. Mechanoreceptors in palps of cockroach.

251. Pringle, J. W. S. *J. Exp. Biol.* 15:467–473, 1938. Proprioception in insects. III. The function of the hair sensilla at the joints.

252. Proske, U. *Exp. Neurol.* 23:187–194,

1969. Vibration receptors in skin of snakes.

253. Pubols, B. H., Jr., P. J. Donovick, and L. M. Pubols. *Brain, Behav. Evol.* 7:360–381, 1973. Opossum trigeminal afferents associated with vibrissae and rhinarial mechanoreceptors.

254. Pumphrey, R. J. *Biol. Rev. Cambridge Philos. Soc.* 15:107–132, 1940. Hearing in insects.

255. Ravizza, R. J. and B. M. Masterton. *J. Neurophysiol.* 35:344–356, 1972. Contribution of neocortex to sound localization in opossum (*Didelphis virginiana*).

256. Ray, R. H. and L. Kruger. *Fed. Proc., Fed. Am. Soc. Exp. Biol.* 42:2536–2541, 1983. Spatial properties of receptive fields of mechanosensitive primary afferent nerve fibers.

257. Reid, K. H. *Science* 172:949–951, 1971. Periodical cicada: Mechanism of sound production.

258. Renaud, D. L. and A. N. Popper. *J. Exp. Biol.* 63:569–585, 1975. Sound localization by the bottlenose porpoise *Tursiops truncatus.*

259. Rheinlaender, J. *J. Comp. Physiol. A* 97A:1–53, 1975. Transmission of acoustic information at three levels in the auditory system of *Decticus verrucivorus* (Tettigoniidae: Orthoptera).

260. Rheinlaender, J. and G. Blätgen. *Physiol. Entomol.* 7:209–218, 1982. The precision of auditory lateralization in the cricket *Gryllus bimaculatus.*

261. Rheinlaender, J. and H. Römer. *J. Comp. Physiol. A* 140A:101–111, 1980. Bilateral coding of sound direction in the CNS of the bushcricket *Tettigonia viridissima* L. (Orthoptera, Tettigoniidae).

262. Rheinlaender, J., H. C. Gerhardt, D. D. Yager, and R. R. Capranica. *J. Comp. Physiol. A* 133A:247–255, 1979. Accuracy of phonotaxis by the green treefrog (*Hyla cinerea*).

263. Rice, W. R. *Auk* 99:403–413, 1982. Acoustical location of prey by the marsh hawk: Adaptation to concealed prey.

264. Roberts, W. M., J. Howard, and A. J.

Hudspeth. *Annu. Rev. Cell Biol.* 4:63–92, 1988. Hair cells: Transduction, tuning, and transmission in the inner ear.

265. Roeder, K. D. *J. Insect Physiol.* 20:55–66, 1974. Responses of the less sensitive acoustic sense cells in the tympanal organs of some noctuid and geometrid moths.

266. Roeder, K. D. and A. E. Treat. *J. Insect Physiol.* 16:1069–1086, 1970. An acoustic sense in some hawkmoths (Choerocampinae).

267. Roeder, K. D. *J. Insect Physiol.* 10:529–546, 1964. Aspects of the noctuid tympanic organ having significance in the avoidance of bats.

268. Römer, H. *Nature (London)* 306:60–62, 1983. Tonotopic organisation of the auditory neuropile in the bushcricket *Tettigonia* viridissima.

269. Römer, H. and J. Rheinlaender. *J. Comp. Physiol. A* 152A:289–296, 1983. Electrical stimulation of the tympanal nerve as a tool for analyzing the responses of auditory interneurons in the locust.

270. Ronacher, B., D. V. Helversen, and O. V. Helversen. *J. Comp. Physiol. A* 158A:363–374, 1986. Routes and stations in the processing of auditory directional information in the CNS of a grasshopper, as revealed by surgical experiments.

271. Rose, G. and R. R. Capranica. *J. Comp. Physiol. A* 154A:211–219, 1984. Processing amplitude-modulated sounds by the auditory midbrain of two species of toads: Matched temporal filters.

272. Rose, G. and R. R. Capranica. *J. Neurophysiol.* 53:446–465, 1985. Sensitivity to amplitude modulated sounds in the anuran auditory nervous system.

273. Rosowski, J. J. and J. C. Saunders. *J. Comp. Physiol. A* 136A:183–190, 1980. Sound transmission through the avian interaural pathways.

274. Rubel, E. W. and T. N. Parks. *J. Comp. Neurol.* 164:411–434, 1975. Organization and development of the brain stem auditory nuclei of the chicken: Tonotopic

organization of n. angularis and n. laminaris.

275. Russell, I. J. and P. M. Sellick. *Nature (London)* 267:858–860, 1977. Tuning properties of cochlear hair cells.

276. Russell, I. J., A. R. Cody, and G. P. Richardson. *Hear. Res.* 22:199–216, 1986. The responses of inner and outer hair cells in the basal turn of the guinea-pig cochlea and in the mouse cochlea grown *in vitro*.

276a. Sachs, F. *Fed. Proc., Fed. Am. Soc. Exp. Biol.* 46:12, 1987. Baroreceptor mechanisms.

277. Sachs, F. *CRC Crit. Rev. Biomed. Eng.* 16(2):141–169, 1988. Mechanical transduction in biological systems.

278. Sachs, M. B., J. M. Sinnott, and R. D. Heinz. *Fed. Proc., Fed. Am. Soc. Exp. Biol.* 37:2329–2335, 1978. Behavioral and physiological studies of hearing in birds.

279. Sachs, M. B., N. K. Woolf, and J. M. Sinnott. In *Comparative Studies of Hearing in Vertebrates* (A. N. Popper and R. R. Fay, eds.), pp. 323–353. Springer, New York, 1980. Response properties of neurons in the avian auditory system: Comparisons with mammalian homologues and consideration of the neural encoding of complex stimuli.

280. Sachs, M. B., E. D. Young, and R. H. Lewis. *Brain Res.* 70:431–447, 1974. Discharge patterns of single fibers in the pigeon auditory nerve.

281. Sakaluk, S. K. and J. J. Belwood. *Anim. Behav.* 32:659–662, 1984. Gecko phonotaxis to cricket calling song: A case of satellite predation.

282. Sales, G. and D. Pye. *Ultrasonic Communication by Animals*. Chapman & Hall, London, 1974.

283. Salmon, M., K. Horch, and G. W. Hyatt. *Mar. Behav. Physiol.* 4:187–194, 1977. Barth's myochordotonal organ as a receptor for auditory and vibrational stimuli in fiddler crabs (*Uca pugilator* and *U. minax*).

284. Sand, O. In *Hearing and Sound Communication in Fishes* (W. N. Tavolga, A. N. Popper, and R. R. Fay, eds.), pp. 459–480. Springer, New York, 1981. The lateral line organ and sound reception.

285. Saunders, J. C. *Hear. Res.* 18:253–268, 1985. Auditory structure and function in the bird middle ear: An evaluation by SEM and capacitive probe.

286. Saunders, J. C., M. E. Schneider, and S. P. Dear. *J. Acoust. Soc. Am.* 78(1):299–311, 1985. The structure and function of actin in hair cells.

287. Scheich, H., G. Langner, and D. Bonke. *J. Comp. Physiol. A* 132A:257–276, 1979. Responsiveness of units in the auditory neostriatum of the guinea fowl (*numida meleagris*) to species-specific calls and synthetic stimuli. II. Discrimination of iambus-like calls.

288. Scheich, H., G. Langner, and R. Koch. *J. Comp. Physiol. A* 117A:245–265, 1977. Coding of narrow-band and wide-band vocalizations in the auditory midbrain nucleus (MLD) of the guinea fowl (*Numida meleagris*).

289. Schildberger, K. *Experientia* 44:408–415, 1988. Behavioral and neuronal mechanisms of cricket phonotaxis.

290. Schlegel, P. *Z. Vergl. Physiol.* 66:45–77, 1970. Spikes and receptor potentials from joint receptors in *Calliphora*.

291. Schmitz, B. *J. Comp. Physiol. A* 156A:165–180, 1985. Phonotaxis in ·*Gryllus campestris*· L. (Orthoptera, Gryllidae). III. Intensity dependence of the behavioural performance and relative importance of tympana and spiracles in directional hearing.

292. Schnitzler, H.-U. In *Localization and Orientation in Biology and Engineering* (M. Varju and H. U. Schnitzler, eds.), pp. 211–224. Springer-Verlag, Berlin, 1984. The performance of bat sonar systems.

293. Schnitzler, H.-U. and O. W. Henson, Jr. In *Animal Sonar Systems* (R.-G. Busnel and J. F. Fish, eds.), pp. 109–182. Plenum, New York, 1980. Performance of airborne animal sonar systems. I. Microchiroptera.

294. Schüh, W. and F. G. Barth. *Behav. Ecol. Sociobiol.* 16:263, 1985. Temporal patterns in the vibratory courtship of a

wandering spider (*Cupiennius salei* keys).

295. Schuijf, A. In *Hearing and Sound Communication in Fishes* (W. N. Tavolga, A. N. Popper, and R. R. Fay, eds.), pp. 267–310. Springer, New York, 1981. Models of acoustic localization.

296. Schuller, G. *Exp. Brain Res.* 34:275–286, 1979. Coding of small sinusoidal frequency and amplitude modulations in the inferior colliculus of "CF-FM" bat, *Rhinolophus ferrumequinum.*

296a. Schuller, G. *J. Comp. Physiol.* 132A:39–46, 1979. Vocalization influences auditory processing in collicular neurons of the CF–FM bat, *Rhinolophus ferrumequinum.*

297. Schuller, G. and G. D. Pollak. *J. Comp. Physiol. A* 132A:47–54, 1979. Disproportionate frequency representation in the inferior colliculus of Doppler-compensating greater horseshoe bats: Evidence for an acoustic fovea.

298. Schwartzkopff, J. *Proc. Int. Ornithol. Congr., 13th, 1962,* pp. 1059–1068, 1963. Morphological and physiological properties of the auditory system in birds.

299. Semple, M. N., L. M. Aitkin, M. B. Calford, J. D. Pettigrew, and D. P. Phillips. *Hear. Res.* 10:203–215, 1983. Spatial receptive fields in the cat inferior colliculus.

300. Seyfarth, E.-A. In *Neurobiology of Arachnids* (F. G. Barth, ed.), pp. 230–248. Springer-Verlag, Berlin, 1985. Spider proprioception: Receptors, reflexes and control of locomotion.

301. Shaw, E. A. G. In *Handbook of Sensory Physiology* (W. D. Keidel and W. D. Neff, eds.), Vol. 5, Part 1, pp. 455–490. Springer-Verlag, Berlin, 1974. The external ear.

302. Shinozawa, T. and M. Kanou. *J. Comp. Physiol. A* 155A:425–493, 1986. Varieties of filiform hairs: Range fractionation by sensory afferents and cercal interneurons of a cricket.

303. Simmons, J. A. *J. Acoust. Soc. Am.* 54:157–173, 1973. The resolution of target range by echolocating bats.

304. Simmons, J. A., W. A. Lavender, B. A. Lavender, C. A. Doroshow, S. W. Kiefer, R. Livingston, A. C. Scallet, and D. E. Crowley. *Science* 186:1130–1132, 1974. Target structure and echo spectral discrimination by echolocating bats.

305. Simmons, J. A., D. J. Howell, and N. Suga. *Am. Sci.* 63:204–215, 1975. Information content of bat sonar echoes.

306. Slepecky, N. and S. C. Chamberlain. *Hear. Res.* 20:245–260, 1985. Immunoelectron microscopic and immunofluorescent localization of cytoskeletal and muscle like contractile proteins in inner ear sensory hair cells.

307. Smith, C. A., M. Konishi, and N. Schuff. *Hear. Res.* 17:237–247, 1985. Structure of the barn owl (*Tyto albo*) inner ear.

308. Smola, U. *Z. Vergl. Physiol.* 70:335–348, 1970. Electrical responses from hair sensilla of locusts.

309. Sobin, A. and A. Flock. *Acta Oto-Laryngol.* 96:407–412, 1983. Immunohistochemical identification and localization of actin and fimbrin in vestibular hair cells in the normal guinea pig and in a strain of the waltzing guinea pig.

310. Spoendlin, H. In *Facts and Models in Hearing* (E. Zwicker and E. Terhardt, eds.), pp. 18–32. Springer-Verlag, New York, 1974. Neuroanatomy of the cochlea.

311. Stephen, R. O. and H. C. Bennet-Clark. *J. Exp. Biol.* 99:279–314, 1982. The anatomical and mechanical basis of stimulation and frequency analysis in the locust ear.

312. Stopp, P. E. and I. C. Whitfield. *J. Physiol. (London)* 158:165–177, 1961. Unit responses from brainstem nuclei in the pigeon.

313. Stout, J. F., C. H. Dehaan, and R. W. McGhee. *J. Comp. Physiol.* 153:509–521, 1983. Attractiveness of the male *Acheta domesticus* calling song to females. I. Dependence on each of the calling song features.

314. Stout, J. F. and F. Huber. *Physiol. Entomol.* 6:199–212, 1981. Responses to

features of the calling song by ascending auditory interneurons in the cricket *Gryllus campestris*.

315. Strelioff, D. and A. Flock. *Hear. Res.* 15:19–28, 1984. Stiffness of sensory cell hair bundles in the isolated guinea pig cochlea.

316. Suga, N. *J. Physiol. (London)* 179:26–53, 1965. Analysis of frequency-modulated sounds by auditory neurons of echolocating bats.

317. Suga, N. In *Dynamic Aspects of Neocortical Function* (G. M. Edelman, W. E. Gall, and W. M. Cowan, eds.), pp. 315–373. Wiley, New York, 1984. The extent to which biosonar information is represented in the bat auditory cortex.

318. Suga, N. and P. H.-S. Jen. *Science* 194:542–544, 1976. Disproportionate tonotopic representation for processing CF-FM sonar signals in the mustache bat's auditory cortex.

319. Suga, N. and W. E. O'Neill. *Science* 206:351–353, 1979. Neural axis representing target range in the auditory cortex of the mustache bat.

320. Suga, N. and P. Schlegel. *Science* 177:82–84, 1972. Neural attenuation of responses to emitted sounds in echolocating bats.

321. Suga, N. and T. Shimozawa. *Science* 183:1211–1213, 1974. Site of neural attenuation of responses to self-vocalized sounds in echolocating bats.

322. Sullivan, W. E. *J. Neurophysiol.* 48:1011–1032, 1982. Neural representation of target distance in auditory cortex of the echolocating bat *Myotis lucifugus*.

323. Sullivan, W. E. and M. Konishi. *J. Neurosci.* 4:1787–1799, 1984. Segregation of stimulus phase and intensity coding in the cochlear nucleus of the barn owl.

324. Takahashi, T., A. Moiseff, and M. Konishi. *J. Neurosci.* 4:1781–1786, 1984. Time and intensity cues are processed independently in the auditory system of the owl.

325. Takasaka, T. and C. A. Smith. *J. Ultrastruct. Res.* 35:20–65, 1971. The structure and innervation of the pigeon's basilar papilla.

326. Tautz, J. and D. C. Sandeman. *J. Exp. Biol.* 88:351–356, 1980. The detection of waterborne vibration by sensory hairs of the chelae of the crayfish.

327. Tautz, J., W. M. Masters, B. Aicher, and H. Markl. *J. Comp. Physiol. A* 144A:533–541, 1981. A new type of water vibration receptor on the crayfish antenna. I. Sensory physiology.

328. Tavolga, W. N. In *Sound Reception in Fish* (A. Schuijf and A. D. Hawkins, eds.), pp. 185–204. Elsevier, Amsterdam, 1976. Recent advances in the study of fish audition.

329. Terhune, J. M. *J. Acoust. Soc. Am.* 56:1862–1865, 1974. Directional hearing of a harbor seal in air and water.

330. Theurich, M., G. Langer, and H. Scheich. *Neurosci. Lett.* 49:81–86, 1984. Infrasound responses in the midbrain of the guinea fowl.

331. Thorson, J., T. Weber, and F. Huber. *J. Comp. Physiol.* 146:361–378, 1982. Auditory behavior of crickets. II. Simplicity of calling song recognition in *Gryllus*, and anomalous phonotaxis at abnormal frequencies.

332. Thurm, U. *Cold Spring Harbor Symp. Quant. Biol.* 30:75–82, 83–94, 1965. An insect mechanoreceptor. I. Fine structure and adequate stimulus. II. Receptor potentials.

333. Tilney, L. G., E. H. Engelman, D. J. DeRosier, and J. C. Saunders. *J. Cell Biol.* 96:822–834, 1983. Actin filaments, stereocilia and hair cells of bird cochlea. II. Packing of actin filaments in the stereocilia and in the cuticular plate and what happens to the organization when the stereocilia are bent; Tilney, L. G., D. J. DeRosier, and M. J. Mulroy. *Ibid.* 86:244–259, 1980. The organization of actin filaments in the stereocilia of cochlear hair cells.

334. Tilney, L. G. and M. S. Tilney. *Hear. Res.* 22:55–77, 1986. Functional organization of the cytoskeleton.

335. Turner, R. G. In *Comparative Aspects of*

Hearing in Vertebrates (A. N. Popper and R. R. Fay, eds.), pp. 205–237. Springer, New York, 1980. Physiology and bioacoustics in reptiles.

336. Turner, R. G., A. A. Muraski, and D. W. Nielsen. *Science* 213:1519–1521, 1981. Cilium length: Influence on neural tonotopic organization.

337. van Bergeijk, W. A. *Contrib. Sens. Physiol.* 2:1–49, 1967. The evolution of vertebrate hearing.

338. Vedel, J. P. *Comp. Biochem. Physiol. A* 80A:151–158, 1985. Cuticular mechanoreception in the antennal flagellum of the rock lobster *Palinurus vulgaris*.

339. von Frisch, K. *Z. Vergl. Physiol.* 25:703–747, 1938. Über die Bedeuting des Sacculus und des Lagena für den Gehörsinn des Fische.

340. Walkowiak, W. *J. Comp. Physiol. A* 155A:57–66, 1984. Neuronal correlates of the recognition of pulsed sound signals in the grass frog.

340a. Weber, H. *Lehrbuch der Entomologie*. Fischer, Jena, 1933.

341. Weiss, T. F. *Annu. Rev. Physiol.* 46:247–259, 1984. Relation of receptor potentials of cochlear hair cells to spike discharges of cochlear neurons.

342. Weiss, T. F., W. T. Peake, A. Ling, and T. Holton. In *Evoked Electrical Activity in the Auditory Nervous System* (R. F. Naunton and C. Fernandez, eds.), pp. 91–112. Academic Press, New York, 1978. Which structures determine frequency selectivity and tonotopic organization of vertebrate cochlear nerve fibers? Evidence from the alligator lizard.

343. Wendler, G., M. Dambach, B. Schmitz, and H. Scharstein. *Naturwissenschaften* 67:99–100, 1980. Analysis of the acoustic orientation behavior in crickets (*Gryllus campestris* L.).

344. Westin, J. *J. Comp. Physiol. A* 133A:97–102, 1979. Responses to wind recorded from the cercal nerve of the cockcrach *Periplanata americana*. I. Response properties of single sensory neurons.

345. Wever, E. G. In *Handbook of Sensory Physiology* (A. D. Keidel and W. D. Neff,

eds.), Vol. 5, Part 1, pp. 423–454. Springer-Verlag, Berlin, 1974. The evolution of vertebrate hearing.

346. Wever, E. G. *The Reptile Ear*. Princeton Univ. Press, Princeton, NJ, 1978.

347. Wiederhold, M. L. *Annu. Rev. Biophys. Bioeng.* 5:39–62, 1976. Mechanosensory transduction in "sensory" and "motile" cilia.

348. Wiederhold, M. L. and N. Y. S. Kiang. *J. Acoust. Soc. Am.* 48:950–965, 1970. Effects of electrical stimulation of the crossed olivocochlear bundle on single auditory nerve fibers in the cat.

348a. Wiersma, C. A., ed. *Physiology of Invertebrate Nervous Systems*. Univ. of Chicago Press, Chicago, IL, 1967.

349. Wiese, K. *J. Comp. Physiol. A* 92A:317–325, 1974. The mechanoreceptive system of prey localization in *Notonecta*. II. The principle of prey localization.

350. Wilczynski, W. and R. R. Capranica. *Prog. Neurobiol.* 22:1–38, 1984. The auditory system of anuran amphibian.

351. Wiley, R. H. and D. G. Richards. In *Acoustic Communication in Birds: Production, Perception and Design Features of Sounds* (D. E. Kroodsma and E. H. Miller, eds.), pp. 132–181. Academic Press, New York, 1982. Adaptations for acoustic communication in birds: Sound transmission and signal detection.

352. Witschi, E. *Z. Naturforsch. B: Anorg. Chem., Org. Chem., Biochem., Biophys., Biol.* 4B:230–242, 1949. The larval ear of the frog and its transformation during metamorphosis.

353. Wohlers, D. W. and F. Huber. *J. Comp. Physiol. A* 146A:161–173, 1982. Processing of sound signals by six types of neurons in the prothoracic ganglion of the cricket, *Gryllus campestris* L.

354. Yodlowski, M. L., M. L. Kreithen, and W. T. Keeton. *Nature (London)* 265:725–726, 1977. Detection of atmospheric infrasound by pigeons.

355. Young, D. *J. Exp. Biol.* 88:407–411, 1980. The calling song of the bladder cicada, *Cystosoma saundersii*: A computer analysis.

356. Young, D. and R. K. Josephson. *J. Comp. Physiol. A* 152A:183–195, 1983. Mechanisms of sound-production and muscle contraction kinetics in cicadas.

357. Zaretsky, M. D. and E. Eibl. *J. Insect Physiol.* 24:87–95, 1978. Carrier frequency-sensitive primary auditory neurons in crickets and their anatomical projection to the central nervous system.

358. Zenner, H. P. *Arch. Oto-Rhino-Laryngol.* 230:81–90, 1981. Cytoskeletal and muscle-like elements in cochlear hair cells.

359. Zhantiev, R. D. *Zool. Zh.* 50:507–514, 1971. Frequency characteristics of tympanal organs in grasshoppers (*Orthoptera, Tettigoniidae*) (in Russian).

360. Zucker, E. and W. I. Welker. *Brain Res.* 12:138–156, 1967. Coding of somatic sensory input by vibrissae neurons in the rat's trigeminal ganglion.

Chapter 6 | Electric Organs and Electroreceptors

Albert S. Feng

The first bioelectric potentials observed by man were the discharges from electric fishes. The electric catfish *Malapterurus* was figured on Egyptian tombs (~2750 B.C.), and the Romans named *Torpedo*, an electric ray (53). The electric eel *Electrophorus* was brought to Europe by South American explorers and examined by Europeans. The shock from these fishes was considered to be similar to lightning or to the electrostatic discharge from a Leyden jar. In 1773, John Walsh showed that the shock is conducted through metals but not through glass or air. Faraday measured the discharge of *Electrophorus* with a galvanometer and remarked that if the nature of the electric discharge were understood, one might "reconvert the electric into the nervous force."

Electric organs present a prime example of evolutionary convergence. Electric organs have apparently evolved at least six times through modification of body tissues (muscles or neurons in various fish species), and their functional properties are adapted to ecological niches (4,

16). Some families of fish, such as marine electric rays, torpedos, stargazers, and freshwater electric eels, emit high voltage electric organ discharges (EODs) that ward off predators or stun prey (11); these fishes are called strongly electric fish. Other families of fish produce weak pulsatile EODS continuously (freshwater mormyrids, gymnotids, gymnarchids, etc.) or sporadically (marine skates) for electrolocation (39) or for communication (46); these are called weakly electric fish and many of them have the capacity both to generate and to sense weak electric current. Weakly electric fish are often referred to as "pulse" or "wave" species depending on the waveform and rhythm of the EOD; a pulse species produces brief EOD pulses separated by long intervals, whereas a wave species elicits a quasisinusoidal EOD.

In all marine electric fishes the discharge consists of postsynaptic potentials (PSPs) that add arithmetically in series. The electrocyte membranes are not electrically excitable but the innervated face is stimulated by acetylcholine (ACh); the noninnervated face has low resistance or shows delayed rectification. The external current is high in low-resistance seawater; the current is low but voltage is high in freshwater species.

Electroreception has evolved at least three times (23). Electroreceptors are present in primitive vertebrates (ancestral to all nonteleost fishes), lost by the intermediate ancestors of holosteans and teleosts, and reappearing at least twice in teleosts (25, 76).

Production of Electric Organ Discharges

Cells that are specialized for generation of EODs are called electrocytes, or electroplaques, irrespective of their embryological origins. In all marine fishes, only one face of each plaque is innervated and it generates postsynaptic potentials. In contrast, both faces of electrocytes of freshwater mormyrids are innervated and excitable. In other freshwater fishes such as gymnotiforms, one or both faces are excitable (16, 17). In general, the noninnervated surface of many electrocytes is highly convoluted, and the increase in membrane area provides for high conductance or capacitance. Electrocytes of freshwater fishes have spike generating membranes, while those of marine species do not.

In strongly electric fishes and some of the weakly electric fishes, the EOD contains a direct current (dc) component. Figure 1 shows the diversity of EOD waveform and rhythm. The EOD waveform is governed by the excitation pattern of the electric organ(s). The rhythm, however, depends on the activity of the pacemaker neurons in the brain that innervate the electric organs.

In the *Torpedo*, a marine elasmobranch, each organ consists of some 45 dorso-ventral columns of electroplaques (700/col-

umn). In each column, the electroplaques are connected in series; the ventral face of each plaque is innervated (nicotinic cholinergic) and generates graded, nonpropagating postsynaptic potentials (epsps). The organ is a rich source of acetylcholone esterase. The dorsal surface is not innervated and has a lower resistance than the ventral face as does the skin overlying the electric organ so that the current can flow along the plaques in a column although only one face is depolarized (Fig. 2). The serial arrangement allows a production of EOD of 20 to 50 (!) V, which in seawater generates a current of several amperes. In marine skates (*Rajidae*), the electric organs are located in the base of the tail (Fig. 2B). The innervated face of each plaque produces an epsp of long duration (50 ms). The noninnervated face shows delayed rectification, thus producing some increase in the external current. Firing of thousands of electroplaques is furthermore synchronized by central and peripheral mechanisms that result in the high voltage and current (16, 17).

The strongly electric eel *Electrophorus* from the Amazon has several electric organs: The main organ has 1000 plaques/column oriented anteroposteriorly. Each discharge contains a train of 3 to 5 (sometimes up to 20) pulses and these pulses can attain 400 V in an open circuit or a current of 1 A in a short circuit. A single plaque has a resting potential of 90 mV (inside negative). Stimulation of the nerve gives a graded epsp, which may elicit a Na^+-activated spike of 60 mV at the posterior surface of the innervated face; hence, the two surfaces in series give a total potential of 150 mV (Fig. 2C). In the strongly electric African catfish *Malapterurus*, a single giant neuron innervates the several million electrocytes on each side of the body. Each electrocyte has a stalk on the innervated face that generates a small brief spike; this is followed by a

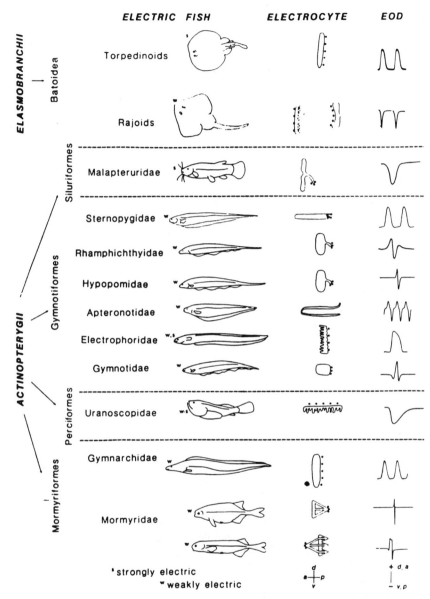

Figure 1. Summary of known groups of electric fish, their electrocytes, and form of electric organ discharges. [From (4).]

large spike of longer duration, sometimes with two peaks, on the noninnervated face (Fig. 2E) (9).

The EOD of strongly electric fish is primarily used to stun prey. The stargazer, *Astroscopus*, lies in the sand with only its eyes exposed; when a small fish approaches, it gives a high-frequency burst before or during opening its mouth, and then after 100 ms it gives a discharge train lasting several seconds (70). The electric organs of *Astroscopus* are modified extraocular muscles. The innervated (dorsal surface) of each electrocyte gives a negative epsp (10–20 mV) that is a cholinergic response. The electric discharge of *Tor-*

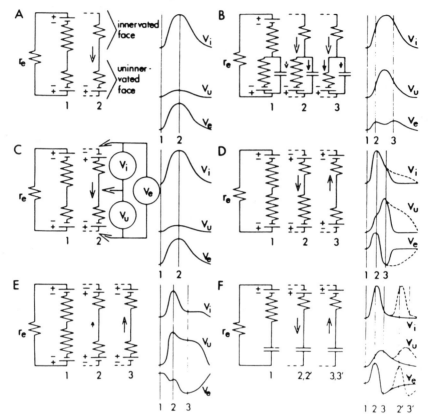

Figure 2. Diagrams of equivalent circuits of different kinds of electrocytes and of recordings across the innervated surface (V_i), across the uninnervated surface (V_u), and across the entire cell (V_e). The successive changes in the membrane properties are shown by numbered branches of the equivalent circuits and their times of occurrence are indicated on the potentials. (A) A strongly electric marine fish; the innervated surface generates an epsp. (B) Electrocyte of a rajid fish; the innervated surface generates an epsp and uninnervated surface shows delayed rectification. (C) Electrocyte of electric eel and some gymnotids; the innervated face generates overshooting spike. (D) Electrocyte of mormyrids and some gymnotids; both faces generate spike, and V_e is diphasic. (E) Electrocyte of electric catfish; the stalk has high threshold and generates a small spike, and V_e is negative on the nonstalk side. (F) Electrocyte of *Gymnarchus*; uninnervated face acts as a series capacity. [From (16).]

pedo starts 80 ms after the first movement toward capture of prey (detected by torpedo's mechanoreceptors); just before "landing" the frequency is 140–290 pulses/s, afterward it is irregular and slow. Prey can be immobilized at a distance of 15 cm (10).

The electrocytes of weakly electric fish produce spikes; different species show different properties of the two surfaces so that spikes may be diphasic or triphasic. In *Gymnotus*, both faces are excitable and each gives a spike response (Fig. 2D). However, the innervated face fires first so that the total response is diphasic. In *Gymnarchus*, an African freshwater fish, the innervated face generates a spike; the noninnervated face has a high capacitance and resistance (Fig. 2F). Current from the innervated face charges the series capacitor that then discharges passively, producing a second peak at a later time. In several species, the electric organ is embryologically derived from neurons instead of muscles.

In *Apteronotus,* spinal nerves form an organ ventral to the vertebral column; many nerve fibers run forward, then loop and pass posteriorly, and end blindly. The EOD in *Apteronotus* is diphasic, and results from synchronous spikes in the nerve fibers; the spikes are superimposed on a dc potential. In some mormyrid fishes, the electrocyte forms a complex stalk that branches and receives rich innervation. The action potential of the noninnervated face is larger and longer than that of the innervated face, so that the second phase of the spikes predominate (78).

In addition to the difference in EOD waveform, the EOD frequency also differs with species, sex, and behavioral state. *Eigenmannia* discharges at 240–600 Hz, while *Apteronotus* fires at ~1200 Hz and *Gymnarchus* has an EOD of 250 Hz. In mormyrids, the basal rate is 1–15 pulses/s, but when actively pursuing prey the rate may increase up to 130/s. The genus *Sternopygus* is sexually dimorphic (44). The EOD of mature female range from 120–240 Hz, which is about one octave higher than the EOD of mature males. Injections of androgens over the course of 2 weeks lower the EOD frequency in either sex by as much as 10% (5, 63). Androgen also influences the tuning characteristics of peripheral electroreceptors, through indirect effects, so that the sensory system matches the motor output at all times (6, 54, 64).

Electroreception

All animals generate low-frequency electrical potentials by asymmetric secretory and neuromuscular actions. Such potentials are produced inadvertently, for example, in respiration or swimming movements; this is in contrast to the EODs, which are generated for specific purposes. There are two major classes of sensory organs that are specialized in detecting minute EODS or other bioelectric potentials (Fig. 2): (1) the ampulla of Lorenzini or similar organs, which are found in almost all nonteleost fishes and in some teleosts (called ampullae of teleosts), as well as in several amphibian species, and (2) tuberous organs, which are found in nonteleosts and teleosts (19, 20, 23, 33–35, 68, 71). Ampullary and tuberous organs differ in morphology as well as function and frequently both occur in the same fish.

Ampullary receptors are diverse in morphology, central projection and sensing mechanism (23, 79), and are specialized to detect low-frequency (0.1–50 Hz) incidental voltages from animate (other than electric organs, e.g., breathing muscle potentials) and inanimate sources. The ampullary organ consists of a short (Fig. 3C) (in freshwater species) or long (Fig. 3D) (in marine species) jelly-filled canal with receptor cells situated in the proximal end of the canal in an enlarged ampullary lumen. The tuberous organ, in contrast, resides in an invagination of the epidermal basement membrane, is typically not associated with a canal, and has epithelial cells separating it from the exterior (79, 80). The tuberous organ is sensitive to voltage fluctuation (such as fish's EOD) at a high frequency (50–2000 Hz) and its firing is in synchrony with the EOD.

The ampullary organs play ubiquitous roles in the wide range of taxa where they occur. Sharks, rays, and skates detect a voltage gradient as small as 0.0005 μV/cm, which is sufficient for detecting the gradient induced by swimming within the earth's magnetic field. This detection

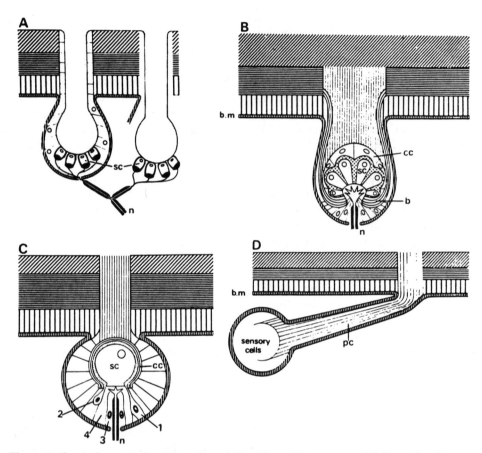

Figure 3. Semischematic drawings of specialized lateral line organs. (*A*) Ampullary (gymnotid and mormyrid), (*B,C,D*) tuberous organs, (*B* and *D*) gymnotid; (*C*) mormyrid fish. (bm) basement membrane. Skin: (a) and (b) layers, vertically hatched; layers (c) and (d) horizontally and obliquely hatched; covering cell (cc), white. (It is still not clearly established whether the multilayered basal part (b) of the perisensory space is formed by the particularly structured covering cells or by special cells of the (c) layer of the skin.) Plug cells in (*B*) and (*C*), dense vertically hatched area. Apical and basal accessory cells in (*A*), white; in (*D*), only plug cells (pc) are indicated. Basal accessory cells in (*B*), white. Differentiated basal accessory cells in (*C*): ramified (1), bottle shaped (2), perinervous (3) and dark supporting (4) elements. Perisensory space (= intraepidermal cavity) in (*B*) and (*C*) as well as canal and ampulla in (*A*), stippled; (n) myelinated nerve fiber (nerve endings in (*A*) black); (sc) sensory cell. [From (79).]

capacity thus provides them a means of using the magnetic cue for orienting in the ocean (52). Furthermore, these animals as well as some teleosts (siluriforms and gymnotiforms) can locate prey by detecting the minute muscle or secretory potential produced by the prey (26, 50, 51,

72). *Scyliorhinus* and *Raja*, can show conditioned circulatory responses to the presentation of muscle potential emanating from a flatfish tens of centimeters away. A dogfish shows behavioral response to a voltage gradient of 0.02 μV/cm, *Raja* to 0.01 μV/cm; these responses are me-

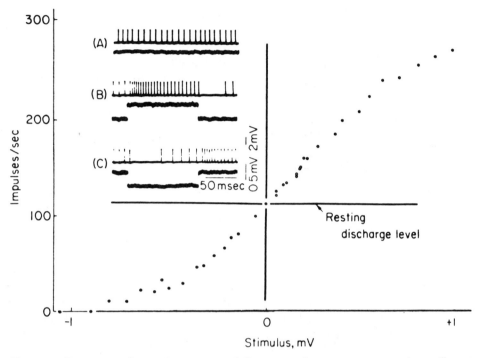

Figure 4. Response of a tonic receptor of *Gymnotus*. Inset upper trace gives afferent impulses, and lower trace gives stimulating potential at opening of receptor. (*A*) Spontaneous discharge; (*B*) anodal stimulation; and (*C*) cathodal stimulation. Graph gives average impulse frequency as a function of stimulation voltage; anodal at right and cathodal at left. [From (16).]

diated by the ampullary organs, since animals respond to the electrical potential when other receptors are excluded (30, 50). Ampullary organs are abundant in the head region. Afferents from the ampullary organs generally fire spontaneously at ~40–100 impulses/s; a cathodal stimulus at the canal opening increases the firing frequency whereas an anodal pulse (hyperpolarizing the inner face of the receptor cell) decreases the frequency (1, 2, 15, 59, 67).

In teleosts, ampullary fibers are, in contrast, excited by anodal current (Fig. 4) (1, 2, 86). While the primary function is electroreception, these organs are also sensitive to temperature, tactile stimuli, CO_2, and salinity (1–3, 21). The firing rate of ampullary fibers also varies with a change in magnetic field (22).

Both ampullary and tuberous (phasic) give generator potentials and chemically activate nerve endings at the base of the receptor cell. In elasmobranchs, the depolarizing receptor potential of the ampullary organ is attributed to an inward Ca^{2+} current at the apical membrane, which is followed by a late outward K^+ current (18, 28, 29). In contrast, in teleosts, the depolarizing receptor potential of the ampullary organ is due to an inward Ca^{2+} current at the basal membrane. Hence, the former is excited by a cathodal current at the pore of the skin, the latter by an anodal current.

Tuberous receptors, found in all weakly electric fishes, are used in: (1) active detection, discrimination, and localization of objects by means of sensing local changes in EOD field intensity (8,

the electrosense (31, 61). An object in the vicinity distorts the fish's EOD and the distortion varies along different parts of the body surface (Fig. 5) providing information about the position of the object relative the the fish (7). In the mormyrid *Gnathonemus* the electric organs of left and right sides are fired alternately; the asymmetry of their EOD fields gives depth perception (27). Weakly electric fish can additonally discriminate objects by detecting differences in resistance and capacitance of objects in their field (32, 62, 74).

Some weakly electric fish produce several forms of EODs according to behavioral contexts, differing in waveform, power spectrum, rate, and amplitude modulations. Various mormyrid and gymnotid species produce different EOD wave forms; individuals recognize conspecific waveforms (41, 47–49, 59, 66).

A pronounced social behavior, the jamming avoidance response (JAR), is observed in many of the wave species when they are presented with a jamming stimulus, that is, sinusoidal stimulus at a slightly higher or lower frequency than the fish's own EOD frequency; the optimal stimulus is a ΔF of 4 Hz (24, 40, 82). The behavior is characterized by a shift in EOD frequency away from the stimulus frequency until the difference is 10–15 Hz (Fig 6). Thus within a social group each fish is able to maintain a private EOD frequency for electrolocation. The behavioral cue for JAR lies in the difference of "contamination" of the fish's own EOD by a foreign EOD (or stimulus) in different parts of the skin. The sign of the ΔF is computed from the information provided by the afferents without internal reference to the pacemaker (40, 42). Pulse species, such as mormyrid fishes, also display echo response to minimize the coincidence of the fish's own EOD with that of its neighbors; this response is characterized by occasional firing 10–12 ms after a

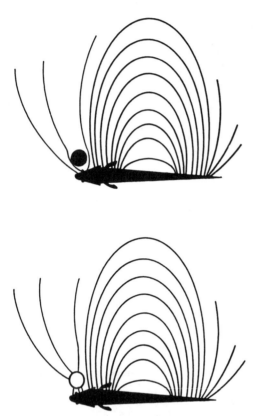

Figure 5. The electric field on one side of an electric fish, shown by current lines, and the distortions due to objects of low conductivity (upper) and high conductivity (lower). [From (60).]

31, 37, 38, 61, 77), or (2) passive electrolocation and reception of frequency modulations of EODs of neighboring fish during social and agonistic encounters (13, 36, 46, 56–58, 65, 84, 85). These fishes live in turbid water and hence have to rely heavily, if not entirely, on nonvisual senses to detect and locate objects. An electrosensory system is shown to be involved in these fishes, and their capacity to detect and locate objects deteriorates rapidly if their EODs are jammed electrically (37, 38). Normally, these fishes maintain a constant postural relationship with surrounding objects, such as plants (37), or the ventral substrate, by

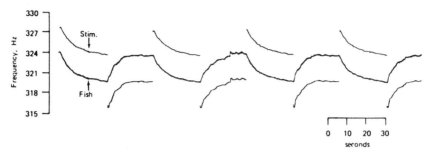

Figure 6. The jamming avoidance response (JAR) of *Eigenmania* stimulated at ~30 dB above threshold with alternately higher and lower stimulus frequencies (clamped) at (ΔF) (difference from the fish = 4 Hz). The alternation occurs automatically about every 25 s. [From (24).]

neighbor's discharge (13, 55). Gymnotiform pulse species fire at a more regular rate than mormyrids. Two gymnotoid fish at not very different repetition rates avoid coincidence by firing at progressively changing phases; each fish scans through advancing or retarding phases relative to the other (23).

The tuberous receptors can be divided functionally into P and T units (75). The P units fire probabilistically and change their firing probability to each stimulus cycle with changing stimulus amplitude. The T units, in contrast, fire one spike that is time-locked to each stimulus cycle and the timing of this spike varies with stimulus strength. These receptors possess V-shaped tuning curves and the best frequency of the receptor matches the predominant spectral frequency of the individual EOD cycle (45, 81). One basis of frequency tuning is electrical resonance of the receptor; an impulse stimulus gives rise to a damped oscillation with a frequency that is highly correlated with the best frequency of the receptor (81, 83). The two types (P and T) of tuberous re-

ceptors have distinct ascending projection patterns (Fig. 7) (42). At the posterior lateral line lobe there are multiple body representations resulting from multiple electroreceptor projections.

Other functional classes of tuberous receptor have been identified (23). Mormyrids have three types of electroreceptors: knollenorgan, ampullary, and mormyromast. Each receptor gives rise to 1–2 somatotopic projections to the electrosensory lateral lilne lobe (ELL), which in turn projects to the lateral torus semicircularis (14). Knollenorgan afferents provide temporal information, which is crucial for sensing the EODs of other fishes during electrocommunication. Such sensing is possible only when a corollary discharge of the EOD motor command is present; when the corollary discharge is removed, knollenorgan afferents respond to the fish's own EOD but not to those of others (12).

The neural bases of electrolocation and JAR have been extensively studied (8, 23, 40, 42). Recordings from the ELL show that there are two functional categories of

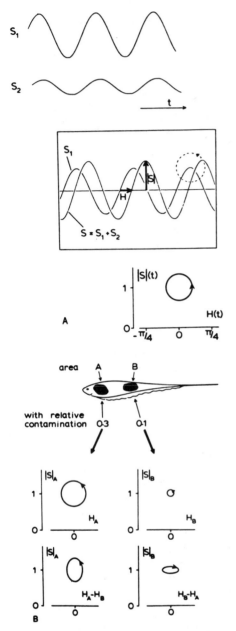

pyramidal neurons in this structure: excitatory (*E*) and inhibitory (*I*) cells. These cells are more sensitive than the primary afferents to small changes in the amplitude of the EOD. More rostrally along the electrosensory pathway, the receptive fields become more complex, sensitivity to movement appears and the range between the fish and object within which neurons respond increases, and in the cerebellum each neuron responds optimally to a specific distance for a given object. Thus there seems to be a hierarchical processing along the pathway to encode the spatial location of objects (8).

The neuronal basis of JAR has also been investigated (40). The fish obtains information about the sign of the difference frequency (ΔF) by simultaneously evaluating the amplitude and phase modulations resulting from the interference of the fish's own EOD by the neighbor's EOD (Fig. 8). Amplitude modulations are primarily encoded by the firing rate of *P*-type tuberous receptors, whereas phase modulation, or timing of zero crossings of the quasisinusoidal signal, is primarily encoded by the time of firing of *T*-type receptor (Fig. 9). In the ELL, the E and I subtype of *P* cells are excited by a rise and fall of stimulus amplitude, respectively. The phase information is processed by a different cell type in the ELL, the spherical cells, which later relay the timing information to layer 6 of the torus semicircularis (Fig. 10). Small toral neurons within this layer compute (compared pairwise) the difference in phase between

Figure 7. (A) Diagrammatic representation of an experiment on curarized *Eigenmannia*. Stimulus (*S*) represents normal electric discharge, S_2 represents a smaller signal of slightly different frequency coming from another fish. Below is simultaneous display of mixed signals. $S_1 + S_2$ yields a sinusoidal signal with momentary amplitude ($|S|$) and phase angle (*H*) modulated according to ΔF, the difference between frequencies of S_1 and S_2. As a result of

the modulation of *S* and *H* the peak of $S_1 + S_2$ rotates. Below is a Lissajous display of *S* versus *H* in a two-dimensional plane. (B) Figure of fish shows two areas of skin where receptors respond differently to *S* and *H*. Upper two Lissajous figures show responses to *S* and *H* of the two areas separately, lower figures show relative responses of the two areas together. [From (40).]

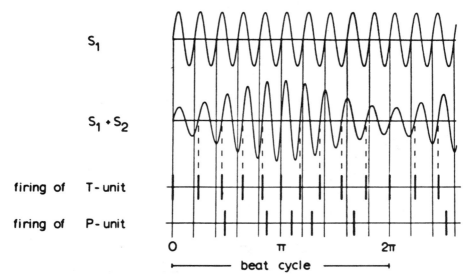

Figure 8. Schematic representation of T and P receptor responses to a beat pattern resulting from the addition of two sine waves, S_1 and S_2, with S_1 being the dominant signal. The T units fire one spike per EOD cycle. Phase locked to zero crossing of this signal and timing of this spike codes the phase of a sinusoidal stimulus. The P units code modulation of amplitude and fire more frequently as current passes through a given amplitude in the positive direction. [From (40).]

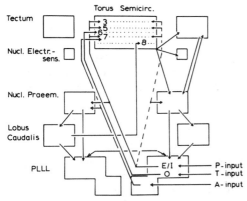

Figure 9. The central neural pathways of electroreception. Primary afferents from ampullary receptors (A input) project to medial portion of the posterior lateral line lobe (PLLL); the tuberous afferents (P and T inputs) project to the central and lateral portions of the PLLL. The T afferents contact spherical cells (O) in PLLL that project to lamina 6 of the torus. The P afferents go to basilar pyramidal cells (E) and nonbasilar pyramidal cells (I), which project to laminae 3, 5, and 7 of torus and to caudal lobe of cerebellum, which in turn projects back to PLL. Most projections of

PLL are contralateral, a few are ipsilateral (dashed line). The lobus caudalis of cerebellum projects to contralateral lamina of torus and torus projects to tectum, nucleus electrosensorius, and nucleus praeeminetialis. [From (42).]

two body areas (Fig. 11); this information is relayed to deeper torus laminae and to the optic tectum (Fig. 9). It is in the torus where the amplitude modulation and phase difference information converges and is used to compute the ΔF (43). These toral cells project to the nucleus electrosensorius complex in diencephalon, which controls the output of prepacemaker nucleus. The prepacemaker nucleus gives the sole input to the pacemaker nucleus in the medulla, which ultimately controls the firing frequency of the EOD such that a negative ΔF elicits an increase, and a positive ΔF a decrease, in pacemaker firing.

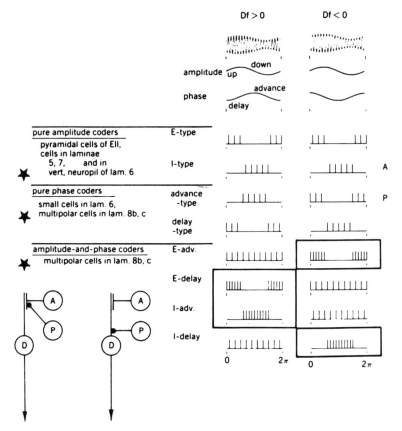

Figure 10. The classification of toral cells as coders of amplitude and phase information contained in beat patterns (top) for positive and negative ΔF values. E delay and I advance cells reach their highest spike densities for positive ΔF, while E advance and I delay cells reach their highest spike densities for negative ΔF values (patterns framed by thick lines). Pure amplitude coders in lamina 5 and in the vertical neuropile of lamina 6, as well as phase and amplitude and phase coders in laminae 8b and 8c, project to the optic tectum (* at the left). The insert in the lower left proposes how an amplitude coder (A) and a phase coder (P) could influence a tectal neuron (D) to become ΔF sign selective. Mechanisms of pre- (left) or postsynaptic (right) inhibition could yield a tectal neuron that is silent for positive ΔF values and active for negative ΔF values. [From (40).]

Detection of Magnetic Fields

The earth's magnetic field in the eastern United States is 0.5 G or 50,000 γ; the lines of the field are vertical at the poles and horizontal at the equator. Orientation with respect to magnetic fields has been observed in many organisms but the mechanism of detection remains obscure.

Some muddwelling bacteria orient in a magnetic field of 0.5 G or higher. When in suspension they normally swim north in the northern hemisphere and reverse their direction if the field is reversed. These bacteria contain crystals of magnetite (Fe_3O_4) (18a, 52a).

Bees use the earth's magnetic field as one of several kinds of sensory input in dance communication and in building comb. On a vertical surface they orient by

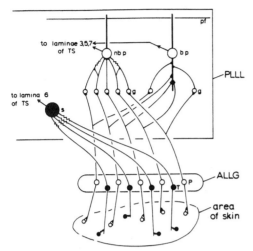

Figure 11. The synaptic organization of primary tuberous electroreceptor afferents in the PLLL. *T*-type afferents (●) form electrical synapses upon spherical cells (s), which in turn project to lamina 6 of the torus (TS) (see Fig. 9). *P*-type afferents (○) form excitatory chemical synapses upon granule cells (g) and upon basilar dendrites of basilar pyramidal cells (bp). Basilar pyramidal cells also receive inhibitory input from more distant granule cells, which in turn are excited by *P*-unit inputs from the periphery of the receptive field. The total *P*-type input to the basilar pyramidal cell thus appears to have an excitatory center and inhibitory surround organization. Nonbasilar pyramidal cells (npb), on the other hand, receive inhibitory input from the nearest granule cells and electrical synapses from more distant granule cells, with the latter inputs being inhibited by more centrally located granule cells. The total *P*-type input to a nonbasilar pyramidal cell thus appears to have an inhibitory center and excitatory surround organization. Both types of pyramidal cells project most heavily to laminae 3, 5, and 7 of the torus (TS) (see Fig. 9). The dorsal dendrites of pyramidal cells are contacted by parallel fibers (pf) that originate in the lobus caudalis of the cerebellum. ALLG is the anterior lateral line nerve ganglion that houses the somata of primary afferent cells. [From (42).]

polarized light, by the angle of the sun, and (on cloudy days) by gravity. On a horizontal field they normally use visual cues but if light is excluded they become disoriented initially. Then, after several days, they dance with respect to the earth's magnetic field. Bees contain many crystals of magnetite in their abdomens. However, when demagnetized, they still orient. It is postulated that magnetite is not used for detection of the earth's field (35a, 54a).

Elasmobranch fishes—sharks, skates, and rays—are sensitive to electric fields as low as 0.01 μV/cm in seawater. They orient in an artificially generated magnetic field when tested in nonmagnetic tanks. It is postulated that these fishes make use of induced electric fields when navigating in ocean currents and that they can use induced currents to detect direction. Detection is by the ampullae of Lorenzini (52a, 52).

The earth's magnetic field has been shown to influence the migration and/or orientation behavior of several species of migratory birds (9b, 35b, 66a, 85a, 85b) and a species of rodent (60b), some migratory fishes (70a, 81b), as well as two species of amphibians (69a, 70b). Homing pigeons normally use visual landmarks for flights. However, their flight may become disoriented if small magnets are attached to them or if they fly over regions where the earth's field is distorted (35b). Deposits of 10^8 crystals of magnetite occur between the olfactory bulb and optic chiasm of homing pigeons (81a). Magnetite crystals have been found between the nasal cavity, the sthmold bone, and the orbit in birds that make long migration flights. Electrical recordings from the pineal gland (29a) and the nucleus of the basal optic root (77a) of pigeons show responses to magnetic fields. In contrast, in the bobolink, magnetic-evoked responses can be found in the trigeminal nerve (9a).

Neurophysiological evidence for detection of magnetic fields has been obtained with the marine opisthobranch mollusc *Tritonia*. In the laboratory these snails orient eastward in the earth's geomagnetic field and move randomly when the field is canceled. Turning in a Y-maze in a magnetic field changes according to the phase of the moon. One neuron in the pedal ganglion shows enhanced electrical activity when a magnetic field is imposed (60a).

Detection of magnetic fields has been proven by behavioral observations. The detection of induced electric fields by elasmobranchs is reasonably established. However, the nature of other detectors is uncertain; several possibilities, such as paramagnetic organic molecules in organelles, have been suggested (54a).

References

1. Akoev, G. N., O. B. Ilyinsky, and P. M. Zadan. *Comp. Biochem. Physiol. A* 53A:201–209, 1976. Physiological properties of electroreceptors of marine skates.

2. Akoev, G. N., O. B. Ilyinsky, and P. M. Zahdan. *J. Comp. Physiol. A* 106A:127–136, 1976. Responses of electroreceptors (ampullae of Lorenzini) of skates to electric and magnetic fields.

3. Akoev, G.N., N. O. Volpe, and G. G. Zhadan. *Comp. Biochem. Physiol. A* 65A:193–201, 1980. Analysis of effects of chemical and thermal stimuli on the ampullae of Lorenzini of the skates.

4. Bass, A. H. In *Electroreception* (T. H. Bullock and W. Heiligenberg, eds.), pp. 13–70. Wiley, New York, 1986. Electric organs revisited: Evolution of a vertebrate communication and orientation organ.

5. Bass, A. H. and C. D. Hopkins. *Science* 220:971–974, 1983. Hormonal control of sexual differentiation: Changes in electric organ discharge waveforms.

6. Bass, A. H. and C. D. Hopkins. *J. Comp. Physiol.* 155A:713–724, 1984. Shifts in frequency tuning in androgen-treated mormyrid fishes.

7. Bastian, J. *J. Comp. Physiol. A* 144A:465–479, 1981. Electrolocation I: An analysis of the effects of moving objects and other electrical stimuli on the electroreceptor activity of *Apteronotus albifrons*.

8. Bastian, J. In *Electroreception* (T. H. Bullock and W. Heiligenberg, eds.), pp. 577–612, 1986. Electrolocation. Behavior, anatomy, and physiology.

9. Bauer, R. *Z. Vergl. Physiol.* 59:371–402, 1968. Electric discharges in *Malapterurus*.

9a. Beason, R. C. and P. Semm. *Neuroscience* 80:229–234, 1987. Magnetic responses of trigeminal nerve of bobolink.

9b. Beason, R. C. and J. E. Nichols. *Nature (London)* 309:151–153, 1984. Magnetic orientation and magnetically sensitive material in a transequitorial migratory bird.

10. Belbenoit, P. *Z. Vergl. Physiol.* 67:205–216, 1970. Electric discharge in *Torpedo*.

11. Belbenoit, P. and R. Bauer. *Mar. Biol.* 17:93–99, 1972. Video recordings of prey capture behaviour and associated electric organ discharge of *Torpedo marmorata* (Chondrichythyes).

12. Bell, C. C. In *Electroreception* (T. H. Bullock and W. Heiligenberg, eds.), pp. 423–452. Wiley, New York, 1986. Electroreception in mormyrid fish. Central physiology.

13. Bell, C. C., J. P. Myers, and C. J. Russell. *J. Comp. Physiol.* 92A:201–228, 1974. Electric organ discharge patterns during dominance-related behavioral displays in *Gnathonemus petersii*.

14. Bell, C. C. and T. Szabo. In *Electroreception* (T. H. Bullock and W. Heiligenberg, eds.), pp. 375–422. Wiley, New York, 1986. Electroreception in mormyrid fish.

15. Bennet, M. V. L. In *Physiological and Biophysical Aspects of Nervous Integration* (F. D. Carlson, ed.), pp. 73–128. Prentice-Hall, Englewood-Cliffs, NJ, 1968. Similarities between electrically and chemically mediated transmission.

16. Bennett, M. V. L. *Annu. Rev. Physiol.* 32:471–528, 1970. Comparative physiology; electric organs.

17. Bennett, M. V. L. *Fish Physiol.* 5:347–391,

1971. Electric organs and electroreception.

18. Bennett, M. V. L. and S. Obara. In *Electroreception* (T. H. Bullock and W. Heiligenberg, eds.), pp. 157–181. Wiley, New York, 1986. Ionic mechanisms and pharmacology of electroreceptors.

18a. Blakemore, R. P. and R. B. Frankel. 1981. *Sci. Am.* 245(6):58–65. Magnetic Navigation in bacteria; *Nature (London)* 286:384–385, 1981. South seeking magnetotatic bacteria in the southern hemisphere.

19. Bodznick, D. and R. L. Boord. In *Electroreception* (T. H. Bullock and W. Heiligenberg, eds.), pp. 225–257. Wiley, New York, 1986. Electroreception in chondrichthyes: Central anatomy and physiology.

20. Braford, M. R., Jr. In *Electroreception* (T. H. Bullock and W. Heiligenberg, eds.), pp. 453–464. Wiley, New York, 1986. African knifefishes: The Xenomystines.

21. Brown, H. R., V. I. Govardovsky, and O. B. Ilyinsky. *Proc. Int. Union Physiol. Sci.* 14:339A, 1980. Thermal sensitivity of Lorenzinian ampullae.

22. Brown, H. R. and O. B. Ilyinsky. *J. Comp. Physiol. A* 126A:333–341, 1978. The ampullae of Lorenzini in the magnetic field.

23. Bullock, T. H. *Annu. Rev. Neurosci.* 5:121–170, 1982. Electroreception.

24. Bullock, T. H., R. H. Hamstra, Jr., and H. Scheich. *J. Comp. Physiol.* 77A:1–22, 23–48, 1972. The jamming avoidance response of high frequency electric fish. I. General features. II. Quantitative aspects.

25. Bullock, T. H. and W. Heiligenberg, eds. *Electroreception.* Wiley, New York, 1986.

26. Butsuk, S. V. and B. I. Bessonov. *J. Comp. Physiol. A* 141A:277–282, 1981. Direct current electric field in some teleost species: Effect of medium salinity.

27. Chichibu, S. *Adv. Physiol. Sci.* 30:165–178, 1981. EOD time series analysis of *Gnathonemus petersii.*

28. Clusin, W. T. and M. V. L. Bennett. *J. Gen. Physiol.* 69:121–143, 1977. Calcium-activated conductance in skate electroreceptors. Current clamp experiments.

29. Clusin, W. T. and M. V. L. Bennett. *J. Gen. Physiol.* 69:145–182, 1977. Calcium-activated conductance in skate electroreceptors. Voltage clamp experiments.

29a. Demaine, C. and P. Semm. *Neurosci. Lett.* 62:119–122, 1985. The avian pineal gland as an independent magnetic sensor.

30. Dijkgraaf, S. and A. J. Kalmijn. *Z. Vergl. Physiol.* 53:187–194, 1966. Function of ampullae of Lorenzini.

31. Feng, A. S. *J. Neurobiol.* 8:429–437, 1977. The role of the electrosensory system in postural control in the weakly electric fish *Eigenmannia virescens.*

32. Feng, A. S. and T. H. Bullock. *J. Exp. Biol.* 66:141–158, 1977. Neuronal mechanisms for object discrimination in the weakly electric fish *Eigenmannia virescens.*

33. Finger, T. E. In *Electroreception* (T. H. Bullock and W. Heiligenberg, eds.), pp. 287–318. Wiley, New York, 1986. Electroreception in catfish.

34. Finger, T. E., C. C. Bell, and C. E. Carr. In *Electroreception* (T. H. Bullock and W. Heiligenberg, eds.), pp. 465–482. Wiley, New York, 1986. Comparisons among electroreceptive teleosts: Why are electrosensory systems so similar?

35. Fritzsch, B. and H. Muñz. In *Electroreception* (T. H. Bullock and W. Heiligenberg, eds.), pp. 483–496. Wiley, New York, 1986. Electroreception in amphibians.

35a. Gould, J. L., J. L. Kirschvink, K. S. Deffeyes, and M. L. Brines. *J. Exp. Biol.* 86:1–8, 1980. Orientation of demagnetized bees.

35b. Gould, J. L. *Nature (London)* 296:205–211, 1988. Orientation and magnetic detection by birds.

36. Hagedorn, M. In *Electroreception* (T. H. Bullock and W. Heiligenberg, eds.), pp. 497–525. Wiley, New York, 1986. The ecology, courtship, and mating of gymnotiform electric fish.

37. Heiligenberg, W. *J. Comp. Physiol. A* 103A:55–67, 1975. Electrolocation and jamming avoidance in the electric fish

Gymnarchus niloticus (Gymnarchidae, Mormyriformes).

38. Heiligenberg, W. *J. Comp. Physiol. A* 109A:357–372, 1976. Electrolocation and jamming avoidance in the mormyrid fish *Brienomyrus*.

39. Heiligenberg, W. In *Studies of Brain Function* (V. Braitenberg, ed.), Vol. 1, pp. 1–85. Springer-Verlag, New York, 1977. Principles of electrolocation and jamming avoidance in electric fish. A Neuroethological approach.

40. Heiligenberg, W. In *Electroreception* (T. H. Bullock and W. Heiligenberg, eds.), pp. 613–650. Wiley, New York, 1986. Jamming avoidance responses: Model systems for neuroethology.

41. Heiligenberg, W. and J. Bastian. *Acta Biol. Venez.* 10:187–203, 1980. Species specificity of electric organ discharges in sympatric gymnotoid fish of the Rio Negro.

42. Heiligenberg, W. and J. Bastian. *Annu. Rev. Physiol.* 46:561–583, 1984. The electric sense of weakly electric fish.

43. Heiligenberg, W. and G. Rose. *J. Comp. Physiol.* 159:311–324, 1986. Gating of sensory information: Joint computations of phase and amplitude data in the midbrain of the electric fish, *Eigenmannia*.

44. Hopkins, C. D. *Science* 176:1035–1037, 1972. Sex differences in electric signalling in an electric fish.

45. Hopkins, C. D. *J. Comp. Physiol.* 111:171–207, 1976. Stimulus filtering and electroreceptors in three species of gymnotoid fish.

46. Hopkins, C. D. In *Fish Neurobiology* (R. G. Northcutt and R. E. David, eds.), Vol. 1, pp. 215–259. Univ. of Michigan Press, Ann Arbor, 1983. Functions and mechanisms in electroreception.

47. Hopkins, C. D. In *Electroreception* (T. H. Bullock and W. Heiligenberg, eds.), pp. 527–576. Wiley, New York, 1986. Behavior of mormyridae.

48. Hopkins, C. D. and A. H. Bass. *Science* 212:85–87, 1981. Temporal coding of species recognition in an electric fish.

49. Hopkins, C. D. and W. Heiligenberg. *Behav. Ecol. Sociobiol.* 3:113–134, 1978. Ev-olutionary designs for electric signals and electroreceptors in gymnotoid fishes of Surinam.

50. Kalmijn, A. J. *J. Exp. Biol.* 55:371–383, 1971. The electric sense of sharks and rays.

51. Kalmijn, A. J. In *Handbook of Sensory Physiology* (A. Fessard, ed.), Vol. 3, Part 3, pp. 147–200. Springer-Verlag, New York, 1974. The detection of electric fields from inanimate and animate sources other than electric organs.

52. Kalmijn, A. J. In *Animal Migration, Navigation and Homing* (K. Schmidt-Koenig and W. T. Keeton, eds.), pp. 347–353. Springer-Verlag, New York, 1978. Experimental evidence of geomagnetic orientation in elasmobranch fishes.

52a. Kalmijn, A. J. In *Electromagnetic Fields and Neurobehavioral Function* (M. E. O'Connor and R. H. Lovely, eds.), pp. 23–45, 1988. Electromagnetic orientation: A relativistic approach.

53. Kellaway, P. *Bull. Hist. Med.* 20:112–137, 1946. Electric fish in history of electrobiology and medicine.

54. Keller, C. H., H. H. Zakon, and D. Y. Sanchez. *J. Comp. Physiol.* 158A:301–310, 1986. Evidence for a direct effect of androgens upon electroreceptor tuning.

54a. Kirschvink, J. L. *Trends Neurosci.* 5:160–167, 1982. Birds, bees. Two magnetisms.

55. Kramer, B. *J. Comp. Physiol. A* 93A:203–236, 1974. Electric organ discharge interactions during interspecific agonistic behavior in freely swimming mormyrid fish. A method to evaluate two or more.

56. Kramer, B. *Behav. Ecol. Sociobiol.* 1:425–446, 1976. The attack frequency of *Gnathonemus petersii* towards electrically silent (denervated) and intact conspecifics, and towards another mormyrid (*Brienomyrus niger*).

57. Kramer, B. *Behav. Ecol. Sociobiol.* 4:61–74, 1978. Spontaneous discharge rhythms and social signalling in the weakly electric fish *Pollimyrus isidori* (Cuvier et Valenciennes) (Mormyridae, Teleostei).

58. Kramer, B. *Behav. Ecol. Sociobiol.* 6:67–79, 1979. Electric and motor responses of the

weakly electric fish *Gnathonemus petersii* (Mormyridae), to playback of social signals.

59. Kramer, B. and G. K. H. Zupanc. *Naturwissenschaften* 73:679–681, 1986. Conditional discrimination of electric waves differing only in form and harmonic content in the electric fish, *Eigenmannia*.

59a. Kuterbach, D. B., B. Walcott, R. J. Reeder, and R. B. Frankel. *Science* 218:695–697, 1982. Iron containing cells in the honey bee.

60. Lissman, H. W. and K. E. Machin. *J. Exp. Biol.* 35:451–486, 1958. The mechanism of object location in *Gymnarchus niloticus* and similar fish.

60a. Lohmann, K. J. and A. O. Willows. *Science* 235:331–334, 1987. Lunar-modulated geomagnetic orientation by a marine mollusk.

60b. Mather, J. G. and R. R. Baker. *Nature (London)* 291:152–155, 1981. Magnetic sense of direction in woodmice for route-based navigation.

61. Meyer, D. L., W. Heiligenberg, and T. H. Bullock. *J. Comp. Physiol. A* 109A:59–68, 1976. The ventral substrate response. A new postural control mechanism in fishes.

62. Meyer, J. H. *J. Comp. Physiol.* 145A:459–470, 1982. Behavioral responses of weakly electric fish to complex impedances.

63. Meyer, J. H. *J. Comp. Physiol.* 153A:29–37, 1983. Steroid influences upon the discharge frequency of a weakly electric fish.

64. Meyer, J. H., H. H. Zakon, and W. Heiligenberg. *J. Comp. Physiol.* 154A:625–631, 1984. Steroid influences upon the sensory system of weakly electric fish: Direct effects upon discharge frequencies with indirect effects upon electroreceptor tuning.

65. Moller, P. *Science* 193:697–699, 1976. Electric signals and schooling behavior in a weakly electric fish, *Marcusenius cyprinoides* L. (Mormyriformes).

66. Moller, P. and J. Serrier. *Proc. Int. Union Physiol. Sci.* 14:591A, 1980. Specificity of

electric organ discharge in social spacing of mormyrid fish.

66a. Moore, F. R. *Biol. Rev.* 62:65–86, 1987. Sunset and the orientation behaviour of migratory birds.

67. Murray, R. W. *J. Physiol.* 180:592–606, 1965. Sensory function of ampullae of Lorenzini.

68. Northcutt, R. G. In *Electroreception* (T. H. Bullock and W. Heiligenberg, eds.), pp. 257–286. Wiley, New York, 1986. Electroreception in nonteleost bony fishes.

69. Obara, S. and M. V. L. Bennett. *J. Gen. Physiol.* 60:534–557, 1972. Mode of operation of ampullae of Lorenzini of the skate, *Raja*.

69a. Phillips, J. B. *Science* 233:765–767, 1986. Two magnetoreception pathways in a migratory salamander.

70. Pickens, P. E. and W. N. MacFarland. *Anim. Behav.* 12:362–367, 1964. Electric discharge and behavior in stargazer.

70a. Quinn, T. P. *J. Comp. Physiol.* 137A:243–248, 1980. Evidence for celestial and magnetic compass orientation in lake migrating sockeye salmon fry.

70b. Rodda, G. H. *J. Comp. Physiol.* 154A:649–658, 1984. The orientation and navigation of juvenile alligators: Evidence of magnetic sensitivity.

71. Ronan, M. In *Electroreception* (T. H. Bullock and W. Heiligenberg, eds.), pp. 209–224. Wiley, New York, 1986. Electroreception in *Cyclostomes*.

72. Roth, A. *J. Comp. Physiol. A* 79A:113–135, 1972. Wo zu dienen die Elektrorezeptoren der Welse?

73. Saunders, J. and J. Bastian. *J. Comp. Physiol. A* 154A:199–209, 1984. The physiology and morphology of two types of electrosensory neurons in the weakly electric fish *Apteronotus leptorhynchus*.

74. Scheich, H. and T. H. Bullock. In *Handbook of Sensory Physiology* (A. Fessard, ed.), Vol. 3, Part 3, pp. 201–256. Springer-Verlag, Berlin, 1974. The detection of electric fields from electric organs.

75. Scheich, H., T. H. Bullock, and R. H. Hamstra, Jr. *J. Neurophysiol.* 36:39–60,

1973. Coding properties of two classes of afferent nerve fibers: High frequency electroreceptors in the electric fish *Eigenmannia*.

76. Scheich, H., G. Langer, C. Tidemann, R. B. Coles, and A. Guppy. *Nature (London)* 319:401–402, 1986. Electroreception and electrolocation in platypus.

77. Schlegel, P. *Biol. Cybernet.* 20:197–212, 1975. Elektroortung bei schwach elektrischen Fischen: Verzerrun gen des elektrischen Feldes von *Gymnotus carapo* und *Gnathonemus petersii* durch Gegenstande und ihre Wirkungen auf Afferente.

77a. Semm, P. and C. Demaine. *J. Comp. Physiol.* 159A:619–625, 1986. Neurophysiological properties of magnetic cells in the pigeon's visual system.

78. Szabo, T. In *Bioelectrogenesis* (C. Chagas and A. P. de Carvalho, eds.), pp. 20–23. Elsevier, Amsterdam, 1961. Peripheral and central components in electroreception.

79. Szabo, T. In *Handbook of Sensory Physiology* (A. Fessard, ed.), Vol. 3, Part 3, pp. 13–58. Springer-Verlag, New York, 1974. Anatomy of the specialized lateral line organs of electroreception.

80. Szamier, R. B. and A. W. Wachtel. *J. Ultrastruct. Res.* 30:450–471, 1970. Ultrastructure of electroreceptor organs.

81. Viancour, T. A. *J. Comp. Physiol. A* 133A:317–327, 328–339, 1979. Electroreceptors of a weakly electric fish I. Characterization of tuberous receptor organ tuning. II. Individually tuned receptor oscillations.

81a. Walcott, C., J. L. Gould, and J. L. Kirschvink. *Science* 205:1027–1029, 1979. Pigeons have magnets.

81b. Walker, M. M., J. L. Kirschvink, S.-B. R. Chang, and A. E. Dizon. *Science* 224:751–753, 1984. A candidate magnetic sense organ in the yellow fin tuna, *Thunnus albacares*.

82. Watanabe, A. and K. Takeda. *J. Exp. Biol.* 40:57–66, 1963. The change of discharge frequency by A.C. stimulus in a weakly electric fish.

83. Watson, D. and J. Bastian. *J. Comp. Physiol. A* 134A:191–202, 1979. Frequency response characteristics of electroreceptors in the weakly electric fish, *Gymnotus carapo*.

84. Westby, G. W. M. *Anim. Behav.* 23:249–260, 1974. Further analysis of the individual discharge characteristics predicting social dominance in the electric fish, *Gymnotus carapo*.

85. Westby, G. W. M. *Sci. Prog. (Oxford)* 69:291–313, 1984. Electroreception and communication in electric fish.

85a. Wiltschko, W. *Trends Neurosci.* 3:140–144, 1980. The earth's magnetic field and bird orientation.

85b. Wiltschko, W. *Naturwissenschaften* 74: S.94, 1987. Pigeon homing. Olfactory experiments with inexperienced birds.

86. Zhadan, P. M., G. G. Zhadan, and O. B. Ilyinsky. *Neurophysiology* 7:225–231, 1976. Functional features of electroreceptors (small pit organs) in the catfish.

7 | Chemoreception

Esmail Meisami

Introduction

Chemoreception is defined as the capacity of organisms to receive and respond to specific chemical stimuli in the environment. Chemosensitivity is present in all the major phyla of bacteria, protozoa, plants and animals, and was one of the first sensory capacities to evolve. In unicellular organisms, a general type of sensitivity to chemical stimuli is present. These stimuli are highly diverse, including food substances, toxic compounds, and molecules involved in intercellular communication. Unicellular organisms respond to these chemical stimuli by moving toward or away from the stimuli.

In cells of complex animals, some of these sensory and motor capacities of unicellular organisms are preserved, or developed further, for example, in such single cells as the white blood cells of vertebrates. With adaptive evolution, various complex sensory systems developed, each responding to a certain group of stimuli. Thus the chemoreceptors of the common chemical sense, in the skin, oral

and nasal mucosa, respond to nonspecific but noxious chemical stimuli in the environment. More specialized internal chemoreceptors, restricted to blood vessels and digestive tract, respond to specific respiratory gases and food molecules. These internal chemoreceptors and their associated reflex responses, which are discussed in other chapters of this book, are essential for proper functioning of the cardiovascular, respiratory, and digestive systems.

As animals evolved and adapted to new and different environments, chemosensory systems diversified and specialized further, developing specialized sensory receptors and separate neural centers. In the higher vertebrates and insects two distinct and special chemosensory systems, with separate receptors, nerve pathways and neural integrative centers, are observed: gustation (taste) and olfaction (smell). The gustatory system deals primarily with food related stimuli and has its receptors mainly in the oral cavity. The olfactory system has receptors in the nasal cavity and deals with a wide variety of chemical stimuli in the environment. Olfactory stimuli carry information about many aspects of the environment, including food, sex, kin, enemy, habitat, and so on. Both olfactory and gustatory systems are linked to important motor pathways that mediate the associated reflexes and higher behavioral responses. Many important behaviors associated with the survival of the individual (e.g., feeding) and species (e.g., reproduction) are intimately regulated by chemical stimuli. In humans, taste and smell also play important roles in pleasures and enjoyments of life (flavors of gourmet foods, aromatic herbs, wines, and perfumes).

Chemoreception in Single Cell Organisms

Many bacteria and protozoa are chemotaxic, that is, show chemically induced orientation and movement. Chemotaxis may be positive or negative, movement toward or away from the source of chemical stimuli (77).

Bacteria

Motile bacteria like *Eschericha coli* and *Bacillus subtilis* use flagella as their motor elements. Counterclockwise rotation of filaments in the flagella causes smooth forward locomotion while clockwise rotation causes random "tumblings" or turns. In a stimulus-free or chemically homogenous environment these bacteria move randomly, swimming a little, then tumbling and turning. Addition of food substances called "attractants" to the medium stops the tumble and turn, and activates counterclockwise rotation of the flagellum, causing the bacteria to swim smoothly. Addition of toxic substances called "repellents," however, activates clockwise flagellar rotation which induces

the tumbling and random turns (4, 120, 124).

The attractants are in general food-related substances such as simple sugars (galactose and ribose) or amino acids (serine, cysteine, alanine, glycine, aspartate, and glutamate. These stimuli act as attractants at the threshold concentrations of 10^{-6}–10^{-8} M) (136) (Fig. 1a). Repellents are inorganic and organic acids, toxic substances (cyanide), uncouplers of oxidative phosphorylation, and even drugs such as local anesthetics and psychoactive chemicals (hydrophobic ionizable molecules) (120). Some amino acids like leucine, isoleucine, and valine act as repellents at threshold concentrations of 10^{-4} M (136).

The bacterial chemotactic responses show specificity, summation, and adaptation. Each stimulus interacts optimally with a particular receptor (specificity); closely related substances showing similar structural features may share the same receptor. The intensity of motor response to different substances is additive so that exposure of the cell to more than one attractant causes corresponding increases in response (summation). When the stimulus remains in the medium at a constant level, the bacterium stops responding (adaptation) Fig. 1b. Then upon reexposure of the cell to the same stimulus but at a higher dose, the response reappears; during adaptation the baseline of the motor response is reset so that the behavior can occur in a wider stimulus range (74, 75).

Bacteria contain membrane proteins that have three domains, periplasmic, plasmic, and cytoplasmic (Fig. 2). The periplasmic domain, situated between the plasma membrane and the cell wall, acts as a chemoreceptor that binds to the chemical stimuli. These chemoreceptors are highly specific for the attractants and there may be many of them (Fig. 1a). As the bacteria swim in the direction of the

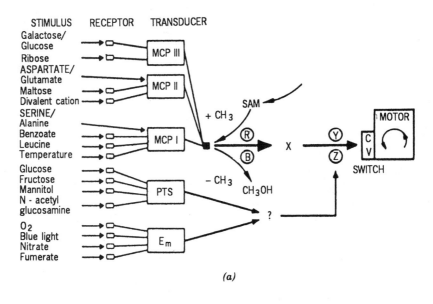

(a)

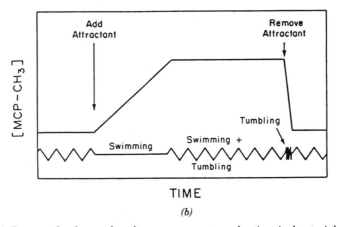

(b)

Figure 1. (*a*) Proposed scheme for chemosensory transduction in bacterial membrane, leading from chemoreception to chemotaxis. Various chemical stimuli bind with receptor molecules in the periplasmic surface; this activates methyl accepting chemotaxic proteins (MCPs); MCPs may be a part of the receptor protein complex. Methylation or demethylation reactions lead to products that act on the bacterial flagellar motor complex, causing negative or positive chemotaxis. SAM, S-adenosylmethionine. R, B, Y and Z, chemotaxic regulatory proteins. [From Van Houten and Preston (164) after Koshland (75).] (*b*) Relationship between methylation and the chemotactic behavior in bacteria. Addition of an attractant substance (e.g., serine) results in methylation of methyl-accepting proteins, triggerring smooth swimming; upon termination of methylation, behavior changes from smooth swimming only to a combination of smooth swimming and tumbling (adaptation); removal of the attractant results in demethylation of MCPs followed by tumbling behavior. [From Ordal (120)]

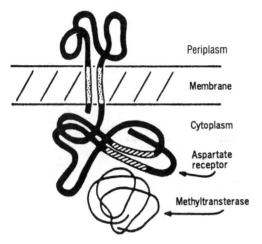

Periplasm

Membrane

Cytoplasm

Aspartate
receptor

Methyltransferase

Figure 2. Various components of the chemo-receptor molecular complex in the bacterial membrane as illustrated by the receptor system for aspartate, which extends into the periplasm. The receptor molecule includes the methyl-accepting chemotaxis protein (MCP) and the cytoplasmic side interacts with a methyltransferase enzyme [From Van Houten and Preston (164).]

attractant source, more attractant–receptor complexes are formed so that the bacteria are more likely to swim smoothly.

The receptor segment is connected to a cytoplasmic signaling protein (cytoplasmic domain), which initiates the intracellular chemical change that activates the switch at the flagellum. These cytoplasmic proteins are capable of being methylated (accept methyl groups) and are therefore called methyl-accepting chemotactic proteins (MCPs) (Fig. 1a). Upon addition of the attractant, MCPs are methylated (4). Methylation of MCPs serves as the signal to induce smooth swimming since swimming persists as long as methylation continues. Removal of the attractant induces tumbling and leads to demethylation (Fig. 1b). Certain regulatory cytoplasmic proteins operate in the adaptation response, in part by selectively methylating or demethylating the MCPs. Other cytoplasmic proteins are

involved in regulation of flagellar switch, promoting clockwise or counterclockwise rotation (75, 120, 124) (Fig. 1b).

The details of the transduction mechanisms operating between methylation of the MCPs and rotation of the flagellum are not fully understood. In *E. coli* transfer of phosphoryl groups among proteins appears to be involved. It has been shown that phosphorylation of some proteins initiates clockwise rotation while dephosphorylation of others leads to counterclockwise motion (124).

Protozoa

Eukaryotic unicells (amoebae and ciliates) show chemoreception and chemotaxis. White blood cells show migratory and other chemosensory responses to wounds and infections. In addition to general responses to food substances like peptides and amino acids, some unicells like *Blepharisma* respond to specific substances such as a mating pheromone. *Paramecium* is a swimmer, utilizing its cilia like oars, to turn, tumble, and swim (77). The frequency of turning and speed of forward movement are studied as quantitative measures of the chemosensory response. Paramecia are attracted to folate and acetate, food-related substances, and even to such signal molecules as cAMP. In the presence of these substances they show fast smooth movement in the up-gradient direction (164).

These chemosensory-induced movements in protists are caused by changes in the membrane potential (132). Attractants cause hyperpolarization of the membrane while repellents cause depolarization. The hyperpolarization response seems to be related to the changes in the level of internal calcium. Hyperpolarization further induces changes in the level of second messengers like cAMP, which in turn trigger the ciliary machinery. Cilia that cover 50% of the surface membrane in *Paramecium* contain high adenylate cyclase activity (164).

Chemoreception in Vertebrates

The vertebrate body cells have different chemoreceptor molecules on their plasma membrane that endow them with a variety of cellular chemosensitivities (13, 49, 170). In special cases, as in white blood cells, this ability is highly developed and extensively used in carrying out the cell's chemical detection, recognition, and response functions. In addition, vertebrates have four basic types of chemosensory systems (123). (1) A common general chemical sense that serves in detection of noxious chemical stimuli; its receptors are concentrated in mucous areas of the skin, particularly in the nasal and oral mucous membranes. (2) An internal chemosensory system that serves in regulation of vital visceral functions; these include the oxygen chemosensors of the carotid and aortic bodies in the vasculature, chemoreceptors of the digestive tract, osmotic and glucoreceptors of the hypothalamus, and the respiratory chemoreceptors in the medulla oblongata. (3) and (4) Special senses of taste and smell that serve in detection and recognition of and responses to food substances and odors (170).

Olfaction and the Olfactory System

The olfactory system is a special sensory system that functions to detect chemical stimuli of relatively low molecular weight that have some adaptive value for the animal. These chemical stimuli are called odorants and their perceptual equivalents are referred to as odors. The odorants may signify food, sex (mates), familiar and unfamiliar individuals, danger, rotting matter, and so on. Odorants carrying specific "signals" that are communicated between individuals of the same or different species, activating stereotyped behavior, are called pheromones. For nonaquatic vertebrates and most insects, odorants are generally volatile substances that diffuse in air before reaching the olfactory receptors. In fish and other aquatic animals the odorants are dissolved in water.

An important characteristic of the olfactory system in all land vertebrates and many aquatic species is that it is able to act as a distant sense, detecting faint traces of substances that have diffused from far-away sources. In contrast, the taste system, because of its lesser sensitivity, requires direct contact of the source material with the taste receptors. The high sensitivity of the olfactory system results from its low thresholds to chemical stimuli, enabling it to detect, in some cases, substances at femtomolar ($10^{-15} M$) concentrations (see Table 1). Thus olfaction and taste both respond to a continuum of chemical stimuli but olfaction is specialized for the low range of concentrations while taste is specialized for the high range. Each system provides the animal with information about different qualities of the stimuli. Also taste stimuli are generally associated with food substances while the range of olfactory stimuli are much wider, providing information not only about food but about sex, reproduction, friend and foe, territory, and so on. In humans and possibly other mammals perception of flavors of many substances like coffee and chocolate is impaired without olfaction (109).

Nasal and Olfactory Cavities

In vertebrates, the access of odorants to the receptors of the olfactory receptor sheet is in general through the external nares, the nasal passages, and the nasal–olfactory cavities. The nasal cavities in general are situated above the oral cavity. The structures of the nose, nares, and nasal cavities show species modifications that are highly adaptive (125).

A nasal cavity communicates directly between the external environment and the internal respiratory or olfactory cavities. The olfactory cavity houses the olfactory receptor sheet (organ). The nasal and olfactory cavities may be nearly identical, sharing the same space, as in humans and most other primates, or they may be set quite apart from, although communicating with, the nasal–respiratory cavity, as in most fish, rodents, and carnivores. In general, the complexity of olfactory cavities seems to increase in proportion with the degree of development of olfactory structures. This is particularly true in air-breathing mammals like carnivores and rodents, in which stimulus delivery to the receptor sheet may play an important role in pretransduction events (113, 125).

FISH. In fish and aquatic amphibians, the nasal cavity is truly an olfactory cavity, having a pit or saclike structure. These sacs are often found as a pair on the dorsal or ventral surface of the snout. In general, in teleost fishes the location of the olfactory pit and the olfactory organ that it houses is on the dorsal surface of the snout with no connection to the gills, while in bottom-feeding elasmobranchs and cyclostomes, the nasal pits are located ventrally on the snout, having connections with the gills and respiratory system. In the lungfishes, olfactory cavities open dorsally on the surface of the snout and ventrally into the oral-pharyngeal cavity (113, 135). Regardless of the location, each sac contains an olfactory organ plus one inlet and one outlet (Fig. 3A). As the animal swims, water enters the sac through the inlet, passes over the lamellae of the olfactory organ, and exits through the outlet (Fig. 3B). Movements of the gills and jaws directly or indirectly aid in the passage of water through the

olfactory pit. The pit may also contain motile cilia that aid in water transport.

The fish olfactory organ forms a rosette with lamellae formed as a result of the folding of the nasal walls and the overlying receptor sheet. The lamellae, which vary widely in number depending on the species, increase the area of contact with the flowing water and its dissolved odorants (Fig. 23), and may therefore be important in determining the degree of olfactory sensitivity (11, 63).

RAT AND RABBIT. In mammals like the rat and rabbit that have well-developed olfactory systems, the anterior nasal cavities form convoluted respiratory turbinates (maxillary turbinals) that aid in warming and moistening the air (Fig. 4). This air can then be shunted in two directions: (1) a lower passage, through the duct of the nasopharynx into the pharynx and lungs; and (2) an upper passage that leads the air into an olfactory cavity with a complex inner wall that is folded into several nasoturbinals or conchae (Figs. 4 and 5). These protrude into the cavity and divide it into even smaller passageways (Fig. 5). The nasopalatine ducts located anteriorly near the vomeronasal organ, permit communication between the nasal and oral cavities (see Fig. 24). The nasal septum that divides the cavity into left and right halves is continuous throughout its length, showing no internasal canals or orifices, except for a small "septal window", located near the beginning of the nasopharyngeal duct; this window may permit flow of air between the two halves (Fig. 5).

The posterior olfactory turbinals increase the surface area of the olfactory epithelium that covers them. The turbinals may also be important in enhancing the dynamics of air flow through the cavity and facilitating interaction of odorants with the receptor sheet. Sniffing movements, performed constantly and skill-

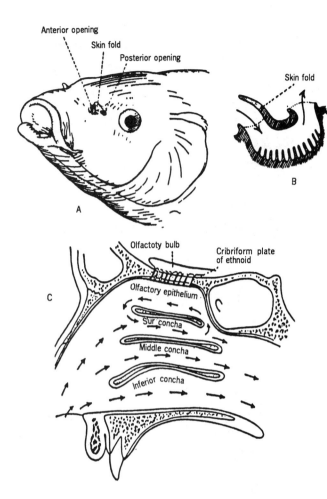

Figure 3. Comparison of naso-olfactory cavities in the fish and the human head. The location of the fish olfactory pit and its inlet (anterior opening) and outlet (posterior opening) are shown in *A*; The direction of water flow and the location of the lamellae in the pit are shown in *B*. The lower diagram shows a mid-sagittal section of the nasal cavities in the human. Arrows indicate possible patterns of air flow through the lower and superior conchae. [Chap. 12 in (135).]

fully by animals like rabbits and dogs, help to promote air flow through the olfactory cavities and increase the intensity of olfactory stimulation (113).

MAN. In humans and higher primates, the left and right nasal cavities are large open chambers connected anteriorly to the nose (external nares) and posteriorly to the pharynx via the internal nares (Fig. 3C). The lateral walls of the cavity form conchae relatively simple, compared to those seen in rodents and carnivores; medially the nasal septum divides the left and right cavity. There are no connections between the left and right nasal cavities, or between the nasal cavity and oral cav-

ity except through the pharynx. The olfactory organ (olfactory mucosa, olfactory epithelium) lies over the upper surface of the nasal septum and in the roof of the superior nasal cavities only, the conchae being devoid of olfactory tissue (Fig. 3C). The inspiratory air entering the nose has the tendency to exit to the lungs via the pharynx, but eddies of air current can reach the olfactory epithelium and stimulate it. Sniffing increases air contact with the epithelium by increasing these eddies.

Main Olfactory System
The main olfactory system (MOS), consisting of the main olfactory organ, olfac-

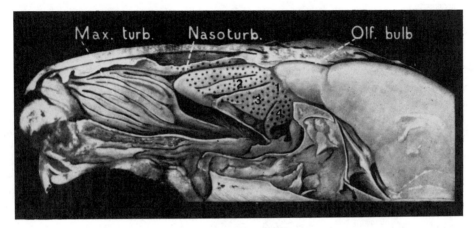

Figure 4. Semidiagramatic drawings of the rabbit head showing the location of the maxillary turbinals (max. turb.), which warm and humidify the inspired air and the nasoturbinals (dotted areas), which house the olfactory receptor sheet over their convoluted surfaces. The numbers 1–4 indicate the various nasoturbinals (conchae). The nasopharynx is a duct running under the nasoturbinals. [From Negus (113) after Le Gros Clark.]

tory bulb, and related olfactory cortical areas, is present in all vertebrate species except Cetacea (whales and dolphins). The size and degree of development of MOS varies markedly between vertebrate species. Species with relatively large MOS are considered to have well-developed olfactory abilities and are called macrosmatic while those with a less developed MOS are called microsmatic (113). The cetaceans that lack an olfactory system entirely are anosmatic. Among the highly macrosmatic animals are nocturnal ground-dwelling mammals (rat and mole), insectivores (hedgehog), and carnivores (dog and cat) (113).

In mammals, MOS is the most complex of the nasal chemosensory systems. The unique neural organization of MOS endows it with high sensitivity, making it the only distance chemical sense, capable of detecting stimuli from far distances and at extremely low concentrations. The MOS is also capable of discriminating between hundreds of different odorants and thus is crucial for cognitive, discriminative and higher associative functions of

odors and olfactory-guided behaviors. The MOS also plays a role in regulation of such vegetative functions as appetite, food search, and food choice; it is essential in feeding behavior of the suckling young (96). The main olfactory system aids in perception of many flavors as well (109).

OLFACTORY EPITHELIUM (MAIN OLFACTORY ORGAN). Olfactory epithelium has a fairly simple and similar structure among all vertebrates (42, 56, 82). It is a pseudostratified epithelium of varying thickness, 100–150 μm. The epithelium consists of three major cell types: the olfactory receptor neurons and the supporting cells comprise the two functional types, and the basal cells constitute a progenitor cell type. The typical histological structure of the olfactory epithelium and the arrangement of olfactory neurons, supporting cells and the basal cells within the epithelium are shown in Figure 6a.

The primary olfactory neurons, containing ciliated dendrites, are the excitable receptive cells, receiving the olfactory

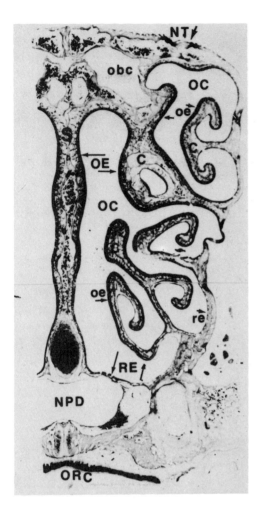

Figure 5. A frontal section through the rat head just in front of the olfactory bulb showing the complexity of the olfactory cavities. Only the left olfactory cavity and the associated structures are shown. Olfactory epithelium (OE,oe) covers the surfaces of the complex nasoturbinals (conchae) except for small portion in the ventral and ventrolateral aspect which is covered by respiratory epithelium (RE,re). NT, nasal tectum; OC, olfactory cavity; obc, olfactory bulb cavity; oe, olfactory epithelium; C, conchae (nasoturbinals); re, respiratory epithelium; NPD, nasopharyngeal duct (nasopharynx); ORC, oral cavity; S, nasal septum. [From Paternostro and Meisami (125a).]

stimuli and conveying the generated electrical signals to the olfactory brain centers. Nonciliated olfactory receptor neurons containing microvilli are also present in the olfactory organs of fishes and in the vomeronasal organ of Jacobson in the nose of certain reptiles and mammals (19, 31, 56). They have also been described in the olfactory epithelium of mammals (103).

The supporting cells act as the neuroglia of the olfactory epithelium, helping to support and nourish the olfactory neurons during their development and functional lifetime. They also secrete some of the constituents of the mucus fluid covering the epithelium. The basal cells occur at the base of the epithelium adjacent to the basal lamina. Cytoplasmic processes of both the olfactory neurons and the supporting cells span the entire thickness of the epithelium, although the nuclei of the supporting cells form a single layer superficially while the nuclei of the olfactory neurons are found between those of the supporting cells and the basal cells (Fig. 6a).

The epithelium is covered at its surface by a mucus fluid layer about 30 μm thick, secreted by the cells of the Bowman's glands, which occur at fairly regular intervals throughout the epithelium (Fig. 6a). The supporting cells are also known to secrete substances into the mucus. This mucus layer, which is rich in mucopolysaccharides and glycoproteins, bathes the cilia of the olfactory neurons and is physiologically important because odorants must dissolve in this fluid layer before interacting with the cilia of the olfactory neurons. The interaction of odorants with the mucus fluid (e.g., solubilization and diffusion) constitute the perireceptor phase of olfactory reception microphysiology (54).

OLFACTORY RECEPTOR NEURON STRUCTURE. Olfactory receptor neurons have a fairly

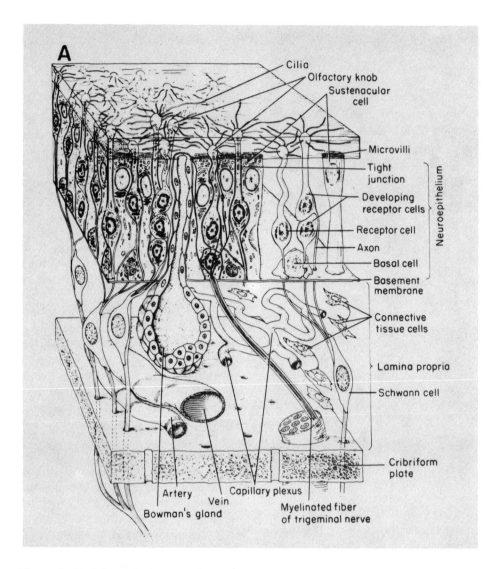

Figure 6. (*a*) Schematic diagram showing the various neural and nonneural structural elements typical of the vertebrate olfactory epithelium. The Bowman's glands are absent in fish. [From Phillips and Fuchs (139), after Greep and Weiss.] (*b*) Ultrastructural details of the single apical dendrite, olfactory knob and the cilia of a typical olfactory receptor neuron. Number of microtubules diminish in distal segments of the olfactory cilia. [From Den Otter (38), after Andres.]

similar structure in different classes of vertebrates (56). Figures 6*b* and 7 depict the structure of a typical mature mammalian olfactory neuron. This neuron has an oval cell body and a single slender dendrite that extends apically toward the na-

sal cavity surface, ending in a bulbous structure called olfactory knob or vesicle. The dendritic shaft varies in length between 20 and 100 μm. In animals like frogs that have a thick olfactory epithelium the dendrites are longer. The length

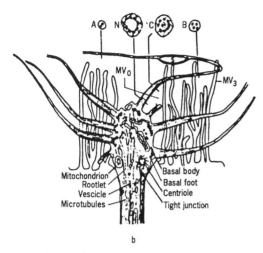

Mitochondrion
Rootlet
Vescicle
Microtubules

Basal body
Basal foot
Centriole
Tight junction

b

Figure 6. (*Continued*)

depends on the position of the cell body within the epithelium, deeper neurons having longer dendrites.

Olfactory knobs give rise to several olfactory cilia which, depending on the species, are 5–20 in number and 30–200 μm long (82, 98, 105). These cilia are bathed in the olfactory mucus that is in contact with the nasal environment. Therefore olfactory neurons, more than any other neuron type, are directly exposed to the external environment and may suffer from its hazardous and toxic effects. The cilia of olfactory neurons lack dynein, the ciliary ATPase, and are therefore nonmotile, in contrast to the rapidly beating cilia of the epithelial cells covering the respiratory surfaces of nasal cavities (98). Olfactory cilia are specialized in that they contain the olfactory receptor molecules, making them the site of olfactory transduction (153). The cilia increase the available receptive surface area of the knob by several hundred folds (53, 82, 98).

The unmyelinated olfactory fibers, upon leaving the olfactory epithelium, form fascicles of about 1000 closely packed fibers. Several fascicles are ensheathed by a single Schwann cell, the plasma membrane of which forms a single mesaxon. These Schwann cells show

similar biochemical properties to brain astrocytes. Although the individual fascicles are far apart (100 μ), the individual fibers within each fascicle are packed close enough (100–150 Å apart) to permit some electrochemical interaction that may lead to synchronization (53, 82, 105).

The small diameter (0.2 μm) of olfactory fibers and the sparcity of voltage gated ion channels make them among the slowest conducting nerve fibers in the vertebrate body (0.2 m/s). In some fish (e.g., salmon, garfish, and carp), frogs and some birds (e.g., pigeon), the olfactory axons form relatively long and distinct nerves before reaching the olfactory bulb. In others, (catfish and goldfish, all mammals, most birds) an olfactory nerve proper is absent; instead, olfactory axons form small and short fascicles that traverse tortuous pathways, pass through the pores of the cribiform plate and immediately enter the bulb. Given the slow conduction velocity of the olfactory fibers, the highly variable fiber lengths may differentially influence the speed of olfactory processing and responsiveness.

Axons of olfactory neurons pass without branching to the olfactory bulb where tens of thousands of them converge to form structures called olfactory glomeruli. The number of glomeruli differs depending on the size of the bulb (2–3 × 10^3 per olfactory bulb in the rodents) (6). Within each glomerulus, the terminal branches of the olfactory axons synapse with the dendritic terminals of the relay neurons (mitral and tufted cells) as well as with those of the interneuronal periglomerular cells. In the adult rabbit that contains about 50 × 10^6 olfactory receptor neurons on each side of the nasal cavities, it is estimated that about 25 × 10^3 olfactory fibers enter a single glomerulus (rabbit bulb contains about 2000 glomeruli) to synapse with 25 mitral cells, creating a convergence ratio of about 1000:1 (6) (Table 2). An extremely high convergence ratio of olfactory neurons to olfactory bulbar relay cells (mitral cells) is a characteristic of the olfactory system of all vertebrates, and is believed to play an important role in the unusually high detection sensitivity of the olfactory system (97, 163).

TRANSMITTER CHEMISTRY AND OLFACTORY MARKER PROTEIN. Olfactory neurons synthesize and release the dipeptide carnosine (β-alanyl-histidine), which may act as a neurotransmitter, at their synaptic terminals in the bulb (91) (Fig. 15b). The neurons contain the enzymes carnosine synthetase and carnosinase, necessary for the synthesis and breakdown of carnosine, respectively. Carnosine is found in several body tissues, chiefly the muscle tissue. In the neural tissue, it is found only in the olfactory neurons and their terminals in the olfactory bulb. Olfactory neurons also contain a unique 19-kD cytoplasmic protein, the olfactory marker protein (OMP), that is expressed by mature neurons only, that is, after the developing neuron has made synaptic contact with the bulb (46). Although OMP is mainly restricted to olfactory tissue and is only associated with mature and functional olfactory neurons, its exact function is not fully understood (92).

DEVELOPMENT, TURNOVER, AND REGENERATION. Olfactory neurons are the only neurons in the nervous system to undergo several turnover cycles during the life of the animal (57, 108). Each cycle consists of the formation of a new olfactory neuroblast from the division of progenitor basal cells, development of a new dendrite and axon, growth of the axon toward the olfactory bulb and establishment of new connections with target cells in the olfactory bulb (Fig. 7). These developmental events take about a week and occur in both young and adult (46). After synaptic function is established, the new olfactory neuron begins its relatively short life (1–2 months) as a mature sensory neuron, relaying

messages from nose to brain. The last stages are aging, degeneration, and death. The olfactory neurons of mice raised in pollution-free cages show longer life spans (66).

If the axons of olfactory neurons are severed, the neurons degenerate (retrograde degeneration) within a few days. The degenerating neurons trigger the basal cells into division and differentiation (57). Of all the nervous system neurons, the olfactory neurons are the ones most exposed to the environment. The nasal cavities inevitably harbor numerous microbial organisms such as bacteria and viruses as well as dust particles and toxic pollutants. The turnover cycle and regenerative ability of the olfactory epithelium is believed to ensure the maintenance of a stable population of olfactory neurons in the face of constant environmental hazards that lead to toxicity, wear and tear, rapid aging, and death of these neurons. It is not known how the regenerated olfactory neurons preserve their specific pattern of connectivity and responsivity to odorants.

ELECTRICAL RESPONSES. Upon stimulation of the olfactory receptor sheet by odorants, and depending on the type of recording electrode used, two types of electrical responses can be observed, one obtained from a population of neurons, the other from single neurons.

Mass Response of Olfactory Neurons or Electroolfactogram (EOG). The electro-olfactogram (EOG) is believed to be the collective response of a relatively large population of receptor neurons and may be the summated responses of the individual olfactory neurons to the applied stimulus (53, 121). The EOG can be recorded by placing an electrode in contact with the surface of the olfactory epithelium. With

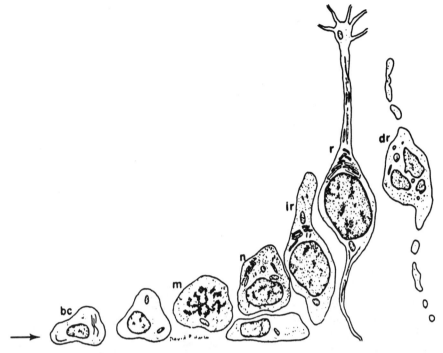

Figure 7. Sketch of the life cycle (development, functional activity, and degeneration) of olfactory receptor neurons. bc, basal cell; m, basal cell undergoing mitosis; n, neuroblast; ir, immature neuron; r, mature, functional receptor neuron; dr, degenerating receptor neuron. Arrow indicates the basal lamina of the epithelium. (Courtesy of P. P. C. Graziadei.)

weak stimuli (a puff of odorous air), a purely negative and monophasic potential is observed. Stronger stimuli result in the appearance of rhythmic oscillatory waves superimposed upon the monophasic potential (Fig. 8). The amplitude of the EOG increases in proportion with the strength of the stimuli. Mathematically, this relationship follows a power function. Since the perceptual intensities of odors also vary according to a power function, these peripheral events may play an important role in determining changes in perceptual intensities (121).

Of the various cellular elements forming the epithelium, the EOG responses are believed to be generated by the olfactory cilia, since their removal eliminates the electrical response. Also if EOG is measured at different depth of the olfactory epithelium, its amplitude diminishes with increasing depth in the mucosa, that is, farther away from the ciliary surface.

Responses of Single Olfactory Neurons. Single olfactory neurons, in the absence of odorants, show a slow and irregular firing. Upon stimulation by odorants, an excitatory response is evoked, consisting of a rapid rise in impulse frequency at the onset of stimulation, followed by a later phase of continuous firing at a reduced rate. The range of impulse frequency is from 1 to 3 impulses/s at threshold levels to 20/s at highest concentrations of odorant (52, 53, 82). Although excitatory responses of olfactory neurons to odorants are the dominant type, some units (~5%) are known to decrease their activity upon stimulation. It is not known whether this

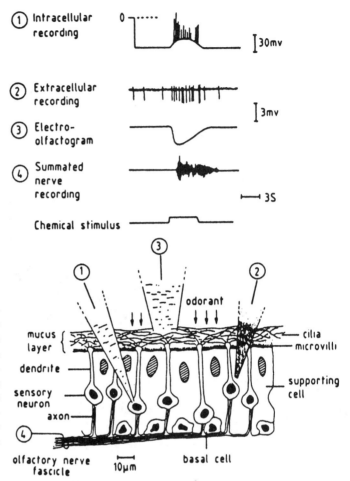

Figure 8. A schematic representation of vertebrate olfactory epithelium and the various electrophysiological responses obtained at the different recording sites and by different procedures. Numbers in circles relate the various types of recording and/or electrical responses to corresponding sites of recording in the epithelium. [From Lancet (82).]

is a true inhibitory response caused by hyperpolarization of the receptor membrane or a secondary response caused by the removal of odorant (52, 70).

Unit responses of frog olfactory receptor neurons are not narrowly specific for particular odorants but show a relatively broad spectrum of responsivity to different odorants (52, 53). However, when the battery of odorants tested is increased, it is found that each neuron is selective to a group of odorants numbering from one to twelve; and no neuron is capable of responding to all the odorants. In turtles, the responses of 20 receptor units to 27 odorants indicated that each unit showed

a unique spectrum, some responding to only one or two, others to as many as fifteen odorants (93) (Fig. 9). The response spectra of different neurons may show much overlap but are never the same. Thus the vertebrate olfactory receptor neurons do not form specific categories based on their response properties. In comparison to insect olfactory receptors the vertebrate neurons are less like the "odor specialists," resembling more closely the "odor generalists" (70, 146) (Figs. 43b and 44).

For each neuron, the response threshold for the different odorants varies markedly. Similarly, various receptor neurons

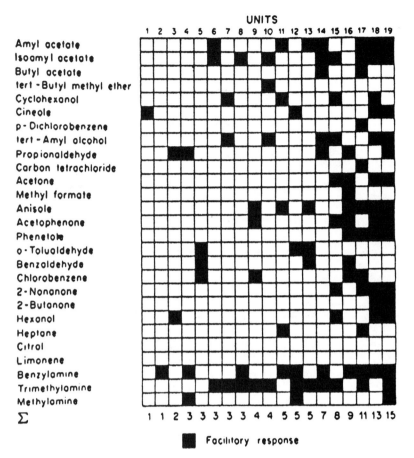

Figure 9. Each turtle olfactory receptor neuron responds to a variety of different odorants, but the response spectra of different units are not alike. This matrix shows at top the responses of 19 turtle receptor units to 27 different odorants. Σ is number of odorants responded to by single units. At the two ends of the spectrum, units 1 and 2 responded to only one odorant each, while units 18 and 19 responded to 13 and 15 odorants, respectively. [From Mathews (93)].

responding to the same odorant do so at different thresholds. Assuming that the stimulus delivery conditions are the same for these units, an explanation for this differential sensitivity may be the existence of different densities of odorant-binding receptor molecules at the ciliary zone, or the existence of different transducing mechanisms, some involving amplification of the response while others involve none (53, 82).

Not only the threshold but also the response patterns of the olfactory neurons to the same odorants show considerable variation from unit to unit, with some units firing very strongly and others very weakly (42, 70, 138). Figure 10 illustrates the responses of 10 different olfactory neurons of the frog to a single odorant, anisol (42, 121). The same unit that responds to one particular odorant with a weak response may respond vigorously to another odorant. This is shown in Figure 11 where the responses of a single frog olfactory neuron to four different odorants (d-citronellol, cyclohexeneone, cycloundecanone, and pinacolone) are represented (138). The unit responded

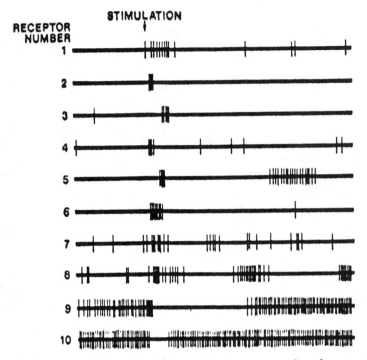

Figure 10. Electrical responses of 10 olfactory receptor neurons in a frog to stimulation with a single odorant (anisol). Each neuron shows a different pattern of response to the same odorant. [From Ottoson (121), after Duchamp et al. (42B).]

with different latencies and duration to the four odorants.

The response of a particular receptor neuron may also vary depending on its immediate past odorant exposure experience. In frog primary olfactory neurons, the response to a given odor varies with time and with the sequence of presentation of other odors. For example, one cell gives a few spikes to butanol, but after musk (which elicits no response itself) the response to butanol is a large burst (121).

Molecular Physiology of Olfaction

ROLE OF PHYSICAL–CHEMICAL PROPERTIES OF ODORANTS. For terrestrial vertebrates, odorants that are capable of stimulating the olfactory neurons of the main olfactory epithelium are highly volatile molecules of relatively small molecular weight (89). High volatility and small size facilitate transport of the odorant via air currents from the source to the nose and through the nasal passages. Various physicochemical properties influence the ability of the odorants to penetrate the epithelial mucus surface layer and interact with the receptor neurons. These properties include the degree of solubility in water and in lipids, the overall size and shape of the molecule, and its surface chemistry (functional groups and charge distribution) (9, 53, 54, 82, 102).

For hydrophilic odorants, water solubility is an important determinant of odorant potency. Substances with a high water solubility tend to be more effective odorants, showing short response latencies and low thresholds. The importance

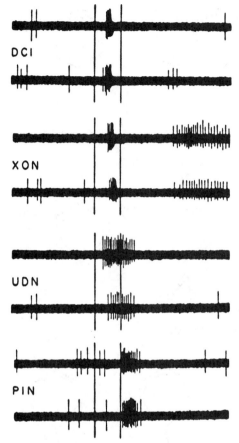

Figure 11. Responses of a single frog olfactory receptor neuron to 2-s pulses of four different odorants: *d*-citronellol (DCL), produces short latency, short lasting, high frequency burst responses; cyclohexanone (XON); produces an early and a late burst; cycloundecanone (UDN), produces long-lasting low frequency discharges; pinacolone (PIN), produces long latency responses. [From Revial et al. (138).]

of water solubility relates to the requirement for the odorants to dissolve in and traverse the surface mucus layer. The olfactory mucus layer, estimated to be about 30 μm thick in the frog, has a mucopolysaccharide scaffolding embedded in a watery medium. Odorants must traverse through the sol phase to reach the receptive sites of olfactory neurons. Thus,

in addition to water solubility, the viscosity and thickness of the mucus are important parameters regulating the diffusion rate of odorants (9, 54).

The diffusion coefficient for a typical odorant in water is estimated to be about 2×10^{-5} cm^2/s, a value nearly three orders of magnitude smaller than that in air. This indicates that odorants encounter a formidable barrier at the air–mucus interface. It takes an average odorant 6 ms to reach a point 5 μm below the mucus surface (tip of cilia) and 40 times longer to reach a depth of 30 μm (shaft of cilia). Traversing a distance of 100 μm requires 2.5 s (54). This is clearly one reason why in a disease condition, when nasal secretion is profuse (rhinorrhea), the ability to smell disappears.

Lipid solubility is also important because some odorants, like alcohols, may not necessarily act by binding with a receptor binding protein; instead they may activate transduction by perturbing the lipid–protein interaction in the receptor membrane (78). For example, stimulatory efficiency of primary alcohols as odorants increases linearly with the chain length, up to eight atoms per chain, decreasing thereafter, as concluded from EOG studies of frog olfactory mucosa. Similar relations have been found for aldehydes and ketones (121). In humans, bees, and dogs, the optimum for some volatile fatty acids is a C_4 compound; for a series of volatile esters, the optimum is at 12 carbon atoms for humans and bees. Possibly lipid solubility limits effectiveness on the side of smaller molecules and molecular size limits it on the high side of the optimum (15, 78, 102, 135).

The nature of the functional group on a molecule may be important in response latency and threshold (15) but does not always critically determine odor quality. Thus nitrobenzene and cyanobenzene both have the odor of bitter almond. However, the same compounds with a 3-

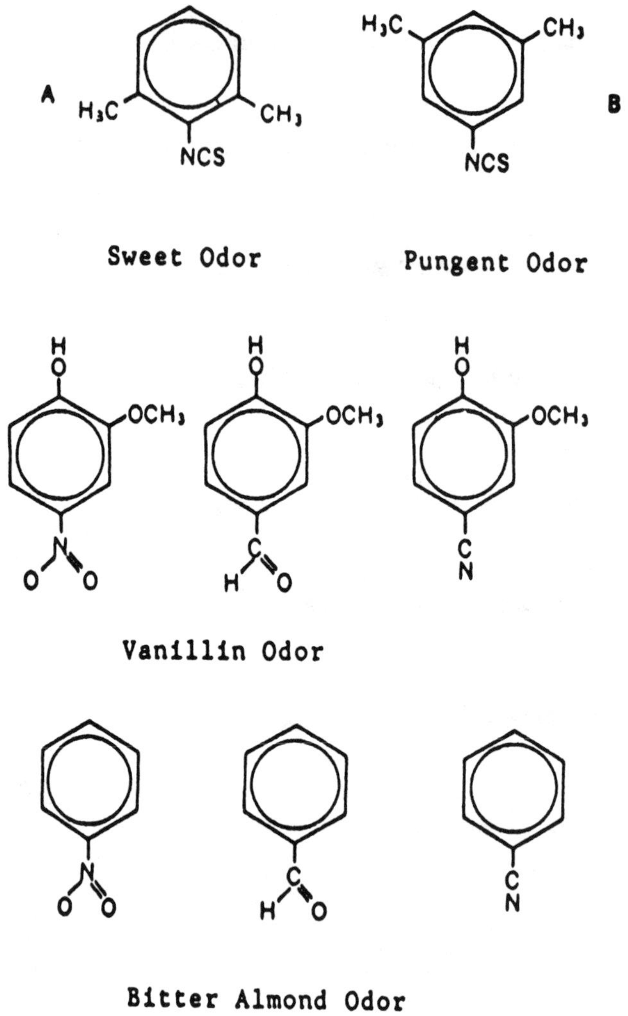

Figure 12. Slight changes in functional group and shape of molecules have marked effects on their properties as odorants. For example, nitrobenzene and cyanobenzene molecules (bottom row) both have a bitter almond odor. The same compounds with a 3-methoxy and 4-hydroxy substitution (middle row) have a vanillin odor. The position of the substitution may also be important. The presence of methyl groups in positions 3 and 5 of this benzene derivative (top row, *B*) gives the molecule a pungent odor while substitution of the same groups at different positions of the ring changes the odor to sweet (*A*). [From Beidler (18).]

methoxy and 4-hydroxy substitution have a vanillin odor (Fig. 12). Position of the substitution may also be important, as shown by the example in Figure 12. Similarly, numerous compounds with differing functional groups have a musky odor (18).

NONTRANSDUCTION OLFACTORY-BINDING PROTEINS. A class of proteins that function in binding and transport of odorants but are not involved in the transduction mechanism are the odorant-binding proteins (OBPs) that are soluble cytoplasmic proteins of about 20 kD secreted by the nasal glands or secretory cells in the epithelium but are not associated with olfactory cilia (127, 153). These OBPs show a high affinity for a large variety of odorants of differing molecular structures and odors and may serve in their transport within the mucus layer and in concentrating the odorants near the ciliary receptor sites or in their removal after transduction. Structurally, OBPs are closely related to the retinal-binding proteins, which bind and transport retinol and vitamin A in the body and in the retina (153).

OLFACTORY RECEPTOR PROTEINS. Olfactory reception is mediated by membrane-bound receptor molecules that bind with high affinity and reversibly to odorants at the surface of receptor neurons and initiate the transduction mechanism. The membrane-bound receptor molecules are thought to be glycoproteins positioned as integral components of olfactory ciliary membranes (82). Based on the densities of intramembranous particles observed in freeze-fractured electron micrographs of olfactory ciliary membranes, the surface density of intramembranous particles (receptor proteins?) is estimated to be 0.9–2.6 $\times$ $10^3/\mu m^2$ in frogs and dogs, and their molar concentration $10^{-5}M$ (98). Based on about 10 cilia per neuron and 200-μm^2 surface area per cilium, the total number of these receptor particles per olfactory neuron would be about 3 $\times$ 10^6 (53, 82).

Each receptor molecule is thought to contain a receptor site, specific for a particular odorant or a class of them. There are two current views regarding the mode of interaction between the odorant and the receptor sites. One concept views the activation of the receptor as a result of the stereochemical fit between the odorant and the receptor site (8). Another considers the activation in the light of interaction of functional groups in the odorant and the receptor site (15).

The number of functional types of these olfactory receptor molecules, as defined by their responsiveness to different odorants, is estimated variously from as low as a dozen to as high as several thousand, depending on which theory of molecular olfaction one follows (153). According to one theory, the peripheral events of olfactory reception–recognition may be analogous to those occurring in the B-lymphocytes of the immune system. Thus there may be thousands of different receptor types, each neuron containing some of these types and each

type responding to a particular specific odorant. Similar to antibody molecules, the variety of the olfactory receptor types can be generated by changing the structure of a variable region of the olfactory receptor molecule that faces the nasal cavity while the constant region stabilizes the receptor within the membrane and helps trigger the intracellular transduction machinery (Fig. 13) (82). According to this theory, some of the olfactory discrimination and recognition processes occur early in the olfactory pathway, at the receptor neuronal level.

Another theory assumes the existence of a much smaller number of types of receptor molecules, each type responding to a class of odorants possessing presumably some common structural features, with each receptor neuron containing one or more of these receptor molecules. According to this view, olfactory discrimination and recognition of the numerous individual odorants (up to 2000 in humans) are thought to occur as a result of the action of neuronal circuits in the olfactory bulb and higher cortical structures, responding to the pattern of afferent input from the various receptor neurons. The second theory is more akin to what exists in other sensory systems (e.g., color vision) and is probably more compatible with the known electrophysiological responses of vertebrate olfactory neurons to odorants (70, 82).

The existence of odorant-binding surface receptor molecules is supported by genetic evidence. Thus, many humans and some animals suffer from specific hereditary anosmias, conditions in which the individual is unable to smell an odorant or a particular class of odorants but can smell other odors. Examples are specific anosmia to odors of *n*-butyl mercaptan (skunk odor), hydrogen cyanide, and freesia scent (7). These conditions are presumed to arise from the absence of a gene(s) coding for specific proteins involved in olfaction, the most likely candidate being olfactory receptor proteins. An interesting example is provided by the ability to smell the steroid androstenone. Nearly one third of humans are unable to smell this substance while in the remaining two thirds, one half report a urinelike smell and the other one half a musky, sandalwood perfumelike smell; identical twins show 100% concordance while dizygotic twins show 60% concordance (153).

In a cilia-enriched fraction of frog olfactory mucosa, a glycoprotein called gp95 has been found that fulfills many of the requirements such as integral disposition, glycoprotein nature, and high concentration in olfactory ciliary membranes. Radiolabeled ligand-binding studies show receptor-binding proteins for amino acids in fish olfactory epithelium. The binding affinity of these receptors correlates with the electrophysiological responsiveness of the receptor neurons (82). Saturable binding sites for camphor have been found in preparations from frog and rat olfactory epithelia. Proteins with high binding affinity for *o*-methylphenols (anisole) and for benzaldehyde (bitter almond) have been purified from dog olfactory mucosa. These proteins are of similar size (60 kD) but differing immunoreactivity (153).

SECOND MESSENGERS AND OLFACTORY TRANSDUCTION. Application of cAMP or its membrane penetrable analog dibutyryl cAMP to the olfactory epithelium can generate EOG responses similar to those produced by odorants (101). Olfactory transduction mechanisms, in many instances, may act by mediation of cAMP or related cyclic nucleotides (82, 87). Cyclic-AMP is formed from ATP by the action of a membrane-bound enzyme adenylate cyclase. Olfactory cilia are highly enriched in adenylate cyclase ac-

tivity (10 times more than brain membranes). In ciliary preparations, adenylate cyclase activity is increased by the addition of odorants, although GTP and a stimulatory G protein must be present (82, 87, 153).

It is postulated that odorants activate a receptor molecule, which in turn activates the adenylate cyclase via mediation of a regulatory G protein. As a result, cAMP is formed, which may act in two ways. Either it acts directly on a cation channel (Na or K) opening it and causing depolarization of the membrane, or cAMP may activate a protein kinase, which in turn causes phosphorylation of a cation channel. There is evidence for both of these pathways (79, 82, 87) (Fig. 13).

The activation of adenylate cyclase by an odorant depends on the type of odorant used. Thus fruity, floral, and minty agents are potent stimulants of adenylate cyclase while putrid odorants do so much less strongly, perhaps because there are fewer receptor molecules for putrid odorants (153). The concentration of odorants required to activate the cyclase is in the same range as that required for producing electrical responses in the receptor neurons. Recent patch clamp studies on frog olfactory epithelium also supports cGMP as an activator of ion channels (82, 87). A main value of second messengers in olfactory transduction is the increased olfactory sensitivity via an amplification cascade of effects. Thus the binding of one odorant molecule to the receptor can produce a thousand molecules of cAMP, which in turn can open at least an equal number of ion channels.

Olfactory Bulb

The olfactory bulb is a layered structure that constitutes the first synaptic and integrative station for processing olfactory signals (Fig. 14). The sensory signals ar-

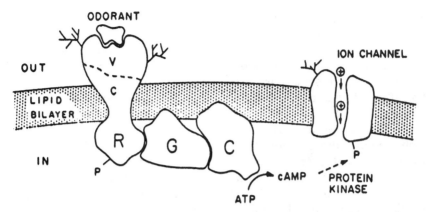

Figure 13. A proposed model for molecular events underlying olfactory transduction. The odorant binds to a binding site in the receptor molecule (R, an integral membrane protein). The receptor's region containing the binding site may be variable (V) (glycoprotein), while the intramembranous region is constant (C). Odorant binding activates a membrane-bound GTP-binding protein (G), which in turn activates the olfactory adenylate cyclase (C), which converts ATP to cAMP. Cyclic-AMP, acting as an intracellular messenger, activates a protein kinase, which in turn phosphorylates (p) an ion channel. This permits the inflow of cations, depolarizing the membrane. Phosphorylation may alter the binding capacity of the receptor molecule eliciting receptor adaptation. [From Lancet (82).]

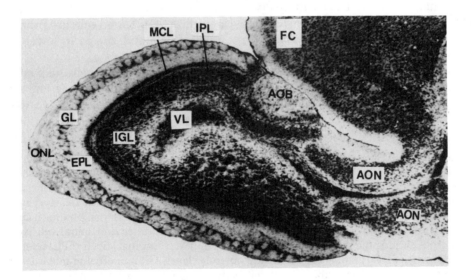

Figure 14. A midsagittal section through the rat olfactory bulb showing the various bulbar layers; the figure also shows some of the anterior olfactory structures and their relation to the forebrain. AOB, accessory olfactory bulb; AON, anterior olfactory nucleus; EPL, external plexiform layer; FC, frontal lobe of the forebrain; GL, glomerular layer; IGL, internal granular layer; IPL, internal plexiform layer; MCL, mitral cell layer; ONL, olfactory nerve layer; VL, ventricular layer. (Meisami, unpublished).

riving from millions of olfactory neurons are transmitted to the principal output neurons of the bulb, the mitral cells, for relay to the higher olfactory center. Before relay, in the bulb, these signals are subjected to a variety of synaptic analysis and integration, carried out mainly by the two classes of inhibitory neurons. These inhibitory neurons are the periglomerular and granular cells that interact with the apical and basal dendrites of the mitral cells (142, 145). The bulb also receives a major centrifugal input from the brain.

NEURONS AND LOCAL CIRCUITRY OF THE OLFACTORY BULB. The basic functional layers of the vertebrate olfactory bulb and their constituent neurons are shown in Figures 15a and 15b; Figure 15b also denotes the interconnectivity and known transmitters of these neurons.

The olfactory fibers from various epithelial regions converge, to form numerous glomeruli on the bulb surface just underneath the olfactory nerve layer (6). Within the glomeruli, the terminal branches of olfactory fibers that release carnosine make excitatory synapses with the terminal branches of the apical dendrites of the mitral cells. Small, mostly

inhibitory, short axon neurons residing in the glomerular layer release dopamine and/or GABA and surround the glomerular neuropil; their processes make synaptic contact with the processes of olfactory axons and mitral cell dendrites. Some of the periglomerular cells exert inhibitory functions within the glomerulus, others between the glomeruli. Both axosomatic and dendrodendritic synapses are found within the glomerulus (Fig. 15). Axons of mitral cells, coursing through the olfactory tract, convey the integrated signals to the olfactory cortical structure (142, 145).

Another cell type with dendrites inside the glomerulus is the tufted cell, which is largely a relay cell communicating with the more anteriorly located olfactory brain structures, for example, the anterior olfactory nucleus, and which is involved in feedback interaction between bulb and brain (Fig. 15). Some of the smaller tufted-like cells may be intrabulbar association neurons (142). Tufted cells are prominent in mammals and sparce in lower vertebrates.

The periglomerular cells help analyze, filter, and integrate the afferent signals that then spread to the cell body and basal

Figure 15. (*a*) Diagram showing the various neuronal cell types of the vertebrate olfactory bulb, their connectivity and interactions (based on Golgi and EM studies); arrows show the directions of impulse movement; insets show various types of axodendritic and dendrodendritic synaptic interactions in the glomerulus (axon terminals of olfactory neurons exciting mitral and periglomerular cell dendrites) and in the external plexiform layer (dendrites of granule cells inhibiting mitral cell basal dendrites and mitral cell basal dendrites exciting the granule dendrites (reciprocally). C, centrifugal fibers; EPL, external plexiform layer; G, granule cell; GLOM, glomerular layer; GRL, internal granular layer; M, mitral cell dendrite; MBL, mitral cell body layer; ON, olfactory nerve layer; PG, periglomerular cell; SA, short axon cells; T_d, deep tufted cells; T_m, middle tufted cells; T_s, superficial tufted cell. [From Shepherd (145), after Getchell and Shepherd.] (*b*) Diagram showing the various synapses and the associated neurotransmitters of the vertebrate olfactory bulb. ACh, acetylcholine; Asp, aspartate; Enk, enkephaline; DA, dopamine; GABA, γ-aminobutyric acid; Glu, glutamate; 5HT, serotonin; LHRH, Luteinizing hormone releasing hormone; NA, norepinephrine; PG cell, periglomerular cell SOM, somatostatin; SP, substance P. [From Shepherd (145), after Halazs and Shepherd (60).]

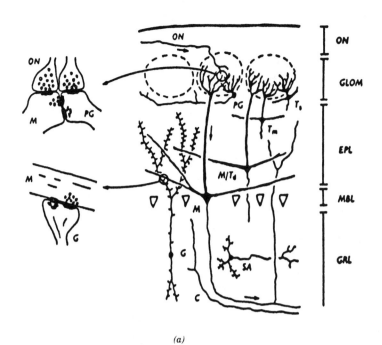

(a)

Olfactory nerve

Carnosine
Marker Protein

Short-axon cell
GABA?

PG cell
GABA
DA
Enk

PG cell

Tufted cell
Glu/Asp
DA
SP

ACh
5HT
Enk
LHRH

Mitral cell
Glu/Asp

ACh
5HT
NA
Enk

Short-axon cell
GABA?
Enk
SOM

Granule cell
GABA
Enk
DA?

ACh
5HT
NA
Enk

ACh
5HT
NA
SP
LHRH

(b)

dendritic tree of the mitral cells. The mitral cells that form a distinct layer of their own are constantly under the influence of inhibitory synapses of granule cells that release GABA as their neurotransmitter (Fig. 15). This interaction, forming a short "local feedback circuit" is carried out via reciprocal dendro-dendritic synapses, between the apical dendrites of the granule cells and the basal dendrites of the mitral cells, both of which ramify within the external plexiform layer (145).

The granule cells are constantly excited by the activity of the centrifugal fibers arising from reticular nuclei of brain, such as the locus coeruleus (adrenergic) and the diagonal band (cholinergic) or from the anterior olfactory nucleus. Activation of the inhibitory granule cells by the brain centers results in suppression of mitral cell activity (51). These interactive relations between the bulb and brain constitute a "long feedback circuit" (142, 145). One function of the local and long feedback circuits is to provide a basis for lateral inhibition between neighboring units of mitral cells, a phenomenon that is of importance in sensory tuning and discrimination. Thus a signal that is relayed by the mitral cells to the brain is never a copy of what it receives from the olfactory neuron, but a highly filtered and refined version, very different from the original message.

COMPARATIVE ASPECTS OF OLFACTORY BULB. The olfactory bulb shows a basically similar structural and presumably functional pattern among all vertebrates, although the relative abundance and development of the various cell types comprising the layers show substantial variation in different species (Fig. 16). Thus while the glomeruli, the mitral cells and the granule cells are the most persistent features, only the macrosmatic mammals show a highly cellular glomerular layer with numerous periglomerular cells; these cells are sparse in lower vertebrates (6, 10).

The basal dendrites of the mitral cells are highly developed in mammals. In fish and amphibia, the basal dendrites are absent or sparse. In mammals, a single unbranched apical dendrite of the mitral cell innervates a single glomerulus, while in other vertebrates from fish to birds, an apical dendrite is branched, innervating several glomeruli; thus each mitral cell is excited by signals from more than one glomerulus. Ultrastructurally, all the synaptic types found in mammalian bulb have been found in lower vertebrates as well (6, 10).

These structural variations may be functionally important. Thus sorting and segregation of olfactory stimuli, and hence their discrimination, may be more effective in mammals that have a more organized and well-laminated olfactory bulb. In contrast, in fish, the structural pattern of the bulb may aid in sensitivity, through increase in convergence between the olfactory fibers and mitral cells. Fish lack tufted cells but have a cell type called "ruffed cell" that may be homologous, serving similar function (48).

Among mammals the size of the olfactory bulb shows great variation (113). Relative to the brain size, the hedgehog has one of the largest and humans one of the smallest. The bulb in most species of dogs and in pigs is large. A variation in size is observed even among related species. Thus fruit eating bats have large olfactory bulbs compared to insectivorous bats (20). The size of the bulb has functional significance as it reflects the number of neural elements in it and is probably proportional to the size of the olfactory receptor sheet (or the number of olfactory neurons).

ELECTRICAL ACTIVITY OF OLFACTORY BULB. The olfactory bulb, like all cortical-type neural

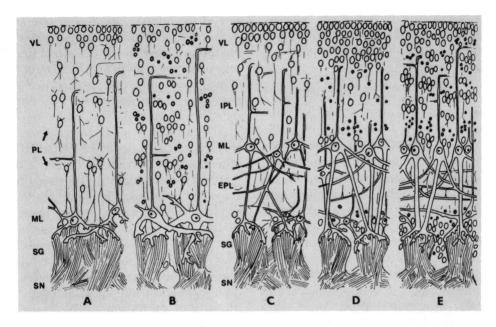

Figure 16. Semidiagramatic representation of the structure of the olfactory bulb in various vertebrates. *A*, lamprey; *B*, elasmobranchs; *C*, amphibian; *D*, reptile; *E*, mammal; parts *A* and *B*, have five layers while *C–E* have six (seven according to other authorities). EPL, external plexiform layer; IPL, internal plexiform layer; ML, mitral cell layer; SG, glomerular layer (stratum glomerulosum); SN, olfactory nerve layer (stratum nervosum); VL, ventricular layer. [From Andres (10).] Note that although olfactory bulb maintains an overall uniformity in structure, important functionally significant differences exist between vertebrate orders. Higher vertebrates (mammals) have: (1) a wider external plexiform layer due to development of basal dendrites of the mitral cells, (2) an internal plexiform layer, (3) increased abundance of the periglomerular cells and internal granule cells and separate layers for these cells (not shown in this diagram), and (4) only one mitral cell apical dendrite per glomerulus.

tissues, shows spontaneous EEG activity. In response to stimulation by odorants, surface recording of electrical activity in the olfactory bulb shows slow potentials (5). These bulb potentials resemble the EOG responses of the epithelium, in time course and general form, and may reflect the collective activity in the glomeruli (121). Oscillatory waves, called "induced waves" that are a form of EEG, are superimposed on these slow potentials (5). The source of these induced waves may be the mitral cells or the granule cells. Odor-induced bulb potentials have been recorded from the olfactory bulbs of all vertebrate classes from fish to mammals (121).

Earlier studies had indicated that all parts of the bulb surface of the rabbit respond to different odorants, but the response latency and amplitude vary depending on location. Thus anterior regions responded maximally to more volatile water soluble odorants like ether while the more posterior regions are the site of maximal responses to less volatile fat soluble odorant (e.g., pentane) (5). Later studies, recording evoked potentials directly from the mitral cell layer, indicated that the different populations of mitral cells show differential responses to various odorants (70, 106).

RESPONSES OF SINGLE MITRAL CELLS. Single cell recordings in the turtle (93), tiger salamander (70), frog (70), and recently in the rat bulb (100, 167) have provided a picture of the responses of individual mitral cells to various odorants. The mitral cells of the olfactory bulb show ongoing spontaneous activity in the absence of stimuli. This spontaneous activity changes upon exposure of olfactory epithelium to the odorants. Each mitral cell may respond to stimulation by one or more odorant. The response may be excitation (increase in activity), inhibition (suppression of activity), or no change at all (70). Figure 17 demonstrates the differential responses of 16 mitral cells in the turtle olfactory bulb to 18 odorants. Figure 18 shows the different patterns of excitation and inhibition in mitral cells of tiger salamander. Comparison of Figure 17 with Figure 9 shows that mitral and olfactory cells share some characteristics in that both cell types have individually specific response spectra to odorants. The main difference between the mitral and receptor cells is in the equal

role of inhibitory responses to odorants in mitral cells; inhibition plays no role in the receptor neurons. Excitation of mitral cells occurs as a result of activation of the olfactory receptors that are sensitive to a particular odorant and to which the mitral cells are connected. The inhibitory response of a mitral cell, however, presumably occurs as a result of excitation of other mitral cells that are adjacent or distant to it by an odorant and the resulting activation of local and long inhibitory circuits through lateral inhibition. Increase in the intensity of odorants usually changes the response pattern of excitatory neurons but not those of the inhibitory ones. In mammals an increase in odor concentration may increase the number of excitatory units recruited (100, 167).

Higher Central Olfactory Structures and Connections

OLFACTORY TRACT. In lower vertebrates, the axons of the mitral cells provide the main output of the olfactory bulb that leaves the bulb via the olfactory tract. In higher vertebrates, where tufted cells are found, these cells also contribute output fibers.

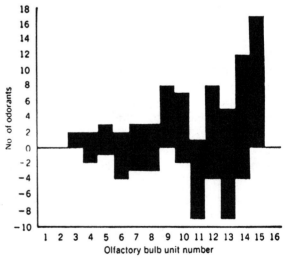

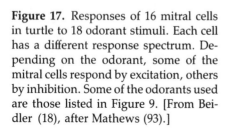

Figure 17. Responses of 16 mitral cells in turtle to 18 odorant stimuli. Each cell has a different response spectrum. Depending on the odorant, some of the mitral cells respond by excitation, others by inhibition. Some of the odorants used are those listed in Figure 9. [From Beidler (18), after Mathews (93).]

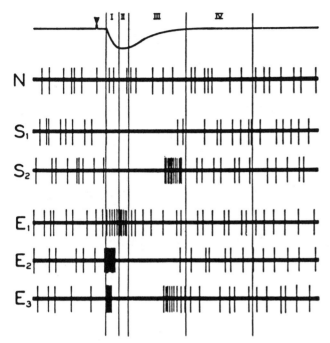

Figure 18. Details of response patterns of single olfactory bulb mitral–tufted cells in the tiger salamander after 1-s pulses of odorant stimulation. Periods of excitation and suppression for each response category are shown in relation to phases of the EOG designated I–IV. Response categories are N, no response (spontaneous activity only); S_1 and S_2, examples of inhibition (suppression of activity); E_1, E_2, and E_3, examples of excitation (increase in activity). [From Kauer (70).]

The length of the olfactory tract varies in different species depending on the location of the bulb with respect to the brain. In humans and other primates the tract forms a distinct and separate nerve bundle. In macrosmatic mammals, where the bulb is positioned close to the brain, the tract is called the lateral olfactory tract due to its course along the side of the olfactory peduncle. In fish the tract has a lateral and a medial component (Fig. 21*b*).

OLFACTORY CORTICAL AREAS. In mammals, tufted cell axons project to the anterior olfactory nucleus and other anteriorly located olfactory structures (142). These fibers provide excitatory input for negative feedback between the brain and the bulb cells. The anterior olfactory nucleus also mediates the excitatory input to the contralateral olfactory bulb and cortical olfactory structures. The ipsilateral cortical structures targeted by the mitral cell axons are the anterior olfactory nucleus, primary olfactory cortex (prepyriform

cortex), and the entorhinal cortex, all part of the basal paleocortex (Fig. 19). In addition to these olfactory structures, the mitral cell axons make contact with the anterior and posterolateral cortical amygdala and the olfactory tubercle. These projections are in general ipsilateral. The axons of the mitral cells traverse a long course along the base of the brain, contributing collateral fibers to these primary and association paleocortical structures. Thus a mitral cell may make contact with numerous pyramidal cells within each cortical structure (e.g., primary olfactory cortex) and with other cortical and limbic structures (133, 142).

The organization of olfactory cortex and bulbar projections to it are known in some details in the rat (59, 133). The major input is from the bulb (Fig. 20*a*). Projecting fibers run superficially under the pia and synapse with the apical dendrites of the most superficial pyramidal cortical neurons (Fig. 20*b*). Individual mitral cells in the bulb project to widely spaced cor-

Figure 19. Right half of this sketch of the ventral view of the rodent brain depicts the primary target structures of the output cells (mitral and tufted) of the main olfactory bulb; the left half depicts targets of the accessory olfactory bulb. Second order connections to the hypothalamic areas are also shown. Acc Olf, Bulb, accessory olfactory bulb; AHp, anterior hippocampus; Ant Olf Nuc, anterior olfactory nucleus; Bed Nuc Stria Term, bed nucleus of stria terrminalis; C1, C2, C3 and M are the nuclei of amygdaloid complex (C1, anterior cortical nucleus, C2, posterolateral cortical nucleus, C3, posteromedial cortical nucleus, M, medial nucleus); Ent, entorhinal area; MPOA-AH, medial preoptic area-anterior hypothalamus; NLOT, nucleus of the lateral olfactory tract; Olf Bulb, main olfactory bulb; Olf Cortex, pyriform cortex (primary olfactory cortex); Olf Tub, olfactory tubercle; NAOT, nucleus of accessory olfactory tract; PMN, premammilary nucleus; VMH, ventromedial nucleus of the hypothalamus. [From Johnston (67a).]

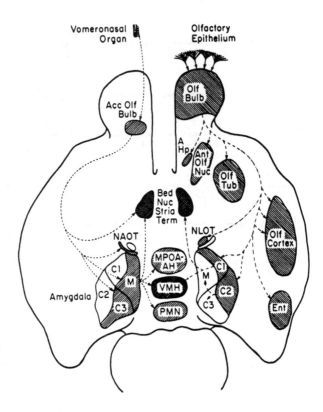

tical areas. There is no clear topographic organization in these bulbar projections. Small areas of the olfactory bulb project to large areas of the olfactory cortex and conversely small areas of the cortex receive fibers from all parts of the olfactory bulb (59, 133).

The olfactory cortex (pyriform cortex) provides a neural substrate for the integration of olfactory information transmitted from the olfactory receptor sheet through the olfactory bulb and for the transfer of this information to other forebrain areas like the frontal cortex, thalamus, and hypothalamus (Fig. 20a). Although spatial organization may be significant in this process, fibers from all parts of the olfactory bulb are distributed diffusely across the cortex with no clear topographic organization. Neurons in the olfactory cortex, like those in the olfactory bulb, are not closely tuned to primary odors or other specific stimuli and respond to a large number of odorants. Intracortical associational fibers provide for interaction between different parts of the olfactory cortex (Fig. 20b). Unlike the neocortex, the olfactory cortex is thin and poorly developed radially; thus cortical processing of olfactory stimuli are mediated by these association projections (133).

Thus spatiotemporal patterns of activity distributed across a large area of the olfactory cortex may be the primary factors in detecting and discriminating odors. According to Haberly (1985), sensory information in the olfactory cortex

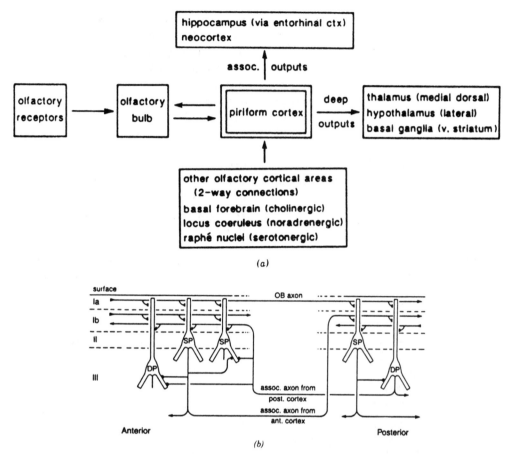

Figure 20. (*a*) Diagram showing the mammalian brain structures which provide input to the piriform cortex (primary olfactory cortex) and those which receive output from it. (*b*) The piriform cortex is a three layered structure containing deep (DP) and superficial (SP) pyramidal neurons which are extensively interconnected. The main input to these neurons is via the output fibers (axons of mitral cells) from the main olfactory bulb. Excitatory intracortical association fibers connect the anterior and posterior parts of the olfactory cortex together. [From Haberly and Bower (59).]

may be in the form of a highly distributed ensemble code; parts of the olfactory cortex particularly the pyriform area may serve as a "content addressable memory," which would detect and discriminate odors on the basis of evoked patterns of activity and would also provide for association of present stimuli with memories of past odors. This model is similar to some of the theoretical neural network models for memory (59).

CONNECTIONS WITH LIMBIC SYSTEM, HYPOTHALAMUS, AND FRONTAL NEOCORTEX. Tertiary fibers from olfactory cortical areas and from the amygdaloid–septal complex make two sets of connections with other brain structures (Figs. 19 and 20a). One set of connections is made with paleocortical structures and with those of the limbic system, especially the hippocampus and hypothalamus. Through these connections the main olfactory system finds access to the memory mechanisms of hippocampus and provides input into the controlling mechanisms of hypothalamus over drives, emotions, and instinctual behaviors (e.g., feeding, rage and aggression, sex and reproduction). A second neocortical set is reached via connections with medial dorsal thalamus; this set aids the olfactory signals to have access to frontal cortex and to other sensory regions served by the neocortical forebrain areas, centers like vision, audition, and especially taste (133).

COMPARATIVE ASPECTS. Early research indicated that olfaction provides the most massive and dominant input into the telencephalon of fish brain. Recent findings show that the olfactory input, while prominent, is less widespread than previously believed (48, 165a). For example, in the goldfish, mitral cell axons, leaving via a medial and a lateral tract, project bilaterally to two distinct zones, one to a ventromedial area and another to a basolateral area of the telencephalon; some fibers also reach the diencephalon, near the hypothalamus or neighboring areas (47, 48, 63) (Fig. 21c).

In the carp, the olfactory nerve consists of two main bundles, medial and lateral. The medial is derived from the more rostral lamellae while the lateral is from the more caudal lamellae. These main branches innervate the lateral and medial aspects of the olfactory bulb, respectively (Fig. 21a). Similarly from the bulb, a lateral and a medial group of fibers emerge. The lateral group innervates the lateral and posterior terminal fields while the medial group innervates the medial and posterior terminal fields of the ipsilateral telencephalon (Fig. 21b). Both groups also send fibers to the contralateral telencephalon via the anterior commissure. Based on this evidence Oka et al. (119) and Hara (63) have hypothesized that in fish there are two distinct, spatially sep-

arate olfactory subsystems that subserve different functions: a lateral olfactory subsystem that processes feeding behavior elicited by substances such as amino acids and a medial subsystem that processes other behaviors, for example, reproductive (63, 119).

The olfactory system, more than any other sensory system, exerts direct control over the structures of the limbic system. In fish and reptiles, in which a true neocortex is lacking, olfactory input provides massive input into the limbic structures of the forebrain, thereby controlling much of the instinctual and vegetative functions of the animals. For this reason many forebrain structures that are currently classified as part of the limbic system were originally considered as parts of the rhinencephalon (smell brain)! This term is now relatively abandoned, especially in relation to higher vertebrates in which olfaction is not always predominant, compared to other senses, and the limbic areas have become fairly isolated from purely olfactory influence.

In humans the importance of olfaction on instinctual behavior and expression of emotions cannot be doubted (45). No other sense can make a face show such extremes of disgust and pleasure in response to a single presentation of unpleasant and pleasant odors, respectively. These innate facial responses are seen even in newborn infants upon presentation of pleasant and unpleasant odorants (96). The intimate contact of the olfactory system with the limbic system and possibly temporal lobe structures may also explain why olfactory memories seem to be so persistent and firmly associated with events occurring in the distant past.

Olfactory Sensitivity
A fox can detect meat–prey odors at a distance of nearly a mile. Humans can detect the odor of a skunk at similar dis-

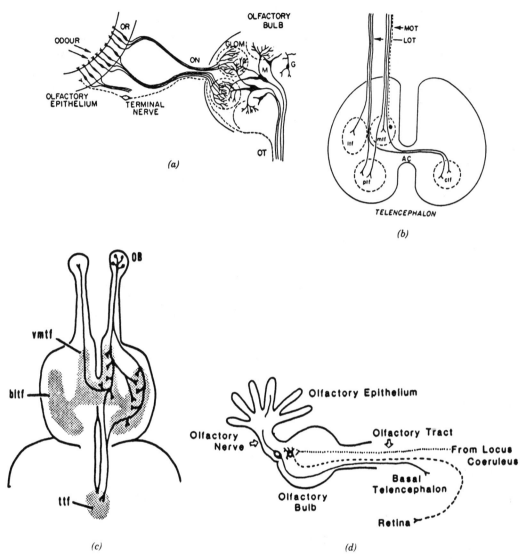

Figure 21. Organization of fish afferent olfactory pathways and terminal fields in the telencephalon. (*a*) Olfactory receptor neurons, the two divisions of the olfactory nerve and the main elements of the olfactory bulb; (*b*) The two divisions of the olfactory tract and their terminal fields in the telencephalon. (AC, anterior commissure; G, granule cell; GLOM, glomerulus; LOT, lateral olfactory tract; M, mitral cells; MOT, medial olfactory tract; ON, olfactory nerves; OR, olfactory receptors; OT, olfactory tract; ctf, contralateral terminal field; ltf, lateral terminal fields; mlf, medial terminal field, ptf, posterior terminal field.) (*c*) Fibers of the olfactory tract from one bulb project to the contralateral bulb, a ventromedial terminal field (vmtf), basolateral terminal field (bltf), and a tuberal terminal field (ttf) (goldfish). (*d*) the relation of the bipolar neurons of the terminal nerve to the olfactory epithelium and bulb in the fish *Carassius*. Note also the multipolar neurons of the nucleus olfactoretinalis that project from the bulb to the retina. The path of the terminal nerve in the fish is also indicated in parts *a* and *b*. [Parts *a* and *b*, Hara (63), after Oka (119); parts *c* and *d*, Finger (48), in part after the works of Meyer and Munz.]

tance, even in calm air (27). Under these conditions, the concentration of the stimuli in the nasal cavity is extremely low, in the nanomolar to picomolar range. Thresholds to several odorants for humans and other animals are listed in Table 1. Thus goldfish olfactory epithelium responds to a pheromonal steroid (17α,20β-dihydroxy-4-pregnene-3-one) at a concentration of 10^{-13} M as shown by EOG recordings (154). The German shepherd threshold for detection of a ketone substance α-ionone (irisone, cedar wood odor) was estimated to be as low 10^{-14} M (107). Even the human, not known for olfactory sensitivity, can smell methyl mercaptan (active odorant in garlic) at a concentration of 40 ng/L of air (about 10^{-10} M) or even more remarkably, an active odorant of thiamin, bis-(2-methyl-3-furyl)disulfide, at a water concentration of 2 parts per 100 trillion! (10^{-14} M) (28). The biological basis for the great sensitivity of the olfactory system may be explained at different levels.

ROLE OF SECOND MESSENGER CASCADE AND CILIA. At the subcellular and molecular levels, olfactory neurons amplify the effect of odorant binding by using a cascade of intracellular second messengers, such as that of c-AMP. Thus the binding of a single odorant with its receptor molecule can result in the opening of hundreds of ion channels, ensuring electrical activation of the olfactory receptor neuron. Another structural specialization of the olfactory neurons that increase their sensitivity is the presence of several long cilia in each olfactory neuron. These cilia that contain the odorant receptor molecules increase the number of odorant-binding sites per neuron by several-hundred folds (53, 82).

Given the structural similarity of the olfactory receptor neurons in different species, it is likely that the response sensitivity of a typical neuron to an odorant is similar in different species. Moulton calculated the sensitivity of a single olfactory neuron to α-ionone (irisone, cedar wood odor) in both humans and dog, and found that in both only one molecule of the odorant per receptor neuron was necessary to cause a minimum perceptible response (107).

ROLE OF NEURON NUMBER AND CONVERGENCE. Another mechanism and possibly the most unique and universal way by which the olfactory system provides for a very high level of detection sensitivity is by having a very large number of primary olfactory neurons that converge to a much smaller number of second-order relay cells (mitral cells in vertebrates) in the olfactory bulb. Although there are wide species differences in the total number of olfactory neurons in the nasal cavity, in most vertebrates, and even in invertebrates, the range is in the order of millions and tens or hundreds of millions (Table 2). Macrosmatic animals have many more olfactory neurons than microsmatic ones. For example, the dog (German shepherd) has nearly a billion receptor neurons, 100 times the number found in humans. Rabbit has 100 million and rat 60 million on both sides of the nasal cavities (97, 105).

However, in all species, the olfactory neurons converge on the glomeruli of the olfactory bulb to form excitatory synapses with the central relay (mitral) cells. The convergence ratio is estimated to be about 1000:1 (olfactory neurons per mitral cell) for the macrosmatic species studied (Table 2). The high convergence means that at extremely low stimulus concentration there will still be enough receptor units activated to excite a single mitral cell. Since each mitral cell is in turn connected to numerous cortical units through axon collaterals, many cortical units will be excited by the activity of a single mitral cell (97, 163). Neurophysiologically, this high convergence permits spatial summation

TABLE 1. Olfactory Thresholds for Some Vertebrate and Invertebrate Species[a]

Odorant	Threshold (M)	
	A. Mammals	
	Humans	Other Mammals
Ammonia	1.8×10^{-6}	
n-Amyl acetate	1.0×10^{-3}	Rat 3.2×10^{-5}
		Rabbit 3.2×10^{-5}
Amyl alcohol	1.1×10^{-8}	
Amyl thioether	5.8×10^{-8}	
Benzaldehyde	4.0×10^{-9}	
Benzol	6.8×10^{-8}	
Bis(2-methyl-3-furyl) disulfide	2.0×10^{-14}	
Butyric acid	1.1×10^{-11}	Dog 1.5×10^{-17}
Caproic acid	3.3×10^{-10}	Dog 6.0×10^{-18}
Chloroform	2.8×10^{-5}	
Citral	6.6×10^{-10}	
Diacetal		Dog 1.7×10^{-18}
Ethanol	5.4×10^{-6}	
Ethyl butyrate	8.6×10^{-6}	
Ethyl ether	1.3×10^{-8}	
Ethyl mercaptan	7.0×10^{-13}	
Eugenol	1.4×10^{-9}	
Iodoform	5.1×10^{-7}	
Ionon	5.0×10^{-13}	Dog 5.1×10^{-17}
Isovalereanic acid		Dog 1.8×10^{-7}
Limonene	7.3×10^{-5}	
Methyl mercaptan	8.3×10^{-11}	
Methyl salicylate	6.6×10^{-5}	
Musk xylene	7.1×10^{-13}	
Phenol *p*-chlorophenol	4.3×10^{-8}	
Propionic acid	6.7×10^{-10}	
Pyridine	5.0×10^{-10}	
Skatol	3.0×10^{-12}	
	B. Birds	
Benzaldehyde		Pigeon 6.0×10^{-7}
Butanoic acid		Vulture 1.0×10^{-6}
Butanol		Pigeon 8.0×10^{-7}
Butanethiol		Pigeon 6.0×10^{-4}
		Magpie 5.0×10^{-4}
Ethanethiol		Pigeon 1.0×10^{-8}
		Magpie 8.0×10^{-4}
Heptane		Pigeon 3.0×10^{-7}
		Chicken 5.0×10^{-7}
Hexane		Pigeon 2.0×10^{-6}
		Chicken 8.0×10^{-7}
		Quail 3.8×10^{-6}
Methanol		Pigeon 8.0×10^{-7}
		Magpie 1.5×10^{-6}
n-amyl acetate		Pigeon 5.0×10^{-7}

TABLE 1. Olfactory Thresholds for Some Vertebrate and Invertebrate Species[a]—Continued

Odorant	Threshold (M)
	B. Birds
n-butyl acetate	Pigeon 2.0×10^{-7}
Pentane	Pigeon 1.8×10^{-5}
	Quail 9.0×10^{-6}
Trimethylamine	Vulture 1.0×10^{-5}
	C. Aquatic Vertebrates
n-Amyl acetate	Tortoise 1.0×10^{-3}
L-Arginine	Catfish (EOG) 1.0×10^{-8}
L-Cysteine	Goldfish (EOG) 1.0×10^{-9}
Eugenol	Minnow 6.0×10^{-14}
Phenylethyl alcohol	Minnow 4.3×10^{-14}
β-Phenylethyl alcohol	Eel 2.8×10^{-18}
β-Phenylpropyl alcohol	Trout 9.9×10^{-9}
17-α, 20-β-Dihydroxy-4-pregnene-3-one	Goldfish 1.0×10^{-13}
Progesterone	Goldfish (EOG) 1.0×10^{-12}
L-serine	Goldfish (EOG) 1.0×10^{-8}
Taurocholic acid	Goldfish (EOG) 1.0×10^{-9}

	D. Invertebrates	
	Insects	Other Invertebrates
ADP		Shrimp 1.0×10^{-5}
AMP		Shrimp 1.0×10^{-7}
Ammonia	Musca 2.0×10^{-6}	
Benzol	Habrobracon 6.4×10^{-6}	
Bombykol	Polyphemus 1.7×10^{-7}	
	Bombyx 2.5×10^{-12}	
Butyric acid	Bee 1.7×10^{-9}	
Caproic acid	Bee 3.6×10^{-10}	
Citral	Bee 1.0×10^{-10}	
Ethanol	Musca 7.2×10^{-3}	
	Habrobracon 1.6×10^{-4}	
Eugenol	Bee 3.3×10^{-11}	
Glycine		Shrimp 1.0×10^{-5}
Ionon	Bee 2.5×10^{-11}	
β-Phenylpropyl alcohol	Bee 3.8×10^{-12}	
Propionic acid	Bee 7.0×10^{-10}	
Skatol	Geotrupes 2.3×10^{-8}	
Taurine		Shrimp 1.0×10^{-5}

[a]Data from various sources cited in the text especially 31, 135, 166, 154. All values are behavioral thresholds unless indicated otherwise [e.g., electrophysiological thresholds obtained by EOG (electroolfactogram)].

TABLE 2. Numbers of Primary Olfactory Neurons and Second-Order Relay Cells and Their Convergence Ratios in Some Vertebrates and Invertebrates

Species[a]	Number of Olfactory Neurons	Number of Mitral or Output Cells	Convergence[b] Ratio
Rat			
Newborn	1×10^6	40×10^3	25
Weanling	10×10^6	40×10^3	250
Adult	32×10^6	40×10^3	800
Rabbit			
Newborn	7×10^6	50×10^3	150
Weanling	35×10^6	50×10^3	700
Adult	50×10^6	50×10^3	1000
Fish (burbot)	11×10^6	10×10^3	1100
Bats			900
Cockroach			
General odorants	200×10^3	250	800
Pheromone system	80×10^3	20	4000

[a]Data sources: *Rat*, Meisami (97), and Paternostro and Meisami (125a); *rabbit*, Meisami, Hudson, and Distel (97a), and Allison and Warwick (6a); *fish*, Geme and Doving (51a); *cockroach*, Boeckh et al. (21a); *bat*, Bhatnagar and Kallen (20).
[b]Convergence ratio is defined as the number of primary olfactory neurons converging on each second-order relay cell and is obtained by dividing the former by the latter.

of the excitatory signals, causing facilitation and amplification of the afferent transmission (43, 163).

In our studies we have also shown that the marked increase in olfactory sensitivity observed in the developing young of mammals, particularly those born immature, like the rat, cat, and dog, may be explained in part by the increase in the number of olfactory receptors neurons and consequently in the ratio of convergence occurring postnatally. In the rat, a minimum of 10-fold increase in convergence ratio is seen during the suckling period alone (96, 97) (Table 2).

Olfactory Discrimination

ROLE OF OLFACTORY EPITHELIUM. If each olfactory neuron has a broad response spectrum to odorants, there must be other factors that enable the brain to recognize information regarding a particular odorant and discriminate it from other odor-

ants. It is possible that collective activity of specific populations of olfactory neurons from the entire mucosa provides the brain olfactory centers with patterned information. The olfactory epithelium participates in this process through temporal and spatial patterning of olfactory signals (94).

Temporal Patterning. As a particular odorant moves across the mucosal sheet, each receptor unit sensitive to that odorant produces its own pattern of response and transmits it to the brain. Thus brain cells connected to receptors for a specific odorant receive a particular set of signals with a distinct temporal order (70). The temporal patterning reflects both activity in the individual units and among the various units across the mucosa that are stimulated at different times. Presumably the brain centers are then tuned to associate this temporal pattern with a particular odorant. Thus a mechanism similar to *across fiber pattern coding*, which is believed to operate for the sense of taste and hearing, can also operate for odor recognition.

Spatial Patterning. Spatial patterning requires that the different groups of receptor cells be clustered in specific parts of the mucosa or that the physical factors governing the distribution of odorants act in such a way that each odorant receives differential distribution across the mucosa.

Evidence for the clustering is that when EOG responses are recorded from small areas of the tiger salamander's mucosa, in response to punctate delivery of odorants, it is found that while all odors activate receptors in all parts of the mucosa, the magnitude of the response differs depending on the location in the mucosa. Thus each odor seems to have an area where maximal response is observed and one where the response is minimal. For example, the maximal response locus for butanol is in the anterior mucosal region, while that for limonene is in the posterior mucosal regions (88, 106). A differential increase in metabolic activity (radioactive 2-deoxyglucose uptake) in response to stimulation by isoamyl acetate occurs between the dorsal and ventral regions of the salamander epithelium (70).

Further evidence for spatial factors is that in the frog epithelium, odorants entering the nasal cavity distribute themselves unequally across the mucosa, based on their physical and chemical properties. Thus the nasal cavity and its olfactory mucosal sheet act like a chromatographic column, adsorbing various compounds at different rates based on their physical and chemical properties. This means that some odorants traverse the length of nasal cavity rapidly while others move slowly (110).

Recording from the two different ol-factory nerves that emerge from the anterior and posterior areas of the frog mucosa showed that a slowly penetrating odorant like butanol produced a more intense response from the anterior areas of mucosa when the odorant was delivered through the external nare (near the anterior mucosal areas). Reversing the flow direction by allowing the odorant to enter through the internal nare caused higher activity in the posterior nerve (110). This suggests that receptors in the anterior and posterior areas have similar sensitivity and their responses depend on the physical factors controlling the amount and kind of odorant.

Spatial patterning of olfactory signals is more possible in animals like reptiles and amphibians with relatively simple olfactory cavities in which the nasal air flow moves passively over the mucosal sheet. In macrosmatic mammals where the olfactory cavities are complex and sequestered from the normal path of air flow through the nasal–respiratory cavities, additional mechanisms such as frequency and depth of sniffing become increasingly more important. These help shunt the air currents to the olfactory cavities and to distribute the odorants across the receptor sheet.

ROLE OF OLFACTORY BULB

Projection Patterns of Olfactory Epithelium to the Bulb. The spatial pattern of projection of the olfactory neurons in the epithelium to the olfactory bulb may be important in olfactory discrimination. In rabbits the projections of the olfactory receptor sheet to the bulb follow a general pattern. Thus dorsal areas of the epithelium project to the dorsal areas of the bulb; similarly, the more anterior and posterior epithelial regions project to the more anterior and posterior regions of the bulb (5). Within this general order, there is no point-to-point correspondence between the receptor sheet and bulb, such as exists between the retina and lateral geniculate body.

Small lesions in the epithelium produce widespread degeneration in the glomerular layer. Lesions show that the epithelium-bulb projection pattern may follow two complimentary systems: a divergent projection system that connects a small area of the epithelium to a wide area of the bulb and a convergent projection system that connects a large area of the epithelium with a small area of the bulb (see 70). This system ensures that receptors located in a small patch of the epithelium have access to a fairly large population of glomeruli and the mitral cells that are connected to them, and that each glomerulus receives fibers from relatively distant areas in the epithelium.

Bulbar Odorant Maps. Central relay neurons increase their metabolic activity when their sensory afferent neurons are stimulated. A permanent record of the increase in metabolic activity can be obtained by measuring the uptake of radioactive 2-deoxyglucose. When rats are exposed to various odorants, it is found that each odorant has a corresponding locus over the olfactory bulb that shows increased activity. These regionally specific loci, which involve mainly the glomeruli, are often observed in the lateral and medial surfaces of the olfactory bulb and may overlap for different odors (144–146). If these loci represent true odor maps, then higher brain centers may read the odor-related information by receiving signals from these distinct bulbar regions. This type of spatial mechanism for odor discrimination has been called the *labeled line pathways* (144). However, in these studies exposure time to odorants is usually fairly long. Thus it is not certain whether these loci represent true odor maps or a secondary effect due to activation of centrifugal fibers or other metabolic factors. Also, it is not clear how the cortex can read these regional bulbar maps since mitral cells projecting from different parts of the bulb do not end in specific olfactory cortical areas but send collaterals to many cortical loci (59, 133).

Bulbar EEG Patterns Upon Odorant Activation. A different mechanism, developed by Freeman, is based on EEG studies from the bulb surface using multielectrode arrays that permit simultaneous recording from numerous loci (51). Exposure to an odorant activates many loci in the bulb, although some loci appear more active than others. Specifically, EEG recordings from 64 chronically implanted electrodes on the olfactory bulbs of rabbits that are trained to discriminate odor and conditioned stimuli show that the odorants induce quality distinctive amplitude patterns of neural activity. The odor-specific information density is inferred to be uniform over the whole main bulb. Excitatory synapses between mitral cells are subject to modification when odorants are paired with unconditioned stimuli, thus forming nerve cell assemblies. Odorant-specific information established by a stimulus locally in the bulbar unit activity is integrated with past experience by an assembly disseminated over the entire bulb. An arbitrary spatial sample on the order of 20% of bulbar EEG activity captures the entire integrated information albeit at lesser resolution than the whole. Thus, to differentiate odorants, the olfactory cortical areas presumably read a spatially dynamic pattern of signals transmitted simultaneously from the entire bulb.

PSYCHOPHYSICAL STUDIES AND PRIMARY ODORS. Moncrieff arrived at 62 general principles relating chemical structures to odor (102). Although intuition assumes that substances with similar odors should have similar structures, there are many compounds with widely differing chemical structures that smell the same. For example, all of the following compounds, although they have differing chemical structures and formulas, have the same camphoraceous odor: d-camphor, hexachlorethane, trinitroacetonitrile, silicononyl alcohol, pentamethyl ethyl alcohol, benzene hexachloride, durene, and tetrachlornaphthalane (8, 27).

There have been several attempts to categorize various odorants into primary groups of odors (130). Amoore (8) conceived of a minimum of seven primary odor groups for humans: camphoraceous, musky, floral, pepperminty, ethereal, putrid, and pungent. Although member compounds within each group have different chemical structures, they share similarity in stereochemical features (shapes) or surface charge distribution. This similarity in shape or charge of odorants permits interaction with the same receptor-binding protein on the olfactory receptor neuron, evoking similar cellular and psychophysical responses.

Minor changes in surface chemistry of odorants may cause marked changes in the odor quality (18). For example, ambergris, a foul-smelling substance regurgitated by the sperm whale, after minor chemical treatment, possibly involving a change of the surface charge from negative to positive, is converted into a pleasant smelling substance that is used in the perfume industry. Further evidence in support of this stereochemical theory comes from studies of individuals with various hereditary anosmias. These individuals may lack the ability to smell one particular group of primary odors (*specific anosmias*) while possessing the ability to smell others (7). These individuals lack the gene necessary for the synthesis of odorant receptor protein. The status of the number of primary odorant groups in

human olfaction is not settled. Animals like dogs and rodents may have a larger repertoire of primary odors than humans while some insects may have a more limited repertoire.

Uses and Adaptations of the Olfactory System in Vertebrates

FISH. Chemoreception is well-developed in fish and the anatomy of olfactory and gustatory structures are extensively described (Figs. 22 and 23). Fish make use of olfaction in their migratory, feeding, social, sexual, and reproductive behavior (63, 66). Sharks are well known for their keen sense of smell, especially for blood. They gather from far distances toward a source of blood. Olfaction is essential for homing behavior in fish, as exemplified by the upstream migration of the *Salmonids* from open oceans to their original breeding regions in the rivers. Two theories, the *olfactory hypothesis* and the *pheromonal hypothesis*, prevail in explaining the homing migration (63, 155). According to the olfactory hypothesis, migrating *Salmonids* imprint to certain distinct odors of the home stream during the early period of residence. As adults they use this olfactory information to locate the home stream. Salmons imprinted to morpholine during their early development, homed for the streams artificially scented with morpholine. The pheromonal hypothesis claims that homing is an inherited response to population-specific pheromone trails released by descending smolts (63, 155).

Olfaction is believed to be involved in the *fright reaction* observed in certain *cyprinid* fishes (e.g., minnow and carp) (129). When the skin of these fish is damaged, alarm substance cells (club cells) are broken, releasing the alarm substance that is identified to be hypoxyamine-3(N)-oxide in the minnow. Nearby conspecifics smell this alarm substance and show a fright reaction (63). The adaptive significance of the alarm reaction may be in dispersing the nearby members of the school after a member is attacked (152).

Olfaction plays an important role in several aspects of the fish reproductive process. Gonads, in addition to releasing sex steroid hormones into the bloodstream, seem to be the source of fish reproductive pheromones that are also steroidlike. Several water-borne steroidal substances (sex-steroid-like compounds and their glucuronide derivatives), acting as pheromones and influencing the olfactory system, are known to regulate fish sexual and reproductive behavior. For example, in the goldfish, $17\alpha,20\beta$-di-*OH*-4-pregnene-3-one, a steroidal hormone that promotes oocyte maturation, is released into water by preovulatory females; this pheromone stimulates the mature males to become sexually aroused and to liberate sperms. F Prostaglandins have also been found to function as fish reproductive pheromones (154).

Olfaction is involved in fish feeding behavior, which is a stereotypic and complex form of behavior. The role of olfaction seems to be in the initial arousal or alerting process. Although various feeding stimulants seem to be species specific, they all share the following chemical properties: low molecular weight, nonvolatile nitrogenous, and amphoteric. Biologically meaningful feeding stimulants may consist of a fingerprintlike mixture or chemical images. Mixtures of certain amino acids may be an example (63).

Adaptive modifications are observed in the anatomy of the fish nasal structures and considerable variation exists in the size of the olfactory receptor organ and central olfactory structures (173). The fish olfactory system is sensitive to dissolved compounds of small molecular weight, although volatility of the stimuli is not a condition of stimulation (162); as in land vertebrates, the dissolved stimuli represent food, mate, habitat, and so on.

In many fishes the floor of the nasal pit gives rise to a central ridge from which many lamellae radiate (Fig. 23). The lamellae are covered with olfactory epithelium resulting in a marked increase in the surface area of olfactory receptor sheet and the contact surface with flowing water and its dissolved odorants. The number of lamellae vary in different species (few in sickleback, 14 in salmon, 25 in the plaice, 30 in slime head, 40 in the gurnard, and 120 in eels and morrays) (63, 113). Because this difference reflects a difference in the total number of olfactory neurons, the number of lamellae may be proportionate to the degree of olfactory sensitivity of the fish. The olfactory organ of the deep-sea fishes like *ceratoid* angler and *cyclothone* spp. show a marked sexual dimorphism, the organ being larger in the male than female (63). In the flatfish, for example, flounder, which feed lopsided on the seafloor, the size of the olfactory organ and the central olfactory structures on the upward-facing side is considerably larger, compared to the downward-facing side (48, 131). It is not known whether this unique sidedness develops during ontogeny as a result of differential use or it is a genetic trait.

The olfactory epithelium in fish differs from the general pattern described in mammals in that it lacks Bowman's glands. Instead, goblet cells are found in the apical region which, along with the supporting cells, secretes a mucus that is thinner than that of the tetrapods. Surface density of olfactory receptor cells varies greatly depending on the species (50,000–500,000 cells/mm^2) while the total number per organ, estimated by counting the number of unmyelinated axons of the olfactory nerve, is between 42,000 and 13 million, depending on the species (31). Two types of bipolar receptor neurons have been recognized, ciliated and mi-

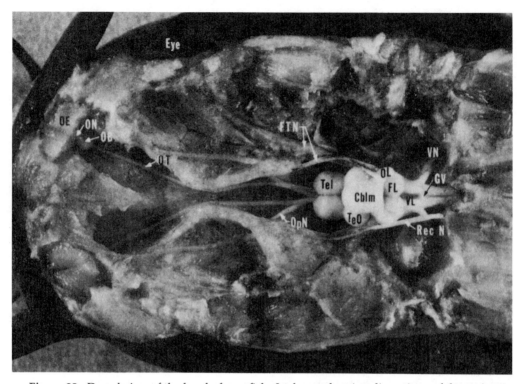

Figure 22. Dorsal view of the head of a catfish, *Ictalurus*, showing dissections of the main sensory organs, nerves and brain regions involved in chemoreception (Dorsal view). Cblm, cerebellum; FL, facial lobe; FTN, facial-trigeminal nerve (mixed nerve to barbels); OB, olfactory bulb; OE, olfactory epithelium; OL, octavolateral area; ON, olfactory nerve; OpN, optic nerve; OT, olfactory tract; Rec N, recurrent nerve of facial nerve that innervates taste buds on fins, flanks, and tail; Tel, telencephalon; TeO, optic tectum; VL, vagal lobe; VN, vagus nerve. [From Finger (48).]

crovillous. Some fish, like the sharks, contain only the microvillous, but most possess both types. The two types may be distributed randomly across the olfactory organ or preferentially segregated in specific regions. As a further elaboration of this regional segregation, it has been proposed that the ciliated neurons of the main olfactory organ and the microvillous receptors of the vomeronasal organ in higher vertebrates may be derived from clustering of these two ancestral receptor types in fish. The ciliated and microvillous types may also be functionally distinct (Fig. 23). In some fish such as the *salmonids*, the ciliated neurons respond to

pheromonal substances like bile salts while the nonciliated ones respond to amino acids. This is, however, not a general rule (31, 63).

Amino acids, fatty acids, nucleotides, and bile salts are known to act as olfactory stimuli in fish, but most studies have dealt with the amino acids. The source of the free dissolved amino acids in the water, present at a concentration of $<10^{-7}$ M, is probably the mucus secretion from the skin of other fish, aquatic invertebrates, or the decaying remnants of other aquatic organisms. They thus act as recognition signals or food, respectively. Generally neutral L-α- amino acids, par-

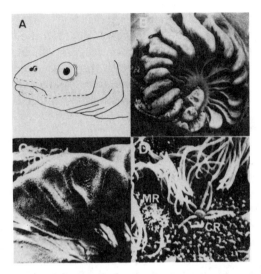

Figure 23. Structural details of the olfactory organ and epithelium in the fish (rainbow trout, *salmo gairdneri*. (A) Position of the nares and olfactory pit; scanning electron micrographs of the olfactory rosette (B); lamella (C), and a surface view of the receptor sheet (D). CR, ciliated receptor cells; MR, microvillar receptor cells. [From Hara (63).]

ticularly those with linear, unbranched, uncharged side chains like methionine, glutamine, cysteine, serine, and alanine, are potent olfactory stimuli, their threshold of activation varying between 10^{-7} and 10^{-9} M, as based on electroolfactogram recording studies (31) (Fig. 36A).

AMPHIBIANS, REPTILES, AND BIRDS. Amphibians and reptiles also use olfaction to search for food and water. Reptiles use smell to find bodies of water at great distances. The use of smell by snakes to track prey and food has been well established, although they may use the vomeronasal organ for some of these behaviors in preference to the main olfactory system. Larvae of frogs can distinguish kin from nonkin by use of olfactory cues. Sea turtles and other reptiles and amphibians may use olfactory cues for homing and orientation (44).

Most species of birds have a poorly developed sense of smell (e.g., the sparrow and finches); in others (e.g., the procellariform species- the sea gull-like Northern fulmar and shearwater) the olfactory system is very prominent both in size and function (14, 168). The procellariform birds and the kiwi, the New Zealand flightless bird, and some of the species of vultures, have keen sense of smell. The Fulmar has a well-developed olfactory epithelium and a large olfactory bulb, containing about 100,000 mitral cells, twice as many as that found in the rat and rabbit (168). Vultures are known to gather around gas leaks from gas lines buried under desert sand, presumably associating the mercaptan odors of the natural gas with those of the rotting bodies. Fulmars appear from far away distances to gather around whaling ships presumably in response to odors of the whale blood and viscera. The kiwi and raven use olfactory means to locate worms placed under the soil (64, 168). Starlings use olfaction to select specific nest materials with odors repellant to nest parasites. There is evidence that the homing behavior in the pigeon and some other birds is in part guided by olfactory cues (166). Nevertheless, the olfactory thresholds of birds to common odorants are in the micromolar range (10^{-5}–10^{-7} M); this is at the higher end of olfactory thresholds for mammals (166).

MAMMALS. Olfaction reaches a peak of usefulness and development in nocturnal ground-dwelling or carnivorous mammals. Many macrosmatic mammals are characterized by their large olfactory organs, cavities, and bulbs. The insectivore hedgehog has the largest olfactory bulb relative to the size of its brain. In mammals, olfaction is used in detecting and discriminating odors of foods and mates, and in behaviors of courtship and copulation, care and rearing of infants, recognition of kin, friends and foes, play and fighting, and social cohesion, in the marking of territories and other individuals (157).

Dogs and pigs also have very large and complex olfactory cavities and bulbs. Because of their ability to track animals, dogs have been used for centuries in hunting. Recently, dogs have been used to track buried humans (under snow or in earthquake rubbles), and to find drugs and contraband in borders and airports (95).

The sense of smell guides much of the feeding behavior in ungulates. Sheep refuse to eat a particular grass, based on smell, while eagerly accepting and devouring another. Even bats, the flying mammals, are macrosmatic. The larger bodied, fruit eating bats have very large olfactory systems that they use to find food sources (20). Finding hidden food, buried under heaps of soil and leaves, which is commonly practiced by dogs, rats, squirrels, and deer mice, is guided by olfaction. Deer mice, in pitch darkness, can find food (conifer seeds) hidden under 5 cm of peat (27).

Even the microsmatic monkeys use their sense of smell continuously and effectively. Not only do they smell each other, but food offered to these animals is often investigated by the nose before being placed in the mouth. In humans, olfaction is important in perceiving and discriminating food flavors. Without smell, the flavors of coffee, chocolate, molasses, and cranberry juice would be indistinguishable (109, 130). In addition to its role in appetite control, stimulation of digestive activity and feeding, olfaction plays an important role in the "good life," in interindividual hedonic relations, social preference and acceptability, and in aesthetics and cosmetics, as evidenced by the widespread use and popularity of aromatic flowers, perfumes, fragrances, and deodorants (45).

Other Nasal Chemosensory Systems

Vertebrate olfaction is typically identified with the main olfactory receptor sheet, the olfactory bulb, and the associated higher olfactory cortical areas. However, among the various vertebrates, particularly in mammals, the nasal cavities contain other chemosensory systems each of which supplies the animal with specific chemosensory information. The best known of these chemosensory systems, some of which are "olfactorylike" are (1) the vomeronasal organ (Jacobson's organ); (2) the chemosensory (nociceptive) receptors of the trigeminal nerve; (3) the sensory branches of the terminal nerve; and (4) the septal organ of Massera. Not all of these are present in every species, and when present, not all necessarily operate during the entire life of the animal (161).

THE VOMERONASAL SYSTEM

Location, Structure and Evolutionary Development. The vomeronasal or Jacobson's organ (VNO) is a pair of structures located anteriorly in the nasal cavity within the septum and embedded in a vomeronasal cartilage, just above the vomer bone in the roof of the mouth (Fig. 24). Vomeronasal organ has a fairly similar structure in various species, but its degree of development shows marked species variation. Fish and birds lack a VNO. Other aquatic vertebrates either lack a VNO or have a poorly developed one. Among the nonaquatic vertebrates, the development of the organ is diminished in arboreal species in contrast to ground-dwelling ones, as illustrated by comparing various lizard species. Snakes have a prominent VNO that they use extensively (55, 61, 113, 172)

The VNO is very prominent in many mammals such as the carnivores (dog and cat), rabbits, and rodents (3). Even the flying bats have a VNO, although it is

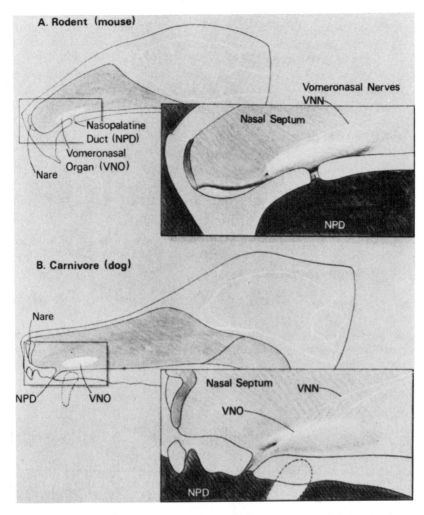

Figure 24. Sketch of the shape and position of the vomeronasal organ in the anterior ventral nasal septum of the mouse and dog. Note the relationship between the vomeronasal organ (VNO), vomeronasal nerve (VNN) and the nasopalatine duct (NPD). [From Wysocki and Meredith (172).]

more developed in the fruit-eating ones, which also have a well-developed main olfactory organ (20). The organ is lacking in aquatic mammals (dolphins and whales) and in certain insectivores (hedgehog) and in most of the higher primates (90, 113). In humans and presumably other higher primates, VNO appears during the fetal stages of development and degenerates before birth (96).

In mammals, each VNO (one on each side) is a cigar-shaped structure (Fig. 24), about 10 mm in length and 1 mm in height. A sensory epithelium covers most of the organ's length and is usually on one side only (Fig. 25a). The organ has a cavity that is blind-ended posteriorly with an opening anteriorly. The opening communicates with a vomeronasal canal, which is in close proximity to the naso-

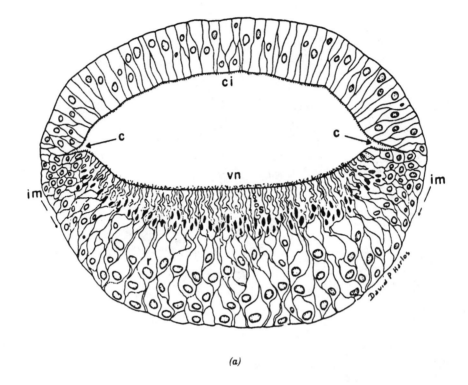

(a)

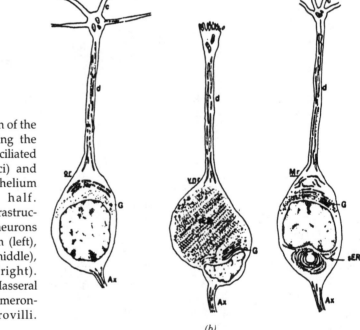

Figure 25. (a) Cross section of the vomeronasal organ showing the cavity in the middle, the ciliated nonsensory epithelium (ci) and microvillous sensory epithelium (vn) at the bottom half. (b) Comparison of the ultrastructure of the chemosensory neurons of the olfactory epithelium (left), vomeronasal epithelium (middle), and Organ of Massera (right). Note that olfactory and Masseral neurons bear cilia while vomeronasal neuron bears microvilli. [From Graziadei (56).]

(b)

palatine canal. This latter small canal permits communication between the oral cavity, the VNO, and possibly the nasal cavity. It is believed that snakes use particular tongue movement and some mammals (goat and horse) use particular lip and facial movements called "flehmen" to place substances directly in the nasopalatine canal or increase access of these chemical stimuli to the VNO (55, 61, 80, 172).

The VNO cavity is filled with fluid, secreted by mucus glands around the organ. Thus chemical stimuli must diffuse through the fluid to contact the VNO receptor surface. A pumping action of the organ increases stimulus access. The organ is surrounded by a rich vascular supply, cavernous sinuses and elastin. Vasoconstriction, directed by sympathetic nerves, contracts the sinuses and causes decreased pressure within the VNO cavity, creating a sucking action, which in turn enhances fluid movement into the cavity and stimulus access (172).

VNO Epithelium and Its Sensory Neurons. The sensory epithelium of the vomeronasal organ is shown in Figure 25*a*. It is a pseudostratified epithelium consisting of bipolar sensory neurons surrounded by supporting cells (158). The ultrastructural features of a typical VNO sensory neuron is shown in Figure 25*b*. This sensory neuron has much in common with the olfactory neurons of the main olfactory organ, except for the lack of cilia at the receptive surface and their replacement with numerous microvilli. This and the VNOs fluid-filled cavity have led to the speculation that VNO neurons are derived phylogenetically from the microvillous olfactory neurons of fish (19).

The microvillous surface of the VNO may be an adaptation to a lesser need for extreme odor sensitivity, compared to the cilia of the olfactory epithelium, which endow the olfactory neurons with increased receptive sites and heightened sensitivity. Thus the VNO, being tuned to more specific fluid-borne, nonvolatile, large molecular weight stimuli, may not require extreme detection sensitivity.

Accessory Olfactory Bulb and Its Higher Connections. The axons of VNO neurons form the bundles of the vomeronasal nerve (Figs. 19, 24 and 26). These run, through the nasal septum, pierce the cribriform plate, bypass the main olfactory bulb medially, and enter into the accessory olfactory bulb (AOB) to make synapses with the dendrites of the mitral cells of this structure (6).

The accessory olfactory bulb, located dorsally in the main olfactory bulb, is prominent in rodents (Fig. 14). The mitral cells of the accessory olfactory bulb, are smaller in size and like those of the main olfactory bulb of fish, reptiles, and birds, make contact with several glomeruli. This is unlike the mammalian main olfactory bulb, where each mitral cell makes contact with only one glomerulus. Other features, such as lack of tufted cells, and paucity of periglomerular cells, are also similar to the main olfactory bulb (OB) of the earlier vertebrates. As shown in Figure 19 axons of the mitral cells of the AOB project to the posteromedial cortical and medial nuclei of amygdala (the vomeronasal amygdala) and the nucleus of accessory olfactory tract (71, 133, 137). Tertiary contacts are made via the stria terminalis or by a ventral pathway with several of the limbic structures and particularly with the reproductive areas of the hypothalamus (i.e., medial preoptic area, bed nucleus of stria terminalis, ventromedial hypothalamus, and mammillary complex), controlling sexual behavior and reproductive functions (171, 172).

General Functions of VNO. The VNO, vomeronasal nerve, the AOB and its higher nerve centers in the amygdala and hypothalamus constitute a separate olfactorylike or nasal chemosensory system, serving in the control of social, sexual, and reproductive behaviors that are under the control of olfactorylike signals.

Experiments such as lesioning of the VNO, sectioning of the vomeronasal nerve, or destruction of the accessory olfactory bulb have been performed in hamsters and snakes. The VNO appears to be involved in regulating the following functions, either wholly or in part: timing of puberty, female cyclicity, reproductive hormonal responses, pregnancy blocks and individual identification, reproductive capacity and maternal behavior, and neonatal behavior (67, 99, 171, 172). In addition, many aspects of scent-marking behavior, aggression, and ultrasonic vocalization involve the VNO system. In snakes, numerous behaviors (feeding, attack, prey trailing and tracking, mating and nesting behaviors) are dependent on vomeronasal input (61, 80).

These functions imply that the VNO provides the sensory detector for those olfactory-guided behaviors that are nonadaptive, preprogrammed, and hardwired. The VNO system is also believed to act as a primer and reinforcer of certain sexual or social behaviors. In contrast to the main olfactory system, the stimuli affecting the VNO system are not believed to be consciously perceived.

Stimuli Acting on the VNO (Pheromones?). The most likely stimuli to activate the VNO may be the interindividual chemical mediators, that is, pheromones. The source of mammalian pheromones is most likely the various skin glands. Compounds found in the urine or feces of mammals (e.g., dogs), which evoke intense olfactory investigation are secreted by the urethral or anal glands and are widely used for territorial markings and identification of individuals and sexes. Vaginal and vulval secretions of the fe-

males are also known to be a source of substances of sexually and reproductively significant information. These pheromonelike compounds and the behaviors that they elicit have been investigated in hamsters and in cows (34, 139, 149).

It is believed that at least some of the mammalian pheromones that stimulate the VNO are proteinaceous substances, nonvolatile, and of high molecular weight. These substances presumably stimulate the VNO by near contact stimulation, as the animals searching for them bring their nostrils very close to the source material–location. Large compounds of more than 10 kD from hamster vaginal fluid placed on anesthetized male hamsters elicit copulatory behaviors from other male hamsters, presumably by stimulating their VNO. Compounds as large as 66 kD have been found to penetrate into the VNO cavity and presumably stimulate the receptors (149).

SEPTAL ORGAN OF MASSERA. The septal organ of Massera is an island of olfactorylike epithelium located on the septum, rostral, and ventral to the main olfactory epithelium (Fig. 26) and is prominent in some species (guinea pig). Thus the air currents

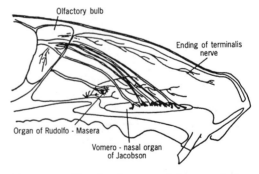

Figure 26. Sketch of some of the nasal chemosensory systems (vomeronasal organ, septal organ of Massera, and terminal nerve) of a rodent. [From Stoddart (157), after Bojsen-Moller and Graziadei.]

containing odorants normally pass by this receptor sheet before reaching the main olfactory epithelium. Because of this location, it was once thought that the organ serves as an early warning–detection system. However, the septal organ is probably not a separate chemosensory system, as it projects diffusely to the main olfactory bulb (56, 126a).

TERMINAL NERVE SYSTEM. The terminal nerve (*nervus terminalis*) connecting the nasal cavity with midline forebrain structures (*lamina terminalis*) is present in all vertebrates including humans; it is well developed in the human fetus and newborn (56, 96). The nerve is macroscopically visible only in the dogfish (56). The cell bodies of the nerve are located in scattered ganglia throughout its course in the nasal septum and on the medial side of the olfactory bulb. The free nerve endings of this nerve innervate the anterior nasal mucosa and the fibers run along with the vomeronasal nerve in the nasal septum, bypassing the main and accessory olfactory bulb, entering the brain, in mammals, with the anterior cerebral artery (172).

Although both autonomic motor and sensory functions have been attributed to the nerve, recent studies have focused on chemosensory functions. Many fibers of the terminal nerve contain LHRH (leuteinizing hormone releasing hormone) and it is now presumed that these neurons, along with other central neurons containing LHRH in the reproductive areas of the hypothalamus (e.g., preoptic area) are part of a network of neurons involved with regulation of some reproductive functions via nasal detection of pheromonal substances (36, 37). The importance of the terminal nerve system in these pheromonal functions is probably critical in species lacking a functioning VNO, such as the human and other primates (172). Similarly, in those marine mammals where the main and accessory olfactory systems are lacking, a functioning terminal nerve may mediate some pheromonal responses.

The path of the terminal nerve fiber in the goldfish brain has been traced (Fig. 21) (165). In male goldfish, electrical stimulation of the terminal nerve, possibly mimicking pheromonal activation, causes sperm release (36). Selective lesions of the terminal nerve in the hamster impairs mating behavior, particularly ejaculation in the presence of receptive females (172). In species of fish where the olfactory bulb is directly in front of the brain, a group of nerve cells (nucleus olfactoretinalis) is found at the bulb–brain border. The cells of this neuronal system, which is believed to be homologous with the terminal nerve, also contain LHRH and make contact with hypothalamus. This nucleus also has a connection with the retina and possibly mediates control over sperm release by visual and pheromonal stimuli (111).

TRIGEMINAL SYSTEM AND GENERAL NASAL CHEMOSENSITIVITY. In addition to the sensory systems designed to detect specific chemical stimuli such as odorants and tastants all vertebrates contain sensory fibers that are specialized to detect highly irritant chemical substances. These substances may or may not have odors or tastes. These specialized nociceptive fibers are a part of the *common chemical sense* system. The fibers of the common chemical sense are in general concentrated in body surface areas where true skin and its appendages are lacking, and the mucus surface is exposed (nasal cavity, oral cavity, cornea, anus, and genital opening), increasing the chance of harm and injury to the tissues (123, 148).

In the vertebrate nasal cavity, the common chemical sense consists of the nociceptive fibers of the trigeminal nerve, specifically the rami of the ophthalmic (ethmoid) and maxillary branches of the trigeminal nerve, the former serving the upper nasal areas while the latter innervates the lower nasal areas. This chemosensory system serves in the detection of noxious gaseous stimuli entering the nose and respiratory system.

The sensory receptors of these nociceptive fibers are free nerve endings that lie just under or in the olfactory and respiratory epithelium of the nasal cavity. These chemosensory fibers contain substance-P, which they presumably release as their neurotransmitter in the brain. Capsacin, the active principle of red pepper, depletes these neurons of their substance-P, causing insensitivity to irritant chemicals (148).

Like the pain fibers of the skin, the nasal nociceptive receptors show a high threshold to chemical stimuli. In cases where the stimuli can be detected by both the olfactory and the nasal trigeminal receptors (many irritant substances like ammonia and ether have distinct odors) the threshold for the trigeminal fibers are severalfold higher. This ensures that the defensive motor responses such as head turning and withdrawal are initiated only if the concentration of the irritant–odorant exceeds tolerable limits.

The perception of the odors of these nonspecific or general substances, which occurs by activation of the main olfactory system, takes place at lower concentrations of the stimuli. The thresholds for activation of trigeminal nociceptive nerve fibers in the rat by several odorants are as follows: phenethyl alcohol, 50 ppm, propionic acid 100 ppm, cyclohexanone 300 ppm, butonaol 400 ppm, and amyl acetate 500 ppm. In pigeons, after sectioning of the olfactory nerves, threshold for detection of amyl acetate is 2.6 log units higher than in intact birds. The operated animals cannot discriminate between amyl and butyl acetate (148).

In humans, responses of the nasal tri-geminal system can be studied in individuals suffering from various anosmias due to lesions of the olfactory system. The degree of unpleasantness of the stimuli is found to be proportional to the intensity of the irritant–odorant. In one study, these anosmic individuals were found to detect 45 of 47 odors presented—albeit at high stimulus concentration, that is, vapor saturation (42a).

Gustation and Gustatory System

Vertebrate taste receptors are found generally in the oral or perioral region and the sense of taste (gustation) serves mainly in detection, selection, or rejection of food related to chemical stimuli. Compared to olfaction, the sensitivity of the gustatory system to various stimuli is low (except for some fishes) and direct contact of a high concentration of dissolved food substances with taste receptors is required for their activation (Table 3). At least four taste qualities are associated with gustatory stimuli: sweet, sour, salty, and bitter; more complex flavors are produced as a result of the perceptual mixing of these basic taste qualities.

Tongue, Gustatory Papillae, and Taste Buds

In the adult vertebrate mouth, taste sensation is mainly, though not exclusively, associated with the tongue. The surface of the tongue contains macroscopic elements called papillae. The filiform papillae contain mechanoreceptors but the remaining types serve in taste sensation. In the human tongue, gustatory papillae have been categorized into three types based on shape: fungiform, circumvallate, and foliate (24). The fungiform type is the most numerous and found mainly in the front and lateral regions of the tongue while the less numerous circumvallate and foliate types are localized to the posterior and posterolateral borders

TABLE 3. Taste Thresholds (behavioral) for Some Vertebrates[a]

	Threshold (M)	
	A. Mammals	
Tastant	Humans	Other Mammals
Glucose	9.6×10^{-2}	
HCL	1.0×10^{-4}	
Monellin	1.0×10^{-7}	
NaCl	2.0×10^{-3}	
Quinine sulfate	1.5×10^{-7}	Rat 7.6×10^{-7}
Saccharin	2.0×10^{-5}	
Strychnine hydrochloride	3.0×10^{-6}	
Sucrose	1.2×10^{-2}	Rat 9.3×10^{-3}
Thaumatin	1.0×10^{-7}	

	B. Aquatic Vertebrates	
Tastant	Minnow	Other Aquatic Vertebrates
Acetic acid	4.88×10^{-6}	
Arabinose	6.51×10^{-5}	
L-Arginine		Catfish (EOG) 9.0×10^{-9}
Fructose	1.63×10^{-5}	
Galactose	1.95×10^{-4}	
Glucose	4.88×10^{-5}	
Lactose	3.91×10^{-4}	
NaCl	4.88×10^{-5}	Frog 1.40×10^{-1}
Quinine hydrochloride	4.07×10^{-8}	
Raffinose	4.07×10^{-6}	
Saccharin	6.51×10^{-7}	
Sucrose	1.22×10^{-5}	

	C. Invertebrates
Tastant	Insects
Glucose	Mosquito 1.1×10^{-1}
HCl	Bee 1.0×10^{-3}
NaCl	Bee 2.4×10^{-1}
	Lepinotarsa 5.0×10^{-3}
Quinine	Bee 8.0×10^{-4}
	Aquatic Beetles 1.6×10^{-6}
Sucrose	Bee 8.0×10^{-3}
	Calliphora 1.0×10^{-2}
	Pyrameis 8.0×10^{-5}
	Danaus 9.8×10^{-6}
	Tabanus 5.0×10^{-3}

Aquatic Invertebrates (thresholds are for single gustatory cells in Crayfish)

L-Alanine	1.0×10^{-6}
L-Glycine	2.5×10^{-5}
L-Histidine	1.0×10^{-5}
L-Methionine	6.3×10^{-4}
L-Serine	3.2×10^{-7}

[a]Data from sources cited in the text especially 14A, 32, 2, 31, 135; all values are behavioral (psychophysical) thresholds unless otherwise indicated.

of the tongue, respectively (Fig. 27). The papillae contain flask-shaped microscopic structures called the taste buds, which occur on the dorsal surface of the fungiform papillae and in the grooves of the circumvallate and foliate types. The taste buds are buried in the epithelial coat of the papillae. The total number of taste buds varies considerably in different vertebrates (Table 4).

The number of taste buds per papilla varies, being very low (1–5) in the fungiform type and very high (>100) in the circumvallate type (18, 24). Taste buds are not exclusively restricted to the tongue. In the adult, they may be found occasionally in the mucosa of cheeks, on the epiglottis, palate, and in the upper one third of the esophagus. In the human neonate and youngs of other mammals the percentage of taste buds in these ex-

TABLE 4. Number of Taste Buds in The Tongue of Different Vertebrates[a]

Species	Number of Taste Buds
A. Fish	
Catfish	180,000[b]
Minnow	8,000
B. Reptiles	
Snake	0
C. Birds	
Parrot	350
Starling	200
Duck	200
Bullfinch	46
Pigeon	37
Chicken	24
D. Mammals	
Calf	25000
Rabbit	17000
Pig	15000
Goat	15000
Human	9000
Bat	800
Kitten	473

[a]Fish data from Caprio (31); other data from Kare (69a).
[b]90% of catfish buds are extra-oral.

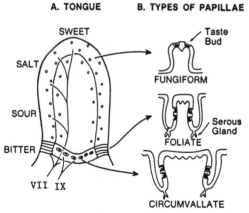

Figure 27. Schematic drawing of a mammalian tongue (*A*) showing the distribution of the three types of taste papillae (A and B), the differential sensitivity of the tongue's surface to the four primary taste groups, (A) and the tongue's gustatory innervation (A). Note that the region of taste sensitivity is around the tongue's edge; also the tongue's anterior two thirds is innervated by taste fibers of the chorda tympani nerve (branch of VII) and the posterior one third by the lingual nerve (branch of IX). [From Shepherd (146).]

tralingual regions is higher (81). The total number of taste buds in the adult human oral cavity is estimated to be about 10,000, 90% of which are on the tongue. Serous glands (von Ebner's glands) occurring at the base of the tongue papillae secrete a fluid that fills the papillar grooves helping to flush out the tastants from the trenches of the buds into the oral cavity (18).

Gustatory Specialization and Sensitivity of the Tongue
Psychophysical and behavioral gustatory studies in the human have indicated that the rich variety of taste sensations (flavors) can be evoked by appropriate com-

bination of substances representing four primary taste sensations, namely, sour, salty, sweet, and bitter (Table 3). It is also widely accepted that the human tongue surface is regionally specialized in regard to these four taste sensations. Thus the lowest threshold (highest sensitivity) for the bitter taste is perceived mainly in the back of the tongue and on the soft palate, that for sweet in the front, and salty and sour sensitivity on the sides of the tongue. There is some overlapping of these regions particularly in regard to sour and salty. However, at high stimulus concentrations all four taste sensations can be evoked from the various tongue regions (24, 33, 130).

Certain anatomic and neural innervation characteristics of the tongue surface may in part underlie the distribution pattern of taste sensitivity. Thus the region of bitter sensation is associated mainly with the taste buds of the circumvallate papillae that occur in the tongue's posterior one third. There is an adaptive advantage in this pattern, since tactile stimulation of this same tongue region evokes gagging, spitting, and vomiting reflexes resulting in food rejection from the mouth. Bitter substances are usually associated with certain plant foods containing alkaloids, diterpenes, and glycosides, substances that are toxic in high concentrations and therefore incompatible with digestion and health. The site of highest sensitivity to sourness is also coincident with the loci of foliate papillae. However, the fungiform papillae that occur all across the tongue's anterior two thirds are not modality specific and seem to respond to the various taste qualities.

The regional gustatory specialization of the tongue may also relate to its pattern of innervation. Gustatory signals are conveyed by afferent fibers in the chorda tympani nerve (a branch of the facial nerve VII cranial) and the lingual branch of the glossopharyngeal nerve (IX cra-

nial). Chorda tympani innervates the anterior two thirds of the tongue, conducting gustatory impulses from all but the circumvallate papillae. The latter are served by the IX nerve (33, 73) (Fig. 27).

Although less widely accepted, there is some evidence for the existence of two other primary tastes, one a metallic taste localized in the middle region of the tongue surface and another a soapy taste in the tongue tip. In rats and rabbits, like insects, taste fibers sensitive to pure water have also been found (147) but the existence of such water sensitivity in humans is unlikely. The "spicy hot" sensations associated with certain foodstuff like pepper are due to the irritative effect of their active ingredient (e.g., capsaicin in the red pepper) on the free nerve ending of the tongue's nociceptive fibers (148, see also below).

Structure of Taste Buds
Vertebrate taste buds are barrel or flask-shaped complex sensory receptor organs that show definite polarity (73, 112) (Fig. 28). Each taste bud has an apical opening, the taste pore, which opens either into the papillar groove (circumvallate and foliate types) or directly into the oral cavity (fungiform). The saliva containing the tastants irrigates the pore and its structures. Each taste bud contains about 50 taste receptor cells along with two other cell types, the supporting cells and the basal cells (18). The endings of the taste nerve fibers enter the taste buds at their basal aspect and terminate on the basolateral sides of the taste cells. As seen in Figures 28 and 29, the basal cells occur at the base of the buds while the supporting cells, like the receptor cells, span the thickness of the bud's epithelium. The receptor cells are also known as the "light" cells and the supporting cells as the "dark" cells, based on their cytological features in ultrastructural sections. The supporting cells, secrete mucus ma-

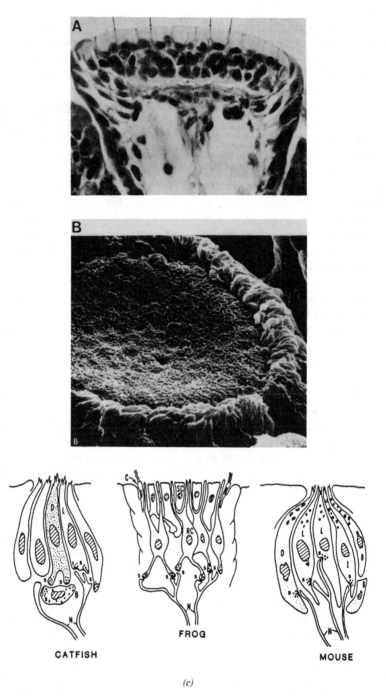

(c)

Figure 28. Taste buds from various vertebrates. (A and B) Microphotograph (A) and scanning electron micrograph of a taste bud from a frog tongue. (Courtesy of P. P. C. Graziadei.) (c) Schematic drawing comparing structure of typical taste buds in a fish, an amphibian, and a mammal. B, basal cell; D, dark cell; I, intermediate cell; L, light cell; N, taste nerve fiber; RC, receptor cell. [From Kinnamon (73).]

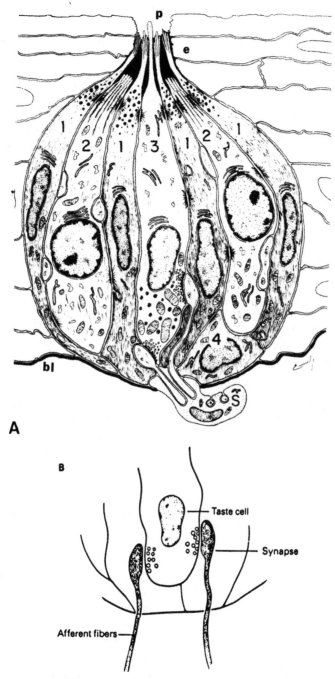

Figure 29. (*A*) Diagram illustrating the typical cell types and cellular specialization of a vertebrate taste bud (rabbit foliate taste bud). 1, type I supporting cell; 2, type II cell; 3, type III receptor cell with long apical microvilli projecting out of the taste pore and synaptic specialization at its basolateral surface; 4, Schwann cell around the innervating taste nerve fiber. (bl, basal lamina; e, epithelial cells around the bud; p, outer taste pore.) (*B*) Schematic drawing of the synaptic area between the branches of the afferent taste fibers and the taste receptor cells. Note that synaptic vesicles are in the receptor cell. [After Murray (112).]

terial into the mouth pore of the taste buds and may interact with the taste cells biochemically and structurally (18, 73, 112)

Turnover of Taste Cells

Thymidine labeling studies have established that the receptor cells of the taste buds turn over continuously, each cycle lasting about 10 days (18). After a functional survival period of a few days, the receptor cell degenerates and is replaced by a new taste cell. The new taste cells may be formed by the mitotic division of the basal or stem cells; the daughter cell differentiates into a new taste cell. Alternatively, taste cells are thought to form from a population of mitotic epithelial cells from outside the taste bud that migrate into the bud, replacing the degenerating cells. These migrating cells may be the source of new supporting cells (18).

The turnover and survival of the taste cells depends on a trophic influence from the taste nerve conveyed by axonal transport. Transection of these nerves interrupts the flow and results in degeneration and death of taste buds and their cells. Supporting cells, which are more numerous, do not show the same rapid turnover rate as the receptor cells but they also receive neural and neurotrophic stimuli, which regulate their secretory activity (130, 150).

Taste Cells: Structure and Cellular Physiology

The taste cells are modified epithelial cells showing some of the characteristics of neurons including polarity. Taste cells have been classified as paraneurons by Fujita. The apical surface of a taste cell is folded into numerous long microvilli that project into the outer taste pore, coming into contact with the tastants dissolved in the saliva. Occluding junctions surround the necks of the taste cells and seal the

intercellular spaces, preventing the flow of ions, water, and tastants in the saliva into the taste buds. Taste cells contain voltage sensitive ion channels in their apical membranes. The potential across this membrane is known to be altered by addition of tastants and by application of electrical currents. These cells can generate potentials resembling receptor potentials (73, 78, 140).

The basal and basolateral aspects of the taste cells contain subcellular specializations that are a mixture of some features found in the absorptive–transportive epithelial cells and neurons. The most prominent is the presence of synaptic-like contacts between the endings of the taste nerve fibers and the basolateral surface of the taste cell. Synaptic vesicles have been observed on the taste cell side but the nature of the chemical transmitter is unknown (Fig. 29b). Typical thickening of the synaptic membranes are observed and these membranes are thought to contain pumps and channels for the flow of ions (73, 159).

The general function of the taste cell is to selectively recognize and bind to specific substances (tastants) at their apical microvillous surface and to transduce the information about the quality and quantity of these chemical stimuli to electrochemical signals. These signals are then transmitted to the taste nerve endings at the basolateral synapse. The microvilli membranes contain the taste receptor molecules or mechanisms that bind with the tastants (159).

The binding of tastant with the receptors can activate one or more of the following membrane events, depending on the specific tastant. Binding can cause depolarization of apical membranes by increasing the inflow of cations or outflow of anions. These electrical events in the apical membrane lead to a transient receptor potential in the taste cell that electrotonically excites the basolateral synapse (72) (Fig. 30*A*). The electrical response of the taste cells are not always depolarizing (122). As shown in Figure 30*B*, hyperpolarizing responses have also been observed (140). Alternatively, receptor binding could activate a G protein, which would in turn activate the enzyme adenylate cyclase, resulting in elevated cAMP levels (Fig. 30*C*). This intracellular second messenger would then activate the basolateral synapses by phosphorylation of kinases or by direct activation of ion channels in the basolateral synaptic region (25, 73, 78).

Involvement of second messengers like cAMP, cGMP, IP3, and the associated biochemical cascades enables the taste cell to modulate the response to the tastant and to amplify the signal (159a). A role for calcium in the taste cell microphysiology is widely accepted (159). The release of the transmitter at the basolateral synapse, as in the case of other chemical synapses, is believed to be preceded by a rise in calcium ion levels in the taste cell. The source of this calcium may be from the intracellular reserves (Fig. 30C). The trigger for the rise may be the receptor potential or the rise in IP3. Activation of the postsynaptic membrane evokes action potentials in the endings of the afferent taste nerve fibers. These action potentials carry the gustatory signals to the brainstem taste centers.

Receptor Physiology of Primary Tastes

SWEET. Among the various primary tastes, the sweet taste provides the best case for a direct relationship between stimulus structure and receptor function (sensation). Compounds that are naturally sweet are few. In general monosaccharides such as glucose, fructose and galactose, or di- and polysaccharides containing these monosaccharides are the best examples of sweet stimuli. The intensity of sweetness is inversely related to the degree to which sugar hydroxyl groups form hydrogen bonds with receptor sites. Thus glucose is sweeter than fructose (18, 159).

The first attempts to purify taste receptor molecules involved those of sugar receptors perhaps because sugar receptors occur in bacterial walls and their biochemistry is well known (75, 164). Initially soluble proteins with high affinity for glucose (sweet sensitive proteins) were extracted from the beef tongue and thought to be the sweet receptors (29, 30). They are found in nongustatory tissues of the tongue as well. They may be equivalent to the soluble odorant-binding proteins and may function in concentrating the tastant near the taste cells.

Sweet sensation may also be produced by stimuli other than carbohydrates. Thus, glycerol, some of the alcohols and ketones, chloroform, beryllium salts, and amides of aspartic acid all taste sweet. Also, a 44-kD protein called miraculin, extracted from the berries of African bush, Miracle Fruit *Synsepalum dulcificum*, and another plant product, a 11-kD protein called Monellin, both taste extremely sweet in humans, at a molar threshold of 10^{-7}. This threshold is about 10^5 times lower than for sucrose (i.e., 10^{-2} M) (18, 85) (Table 3). Miraculin make acidic compounds taste sweet. A third sweet protein, thaumatin, has been investigated in terms of molecular biology. A 28 amino acid flexible sequence of the molecule is responsible for the sweetness; also the mRNA of thaumatin has been isolated and through recombinant DNA biotechnology, its large scale microbial synthesis for commercial use is underway (18, 44a).

The dipeptide L-aspartyl-L-phenylalanine methyl ester (aspartame), which is the active ingredient in NutraSweet is also extremely sweet at low concentration. The aspartame, saccharin, and other artificial noncalorigenic sweeteners form the base of a major food industry in mod-

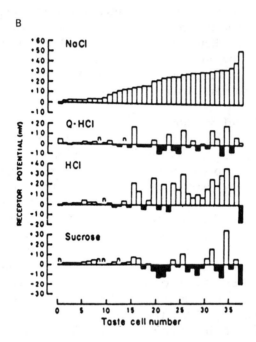

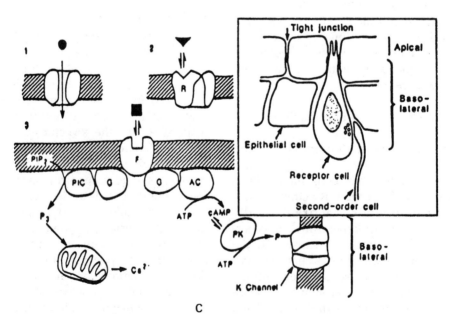

ern times. Several types of membrane receptor molecules with high affinities for glucose have been isolated (30). It is not known whether the peptidergic or proteinacious sweet tastants bind with the same glucose receptors or with other membrane elements in the microvilli. There are also substances that specifically abolish the sweet sensation. An example is gymnemic acid extracted from the leaves of *Gymnema sylvestre*, a native vine of India. Application of this compound to the tongue abolishes sweet sensitivity for 30 min (18, 59).

SOUR AND SALTY. The acidic chemicals capable of ionizing and releasing protons in the dissolved state evoke the sensation of sour. In the case of inorganic acids, protons alone act as the stimuli. For organic acids the anions may also play a role as the size and chemical nature of the anion binding with the receptor membrane can influence the number of protons that bind with this membrane and thus affect the degree of sourness. Also equimolar concentrations of different acids at similar pH will not evoke the same degree of sour taste sensation. Thus, to produce a sensation equal to a given concentration (e.g., 2.2 mM) of sulfuric acid, about 2 times as much hydrochloric acid, 3 times maleic acid, 5 times glutaric acid, 7 times lactic acid, 30 times acetic acid, and 70 times as much butyric acid are required. In general the ionic strength of the acid solution and the electrical charge of the receptor membrane seem to be the determining factor (18).

The salty sensation in its purest form

Figure 30. Receptor potentials of taste cells. (A) A single taste receptor cell may be sensitive to stimuli representing one or more of the four primary taste groups (sweet, salty, bitter, and sour). This hamster taste receptor cell responds to all four tastant solutions (sucrose, salt, quinone, and acid) applied to the tongue by producing depolarizing receptor potentials of differing amplitudes. [From Kimura and Beidler (72).]. (B) Differential responses of 38 rat taste cells to stimulation by salt, quinone, acid, and sucrose, as measured by changes in the receptor potential. Responses of each cell varies with regard to differential sensitivity to the stimulus type and the amplitude and direction of the receptor potentials. While all of the salt sensitive receptors show depolarizing response (open bars), nearly one half of the sweet and bitter sensitive receptors and a few of the acid sensitive receptors show hyperpolarizing responses (closed bars). n, no response. [From Sato and Beidler (140).] (C) Proposed scheme of molecular events involved in activation of taste receptor cells by salty, sour, sweet and bitter substances. (1) Salt stimuli (cations) may depolarize the cell by entry through passive (ungated) ion channels in the apical membrane. Acid (sour) stimuli may inactive voltage dependent K^+ channels in the apical membrane. (2) Amino acids bind to membrane receptors (R); this allosterically modifies the receptor structure to induce opening of ion channels. (3) Sweet and bitter stimuli may bind to membrane receptors (F), activating different types of G proteins and initiating effects involving cascades of intracellular second messengers (cAMP, IP_3, Ca^{2+}). Two pathways are known: In the first, G protein activates adenylate cyclase (AC) resulting in the formation of cAMP from ATP. The cAMP activates a protein kinase (PK), which then blocks by phosphorylation of a K^+ channel in the basolateral membrane (synapse); in a second pathway, another G protein activates a membrane enzyme, phosphoinositidase C (PIC) which forms inositol triphosphate (P_3) from hydrolysis of membrane phospholipid phosphotidylinositol-biphosphate (PIP_2); P_3 can release Ca^{2+} from intracellular sources such as mitochondria; Ca^{2+} then may act on basolateral synapses. [From Teeter and Gold (159a).]

is produced by NaCl. However, many other compounds can produce salty sensation. The cation plays the most important role although anions are also important, particularly in regard to their size (hydration diameter): Salts with smaller anion hydration diameter produce a higher taste response. Thus equimolar amounts of sodium chloride produces markedly higher activity in the taste nerve than sodium acetate (Fig. 31). The diuretic compound amiloride that inhibits sodium transport, selectively abolishes the salty taste (159).

BITTER. Although bitterness is such a unique taste sensation subjectively, it may be evoked by a variety of diverse and structurally unrelated compounds. Indeed no single structural feature seems to determine bitterness (18, 85, 159). The main category of bitter tasting compounds are the many plant secondary metabolites (alkaloids, diterpenes, and glycosides). A well-known example is quinine. Interestingly, many familiar and favorite substances like caffeine and theophylline (xanthines alkaloids in coffee and tea, respectively), which have a pleasant flavor when appropriately prepared in solutions or in mixture with other foodstuff, are intensely bitter in their raw forms.

Some amino acids and dipeptides like L-leu-Gly, L-lys-L-Ala, and Gly-L-Try have bitter tastes. Many sweet substances can be made to taste bitter by slight changes in molecular structure (18, 159).

Although bitter compounds have a high lipid solubility and are believed to penetrate deep into the apical microvillous membranes, or enter the receptor cell, there is indirect evidence for a protein receptor or receptor mechanism for these compounds. This evidence comes chiefly from observation of people who are genetically nontasters to bitter compounds containing the N—C≡S group, like thiocarbamides, for example, phenylthiocarbamide (PTC) (= phenylthiourea). The sensitivity to this bitter taste is mediated by a single Mandelian dominant gene (84, 85). The nontasters (30% of Caucasians) inherit the recessive component and presumably lack the gene product (a protein) involved in bitter reception. The ability to taste caffeine is also reduced but not as much.

Application of bitter stimuli to taste cells evokes a graded depolarization of the cell membrane. This response occurs probably due to outward flow of chloride ions across the apical membrane. The high lipid solubility of bitter compounds permits their entry into the taste cells.

Figure 31. Effects of the increase in concentration of the taste stimulus (sodium salts of different acids) and diameter of the anion group on integrated responses of the taste nerve (chorda tympani). *Note:* (1) the response increases with increasing stimulus concentration (up to $0.5\,M$); and (2) peak response of the nerve is reduced in proportion to the increase in diameter of the anion group. [From Beidler (17).]

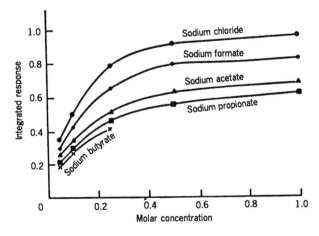

Most bitter compounds, like caffeine, under *in vitro* conditions, inhibit the enzyme phosphodiesterase, which normally inactivates cAMP (73). Thus bitter compounds may exert some of their effects by inhibiting phosphodiesterase, resulting in elevated levels of cAMP. The exact role and cellular physiology of cAMP in bitter reception remains to be established.

Central Gustatory Pathways

Taste buds from the anterior two thirds of the tongue give rise to the chorda tympani nerve and those from the posterior one third to the lingual nerve. Chorda tympani join the facial (VII cranial) and the lingual joins the glossopharyngeal nerve (IX cranial). Taste fibers from the pharyngeal region join the vagus nerve (33). After entering the medulla at their respective locations, the central processes of the taste fibers enter the gustatory nucleus of the solitary tract, located at the rostral portion of the solitary tract nuclear complex. The gustatory nucleus of medulla is the first synaptic station for the taste fibers and a major sensory–motor relay center (26). The caudal portion of the solitary tract nuclear complex receives afferent information from the gut and other visceral organs (115). From the gustatory nucleus, three sets of connections arise that subserve the three different gustatory related neural functions: (1) gustatory reflexive, (2) affective responses and drives, and (3) higher aspects of gustatory discrimination and perception (33, 114, 130). These projections are shown in Figure 32.

In the first set of projections, second-order neurons from the gustatory center innervate the medullary and other brainstem motor nuclei of vagus and salivary nerves. Some of these connections may be via small short axon neurons. These short connections, which are restricted to the medullary and upper brainstem regions, enable activation of taste stimulated reflex swallowing, salivation, and gastric secretion. The flow of saliva is essential for taste as it helps dissolve food substances. Additionally, many taste stimuli increase secretion of gastric juices (115). Other reflexive connections from bitter and salty fibers activate brainstem motor nuclei regulating gagging, spitting, and vomiting reflexes, which aid in rejection of undesirable food from the mouth.

Another set of second-order fibers arising from the medullary gustatory nucleus projects to the hypothalamus and limbic system via relays in the pons (parabrachial nucleus). These connections serve in the affective aspects of taste sensations (e.g., acceptibility and pleasure of foods) and unconscious drives for foods of particular tastes (e.g., appetite for sweet foods in humans and salt appetite in rats).

In primates (old world monkeys, apes, and humans), a third and the most massive set of second-order cells project, from the gustatory nucleus, directly to the ventral posteromedial nucleus of thalamus, which integrates and relays all afferent signals relating to the tongue (130). In the rat and subprimate mammals, the projecting fibers in this category, synapse first in the parabrachial pontine nucleus, before moving on to the thalamic taste center (47, 130) (cf. Fig. 32*a* and *b*). From the thalamic nucleus, higher order thalamic fibers project to the taste area of the cerebral cortex, which is located at the bottom of the somatic sensory cortex below and adjacent to the tongue area. In the rat only one such area has been found. In primates two adjacent taste areas have been found, one in the ventral aspects of the postcentral gyrus and a smaller second area in the insular cortex (33, 114).

The cells of neocortical gustatory centers deal with localization of the tastant source on the tongue, taste discrimination and cognitive aspects of gustation (174). Lesions in the cortical taste areas impair

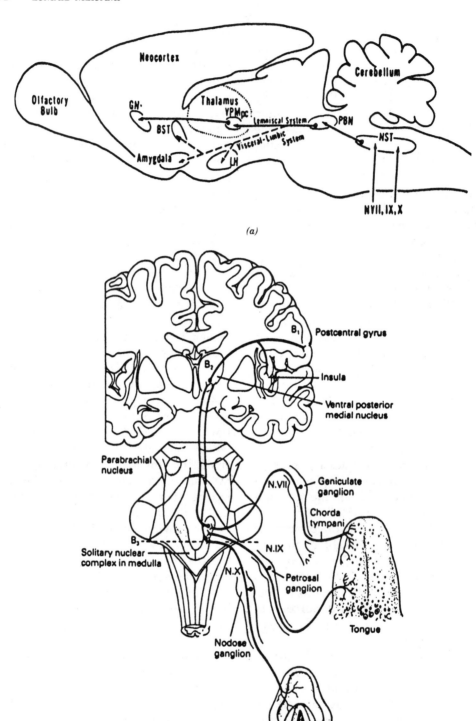

(a)

(b)

taste discrimination and sensibility. These higher sensory centers for gustation are adjacent to the higher somatic sensory centers for tongue serving in tactile and thermal discrimination and cognition. It is a familiar experience that gustatory sensations are intimately associated with, and are altered by, the tactile and thermal nature of gustatory (food) stimuli. Interestingly, in contrast to the somatic sensory projections from the tongue to the sensory cortex that is crossed, the gustatory projections from the tongue to the taste cortex are essentially uncrossed so that the taste sensation from each half of the tongue is initially perceived by the taste centers in the ipsilateral cerebral hemisphere (33, 130).

Mechanisms of Taste Discrimination
How does the information concerning the quality of the various tastants become encoded by the gustatory receptor cells, taste buds, and taste fibers? How is this information communicated to the neurons of brain taste centers? Since different parts of the tongue are innervated by different taste nerves, and since the tongue surface shows some regional specialization for various taste sensations, it was expected that different taste nerve fibers would be specific for each of the taste primaries and the taste cells belonging to taste buds from different tongue regions would show taste specificity (94). Psychophysical studies in humans, involving application of a stimulus to single taste papilla, initially provided some support for this type of "labeled lines" theory of gustatory coding (33, 85).

RESPONSES OF SINGLE TASTE FIBERS TO TASTANTS. Recording from single taste fibers of mammals, pioneered by Pfaffmann, revealed that they were not specific for any particular taste primary. Each single fiber conducted action potentials in response to stimulation of the tongue surface by compounds representing each of the four tastant types: sour, sweet, salty, and bitter (128). A typical demonstration of this polymodal behavior of single taste fibers is shown in Figure 33. It can be seen that of the 50 rat taste fibers tested, many respond to the representative substances of all four taste primaries, NaCl, sucrose, HCl, and quinine.

Two questions arise: (1) What makes the single taste fibers respond nonspecifically? Is it the polymodal responsiveness of the taste cells or their particular pattern of connectivity to the taste fibers? (2) How do the brain cells decipher specific information about the different tastants when taste fibers seem to respond nonspecifically?

An understanding of the pattern of innervation of taste buds and cells by taste nerves is important. Each taste nerve fiber ramifies extensively before it enters a taste bud and after entry it ramifies again to innervate several taste cells. Each taste bud is innervated by several taste nerve fibers (18, 73). Thus each taste nerve fiber carries signals from numerous taste cells and from several taste buds and each taste bud sends signals to the brain via several different taste fibers.

Here again two possibilities exist: (1) That the taste receptor cells, innervated by the branches of the polymodal taste

---

Figure 32. Central gustatory pathways in subprimate mammals (*a*) and in primates (*b*). BST, bed nucleus of striae terminalis; GN, gustatory neocortex; LH, lateral hypothalamus; NST, nucleus of solitary tract; PBN, parabrachial nucleus (pontine taste area); VPMpc, ventral posteromedial nucleus of the thalamus, *pars parvocellularis*. Note that the primate pontine parabrachial nucleus also receives input from the medullary centers. [Part *a*, Finger (47); Part *b*, Castelluci (33).]

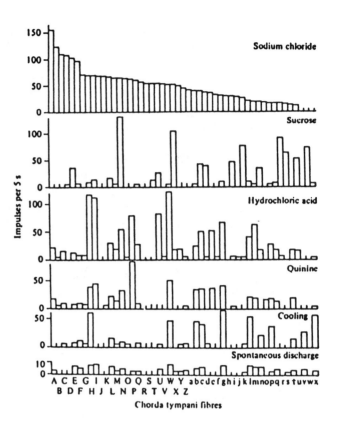

Figure 33. The polymodal responses of 50 single taste fibers (rat, *chorda tympani*) to the four primary taste stimuli (sweet, bitter, salty, and sour); bottom profiles indicate the spontaneous discharges from fibers and the response to cooling. (Letters A–Z, and a–y indicate individual fibers and are arranged in order of their responsiveness to 0.1 M NaCl). *Note:* (1) Single fibers do not show specific sensitivity to a particular taste stimuli but each has its own characteristic response spectrum. [From Ogawa et al. (118).]

fibers, are also nonspecific. (2) That the receptor cells are specific but a single fiber innervates different receptor cell types, each sensitive to a specific taste primary, resulting in each fiber carrying messages from all the different taste cells.

Recording from single taste cells in the rat and hamster showed results in favor of the first possibility. Thus in response to application of NaCl (salty), quinine (bitter), sucrose (sweet), and HCl (sour), each receptor cell responded to all these stimuli by a depolarizing receptor potential (Fig. 30A) and the amplitude of the graded potential was proportional to the intensity of the chemical stimulus (72, 140). Thus the polymodal responses of the single nerve fibers reflected the polymodal responses of the single receptor cells to which they were connected.

Since each single taste fiber increases activity in response to all tastants, then brain cells in the taste centers could not decipher information about the specific taste qualities by tuning in to individual fibers. Pfaffmann reasoned that this information must somehow be deciphered by comparing the response pattern (characterized by different firing rate) of the individual members of a population of taste fibers activated by the particular tastant or combination of tastants. Pfaffmann called his theory of sensory communication in the gustatory system, the *across fiber pattern theory* (33, 128, 143). This type of signal communication and analysis is now found to be operating in other sensory systems as well, aiding in frequency discrimination in the auditory system, color perception in the visual system and as we have already noted in olfactory discrimination.

Further quantitative recordings from many single taste nerve fibers in response to a large variety of tastants have shown that although broadly tuned, taste fibers show some of the general characteristics of "labeled lines" as well (85, 128, 143). Frank recorded from 80 hamster taste fibers and showed that although all fibers respond to all primary taste substances, each fiber had a best response (maximal impulse rate) to one particular taste primary (50). Thus the taste fibers may be grouped into specific subpopulations (bitter-best, sour-best, etc.) the members of which would show similar stimulus preference and frequencies of impulse propagation for their preferred stimulus. Thus brain centers, although receiving signals from all afferent fibers, can decipher the specific information about the particular taste involved, by tuning to all those fibers carrying the fastest rates of impulses. This requires that each fiber would have a repeatable and characteristic response pattern upon the presentation of the same stimulus or mixtures of stimuli (143).

Responses of central gustatory neurons to tastants have also been studied and elements of both labled lines and across neuron coding have been found at that level as well. Gustatory neurons in the hindbrain show differential sensitivity to taste stimuli, even though their tunning spectrum is relatively broad. Thus, as shown in Figure 34 for the hamster, some of the neurons of the nucleus of the solitary tract respond best to sugars, others to salts, and so on. Neurons of similar chemosensitivity are clustered together (151, 160).

GUSTATION AND AFFECTIVE RESPONSES. A characteristic attribute of odors and tastes are their affective qualities. Many odorants and tastants, although very different in sensory perceptual qualities, may appear either pleasant or unpleasant. These affective qualities influence food choice and preference. Interestingly, in humans the degree of pleasantness or unpleasantness of a tastant depends on the stimulus strength, that is, tastant concentration (58). Thus sour and salty substances taste pleasant at low concentrations becoming more and more unpleasant with increasing stimulus concentrations. Similarly the generally unpleasant bitter substances taste slightly pleasant at very low concentrations, explaining the use of bitter spices as food flavorizers.

Gustatory preferences and aversions can be formed very rapidly and may be long term and even permanent. Humans form aversion to foods consumed accidentally with a toxic and nauseous substance that has made them sick. Even frogs, not known for good learning ability, require a single exposure to a toxic bitter foodstuff to develop a permanent aversion toward that particular food.

GUSTATION AND DIETARY CONTROL. Many animals use their gustatory system to control their diet. Rats, made salt deficient by adrenalectomy, drink salty water in preference to tap water (76). Similarly, parathyroidectomized rats preferentially choose water with added calcium chloride. Animals depleted of blood sugar by constant injection of insulin develop an intense appetite for sweet foods (58).

General (Nociceptive) Chemoreception in the Oral Cavity
Like the nasal mucosa, the oral mucosa including that of the tongue, is richly innervated by free nerve endings belonging to the trigeminal nerve (V cranial). These endings convey thermal and nociceptive sensations from the oral region and in doing so influence gustatory responses and preferences. Among the protective reflexive responses elicited by the activation of these fibers are increased salivation, vasodilation, swallowing and nasolacrimal secretions. Increased sweat-

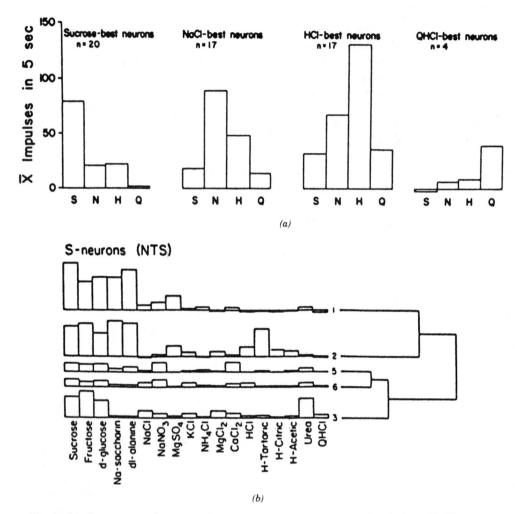

Figure 34. Responses of neurons in gustatory centers to taste stimulation. (*a*) Neurons of the gustatory nucleus of solitary tract (NST) respond to all taste stimuli but not to the same extent; some neurons respond maximally to sucrose (sweet-best), others are salt-best, bitter-best, or sour-best. (*b*) Stimulation of sucrose-best neurons with a variety of different tastants (applied to the tongue) reveals their differential sensitivity. [Parts *a* and *b*, Travers et al. (160), after Smith et al. (151).]

ing in the head and neck is also observed (gustatory sweating). Increased salivation and vasocontractile and secretory changes in the papillary regions may influence gustatory responses. However, these trigeminal fibers do not influence detection of the gustatory stimuli and thresholds. Humans whose tongues' free nerve ending are desensitized with cap-

saicin treatment can detect quinine, glucose and sodium chloride, and menthol at the same thresholds as before the desensitization (148).

The maxillary and mandibular branches of the trigeminal nerve innervate the oral region. The relevant subdivisions of these nerves include the lingual nerve (for tongue and mouth floor), the nasopalatine nerve (for soft palate), and buccal nerve (for cheeks). Even the taste papillae are innervated by the free nerve endings. Nearly 75% of the innervation of fungiform papillae is due to trigeminal fibers.

Free nerve endings are much less sensitive compared to taste cells, responding to tastant solutions at a concentration 1000 times higher (e.g., 1.5 M NaCl). These trigeminal fibers respond especially to pungent stimuli like those in the "hot" spices (pepper, horseradish, and ginger). The active ingredients of pungent spices and vegetables are capsaicin (chili pepper), piperine (black pepper), 6-gingerol (ginger), β-butenyl isothiocyanate (mustard), and eugenol (cloves). Although these pungent compounds are aversive, humans develop a preference for these as indicated by wide interest in chilies and other hot spices in the diet. Animals do not develop such a preference (148).

Gustation in Fish

Fish live and feed in aquatic environments. This necessitates the development of specific chemosensory systems that provide information about the quality and quantity of foodstuff and related chemosensory signals in the immediate environment (116). Peripheral taste organs and a neural gustatory system exist in all fish species although their relative development shows considerable species variation. Thus catfish have a large and markedly developed gustatory system, compared to sicklebacks that have a small one. Also, compared to the land vertebrates, the fish taste system shows marked differences in the number and bodily distribution of the peripheral taste organs, the type of taste stimuli that evoke the best response, and the relative sensitivity to tastants (31, 47, 48).

Although fish are capable of taste, it is not clear whether the taste qualities found in the land vertebrates (e.g., four taste primaries in humans) are also perceived by fish. Such classification may not even be valid for aquatic animals. Sensitivity for amino acids is widely distributed. Responses to compounds representing the four taste primaries have been observed in fish (149).

FISH TASTE BUD STRUCTURE AND DISTRIBUTION. In fish, as in land vertebrates, the taste cells occur mainly in taste buds. The general shape and structure of taste buds in fish and their turnover characteristics are remarkably similar to those of the typical terrestrial vertebrate described above (31, 73, 112). However, the apical taste pores are nonexistent or less prominent and a thin mucus layer covers the bud. Also, the fish taste cells contain large club shaped microvilli, up to 3 μm long, which pierce through the thin mucus into the oral cavity (Fig. 28c). Thus gustatory stimuli may have more direct access to the taste cells in fish than in mammals.

In contrast to land vertebrates, where the taste buds are restricted mainly to the tongue, in fish the taste buds can occur in other areas including the pharyngeal compartment of the oropharyngeal cavity and in the external body surface, especially in the head region and its appendages like the barbels (31) (Fig. 35).

Different fish vary considerably in regard to total number of taste buds and their bodily distribution. For example, the Japanese minnow that reaches a length of 6 cm has a total of 8000 taste buds, 90% of which are intraoral. The catfish with a body length of about 25 cm has more than 180,000 buds, 90% of which are distributed on the external body surface (31).

FISH GUSTATORY SENSITIVITY AND DISCRIMINATION. In contrast to the olfactory system, the large numbers of taste buds in the catfish or even larger fish, are not associated with higher gustatory sensitivity. Minnow, catfish, and eel all have similar gustatory thresholds to amino acids as

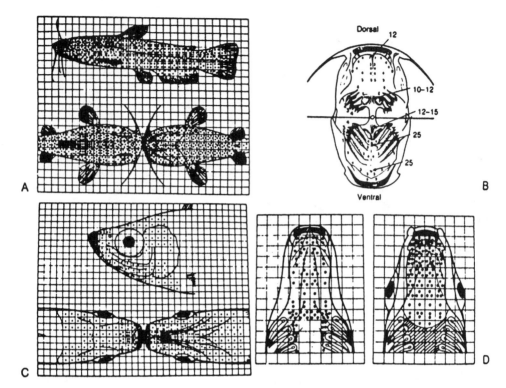

Figure 35. Distribution of oral and extraoral taste buds in the yellow bullhead catfish (*Icxtalurus natalis*, 25 cm long) and Japanese minnow (*Pseudorasbora parava*, 6 cm long). Each sq = 1 mm². (*A*) Distribution of taste buds on the body surface of catfish. Each dot = 1 taste bud; solid black = >15 buds; top, lateral view; bottom, left and right are the dorsal and ventral views. (*B*) Distribution of oral taste buds in the catfish; arrows and numbers indicate high density sites. (*C*) Distribution of taste buds in the minnow; top, lateral view; bottom, left and right are the dorsal and ventral views. (*D*) Distribution of oral taste buds in the minnow; left and right are the dorsal and ventral views; each dot = 15 buds; oblique lines on lips and palatal organ = 140 buds. *Note:* Ninty percent of the catfish taste taste buds occur extraorally, on the head and body surface, compared to 15% in the minnow. [From Caprio (31), after Atema (catfish) and Kyohara (minnow).]

measured by recording from gustatory nerves, although they vary markedly in their total taste bud number (31). Eels have no external taste buds. Nevertheless, the larger number of taste buds in their head and body surface and their wide distribution increase the ability of the fish to localize food sources more efficiently. In particular, the fish with large body size may use gradients of chemical cues, detected by external taste buds, to generate appropriate behavioral re-

sponses such as turning (31). The existence of extraoral taste receptors in some fish and also in aquatic invertebrates have led to the concept that these animals may have a dual taste system. The extraoral taste helps in detection and capture of the foodstuff while the intraoral taste system helps in its recognition and discrimination, as proper and palatable food (11).

Fish respond to sweet, salty, sour, and bitter substances. Many fish spit out undesirable foodstuff, indicating their sensitivity to bitter compounds. Of all the chemical gustatory stimuli for the fish, the L-α-amino acids seem to be the most potent (Table 3). These relatively water soluble compounds may signal foodstuff from decomposing biomass; the amino acids may also be used for conspecific recognition of skin mucus. The gustatory receptor cells show high sensitivity (low thresholds) to amino acids. In the catfish, the threshold to evoke an electrophysiological response to amino acids by the gustatory system is in the range of 10^{-7}–$10^{-9} M$ (Fig. 36). This high sensitivity is not necessarily observed in all fish. It is interesting that the behavioral–psychophysical gustatory threshold of humans to various sweet proteins is also in the range of $10^{-7} M$ (18).

The high sensitivity of fish gustatory receptors to amino acids rivals the ability of the olfactory system, which is thought to be the most sensitive of the chemoreceptor systems. This creates a problem of definition, since it was generally thought that, compared to olfaction, a lower sensitivity was a characteristic of gustation whose stimuli required direct contact and high concentration for activation. Clearly fish gustation is an exception to the rule. However, there is a qualitative difference between the fish olfactory and gustatory systems in that, although both systems respond to all amino acids, each system has a differential preference for a different type of amino acid side chain. Thus olfactory receptors respond to neutral amino acids with long, linear, uncharged, and unbranched side chains, while the gustatory system responds best to amino acids with short side chains (alanine, glycine, or serine) (31).

Biochemical purification and ligand-binding studies have shown the existence of receptor proteins, with high affinity for amino acids, associated with the taste buds of catfish (29, 30). Electrophysiological cross-adaptation studies point to the

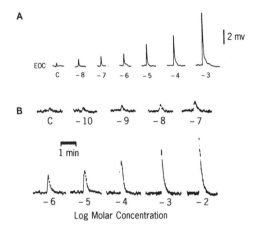

Figure 36. Comparison of responses of the catfish olfactory and gustatory system to stimulation by increasing concentrations of amino acids. (*A*) Olfactory responses (EOG, from channel catfish *Ictalurus punctatus*) to L-arginine; (*B*) Integrated taste nerve responses (facial nerve to the barbel taste buds) in the spotted bullhead catfish *Ictalurus serracanthus* to L-alanine; c, response to control well water. *Note:* (1) Both the olfactory and gustatory system show high sensitivity (low thresholds, $10^{-8} M$) to amino acids; (2) the response magnitude increases with increasing stimulus concentration. [From Caprio (31).]

existence of four general categories of receptor sites: (1) for acidic amino acids, (2) for basic amino acids, (3) for neutral amino acids with long hydrophobic side chains, and (4) for neutral amino acids with long hydrophilic side chains. All of these receptor sites may or may not occur on the same receptor cell (31).

GUSTATORY NERVES. The same three cranial nerves (facial, glossopharyngeal, and vagus) that carry afferent taste fibers in land vertebrates do so in fish. The extraoral taste buds of the head region including those from the appendages (e.g., catfish barbels) are innervated by the branches of the facial nerve. The taste buds in more caudal body surfaces are innervated by the glossopharyngeal and vagus nerves. Also, in contrast to land vertebrates, where the vagal share of taste fibers is small, in fish this is highly enlarged. Instead, the relative number of taste fibers in the glossopharyngeal nerve is small, possibly due to the lesser importance of the tongue in fish gustation and the reduced gustatory innervation of this organ (47, 48) (Fig. 37).

GUSTATORY CENTERS. The general pattern of organization of the gustatory system in the fish brain is similar to that of the land vertebrates. Differences are mainly in: (1) the relative development of specialized peripheral taste organs such as the taste buds in the barbels of the catfish and (2) the organization of telencephalon, which in the fish lacks a neocortex. Figure 37 illustrates the location and connectivity of the gustatory centers in a fish brain. The central processes of the taste nerve fibers terminate in an orderly manner in the gustatory nucleus located in the dorsal medulla of fish brainstem. The taste fibers in the facial nerve innervate the rostral

part of this nucleus, those in the vagus innervate the caudal part and the glossopharyngeal gustatory fibers innervate the middle zone. The gustatory nucleus accordingly has three lobes: two prominent facial and vagal lobes and a minor glossopharyngeal lobe (47, 48).

The fish gustatory nucleus is more prominent than its mammalian counterpart. In fish with highly specialized gustatory appendages in the head (e.g., barbels of the catfish) the facial lobe is very large and subdivided into lobules each representing a single barbel; in those with gustatory appendages in the oropharyngeal cavity, for example, the spinal epibranchial organ of the *Heterotis*, the vagal lobe is well developed, and contains a matching spiral neural lobulation (48).

From the medullary gustatory lobe, second-order cells project to a pontine gustatory nucleus where two higher order ascending projection systems arise: one innervates the lateral hypothalamic area and the other visceral centers of the inferior lobe; the second projection terminates, via the thalamic gustatory

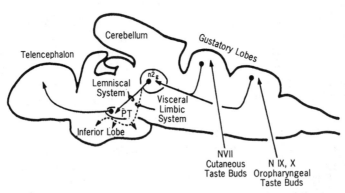

Figure 37. Semischematic diagram of the gustatory pathways and centers in the fish brain (catfish). n2g, secondary gustatory nucleus (in pons); PT, posterior thalamic nucleus. *Note:* (1) The highly developed facial and vagal lobes of the gustatory nucleus of medulla to which taste nerves (VII, IX, and X) project; (2) similarity to the mammalian gustatory brain system (cf. with Fig. 32a); (3) fish inferior lobe is equivalent to the hypothalamus. [From Finger (47).]

complex, in discrete areas of the telencephalon. As fish do not have a neocortex, these telencephalic projection fields are the equivalent of the neocortical taste areas of mammals.

Swallowing, spitting and food rejection and other immediate reflex responses are mediated by connections to the oropharygneal motor nuclei and those in the reticular formation. Drive and appetite-related responses are carried out by the hypothalamic connections. The higher order cognitive aspects of gustatory functions, requiring learning and memory and integration of gustation with olfaction and somatosensation, are carried out by the thalamic and telencephalic centers. Older papers indicated the fish forebrain to be both gustatory and olfactory. Recent research shows that the primary projection fields of higher order olfactory and gustatory fiber are more discrete than previously thought (47, 48). The remaining parts of the fish telencephalon are involved either in higher integrative functions related to olfaction and taste or in other aspects of sensory and motor functions.

Chemoreception in Invertebrates

Chemoreception is well developed in invertebrates, particularly in insects where differentiation of various general and special types of chemosensation is possible. Insects have not only distinct olfactory and gustatory systems, but also are capable of general chemosensitivity as indicated by their responsiveness to irritant substances like ammonia or chlorine fumes. The sensory receptors for these noxious substances are distributed in many different parts of the body and resemble the common chemical sense described in vertebrates. In termites these receptors are abundant in the legs. In the cricket *Gryllus,* the receptors of the cerci respond to these irritants (169).

Olfaction in Insects

Insects have a well-developed olfactory system that shows great sensitivity and discriminative power. Insects respond to numerous odorants associated with plant and animal sources (126). Various alcohols and esters, aldehydes, and some volatile fatty acids act as particularly strong and effective odorants. In the honeybee the behavioral thresholds for olfactory detection of propionic acid, eugenol, phenylpropyl alcohol, and citral are low (10^{-10} M) and in the same range as found for humans (Table 1). A honeybee can be trained to distinguish the essence of orange from 43 other ethereal oils (135, 169).

Pheromonal Communication in Insects

Olfactory communication plays a major role in the life of all insects, in particular the social insects like ants and honeybees. The bodies of social insects contain numerous exocrine glands many of which release, into the environment, volatile substances called pheromones that act as olfactory signals (Fig. 38). Chemically, insect pheromones are often alcohols or aldehydes, contain from 5 to 20 carbon atoms, and range in molecular weight from 100 to 300 D (21). Pheromones are powerful regulators of reproductive and social behaviors. Among the specific social behaviors known to be regulated by pheromones are alarm, orientation, swarming, defense, fighting, food source tracking and localization, and attraction and assembly of different members of the colony. Study of the sex pheromones and the olfactory system of moths like the silk and gypsy moths (141) and *Manduca* (165), as well as the cockroach (23), has contributed significantly to our knowl-

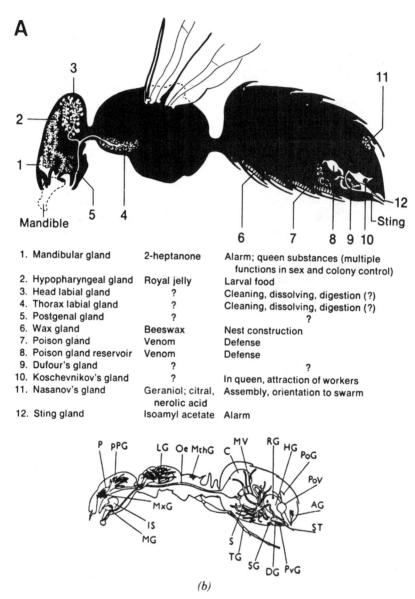

A

1. Mandibular gland — 2-heptanone — Alarm; queen substances (multiple functions in sex and colony control)
2. Hypopharyngeal gland — Royal jelly — Larval food
3. Head labial gland — ? — Cleaning, dissolving, digestion (?)
4. Thorax labial gland — ? — Cleaning, dissolving, digestion (?)
5. Postgenal gland — ? — ?
6. Wax gland — Beeswax — Nest construction
7. Poison gland — Venom — Defense
8. Poison gland reservoir — Venom — Defense
9. Dufour's gland — ? — ?
10. Koschevnikov's gland — ? — In queen, attraction of workers
11. Nasanov's gland — Geraniol; citral, nerolic acid — Assembly, orientation to swarm
12. Sting gland — Isoamyl acetate — Alarm

(b)

Figure 38. Diagrams of the body of a honeybee (*a*) and an ant (*b*) showing the location of the various exocrine glands some of which have pheromonal secretory functions. Abbreviations for ant glands: AG, anal gland; DG, Dufour gland; LG, labial gland, MthG, metapleural gland; MxG, maxillary gland; MG, mandibular gland; PvG, Pavan gland; PoG, poison gland; RG, rectal gland; SG, sternal gland; TG, tibial gland; the remaining structures indicated in part *a* are digestive. [Part *a*, Shepherd (146), after Wilson; part *b*, Attygalle and Morgan (13a), after Dumpert.]

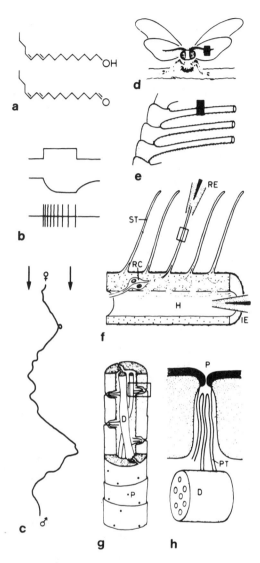

edge of the physiology and neurobiology of insect olfaction.

Receptive female moths emit a sex pheromone that attracts the male moth. The first of such pheromones to be identified was bombykol, emitted from the female silk moth *Bombyx mori*. This pheromone elicits orientation towards the female, fluttering and movement. Bombykol is a C_{16} double unsaturated alcohol (*trans*-10, *cis*-12-hexadecadien-1-ol) (141). A small amount of bombykol on an inert object evokes similar orientating and other behavioral responses in the male moth. The aldehyde form of this substance, bombykal, in contrast to the alcohol, suppresses the fluttering response (Fig. 39). Other insects may require a more complex repertoire of pheromones to carry out the same behaviors. The cabbage looper moth (*Trichoplusia ni*) uses seven behaviorally relevant pheromones, 7-dodecenyl acetate and other closely related compounds, to regulate attraction and flight toward the female (117, 141, 146).

Insect sex pheromones are highly potent chemical messengers. Pheromones from the female silk or gypsy moths can attract males from a distance of 1 mile. The males show an extremely low behavioral threshold to these substances (10^{-13} g/L of air). To obtain a behavioral response under threshold conditions, only 40 of the insect's 40,000 olfactory receptor neurons need be activated by one molecule of bombykol per second. The indi-

Figure 39. Mechanisms of pheromone communication in the silk moth, *Bombyx*. (*a*) Bombykol (top) and bombykal (bottom), the two pheromones, each acts as specific stimulus for one of the two male odor receptor cells (see RC in *f*) in a hair sensillum. (*b*) Schematized extracellular responses: odor puff (top), generator potential (middle), nerve impulses (bottom); recorded as in *f*. (*c*) Upwind path of the flightless male moth to its calling female or to pure bombykol. Parts *d–h* are progressive enlargements of the male antennal structures; (*d*) male with comb shaped antenna; wingspan 40 mm; (*e*) section of antennal stem with branches and hair sensilla; (*f*) opened branch with hairs and receptor cells [(ST) sensillum trichodeum]; RC, receptor cell; H, hemolymph; RE, recording electrode; IE, indifferent electrode; (*g*) section of a hair with pores (P), pore tubules and two dendrites (D); (*h*) epi- and endocuticle with pores tubules (PT) and a dendrite with microtubules. [From Schneider (141).]

vidual receptor neurons are so sensitive that a single molecule of the pheromone is capable of eliciting action potentials in the nerve fiber (68, 141). However, insects, like vertebrates, show a very high convergence ratio of olfactory neurons to central relay neuron, which markedly enhances the sensitivity of the olfactory system (1, 2, 23).

Peripheral Olfactory System:
Antennae and Sensilla

The insects' two flagellar antennae, which may be branched or unbranched, are complex sensory organs. Early studies indicated that when the antennae of bees were removed, the insects failed to respond to odors; this indicated that these appendages function as the insect's olfactory organ (169). Each antenna contains numerous hairs or sensilla (Figs. 40 and 41). These receptor organs are of several morphological types. Each type is specialized for one of the olfactory, thermal, or mechanoreceptor functions. The sensilla with numerous surface pores are olfactory, each containing two or more olfactory receptor neurons. In certain insects the olfactory sensilla of the antennae are peg shaped; in rarer cases of insects with no antennae, the olfactory sensilla are in the tarsi or other body parts (1, 169, 175).

Various insects contain between 20,000 and 50,000 olfactory sensilla on their two antennae. Some of the olfactory sensilla function in sex pheromone detection, others in detection of plant odorants. In the male *Manduca*, nearly 40% of the olfactory sensilla are male specific, each housing the dendrites of two male specific olfactory neurons, specialized to detect only the female sex pheromones (23, 35, 65). In these insects as well as in the cockroach the antennae are longer in the males. The total number of olfactory receptor neurons per animal range from 40,000 to 300,000, depending on the species (175).

Structure of Insect Olfactory Sensillum

The typical structure of an olfactory sensillum is schematically shown in Figure 40B. Each sensillum contains one or more olfactory neurons. Each neuron has a branched dendrite that is supported by a ciliary base. The branches ramify within an extracellular fluid (lymph) that is secreted by the nonsensory sensillar cells and fills the sensillar core. This lymph has a specific ionic composition and contains specialized proteins, which bind with pheromones and odorants and aid in their transport or inactivation (69).

The cuticle (outer surface) of each olfactory sensillum has numerous tiny pores of about 10 nm in diameter (up to 15,000/hair, density, 1–100/μm^2), which open inward into the fluid (lymph) filled space bathing the receptor dendrites. Odorants present in air currents are adsorbed on the cuticular surface, which is covered by a sticky substance. They then diffuse through the pore and across the fluid layer before binding to the receptor-binding proteins of the dendrite membrane. Transport through the extracellular fluid may also be aided by soluble odorant-binding proteins. In pheromone-sensitive sensilla, the pores connect with specialized tubules, which in turn connect to the dendritic membrane. These tubules facilitate the transport of pheromone molecules. After interaction with the receptor molecules, the pheromone must be removed to eliminate its continuous effect. This inactivation of the pheromone is believed to be carried out by pheromone-binding proteins in the extracellular fluid and by enzymatic inactivation. Degrading esterase enzymes have been found in the antennae (69).

Olfactory Transduction and
Electro-antennogram

Stimulation of olfactory receptor neurons by a minute amount of the pheromone

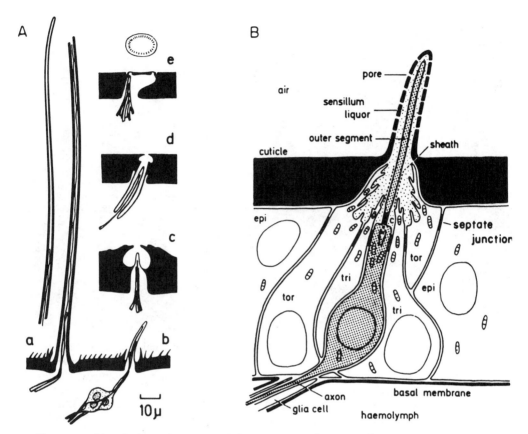

Figure 40. Morphological types and the structure of insect olfactory sensilla. (*A*) The different morphological types of olfactory sensilla in different insect species: (a) long hairlike sensillum trichodeum, (b) short hairlike sensillum basiconicum, (c) pit organ or sensillum coeloconicum, (d) sunken peg or sensillum ampullaceum, and (e) platelike sensillum placodeum. (*B*) The fine structure of a hairlike sensillum showing only one of the several olfactory sensory neurons (shaded) that typically innervate these sensilla; tri, tor, trichogen, and tormogen cells (supportive, secretory cells); epi, epithelial cells. [From Ache (1), after Kaissling.]

causes small negative potentials, called elementary receptor potentials. These potentials are about 0.5 mV in amplitude and 10–50 ms in duration and may represent opening of single ion channels in the receptor membrane. In the presence of stronger stimuli, these small graded potentials summate to produce larger receptor potentials (20–30 mV), which electrotonically spread through the dendrite and cell body and produce action potentials in the nerve fibers (69). Typical extracellular recordings of these elementary receptor potentials and nerve impulses in response to pheromone stimulation are shown in Figure 42*b*. The summated responses of several sensilla may be recorded from the whole antennae and are called the electroantennogram (EAG) (Fig. 42*a*).

Intensity Coding
Insect olfactory neurons respond to a wide range of stimulus intensities cov-

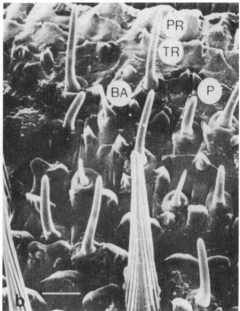

Figure 41. (*a*) Microphotograph of a distal segment of the antennal flagellum of a male cockroach, *Periplaneta americana* with sensilla of various types (bar, 50 μm); (*b*) Enlarged distal portion of the segment in (*a*), showing morphologically and physiologically identified sensilla: BA, cells in this dwA-type sensillum respond to compounds such as butyric acid; P, swA-type sensillum with cells responding to butanol; PR and TR, swB-type sensillum with receptors for pheromone (PR) and terpenoid (TR) compounds. [From Boekh and Ernst (23), after Sass.]

ering nearly three or more orders of magnitude. To code for increasing amounts of stimulus the receptors change the amplitude and duration of their receptor potentials and the frequency of nerve discharges. The amplitude of receptor potential increases in proportion to the logarithm of the stimulus intensity until saturation. The change in the pattern of action potentials to increasing stimulus intensity is more complex. In *Antheraea polyphemus*, in response to the increasing concentration of the pheromone (11Z, 6E)-hexadecadienyl acetate, it is found that at low stimulus intensities, that is, pheromone concentration (10^{-4} μg), the response pattern is tonic, becoming tonic-phasic at medium concentrations (10^{-3}

μg) and purely phasic at higher (10^{-1} μg) concentrations (23, 69) (Fig. 42*b*).

Functional Types of Insect Olfactory Neurons

Based on their patterns of responsivity to various pheromones and general odorants, the insect olfactory neurons may be divided into four categories: narrow specialists, group specialists, group generalists, and broad generalists (146) (Figs. 43 and 44). It is not known whether all of these types occur in every insect. Narrow specialists such as those that respond to sex pheromones, show a highly specific and narrow spectrum, responding only to a given compound. In the silk moth, one type of such receptor neuron responds to

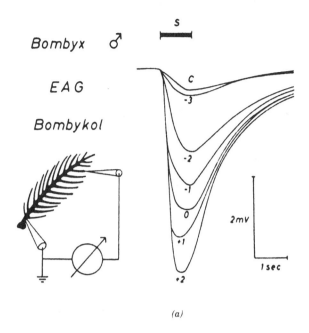

(a)

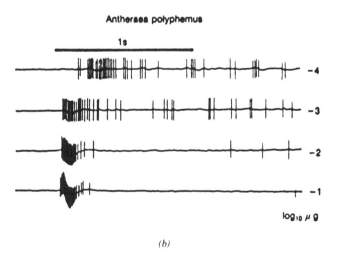

(b)

Figure 42. (a) The electroantennogram (EAG) from isolated antennae of a male silkworm *Bombyx* in response to stimulation by different amounts of bombykol; s, duration of the stimulus; *Note:* The amplitude of the EAG, believed to be the sum of receptor potentials, increases with increasing amount of the pheromone; the recording arrangement is shown at left. [From Prosser (135), after Schneider.] (b) The effect of an increase in the stimulus intensity (amount of pheromone applied) on discharge rates of fibers from an olfactory sensillum of the insect *Antheraea polyphemus*. The pheromone (11Z,6E)-hexadecadienyl acetate) is applied for 1 s (bar). *Note:* (1) Response is tonic at low stimulus intensities ($<10^{-4}$ μg, not shown), phasic-tonic at medium intensities (10^{-3} μg), and purely phasic at higher intensities (10^{-2} μg). [From Kaissling (69).]

bombyk*ol*, the alcohol form of the sex attractant, which causes fluttering, while another neuron responds to bombyk*al*, the aldehyde form, which suppresses the fluttering. These narrowly tuned receptor neurons are the best examples of "labeled lines" in sensory communication, as they carry highly specific information to the brain.

The group specialists respond to a group of closely related compounds such as alcohols or esters associated with odors

Figure 43. (*a*) Responses of receptor neurons (right curve), antennal lobe projection neurons (left curve) and behavioral (wing raising) responses (middle curve) of the male cockroach, *Periplaneta americana*, to different amounts of a cockroach pheromone (periplanone B). *Note:* Central and behavioral responses can be observed at concentrations in which individual receptor neurons show little response. (*b*) Various insect olfactory receptors show differential sensitivity to odorants, some narrow and some wider. This scheme shows the responses of 16 antennal olfactory receptor cells of a male cockroach, *Periplaneta americana*, to various odors: (1) banana; (2) orange; (3) well-hung meat; (4) female lure; (5) *n*-hexanol; (6) *n*-octanol; (7) *n*-decanol; and (8) *n*-hexanoic acid. Filled circles, 50–100% maximal response; stippled circles, 25–50% maximal response; open circles, 0–25% maximal response. [From Boeckh and Ernst (23), after data of Sass and Selzer.]

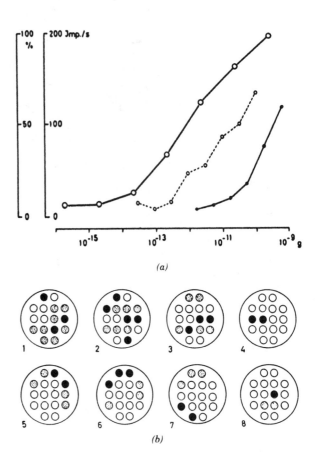

(*a*)

(*b*)

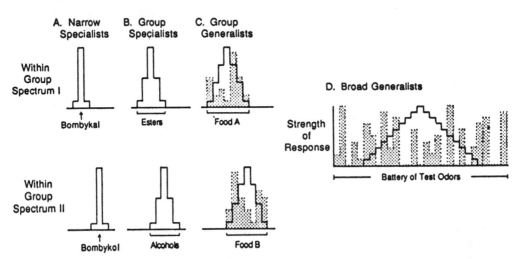

Figure 44. Scheme showing a classification of insect olfactory receptor neurons into four functional categories based on their differential responsiveness to odorants; (*A*) narrow specialists, (*B*) group specialists, (*C*) group generalists, and (*D*) broad generalists. [From Shepherd (146), after Boeckh.]

of meats and fruits. A neuron in this category responds identically to a series of alcohols while another responds to a series of esters. These cells have been found in locusts and cockroaches (Fig. 43). Receptor neurons in the category of group generalists are similar to the group specialists in that they too respond to a single group of compounds like the alcohols, but the response spectra of each cell is different. Such cells have been found in honeybees (146).

The fourth group of olfactory receptor neurons, which is found in some moths, function as broad generalists, because they respond to a variety of compounds. The neural discriminatory mechanisms involved for the group generalist and broad generalist types must be more complicated since the labeled line coding system cannot operate for them. Presumably the insect brain, like the taste and olfactory brain centers of vertebrates, uses some type of "across neurons coding" system to decipher the complex pattern of information arriving from the group generalists and broad generalists receptor types.

Antennal Lobe and Higher Olfactory Brain Centers

After leaving the sensilla, the axons of olfactory cells form the antennal nerve, which projects ipsilaterally to the antennal lobe of the deutocerebrum in the insect brain (23, 35, 65) (Fig. 45). The projecting fibers converge to form glomeruli within the lobes. The number of these glomeruli, which are about 50–100 μm in diameter, differ depending on the species, 60 in *Manduca*, 125 in the cockroach, and several thousand in the locust (1). The glomeruli help segregate the input and analyze the olfactory signals. At least some of the glomeruli are functionally distinct and highly specialized. The male moth and cockroach contain a special type of glomerulus not found in the female. This male specific brain structure

called the macroglomerular complex receives its input mainly from the male specific pheromone sensitive receptor neurons. If the male antennae are transplanted in the head of the developing female *Manduca*, the recipient insect antennal lobe will develop the male specific macroglomerulus (35, 65) (Fig. 45).

The antennal lobes contain both relay and intrinsic neurons. In *Manduca* and cockroach the number of these neurons is about 1000, the intrinsic neurons comprising the majority. The terminals of olfactory neurons make chemical synapses with both the relay and intrinsic neurons within the glomeruli. As the number of olfactory neurons exceeds that of the relay neurons by many-fold, there occurs a convergence of sensory input onto the central relay cells. As for vertebrates, this convergence is an important neural basis for the high sensitivity of the olfactory system. In the American cockroach, this convergence ratio is estimated to be in the order of 1000:1 for receptor neurons involved in food related plant odorants and 4000:1 for the sex pheromone receptors (1, 21a, Table 2). Thus in terms of this aspect of sensitivity, insects are equal or better than the macrosmatic mammals (23, 35, 65).

The local interneurons are axonless amacrine cells. Each interneuron ramifies its dendrites within all the glomeruli of the antennal lobe while output neurons ramify their dendrites within single glomeruli (Fig. 46a). The electrophysiological responses of local interneurons to application of odorants to sensilla have been studied. In *Manduca* the interneurons respond to a broad spectrum of odorants but the response pattern for each odorant is specific (Fig. 46b).

The axons of the relay (output) neurons constitute the central olfactory tract, which carries the output of antennal lobe to higher areas of the insect brain, mainly the corpora pedunculata (mushroom

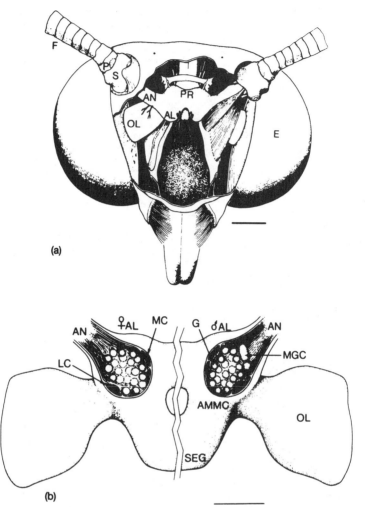

Figure 45. (*a*) Cut-away frontal drawing of a moth *Manduca sexta* head showing the major olfactory structures. (*b*) Cut-away drawing of the antennal lobe of the olfactory lobe in the male (right half) and female (left half), showing the olfactory glomeruli. Note the presence of the macroglomerular complex in the male and its absence in the female. AN, antennal nerve; AL, antennal lobe; E, compound eye; F, flagellum (antenna); G, glomeruli; LC and MC, lateral and medial clusters of neurons; MGC, macroglomerular complex; OL, olfactory lobe; PR, protocerebrum; (P, pedicle and S, scape, basal structures of the flagellum). [From Hildebrand and Montague (65).]

body) and the lateral lobes of protocerebrum (Fig. 47). From these structures neural commands are sent to the motor centers of the insect nervous system to elicit the behaviors associated with the pheromones and odorants.

Synaptic Neurotransmitters
The primary olfactory neurons in *Manduca* and many other insects are cholinergic, releasing the neurotransmitter acetylcholine at their terminals in the antennal lobes; the cholinergic receptor type

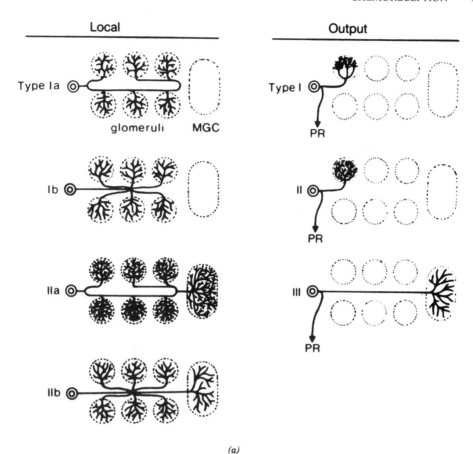

Local

Output

Type Ia

glomeruli   MGC

Ib

IIa

IIb

Type I

PR

II

PR

III

PR

(a)

Figure 46. (*a*) Schematic diagram of the various types of local neurons and output neurons in the antennal lobe of the male moth *Manduca sexta*. MGC, macroglomerular complex; PR, protocerebrum. *Note:* (1) Dendrites of each output neuron innervates only one glomerulus while those of local neurons innervate many; (2) types 1a and 1b of local neurons send dendrites into all the ordinary glomeruli but not into the MGC, while types IIa and IIb innervate both the ordinary and the male specific MGC; type III innervates only MGC. (*b*) Responses of a local neuron in the antennal lobe of the *Manduca* to stimulation of the antennae by odorants and pheromones. This neuron, like the type IIa local neurons (Fig. 46a), is in contact with both the ordinary glomeruli and MGC. *Note:* While the neuron responds to both ordinary odorants (hexenal) and sex pheromones (bombykal), it does so with differential patterns; it also does not respond to tobacco odors. [Part *a*, from Hildebrand and Montague (65); Part *b*, from Christensen and Hildebrand (35).]

mediating this action is nicotinic. Some of the antennal lobe intrinsic neurons may be cholinergic also. Some of the inhibitory interneurons of the antennal lobe release GABA as their neurotransmitter. γ-Aminoleutyric acid is also the neurotrans-

mitter for some of the fibers connecting the antennal lobe to the protocerebrum. Serotonin, histamine, and tyramine, as well as some of the neuropeptides and neurohormones like substance-P, locust adipokinetic hormone, FMRFamide, cor-

B

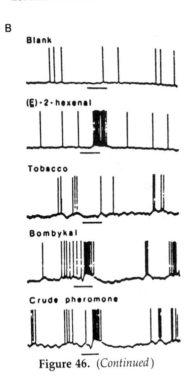

Blank

(E)-2-hexenal

Tobacco

Bombykal

Crude pheromone

Figure 46. (*Continued*)

ticotropin-releasing hormone, and cholecystokinin have also been found associated with some of the antennal lobe neurons (35).

Gustation in Insects

Insects have distinct taste organs and a well-developed gustatory system that can respond to, and discriminate among a relatively wide variety of gustatory stimuli. As with vertebrates the insect gustatory stimuli are chiefly dietary substances, although this restriction is narrower and more restrictive in the insects. Also unlike vertebrates, the insect gustatory responses, even taste thresholds, vary with the animal's nutritional and metabolic state. As in vertebrates, the gustatory thresholds are generally high, compared to olfactory thresholds (Table 3).

Gustatory Sense Organs

Insect gustatory receptor neurons occur, like the olfactory neurons, in specialized sensory structures called taste sensilla. The gustatory sensilla occur commonly as peg- or hairlike organs in different body regions; the distribution varies depending on the species. In flies, butterflies and moths, the most common locations are the mouth parts, inner mouth wall and the tarsi, the terminal segments of the jointed walking legs. In other insects, the taste sensilla are also encountered on the ovipositors (e.g., crickets and Ichneumonid flies), the antennae (bees and ants), and other less specialized surface regions of the body (135, 169, 175).

A gustatory sensillum has a thin outer wall with a single terminal pore of about 2-μm diameter that is filled with a viscous fluid (104). Each sensillum has several chemosensory receptor neurons clustered at its base. The long dendritic processes of these neurons, being modified cilia, project into the cuticular sheath toward the sensillar tip (40). Gustatory stimuli penetrate the sensillum through this pore and bind with receptor proteins on the surface of the receptive zone of the dendritic process. The binding results in gustatory transduction; it depolarizes the membranes and initiates action potentials in the sensory fibers. The structure of an insect taste sensillum is shown in Figure 48. Although structurally alike, gustatory sensilla may differ in their sensitivity to different tastant types, depending on their location in the insect body (169, 175).

A characteristic of gustatory sensilla is that in addition to the taste sensitive neurons, they invariably contain a mechanoreceptor neuron. The combined presence of gustatory and mechanoreceptor elements within the taste sensilla may be one reason why gustation is also called

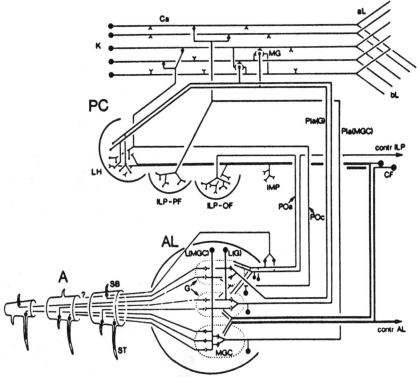

Figure 47. Schematic diagram of olfactory pathways in the brain of the adult male moth *Manduca sexta*. Axons of pheromone receptor cells in sensilla trichodea (ST) on the antenna (A) terminate in the male specific macroglomerular complex (MGC) of the antennal lobe (AL). Receptors sensitive to general odorants project to ordinary glomeruli (G). Receptor neurons synapse on two types of local interneurons, one with dendritic branches in ordinary G and MGC [L(MGC)], and another with branches in ordinary Gs only [L(G)]. AL relay neurons synapse with local neurons and project to olfactory areas in the protocerebrum (PC). Projection areas include calyces (Ca) of the mushroom body, the lateral horn (LH) of the protocerebrum, olfactory foci in the inferior lateral protocerebrum (ILP-OF), and a pheromone focus in the inferior lateral protocerebrum (ILP-PF). (From Homberg, Christensen, and Hildebrand (66a).]

contact chemoreception, in contrast to olfaction, which requires no direct contact of the stimulus mass. In addition, organs containing taste sensilla are often motile, performing mechanical functions in association with food retrieval and ingestion. A mechanoreceptor can provide positional information (83, 175).

Taste Discrimination by Insects
Insects respond not only to food-related stimuli but they also discriminate among a wide variety of gustatory stimuli. Some insect gustatory receptors also respond to water. Many of the insect behavioral responses to typical tastants resemble those shown by land vertebrates under similar conditions. For example, when the tarsi of flies are placed in contact with sweetened water, the proboscis is extended to suck the solution; this response does not occur with water alone. Addition of bitter tasting quinine to the sweetened water prevents the proboscis extension (40, 169).

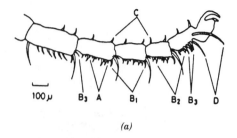

(a)

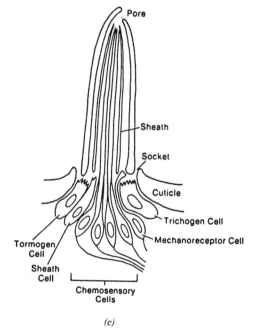

(b)

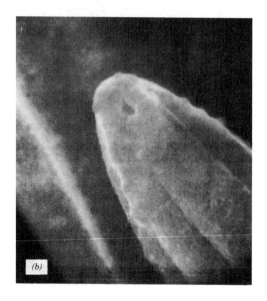

Figure 48. Insect gustatory sensilla (taste hairs). (a) Side view of the tarsus in a blowfly's prothoracic leg, showing the fine taste hairs (contact chemosensilla); the larger mechanoreceptor hairs have been removed. [From Morita and Shiraishi (104).] (b) Microphotograph of a labellar taste hair in the blowfly *Phormica regina*. Note the apical taste pore (courtesy of V. G. Dethier). (c) Schematic representation of the fine structure of insect taste hair. Note the apical pore, the cluster of chemosensory cells and the single mechanoreceptor cell. [From Dethier (40).]

(c)

Caterpillars make vigorous movements of the mouth in response to salt, acids, or bitter compounds. Bees reject honey treated with quinine or salt. As in the human, the intensity of response to sourness is proportional to the degree of acidity (proton release). Although the insect as a whole shows sensitivity to all four taste primaries, behavioral evidence indicates that the insect's various taste organs, for example, tarsi and labellar organs, are differentially sensitive to the various tastant categories. Cell physiology of insect gustation has been focused mostly on the sugar and salt receptors (104, 169, 175).

Insect Gustatory Thresholds and Sensitivity
In addition to behavioral experiments that show the capacity of insects to differentially recognize and respond to the four primary taste qualities, insects show different behavioral thresholds to these gustatory stimuli (Table 3). The taste thresholds are in the same range as those found for humans. Comparison of these insect taste thresholds to olfactory thresholds in insects (Table 1) reveals another similarity with the vertebrates, namely, that gustatory thresholds are higher than olfactory thresholds by more than a thousandfold.

Since the insect gustatory and olfactory receptor cells and sensilla are very similar and thresholds of direct responses from the sensilla are in the same order, the lower behavioral sensitivity of the gustatory system, in both insects and higher vertebrates can only be explained by the difference in central neural factors; chief among these is the absence of a high convergence ratio of the primary sensory afferents on the second-order central relay cells in the gustatory system. These ratios are very high in the olfactory system, in the order of 1000:1 (1, see also above) (Table 2).

In contrast to humans and other vertebrates, insect gustatory thresholds decrease markedly if the animal is starved (169); this indicates a relationship between nutrition, metabolism, and taste sensitivity. This is true especially for sweet substances. Such an ability endows the insect with a special adaptive advantage so that when the insect is in a state of dietary deficiency, even sources with a poor supply of nutrients will be selected and ingested.

Gustation and Feeding in Insects
The sweet taste of nutritionally rich foodstuffs make these foods highly acceptable for insects. Although best exemplified by the honeybee's appetite for the sweet sap of flowers and plants, many insects have a well-developed capacity to detect, choose, and ingest sweet tasting stimuli (62). However, not all the substances that appear sweet to humans taste "sweet" or acceptable to the bee. For example, pentoses (e.g., arabinose), sugar alcohols (e.g., mannitol), and many true hexoses are apparently tasteless to the bee. These "tasteless" sweet substances are absent from the bee's natural diet or habitat (169).

Although deficient peripheral mechanisms cannot be ruled out, the lack of response by the bee to the sweet tastants may be due to absence of gustatory conditioning. Early gustatory conditioning plays an important role in the development of chemosensory preference in many insects, as it does for vertebrates. For example, many insects only oviposit on those plants on which they feed as larvae.

In addition to its use in selection and ingestion of specific food substances, gustation also partly determines the insect's choice of a specific host plant. Phytophagus insects like locusts do not eat all plants indiscriminately. Instead they feed on a family of related plants, often of one genus. This choice occurs in response to specific nonnutritional gustatory stimuli, which commonly occur in all members of the genus. These common tastants are thought to be secondary metabolites (glycosides and other plant alkaloids), which act as attractants or repellents, making one group of plants highly tasty and acceptable and others unacceptable. Thus potato beetles and tobacco hornworm feed only on Solanaceae. Among Solanaceae, the beetles feed on the potato but avoid the tobacco and tomato leaves. This behavior is believed to be associated with the nicotine in the tobacco and the glycoside tomatin in the tomato, which are bitter substances. The opposite behavior is seen in tobacco hornworm, which avoids the potato and feeds on the tobacco and tomato leaves (41, 135).

Functional Responses of Insect Taste Neurons

Recording from single taste receptor cells in the blowfly, in response to presentation of various gustatory stimuli at different conditions (single tastant, mixtures, and low and high concentration of stimuli) yields results comparable with those found in vertebrates. For example, at low stimulus concentration each sensillar taste receptor neuron seems to be preferentially tuned to a particular gustatory stimulus. In the labellar sensilla that occur at the proboscis tip, of the four chemosensory neurons, one showed its best sensitivity to water, the second to sugar, and the third and fourth neurons showed sensitivity to the cation and anion of the salt. The fifth neurons was found to be mechanosensory (40, 146) (Fig. 49).

When, however, these stimuli are presented at higher concentration, all four sensillar neurons show sensitivity, responding to various tastants regardless of their quality. The rate of response as measured by impulse frequency is different, depending on the stimulus type (40, 104).

The multimodal character of the gustatory receptor neurons is further shown when natural mixtures of gustatory stimuli like beer, meat, fish, honey, and apple are presented. Under these conditions, all sensillar fibers respond, but the response pattern (discharge frequency, and other characteristic) is different for each of the complex stimuli (146) (Fig. 49).

Thus the blowfly's labellar taste neurons behave in a manner similar to the taste cells and taste fibers of the vertebrate gustatory system, showing the characteristic features of both "labeled lines" and "across neurons coding" forms of sensory communication. The insect's central neurons receive these different action spectra of fibers, which are determined by the stimulus concentration and mixture interaction, as well as the animal's nutritional and metabolic state, selecting and integrating the appropriate information.

Chemoreception in Invertebrates Except Insects

Chemoreception is well developed in non-insect invertebrates as well. In other aquatic invertebrates, distinction between olfactory and gustatory stimuli and receptor types is not always clear, although the existence of two different chemosensory systems, each with different behavioral functions, is generally accepted (2, 11, 83).

Structure of Sense Organs

Invertebrates show a wide variety of chemoreceptor organs some of which may be olfactory, others gustatory (Fig. 50). However, the receptor cells within these organs show much structural similarity. In annelids like the earthworms (Fig. 50a) (oligochaetes) and marine worms (polychaetes) (Fig. 50b) and leeches (hirudina), chemosensory cells are multiciliate spindle-shaped cells that terminate distally in microvilli and cilia,

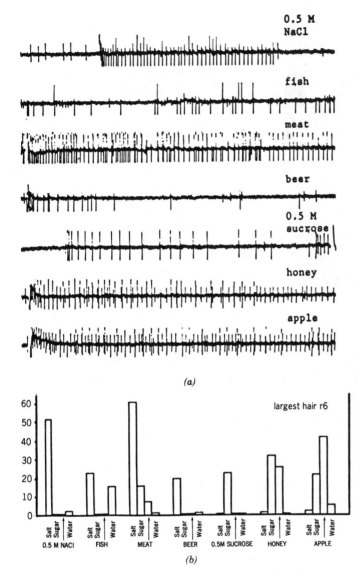

**0.5 M
NaCl**

fish

meat

beer

**0.5 M
sucrose**

honey

apple

(a)

(b)

Figure 49. Responses of insect taste hairs to taste stimuli. (*a*) Electrical responses of a labellar hair of the blowfly to stimulation by simple tastants (NaCl and sucrose) and by some complex food substances. *Note:* The sensillum responds to all these stimuli but with different discharge rates. (*b*) The responses of the labellar sensillum is due to the collective activity of the several receptor cells, a salt-best cell, a sugar-best cell, a water cell, and an undefined cell. The histogram shows the responses of individual cells of the sensillum during the first second of exposure to the substances in (*a*); open bars represent (from left to right) activity in salt, sugar, water cells, and undefined cells indicated by arrows; Y axis: number of impulses per first second of stimulation. [From Dethier (40).]

which project beyond the cuticle. Clusters of these cilia occur in small mounds called sensilla. In the leech, about 150 sensilla line the dorsal edge of the mouth. Ablation of these sensilla eliminates feeding, which is normally elicited by blood. In marine polychaetes sensilla occur in cephalic appendages, on the parapodial cirri and on the anal cirri. Parapodial cirri contain sex pheromone receptors as indicated by their swelling during sexual maturation. Many worms also have, on their heads, an elaborate chemosensory structure known as the nuchal organ, that functions as an olfactory organ (1).

In molluscs, like snails, clams, squids, and octopus, chemosensory cells are bipolar neurons with a ciliated dendrite that projects to the surface. In gastropods, three different appendages are thought to be chemosensory: cephalic tentacles, rhinophores, and osphradium (Fig. 50c).

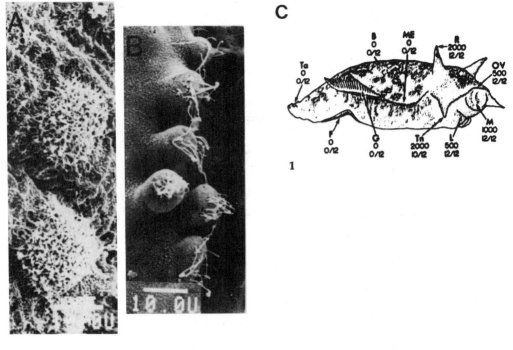

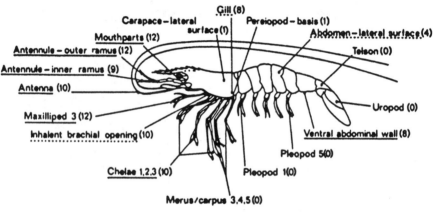

(d)

Figure 50. Sites of chemosensitivity or chemosensory organs in some non-insect invertebrates. (*a*) Earthworm: Ciliary chemosensory mounts from the prostomium of oligochaete *Allolobophora longa;* (*b*) Ciliated chemosensory organs from the ventral cirrus of polychaete *Spinther miniaceus;* (*c*) sea slug, *Pleurobranchaea;* [B, back; F, foot; G, gill; OV, oral veil; M, mouth; ME, mantle edge; R, rhinophore; Ta, tail; Tn, tentacle; the numbers immediately below each letter are the density of chemoreceptors and the ratio below that is the number of trials out of 12 in which an extract of squid elicited feeding behavior. (*d*) Crustacean, *Penaeus merguiensis*, numbers indicate positive responses to meat extract in 12 trials; solid underlines = strong responses; dashed underlines = weak responses. [Parts *a* and *b*, from Laverack (83); Part *c*, from Ache (1), after Davis and Matera; Part *d*, from Ache (2), after Hindley.]

The osphradium occurs in many gastropods except terrestrial snails and slugs. In clams, the edges of the mantle, and the tip of the siphon, are believed to be chemosensory. In cephalopods, the tentacles are chemosensory. Cephalopods also contain olfactory pits, flat disks of tissue located on either side of the head in front of the mantle cavity (1).

In all arthropods (insects, crabs, lobsters, shrimps, and spiders) chemosensitivity is associated with cuticullar sensilla, particularly hairlike setae. Crustaceans have two pairs of chemosensory organs, the antennules, which are olfactory (83). Taste organs are located on the terminal segments of the walking legs (dactylopodites or dactyls) and on the mouth parts. The sensory pore, a complex of three structures that occurs near the margin of the compound eye in the stalked eye crustaceans is also chemosensory. The olfactory sense organs of the antennule are called aesthetases. Aesthetases have thin cuticular walls that are permeable to dyes along their length. The thick walled sensilla are permeable only at their tips and function in taste (2, 13, 83). Figure 50d shows the chemosensitive sites of a typical crustacean. The ultrastructural features of a crustacean olfactory aesthetases as well as those in taste organ (fringed seta) are shown in Figure 51. The tarsi of spiders have taste sensilla that respond to sugars and salts. Spiders also have in their distal tarsi, structures called tarsal organs with many sensilla clustered in a capsule. The tarsal organs respond to odorants.

In snails and lobsters and other arthropods, the olfactory glomeruli have been found in the olfactory lobes, to which the nerve fibers from the antennule project (Fig. 52). In other molluscs and arthropods, one way to ascertain the olfactory nature of a chemoreceptor organ is to trace the thin fibers from the organ in question into the brain; the fibers, upon arrival within the deutocerebrum, converge to form characteristic olfactory glomeruli, as in the antennal lobes of the insects.

Functional Properties of Sense Organs
Chemosensory organs of aquatic invertebrates respond to amino acids and peptides, quaternary ammonium compounds (e.g., betaine), nucleotides (ATP, ADP, and AMP), nucleosides (e.g., inosine), and organic acids (e.g., lactate) (32). Molecules that stimulate the taste receptors are food related. Many of the molecules that stimulate the olfactory system are known or suspected to be neuroactive (e.g., GABA) and may function as pheromone-like signal molecules.

SELECTIVE RESPONSES OF CHEMORECEPTORS Invertebrate chemoreceptors may be narrowly tuned to a single type of chemical stimulus (39). Thus, as shown in Figure 53a, a glutamate sensitive chemoreceptor in the lobster shows activity only when stimulated by L-glutamate but is basically insensitive towards taurine, betaine, L-pyroglutamate, ammonium chloride, glycine, and L-arginine (11). Other chemoreceptor neurons are more broadly tuned but are still differentially sensitive (1) as shown by the example in Figure 53c, which shows the electrophysiological responses of some chemoreceptor neurons in a freshwater crayfish walking leg. These neurons show differential sensitivity to different types of amino acids, as shown by differences in their discharge rates. The sensitivity is in the order: serine > alanine > histidine > glycine > methionine. The threshold response to serine is at $10^{-6}\,M$ and to methionine at $10^{-3}\,M$. When the concentration of a particular amino acid in the stimulating medium is increased, the frequency of impulse propagation in the corresponding chemosensory nerve also increases (Fig. 53b).

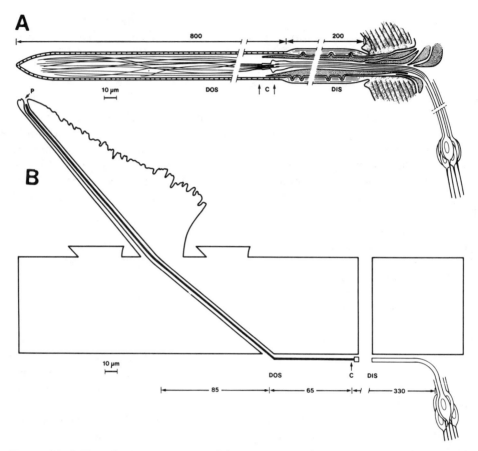

Figure 51. Schematic representation of fine structure of some noninsect invertebrate chemosensory organs. (*A*) Lobster olfactory organ (aesthetasc sensillum) on the antennule of *Homarus americanus:* this sensillum receives ciliated dendrites (one shown greatly enlarged) from some 400 receptor cells. (*B*) Crayfish gustatory organ (fringed seta) on the leg of *Austropotamobius torrentium.* Chemical stimuli enter through the subapical pore. DOS, outer ciliary segment of sensory dendrite; DIS, inner segment of sensory dendrite; c, ciliary junction; p, pore at the tip of the seta. [Parts a and b, Courtesy of J. Atema.]

LOBSTER SEX PHEROMONES. Aquatic invertebrates use pheromones for courtship and sexual communication. Female lobsters (*Homarus americanus*) release sex pheromones near the entrance of the male shelter to encourage the male to initiate selective cohabitation and mating in his shelter. Male lobsters also have sex pheromones that are excreted in their urines; males fan the pheromone rich shelter water with their pleopods, blowing strong currents (odor plumes) to the environment that advertises their sexual readiness (12, 13).

Comparison of Olfaction and Taste

Both olfactory and gustatory stimuli are chemical in nature and share much in common in their properties and effects. For example, both odorants and tastants must dissolve in a watery mucus before

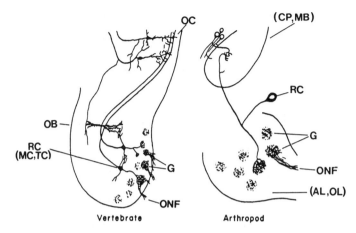

Figure 52. A sketch to emphasize the similarity of the general features of the central olfactory pathway in arthropods and vertebrates. In both, olfactory receptor neurons send their fibers (ONF) to the olfactory bulb (OB) [in vertebrates] or antennal/olfactory lobe (AL, OL) [in invertebrates] to converge onto relay cells (RC) like mitral cells (MC) and tufted cells (TC) of the vertebrates and output cells of the invertebrates; convergence occurs in glomeruli (G). Fibers of output cells project to higher olfactory centers like olfactory cortex (OC) in vertebrates and corpora pedunculata (CP) and mushroom body (MB) in anthropods. [Modified from Ache (2).]

they can bind with their respective receptive surfaces. Also, in many instances taste sensations are influenced by olfactory stimuli. In higher land vertebrates and insects the distinction between taste and smell is clear; however, in aquatic animals, particularly the arthropods, the distinctions are sometimes difficult to decipher. The following general criteria summarize some of the differences between the two systems:

1. Olfactory receptors of terrestrial vertebrates respond to volatile stimuli of small size and diverse chemical nature and source. In aquatic vertebrates and invertebrates volatility is not a criterion for odorants as these present themselves, in the dissolved form, in the aquatic habitat. However, the molecular size of these dissolved odorants is in the small range. Gustatory stimuli may be of varying size and nature but are associated mainly with food substances in the fluid or solid

forms. In fish and aquatic invertebrates, olfactory receptors also respond to food related compounds like amino acids.

2. Sensitivity of the olfactory system to odorants, as indicated by extremely low olfactory thresholds, is usually very high, more than a million times higher than the gustatory system. The olfactory system is therefore traditionally considered as the "distance" chemical sense, permitting long-distance communication, while the gustatory system as the "contact" chemical sense. Only the fish gustatory sensitivity to amino acids is in the same range as that for olfactory receptors (10^{-6} M) and can thus also serve as a distance sense.

3. In insects and aquatic arthropods, the general microscopic appearance of the gustatory and olfactory receptor elements (sensilla) are fairly similar but ultrastructurally the two show distinctive differences.

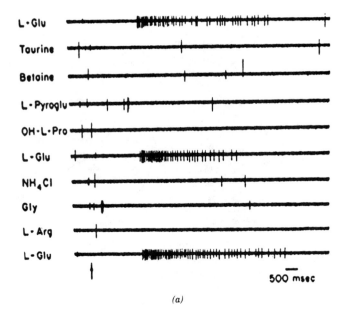

(a)

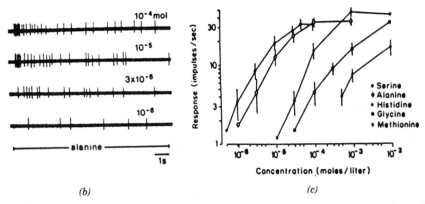

(b)   (c)

Figure 53. (a) Response specificity of a single chemoreceptor cell in a gustatory sensillum of the lobster's walking leg to amino acids. This cell is a narrow specialist showing exclusive responsiveness to L-glutamate and lack of response to other structurally similar amino acids or amine compounds. (b) Increase in the concentration of L-alanine in the stimulation medium results in increase in the rate of discharge by this alanine-sensitive gustatory chemoreceptor cell in the crayfish walking leg. (c) Dose-response plots of the maximum discharge frequency of amino acid sensitive neurons in the dactyls of freshwater crayfish (*Austropotamobius torrentium*) to five amino acids. Each cell responds to all five amino acids but there is differential sensitivity to various amino acids in the following order: serine > alanine > histidine > glycine > methionine. [Part *a*, from Atema (11), after Derby; Parts *b* and *c*, from Ache (1), after Bauer, Dudel, and Hatt (14b).]

4. Gustatory receptor elements, but not the olfactory elements, are often accompanied by mechanoreceptor and other receptor elements of the somatic sensory system. The gustatory rich vertebrate tongue is also well endowed with other types of somatosensory receptors. Even in insects, gustatory elements contain a distinct mechanoreceptor cell. Gustatory perception may be linked to somatosensory perception.

5. Gustatory stimuli must first come in direct contact with the receptor surface before stimulation can occur. Contact may provide higher stimulus concentration for activation of gustatory stimuli; contact may also be necessary for activation of the accompanying mechanoreceptor, thermoreceptor, and nociceptor elements.

6. Olfactory and gustatory receptors are usually housed on different organs or in different cavities. For example, olfactory receptors are usually found in the insect antennae and the vertebrate nasal cavities. Gustatory receptors are found in the insect's tarsi and labellar organs and the vertebrate tongue and oral cavity.

7. In higher animals, olfactory and olfactorylike signals are carried by the fibers of the first cranial nerve and other associated nerves of the nasal region (terminal and vomeronasal). These messages are processed primarily by centers in the paleocortical areas. Gustatory messages travel by the VII, IX, and X nerve to the brain and are processed by neocortical centers. One prominent structural feature of the olfactory brain in vertebrate and invertebrate brain alike is the presence of olfactory glomeruli, formed by the convergence of a large number of the primary afferent olfactory fibers.

8. Olfactory and gustatory discrimination have certain mechanisms in common but they also differ in many regards. The existence of at least four primary

tastes in terrestrial vertebrates is generally accepted and it is thought that various flavors are produced as a result of sensory and perceptual mixing of these primary taste qualities. But the existence of primary odors appears to be a more difficult problem. If primary odor qualities exist, their number is higher than the primary tastes. It is at least seven.

References

1. Ache, B. W., in *Neurolobiology of Taste and Smell* (T. E. Finger and W. L. Silver, eds.), pp. 39–46. Wiley, New York, 1987. Chemoreception in invertebrates.

2. Ache, B. W., in *Sensory Biology of Aquatic Animals* (J. Atema, R. R. Fay, A. N. Popper, and W. N. Tavolga, eds.), pp. 387–401. Springer-Verlag, New York, 1988; in *Biology of Crustacea* (H. Atwook and D. C. Sanderman, eds.), Vol. 3, pp. 369–398. Academic Press, New York, 1982. Chemosensory integration in aquatic invertebrates/crustacean chemoreception.

3. Adams, D. R., and Weikamp, M. D., *J. Anat.* **138**:771–787, 1984. Canine vomeronasal organ.

4. Adler, J., *Science* **166**:1588–1597, 1969. Bacterial chemoreceptors.

5. Adrian, E. D., *J. Laryngol. Otol.* **70**:1–14, 1956; *Acta Physiol. Scand.* **29**:4–14, 1953; *Br. Med. Bull.* **6**:330–331, 1950. Olfactory discrimination by the mammalian olfactory epithelium.

6. Allison, A. C., *Biol. Rev. Cambridge Philos. Soc.* **28**:195–244, 1953. Vertebrate olfactory morphology.

6a. Allison, A. C. and Warwick, R. T. T. *Brain* **72**:186–197, 1949. Number of neurons in rabbit olfactory system.

7. Amoore, J. E., in *Handbook of Sensory Physiology* (L. M. Beidler, ed.), Vol. 4, Part 1, pp. 245–256. Springer, Berlin, 1971. Olfactory genetics and anosmia.

8. Amoore, J., in *Fragrance Chemistry: The Science of the Sense of Smell* (E. T. Theimer, ed.), pp. 27–76. Academic Press, New York, 1982; *Molecular Basis of Odor*.

Thomas, Springfield, IL, 1979; Amoore, J. E., Johnston, J. W., and Rubin, M., *Sci. Am.* **210**:42–49, 1964. Odor classification/stereochemical theory of odor.

9. Amoore, J. E., and Buttery, R. G., *Chem. Senses Flavor* **3**:57–71, 1978. Partition coefficients and comparative olfactometry.

10. Andres, K. H., in *Taste and Smell in Vertebrates* (G. E. W. Wolstenholme and J. Knight, eds.), pp. 177–193. Churchill, London, 1970. Comparative structure of the olfactory bulb.

11. Atema, J., *Oceanus* **23**:4–18, 1980. Smelling and tasting underwater.

12. Atema, J., *Can. J. Fish. Aquat. Sci.* **43**:2283–1190, 1986. Sexual selection and chemical communication in the Lobster.

13. Atema, J., Fay, R. R., Popper, A. N., and Tavolga, W. N. (eds.), *Sensory Biology of Aquatic Animals.* Springer-Verlag, New York, 1988.

13a. Attygalle, A. B. and Morgan, E. D. *Adv. Insect Physiol.* **18**:1–30, 1985. Ant pheromones.

14. Bang, B. G., and Cobb, S., *Auk* **85**:55–61, 1968; Bang, B. G., *Acta Anat.* **79**(Suppl.):1–76, 1971. Size of olfactory bulb/functional anatomy of olfactory system in different species of birds.

14a. Bardach, J. E., and Atema, J., in *Handbook of Sensory Physiology* (L. M. Beidler, ed.), Vol. 4, Part 2, pp. 293–336. Springer, Berlin, 1971. The sense of taste in fishes.

14b. Bauer, U., Dudel, J., and H. Hatt. *J. Comp. Physiol.* **144**:67–74, 1981. Responses of single taste receptors in crayfish leg to amino acids.

15. Beets, M. G. J., in *Handbook of Sensory Physiology* (L. M. Beidler, ed.), Vol. 4, Part 1, pp. 257–321. Springer, Berlin, 1971. Olfactory response and molecular structure.

16. Beidler, L. M. (ed.), *Handbook of Sensory Physiology*, Vol. 4, Part 1. Springer, Berlin, 1971.

17. Beidler, L. M. (ed.), *Handbook of Sensory Physiology*, Vol. 4, Part 2. Springer, Berlin, 1971.

18. Beidler, L. M., in *Medical Physiology* (V. B. Mountcastle, ed.), 14th ed., Vol. 1, pp. 586–602, Mosby, St. Louis, MO, 1980. Physiology of taste and olfaction; in *Neurobiology of Taste and Smell* (T. E. Finger and W. L. Silver, eds.), pp. 423–437. Wiley, New York, 1987. Future trends in taste research.

19. Bertmar, G., *Evolution (Lawrence, Kans.)* **35**:359–366, 1981. Evolution of vomeronasal organs in vertebrates.

20. Bhatnagar, K. P., and Kallen, F. C., *Acta Anat.* **91**:272–282, 1975; *J. Morphol.* **142**:71–90, 1974. Olfactory organ and bulb size in relation to diet and acuity in bats.

21. Bjostad, L. B., *Chem. Senses* **14**:411–420, 1989. Biochemistry of insect sex pheromones.

21a. Boeckh, J. et al. *J. Insect Physiol.* **30**:15–26, 1984. Receptor number and convergence in insect olfactory system.

22. Boeckh, J., in *Sense Organs* (M. S. Laverack and D. J. Cosens, eds.), pp. 87–99. Blackie, Glasgow, 1981. Structure and function of chemoreceptors.

23. Boeck, J., and Ernst, K.-D., *J. Comp. Physiol. A* **161**:549–565, 1987. Single unit analysis in insect olfactory system; Boeckh, J., Ernst, K.-D., and Selsam, P., *Ann. N.Y. Acad. Sci.* **510**:39–43, 1987. Overview of cockroach olfactory pathways.

24. Bradley, R. M., in *Handbook of Sensory Physiology* (L. M. Beidler, ed.), Vol. 4, Part 2, pp. 1–30; Springer, Berlin, 1971. Tongue topography.

25. Bruch, R. C., Kalinoski, D. L., and Kare, M. R., *Annu. Rev. Nutr.* **8**:21–42, 1988. Biochemistry of vertebrate olfactory and gustatory reception.

26. Burton, H., and Benjamin, R. M., in *Handbook of Sensory Physiology* (L. M. Beidler, ed.), Vol. 4, Part 2, pp. 148–164. Springer, Berlin, 1971. Central gustatory projections.

27. Burton, R. *The Language of Smell.* Routledge & Kegan Paul, London, 1976.

28. Buttery, R. G., Haddon, W. F., Seifert, R. M., and Turnbaugh, J. G., *J. Agric. Food Chem.* Vol. 32, pp. 674–676, 1984.

Olfactory threshold of humans to thiamine odor.

29. Cagan, R. H., in *Biochemistry of Taste and Olfaction* (R. H. Cagan and M. K. Kare, eds.), pp. 175–203. Academic Press, New York, 1981. Recognition of taste stimuli at the initial binding interaction.

30. Cagan, R. H., and Kare, M. K. (eds.), *Biochemistry of Taste and Olfaction.* Academic Press, New York, 1981.

31. Caprio, J., in *Sensory Biology of Aquatic Animals* (J. Atema, R. R. Fay, A. N. Popper, and W. N. Tavolga, eds.), pp. 313–338. Springer-Verlag, New York, 1988. Chemoreception in fishes.

32. Carr, W. E. S., in *Sensory Biology of Aquatic Animals* (J. Atema, R. R. Fay, A. N. Popper, and W. N. Tavolga, eds.), pp. 3–28. Springer-Verlag, New York, 1988. Chemical stimuli in the aquatic environment.

33. Castelluci, V. F., in *Principles of Neural Science* (E. Kandel and J. H. Schwartz, eds.), 2nd ed., Chapter 32. Am. Elsevier, New York, 1985. Vertebrate taste and olfaction.

34. Clancy, A. N., Macrides, F., Singer, A. G., and Agosta, W. C., *Physiol. Behav.* **33:**653–660, 1984. Vomeronasal role in the male hamster response to a high molecular weight vaginal pheromone.

35. Christensen, T. A., and Hildebrand, J. G., in *Arthropod Brain* (A. P. Gupta, ed.), pp. 457–484. Wiley, New York, 1987. Organization of olfactory pathways in the moths brain.

36. Demski, L. S., *Am. Zool.* **24:**809–830, 1984. Evolution of LHRH system and terminal nerve in the vertebrate reproductive brain.

37. Demski, L. S., and Northcutt, R. G., *Science* **220:**435–437, 1983. Terminal nerve chemosensory function in vertebrates.

38. Den Otter, C. J., in *Sense Organs* (M. S. Laverack and D. J. Cosens, eds.), pp. 186–215. Blackie, Glasgow, 1981. Mechanisms of transduction in chemoreceptors.

39. Derby, C. D., and Atema, J., in *Sensory Biology of Aquatic Animals* (J. Atema, R.

R. Fay, A. N. Popper, and W. N. Tavolga, eds.), pp. 365–386. Springer-Verlag, New York, 1988. Chemoreceptor physiology in crustaceans.

40. Dethier, V. G., *The Hungry Fly: A Physiological Study of Behavior Associated with Feeding.* Harvard Univ. Press, Cambridge, MA, 1976.

41. Dethier, V. G., *Entomol. Exp. Appl.* **31:**49–56, 1982. Insect gustation/mechanisms of host-plant recognition.

42. Dodd, G. H., and Squirrell, D. J., in *Olfaction in Mammals* (D. M. Stoddart, ed.), pp. 35–56. Academic Press, London, 1980. Structure/function in the mammalian olfactory system.

42a. Doty, R. L., W. E. Brugger, P. C. Jurs, M. A. Orndorff, P. J. Snyder, and L. D. Lowry, *Physiol. Behav.* **20:**175–187, 1978. Odorant discrimination by trigeminal system in anosmic humans.

42b. Duchamp, A., Revial, M. F., Holley, A., and MacLeod, P. *Chem. Senses Flavor* **1:**213–233, 1974. Responses of frog receptors to odorants.

43. Duchamp-Viret, P., Duchamp, A., and Vigouroux, M., *J. Neurophysiol.* **61:**1085–1094, 1989. Amplifying role of convergence in olfactory system.

44. Duval, D., Müller-Schwarze, D., and Silverstein, R. M. (eds.), *Chemical Signals in Vertebrates*, Vol. 4. Plenum, New York, 1985.

44a. Edens, L., and Van der Well, H., *Trends Biotechnol.* **3:**61–63, 1985.

45. Engen, T., *Am. J. Otolaryngol.* **4:**250–251, 1983. Uses of olfaction in the human.

46. Farbman, A. I., *Chem. Senses* **11:**3–18, 1986. Development of mammalian olfactory receptor cells.

47. Finger, T. E., in *Neurobiology of Taste and Smell* (T. E. Finger and W. L. Silver, eds.), pp. 331–354. Wiley, New York, 1987. Gustatory nuclei and pathways in the fish and mammalian CNS; in *Fish Neurobiology* (R. G. Northcutt and R. E. Davis, eds.), pp. 285–310. Univ. of Michigan Press, Ann Arbor, 1984. The gustatory system in teleost fish.

48. Finger, T. E., in *Sensory Biology of Aquatic*

Animals (J. Atema, R. R. Fay, A. N. Popper, and W. N. Tavolga, eds.), Springer-Verlag, New York, 1988. pp. 339–364. Fish chemosensory brain.

49. Finger, T. E., and Silver, W. L. (eds.), *Neurobiology of Taste and Smell*. Wiley, New York, 1987.

50. Frank, M., *J. Gen. Physiol.* **61**:588–618, 1973. Hamster afferent taste nerve responses.

51. Freeman, W. J., and Skarda, C. A., *Brain Res. Rev.* **10**:147–175, 1985. Odor induced spatial EEG patterns in the mammalian olfactory bulb; Freeman, W. J., and Schneider, W., *Psychophysiology* **19**:44–46, 1982. Olfactory bulb EEG spatial patterns and conditioning.

51a. Geme, G., and Doving, K. D. *Amer. J. Anat.* **26**:457–463, 1969. Receptor number and convergence in the fish olfactory system.

52. Gesteland, R. C., Lettvin, J. Y., and Pitts, W. H., *J. Physiol.* (*London*) **181**:525–559, 1965. Responses of frog olfactory receptors to odorants; Gesteland, R. C., *Experientia* **42**:287–291, 1986. Olfactory receptor cells as analyzers and filters.

53. Getchell, T. V., *Physiol. Rev.* **66**:772–818, 1986. Functional properties of vertebrate olfactory receptor neurons.

54. Getchell, T. V., Margolis, F. L., and Getchell, M. L., *Prog. Neurobiol.* **23**:317–345, 1984. Perireceptor and receptor events in vertebrate olfaction.

55. Gillingham, J. C., and Clark, D. L., *Can. J. Zool.* **59**:1651–1657, 1981. Snake tongue flicking and vomeronasal function.

56. Graziadei, P. P. C., in *Handbook of Sensory Phsyiology* (L. M. Beidler, ed.), Part 1, pp. 27–58. Springer, Berlin, 1971; in *Essays on the Nervous System* (R. Bellairs and E. G. Gray, eds.), pp. 191–222. Oxford Univ. Press, London and New York, 1974. Evolution and structure of vertebrate olfactory epithelium; in *Chemical Signals in Vertebrates* (D. Müller-Schwarze and M. M. Mozell, eds.), pp. 435–454. Plenum, New York, 1977. Functional anatomy of the mammalian chemoreceptor system.

57. Graziadei, P. P. C., and Monti-Graziadei, G. A., in *Handbook of Sensory Physiology* (M. Jacobson, ed.), Vol. 9, pp. 55–83. Springer-Verlag, Berlin, 1978; *Ann. N.Y. Acad. Sci.* **457**:127–142, 1985. Neurogenesis and continuous nerve cell renewal in the olfactory system.

58. Guyton, A. C., *Textbook of Medical Phsyiology*, 6th ed., Chapter 62. Saunders, New York, 1981. Physiology and psychophysiology of taste and smell.

59. Haberly, L. B., *Chem. Senses* **10**:219–238, 1985; Haberley, L. B., and Bower, J. M., *Trends Neurosci.* **12**:258–264, 1989; Implications of neuronal circuitry in the olfactory cortex for memory.

60. Halazs, N., and Shepherd, G. M., *Neuroscience* **10**:579–619, 1983. Neurochemistry of vertebrate olfactory bulb.

61. Halpern, M., *Annu. Rev. Neurosci.* **10**:325–362, 1987. Organization and functions of vomeronasal system.

62. Hansen, K., and Wieczorek, H., in *Biochemistry of Taste and Olfaction* (R. H. Cagan and M. K. Kare, eds.), pp. 139–162. Academic Press, New York, 1981. Sugar receptors in insects.

63. Hara, T. J., *Prog. Neurobiol.* **5**:271–335, 1975; in *The Behavior of Teleost Fishes* (T. Pitcher, ed.), pp. 152–176. Johns Hopkins Univ. Press, Baltimore, Maryland, 1986. Olfactory system in fishes; olfaction in fish behavior.

64. Harriman, A. R., and Berger, R. H., *Physiol. Behav.* **36**:257–262, 1985. Olfactory acuity in the common raven.

65. Hildebrand, J. G., and Montague, R. A., in *Mechanisms of Insect Olfaction* (T. L. Payne, M. C. Birch, and C. E. J. Kennedy, eds.), pp. 279–288. Oxford Univ. Press (Clarendon), London and New York, 1986. Olfactory pathways of insect brain *Manduca sexta*.

66. Hinds, J. W., and McNelly N. A., *Anat. Rec.* **210**:375–383, 1984. Prolonged survival of olfactory neurons in mice reared in polution-free environment.

66a. Homberg, U., Christensen, T. A., and Hildebrand, J. G. *Annu. Rev. Entomol.* **34**:477–501, 1989. Insect olfactory brain and neural pathways.

67. Johns, M. A., in *Chemical Signals* (D. Müller-Schwarze and R. Silversteine, eds.), Vol. 2, pp. 341–364. Plenum, New York, 1980. Vomeronasal functions in mammalian reproduction.

67a. Johnston, R. E., in *Taste, Olfaction and the Central Nervous System* (D. W. Pfaff, ed.), pp. 322–346. Rockefeller University Press, New York, 1985.

68. Kaissling, K. E., in *Handbook of Sensory Physiology* (L. M. Beidler, ed.), Vol. 4, Part 1, pp. 351–431. Springer, Berlin, 1971. General mechanisms of insect olfaction.

69. Kaissling, K. E., *Annu. Rev. Neurosci.* **9:**121–145, 1986. Chemoelectrical transduction in insect olfactory receptors.

69a. Kare, M., in *Handbook of Sensory Physiology* (L. M. Beidler, ed.), Vol. 4, Part 2, pp. 278–292. Springer, Berlin, 1971. Comparative aspects of taste.

70. Kauer, J. S., in *Neurobiology of Taste and Smell* (T. E. Finger and W. L. Silver, eds.), pp. 205–231. Wiley, New York, 1987. Coding in the olfactory system.

71. Keverne, E. B., *Trends Neurosci.* **1:**32–34, 1978. Olfaction and taste-dual systems for sensory processing.

72. Kimura, K., and Beidler, L. M., *J. Cell. Comp. Physiol.* **58:**131–139, 1961. Taste receptor potentials in rat and hamster.

73. Kinnamon, J. C., in *Neurobiology of Taste and Smell* (T. E. Finger and W. L. Silver, eds.), pp. 277–298. Wiley, New York, 1987. Organization and innervation of taste buds.

74. Kleene, S. J., *Experientia* **42:**241–250, 1986. Bacterial chemotaxis in vertebrate olfaction.

75. Koshland, D. E., Jr., *Annu. Rev. Neurosci.* **3:**43–74, 1980. Bacterial chemotaxis in relation to neurobiology.

76. Kosten, T., and Contreas, R. J., *Behav. Neurosci.* **99:**734–741, 1985. Effect of adrenalectomy on rat gustatory nerve responses.

77. Kung, C., and Saimi, Y., *Annu. Rev. Physiol.* **44:**519–534, 1982. The physiological basis of taxis in Paramecium.

78. Kurihara, K., Yoshii, K., and Kashiwayanagi, M., *Comp. Biochem. Physiol.* *A* **85:**1–22, 1986. Transduction mechanisms in chemoreception.

79. Labarca, P., and Bacigalupo, J., *J. Bioenerg. Biomembr.* **20:**551–569, 1988. Ion channels in chemosensory olfactory neurons.

80. Ladewig, J., and Hart, B. L., in *Olfaction and Endocrine Regulation* (W. Breipohl, ed.), pp. 237–247. IRL Press, London, 1982. Flehmen and vomeronasal function; Hart, B. L., in *Chemical Signals in Vertebrates* (D. Müller-Schwarze and R. Silverstein, eds.), Vol. 3, pp. 87–103. Plenum, New York, 1983. Flehmen behavior and vomeronasal function.

81. Lalonde, E. R., and Eglitis, J. A., *Anat. Rec.* **140:**91–95, 1961. Extralingual taste buds in the human newborn.

82. Lancet, D., *Annu. Rev. Neurosci.* **9:**329–355, 1986. Vertebrate olfactory reception.

83. Laverack, M. S., in *Sensory Biology of Aquatic Animals* (J. Atema, R. R. Fay, A. N. Popper, and W. N. Tavolga, eds.), pp. 287–312. Springer-Verlag, New York, 1988. Chemoreceptor diversity in aquatic animals.

84. Lawless, H. T. *Chem. Senses* 5:247–256, 1980. Taste sensitivity to phenylthiocarbamides.

85. Lawless, H. T. in *Neurobiology of Taste and Smell* (T. E. Finger and W. L. Silver, eds.), pp. 401–420. Wiley, New York, 1987. Gustatory psychophysics.

86. Little, E. E., in *Fish Neurobiology* (R. G. Northcutt and R. E. Davis, eds.), Vol. 1, pp. 351–377. Univ. of Michigan Press, Ann Arbor, 1984. Behavioral functions of olfaction and taste in fish.

87. Lowe, G., Nakamura, T., and Gold, G. H., *Proc. Natl. Acad. Sci. U.S.A.* 86: 5641–5645, 1989. Adenylate cyclase and olfactory transduction.

88. Mackay-Sim, A., Shamnan, P., and Moulton, D. G., *J. Neurophysiol.* **48:**584–596, 1982. Topography of coding in the salamander olfactory epithelium.

89. Macleod, A. J., in *Olfaction in Mammals* (D. M. Stoddart, ed.), pp. 13–34. Academic Press, New York, 1980. Chemistry of odours.

90. Maier W., in *Evolutionary Biology of the New World Monkeys and Continental Drift* (R. L. Ciochon and A. B. Chairelli, eds.), pp. 219–241. Plenum, New York, 1980. Nasal structures in old and new world primates.

91. Margolis, F. L., in *Biochemistry of Taste and Olfaction* (R. H. Cagan and M. K. Kare, eds.), pp. 369–394. Academic Press, New York, 1981. Neurotransmitters of the mammalian olfactory bulb.

92. Margolis, F. L. *Trends Neurosci.* **8:**542–546, 1985. Functions and molecular biology of olfactory marker protein.

93. Mathews, D. F., *J. Gen. Physiol.* **60:**160–180, 1972. Response patterns of single neurons in the tortoise olfactory epithelium and olfactory bulb.

94. McBurney, D. H., in *Handbook of Physiology* (I. Darian-Smith, ed.), Vol. 3, pp. 1067–1086. Am. Physiol. Soc., Bethesda, MD, 1984. Taste and olfaction: Sensory discrimination.

95. McCartney, W., *Olfaction and Odours.* Springer-Verlag, Berlin, 1968.

96. Meisami, E., in *Fetal and Neonatal Development* (C. T. Jones, ed.), pp. 195–203. Perinatology Press, Ithaca, New York, 1988. Olfactory neural development and uses in the neonate (mammals): in *Handbook of Human Growth and Developmental Biology* (E. Meisami and P. S. Timiras, eds.), Vol. 1, Part B, pp. 33–62. CRC Press, Boca Raton, FL, 1988. Olfactory development in the human.

97. Meisami, E., *Dev. Brain Res.* **46:**9–19, 1989. Neuron number, convergence and olfactory sensitivity during development (rat).

97a. Meisami, E., Hudson, R., and Distel, H. *Cell Tissue Res.* 1990 in press. Growth of rabbit olfactory epithelium bulb.

98. Menco, B. P. M., *Cell Tissue Res.* **211:**5–29, 1980. Quantification of vertebrate olfactory cilia; in *Nasal Tumors in Animals and Man* (G. Reznik and S. F. Stinson, eds.), Vol. 1, Chapter 3. CRC Press, Boca Raton, FL, 1983. Ultrastructure of olfactory cilia.

99. Meredith, M., in *Pheromones and Reproduction in Mammals* (J. G. Vandenbergh, ed.), pp. 199–252. Academic Press, New York, 1983. Sensory physiology of pheromone communication; in *Chemical Signals* (D. Müller-Schwarze and R. M. Silversteine, eds.), Vol. 2, pp. 303–326. Plenum, New York, 1980. Vomeronasal system in mammals especially hamster.

100. Meredith, M., *J. Neurophysiol.* **56:**572–597, 1986. Patterned response to odor intensity in mammalian olfactory bulb.

101. Minor, A. V., and Sakina, N. L., *Neirofiziologiya* **5:**415–422, 1973. cAMP in olfactory reception.

102. Moncrieff, R. W., *The Chemical Senses.* CRC Press, Cleveland, OH, 1967.

103. Moran, D. T., Rowley, J. C., and Jasek, B. W., *Brain Res.* **253:**39–46, 1982. Microvillar cell in human olfactory epithelium.

104. Morita, H., and Shiraishi, A., in *Comprehensive Insect Physiology, Biochemistry and Pharmacology* (G. A. Kerkut and L. I. Gilbert, eds.), Vol. 6, pp. 133–166. Pergamon, Oxford, 1985. Insect chemoreception physiology/gustation.

105. Moulton, D. G., and Beidler, L. M., *Physiol. Rev.* **47:**1–52, 1967. Structure and function in the vertebrate peripheral olfactory system.

106. Moulton, D. G., *Physiol. Rev.* **56:**578–593, 1972. Spatial patterning of response to odors in the peripheral olfactory system.

107. Moulton, D. G., in *Chemical Signals in Vertebrates* (E. D. Müller-Schwartz and M. M. Mozell, eds.), pp. 455–464. Plenum, New York, 1977. Odor thresholds/sensitivity in the dog.

108. Moulton, D. G., Celebi, G., and Fink, R. P., in *Taste and Smell in Vertebrates* (G. E. W. Wolstenholme and J. Knight, eds.), pp. 227–250. Churchill, London, 1970. Renewal of olfactory neurons and sensitivity to odors in mammals.

109. Mozell, M. M., Smith, B. P., Sullivan, R. J., and Swindler, P., *Arch. Otolaryngol.* **80:**367–373, 1969. Nasal chemoreception and flavor identification.

110. Mozell, M. M., *J. Gen. Physiol.* **50:**25–41, 1966; **56:**46–63, 1970; Mozell, M. M., and Honnung, D. E., in *Taste, Olfaction*

and the Central Nervous System (D. W. Pfaff, ed.), pp. 253–279. Rockefeller Univ. Press, New York, 1985. Spatiotemporal mechanisms and chromatographic model of odorant discrimination in the olfactory periphery.

111. Munz, H., Claas, B., and Jennes, L., *Brain Res.* **221:**1–13, 1981. LHRH systems in the brain of Platyfish.

112. Murray, R. G., and Murray, A., in *Taste and Smell in Vertebrates* (G. E. W. Wolstenholme and J. Knight, eds.), pp. 3–30. Churchill, London, 1970; Murray, R. G., in *The Ultrastructure of Sensory Organs* (J. Friedmann, ed.), pp. 1–81. Am. Elsevier, New York, 1973. Taste bud structure.

113. Negus, V., *The Comparative Anatomy and Physiology of the Nose and Paranasal Sinuses.* Churchill-Livingstone, Edinburgh, 1958.

114. Norgren, R., in *Handbook of Physiology* (I. Darian-Smith, ed.), Vol. 3, pp. 1087–1128. Physiol. Soc., Bethesda, MD, 1984. Central neural mechanisms of taste.

115. Norgren, R., *Chem. Senses* **10:**143–161, 1985. Taste and the autonomic nervous system.

116. Northcutt, R. G., and Davis, R. E. (eds.), *Fish neurobiology: Vol. 1. Brain stem and sense organs.* Univ. of Michigan Press, Ann Arbor, 1984.

117. O'Connel, R. J. O., *Experientia* **42:**232–241, 1986. Chemical communication in invertebrates.

118. Ogawa, H., Sato, M., and Yamashita, S., *J. Physiol. (London)* **199:**223–240, 1968. Polymodal responses of rat taste nerve fibers.

119. Oka, Y., Ichikawaa, M., and Ueda, K., in *Chemoreception in Fishes* (T. J. Hara, ed.), pp. 61–75. Elsevier, Amsterdam, 1982. Synaptic organization and central projections of the olfactory bulb in fishes.

120. Ordal, G. W., *BioScience* **30:**408–410, 1980; *CRC Crit. Rev. Microbiol.* **12:**95–130, 1985. Biochemistry of bacterial chemotaxis.

121. Ottoson, D., in *Handbook of Sensory Phys-iology* (L. M. Beidler, ed.), Vol. 4, Part 1, pp. 95–131. Springer, Berlin, 1971. The electro-olfactogram; *Physiology of the Nervous System*, Chapters 28 and 29. Oxford Univ. Press, New York, 1983.

122. Ozeki, M., and Sato, M., *Comp. Biochem. Physiol. A* **41,** 391–407, 1972. Response of rat gustatory cells to four taste qualities.

123. Parker, G. H. *Smell, Taste and Allied Senses in the Vertebrates.* Lippincott, Philadelphia, PA, 1922.

124. Parkinson, J. S., *Cell (Cambridge, Mass.)* **53:**1–2, 1988. Protein phosphorylation in bacterial chemotaxis.

125. Parsons, T. S., in *Handbook of Sensory Physiology* (L. M. Beidler, ed.), Vol. 4, Part 1, pp. 1–26. Springer, Berlin, 1971. Comparative Anatomy of Nasal Structures.

125a. Paternostro, M. and Meisami, E. *Inter. J. Develop. Neurosci.* **7:**243–255, 1989. Growth of olfactory epithelium and its hormonal control in the rat.

126. Payne, T. L., Birch, M. C., and Kennedy, C. E. J., eds., *Mechanisms of Insect Olfaction.* Oxford Univ. Press (Clarendon), London and New York, 1986.

126a. Pedersen, P. E., and T. E. Bensen, *J. Comp. Neurol.* **252:**555–562, 1986. Projections of Masseral organ to the main olfactory bulb (guinea pig).

127. Pevsner, J. P., Sklar, B., and Snyder, S. H., *Proc. Natl. Acad. Sci. U.S.A.* **83:**4942–4946, 1986. Odorant binding protein: Localization to nasal glands and secretions.

128. Pfaffmann, C., *J. Neurophysiol.* **18:**429–440, 1955. Gustatory nerve impulses in mammals: Pfaffmann, C., Frank, M., Bartoshuk, I. M., and Snell, T. C., *Prog. Psychobiol. Physiol. Psychol.* **6:**1–27, 1976. Gustatory coding by the squirrel monkey chorda tympani nerve.

129. Pfeiffer, W., in *Pheromones* (M. C. Birch, ed.), pp. 269–296. Elsevier/North-Holland, Amsterdam, 1974. Pheromones in fish and amphibia.

130. Philips, J. O., and Fuchs, A. F., in *Textbooks of Physiology* (H. D. Patton, A. F. Fuchs, B. Hille, A. M. Scher, and R.

Steiner, eds.), 21st ed., Vol. 1. Saunders, Philadelphia, PA, 1989. Physiology and neurobiology of gustation and olfaction.

131. Prasada-Rao, P. D., and Finger, T. E., *J. Comp. Neurol.* **225**:492–510, 1984. Asymmetry of olfactory structure in the flatfish.

132. Preston, R. R., and van Houten, J. L., *J. Comp. Physiol.* A **160**:525–536, 1987. Hyperpolarizing responses to acetate and folate in *Paramecium*; 537–542. Localization of chemoreceptive sites in *Paramecium tetraurelia*.

133. Price, J. L., in *Neurobiology of Taste and Smell* (T. E. Finger and W. L. Silver, eds.), pp. 179–205. Wiley, New York, 1987. Central neural structures of the olfactory systems of mammals.

134. Price, S., *Chem. Senses* 8:341–354, 1984. Mechanisms of stimulation of olfactory neurons.

135. Prosser, C. L., in *Comparative Animal Physiology* (C. L. Prosser, ed.), 3rd ed., Chapter 13. Saunders, Philadelphia, PA, 1973; Chemoreception; in *Comparative Animal Physiology* (C. L. Prosser and F. A. Brown, Jr., eds.), 2nd ed., Chapter 11. Saunders, Philadelphia, PA, 1961. Chemoreception.

136. Prosser, C. L., *Adaptational Biology: Molecules to Organisms*, pp. 520–522. Wiley, New York, 1986.

137. Quay, W., in *Chemical Signals in Vertebrates* (D. Müller-Schwarze and R. M. Silverstein, eds.), pp. 105–118. Plenum, New York, 1983. Olfactory pathways in neuroendocrine responses.

138. Revial, M. F., Sicard, G., Duchamp, A., and Holley, A., *Chem. Senses* 7:175–190, 1982. Odor discrimination in the frog olfactory receptor cells.

139. Rivard, G., and Klemm, W. R., *Chem. Senses* 14:273–279, 1989. Estrous pheromones in body fluids (cows).

140. Sato, T., *Prog. Sens. Physiol.* 6:1–37, 1986. Receptor potential in rat taste cells; Sato, T., and Beidler, L. M., *Comp. Biochem. Physiol.* A 75:131–137, 1983. Depolarizing and hyperpolarizing receptor potentials of taste cells in the rat.

141. Schneider, D., *Science* **163**:1031–7, 1969. Insect olfaction/chemical communication; in *Foundations of Sensory Sciences* (W. W. Dawson and E. J. Enoch, eds.), pp. 381–418. Springer-Verlag, Heidleberg, 1984. Insect olfaction—review of own research; in *Encyclopedia of Neurosciences*, (G. Adelman ed.), Vol. 2, pp. 877–879, Birkhaueser, Boston, 1987. Insect olfaction.

142. Scott, J. W., *Experientia* **42**:223–232, 1986; Scott, J. W., and Harrison, T. A., in *Neurobiology of Taste and Smell* (T. E. Finger and W. L. Silver, eds.), pp. 151–178. Wiley, New York, 1987. Functional anatomy of mammalian olfactory bulb and central olfactory pathways.

143. Scott, T. R., and Chang, F. C. T., *Chem. Senses* **8**:297–314, 1984. Gustatory neural coding.

144. Shepherd, G. M., in *Taste, Olfaction and the Central Nervous System* (D. F. Pfaff, ed.), pp. 307–321. Rockefeller Univ. Press, New York, 1985. Responses of olfactory bulb to odors; labeled-lines in the olfactory pathways.

145. Shepherd, G. M., in *The Neurosciences, Fourth Study Program* (F. O. Schmitt and F. G. Worden, eds.), pp. 129–143. MIT Press, Cambridge, MA, 1979; in *Encyclopedia of Neurosciences* (G. Adelman, ed.), Vol. 2, pp. 879–881. Birkhaueser, Boston, MA, 1987. Olfactory bulb neurobiology.

146. Shepherd, G. M., *Neurobiology*, 2nd ed., Chap 11, Oxford Univ. Press, New York, 1988. Overview of invertebrate and vertebrate chemical senses.

147. Shingai, T., *Jpn. J. Physiol.* **30**:305–307, 1980. Water taste fibers in the rat.

148. Silver, W. L., in *Neurobiology of Taste and Smell* (T. E. Finger and W. L. Silver, eds.), pp. 65–87. Wiley, New York, 1987. The common chemical sense in mammals and humans.

149. Singer, A. G., Clancy, A. N., Macrides, F., and Agosta, W. C., *Physiol. Behav.* **33**:645–651, 1984. Responses of male hamster to high molecular weight vaginal pheromone; in *Chemical Signals* (D. Müller-Schwarze and R. M. Silverstein,

eds.), Vol. 2, pp. 365–375. Plenum, New York, 1980. Hamster reproductive pheromones.

150. Sloan, H. E., Hughsess, E., and Oakley, B., *J. Neurosci.* **3:**117–123, 1983. Role of axonal transport in taste responses and taste bud renewal.

151. Smith, D. V., Van Bushkirk, R. L., Travers, J. B., and Bievber, S. L., *J. Neurophysiol.* **50:**522–540, 1983. Gustatory neuron types in hamster brainstem.

152. Smith, R. J. F., in *Chemical Signals in Vertebrates* (D. Müller-Schwarz and M. M. Mozell, eds.), pp. 303–320. Plenum, New York, 1977. The alarm substance of fish.

153. Snyder, S. H., Sklar, P. B., and Pevzner, J., *J. Biol. Chem.* **263:**13971–13974, 1988. Molecular mechanisms of olfaction.

154. Sorensen, P. W., Hara, T. J., and Stacey, N. E., *J. Comp. Physiol. A* **160:**305–314, 1987. Extreme olfactory sensitivity of goldfish to a steroidal pheromone; Sorensen, P. W., Hara, T. J., Stacey, N. E., and Goetz, F. W., *Biol. Reprod.* **39:**1039–1050, 1988. Prostaglandins as female sex pheromone in goldfish.

155. Stabell, O. B., *Biol. Rev. Cambridge Philos. Soc.* **59:**333–338, 1984. Homing and olfaction in salmonids.

156. Stoddart, D. M., (ed.), *Olfaction in Mammals*, Symp. Zool. Soc. London, No. 45. Academic Press, London, 1980.

157. Stoddart, D. M., in *Olfaction in Mammals* (D. M. Stoddart, ed.), Chapter 1. Academic Press, London, 1980. Evolutionary biology of mammalian olfaction.

158. Taniguchi, K., and Mochizukin, K., *Jpn. J. Vet. Sci.* **45:**67–76, 1983. Vomeronasal organ morphology in rats, mice and rabbits.

159. Teeter, J. H., and Brand, J. G., in *Neurobiology of Taste and Smell* (T. E. Finger and W. L. Silver, eds.) pp. 299–330. Wiley, New York, 1987. Peripheral mechanisms of gustation.

159a. Teeter, J. H. and Gold, G. *Nature* **331:**298–299, 1988. Molecular physiology of taste reception.

160. Travers, J. B., Travers, S. P., and Nor-gren, R., *Annu. Rev. Neurosci.* **10:**595–632, 1987. Gustatory neural processing in the hindbrain.

161. Tucker, D., in *Handbook of Sensory Physiology* (L. M. Beidler, ed.), Vol. 4, Part 1, pp. 151–181. Springer, Berlin, 1971. Non-olfactory nasal chemosensation.

162. Tucker, D., in *Fish Neurobiology* (R. G. Northcutt and R. E. Davis, eds.), Vol. 1, pp. 311–351. Univ. of Michigan Press, Ann Arbor, 1984. Fish chemoreception: Peripheral anatomy and physiology.

163. Van Drongelen, W., Holley, A., and Doving, K. B., *J. Theor. Biol.* **71:**39–48, 1978. Convergence in the vertebrate olfactory system.

164. Van Houten, J., and Preston, R. R., *Ann. N.Y. Acad. Sci.* **510:**16–22, 1987. in *Neurobiology of Taste and Smell* (T. E. Finger and W. L. Silver, eds.), pp. 11–38. Wiley, New York, 1987. Chemoreception in single-celled and eukaryotic organisms.

165. von Bartheld, C. S., and Meyer, D. L., *Cell Tissue Res.* **246:**63–70, 1986. Nervus terminalis in the goldfish brain.

165a. von Bartheld, C. S., D. L. Meyer, E. Fiebig, and Ebbesson, S. O. E. *Cell Tissue Res.* **238:**475–487, 1984. Central olfactory connections in the fish.

166. Waldvogel, J. A., *Curr. Ornithol.* **6:**269–321, 1989. Olfactory orientation and homing by birds.

167. Wellis, D. P., Scott, J. W., and Harris, T. A., *J. Neurophysiol.* **61:**1161–1177, 1989. Odorant discrimination by single olfactory bulb neurons (rat).

168. Wenzel, B. M., in *Handbook of Sensory Physiology* (L. M. Beidler, ed.), Vol. 4, Part 1, pp. 432–448. Springer, Berlin, 1971. Olfaction in birds; in *Physiology and behavior of the pigeon* (M. Abs, ed.), pp. 149–167. Academic Press, London, 1983. Chemical senses in the pigeon; in *Chemical Signals in Vertebrates* (D. Duval, D. Müller-Schwarze, and R. M. Silverstein, eds.), pp. 357–368. Plenum, New York, 1985. Olfaction in procellariform birds; Wenzel, B. M., and Meisami, E., *Ann. N.Y. Acad. Sci.* **510:**200–202, 1987.

Mitral cell number in fulmar and pigeon.

169. Wigglesworth, V. B., *The Principles of Insect Physiology*, 7th ed., pp. 279–291. Chapman & Hall, London, 1972.

170. Wolstenholme, G. E. W., and Knight, J. (eds.), *Taste and Smell in Vertebrates*. Churchill, London, 1970.

171. Wysocki, C. J., *Neurosci. Biobehav. Rev.* **3**:301–341, 1979. Vomeronasal system in mammalian reproduction.

172. Wysocki, C. J., and Meredith, M., in *Neurobiology of Taste and Smell* (T. E. Finger and W. L. Silver, eds), Chapter 6.

Wiley, New York, 1987. Physiology and neurobiology of the vomeronasal system.

173. Yamamoto, M., in *Chemoreception in Fishes* (T. J. Hara, ed.), pp. 39–59. Am. Elsevier, New York, 1982. Comparative morphology of fish olfactory organs.

174. Yamamoto, T., *Prog. Neurobiol.* **23**:273–315, 1985. Taste responses of cortical neurons.

175. Zacharuk, R. Y., in *Comprehensive Insect Physiology, Biochemistry and Pharmacology* (G. A. Kerkut and L. I. Gilbert, eds.), Vol. 6, pp. 1–63. Pergamon, Oxford, 1985. Insect antennae and sensilla.

Chapter 8

Circadian Rhythms: The Physiology of Biological Timing

J. Woodland Hastings, Benjamin Rusak, and Ziad Boulos

Biological Rhythmicity

J. W. Hastings

Circadian Rhythms: Endogenous versus Exogenous Theories

Circadian rhythmicity has been documented as a prominent and pervasive feature of animal and plant physiology (16, 74, 147, 387, 563). Such rhythms have been reported in all major eukaryotic taxa, and at levels of biological complexity ranging from unicells to mammals. Although circadian rhythms have not generally been found to occur in prokaryotes (230, 563), there have been several reports of rhythmicity having circadian properties in nitrogen fixing cyanobacteria (210, 382, 564). The existence of such rhythms in bacteria thus shatters the long standing generalization that circadian rhythms do not occur in prokaryotes! The evolutionary significance of this could also be substantial, especially with regard to the hypothesis of the eukaryotic origin of circadian rhythmicity (see the section on evolution and adaptive significance).

Circadian rhythmicity is based on an endogenous biological mechanism with clocklike properties. Under light–dark (LD) cycles, circadian systems exhibit daily rhythms in many different functions, each with a given stable phase relation (referred to as ψ) to the daily LD cycle. Thus, different functions may exhibit maxima at different times of day. The autonomy of the mechanism is deduced from the fact that if organisms are maintained in the laboratory under conditions of constant light or dark and constant temperature, daily cycles of activity and rest continue essentially unabated and for indefinitely long periods (Fig. 1). Moreover, while the periods are generally close to 24 h, they are not exactly so, thus the term circadian (*circa*, about; *dian*, 1 day) (222). Figure 2, in which the sequential days of records are plotted beneath one another, highlights this point. The free-running period may be either longer or shorter than 24 h in different species or individuals, and even in a given organism it may differ somewhat depending on environmental conditions.

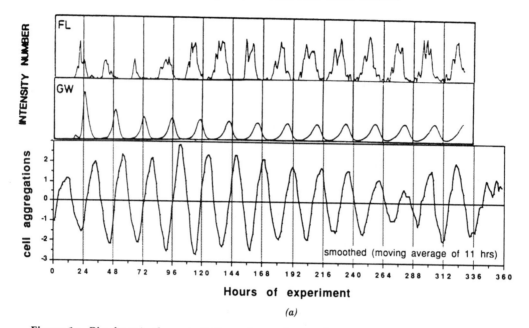

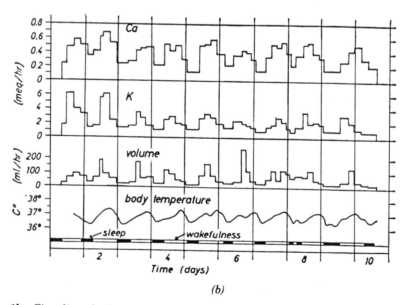

(a)

Figure 1a. Rhythms in the unicell *Gonyaulax* under conditions of constant dim light and temperature. *Upper panels:* bioluminescence flashing (FL) and glow rhythms (GW), peaking in the middle and towards the end of the night phase, respectively (68). *Lower panel:* aggregation rhythm, maximal in the day phase (489). Other rhythms (cell division, photosynthesis) peak at other times.

(b)

Figure 1b. Circadian rhythms of urine excretion: amounts of calcium and potassium; volume; body temperature; and sleep–wake cycle in a human subject kept without timepiece in complete isolation from the outside world for 24 days (12).

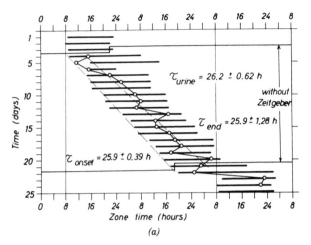

Figure 2a. Circadian rhythm of activity and urine excretion in a human subject kept for 3 days under normal living conditions, then for 18 days in isolation, and finally again under normal conditions. *Black bars,* times of being awake; *circles,* maxima of urine excretion; τ, mean values of period for onset and end of activity and for urine maxima (12).

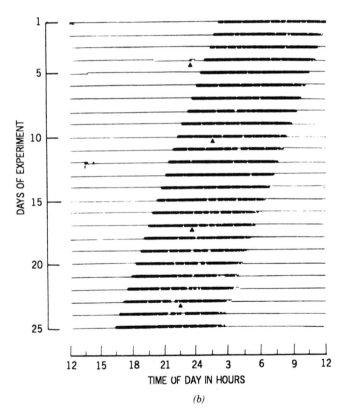

Figure 2b. Activity graph of one flying squirrel in a recording wheel cage in continuous darkness at 20°C (127).

However, the endogenous hypothesis is not accepted by all workers; an exogenous theory holds that there exist unspecified "subtle cyclic geophysical factors," which are available to organisms as time cues, even under so called constant laboratory conditions (69). However, many counterarguments have been advanced (153). Also, experiments have shown that circadian rhythms persist nor-

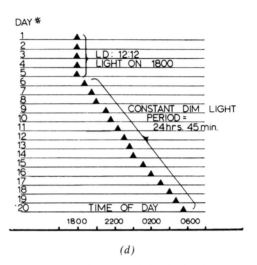

(d)

Figure 2d. Peak times of the bioluminescent glow in a culture maintained first in LD and then in LL at 24°C (237).

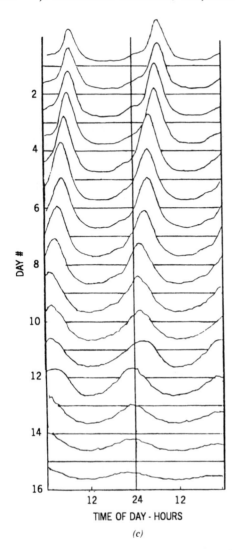

(c)

Figure 2c. Bioluminescence glow recorded for 16 days from *Gonyaulax* cells kept in light–light (LL) at 19°C. Rhythm damping may be due to desynchronization between individual cells (240).

mally in organisms maintained at the South Pole (226), as well as in earth-orbiting satellites (556), suggesting that exogenous timing factors are not necessary, a view to which most workers now subscribe (230).

Infradian and Ultradian Rhythms

There exist a multiplicity of different biological rhythms, with periods ranging

from milliseconds to years (336, 387), and it would seem difficult to accommodate all of these within the framework of an exogenous hypothesis. Over four hundred cellular oscillations have been listed by Rapp (462a). Although lower frequency (longer period) rhythms might depend in some way on higher frequency oscillations, no plausible biological mechanism for this has been suggested.

Rhythms with periods other than circadian may be grouped according to whether the period is shorter (ultradian) or longer (infradian) than 1 day (Table 1). This classification in itself implies neither a mechanistic nor functional distinction between the two, nor a commonality of mechanisms within or between any of the groups. For many ultradian and infradian biological rhythms the period and presumed functional significance are apparently not related to the frequency of any known environmental phenomenon. These range from high-frequency heart beat, brain waves, and oscillatory nerve circuits (83, 272), to the 4–5-day long estrous cycle of rodents and the several day long but circadian gated cell division cycles (146, 336). So far as is known, most

TABLE 1. Rhythms and Cycles with Periods Shorter (ultra) and Longer (infra) Than 24 Hours

Ultradian		Infradian	
Sperm flagellar beating	0.01 s	Estrous	4–5 d
Brain waves	0.1 s	Semilunar	14.8 d
Cardiac	1.0 s		
Respiratory (human)	5.0 s	Menstrual	28 d
Wing beat (*Drosophila* courtship)	1.0 min	Lunar	29.5 d
Glycolytic oscillations	0.5–5 min	Annual	365 d
Cell cycles	20 min–h		
Tidal	12.6 h		

of these cycles do not themselves exhibit the key "clocklike" features of circadian rhythms. However, the minute-long courtship-related wing beating cycle in fruit flies does appear to be related; it exhibits temperature compensation and regulation by the *per* gene (321).

There may also be a relationship between circadian rhythms and some of the longer period cycles. For example, the timing of the 4–5-day estrous cycle in rodents, studied in hamsters and rats, is clearly timed (or gated) by the circadian clock (180). Cell division may be similarly gated in a number of species (146, 148, 258, 334): Cells sufficiently mature to divide, do so at a given circadian time or time of day; those not yet ready are obliged to wait for a subsequent window about $n \times 24$ h later.

Circannual, Circalunar, and Circatidal Rhythms

A more meaningful classification, as elaborated by Sweeney (563), relates to the distinction between circa rhythms, namely those that match environmental periodicities and those that do not. Many infradian and ultradian rhythms have no environmental correlate, as mentioned above. In the cases of circannual, circalunar, and circatidal rhythms, however, the period of the particular persistent endogenous rhythmicity under constant

conditions corresponds approximately to the given environmental periodicity (Table 1), which also relates to the evolutionary origin and functional importance of each.

An exception to this nomenclature is the use of the term circaseptan to designate rhythms with a period of ~1 week (103). In this case it is not proposed or believed that there is an environmental correlate related to the biological rhythmicity. Indeed, the statistical significance of such a frequency in biological cycles has been questioned (157).

The status of knowledge concerning circannual rhythms has been summarized recently (219). Annual rhythmicity under constant laboratory conditions is documented in many species, as illustrated in Fig. 3a with the ground squirrel for 47 months. Although normally synchronized, these free-running periods are considerably less than 12 months, and not greatly affected by temperature (47, 396).

Entrainment has not been extensively studied, but there is clear evidence in both birds and mammals (242), that progressive (and cyclic) changes in photoperiod fraction (equal to those that occur at 40° latitude) can entrain. Indeed the range of entrainment with cycles shorter than 12 months is remarkable (202, 219); periods as short as 1.5 months appear to entrain. However, the free-running

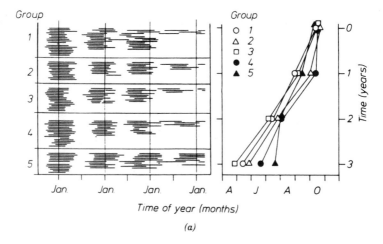

(a)

Figure 3a. Circannual rhythms of hibernation in five groups of golden-mantled ground squirrels (*Spermophilus lateralis*) kept for 47 months in 3°C and under different constant photoperiods. *Group 1* physiologically dark, animals surgically blinded. *Group 2* LL (500 lux). *Group 3* LL (500 lux), animals surgically blinded. *Group 4* LL (20 lux). *Group 5* LD 12:12 (200:0 lux). *Left diagram: Black bars* arranged in horizintal rows represent hibernation periods of individual animals. *Right diagram: Symbols connected by lines* indicate mean dates at which the animals of the various groups entered hibernation in successive years of the experiment (429).

periods following release into constant conditions after entrainment were not reported. Whether or not cycles in ambient temperature, an important feature of the yearly environmental cycle, can entrain, has not been resolved, but some experiments suggest that they can.

Whether or not the circadian system interacts with the circannual one is not known, but experiments intended to test the possibility suggest that it is not. Ground squirrels with suprachiasmatic nucleus (SCN) lesions exhibited no circadian rhythmicity, but circannual rhythms persisted (120). However, since photoperiodic time measurement is mediated in many organisms by circadian clocks and since photoperiodic cycles can evidently entrain circannual rhythms, a connection still seems possible. If so, the relation is not a simple one. For example,

circannual rhythms continue after pinealectomy in many mammals and birds, but in ferrets and sheep pinealectomy

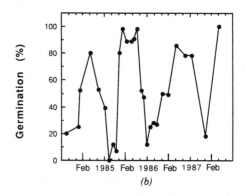

(b)

Figure 3b. Germination of encysted cells of *Gonyaulax tamarensis* collected in 1984 and maintained in the dark at 4°C thereafter. Samples were removed at the times indicated and tested for germination (7).

Dinoflagellate

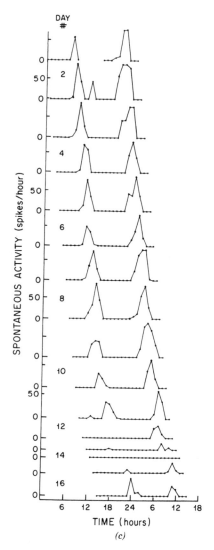

Figure 3c. Tide-associated activity rhythms (ordinate) of a fiddler crab (*Uca pugnax*) maintained in constant illumination (22 lux) and 22°C. The peaks for the first few days centered on the times of low tide at the site of collection; this crab exhibited the persistent rhythmicity for 58 days (from 426a).

prevents circannual entrainment without abolishing the yearly rhythm.

Circannual rhythmicity need not involve higher levels of organization, such as neural and/or endocrine systems. For example, a circannual rhythm of germi-

nation of dormant cysts of the unicellular marine dinoflagellate *Gonyaulax tamarensis* (7) (Fig. 3b).

Lunar and Tidal Rhythms
(129, 405, 408)
Environmental cycles due to the moon include moonlight and tides, both of which may have profound behavioral and physiological effects in the appropriate environments. Moonlight is reflected sunlight, spectrally shifted slightly to the red. The *lunar day* corresponds to the 24.8-h period between two consecutive moon culminations; the lunar month is the 29.53-day interval between new moons.

Circadian-controlled activity may be modulated (exogenously) by moonlight (163), such that a lunar monthly rhythm emerges under natural light conditions. There are also indications of an endogenously controlled lunar monthly rhythm in animals maintained in the laboratory in the dark (408).

Circatidal rhythms associated with semidiurnal tides having a period of about 12.4 h occur in many marine organisms, and persist in the laboratory under constant conditions (Fig. 3c; 249, 426a). The principal zeitgeber (time giver) for entrainment of such rhythms appears to be mechanical stimulation, as generated by tidal and wave action (152); temperature and light cycles may also contribute in lesser but still-poorly understood ways. Like the circadian mechanism, circalunar rhythms appear to be temperature compensated for period and manipulatable by certain drugs.

Circalunar rhythms might be based on a circadian-type clock with a bimodal expression, or to a separate circa-12.4-h clock. Palmer (426) has proposed a third explanation: The two peaks correspond to two independent "circalunidian" clocks. This proposal is based on the observations that the two peaks may concurrently exhibit different free-running periods and that one may "split" or vanish independently of the other.

Key Characteristics of Circadian Rhythms

Four key features are associated with circadian rhythmicity: (1) Such rhythms persist under constant conditions with a circa-24-h period, the so-called natural- or free-running period (16). (2) The length of the free-running period is temperature compensated; it is only slightly different at different constant temperatures within a physiological range. (3) Rhythmicity is not expressed in some organisms under certain environmental conditions, including bright light, low temperature, and anoxia; rhythmicity may thus be considered to be conditional upon certain environmental factors (413). (4) Circadian rhythms are entrained by environmental cycles having periods of about 24 h. Light, but also other factors, are effective as zeitgebers (442).

Persistence: Free-Running Period

In the natural environment, daily rhythms have precise 24-h periods due to entrainment by the LD cycle. However, if animals are transferred in the laboratory to constant conditions of light (or continuous dark) and constant temperature, their daily cycles (e.g., of activity and rest) continue, essentially unabated for weeks, months, or years (Fig. 4). Such free-running circadian rhythms may be considered analogous to a free-swinging pendulum or an electronic oscillator, whose periods are related to the properties of the particular systems, and not to any exogenous environmental timer. However, the free-running period of a circadian rhythm may be longer or shorter than 24 h; Its value is also dependent on environmental conditions.

A rhythm cannot be called circadian unless it meets the criterion of persistence. A number of biological processes do vary in *direct* response to the environmental cycles; the judgment as to whether a rhythm is circadian should be reserved until the appropriate studies have been made.

Temperature Compensation

A second and very distinctive and diagnostic feature of a circadian rhythm is the effect of temperature on period. Biological processes are generally accelerated by a factor of 2 to 3 with a 10°C temperature increase (within a permissive range), but the circadian frequency is only slightly changed at different temperatures. Some circadian oscillations are slightly speeded up by increases in temperature, while others are actually slower at higher temperatures (Fig. 5). This suggests that the biochemical mechanism involves temperature compensation, with possible slight inaccuracies, usually resulting in either under- or overcompensation (238). This was modeled by two reactions $A \rightarrow B$ and $C \rightarrow D$; where B controls the period; the reaction $A \rightarrow B$ is inhibited by D, produced in the reaction $C \rightarrow D$. Both reactions are inherently temperature dependent in a normal way, but the rates of $A \rightarrow B$ becomes effectively constant with temperature because of the variation in amounts of inhibition at different temperatures.

The endogenous biological clock hypothesis capitalizes on these facts by observing that selection pressure in the evolution of a functional biological clock would be expected to operate to match only *approximately* the period of an exogenous oscillation to that of the day— since the daily dawn and dusk resetting (entraining) signals would normally operate to correct for small inaccuracies.

In some cases, however, temperature compensation is apparently almost perfect. For example, the Q_{10} for the *Droso-*

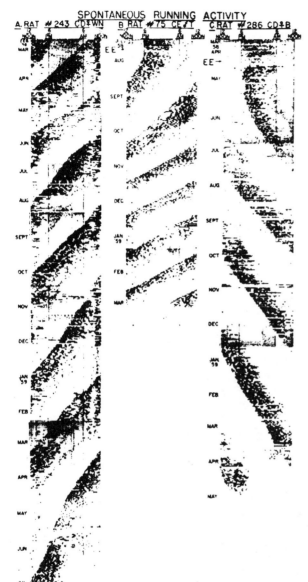

Figure 4a. Long-term activity rhythms of three rats. (A) A wild rat; records shown for 18 months after blinding. (B) and (C) Domesticated Norway rats; records shown both before and after blinding, after which rhythm began to free-run (480).

phila eclosion rhythm was estimated to be 1.01 (Fig. 5; 437). Indeed, the origins of the exogenous theory of circadian control can be traced to a case of such near-exact temperature compensation. In the rhythm of color change in the crab Uca the period appeared to be *completely* tem-perature independent, suggesting an exogenous timer (70).

There are examples of rhythms with periods in the vicinity of 24 h that lack temperature compensation. The marine plasmodial rhizopod, *Thalassomyxa australis,* displays a 22–24-h rhythm in mo-

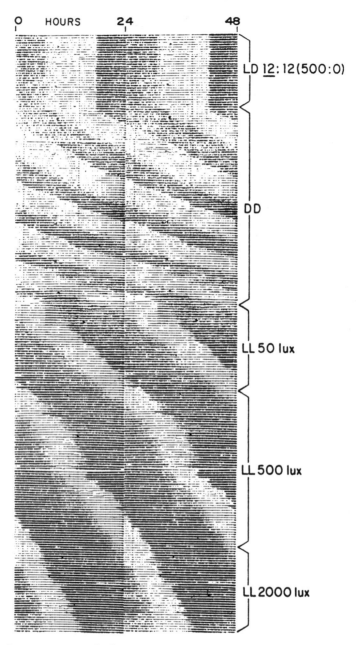

Figure 4b. Long-term record of activity in a blinded bird which, though still entrainable by the LD cycle (*top*), persists with a free-running period under both DD and LL. The period in LL is dependent on light intensity (Aschoff's rule) (372).

bile and resting phases whose Q_{10} is about 2.7 (524). But this rhythm is also not synchronized by environmental factors (light, temperature, or turbulence) and hence does not qualify on that basis as circadian; the authors suggest that this rhythm may represent a primordial "Ur-uhr" type of rhythmicity. Nontemperature compensated and nonentrainable rhythms have also been described in a number of fungi.

Although temperature compensation

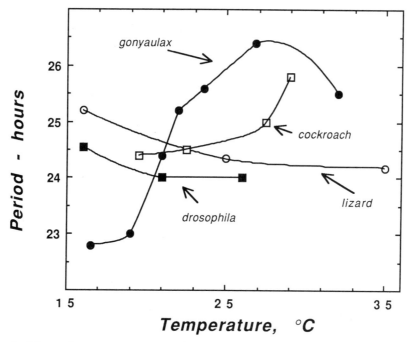

Figure 5. Effect of temperature upon the *period* of circadian rhythms in four species, illustrating that the Q_{10} for rate may be either very close to (*Drosophila*) (437), greater than (lizard and cockroach) (74), or less than 1.0 (*Gonyaulax*) (238).

is characteristic of, it is not unique to circadian rhythms. For example, the period of the ultradian rhythms of respiration and cellular protein content in *Acanthamoeba castellanii* ($\tau = 1$–1.5 h) remains approximately the same at temperatures of 20–30°C (335). Similarly, the ultradian rhythm of tyrosine transaminase activity in *Tetrahymena thermophila* ($\tau = 4$–5 h) has a Q_{10} close to 1 at temperatures between 10–30°C (378). In the hedgehog, *Erinaceus europeanus*, the NREM–REM (rapid eye movement) sleep cycle has an average duration of 17 min at a body temperature of 34°C, and each cycle is followed by a brief arousal. During hibernation at a body temperature of 13°C the electrocorticogram disappears, but the animals continue to show periodic arousals at intervals of about 19 min (583). The arousal cycle thus constitutes a temperature-compensated endogenous ultradian rhythm with a Q_{10} close to 1.

Ultradian rhythms may share some key mechanistic features with circadian rhythms. A case in point is the courtship rhythm in *Drosophila*, involving wing beating with a period of about 50 s (321). This period, like the circadian rhythm of eclosion, is temperature compensated, and mutants with long or short circadian periods (see the section on biochemistry and genetics), exhibit correspondingly altered periods in the ultradian rhythm. Thus these two rhythms, having very different periods, share both physiological and genetic controls.

Some evidence suggests that the temperature compensation mechanism is not an inherent feature of the oscillating mechanism. Certain *Neurospora* mutants lack temperature compensation in the otherwise phenotypically normal circadian rhythm of conidiation during growth (351). Also, temperature compensation is better in organisms from higher latitudes than in those from tropical areas where temperatures fluctuate less (30, 74). Finally, the importance of temperature compensation would be expected to be greater in poikilotherms and heterotherms than in homeotherms. Experiments suggest that induced hypothermia causes a greater reduction in the rate of the circadian clock in the homeothermic rat than in the heterothermic hamster (195, 196). Opposite results were obtained in two neotropical bat species: temperature dependence of circadian period length was greater in *Molossus ater*, a heterotherm, than in the homeotherm *Phyllostomus discolor* (164). It should be noted, however, that period length was reported as a function of ambient, not body temperature, and the latter may have differed in the two species.

Conditional Arhythmicity

The fact that the circadian clock is temperature compensated has somewhat overshadowed another very interesting and mechanistically important effect of temperature: Expression of rhythmicity may be conditional upon temperature. This has been found in several different organisms including *Drosophila* and cockroaches, and analyzed in detail in *Gonyaulax* (413). Below a certain critical temperature—this may still be within the physiological range—the rhythmicity is not expressed, and it is thus conditional upon temperature, in that it occurs at certain "permissive" temperature levels but not at others. Rhythmicity may also be conditional upon other environmental factors, such as light intensity and oxygen concentration.

In classical temperature conditional mutants, the phenotype is typically due to the thermal denaturation of a protein and the phenotype is expressed at higher but not lower temperatures. In temperature conditional circadian expression,

however, it is the lower temperature range that is viewed as nonpermissive. Moreover, only the rhythmic expression of a function is lost, not the function itself. But, as with conditional mutations, the critical temperature at which the transition occurs is very sharp; in *Gonyaulax* normal rhythmicity may occur at 14°C while the system is arhythmic at 12.5°C (413). The identity of the critically temperature-dependent molecular species responsible for rhythmic expression is not known. It could be a protein that inhibits or blocks the rhythm and is denatured above 12.5°C. Alternatively, a protein could be required for rhythmicity and this protein could be *cold sensitive* and inactivated below 14°C, or another molecular species might be responsible for rhythmicity, for example, a membrane lipid and its temperature-dependent phase transition.

The actual mechanism of the arhythmicity is not known. The individual oscillatory components (be they whole cell or subcellular oscillators) could become asynchronous and thus the organism would exhibit apparent arhythmicity. Alternatively, arhythmicity might represent the dampening or absence of the oscillation itself. These two possibilities have also been suggested as explanations for the ability of constant bright light to dampen the expression of rhythmicity. While neither can be rigorously excluded without knowing the identity of the oscillator, it has been shown in *Gonyaulax* that individual cells do exhibit loss of rhythmicity in constant bright light (559), although subsequent studies with populations showed that rhythmicity does indeed continue for some time under bright light (562).

A striking feature of low-temperature arhythmicity is that the rhythm resumes following a return to the higher permissive temperature, with no need for any other signal. The phase of the rhythm after the return is determined by the time of transfer to the higher temperature (Fig. 6). In simple terms, the low-temperature treatment may be viewed as "stopping" or "holding" the oscillation at a specific phase under nonpermissive conditions, and then simply allowing it to start up again.

Is the clock really stopped, or does it continue all the while and experience a reset by the temperature shift? The distinction between holding and shifting an oscillating system may not be simple. Consider that the fundamental mechanism can be described as a limit cycle (see the section on biochemistry and genetics). It is not readily possible to distinguish between (a) holding some point on the cycle from a (b) continued oscillation in a very small region of the normal phase domain, embracing isochrons of only a few minutes on the trajectory under permissive conditions (430). In the latter case, the oscillation would continue, but the amplitude of overt rhythmicity would be small, possibly undetectable, and following a shift to permissive conditions the oscillation would be reset at the phase corresponding to the isochron intersecting the region of the small oscillation.

This reset phase corresponds to circadian time (CT) 12 for all known cases of conditional arhythmicity, whether due to low temperature, bright light (238, 238a, 413), or anaerobiosis (437, 441). Indeed, under constant bright (white) light, rhythmic expression is reported to be lost in many organisms, and for some systems even moderate light intensities are effective. Nocturnal animals are generally more sensitive in this respect. Upon return to permissive conditions, for example, darkness or dim light, rhythmicity resumes, with the new phase corresponding to CT12.

Light differs from temperature in causing arhythmicity because there are intermediate effects at intermediate light intensities; as light intensity is increased the amplitude of the expressed rhythm is

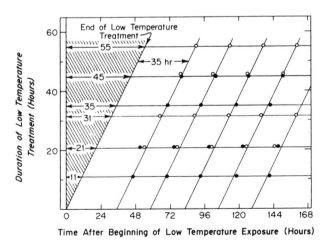

Figure 6. The new phase of the *Gonyaulax* glow rhythm is determined by the time at which cells are returned to 21°C following extended 11°C exposures. Cultures were transferred from LD 12:12 to constant dim light (170 ft-c) at the end of a light period. Either 8 h later (○) or 18 h later (●), the temperature was lowered from 21 to 11°C (with no change in the intensity of illumination) and maintained for various durations as indicated. After returning to 21°C, again with no change in the illumination, the glow rhythm was monitored. The times of the glow maxima following a given treatment are plotted along a horizontal line (413).

commonly decreased until at higher intensities the rhythmicity is abolished. This was studied in detail in the mosquito (430–432) and modeled by a series of smaller and smaller limit cycles at higher and higher intensities such that, as described above for temperature, the arhythmic condition involves a continued oscillation but in a very small region of the normal phase domain, again embracing isochrons in the region of CT12 corresponding to only a small fraction of the 360° day. Thus a shift back to constant darkness resets to that isochron irrespective of the actual phase of the oscillation at the time of the shift back. Observations in *Gonyaulax* (562) are consistent with this interpretation.

Entrainment

With most animals maintained under constant laboratory conditions, the free-running circadian period of whatever function is being measured is only approximately 24 h; under natural conditions, however, it is exactly 24 h, matching the 24-h LD cycle (Figs. 2 and 4). This frequency control is referred to as entrainment and is analogous to resetting an inaccurate watch each day. If the circadian period is longer than 24 h, then the phase must be advanced each day, and conversely delayed if it is shorter. The mechanism whereby this is achieved will be considered in the following section.

Effects of Light
Z. Boulos

Of the various geophysical cycles engendered by the daily rotation of the earth around its axis, that of light and darkness is the least subject to day-to-day variability and is therefore the most reliable index of local time (442). Thus it is not surprising that the LD cycle is the most universally effective and the primary entraining

agent, or zeitgeber, for almost all circadian rhythms. This is true not only for functions such as photosynthesis, where the selective advantage conferred by circadian timing derives directly from the daily alternation of light and darkness, but also of functions where circadian rhythmicity may have evolved as an adaptation to other daily cycles in the environment, for example, temperature or humidity.

Two conceptually different effects of light on circadian rhythms can be distinguished: Discrete or phasic effects, such as may be exerted by brief pulses or single steps of light or darkness, and continuous or tonic effects, best illustrated by the systematic influence of LL intensity on τ (114). By analogy, tapping a pendulum in motion is a discrete effect, whereas altering the air resistance exerts a continuous effect. Light and darkness (as well as other environmental factors) can also directly affect the behavior and physiology of many animals, either suppressing (positive masking) or stimulating activity (negative masking), independently of any effects they may have on the endogenous pacemaker responsible for the generation of circadian rhythms in that activity. Most overt rhythms observed in nature probably reflect the combined influences of such endogenous and exogenous factors.

Phase Shifting by Single Light Pulses: Phase Response Curves

Depending on the phase of a circadian rhythm at which it is presented, a single light pulse, which may be as brief as 0.5 ms (282), can phase advance, phase delay, or have no effect on that rhythm. The relationship between the phase of light exposure and the size and direction of the ensuing phase shift is represented graphically in the form of a phase response curve (PRC) (238a). In general, light pulses presented in otherwise constant darkness around the beginning of subjective night (i.e., near the onset of the daily activity phase for a nocturnal animal and the rest phase for a diurnal animal) cause phase delays, while pulses presented near the end of subjective night cause phase advances. Light pulses presented during much of subjective day have little or no effect (Figs. 7 and 8).

In a number of species, the final steady-state phase after exposure to a light pulse is achieved gradually, over one or more transient cycles (449). Both advancing and delaying transients have been observed, depending on the phase of the rhythm at the time of light presentation. In *Drosophila*, as well as in hamsters (Fig. 7), advance phase shifts take place over many more transient cycles (3–4 cycles in *Drosophila* and up to 10 cycles in hamsters) than do delay phase shifts, which are completed within one or two cycles. In *Drosophila*, however, the phase and shape of a PRC for light pulses obtained shortly after presentation of a separate advance-inducing light pulse indicate that the phase advance of the pacemaker is completed within one circadian cycle, if not less (441). The occurrence of transients in this organism may therefore reflect the motion of a slave oscillator as it gradually regains its steady-state phase relative to the pacemaker, which has been reset instantaneously (442, 449). In the hamster, similar experiments showed that the pacemaker had already advanced by the first day after presentation of the initial light pulse, although the PRC had still not regained its original shape by that time (442). The hamster data are interpreted in the context of a different model, one involving two mutually coupled oscillators.

Two different types of PRC can be distinguished (442, 613): in weak or Type 1 PRCs, the maximum phase shifts are typically of the order of a few hours, and there is a more or less gradual transition

between the advance and delay regions of the curve (Fig. 8). Furthermore, when the phase of the shifted rhythm (new phase) is plotted as a function of the phase at light exposure (old phase), the average slope of the resulting phase transition curve is equal to 1 (hence Type 1). In some species, however, increasing the strength of the light pulse (by increasing its intensity or duration) can produce phase shifts of up to 12 h, and there is an apparent discontinuity in the PRC, with a sudden jump from maximum phase delays to maximum phase advances (Fig. 8). The direction of these large shifts is often ambiguous, since a phase shift of about 12 h may be construed either as a phase advance or a phase delay. Such PRCs are known as strong or Type 0 PRCs, as the average slope of the corresponding phase transition curve is equal to 0.

Phase-dependent phase shifts also occur in response to dark pulses presented in LL, and to step transitions from LL to DD and vice versa. Dark pulse PRCs tend to be mirror images of light pulse PRCs, although there is often a sizable displacement between the advance region of one and the delay region of the other; the minimum effective duration may also be longer for dark than for light pulses (56, 68, 305, 404, 551, 619). The phase shifts caused by step transitions are generally more variable (both within and between species) than those caused by single pulses, and the phase dependence of the response is not always clear, but there is some tendency for a mirror-image relation between the PRCs for LL–DD and DD–LL transitions (4, 13).

The PRC is an intrinsic property of the circadian pacemaker, and the steady-state response to light pulses or other stimuli is therefore a direct measure of the status of the pacemaker itself. As will be seen in the following sections, the phase response relationship depicted by the PRC can account for much of the behavior of

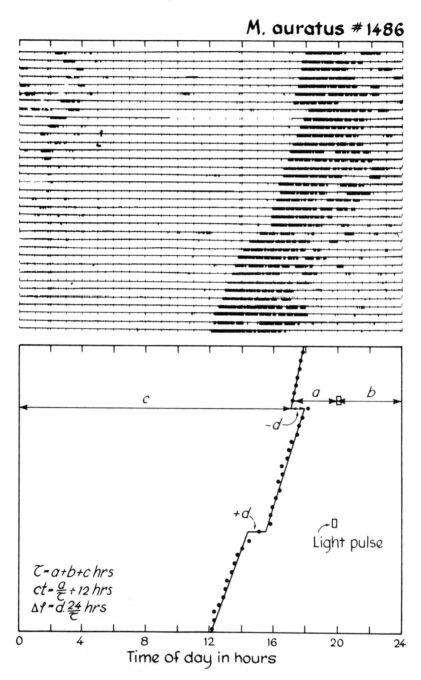

Figure 7. Phase shifts of the circadian activity rhythm of a hamster caused by 15-min light pulses. *Upper panel:* raw data. *Lower panel:* computation of free-running period (τ), phase in circadian time (ct) at which the light pulse was given, and phase shift (ΔΦ) produced by that pulse. The first pulse caused an immediate phase delay, while the second caused a gradual phase advance that took place over 3–4 transient cycles (118).

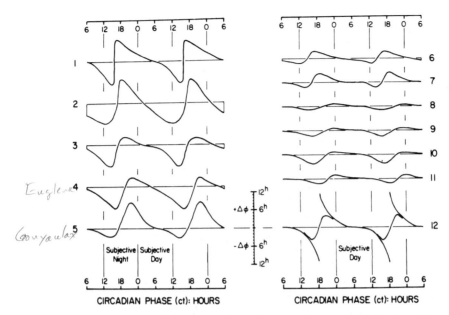

Figure 8. Phase response curves for light pulses in various unicellular organisms, plants, insects, and vertebrates: (1) *Sarcophaga*, (2) *Coleus*, (3) *Leucophaea*, (4) *Euglena*, (5) *Gonyaulax*, (6) *Anopheles*, (7) *Mescoricetus*, (8) *Peromyscus leucopus*, (9) *Peromyscus maniculatus*, (10) *Mus*, (11) *Taphozus*; and (12) *Drosophila*. Note that two separate PRCs are shown for *Drosophila*, a weak or Type 1 PRC, obtained with 1-ms pulses, and a strong or Type 0 PRC, obtained with 15-min pulses. All other PRCs are Type 1. See Pittendrigh (442) for pulse durations and intensities (442).

circadian rhythms under both cyclic and constant lighting conditions.

Mechanisms of Entrainment

Circadian rhythms recorded in constant conditions free run at periods that only approximate 24 h. When entrained to daily LD cycles, however, the rhythms assume an exact 24-h period and a stable phase relative to the entraining cycle. Two mechanisms have been proposed to account for entrainment. The PRC figures prominently in both.

Discrete or Nonparametric Entrainment

In what is perhaps the most widely accepted view, entrainment by LD cycles is accounted for by discrete, daily phase shifts, caused by periodic light exposure, which correct for the difference between the free-running period, τ, recorded in

DD, and the period, T, of the LD cycle. The light portion of the LD cycle need not be very long, and daily exposure to light pulses lasting only a few minutes or seconds are sufficient to entrain the rhythms of many animals. Indeed, this ability of brief daily light pulses to entrain circadian rhythms is central to the nonparametric entrainment model, which predicts that steady-state entrainment occurs when the periodic light pulses coincide with that phase of the rhythm at which a single light pulse (of the same intensity and duration) causes a phase shift equal to $(\tau - T)$. Results obtained in *Drosophila* (440) and in rodents (446) using LD cycles with 15- or 60-min light segments and with different period lengths show a fairly close agreement between the observed phase angle of entrainment, ψ, and the phase angle predicted on the basis of τ and the PRC.

In a Type I PRC, a phase shift of a given size and direction can result from light pulses delivered at either of two different phases, on the positively and negatively sloping portions of the PRC, respectively. Only at the latter phase, however, can stable entrainment be achieved, and only if the slope of the PRC at that phase is less than 2.0 (440). This is because at that phase, any change in the phase angle of entrainment will, in the subsequent cycle, lead to a shift in the opposite direction, and the original phase angle is thereby restored. In contrast, if the light pulse falls on the positively sloped part of the PRC, a change in a given direction will be followed by a shift in the same direction, and this will continue until a new phase angle is reached, on the negatively sloping part of the PRC. Thus, the PRC provides a mechanism for achieving and maintaining a stable phase relative to the entraining cycle.

Light cycles with periods very different from 24 h fail to entrain circadian rhythms (21). According to the nonparametric model, the limits of entrainment are determined by the maximum phase shifts measured in the PRC. In *Drosophila*, the maximum usable phase advances and phase delays caused by 15-min light pulses (corresponding to points on the PRC where the slope is between 0 and -2.0) are both about 6 h, and the eclosion rhythm can in fact be entrained by LD cycles (15-min light pulse/cycle) with periods between 18 and 30 h (440). In nocturnal rodents (*Mesocricetus* and *Peromyscus*), the maximum phase shifts are smaller than in *Drosophilia* and the range of entrainment is correspondingly reduced. However, individual animals sometimes fail to entrain even when the difference between τ and T is within the range of phase shifts seen in the PRC (446). Unlike the case in *Drosophila*, where the tests of entrainment involved entire populations, in rodents each animal is

tested individually and its entrainment performance compared with the population PRC. Thus, the occasional failures of entrainment could be due to inter- or intraindividual variability in τ and PRC shape (446).

When entrained by natural LD cycles, animals are likely to be exposed to light during both advance and delay portions of their PRC, at dawn and dusk, respectively. This can therefore be simulated more realistically with skeleton photoperiods consisting of two brief light pulses per cycle, rather than only one. In this case, the nonparametric model predicts—and data from *Drosophila* (448) and nocturnal rodents (446) confirm—that the net phase shift caused by the two pulses (determined once again from the PRC) is equal to $(\tau - T)$. In general, entrainment is more stable under two-pulse than under one-pulse cycles.

There is, however, an important difference between the effects of skeleton and complete photoperiods. When the two light pulses that constitute the skeleton are approximately 12 h apart, either one can serve as the dawn signal, depending on the initial phase relations between the light pulses and the rhythm. However, if the interval between the dawn and dusk signals is increased beyond a certain limit (simulating an increase in daylength and a corresponding decrease in nighttime duration), the entrained rhythm shows a sudden phase jump such that the former dawn signal is now treated as dusk, and vice versa. There is, in effect, a minimum tolerable night (442) of about 10 h in *Drosophila* (448), and about 6 h in *Mesocricetus* and *Peromyscus* (446). These phase jumps are predicted by the nonparametric model (442), although the minimum tolerable night in the rodents is somewhat shorter than expected from the model. Since phase jumps do not occur under complete photoperiods, this suggests that entrainment normally involves a continuous action of light in addition to any discrete phase-shifting effects.

Continuous or Parametric Entrainment

The fact that circadian rhythms can entrain to sinusoidal light cycles (558) suggests that entrainment can also be achieved through a continuous action of light. According to parametric entrainment models (9, 558), sinusoidal light cycles exert a continuous effect on the period of the circadian pacemaker, alternately slowing it down and speeding it up as the brighter portion of the cycle falls on the delay and advance regions of the PRC, respectively. Entrainment is thus achieved when the change in τ caused by the light cycle is equal to (τ − T). This process is best illustrated in cases where the sinusoidal zeitgeber is too weak to entrain a circadian rhythm but strong enough to affect its period, an effect known as relative coordination or oscillatory free-run (Fig. 9).

It is not yet clear whether discrete and continuous entrainment represent two separate mechanisms, or whether they are different manifestations of the same underlying process. The two certainly are not mutually exclusive, but their relative contributions under natural conditions are likely to depend on the activity pattern of the organism. Both may play a role in diurnally active animals, which are normally exposed to light over a substantial portion of the day. In contrast, many nocturnal animals spend most of the daylight hours in light-excluding shelters, emerging at dusk and regaining their shelters at dawn. Such animals are only exposed to light for brief durations, and entrainment would therefore necessarily be achieved by discrete phase shifts as observed under one- or two-pulse light cycles.

A remarkable example of this process is provided in a recent study of entrainment in the nocturnal flying squirrel, *Glaucomys volans* (130). The animals were kept in a simulated den environment consisting of a light-excluding nest box connected by a narrow exit tunnel to an exposed activity box containing a running wheel. When the activity box was illuminated by daily light cycles (LD 12:12 or

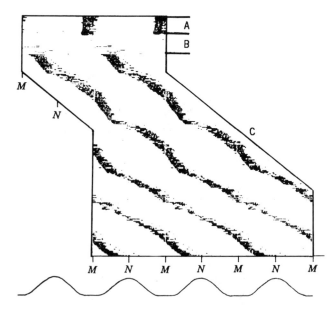

Figure 9. Activity record of a hamster exposed to (*A*) chopped sinusoidal light cycle (max = 186 lux, min = 83 lux); (*B*) constant light and sinusoidal temperature cycle (max = 25.5°C; min = 14.5°C; and (*C*) chopped sinusoidal light cycle (max = 186–147 lux, min = 88–68 lux). The activity rhythm is entrained in (*A*), but shows relative coordination in (*C*) (558).

14:10), the animals showed an entrained pattern of wheel-running activity, which took place exclusively during the hours of darkness. However, whereas in the absence of a nest box, activity onsets took place at the same time each day, in the simulated den environment they showed a characteristic zigzag pattern, occurring a little earlier each day for several days, then showing a single abrupt phase delay (Fig. 10).

Continuous monitoring of behavior revealed that the hours of light were spent in the nest box, where the squirrels alternated between short sleep episodes and brief bouts of grooming, and only on very rare occasions did they emerge from the tunnel. Arousal from the last sleep episode of the daylight hours took place around the time of lights-off, whereupon the animals proceeded to a porthole at the end of the exit tunnel. If the lights were off, they immediately exited the tunnel and began to run in the wheel. If, on the other hand, the lights were still on, they returned to the nest box for an additional brief sleep episode, emerging again to run in the wheel at a later time than in preceding days. Thus, the animals saw very little light, in some cases none at all, for several consecutive days, during which time their activity rhythm free-ran with a period slightly less than 24 h. Eventually, the light-sampling behavior that preceded running activity exposed the squirrels to light for a few minutes, causing an instantaneous phase delay of the activity rhythm. Similar light-sampling behavior has been observed in bats and other nocturnal animals, but only in the flying squirrel has the relation of this behavior to the entrainment process been studied in such detail.

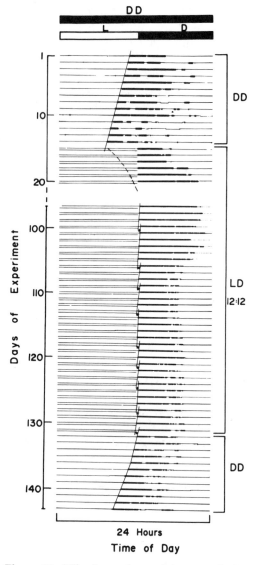

Figure 10. Wheel-running activity record of a flying squirrel kept in a simulated den cage. Schedule includes DD free-run, entrainment to LD 12:12 cycle, and final DD free-run. Sloping lines fitted to daily activity onsets during LD entrainment emphasize zigzag pattern of light sampling and phase resetting (130).

Free-Running Period in DD and LL

Species and Individual Differences

The periods of circadian rhythms free-running in DD show species-character-

istic means and ranges (15). Only rarely is τ exactly equal to 24 h, and the question arises as to whether this deviation from strict 24-h periodicity is merely an inaccuracy, tolerable as long as τ is close enough to 24 h to allow entrainment, or whether it has functional significance.

Under natural conditions, the vast majority of animals are regularly exposed to and entrained by external zeitgebers, and it is therefore probable that τ was not selected for directly, but only through its effect on ψ, the phase angle of entrainment. Clearly, the more τ deviates from 24 h, the larger the phase shift necessary for entrainment to daily LD cycles. From the shape of the PRC, it is also clear that a larger phase shift requires that the daily light signal fall on a steeper segment of the PRC, where any variability in τ can be corrected for by relatively small adjustments in ψ. In contrast, when τ is very close to 24 h, little or no daily phase shift is required, and the entraining light signal would fall on the more insensitive and flatter segment of the PRC, which often spans several hours of an animal's subjective day. In this case, changes in τ would require larger adjustments in ψ for entrainment to be maintained. Thus, variability in τ translates into greater instability in ψ the closer τ is to 24 h (442, 446).

Data relevant to this issue were obtained by Pittendrigh and Daan (446), who compared mean and standard deviations of τ in DD, in four nocturnal rodent species. As seen in Table 2, the variability of τ was smallest in species with a mean τ close to 24 h, and the more τ deviated from 24 h, the greater was its variability. A similar relationship was also demonstrated for individual animals within these species (446), as well as in diurnal rodents (314), birds and humans (28), and insects (190). Apparently then, selection pressure for the stabilization of ψ has favored either a strict homeostasis of τ in cases where τ is close to 24 h, or a systematic deviation of τ from 24 h, with a concomitant relaxation in the requirement for τ stability (442, 446).

A similar argument accounts for the greater stability of ψ under two-pulse rather than under one-pulse light cycles.

As an example, if $\tau = 24.5$ h, entrainment by a single daily light pulse requires that the pulse cause a 30-min phase advance. Under a two-pulse skeleton photoperiod, one of the pulses will fall on the delay region of the PRC causing, say, a 15-min phase delay. In that case, the second pulse must now cause a 45-min rather than a 30-min phase advance, and must therefore fall on a steeper portion of the PRC (446).

The discussion thus far has centered on the possible significance of τ being different from 24 h, with no consideration given to the sign of that difference. In general, τ tends to be less than 24 h in day-active animals and greater than 24 h in night-active animals (15, 119). Pittendrigh and Daan (446) suggest that this trend may be functionally significant in contributing to the conservation of ψ under seasonally changing photoperiods, based on the assumption that it is advantageous that daily activity onset maintain a nearly constant phase relative to dusk in nocturnal animals, and relative to dawn in diurnal animals. Computer simulations of the behavior of model pacemakers with different τ values and PRC shapes under a range of skeleton photoperiods indicate that the seasonal conservation of ψ is dependent on τ, PRC amplitude, and PRC asymmetry. Maximal conservation of ψ relative to the morning light pulse (optimal diurnal strategy) is achieved when τ is longer than 24 h and the slope of the advance segment of the PRC is greater than that of the delay, while maximal conservation of ψ relative to the evening light pulse (optimal nocturnal strategy) is achieved with τ shorter than 24 h and a steeper delay than advance segment of the PRC.

In nocturnal rodents, there is in fact a close relationship between τ and the relative sizes (and slopes) of the delay and advance regions of the PRC, such that the shorter is τ, the greater the difference between the area under the delay segment of the PRC, or D, and the area under the advance segment, or A (118). This interdependence of τ and PRC shape was found in both inter- and intraspecific comparisons, but its functional significance may differ in the two cases. At the species level, a short τ and a large difference between D and A contribute independently to the stabilization of ψ. Their effects are therefore additive and complementary. Within a species, however, differences in τ would lead to differences in ψ unless offset by changes in $(D - A)$. In this case, a larger difference between D and A compensates for a shorter τ, resulting in a closer synchronization of activity times among members of a species (446).

Effects of LL Intensity

Aschoff's rule states that as LL intensity increases, τ lengthens in nocturnal organisms and shortens in diurnal oganisms. This generalization has several exceptions, particularly among arthropods and diurnal mammals, but most animals do conform to Aschoff's rule (Fig. 11) (15).

The effects of LL intensity on τ may be related to the relative sizes of the advance and delay regions of the PRC (126). More specifically, if D is greater than A, LL would be expected to have a net delaying effect, that is, a lengthening of τ relative to DD, whereas if A is greater than D, τ should be shorter in LL than in DD. Given the rapidity of phase resetting by light, however, one would expect that as the pacemaker goes through its delay-sensitive phase, its motion, or angular velocity, will be slowed, while during the advance-sensitive phase, it will be accelerated. A transformation of the PRC, which takes into account these changes in angular velocity caused by LL at dif-

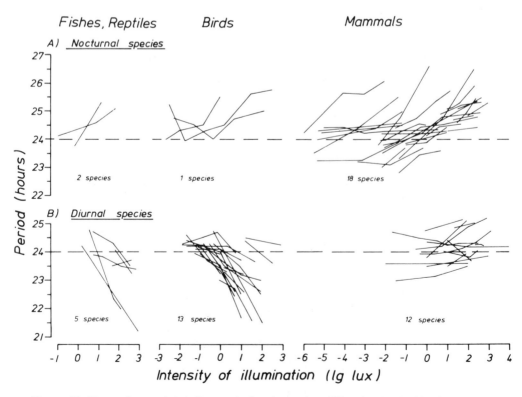

Figure 11. Dependence of circadian period on intensity of illumination in LL. For species and reference sources, see Aschoff (15) (15a).

ferent circadian phases, is known as the velocity response curve (VRC), which differs from the PRC in having an expanded delay and a compressed advance region (119, 558). The VRC could then account for cases where τ is longer in LL than in DD, even though the PRC shows a larger advance than delay region, as found, for example, in hamsters (119).

The VRC also provides one of several possible mechanisms for the loss of circadian rhythmicity commonly observed under bright LL. As LL intensity increases, the phase delay resulting from exposure to light in early subjective night for a given duration, for example, 15 min, will also increase. If the size of that phase delay reaches 15 min, the pacemaker will remain at that phase indefinitely and its motion will therefore stop (119). This interpretation is compatible with the fact that nocturnal mammals, which are expected to show larger phase delays than diurnal mammals, tend to lose circadian rhythmicity at lower LL intensities (15).

The VRC is a theoretical construct, and may not be measurable experimentally. There is, however, one prediction that can reasonably be made on the basis of the PRC itself, namely, that the greater the difference $(D - A)$, the more positive is the slope of the function relating τ to LL intensity. A comparison of PRC shape and τ values in four nocturnal rodent species confirms this prediction (119).

The results described in this and the preceding section suggest that the dependence of τ on LL intensity, itself of no functional importance, is a consequence of the asymmetry between the advance and delay regions of the PRC, which does

have functional significance in promoting stable entrainment patterns (119).

Precision and History Dependence
As seen earlier, the day-to-day variability in τ under constant conditions depends on the value of τ itself, and the closer τ is to 24 h, the smaller its variability. From the data shown in Table 2, it is clear that even in the species with the greatest variability, the standard deviation of τ is still remarkably small, representing only 1.3% of the value of τ. Pittendrigh and Daan (445) have argued convincingly that the precision of the circadian pacemaker is even greater than these measurements— obtained from the daily onsets of overt wheel-running activity—would indicate. In all four species of nocturnal rodents that they studied, they found a significant negative serial correlation between successive activity onsets, such that a deviation from the mean τ value on one day tended to be followed by a change in the opposite direction on the next. This implies the presence of a reference signal, presumably originating from the pacemaker itself, which is less variable than the observed activity onsets.

Pittendrigh and Daan used the computed serial correlation coefficients to estimate the relative contributions of pacemaker and nonpacemaker variability; they found that in mice, for example, the precision of the pacemaker is about twice as high as that of the overt activity rhythm, and furthermore that intraspecific differences in the variability of activity onsets are attributable primarily to nonpacemaker processes.

What are these nonpacemaker pro-

TABLE 2. Variation in τ in Constant Darkness. All values are in hours (h) (from 446).

	Mean τ	95% Range	Mean SD of τ
Mesocricetus auratus	24.07	23.75–24.35	0.08
Peromyscus leucopus	24.01	23.60–24.60	0.09
Mus musculus	23.50	23.70–24.50	0.15
Peromyscus maniculatus	23.36	22.15–24.65	0.30

cesses? Pittendrigh and Daan suggested that the circadian pacemaker initiates a sequence of events that culminate in the onset of overt activity, and introduced the term "wake-up time" to refer to the duration of that sequence. Although the term was only used as a descriptive label, it is at least plausible that the pacemaker generates a wake-up signal that is followed by arousal from sleep and this, in turn, leads to activity onset. This would suggest that wake onset is under more direct pacemaker control than activity onset and should therefore be less variable. Recently, however, the opposite result was obtained in the mouse (605). It is likely that in a natural environment, precise timing of wake onset inside a sheltered nest has less survival value than precise timing of foraging activity outside it (605).

In addition to day-to-day variability, τ also displays a longer term lability, some of which is attributable to the lighting conditions experienced by the animal prior to being released into DD [Table 3 (445)]. The most consistent effects are caused by entrainment to LD cycles with different periods. In mammals, birds, and arthropods, the changes in τ are always in the direction of T, such that τ is lengthened when $T > \tau$, and shortened when $T < \tau$. These aftereffects of entrainment eventually decay and τ returns to a value closer to the species average, but the process can take weeks or months (445).

The length of the photofraction in a daily LD cycle has also been reported to affect the subsequent τ, but the direction of these aftereffects varies in different species, even closely related ones. Among diurnal birds, for example, τ is longer following entrainment to long photoperiods in some species, and shorter in others. Long-lasting changes in τ can accompany phase shifts induced by single light pulses in DD. In general, τ tends to be shorter after phase advances and longer after phase delays. In some animals, τ is also lengthened by prior exposure to LL (445).

There is some evidence that the changes in τ induced by prior lighting conditions are accompanied by changes in the shape of the PRC. Individual mice exposed to 15-min light pulses at CT15 show larger phase delays after entrainment to $T = 22$ h (which caused a shortening of τ) than after entrainment to $T = 26$ h (which caused a lengthening of τ). This suggests that the interdependence between τ and PRC shape demonstrated at the species level and among members of a species may also hold within the same individual (118). Substantial differences in the PRC for 4-h light pulses were obtained following entrainment to long (LD 16:8) and short photoperiods (LD 8:16) in the house sparrow, *Passer domesticus*. In this species, however, τ is shorter after long than after short photoperiods (265), but the PRC shows larger advances and a complete absence of delays after long photoperiods (44).

Spectal Sensitivity: Photoreceptors and Action Spectra

In view of the central importance of light in circadian entrainment, it is surprising that relatively few studies have examined the effects of the spectral composition of light on circadian processes. The limited data available suggest important species differences in the pigments, receptors, and pathways involved.

Phase Shifting and Entrainment

An action spectrum for phase advances of the luminescence rhythm in *Gonyaulax* induced by 3-h light pulses showed peaks at 475 and 650 nm, while light at 550 nm was ineffective (239). While this spectrum is similar to the action spectrum for photosynthesis, it is not identical. Ehret (149) reported similar wavelength effects with a chlorellaless strain of *Paramecium bursaria,* but did not speculate on the possible receptor in the red absorbing region.

In *Drosophila,* action spectra for delay and advance phase shifts induced by 15-min light pulses were similar, both peaking between 420 and 480 nm (186). A more detailed action spectrum, but for phase delays only, suggests that the photoreceptive pigment is a flavin (306). *Drosophila* grown on a carotenoid-free diet exhibit a visual sensitivity loss of three orders of magnitude, but no impairment of light entrainment of the circadian system (623). A comparable distinction exists in *Chlamydomonas:* Phototaxis exhibits a rhodopsin-like action spectrum with a single peak around 500 nm (184), but the action spectrum for phase shifting of the *rhythm* of phototaxis shows two peaks, at 470 and 660 nm; yellow-green light (550 nm) is completely ineffective. Under some conditions, the administration of light pulses at a fixed circadian phase caused phase advances in the blue region and phase delays in the red (241).

In the cockroach, *Periplaneta americana,* which has two photoreceptor types in the ultraviolet and blue-green regions, the threshold for circadian entrainment was found to be lowest at 495 nm, indicating a predominant role for the blue-green receptor (395). The action spectrum for initiation of the circadian rhythm of egg hatching with brief light pulses in the moth, *Pectinophora gossypiella,* shows peak sensitivity between 390 and 480 nm, with a rapid drop at longer wavelengths and a more gradual decrease at shorter wavelengths (72).

Vertebrates other than mammals have extraretinal input to the circadian system, as do many invertebrates (cf. the section on neural control of circadian rhythmicity). In blinded sparrows, the brain itself is the photoreceptive organ and, with

feathers removed, moonlight-level LD cycles suffice to entrain (372); extraretinal photoreception in quail brain exhibits a rhodopsin-like spectral sensitivity (185). The lake chub, *Couesius plumbeus,* exhibits peak sensitivity for entrainment at 538–568 nm, but this is shifted to the red in blinded and blinded-pinealectomized fishes, possibly because longer wavelengths penetrate the tissues more readily (290).

In mammals, action spectra for entrainment in pocket mice, *Perognathus penicillatus* (201), and rats (359) were found to resemble the spectral sensitivity curve for rhodopsin, with a single peak in the blue-green region. Similar results were obtained in a more detailed study of the phase advancing effects of 15-min pulses in the hamster (568). In the flying squirrel, *G. volans,* PRCs for 15-min pulses of monochromatic red (620 nm) and green (500 nm) light were similar and resembled the PRC for white light pulses, except that the amplitude of the green light PRC exceeded that of the red light (131). Since the circadian effects of light in mammals are mediated exclusively by the retina, these results suggest the involvement of a single class of retinal photoreceptors with a rhodopsin-like pigment. The transduction process mediating these effects, however, differs from that mediating visual image formation: The threshold of the phase shift response is considerably higher, and the reciprocal relation between light intensity and duration is maintained for up to 45 min (568).

Results suggesting the involvement of two photoreceptors were obtained in the nocturnal cave-dwelling bat, *Hipposidereos speoris.* In initial studies (283), different PRCs were obtained with fluorescent and incandescent light sources, and this led to the determination of action spectra with monochromatic light pulses (284). Phase advances with 15-min light pulses were greatest at 520 nm, but phase delays were maximal at 430 nm, the shortest wavelength used in the study. The effects of 2.77-h pulses were even more surprising. With irradiance held constant, light pulses at 520 nm caused phase advances at each of four circadian phases, while pulses at 430 nm caused only phase delays.

Free-Running Period: Aschoff's Rule
The effects of the spectral composition of LL on τ have been almost completely ignored thus far. Indeed, the only systematic study was performed recently in *Gonyaulax* (487). Here again, examination of the effects of specific wavelengths was prompted by the chance observation that the effect of light intensity on τ was different with fluorescent and incandescent lights. The results showed that with blue light (400–500 nm), τ shortens as LL intensity increases, while with red light (>600 nm) τ lengthens. At intermediate wavelengths, or with a combination of blue and red, there is a smaller or no effect of intensity on τ (Fig. 12). These findings are consistent with the earlier results in *Gonyaulax* (239) showing two distinct

peaks in the action spectrum for phase shifting, one in the blue region (475 nm) and the other in the red region (650 nm).

In summary, there is evidence in at least some species that (1) the pigments, receptors, and pathways that mediate the circadian effects of light may differ from those mediating other light-dependent functions (photosynthesis, vision, and phototaxis). (2) More than one pigment, receptor, or pathway may be involved, and these may affect circadian rhythms in a qualitatively different manner.

Entrainment in Natural Daylight: Effects of Season and Latitude

Except at the equator, the daily illumination cycle undergoes seasonal changes in such characteristics as photoperiod, light intensity and twilight duration, the extent of which is greater the higher the latitude (Fig. 13). These changes are reflected in the timing of locomotor activity and other rhythmic functions. In some organisms, there is also evidence of latitude-dependent genetically determined differences in circadian parameters.

Activity Duration, Phase, and Period
As the photoperiod lengthens, the duration of daily activity increases in diurnal animals and decreases in nocturnal animals (Fig. 14a). A systematic study by Daan and Aschoff (116), involving eight species of birds and mammals, indicates that the changes in activity duration are greater in birds than in mammals, but in both cases there are limits beyond which further increases and decreases in photoperiod have no effect. As a result, the relationship between these two variables is described by an S-shaped function, the difference between activity duration and photoperiod being greatest under very long and very short days (Fig. 15). This generalization applies to several other mammals and birds, as well as to some

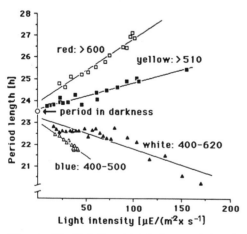

Figure 12. Relationship between intensity and circadian period length for four different spectral bands (in nm) of constant light in *Gonyaulax* (487).

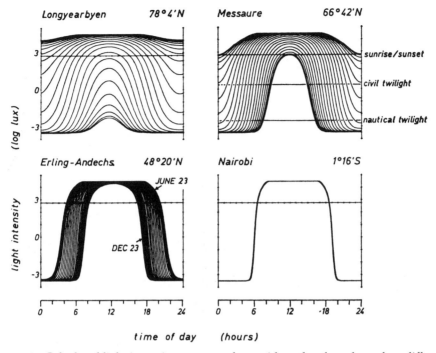

Figure 13. Calculated light intensity curves at the zenith under clear sky at four different latitudes. The uppermost curves represent June 23, the lowermost December 23. The curves in between are for semimonthly midpoints. The data for Messaure include light intensities at sunrise–sunset, civil twilight, and nautical twilight (sun altitude = 0°, −6°, and −12°, respectively) (116).

fishes, as studied under field conditions and in caged animals held under either natural or artificial LD cycles (116, 178, 295).

It should be noted that most of the data from higher latitudes were obtained in animals that normally inhabit more temperate regions, at least for some portion of the year. There is evidence that some year round residents of subarctic regions show greater flexibility in activity duration. An example is the reindeer, *Rangifer tarandus*, whose diurnal activity lengthens to about 22 h in midsummer (162). In the most extreme cases, activity takes place intermittently throughout the continuous daylight of midsummer. Examples include voles (428, 557), ptarmigans (546, 607), several subarctic diurnal fishes (Fig. 14b) (401), and the aquatic larvae of

some flying insects (400). When studied under similar conditions, animals imported from lower latitudes maintain circadian rhythmicity with a more limited activity phase. This suggests that the greater flexibility seen in subarctic species may, at least in some cases, represent a specific adaptation to conditions at higher latitudes, allowing these animals to maximize food intake in summer in preparation for the low illumination and relative scarcity of food resources during the winter months (400, 546).

Examples of extreme inflexibility in activity duration have also been reported. In the tropical cave-dwelling bat, *Taphozous melanopogon*, the times of emergence from and return to the roost, and consequently the duration of activity, were found to remain relatively constant

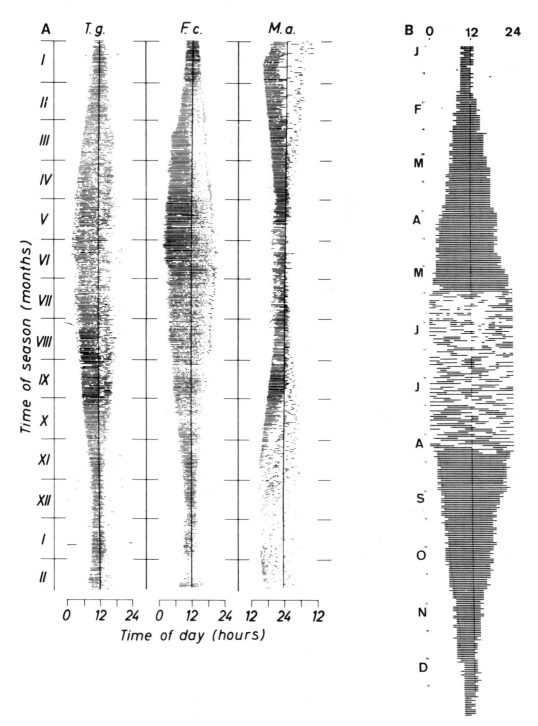

Figure 14. Seasonal changes in activity in animals exposed to natural light conditions at the Arctic Circle (Messaure, 66° N). (*A*) Tree shrew (*Tupaia glis*, T.g.), chaffinch (*Fringilla coelebs*, F.c.), and hamster (*Mesocricetus auratus*, M.c.) (27). (*B*) Fish species, *Coregonus lavaretus* (400).

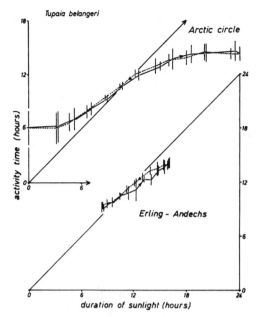

Figure 15. Activity duration (semimonthly means and standard deviations) as a function of daylength (sunlight duration) in tree shrews, *Tupaia belangeri*, at the Arctic Circle (Messaure, 66° N) and in Southern Germany (Erling-Andechs, 48° N). *Solid lines:* January–June; *dashed lines:* July–December. Arrows indicate passage of time (116).

throughout the year (550). These results were obtained at a latitude of only 10° N, but even at that latitude the seasonal variation in photoperiod is close to 2 h, and the activity of another nocturnal bat, *H. speoris*, studied in the same location, closely parallels nighttime duration (346). A similar independence of activity duration from photoperiod was observed in the diurnal ground squirrel, *Ammospermophilus leucurus*, at a 40° N latitude (301).

The seasonal changes in activity duration are the result of changes in the timing of activity onsets and ends. As the days get longer, the activity of diurnal animals starts earlier and ends later in the day. At the same time, however, the phase angle of activity onset relative to sunrise (ψ-onset) decreases (i.e., activity starts later relative to sunrise) while that of activity end relative to sunset (ψ-end) increases. The same holds true for the activity of nocturnal animals in relation to nighttime duration. Maximum ψ-onset and minimum ψ-end values at high latitudes are thus reached in midwinter in diurnal birds and mammals, and in midsummer in nocturnal animals. In chaffinches and other passerine birds, a further increase in ψ-onset, related to reproductive state, takes place during the spring breeding season and is superimposed on the otherwise typical seasonal pattern (Fig. 14*a*) (116).

An alternative phase angle measure, ψ-midpoint, is the difference between the midpoint of activity and either midday for diurnal animals, or midnight for nocturnal animals. In the level-and-threshold model of Aschoff and Wever (26), the daily onset and end of activity are represented, respectively by the times at which the rising and falling portions of a sinusoidal circadian oscillation intersect a horizontal threshold. It follows that any change in the level of the oscillation (i.e., a vertical displacement of the circadian sinusoid relative to the threshold) will alter ψ-onset and ψ-end. Since ψ-midpoint is not affected by such changes in level, this makes it a more accurate index of the phase angle of entrainment.

Systematic seasonal variations in ψ-midpoint have been documented in a number of mammals and birds, at least at high latitudes. They are considerably smaller than the seasonal variations in ψ-onset and ψ-end, which tend to change in opposite directions and thus cancel each other out (116). They also appear to be species specific and are therefore difficult to account for by any of the general properties of circadian oscillations. One such property is the dependence of phase angle of entrainment on circadian period. Although the natural period, τ, cannot be measured directly in entrained conditions, the assumption has been made that it is affected by the average light intensity over the entire LD cycle in the same way as by LL intensity (14). Since average light intensity increases with a lengthening of photoperiod, and in view of Aschoff's rule on the effects of LL intensity on τ, this has led to the generalization, known as the Seasonal rule (11), that a lengthening of photoperiod leads to an increase in ψ-midpoint in diurnal animals and a decrease in nocturnal animals. None of the five bird species studied by Daan and Aschoff, however, showed the predicted midsummer maximum and midwinter minimum in ψ-onset. A midsummer maximum was obtained in the diurnal tree shrew, but in this as in other diurnal primates, τ-increases with increasing LL intensity and maximum ψ-onset would therefore have been expected in midwinter. Only the remaining two mammalian species, both nocturnal, showed the seasonal pattern predicted by the Seasonal rule.

Tree shrews and hamsters show secondary ψ-midpoint maxima at the Arctic Circle in midwinter and midsummer, respectively. This seasonal pattern may be related to twilight duration, which also shows two yearly peaks at the summer and winter solstices, and secondary maxima are in fact predicted by a quantitative formulation by Aschoff and Wever (609). In nocturnal kangaroo rats, *Dipodomys* (300), and in the diurnal fish, *Couesius plumbeus* (295), ψ-onset shows two yearly maxima and minima (at less extreme latitudes), while ψ-end shows only one, suggesting that the former is dependent on twilight duration and the latter on daylength. Several other species, however, show little or no effect of twilight duration on ψ, and a comparison of activity rhythms at Erling-Andechs (48° N) and at the Arctic Circle around the equinoxes (when photoperiod is the same but twilight is longer at the higher latitude) showed no consistent differences in ψ-midpoint (116).

Subsequent results (113) indicate that the assumption that average light intensity in LD has a similar effect on τ (and thereby on ψ-midpoint) as LL intensity may be unrealistic. Finches were studied over the course of a year at Erling-Andechs under natural daylight; during half of each month, light intensity was reduced uniformly over the entire LD cycle by about 2 log units. The reduction in light intensity delayed activity onset and advanced activity end, but the change in ψ-end exceeded that in ψ-onset. As a result, ψ-midpoint was more positive in the low intensity condition, contrary to the predictions of the level-and-threshold model.

There is some evidence of seasonal variations in circadian period measured in constant conditions following exposure to natural daylight. In starlings (217) and in some finches (455), τ is slightly shorter during the spring breeding season than at other times. In both cases, this difference reflects a seasonal variation in the sensitivity of the circadian system to light: In starlings, seasonal differences in τ occurred only in LL, not in DD, while in finches, the slope of the function relating period length to LL intensity was steeper in spring and summer than in autumn and winter. There is also a separate seasonal variation in activity duration, measurable both in LL and DD, which is caused by endocrine changes reflecting the birds' reproductive state (217, 455). One or both of these effects may be responsible for the advanced activity onset and longer activity duration seen in many passerine birds in spring.

Among mammals, seasonal changes in τ were not found either in the diurnal ground squirrel, *A. leucurus*, or the nocturnal kangaroo rat, *Dipodomys merriami*, at either of two LL intensity levels (301). However, seasonal variations were observed in other mammals which, unlike *A. leucurus* and *D. merriami*, are hibernators. In the little brown bat, *Myotis lu-*

cifugus, τ measured in DD at constant but low temperatures (3–10°C) was longer by an average of 2.5 h in animals captured in winter than in animals captured in summer (371). Golden-mantled ground squirrels, *Citellus lateralis*, show a decrease in τ in spring when kept in LL for over a year (625), and a phase advance of activity onset at the same time of year when kept under a constant LD 14:10 cycle for similar durations, probably as a result of a change in τ (326). These changes are accompanied by changes in body weight and in reproductive state and represent an endogenous circannual variation in circadian parameters.

Clear seasonal variations in τ were also obtained in DD in two fish species, the lake chub, *C. plumbeus* (288), and the burbot, *Lota lota* (291), with maximum period lengths in summer. In *C. plumbeus*, a similar effect was obtained following exposure to artificial LD cycles with seasonally appropriate photoperiods (for 54° N latitude), but only if the LD cycle included twilight transitions (295). Since twilight transitions were also necessary for the sigmoidal relation between photoperiod and activity duration in this species, this suggests that the changes in τ may represent aftereffects of, rather than being responsible for, the seasonal variations in the phase and duration of activity (295).

Precision
Circadian activity rhythms also show seasonal differences in the day-to-day variability of activity onsets and ends, as measured by the standard deviations of ψ-onset and ψ-end. Both mammals and birds show the least variability (maximum precision) around the equinoxes, when activity onsets and ends fall during civil twilight. This is the time of day when light intensity is changing most rapidly, and, indeed, the precision of activity onset and end generally shows a positive correlation with the rate of change of light intensity. In most cases, this correlation is higher

for the end of activity than for activity onset (116).

Daan and Aschoff (116) also found that the time of day at which a mechanical twilight sensor registers a fixed light intensity (either 0.35 or 4.5 lux) shows the same seasonal pattern as activity, with the least day-to-day variability occurring at times of rapidly changing light intensity. At such times, differences in cloud cover are likely to have less of an effect on the time at which a fixed light intensity is reached than when light intensity is changing more slowly. These observations suggest that seasonal changes in the precision of activity timing represent primarily a direct action of light on an animal's activity. The fact that the precision of activity end is more highly correlated with the rate of change of light intensity than is the precision of activity onset is evidence that activity onset is more dependent on an endogenous circadian oscillation and less on the direct, exogenous action of light than is the end of activity.

Seasonal Inversion of Activity Pattern
Perhaps the most dramatic effect of season at high latitudes consists of an inversion from a diurnal activity pattern in winter to a nocturnal pattern in summer. This phenomenon has been studied most extensively in fishes (161, 400, 401), but it also occurs in voles (166) and in the bat, *Myotis mystacinus* (416). In the last species, the seasonal inversion is accomplished by a phase delay of the activity rhythm in spring and a phase advance in autumn (Fig. 16a), while in brown trout, *Salmo trutta*, and Atlantic salmon, *S. salar*, the dawn and dusk components of what is essentially a crepuscular pattern with maximal activity centered around midday gradually separate in late winter, until they merge again in late spring, with maximal activity now centered around midnight (Fig. 16b; 160). In other fishes, as well as in voles, the diurnal and nocturnal patterns are often separated by transi-

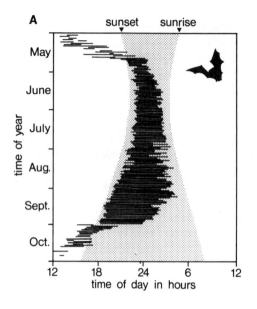

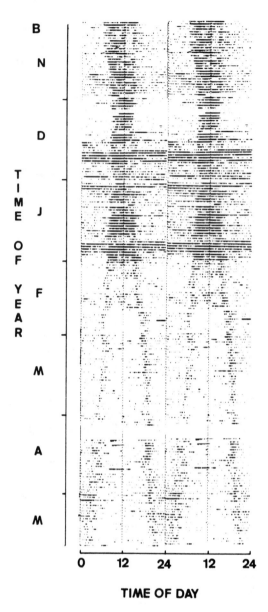

Figure 16. Seasonal inversion of activity patterns. (A) Daily hunting times in a Finnish colony of bats, *Myotis mystacinus*. Original record from Nyholm (416), as modified in Daan (115). (B) Activity of a group of eight brown trout parr, *Salmo trutta* (160).

tional periods in spring and autumn during which the animals may be equally active by day and night. This transitional arhythmicity is distinct from the arhythmicity observed in many species in midsummer at high latitudes, the latter being a consequence of the reduction in the amplitude of the daily light intensity cycle.

In fishes, the seasonal inversion is only observed at latitudes higher than about 60° N. Thus, populations of the sculpin, *Cottus poecilopus*, native to the region of Lund (55° N) maintain a nocturnal pattern year round, whereas populations of the same species from Messaure (66° N) switch to a diurnal pattern in autumn. Sculpins transported from Lund to Messaure in August became diurnally active in September–October (399), while sculpins transported from Messaure to Lund in November remained diurnally active until March–April, when they became

nocturnal again (8). In both *C. poecilopus* and *L. lota*, seasonal inversions take place at the appropriate times when the fishes are maintained under constant LD 12:12 cycles for an entire year, but only if light intensity does not exceed about 50 lux (399, 401). These results are evidence that seasonal inversions in these species reflect an endogenous circannual variation that depends on the presence of appropriate lighting conditions for its expression as well as for precise timing.

The functional significance of seasonal inversions is not fully understood. In some cases, they may reflect seasonal changes in the availability and activity patterns of invertebrate prey species (160, 400, 401, 416), while in others, they may be related to seasonal changes in reproductive state and migratory behavior (401). A relation between activity pattern and migration is well established in several birds that are normally diurnally active but migrate at night. At the times of the spring and fall migrations, caged birds exhibit migratory restlessness (Zugunruhe), with a large increase in activity at night, sometimes exceeding that occurring by day (e.g., 174, 215). That this nocturnal activity is under circadian control is made clear by the demonstration that, when placed in dim LL while showing nocturnal activity in spring, white-throated sparrows, *Zonotrichia albicollis*, start to free-run from the dark phase of the prior LD cycle, whereas birds in post-nuptial molt, showing only diurnal activity, start to free-run from the prior light phase (362).

Other examples of activity pattern inversion are observed in insects. Although these inversions occur at particular times of the year, they are unique events related to specific developmental stages. Thus, the aquatic nymphs of *Leptophlebia* species are nocturnally active in March and April (at a latitude of 63° N), but switch to diurnal activity in May, apparently in preparation for emergence, which takes place only by day (529). A rapid switch from a diurnal (nymphal) to a nocturnal (adult) activity pattern takes place in the cricket, *Gryllus bimaculatus*, a few days after the final imaginal moult. This inversion also occurs in DD (582).

Zeitgebers at High Latitudes
As seen earlier, the locomotor activity patterns of several animals, particularly high northern species, is often arhythmic during the continuous daylight of midsummer at high latitudes; some animals have also been reported to show free-running circadian rhythms in natural daylight (167, 398), although in some cases this may represent the transition between diurnal and nocturnal patterns (318). Many other species, however, maintain an entrained circadian pattern under midsummer daylight conditions, with clearly demarcated rest and activity phases, suggesting that the amplitude of the daily cycle of light intensity, although much reduced at that time of year, is nevertheless sufficient to entrain these rhythms. Indeed, the L:D intensity ratio in midsummer is still around 100:1 at 66° N latitude (116, 400), and 23:1 at 70° N (414, 557), and laboratory studies have demonstrated entrainment with considerably smaller ratios (165, 317, 557, 608). However, the strength of the LD Zeitgeber is also reduced under very long photoperiods, and when overall light intensity levels are high (9, 128, 165, 317, 608), both conditions existing in the arctic midsummer. Furthermore, Krüll (316) has pointed out that the usual light intensity measurements do not take into account the extreme angle of incidence of the sun's rays at high latitudes. When the appropriate conversion is applied, the daily L:D intensity ratio at 78° N is only in the order of 1.4:1, and even less on cloudy days (316). The daily temperature cycle is also much reduced in amplitude at these latitudes, and is therefore unlikely to be effective as a Zeitgeber (471).

Krüll (316, 318) recorded the activity of green finches, *Carduelis chloris*, and snow buntings, *Plectrophenax nivalis*, in Spitsbergen (78° N) in midsummer. The former received direct sunlight 24 h a day, while the latter were studied in a narrow valley which, because of its orientation, receives direct sunlight for only 2 h a day around midnight. The activity patterns of both species included a period of relative inactivity centered around midnight, suggesting entrainment by a Zeitgeber other than light intensity. The most likely candidate appears to be a daily cycle in the spectral composition of light. Recordings at a latitude of 59° N in July indicate a shift toward the red end of the spectrum at twilight, and toward the blue end at night (414). These changes are also reflected in color temperature, which shows a clear daily oscillation ranging between about 3000 and 7000°K in Spitsbergen (316). Laboratory studies have shown that daily color temperature cycles of smaller amplitude are sufficient to entrain the circadian activity rhythms of zebra finches (134) and chaffinches (318).

One additional daily cue in the arctic is provided by the position of the sun. That this cue can also be an effective Zeitgeber was demonstrated in several bird species exposed to a constantly lit incandescent bulb that circled continuously over the cages, completing one rotation every 24 h (315). The fact that the birds showed at least temporary entrainment under these conditions suggests that they made use of the position of the light source in relation to some fixed aspects of their environment.

Genetic Adaptations to Latitude

Some of the results reviewed in preceding sections suggest the possibility of genetic variation in circadian properties related to latitude. More conclusive evidence of such adaptations was obtained in *Drosophila* species. One study (6) compared the circadian oviposition rhythms of 15 strains of *D. melanogaster* from latitudes ranging between 0 and 60° N, all reared and studied under LD 12:12. The proportion of eggs layed during the dark phase of the LD cycle and the relative size of the major nocturnal oviposition peak both decreased with decreasing latitude. A similar comparison of 57 strains of *D. littoralis* (325) showed a latitude depen-

dence of several parameters of the circadian pupal eclosion rhythm, as well as in the critical daylength for photoperiodic adult diapause (Fig. 17). Differences were also seen in the shapes and phases of PRCs for light pulses in two strains of *D. littoralis* from different latitudes (325), as well as in those of two Japanese strains of *D. auraria* (451). In the latter species, a similar latitude dependence was found for phase angle of entrainment (in LD 14:10) and for critical daylength, but unlike the case in *D. littoralis*, τ was positively correlated with latitude (451). The relationships between the various circadian parameters that show latitude dependence are not completely understood. In *D. littoralis*, for example, τ is correlated with ψ, even when the effects of latitude are partialled out, but the variation in τ does not fully account for that in ψ (325).

Functionally, the role of the latitudinal variation in ψ is most probably to ensure that in each population, eclosion will occur at dawn, when relative humidity is highest, regardless of the duration of the photoperiod at that time of year and at that latitude. The latitude dependence of critical daylength, on the other hand, may provide the necessary balance between maximizing the duration of the reproductive season while still ensuring that the flies are in diapause at the onset of winter (325).

The Splitting Phenomenon

When hamsters are kept in LL for extended durations, a certain proportion shows a splitting of their circadian activity rhythms into two components, which free-run temporarily at different frequencies. The two components generally resynchronize upon reaching a 180° phase relation and assume a common period slightly shorter than that observed prior to splitting (Fig. 18) (441, 447); the split components merge again to form a single daily activity band following transfer to

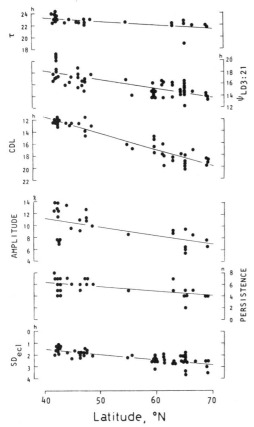

Figure 17. Parameters of the circadian pupal eclosion rhythm and photoperiodic adult diapause in strains of *D. littoralis* from different latitudes. *From top to bottom:* free-running period in DD, phase of entrainment under LD 3:21, critical daylength for adult diapause, amplitude of free-running rhythm, persistence of free-running rhythm, and amplitude of entrained rhythm under LD 3:21 (325).

(*Sturnus vulgaris*), however, a strongly bimodal activity pattern suggestive of splitting can sometimes accompany a unimodal feeding rhythm (191).

The factors responsible for the development of splitting are not fully under-

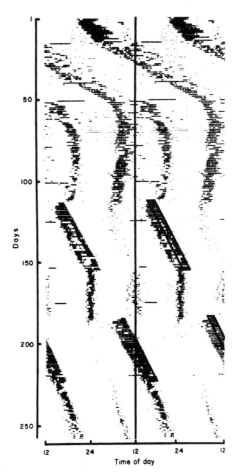

Figure 18. Activity record of a hamster showing spontaneous splitting of its activity rhythm in LL. The hamster was then exposed twice to LD cycles with 2-h dark segments and with a period T = 24.23 h. The 2-h dark segment (enclosed by solid lines on the right half of the plot) was timed to overlap with one split component during the first exposure and with the other component during the second exposure. On both occasions, the rhythm remained split and both split components entrained to the LD cycle (55).

DD (141). Similar effects have been observed in several other vertebrate species (58, 216, 256, 293, 455a, 557, 592). In rats, nearly identical patterns of splitting were seen in concurrently recorded circadian rhythms of drinking, feeding and electrical brain self-stimulation (58), and in locomotor activity and drinking (521) and locomotor activity and body temperature (434) in hamsters. In starlings

stood. In hamsters, splitting occurs spontaneously in relatively bright LL, but it is sometimes triggered by exposure to dark pulses (56). The proportion of hamsters showing split rhythms has also been found to depend on age (391) and, in females, on ovarian hormone levels (392). In the diurnal tree shrew, *Tupaia glis*, splitting results from a reduction, and merging of the two components follows an increase, in LL intensity (Fig. 19; 256). In starlings, splitting can be induced by continuous administration of testoster-

one (216), while in the lizard, *Sceloporus olivaceous*, it occasionally follows pinealectomy (592).

The splitting phenomenon is evidence that the pacemaker underlying circadian rhythms of locomotor activity and other functions in several species consists of two mutually coupled oscillators, or populations of oscillators. In a model of pacemaker structure in nocturnal rodents (441, 442, 447), light intensity is assumed to differentially affect the periods of the two oscillators; as a result, one oscillator

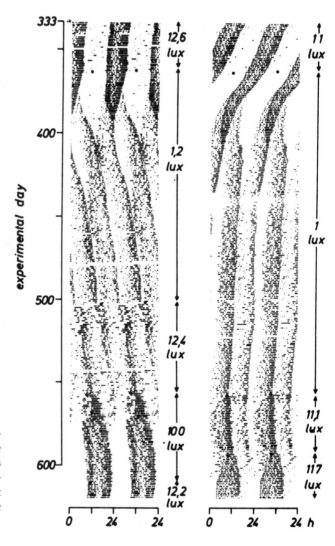

Figure 19. Splitting of the circadian activity rhythms of two tree shrews, *Tupaia glis*. Splitting was induced by a reduction in LL intensity, and merging of the split components followed an increase in LL intensity (256).

(E) controls the onset of activity in the evening and is synchronized by dusk, while the other (M) controls the end of activity in the morning and is synchronized by dawn. Under normal (i.e., nonsplit) conditions, E and M are mutually coupled, with E phase-leading M by several hours. The phase relation between E and M, however, is somewhat flexible, its precise value depending on the lighting conditions to which the animal is exposed, and this phase relation, in turn, determines the duration of daily activity and the free-running period of the compound pacemaker. A second stable phase relation (or range of phase relations) is also possible, as indicated by the fact that the two split components often free-run in synchrony when they are approximately 180° apart.

The dual-oscillator model can account for a wide variety of circadian phenomena. Thus, the coupling of E and M to dusk and dawn, respectively, accounts for the changes in activity duration in response to changes in photoperiod, while the mutual coupling between the two oscillators accounts for the sigmoidal relationship between activity duration and photoperiod. The phase displacement between E and M also implies that a pulse of light or darkness will fall on different phases of the two oscillators, thereby causing unequal phase shifts; the transients that can follow such pulses can therefore be viewed as reflecting the motion of one or both oscillators as they regain their former phase relation. In addition, the flexibility of the phase relation between E and M and its effect on pacemaker period provides an explanation for the aftereffects of various lighting conditions on τ. Several of these features have been successfully simulated in a quantitative version of the dual-oscillator model (117). Other possible applications of the model include the seasonal inversion of activity patterns at high latitudes, and the nocturnal migratory patterns of

otherwise diurnally active birds. Finally, a pacemaker consisting of two coupled oscillators provides a possible mechanism for an internal coincidence model of photoperiodic time measurement (447).

Several studies have examined the properties of the two oscillators in hamsters by comparing the responses of split activity components to dark pulses and to LD cycles. Single 6-h dark pulses (57, 327), as well as composite dark–light–dark pulses (142), were found to cause concurrent but unequal, in some cases opposite, phase shifts of the two components, each of which showed a separate, bidirectional PRC. The PRCs of the two components were qualitatively similar, and resembled the PRC for 6-h dark pulses in animals showing nonsplit activity rhythms. Either or both components also entrained to LD cycles with 2-h dark segments (Fig. 19) and did so in a qualitatively similar manner (55). Other studies have shown that in female hamsters, both split components are preceded daily by a surge in serum LH levels, and either or both components can be associated with the onset of lordosis (589). Thus, the available evidence suggests that the two oscillators presumed to underlie split rhythms are equivalent, differing only in their intrinsic periods, as seen during the initial stages of splitting.

Effects of Temperature and Other Physical Factors
Z. Boulos

Next to light, temperature is undoubtedly the most important environmental factor in the regulation of circadian rhythmicity, both constant and changing temperature levels having a number of different effects on circadian parameters.

Constant Temperature Levels

Two effects of constant temperature levels were described in the section on key characteristics of circadian rhythms; the presence or absence of circadian rhythmicity conditional on ambient temperature, and the relative insensitivity of τ to temperature. The latter, however, is only true in comparison with the effects of temperature on other metabolic processes, as temperature-related differences in τ are generally only slightly smaller than those resulting from changes in LL intensity (cf. Figs. 5 and 11).

The available data indicate that, in both homeotherms and poikilotherms, the effect of ambient temperature on τ is usually in the same direction as that of LL intensity (15). Thus, as temperature increases, τ generally lengthens in nocturnal animals (and in diurnal primates) and shortens in diurnal animals. In several species, however, the temperature dependence of τ is found at some LL intensities and not others (15, 579), while in a few cases, temperature has no consistent effect at any LL intensity (22).

A change in τ would be expected to alter the phase of an entrained rhythm relative to the entraining cycle. This prediction is borne out in the nocturnal dormouse, *Glis glis*, where a lowering of ambient temperature shortens τ in LL and advances the phase of the activity rhythm in LD 12:12; similarly in the diurnal chaffinch, *Fringilla coelebs*, lowering temperature lengthens τ and delays the rhythm (453, 454).

Many animals tend to avoid temperature extremes if they have the opportunity to do so. Diurnal animals, for example, may delay their daily emergence and activity onset on cold mornings, or retreat to shelters around midday on very warm days. Often, these are direct responses to temperature fluctuations, but in several poikilotherms, similar changes in phase and activity pattern can be observed under constant levels of ambient temperature. The locomotor activity of the common water snake, *Natrix sipedon*,

in LD 12:12 changes from predominantly diurnal at 17.8°C to predominantly nocturnal at 33.9°C (246). In garter snakes, *Thamnophis radix*, the activity rhythm shows a single major peak in late morning at low temperatures, two daily peaks at dawn and in late afternoon at intermediate temperatures, and a single major peak at dawn at temperatures of 31–35°C, (246, 247). Similar changes in phase and pattern are seen in turtles under both long and short photoperiods (Fig. 20; 208). Females of several nocturnal insect species show a phase advance of their circadian rhythm of "calling" behavior (sex pheromone release) relative to the LD cycle when exposed to constant low temperatures, and a similar shift is seen in the flight activity of males (82, 508, 533). In the silkmoth, *Antherea pernyi*, a similar, permanent shift in adult activity rhythms follows exposure of the pupae to low temperatures (587). Emergence of the midge, *Chironomus thumni*, from 9–12°C water takes place during the day, but is delayed until after dusk in 16–25°C water. This shift is independent of photoperiod, and takes place in DD as well (319). Under naturally fluctuating levels of ambient temperature, the result of these changes in circadian phase and pattern is invariably a shift toward the cooler hours of the day or night in warm weather, and toward the warmer hours in cold weather.

Temperature Cycles, Pulses, and Steps

Daily temperature cycles readily entrain the circadian rhythms of plants and poikilothermic animals, and can prevent or reverse the loss of overt rhythmicity that is sometimes observed in constant conditions (563a). In several insects, the incidence of diapause is sensitive to the duration of the warm portion of the entraining temperature cycle (thermoperiod), just as it is to photoperiod (510). Poikilotherms also show phase-depen-

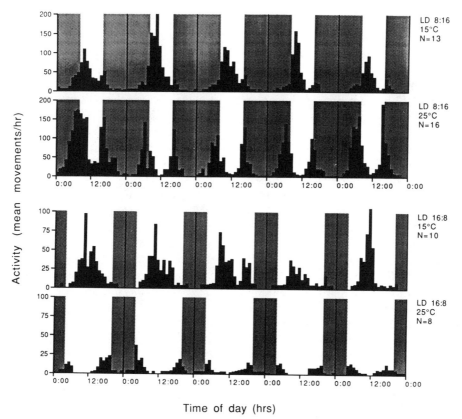

Figure 20. Daily activity patterns of spotted turtles, *Clemmys guttata*, during five consecutive days at two constant temperature levels (15 and 25°C) and two photoperiods (LD 8:16 and LD 16:8). Under both photoperiods, the activity pattern shows one major peak during the light portion of the LD cycle at 15°C, and two peaks around the times of lights-on and lights-off at 25°C (redrawn from 208).

dent phase shifts of their circadian activity rhythms in response to upward or downward temperature steps, and to high- and low-temperature pulses (HTPs and LTPs, respectively).

In general, temperature steps-up cause only phase advances and steps-down only phase delays (though there may be exceptions, cf. 190), the sizes of these phase shifts depending on the phase at which the temperature step is applied (624). Temperature pulses, on the other hand, cause phase advances as well as phase delays (90, 190, 485, 624). The PRCs for HTPs and LTPs are roughly mir-

ror images although the phase shifts induced by the two procedures are not necessarily equal. In *Drosophila pseudoobscura* HTPs cause phase advances when administered in subjective day and phase delays in subjective night, while LTPs have the opposite effect. Depending on their duration and phase of administration, temperature pulses can also attenuate the amplitude of the rhythm, or cause arhythmicity.

A comparison of the responses of the *Drosophila* eclosion rhythm to temperature pulses, steps, and cycles (624) shows that the phase shift generated by a 12-h

HTP or LTP corresponds closely to the sum of the phase shifts generated by its constituent temperature steps. It was also possible to accurately predict the phase angle of entrainment to a temperature cycle with a 12-h thermoperiod based on the PRC for single 12-h pulses. The phase shifting effects of pulses of other durations, estimated by adding the effects of the appropriate LTPs and HPTs, similarly predicted the phase angle of entrainment to daily temperature cycles with thermoperiods ranging between 2 and 21 h. These results clearly indicate that entrainment by temperature cycles can be accounted for by discrete, daily phase shifts caused by periodic exposure to high or low temperatures, in a manner analogous to entrainment by LD cycles. Circadian rhythms entrained to sinusoidal temperature cycles (190, 225, 485) can persist during continuous action of temperature.

The phase shifts induced by temperature steps and pulses described thus far are permanent shifts, recorded in the steady state. In *Drosophila*, however, the first eclosion peak after a 10°C step-down may be phase delayed by as much as 12 h (437), but the eclosion peak advances gradually over the next few cycles, and the final steady-state delay is consequently much reduced (under 3 h for an 8°C step-down) (624). This transient behavior has been accounted for in terms of a two-oscillator model (443, 444), which assumes that the gating of eclosion in *Drosophila* is directly determined by a temperature-sensitive slave oscillator (the B oscillator), which is in turn driven by a temperature-compensated and light-entrainable pacemaker (the A oscillator). The large phase delays observed immediately after the temperature drop are thus attributed to a slowing of the temperature-sensitive B oscillator, which is then reentrained by the A oscillator. The model also assumes that B exerts some feedback effect on A, hence the (relatively small) phase delay in the steady state.

There is, in fact, more direct evidence for the existence of separate light- and temperature-entrainable oscillators in the cricket, *Teleogryllus commodus*. Severing the optic lobes in this species abolishes the rhythm of stridulation in any lighting condition, but the rhythm can be reinstated and entrained by daily temperature cycles (472).

The role of temperature cycles in natural conditions and their interaction with light cycles have not received sufficient attention, but the available evidence suggests a range of possible effects. Diapausing prepupae of the leafcutter bee, *Megachile rotundata*, are not normally exposed to LD cycles; their emergence is unresponsive to LD cycles, and, under constant temperature levels, is distributed over the entire 24-h day. Exposure to a single 24-h low-temperature pulse, however, results in the immediate synchronization of the population and a clear free-running rhythm ensues (590), suggesting that temperature may be the primary, if not the only, Zeitgeber at this stage of development. The Queensland fruit fly, *Dacus tryoni*, is similarly unresponsive to light in its pupal stage. In the absence of temperature variations, the phase of its eclosion rhythm is determined by the LD cycle experienced in prior developmental stages, and the effect of even small daily temperature oscillations is to limit the spread of eclosion to fewer hours of the day (37). Other poikilotherms, when exposed to concurrent light and temperature cycles in different phase relations to one another (444) or having different periods (173), adopt a compromise phase that may be influenced more by one or the other cycle, depending on the species. Thus, the locomotor activity rhythm of the diurnal heliothermic lizard, *Uta stansburiana*, is more strongly dependent on the temperature cycle and less on the light cycle than that of the nocturnal gekkonid, *Coleonyx variegatus* (173). These differences appear to be related to the different thermoregulatory strategies of diurnal and nocturnal reptiles.

Very few studies have examined the effects of temperature cycles and pulses in heterothermic animals. The little pocket mouse, *Perognathus longimembris*, shows a very high sensitivity to such treatments, rivaling that of many poikilotherms. Indeed, a temperature cycle of only 1.5°C entrained the circadian rhythm of arousal from daily torpor in 2 of 3 animals, and all animals exposed to a 3°C cycle were entrained. The PRC for 10°C high-temperature pulses in these animals shows phase delays in the first half of subjective night and phase advances in the second; the phase shifts appear to be determined by the rise in temperature rather than by the duration of the pulse (332). A 10°C temperature cycle entrained the activity rhythms of only 6 of 12 heterothermic bats, *Molossus ater*, but the same treatment failed to entrain the rhythms of the homeothermic species, *Phyllostomus discolor* (164). Results from several other mammals and birds (22, 185a, 225, 554, 579) indicate that temperature is only a weak Zeitgeber in homeotherms. However, even temperature cycles too weak to entrain a rhythm can affect its phase of entrainment to a LD cycle (580).

Other Factors

Changes in relative humidity affect the behavior of many animals, particularly those with more permeable skins or integuments. Thus, rain can lead to the emergence of amphibians, earthworms and slugs, and, indirectly, to that of their predators. The snake, *Natrix natrix*, often hunts after rain in summer, even when ambient temperature is below optimum levels, while the usually nocturnal-crepuscular *Anguis fragilis* can be seen hunting during the day after rainfall (505). The flight activity pattern of mosquitoes (*Anopheles*) in the field parallels the daily curve of relative humidity more closely that that of light or temperature (436). In general, there is little evidence that any of these effects are mediated by a circadian timing mechanism.

It is very likely, however, that daily humidity cycles played an important role as an ultimate factor in the selection of daily activity and rest phases in many organisms. In *Drosophila*, the timing of adult eclosion is widely believed to have evolved to coincide with the time of day when relative humidity levels are highest (i.e., at dawn); the flies are highly susceptible to desiccation in the first few hours after emergence (438). Among terrestrial arthropods, species that lack an epicuticular wax layer (e.g., woodlice, centipedes, and millipedes) tend to be nocturnal (98), and in some groups the degree of nocturnality, determined under laboratory conditions, is correlated with the rate of water loss by transpiration (97).

A few other physical factors can entrain the circadian rhythms of some animals. These include daily cycles of conspecific sounds in birds (214, 374, 465), of mechanical noise in birds (338) and hamsters (376), and cycles of atmospheric pressure (243), and electrostatic and electromagnetic fields (136, 610). Little is known about the mechanisms involved.

Effects of Biotic Factors

B. Rusak

Food Availability

The Time Sense of Bees
Forel's observations (183) that honeybees would arrive at his outdoor breakfast ta-

ble every morning at the same time, even on days when no food was served, were the first indications that bees possess a memory for time, or time sense. Subsequently, a procedure was developed for the experimental study of this phenomenon, by offering bees a sugar water solution for a limited time each day, and noting the times of their arrivals at the feeding place (40). The results of several studies, which include training and testing bees in underground salt mines, and training bees in one location then transporting them to another location several time zones away for testing, showed conclusively that the bee's memory for time was not dependent on exogenous cues but rather was based on an endogenous circadian timing system. Thus, bees can be trained to visit a food source at 24-h or near-24-h intervals, but not if the training interval lies outside the normal circadian range (e.g., 19 or 48 h) (39, 40). They can also be trained to arrive at two or more feeding times each day, and to learn to associate specific food sources with particular times of day (204, 308). In nature, these abilities enable bees to time their visits to different flowers to coincide with the daily times of opening of these flowers, or the times of maximum nectar production.

The circadian mechanism underlying the time sense of bees represents an example of a continuously consulted clock, as it allows them to associate any phase of the circadian cycle with the availability of food (438). A similar mechanism underlies the time-compensated sun orientation ability of birds and other animals (257). In both cases, the clock can be entrained by daily LD cycles and free-runs in constant conditions.

Bees show circadian rhythms of flight activity and foraging behavior even when food is continuously available, and these rhythms are entrainable by LD cycles and

free-run in constant conditions (42, 534, 535). Feeding schedules with periods of 22–25 h are effective in entraining the circadian activity rhythms of honeybee colonies (Fig. 21) (187). These results suggest that both the time sense and activity rhythms of bees may be controlled by the same circadian clock. Similar though more limited results have been obtained in birds (1, 60).

Anticipation of Daily Feedings in Other Animals

Rats and mice, allowed access to food for a limited duration at regular daily phases, gradually come to show a sharp increase in locomotor activity or in food motivated lever pressing immediately preceding the scheduled feeding time, and this is accompanied by similar anticipatory rises in body temperature, corticosterone secretion, gastrointestinal motility, and in the activities of various intestinal and hepatic enzymes (59, 101). Anticipatory activity in rats fails to develop when food is made available at 18–19- or 29–30-h intervals, but it does occur under feeding schedules with periods in the circadian range (59). Within this range, the phase angle of anticipatory activity onset (and, therefore, the duration of anticipatory activity) shows the same relation to the period of the feeding schedule as that between the phase angle of a light-entrained rhythm and the period of the entraining LD cycle (29, 261). Anticipatory activity also shows gradual resynchronization following a shift in the daily feeding time (541). These results are clear evidence that the ability to anticipate cyclic food availability depends on a circadian timing mechanism.

The food-anticipatory mechanism is separate from that which underlies free-running and light-entrained rhythms. Thus, food-anticipatory increases in locomotor activity and other functions in rats kept in LD can occur concurrently

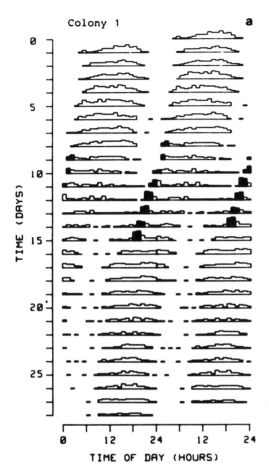

Figure 21. Double plotted activity records of two bee colonies. Hourly values of activity at the hive entrance are plotted in the form of histograms. One ordinate unit corresponds to 1500 passages through the hive entrance. From day 1 to day 8 sugar water was available constantly outside the hive. From days 9–15 food was present for 2 h with an interfeeding interval of 23 h (feeding time indicated by black columns). From day 16 to the end of the experiment food was again available ad libitum. Colony activity was entrained by the feeding schedule and free-ran from an appropriate phase after release from entrainment (187).

with the normal nocturnal rise in these functions. More importantly, when studied in constant lighting conditions, rats and mice often show both an anticipatory component synchronized by the feeding schedule as well as a free-running component. Occasionally, a free-running component may not be visible during exposure to the feeding schedule, but free-running rhythms reappear upon reinstatement of ad libitum feeding, and do so from an initial phase predicted by extrapolation of the prerestriction free-run, not the phase of the prior feeding time; this indicates that although the free-running rhythm may be masked during food restriction, it is not entrained by it (60). The food-anticipatory component generally disappears immediately when free-feeding is restored, but it can be seen to free-run for several cycles if the animal is instead starved for 1–5 days (59). Furthermore, if after exposure to restricted daily feedings, rats are first allowed free access to food and then food deprived for 48 h, they show a resurgence of activity at roughly the same circadian phase relative to their LD entrained or free-running rhythms as that seen just before termination of the feeding schedule. This memory for feeding time can be demonstrated repeatedly for several weeks with no further experience of temporally restricted feedings (100, 490). Finally, rats continue to show food-anticipatory activity after receiving lesions of the SCN [the suprachiasmatic nucleus—the putative light-entrainable pacemaker (see the section on neural control of circadian rhythmicity)], which abolish their free-running rhythms. This anticipatory activity is also limited to feeding intervals in the circadian range, reentrains gradually after a shift in feeding time, and persists for several cycles after complete food deprivation (Fig. 22) (60, 540, 541, 543).

The presence of food-anticipatory ac-

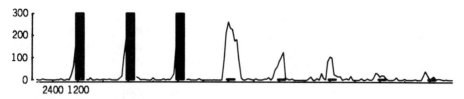

Figure 22. Lever pressing in a suprachiasmatic-lesioned rat on a restricted food availability schedule. *Black bar on days 1–3,* time of food availability; *small bar on days 4–8,* predicted time when food would have been available. On *days 4–8* no food was available. Food-related lever pressing increases on a circadian schedule in anticipation of the feeding time; the rhythm persists with a circadian period in the absence of food (60).

tivity concurrently with free-running or LD-entrained rhythms is not unique to rats and mice; it is also seen in squirrel monkeys (23, 61), marsupials (418), weasels (621), and rabbits (276). The fact that these animals include omnivores, carnivores, and herbivores indicates that food anticipation is not dependent on diet or feeding habits.

Daily feeding schedules are not entirely without effect on the circadian pacemaker underlying free-running and light-entrainable rhythms. In rats and monkeys, the free-running rhythms often exhibit relative coordination as they intersect the scheduled feeding time, and, in a few cases, have even been found to entrain to the scheduled feeding cycle. Such entrainment, however, only occurs when the period of the free-running rhythm is sufficiently close to that of the feeding schedule (23, 61, 542). These results suggest that the food-entrainable oscillator is weakly coupled to the light-entrainable oscillator, and that by entraining the former, daily feeding schedules can exert some effect on the latter (61, 542). One recent exception was reported in a study involving two inbred strains of mice (2). Daily feeding schedules elicited anticipatory activity in both strains in dim LL, but while the free-running rhythms of C57BL mice were not af-

fected by the schedules, those of CS mice were consistently entrained; the presumed genetic basis for this difference remains unidentified.

Social Interactions

Reproductive Synchronization

The circadian system is known to be involved in the regulation of several longer period rhythms, including the estrous cycle of female rodents and seasonal reproductive cycles (502). Evidence that social cues can affect each of these cycles might imply a role for social stimuli in affecting the circadian clock. The estrous cycles of female mice can be synchronized by exposure to pheromones from a male mouse (612). Female rats that are housed together show a nonrandom degree of synchrony among their estrous cycles (356), just as women may show relative synchrony among their menstrual cycles when housed together (355). Female hamsters may also experience social synchronization of estrous cycles, and entrainment to a dominant female was suggested in one study (227).

In rodents, a pheromonal mechanism may affect estrous cycles through an affect on the circadian clock, since the luteinizing hormone signal that triggers ovulation and estrus is normally controlled by a circadian clock. In primates, there is no evidence of circadian clock regulation of menstrual cycles (502), so there is no implication that pheromonal or other social cues operate on menstrual cycles through a circadian clock mechanism.

Maternal Synchronization

Mammalian mothers expose their offspring to a variety of rhythmic cues both prenatally and postnatally. These cues appear to convey information to the young about both circadian time and seasonal daylength changes in some species. Rhythms of pineal N-acetyltransferase in rat pups are synchronized by cues from their mothers both pre- and postnatally (132), as are behavioral rhythms (121). Female rats and Syrian hamsters convey a signal to their young early in pregnancy that synchronizes their circadian clocks *in utero* (121a, 474). Female rodents are also effective in entraining the physiological and behavioral rhythms of their pups postnatally (254, 599). The mother's rhythmic presence in the nest appears to provide an adequate synchronizing cue for the young (599).

The photoperiod to which pregnant dams are exposed also affects the subsequent postnatal reproductive development of rodents. Voles (*Microtus*) whose mothers were exposed to a decreasing photoperiod during gestation show delayed reproductive development compared to those with mothers exposed to increasing photoperiods even when both groups are tested in identical postnatal environments (Fig. 23) (263). Similar transduction of prenatal photoperiodic effects also occurs in female Djungarian hamsters, whose photoperiodic experience influences their pups' subsequent sensitivity to day length (545). The mechanisms underlying these effects are not known, but may include trans-placental endocrine signals that convey photoperiodic information to pups *in utero*.

Social Synchronizers and Resource Partitioning

There are a wealth of anecdotal reports and incidental observations of social effects on circadian rhythms (496), and experimental studies have also yielded evidence for social synchronization in several species. Social cues appear to provide synchronizing information for isolated humans (610), and house sparrows can be entrained by the recorded sounds of conspecifics (374, 465). In natural and seminatural environments, social cues can entrain activity rhythms in bats (Fig. 24) (347), and other species (496). Singly housed white suckers (*Catostomus commersoni*) show short free-running periods while schools of these fishes show longer periods (292).

In other studies, salient social stimuli that had strong behavioral effects failed to influence the circadian systems of crickets or hamsters. Female crickets exposed to male calling songs responded with increased activity, but did not synchronize their circadian clocks (337). Similarly, highly arousing social contacts affected overt behavior without synchronizing the circadian clocks of Syrian hamsters (123). Other studies have demonstrated, however, that brief periods of social interaction between hamsters can produce substantial phase shifts in their free-running activity rhythms and can entrain them when repeated daily (397).

Animals can affect the activity patterns of other individuals not only by coordinating to a common phase, but also by competitive temporal exclusion. Such exclusion has been reported to occur intraspecifically between dominant and subdominant rats (80), as well as interspecifically. Kenagy (299) has documented aggressive interactions among

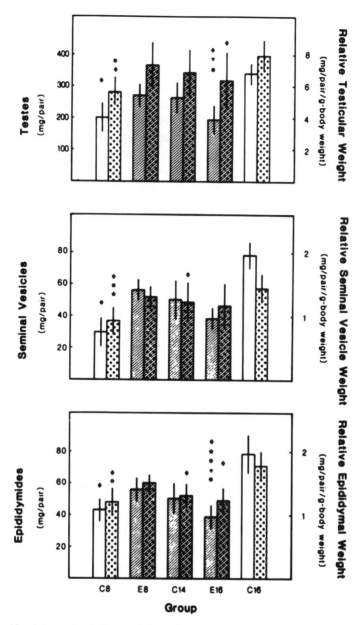

Figure 23. Absolute and relative weights of testes, seminal vesicles and epididymes in 5 groups of voles (*Microtus montanus*) at 74 days of age. Groups C8, C14, and C16 born to females exposed to daily light cycles of 8, 14 and 16 hours; pups remained at these photoperiods. Groups E8 and E16 were born to females exposed during pregnancy to LD 8:16 and LD 16:8 and the pups transferred to LD 14:10 at birth. Reproductive organs largest from long-day periods of gestation and maintenance, smallest in voles gestated and maintained on short photoperiods. Photoperiod during gestation had less effect than after birth. [Modified from Horton (262a)].

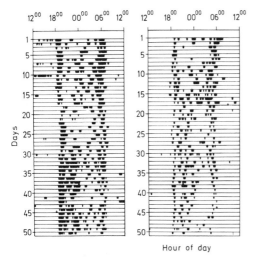

Figure 24a. The flight activity patterns of three captive bats for 40 days in one case and 50 days in the other case recorded 40 m inside a narrow "true cave" in Madurai. The bats could fly within the flight activity cages and the movements of the cages were recorded directly. Activity bouts are indicated by black bars. The activity–rest data are schematized from original data and presented one below the other for successive days. Despite being in constant darkness, the bats are entrained to a 24-h period by the activity of their free-living conspecifics.

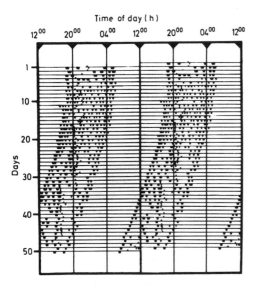

Figure 24b. A record similar to Fig. 24*a* from a solitary male bat caged without access to information from conspecifics (347).

individuals of two species of kangaroo rats (genus *Dipodomys*) that live sympatrically, which result in one species being excluded from favored feeding sites around dusk, an apparently preferred feeding time. There are abundant anecdotal reports of changes in activity patterns of species that face environmental encroachment by humans (e.g., 213) but few systematic studies of these changes.

Temporal niche sharing has been described as a useful mechanism for reducing interspecific competition. On a long time scale, evolutionary forces have resulted in divergence of activity types among species that permit effective sharing of temporally distributed resources; for example, among insectivorous diurnal birds and nocturnal bats, and among diurnal and nocturnal coral reef fishes. Evolution of temporal partitioning to share spatial niches would involve selection over a long interval, but another possible mechanism could involve immediate behavioral responses rather than selection. Activity patterns might differ stably among individuals in a population or might be plastic within individuals. Such plasticity in a population has been documented in cotton rats (*Sigmodon hispidus*) (281), but there is no substantial evidence that this plasticity has functional significance in natural populations.

Feedback Effects of Behavior

It has long been a standard view that circadian clocks are not influenced by feedback from the rhythms they regulate. Recent studies have indicated, however, that changes in activity levels (a rhythm regulated by the circadian pacemaker) can phase shift and entrain circadian clocks. Hamsters that are exposed to a variety of environmental events that increase their activity level (social contact, confinement to an activity wheel, and exploration of a novel environment) may show phase shifts in their activity rhythms and ultimately entrainment of the circadian pace-maker if the stimuli are presented on a daily basis (Fig. 25) (397, 466, 503). Mrosovsky (397) has suggested that a number of different environmental and pharmacological stimuli might act by affecting a common mediating variable, namely, level of arousal or activation. Evidence that temperature cycles can entrain cir-cadian activity rhythms in squirrel monkeys has also been interpreted as possibly reflecting a direct effect of temperature on activity levels, which in turn influence the circadian pacemaker (22). This indirect feedback loop has been proposed to account for phase shifting in rodents in response to social cues, food restriction schedules, benzodiazepines, and dark

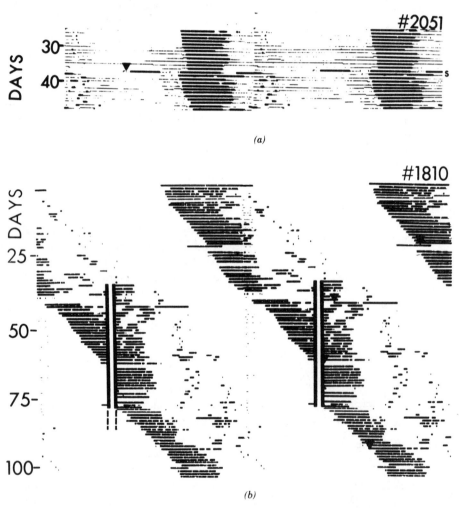

Figure 25. (*a*) Phase shifts in free-running activity of hamsters in response to social interactions (s) at times indicated by arrow and horizontal lines. (*b*) Effects of periodic social interaction at times indicated by solid black bars. Broken vertical lines indicate social interactions for other animals in the room. [From Mrosovsky (397)].

pulses (467). It seems unlikely, however, that such cues act in all species via their influence on arousal mechanisms since dark pulses, for example, are effective phase-shifting agents in cultured chick pineal cells (619), and in *Paramecium* (278).

Circadian Control of Physiological Functions: Thermoregulation and Energy Metabolism
Z. Boulos

Very few physiological functions lie entirely outside the control of the circadian timing system. This section examines the role of circadian timing as it pertains to two representative functions: thermoregulation and energy metabolism. Circadian rhythms in body temperature are characteristic of most endotherms and of many ectothermic animals as well, while rhythms in energy metabolism occur throughout the animal kingdom. The two functions are closely related: In endotherms, the costs of thermoregulation represent a substantial proportion of an animal's daily energy budget, while in ectotherms, metabolism is strongly dependent on body temperature.

Vertebrate Endotherms: Mammals and Birds

In general, body temperature and metabolism are influenced by an animal's behavior, particularly locomotor activity, feeding, and sleep. These behaviors also display circadian rhythmicity, with daily increases in locomotor activity and feeding and a decrease in sleep coinciding with the times of high body temperature and metabolic rate. Such activities, therefore, often contribute to the daily variations in temperature and metabolism, but they do not account for them completely. Daily rhythms of body temperature have been recorded in fasted human subjects during complete bed rest, and in subjects kept awake for several days (387). Fur-

thermore, human subjects studied under constant conditions often exhibit internal desynchronization, with body temperature and sleep–wake rhythms showing radically different free-running periods (610).

In rats (248) and gerbils (458), equal amounts of activity measured during the day and at night are accompanied by equal increases in O_2 consumption. Activity-induced increases in body temperature vary in different species and experimental conditions (125, 182, 259, 393), but in most cases, the activity-induced changes are superimposed on circadian oscillations in resting levels of O_2 uptake and body temperature.

Body temperature and metabolism are highly correlated, but this correlation changes with time of day. In birds as well as humans, the correlation is higher during the daily rest phase than during the active phase (20).

Range of Circadian Oscillation
Recent compilations of data from a large number of species indicate that, in general, the daily ranges of the circadian rhythms of body temperature and O_2 consumption decrease with increasing body weight (17, 302). In the case of body temperature, this relation was obtained in passerine and nonpasserine birds, and in primates and nonprimate mammals weighing up to 10 kg (17). Interestingly, the temperature recorded during the animals' active phase remains more or less constant within each group across all weight classes and shows a relatively small difference between groups (passerine birds: $42.52 \pm 0.75°C$, nonpasserine birds: $41.04 \pm 1.03°C$, nonprimate mammals: $38.36 \pm 0.42°C$), while the temperature recorded during the inactive phase increases with increasing body weight (17). These results suggest that there is an optimal body temperature, maintained during an animal's active phase, that is

very similar across a wide variety of endotherms of all sizes, and that the relatively greater cost of thermoregulation in small animals, resulting from a greater surface-to-mass ratio, is offset in part by a more substantial lowering of body temperature during the inactive phase.

Stability and Flexibility in the Range of Circadian Oscillations

Homeothermic mammals and birds are able to maintain daily (or seasonal) variations in core temperature within fairly narrow limits (arbitrarily defined as $\pm 2°C$; 48), despite much wider fluctuations in ambient temperature, and most show little or no change in the range of their circadian rhythms of body temperature and O_2 consumption as a function of ambient temperature (17).

Some animals, however, when faced with unfavorable climatic conditions and/or inadequate supplies of food or water, display considerable flexibility in the range of their circadian temperature oscillation. For example, camels (Camelus dromadarius) in the Algerian Sahara in summer show a diurnal rhythm of rectal temperature with a range of about 2°C, but when deprived of water, they show higher daily temperature maxima and lower minima, resulting in a widening of the daily range to 6°C. The increase in heat storage by day allows water conservation by decreasing evaporative water loss, and by reducing the temperature gradient between the animal and its environment, thereby decreasing heat gain (515). An increase in the range of the circadian rhythm of body temperature also occurs in response to food deprivation in several larger bird species that do not normally exhibit daily torpor (112, 207). In this case, the increase in the daily range is the result of a regulated reduction in nighttime temperatures proportional to body weight loss (207).

The widest daily temperature ranges, however, are those of animals showing daily episodes of torpor, during which body temperature can fall by as much as 20°C below euthermic levels. The reduction in metabolic rate that accompanies the daily fall in body temperature results in substantial energy savings (264), thereby increasing an animal's chances for survival during periods of food scarcity or at low ambient temperatures. The distinction between daily torpor and hibernation is not always clear, as many animals display both patterns. In general, however, body temperature is maintained at higher levels in daily heterotherms (minimum levels about 15°C) (264) than in hibernators, and a given reduction in body temperature (e.g., to 20°C) is accompanied by a smaller reduction in metabolic rate (194). The regularity and duration of torpor can vary depending on the species, on ambient temperature, on photoperiod, and on food availability, but it generally occurs during the animal's inactive phase and spontaneous arousal from torpor in LD-entrained animals often occurs at a fixed time of day (588). This suggests an important role for the circadian timing system in the regulation of daily torpor, as does the recent demonstration that lesions of the suprachiasmatic nuclei in the Djungarian hamster, Phodopus sungorus, which abolish circadian activity and temperature rhythms, also prevent the occurrence of daily torpor (494).

Ecological Considerations

Several mammalian species show little or no circadian variation in energy metabolism or body temperature; the absence of circadian rhythmicity in these animals tends to be associated with specific ecological characteristics. Two broad categories can be distinguished, (1) animals whose physical environment (light, temperature, and humidity) is relatively constant and who may therefore derive little benefit from a circadian pattern of activity and metabolism, and (2) animals with very high metabolic rates, or herbivores whose diet is high in bulk and low in energy content, necessitating round-the-clock feeding activity to satisfy daily energy requirements.

A number of species belonging to these categories were included in a recent comparison of resting O_2 consumption rates in 15 rodent and 3 insectivore species, all studied at thermally neutral temperatures in DD (302). Little or no difference between day and night levels was found in the continuously fossorial pocket gophers, *Thomomys talpoides*, and moles, *Scapanus orarius* and *S. townsendii* (constant physical environment), in the voles, *Microtus longicaudis* and *M. townsendii* (herbivorous high bulk diet), and in the masked shrew, *Sorex cinereus* (high metabolic rate).

In a study involving 19 mammalian species (marsupials, Chiroptera, and primates), diurnal rhythms of body temperature were found in all but two species, *Wallabia dorsalis* and *Macropus major* (393). Both are marsupials in the family Macropodidae whose members include many grazers and browsers with ruminant-like bacterial digestion that enables them to survive on a high-bulk, low-energy diet (601).

The shelters used by hibernating animals often exclude diurnal variations in light and temperature, and it has been argued that there is no functional requirement for a daily rhythm of metabolism in hibernating animals (302). No persisting circadian variations of O_2 consumption were found in hibernating dormice, *Eliomys quercinus* (296) and *G. glis* (297), ground squirrels, *Citellus citellus* (296), and marmots, *Marmota marmota* (298), nor of body temperature in hibernating little pocket mice, *P. longimembris* (36). In the golden-mantled ground squirrel, *Spermophilus saturatus*, a crepuscular rhythm

in O_2 consumption was seen during the posthibernation season but not during the hibernation season, despite the fact that both measurements were made at a thermally neutral temperature (302). Daily rhythmicity was also absent in the 13-lined ground squirrel, *Spermophilus tridecemlineatus*, during winter hibernation as well as during the preceding fall, when the animals were still behaviorally active (513). However, free-running circadian rhythms of body temperature have been recorded for several days in DD in hibernating bats, *Myotis lucifugus* and *Eptesicus fuscus* (370, 371). Circadian rhythms of CO_2 production were also seen in another bat, *Myotis myotis*, but only during the first few days of dormancy (452).

Heat Production and Heat Loss
The circadian rhythm of body temperature is necessarily the result of rhythmic changes in the difference between heat production and heat loss. Heat production, generally measured as the rate of resting O_2 consumption, shows a circadian oscillation in phase with the oscillation in deep body temperature. An effective means of heat loss regulation is provided by control over cutaneous blood circulation, particularly in poorly insulated and more exposed parts of the body. Changes in blood flow to and from these areas are reflected in changes in skin temperature, which provide a convenient index of heat loss (626).

Skin temperatures in such areas as the tail and foot in squirrel monkeys (189), the ankle in pig-tailed macaques (552), and the foot in pigeons (205), show prominent daily rhythms with high levels near the end of the animal's active phase, when deep body temperature is decreasing, and low levels near the end of the inactive phase, when body temperature is rising. Skin temperature of the hands and feet of human subjects shows a similar rhythm, but that of the head, thorax, and abdomen shows a much lower amplitude rhythm, in phase with the rhythm of deep body temperature (19, 578). As a result, mean skin temperature exhibits a daily rhythm in phase with that at the extremities, but with a smaller amplitude. Comparable results are obtained when internal conductances (core to skin) are measured directly using heat flow disks (19). These observations indicate that at least part of the daily rise and fall in deep body temperature is due to rhythmic shifts in blood flow to and from the skin of the extremities.

A more complete picture of the role of heat loss in humans is obtained by examining the daily distribution of total conductance (core to environment) and of total (nonevaporative) heat loss. Both measures show daily rhythms with amplitudes similar to that of the heat production rhythm, but with phase delays of 2.2 h for total conductance and 1.4 h for heat loss (Fig. 26). The small phase displacement between the rhythms of heat loss and heat production is sufficient to account for the rhythm in deep body temperature (18, 19). It is estimated that the diurnal heat loss rhythm accounts for 75% of the daily range of the core temperature oscillation, while the heat production rhythm accounts for the remaining 25% (19). It should be kept in mind, however, that these results were obtained at thermal neutrality, and that the relative contributions of the various thermoregulatory mechanisms are likely to differ at different ambient temperatures (19). Most of these studies fail to take into account evaporative heat loss, which also displays daily variations (578).

Responses to Thermal Challenges
Diurnal oscillations are also evident in thermoregulatory responses to thermal stimuli. In humans, immersion of the hand or foot in a cold bath during the

rising phase of the body temperature rhythm leads to an earlier onset of shivering and longer rewarming times than during the falling phase (252). Body immersion in 30°C water leads to a decrease in deep body temperature in the evening, but is followed by a small increase in the morning (77), indicating greater cold defense efficiency during the rising phase of the body temperature rhythm. Conversely, sweating and peripheral vasodilation in response to heat loads have lower thresholds and latencies during the descending phase of the temperature rhythm than during the ascending phase (252, 544, 606). These changes in thermoregulatory responses are accompanied by rhythmic changes in the subjective sensations evoked by cold and warm stimuli (77).

Similar results are seen in animals. Pigeons show diurnal changes in shivering and panting thresholds in response to spinal cooling and warming, respectively (286). Rats subjected to intraperitoneal heat loads show a lower threshold for tail skin vasodilation during the day than at night, as well as steeper curves relating evaporative heat loss, nonevaporative heat loss, and thermal conductance, to hypothalamic temperature (522).

There is some evidence of impaired capacity for thermoregulation in squirrel monkeys, *Saimiri sciureus*, kept in bright LL (188). A 6-h drop in ambient temperature from 28 to 20°C caused a significant decrease in body temperature in LL, but had little effect in LD-entrained animals (Fig. 27). The impairment may have been due to altered phase relations between rhythmic components of the thermoregulatory system, particularly heat production and heat loss.

Taken together, these data indicate that the circadian rhythm of body temperature is a consequence of rhythmic changes in one or more central reference temperatures, or set points (18), and pos-

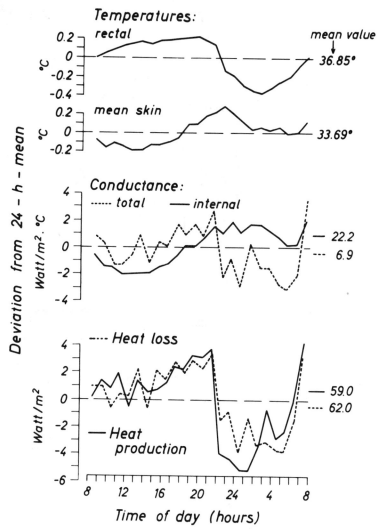

Figure 26. Circadian rhythms of rectal and mean skin temperature (*top*), internal and total conductance (*middle*), and heat production and heat loss (*bottom*), averaged from eight sites of the body surface. Means from eight female subjects recorded twice at an ambient temperature of 28°C (18).

sibly in the sensitivity or efficiency of thermoregulatory effector mechanisms as well.

Vertebrate Ectotherms: Reptiles, Amphibians, and Fishes

Although body temperature in ectotherms is largely dependent on heat transfer from the environment, many

such animals actively regulate their body temperatures, often with a high degree of precision. Thermoregulation is achieved primarily by behavioral means, the most important being the selection of favorable microclimates, but many ectotherms also use physiological mechanisms for controlling the rates of heat gain and heat loss.

Temperate zone ectotherms are often

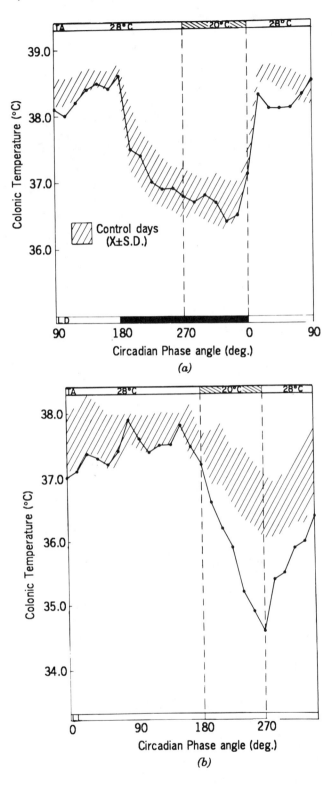

Figure 27. Effects of 6-h cold exposure on colonic temperature in squirrel monkeys during entrainment to LD 12:12 cycles (*a*) and under LL (*b*). Ambient temperature on the day of cold exposure is indicated at the top of each graph. The shading represents the mean ± standard deviation of the previous three cycles at an ambient temperature of 28°C (188).

exposed to wide fluctuations in ambient temperature on a daily (and seasonal) basis, and may therefore be expected to show parallel diurnal variations in body temperature. However, the daily rhythm is rarely a passive reflection of macroclimatic cycles, and many ectotherms exhibit diurnal changes in body temperature in the absence of a daily cycle in ambient temperature.

Reptiles
Behavioral activity in reptiles and other ectotherms is generally confined to a species-specific range of body temperatures. Data from 161 species, summarized by Brattström (64), indicate that activity temperature ranges are related to whether the animal is primarily diurnal or nocturnal, terrestrial or aquatic, and to the extent to which it is heliothermic, obtaining heat from solar radiation, or thigmothermic, obtaining heat by conductance from the substrate.

The means of achieving and maintaining activity temperatures differ. A diurnal, heliothermic, terrestrial reptile will raise its temperature in the morning by basking, orienting its body to maximize heat gain from solar radiation. Both diurnal and nocturnal animals may also flatten their bodies against the substrate to obtain heat by conduction, while semiaquatic reptiles can take advantage of temperature differentials between air and water to raise body temperature. Once activity temperature is reached, the animal can engage in foraging and other behaviors while maintaining its temperature within the activity range by shuttling between sun and shade, and by changes in posture and orientation. With a continued rise in ambient temperature, some reptiles increase the rate of evaporative cooling by gaping or panting. If temperatures rise beyond a certain level, most reptiles retreat to the cooler temperatures of their burrows or shelters,

emerging again later in the day when environmental temperatures have dropped to more suitable levels.

Some reptiles also make use of physiological mechanisms that allow faster warming than cooling (626). In *Lacerta viridis*, this difference is consistently observed only during the day, not at night (481). In the larger reptiles, including the Galapagos tortoise (342) and some crocodilians (527), a low surface-to-mass ratio and high thermal inertia, in combination with circulatory changes, enable the animals to maintain a relatively constant body temperature, with dirunal variations of only a few degrees despite much wide fluctuations in ambient temperature.

An example of the daily course of body temperature and of thermoregulatory behavior in an unstrained lace monitor, *Varanus varius*, is shown in Fig. 28. The data illustrate the stability of daytime body temperature, achieved in spite of substantial fluctuations in air temperature (539).

Several lines of evidence point to the involvement of circadian mechanisms in reptilian thermoregulation. First, many, though not all species studied in thermal gradients show daily rhythms in temperature preference in the absence of daily cycles in ambient temperature (41, 267). In *Sceloporus occidentalis* (105) and *Klauberina riversiana* (469), temperature preference rhythms free-run for several days under constant lighting conditions. In *L. viridis*, the rhythm in selected body temperature shows photoperiod-dependent seasonal changes in amplitude. Body temperature is maintained at a high level with only small day–night differences in summer; the amplitude of the rhythm increases throughout the fall, but the rhythm is abolished during winter dormancy (482). A number of physiological and behavioral processes in reptiles and other ectotherms are optimized at or near

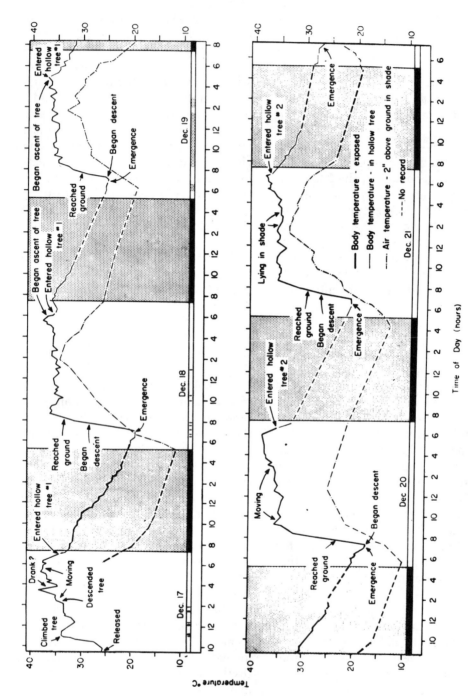

Figure 28. Telemetered abdominal temperature and behavioral activities of a lace monitor, *Varanus varius*, in its natural habitat over five consecutive days. Stippling in the horizontal bar below the dates 17–19 December represent times of light cloud cover (539).

the preferred temperatures, especially the higher levels selected at certain times of day in animals displaying daily rhythmicity (66, 124). Digestion is particularly sensitive to temperature, and reptiles often seek a warm environment after feeding (468). Concurrent telemetric recordings of gastrointestinal contractions and body temperature in both carnivorous (*Varanus flavescens*) and herbivorous (*Ctenosaura pectinata*) reptiles show parallel diurnal rhythms in these two variables (343).

Second, daily rhythms are found in critical thermal maxima in the Eastern painted turtle, *Chrysemys picta,* under LD (312), and in critical thermal minima in the lizard, *Lacerta sicula,* in LD as well as LL (536). A circadian rhythm in panting threshold was also recorded in *Amphibolurus muricatus* in LL (93).

Finally, the circadian system is involved in the timing of daily emergence from nocturnal burrows, as shown, for example, in horned lizards (244). When exposed to daily cycles of access to radiant heat and light, the animals had regular emergence times that anticipated heat and light onset. Furthermore, emergence times were the same under environmental temperatures (and consequently body temperatures) of 18 and 27°C, both of which are substantially lower than normal activity temperatures (34–38°C). Regular daily emergence times can also be seen in the data from the lace monitor in Fig. 28. Over the 5-day recording interval, emergence times varied by no more than 25 min, despite body temperature at these times ranging from 17.8 to 27.0°C (539). A circadian rhythm of emergence allows ectothermic reptiles to use safe nocturnal shelters without loss of valuable activity time in the morning (244).

Several crocodilian species show regular daily movements into the water in the evening and onto land in the morning (527). Since air temperature is generally higher than water temperature by day and lower by night, these movements enable the animals to maintain higher body temperatures than they would otherwise. However, American alligators, *Alligator mississippiensis,* move onto land at dawn when air temperature is still lower than water temperature (and when the effects of solar radiation are still negligible), and into the water at dusk when air temperature is still higher (324). These and other results strongly suggest that the daily movements reflect an endogenous circadian rhythm, rather than being direct responses to temperature.

Oxygen consumption is characterized by a daily rhythm in many diurnal lizards (106, 274, 352, 483, 531, 615) and in snakes, both terrestrial (251) and aquatic (209). The rhythm is often endogenously generated, but there are exceptions. In a comparison of three *Lacerta* species, Cragg (106) found that the rhythm persisted in LL and/or DD in *L. sicula* and *L. viridis* but not in *L. vivipara.* The rhythm shown by the latter species thus appears to be entirely exogenous. One of the five species of xantusiid lizards studied by Mautz (352), the cave-dwelling *Lepidophyma smithii,* showed little or no daily rhythmicity, despite the fact that the lizards were under LD. In its natural cave environment, this troglophile appears to be equally active by day and night.

It is clear that O_2 consumption patterns reflect daily changes in motor activity. It remains possible, however, that in some reptiles, there is also a circadian rhythm in resting metabolic rate, which contributes to the daily rhythm observed in active animals.

Amphibians

Amphibians differ from reptiles in their thermal relations, mostly because of differences in moisture requirements. Unlike reptiles, amphibians have highly permeable skins through which they can both take up and lose water. While this allows terrestrial species to make efficient use of evaporative cooling as a thermoregulatory mechanism, it also makes them more vulnerable to water loss. In addition, the skin is used for respiration and must remain moist for this to proceed effectively (65). Thus, in comparison with reptiles, amphibians tend to be more nocturnal (although even nocturnal forms often engage in daytime basking), to have wider activity temperature ranges with lower minima, and to display less temperature preference.

As seen in Fig. 29, the daily activity pattern of the toad, *Bufo boreas*, changes drastically with the seasons. Diurnal rhythms of body temperature recorded in early March, late April, and mid-June showed considerable differences in daily range (5, 20, and 10°C, respectively). Body temperatures were always highest during the day, but this was achieved by different means at different times of the year: In March and April this was achieved by basking during the day and in June by emerging only at night when air temperature was lower than burrow temperature. When studied in thermal gradients, however, *B. boreas* showed little evidence of daily rhythmicity in temperature preference (528). Other terrestrial amphibians (96, 331) also show little or no diurnal change in temperature preference, but the strictly aquatic mudpuppy, *Necturus maculosus*, shows a clear rhythm with nocturnal temperatures exceeding diurnal values by up to 6°C (268). In *Rana pipiens* tadpoles the diurnal temperature preference pattern is bimodal with one peak in the first half of the nightly dark phase and one near the middle of the light (87). Aquatic amphibians are not faced with the same problems of water regulation that terrestrial forms have to contend with, and may therefore have greater latitude in selecting physiologically optimal temperatures (268).

Daily rhythms in thermal tolerance

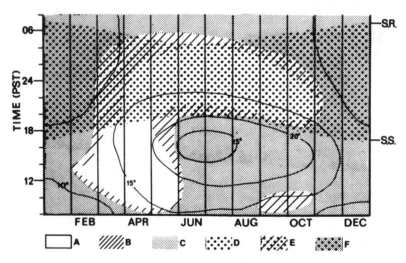

Figure 29. Seasonal changes in activity pattern and microhabitat selection of 14 toads, *Bufo boreas*, observed in an outdoor enclosure, and their relations to air temperature and photoperiod. SR and SS indicate sunrise and sunset, respectively. A, Toads completely emerged during daylight; B, toads at burrow entrance or shuttling in and out of the burrow during daylight; C, toads in burrow during daylight; D, E, F as in A, B, C above, respectively, but during the night (528).

(critical thermal maxima) are consistently found in several amphibians, both in LD and DD (267). Day–night differences, however, are only in the order of 1–2°C, and it is not clear whether these rhythms have any functional value or whether they are merely a by-product of rhythmic changes in other processes.

The absence of consistent daily activity patterns in many amphibians appears to be reflected in their metabolic rates. Even in LD, daily rhythms of O_2 consumption in *Rana temporaria* (287), *R. pipiens* (211) and *R. esculenta* (297a) show considerable variability and fail to appear under certain acclimation conditions. In the last species, as well as in *Bufo marinus* (266), the rhythms are primarily, if not entirely, exogenous, as they are not consistently found in constant conditions. The daily O_2 consumption pattern of the salamander, *Notophthalmus* (*Triturus*) *viridescens*, was also found to be very variable in dim LL, and a daily rhythm could only be demonstrated statistically, after averaging several days of data (71).

Fishes
The daily distribution of temperature preference and of thermal tolerance has been studied in a number of fish species in LD (267, 294, 479). In species that exhibit daily rhythmicity, higher temperatures usually coincide with the animals' active phase, but temperature selection is not always strictly correlated with locomotor activity. Goldfish, *Carassius auratus*, showed a bimodal activity pattern, with peaks in early and late portions of the daily dark segment, while temperature preference had a unimodal rhythm with maximum levels near the end of darkness. The bowfin, *Amia calva*, also displayed a diurnal rhythm in temperature preference, but its locomotor activity was arhythmic (479).

Thermoregulation in fishes can be achieved by taking advantage of the ther-mal stratification that often exists in water bodies and of temperature differentials between shallow and deeper waters. The role of thermoregulation in the physiology and ecology of lake-dwelling sockeye salmon, *Oncorhynchus nerka*, was examined in some detail by Brett (66). Physiological studies indicated that several processes, including metabolism, growth, food intake, blood circulation and performance, were optimal at temperatures in the region of 15°C. This was also the preferred temperature of the salmon, as determined in temperature selection experiments. Brett next examined the diurnal patterns of behavior exhibited by *O. nerka* in lake Babine. Results obtained in midsummer (Fig. 30) show that the fishes spend the day at a depth of 30–45 m (5°C), rising at dusk to feed near the surface (17°C). The middle of the night is often spent at depths of 10–12 m, but the fishes rise back to the surface to feed at dawn, before descending again to the daytime depths.

By spending the day near the bottom of the lake, the salmon can avoid predators and at the same time conserve energy as a result of the low temperatures found at that depth. The latter may be the more significant advantage, as the regular vertical migration stops in the fall at the time of thermal stratification turnover. By rising to the surface at night, the salmon are able to feed at near optimal temperatures. Since most of the zooplankton on which the salmon feed, remain concentrated near the surface (only two of the eight major species display diurnal vertical migration, and do so in opposite directions), this suggests again that the primary function of the vertical migration rhythm is thermoregulatory.

Daily movements of a somewhat different nature are seen in desert pupfish, *Cyprinodon macularius*, in shore pools of the Salton Sea, a saline lake in California (34). These fishes spend the night in the

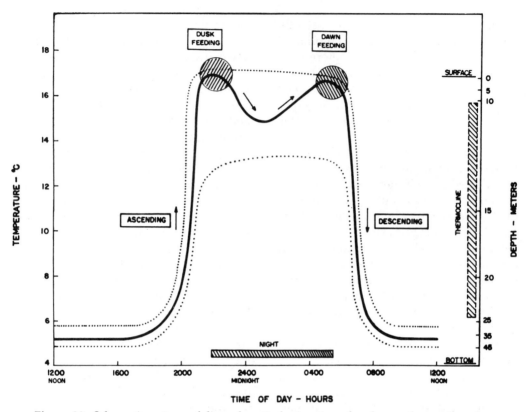

Figure 30. Schematic pattern of diurnal vertical migration of sockeye salmon, *Oncorhynchus nerka*, in midsummer with corresponding water temperatures. Feeding near the surface takes place over a 1–2 h period at dusk and dawn. Dotted lines indicate the general limits of thermal experience (66).

shallowest, and therefore the coolest, areas of the pools. At dawn, they move to the deepest and warmest areas, then into progressively shallower waters as these warm up during the day. If the temperature in the shallow areas rises above 36.5°C, as it does in midafternoon in summer, they move to deeper waters, returning to the shallows in the early evening. In winter, the morning movements to shallow waters are delayed by several hours, and, since water temperature stays well below the 36.5°C maximum, the fishes do not withdraw to deeper waters in midafternoon. In both seasons, the result of these movements is a diurnal rhythm in water (and body) temperature, with a daily range of over 10°C.

Oxygen consumption exhibits daily rhythmicity in several marine and freshwater teleost fishes and in some elasmobranchs (139, 345, 506). In a few species, the rhythms were found to persist under constant light or darkness (492). Clausen (95) found daily rhythms in five freshwater species in 24-h recordings under constant conditions, but three additional species failed to show consistent rhythmicity. The former species all inhabit relatively quiet waters, while the latter are found in rapidly flowing waters, which may require active swimming both day and night.

Invertebrates

Many invertebrates, particularly among arthropods, actively regulate body temperature using a remarkable diversity of behavioral and, in some cases, physiological mechanisms (354). As in vertebrate ectotherms, behavioral means of thermoregulation include the selection of suitable microclimates, changes in posture and orientation, and modification of diurnal activity patterns. Honeybees and other social insects also use behavioral (and physiological) means to control nest temperature.

The temperature preferences of numerous invertebrates have been studied in thermal gradients, but few authors have looked for possible diurnal variations. A daily rhythm was seen in the nocturnally active crayfish, *Orconectes immunis*, with mean night-time temperatures exceeding daytime levels by about 4°C (107), but not in the lobster, *Homarus americanus* (477). In the pink shrimp, *Penaeus duorarum*, another nocturnal species, temperatures selected at night were only slightly higher than daytime levels, but they were maintained within considerably narrower limits, indicating greater thermoregulatory precision during the active phase (478). Remmert (470) examined the diurnal distribution of locomotor activity and of temperature selection in two insects, the diurnally active *Cicindela campestris* and the nocturnal *Gryllus domesticus*. Both showed daily activity rhythms, but only the former species showed a rhythm in temperature selection. A daily rhythm associated with brood development was observed in *Camponotus mus* (457). This ant transports its brood to progressively warmer parts of a thermal gradient during the day, and back to cooler parts at night. There is evidence that in some ant species, exposure to daily temperature cycles is actually necessary for brood development in the prewinter stage, and beneficial in other stages (76).

Insects from several orders can raise body temperature by physiological means, primarily by activation of flight muscles resulting in shivering or low-am-

plitude wing vibrations (626). Many use endothermic warm-up to attain the relatively high wing muscle temperatures that are often necessary for flight, particularly in larger insects, and for stridulation in some Orthoptera. In addition, some insects will maintain thoracic temperature at a high and relatively constant level for extended durations. The ability to reach and maintain flight temperature by endothermic means enables insects to engage in such activities as foraging and reproductive behavior at times of the day (or year) when solar radiation and ambient temperature (T_a) are too low. Thus, male rain beetles (*Pleocoma*), which normally engage in mate-searching behavior at night in winter, maintain a high thoracic temperature, in some cases for over 4 h, while walking in a respirometer chamber (388).

Several nocturnal moths also show endothermic warm-up and maintain a relatively constant thoracic temperature while in flight. When studied in the laboratory, male American silkmoths, *Hyalophora cecropia*, show distinctive nocturnal-crepuscular patterns of warm-up and flight activity. These patterns are unaffected by changes in T_a, and there is evidence that the daily activity rhythm, at least in males, may be endogenous. These moths do not feed as adults; their daily activity patterns probably reflect the timing of reproductive behavior in their natural environment (229).

The capacity for endothermic warming allows some diurnal insects, including honeybees, bumblebees, wasps, and some of their syrphid fly mimics, to forage at dawn or on overcast or rainy days when T_a values are low (354, 389). In other diurnal species, the ability to maintain an elevated body temperature while in flight increases with body size, due to the more rapid heat loss incurred by smaller animals. This allows males of the larger dragonfly species of the genus *Micrathyria* (353), and larger male hoverflies,

Syrphus ribesii (197), to arrive at breeding congregations earlier in the morning, thereby gaining a competitive advantage over their smaller congeners or conspecifics.

Circadian rhythms of thermoregulation and energy metabolism have been studied extensively in the honeybee, *Apis mellifera*. Honeybee workers maintained in isolation or in small groups show prominent circadian rhythms of O_2 uptake in DD and at constant T_a values (Fig. 31) (250, 547). Oxygen consumption is low and stable during subjective night, and it increases with increasing T_a values with a $Q_{10} = 2.2$ (250). In contrast, O_2 consumption during subjective day is high and more variable, and it increases with decreasing T_a, indicating active thermoregulation (79, 250). Indeed, thoracic temperature in workers is maintained between 34 and 36°C during the day at T_a values of 15–40°C (79). Thus, the honeybee worker behaves like a typical daily heterotherm, maintaining body temperature (T_b) at a high and relatively constant level at one phase of its circadian cycle, and allowing it to drop to ambient levels at the opposite phase. The high T_b values achieved by day are accompanied by, and are apparently the result of, an increase in locomotor activity (549).

When they are part of a colony and have access to a nest or hive, however, honeybees adopt a different thermoregulatory strategy, and the colony as a whole behaves more like a typical homeotherm. At low T_a, the bees form clusters that expand and contract in response to increases and decreases in T_a. Heat is

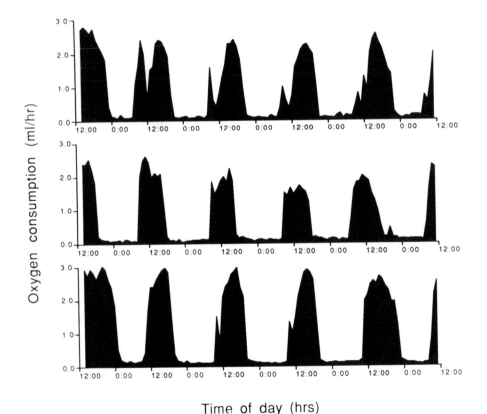

Time of day (hrs)

Figure 31. Oxygen consumption of three isolated honeybees recorded in constant darkness (redrawn from 547).

generated by the bees that occupy the central core of the cluster, and is transmitted to the bees that form the insulating peripheral shell (525). Long-term recordings of cluster temperature in winter show that this is kept above 15°C despite outside temperatures as low as −10°C (253). Concurrent temperature recordings within the core and shell of the cluster indicate that heat production by the core increases when shell temperature goes down. As a result, core and shell temperatures may show opposite daily patterns, with high core and low shell temperatures at night, and low core and high shell temperatures by day (532). Nest temperatures are also controlled, with varying degrees of precision, in other social insects (253).

Circadian rhythms of O_2 uptake have been recorded in DD in several other insect species, both diurnal and nocturnal (38, 92, 473, 548). Daily rhythms in O_2 consumption have also been recorded from whole colonies of the termite, *Cubitermes exiguus*, maintained inside their nests (245). The rhythms were still evident when the nest was broken into small fragments, but only if the colony had access to at least one fragment. A piece of the nest apparently acts as a stimulus for synchronized activity by the termite population. Other invertebrates showing diurnal or circadian rhythmicity include earthworms, *Lumbricus terrestris* (459), crayfish, *Orconectes clypeatus* (179), amphipods, *Pontoporeia* spp. (88), and seapens, *Cavernularia obesa* (390). Some marine crustaceans (89) and molluscs (507) show both diurnal and tidal rhythmicity.

Neural Control of Circadian Rhythmicity

B. Rusak

This section reviews information on the neural systems that function as the pacemakers and entrainment routes in several well-studied species.

Lesion Studies

The putative circadian pacemaker in mammals was initially identified in studies of a previously controversial ganglion cell projection from the retina directly to the suprachiasmatic nucleus (SCN) of the anterior hypothalamus (the retinohypothalamic tract; RHT; 279, 383; Fig. 32). Ablation of the SCN in rodents led to the loss of circadian rhythms of endocrine, behavioral and other physiological functions, and disruption of photic entrainment of these functions (365, 495, 498, 502; Fig. 33). These results led to the hypothesis that the SCN serves as the pacemaker of the mammalian circadian system and the RHT as its photic entrainment route. Ablation studies are, however, inherently ambiguous for several reasons. First, the loss of rhythmicity may result from damage to systems other than the pacemaker; for example, a mechanism coupling the pacemaker to motor systems. Second, lesion effects may be situation dependent, and surviving functional capacities may only be revealed under a limited set of conditions (see the section on effects of temperature and other physical factors). Finally, rhythmicity might fall below statistical detection threshold even when the pacemaker has not been destroyed. Several statistical techniques for assessing rhythmicity are used, but each is subject to criticism, and none can be considered universally appropriate for the many different types of rhythms that have been studied.

The SCN as Oscillator

The first requirement for a putative circadian pacemaker is that it be capable of independently generating a circadian rhythm; that is, that it be a self-sustaining oscillator. A second is that its ouptut regulate the expression of one or more overt circadian rhythms. Several lines of evidence provide convincing evidence that the SCN satisfies both of these criteria.

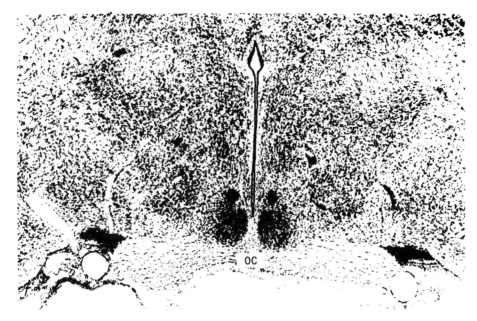

Figure 32. Frontal section through the anterior hypothalamus of a rat showing the location of the SCN (arrows) with their characteristic closely compacted, small cells lying just above the optic chiasm (OC). Cresyl violet stain, ×64 (383a).

Figure 33. Running wheel activity record of a hamster which received an SCN lesion on the day indicated by horizontal line. The LD cycle indicated above the record was maintained until the day marked ΔΦ, at which time it was phase delayed by 4 h. The animal was subsequently kept in constant dim illumination. The SCN ablation severely disrupted normal photic entrainment and prevented the expression of a free-running circadian rhythm in constant conditions (495).

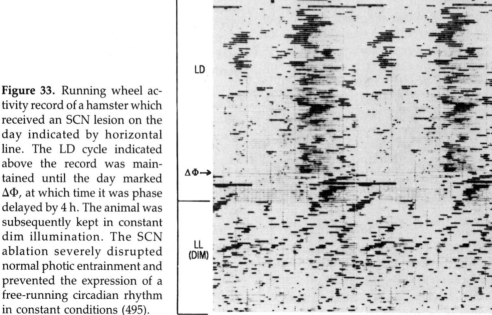

The uptake of ^{14}C-2-deoxyglucose (2-DG), widely used as a marker of neuronal metabolic activity, exhibits a daily rhythm in the SCN, with highest levels during the light phase in LD cycles, and during the subjective day in constant conditions (516). A few other structures show rhythms of glucose utilization that are somewhat different from the SCN rhythms (491). The conclusion that SCN neurons are most active during the subjective day is confirmed by evidence from *in vivo* recordings of multiple-unit neural activity in the SCN (268a). These rhythms persist in an island formed by a knife cut around the SCN, ruling out the need for direct neural input from outside the hypothalamus to sustain SCN rhythmicity. The daytime peaks of 2-DG uptake and of neural activity are observed in both nocturnal and diurnal species (475, 509). It appears that species differences in the timing of activity do not reflect oppositely phased oscillations in their pacemakers, but must reflect a difference in the coupling of effector systems to the pacemaker rhythm.

Convincing evidence for the intrinsic origin of rhythmicity in the SCN came from studies of neural activity in hypothalamic tissue maintained in constant conditions *in vitro*. The SCN generates a circadian rhythm of increasing and decreasing neural activity in slice preparations from both rats and hamsters (365), and the rhythm has been reported to persist for at least three cycles (Fig. 34) (456). The highest level of activity is found during the subjective day as measured by both neural firing rates and 2-DG uptake in the SCN slice (410). These studies together demonstrate convincingly that the SCN functions as a self-sustained oscillator.

There is little information on the cellular mechanisms involved in rhythm generation in the SCN. Recent evidence indicates that the mRNA for vasopressin is expressed rhythmically in the SCN (591). Protein synthesis in the SCN, as measured by lysine incorporation, shows a decline late in the dark phase (596), but there are no differences in leucine incorporation at single time points sampled during L and D (199).

The Entrainment Route

The autoradiographic evidence for a retinal projection to the hypothalamus of rats has been confirmed using a variety of other techniques (433), and in many mammalian and avian species (85, 102, 383), but the relation of the hypothalamic retinorecipient area in birds to the mammalian SCN remains uncertain (415). Studies using choleratoxin-conjugated HRP as a sensitive tracer in rodents have indicated a broad retinorecipient area in the anterior hypothalamus, which is centered on the SCN, but includes nearby structures (279).

In rats and hamsters, the direct retinal projection to the SCN is overlapped by a projection (the geniculohypothalamic tract or GHT) from a region of the lateral geniculate nuclei (LGN) that is itself retinorecipient. This area includes the intergeniculate leaflet and portions of the ventral LGN that contain cells immunoreactive for neuropeptide Y (NPY) (81, 234). This NPY-containing projection to the SCN has also been implicated in rhythm entrainment. NPY injections into the SCN can phase shift circadian activity rhythms (5). Activation of GHT neurons by electrical stimulation (504) or injection of the excitotoxic amino acid N-methyl aspartate into the geniculate area (280) also cause phase shifts of activity rhythms. Destruction of the GHT in hamsters causes reduced sensitivity to the phase-advancing effects of photic stimuli (231, 435), and alters the responses of hamster rhythms to constant illumination (231, 232, 435).

Cholinergic mechanisms have been

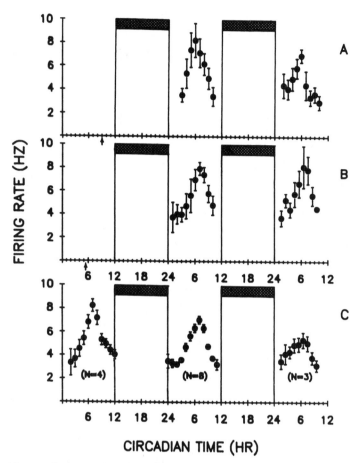

Figure 34. Neuronal electrical activity rhythm continues for multiple cycles *in vitro*. A and B, plotted are the 2-h means ± SEM of all single units recorded on days 2 and 3 in single experiments. The running averages shown (using 1-h lags) were chosen on the basis that they illustrate the most symmetrical pattern of activity around the peak. In the experiment shown in (*A*) electrical activity peaked at CT 6 on both days 2 and 3. In (*B*) the rhythm peaked at CT 7 on day 2 and CT 6.5 on day 3. (*C*) Summed data from all control experiments, showing the stability of the electrical activity rhythm during the first 3 d *in vitro*. N = the number of experiments contributing to the data for each day; *horizontal bars* indicate subjective night for the SCN, the time of lights-off in the donor colony; and *arrows* indicate the time of slice preparation (456).

suggested to play a role in the entrainment route (365, 499). The pharmacological effects of cholinergic agents applied to the SCN are somewhat inconsistent, perhaps because of technical differences among studies (313, 519). Blockade of cholinergic transmission has failed to prevent retinal input to the SCN (78, 520),

suggesting that acetylcholine is not the transmitter in the RHT.

Despite the fact that effects of cholinergics on rhythms appear to involve a nicotinic receptor, no nicotine or acetylcholine-binding sites have been identified in the rat SCN (94). The only cholinergic receptor type found in the rat

SCN is that which binds α-bungarotoxin (BTX; 380). It is unclear which endogenous ligand in the CNS affects this receptor, or whether it can be considered a "nicotinic"-type receptor. The BTX binding in the SCN of female rodents is greatly reduced by ovariectomy (380), which also reduces neurophysiological responses in the SCN to cholinergic agents (313). Yet, ovariectomy does not prevent phase shifts to intraventricular carbachol injections, so the BTX binding sites do not appear to be necessary to mediate phase-shifting effects of carbachol (381). It remains to be determined where and on what receptor-type cholinergic agents act to influence circadian rhythms (365, 499).

Other studies have implicated an excitatory amino acid in the entrainment pathway (78, 333), and glutamate injected into the SCN causes phase-dependent phase shifts in hamsters (368). The phase-dependence of these shifts does not resemble that of light pulses, so this finding is ambiguous with respect to the role of glutamate in entrainment. The ability of drugs acting on the γ-aminobutyrate (GABA)–benzodiazepine receptor complex to block the effects of light pulses on rhythms and to generate phase shifts in darkness suggests a role for GABA-ergic SCN neurons in entrainment (461). Drugs acting on both $GABA_A$ and $GABA_B$ receptors have been found effective, so the pharmacological mechanisms involved remain uncertain. It has been suggested that these mechanisms act by modulating the release or the postsynaptic effectiveness of the transmitter in the RHT (461). Whatever the transmitter(s) involved in entrainment, the process appears to depend on classical Na^+-dependent action potentials. Treatment of the SCN with tetrodotoxin (TTX) prevents the generation of such action potentials, and also blocks sensitivity of the circadian system to shifts in environmental lighting that occur during the TTX infusion (517).

The SCN cells in cats, rats, hamsters, and squirrels share a characteristic pattern of responsiveness to retinal illumination, with modest variations among species (366, 369). Similar responses are recorded from geniculate cells that give rise to the GHT in rats and hamsters (233). These cells show tonic changes in firing rates in response to tonic changes in illumination intensity. The thresholds for responsiveness to light are quite high and in hamsters fall roughly in the range of natural twilight intensities. They track environmental illumination faithfully, and cells activated by light intensity increases may maintain elevated firing rates as long as the light remains on (at least up to 60 min) (Fig. 35) (233). This observation is consistent with the finding that the circadian system can integrate illumination intensity over very long intervals in determining the amplitude of phase shift generated (568).

In the nocturnal rodents tested, about 75% of cells that are photically responsive are activated by increasing light intensities, but in a diurnal squirrel (*Spermophilus tridecemlineatus*) slightly more of the photically responsive cells were suppressed than activated by increasing light intensity (369). Given the few species studied and the small numbers of cells tested, it is uncertain whether this apparent difference between diurnal and nocturnal species will prove to be general. It also remains to be established how such differences can be related to behavioral differences in the responses of nocturnal and diurnal species to light (16).

The SCN as Pacemaker

One approach to assessing the SCN pacemaker hypothesis is to manipulate the activity of the SCN and determine the consequences for overt rhythms. Electrical stimulation in the SCN, but not outside it, produced phase-dependent shifts

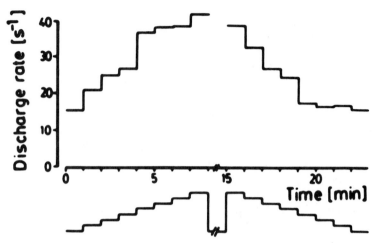

Figure 35. Firing rate record of an SCN cell from a Syrian hamster exposed to "staircase" increases and decreases in ocular illumination in the pattern indicated below the figure. Firing rates tracked changes in the log light intensity in a near-linear fashion (366).

of rhythms of activity in hamsters and of feeding in rats; the phase-dependent effects were similar to those of environmental light pulses, suggesting that photic input might act as electrical stimulation does, by altering the neural activity of the SCN (500).

Another line of evidence supporting a pacemaker function for the SCN is the restoration of circadian rhythmicity when fetal hypothalamic tissue including the SCN is transplanted to the brains of SCN-ablated rats and hamsters (Fig. 36) (328, 512). It is unclear to what extent specific subtypes of SCN neurons must survive in order to restore rhythmicity, and whether and to what degree neurons from the implant must innervate the host tissue. These studies have served, however, to reinforce the idea that the SCN serves as a pacemaker for the system. A finding that is critical to supporting this hypothesis is that the implant conveys the period of the donor's SCN to the recipient nervous system (462, 530). This result implies that the implant is actually regulating the host circadian system and

not merely allowing the expression of host rhythmicity.

The intrinsic mechanism of the SCN pacemaker remains a mystery. The SCN of rats contain about 24,000 neurons, with a rich variety of neurotransmitters, neuropeptides and synaptic relations among neuronal types (212, 597). The major efferent pathways include VIP/PHI immunoreactive neurons and vasopressinergic neurons (603), but the roles of each of these classes of cells is uncertain.

Since TTX infusions into the SCN disrupt overt rhythmicity, Na^+-dependent action potentials appear to be necessary to convey rhythmic information from the SCN to effectors (517). But the behavioral evidence indicates that the SCN pacemaker runs undisturbed through the treatment, so its intrinsic timekeeping does not depend on a TTX-sensitive process (517).

The SCN cells show rhythms in their responsiveness to external inputs as well as in their spontaneous activity. Responsiveness of SCN neurons to serotonin (348), neuropeptide Y (350, 518a) and me-

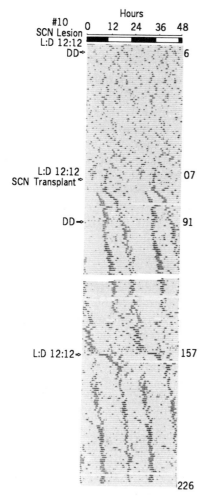

Figure 36. Double-plotted locomotor activity records of a male SCN-lesioned hamster in which fetal SCN grafts restored free-running rhythms. The animal showed a "split" pattern of locomotor activity, following the transplant and did not become entrained to the imposed lighting cycle (328).

latonin (86, 349, 501), varies with circadian phase, and the uptake of serotonin by the SCN (377) depends on time of day. Rhythms of electrical activity or vasopressin secretion recorded *in vitro* from the SCN can be phase-shifted depending on the time of application of pulses of

high potassium or cyclic AMP (143, 198, 456).

The neurophysiology of the SCN has been reviewed recently (365). Several patterns of neural firing have been described but it is uncertain whether these represent different cell types or different states of one cell type (577; see 365). Intracellular recordings show that SCN neurons have a resting membrane potential of about -60 mV. Spontaneous or current-evoked action potentials are followed by after-hyperpolarizations, while synaptically evoked action potentials are often followed by "fast potentials"—small depolarizations that occasionally lead to a second action potential (611). An extracellular recording study had reported that SCN cells that project to the periventricular zone generate a strong recurrent inhibition after being antidromically activated (313).

The SCN appears to function as part of a complex hierarchy of oscillators that interact with each other. The SCN may entrain lower order oscillators elsewhere in the nervous system (498), or may gate the overt expression of such oscillators (122). The existence of self-sustaining circadian oscillators outside the SCN is suggested by the patterns of activity generated after SCN ablation in hamsters (495). The clearest evidence is the presence of a circadian mechanism sensitive to entrainment by temporally restricted food availability after SCN ablation in rats (60, 543; see the section on the effects of biotic factors), and the existence of a rhythm of photic sensitivity after SCN ablation (575). Recent reports have described a restoration of clear circadian rhythms in SCN-ablated rats treated chronically with methamphetamine and kept in either LD cycles or constant conditions (260). The restored rhythms cannot be entrained by photic cues; they could represent the output of the food-entrainable oscillator, since methamphetamine is an anorexigenic drug and treated rats would be effectively chronically food deprived.

Birds

Entrainment
As with mammals, the physiological analysis of the avian circadian system began with ablation studies aimed at identifying the photic entrainment route. Ablation of structures with an obvious photoreceptive function, the eyes and the pineal gland, in house sparrows (*Passer domesticus*) and chickens (*Gallus*) did not prevent photic entrainment of activity rhythms (193, 341, 372). Subsequent studies demonstrated that an encephalic photoreceptor mediates these photic effects in blinded, pinealectomized birds (373). This finding is consistent with earlier demonstrations that measurement of seasonal changes in daylength (which typically involves the circadian system) is mediated by an encephalic photoreceptor (42a).

A variety of attempts to identify the sites at which light exerts direct effects on the brain in birds have used functional criteria such as entrainment or reproductive responses to focal light exposure in the brain. These studies have suggested hypothalamic and other diencephalic sites of action, but they have always suffered from an inability to control the extent of light spread within the brain. A recent study used a monoclonal antibody to a photosensitive protein (opsin) to identify brain sites with the capacity to respond directly to light. Cells immunoreactive to the antibody were identified in the hypothalamus and in the septum (523). These areas are, therefore, good candidate targets for the entrainment effects of light mediated by direct brain photoreception.

The sufficiency of encephalic photoreceptors for entrainment does not imply that other photoreceptors do not affect the circadian system. The threshold for entrainment has been reported to be lower in sighted than in blinded house sparrows, and blinded sparrows maintain activity rhythms during exposure to an intensity of LL that renders sighted birds arhythmic (372). The phase of entrainment to a short photoperiod is different in intact and pinealectomized birds (322). It remains unclear to what degree the inputs from retinal, pineal, and encephalic photoreceptors serve separate or overlapping functions in mediating the various effects of light on circadian systems.

The Pineal Gland
Although pinealectomy did not prevent entrainment to lighting cycles in sparrows, it did lead to a loss of circadian activity rhythms in constant darkness (193) (Fig. 37). The pattern of rhythm loss after environmental entraining cues were removed suggested the existence of oscillators outside the pineal gland that became dissociated after pinealectomy. This observation led to the hypothesis that the pineal acts as a driving pacemaker for a family of lower order oscillators (192).

Support for this hypothesis came from transplantation studies, which demonstrated that a pineal gland implanted into the anterior chamber of the eye of a pinealectomized sparrow could restore circadian activity rhythms (192, 622). More importantly, the phase of the restored rhythm was dictated by the lighting cycle to which the donor bird had been entrained (622), indicating that the pineal actually set the phase of the circadian system, as a driving oscillator would be expected to do. Further support came from studies of isolated pineal glands of chickens, which continued to exhibit robust, light-entrained rhythms of production of the hormone melatonin under a LD cycle. In continuous darkness, however, the rhythm damped out after a few cycles (45,

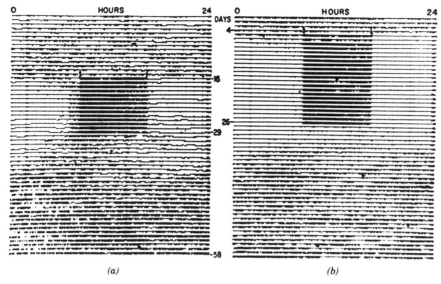

(a)   (b)

Figure 37 (A) and (B) show entrainment patterns of two pinealectomized sparrows. The beginning and end of the daily light period are marked with arrows. The dense black bars during the light fraction indicate intense perching activity. In (A), days 1–15 demonstrate arhythmic activity in DD. On days 16–29, the bird received 8 h of light followed by 16 h of darkness per 24-h period and from days 39–59, the bird was once again in DD. A positive phase angle of about 1 h is evident after the third day of the light cycle. After the light cycle was discontinued, about 8 days were required for this bird to re-establish an arhythmic pattern. In (B), days 1–4 show arhythmic activity in DD, on days 5–26 the bird is on LD 8:16 and on days 27–59 the bird was in DD. Notice the "decay" of rhythmicity on days 27–33, illustrating the transition to arhythmicity more clearly than does (A). The pattern of this decay, with activity onsets occurring earlier and activity terminating later each day is characteristic of pinealectomized birds released in DD from LD entrainment (193; copyright 1968 by AAAS).

567). Similar rhythms could be generated by small pieces of a pineal gland or even by a few dissociated cells in culture, including probably only a single pinealocyte (Fig. 38; 486, 566). It appears that individual pinealocytes contain both an endogenous circadian oscillator and a photic entrainment route. This finding suggests that the damping of the rhythm *in vitro* reflects the loss of synchrony among a population of self-sustained, weakly coupled oscillators within the gland.

While *in vitro* rhythms indicate an oscillatory capacity, the damping of the rhythm after a few cycles is potentially inconsistent with a pacemaker role for the pineal gland. A pacemaker must be capable of sustaining rhythmicity in constant conditions, since intact birds show behavioral rhythmicity in DD for an indefinite length of time. In addition, pinealectomy in chickens (341) or quail (*Coturnix coturnix*; 526) did not prevent the expression of circadian activity rhythms in DD, while pinealectomized starlings (*Sturnus vulgaris*) showed disrupted rhythms but rarely became permanently arhythmic (218).

The Hypothalamus
Gwinner (218) suggested that the species differences with regard to pineal function could be attributed to the degree of mu-

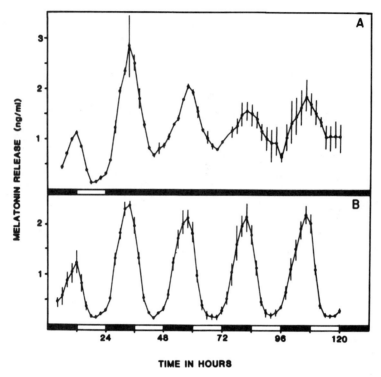

Figure 38. Rhythmic release of melatonin from dispersed chick pineal cells. The *bar* at the bottom of each panel indicates the LD cycle. Flow rate was 0.5 mL/h. (*A*) Circadian oscillation of melatonin release. *Points* and *bars* represent mean and range of duplicate cell chambers exposed to LD 12:12 for one cycle followed by constant darkness for 4 d. (*B*) Diurnal rhythm of melatonin release. *Points* and *bars* represent mean ± SEM of three replicate cell chambers exposed to LD 12:12 for the duration of the experiment (486).

tual coupling among extra-pineal oscillators that influence behavioral rhythmicity. Birds with strongly coupled oscillators could maintain a freerunning rhythm after pinealectomy; those with weakly coupled oscillators depended on the robust signal from the rhythm of pineal melatonin secretion to maintain their mutual phase synchrony. Support for the existence of extrapineal oscillators came from studies showing that hypothalamic lesions in the suprachiasmatic area of house sparrows also prevented the expression of circadian activity rhythms in DD, while permitting entrainment to LD cycles (565). Similar lesions disrupt rhythmicity in other birds (144, 526).

The sustained rhythm that emerges after pineal implantation can be attributed to a continued mutual synchronization of two oscillator populations in the hypothalamus and pineal, neither of which can maintain internal synchrony on its own (84, 497). The mammalian SCN shows a strong daily rhythm of neural activity (268a), and SCN activity appears to inhibit pineal melatonin secretion (411). If one applies what is known about the mammalian SCN and pineal to the avian system, one can hypothesize that the declining activity of SCN neurons as the dark phase approaches disinhibits pineal melatonin secretion; the consequent rise in melatonin levels further reduces SCN activity, since melatonin is known to suppress SCN neural activity in rats and hamsters (86, 349, 501). As melatonin levels fall in the morning, especially in response to the strong suppressive effects of light, the spontaneous increase in SCN activity would be reinforced by the loss of melatonin inhibition. This hypothetical neuroendocrine loop, combined with the assumption of differential coupling strengths among the oscillators in each structure can account for the data on the regulation of avian activity rhythms by the pineal gland and hypothalamus.

The Eyes

The eyes of a number of vertebrates have been shown to contain the enzymatic machinery necessary for the production of melatonin (46, 194a). Coupled with evidence that in some vertebrates the eyes contain a competent circadian pacemaker (43a), this finding prompted an evaluation of whether the eyes of birds are involved in circadian behavioral organization.

Quail, whose activity rhythms are virtually unaffected by pinealectomy, show some disruptions of free-running rhythms after bilateral blinding. If blinding is combined with pinealectomy, however, almost all quail lose their free-running activity rhythms (595). Blinding also affects the expression of circadian activity rhythms in pigeons (145), although not in sparrows (373) or chickens (341). Whether pinealectomy disturbs the integrity of the circadian system in birds may depend on the degree to which the eyes contribute to circadian organization. In at least some avian species, it appears that the pacemaker function is subserved by a complex of interacting oscillators in the hypothalamus, eyes and pineal, rather than by a single discrete pacemaker structure.

The SCN, pineal, and eyes are connected in several ways among vertebrates: The mammalian pineal depends on a signal from the SCN to generate its daily melatonin rhythm (384), and melatonin injections can entrain activity rhythms of both mammals and birds (220, 464), presumably by affecting the SCN (349, 501). The SCN is one of several structures that bind melatonin, as indicated by early studies of hypothalamic melatonin content (73), and by recent autoradiographic studies (598). The eyes of mammals provide the only photic input to the circadian pacemaker in the SCN (407), and therefore to the mammalian pineal gland, and the SCN may in turn project to the retina (54).

The Complexity of the Avian System

The avian circadian system is characterized by complexity in the structure of the pacemaker, and in the variety of available entrainment routes. An additional complexity is suggested by studies of the effects of manipulations on more than one overt rhythm. Starlings, which lose their circadian activity rhythms in LL, may still show robust circadian rhythms in feeding (191). Similarly, pinealectomy renders perch-hopping activity arhythmic in some starlings, but feeding may still show a strong circadian rhythm (221).

Since activity has traditionally been used as an assay of the sparrow circadian system, these studies raise important questions about the generality of the models that have so far been developed. It is possible that feeding and activity are controlled by separate pacemakers, but given that the activity pacemaker is itself known to be composed of several interacting elements, alternative explanations should be considered before multiplying the number of necessary oscillators in the system. The differences between these behaviors may reflect differences in the way they are measured or in the motivational states that promote their performance. Alternatively, activity may be particularly sensitive to the positive and negative masking effects of light on behavior (24), and to changes in photic sensitivity that may follow pinealectomy (497). More rhythmic functions will have to be studied in detail in birds and linked to pacemaker and entrainment mechanisms before these complexities can be resolved.

Lower Vertebrates

Entrainment

In reptiles, amphibia, and fishes that have been studied, blinding by optic nerve transection or enucleation does not prevent entrainment to photic cycles. Most lower vertebrates have a structure known generically as the parapineal organ or frontal organ, which is attached to the pineal organ. The parapineal is clearly photoreceptive and seems an attractive choice as the extraocular photoreceptor for the circadian system.

However, removal of the parapineal (and/or pineal) in lizards and fishes, as well as some amphibians does not prevent photic entrainment (3, 133). Thus, the pineal complex may contribute to entrainment to varying degrees among lower vertebrates, but it does not generally appear to be the primary site of extrocular entrainment.

Rhythm Generation

The amphibians that have been studied do not show very robust free-running behavioral rhythms (3), so the analysis of their control mechanisms has been stymied. The frog retina has been shown to contain a self-sustaining oscillator (43a), but its contribution to the regulation of other rhythms has not been documented.

The pineal gland appears to play a role in fishes and reptiles that is similar to that played by the avian pineal; that is, its rhythmic secretion of melatonin contributes to the integrity of the oscillators mak-

ing up the circadian pacemaker system (594). In several fish species, ablation of the pineal gland under constant conditions may cause splitting of activity rhythms, a general loss of rhythm stability and altered free-running periods (289, 293). In two lizard species pinealectomy also caused changed periods, arhythmicity or splitting of activity rhythms (592, 593). One study found that the pineal gland of the lizard *Anolis carolinensis* isolated *in vitro* maintains a robust rhythm of melatonin secretion for up to 10 days under constant conditions (375).

The site at which melatonin secreted by a pineal oscillator might act to help maintain the integrity of circadian organization is not known. The suprachiasmatic region of the hypothalamus seems a likely target, based on the evidence obtained in birds and mammals. This suggestion is supported by evidence that the SCN of fishes receives a binocular retinal projection, as in other vertebrates (537). In goldfish (*Carassius auratus*), this projection appears to be specialized for detection of changes in environmental illumination, since it is formed strictly by ganglion cells in the ventral retina, which are positioned to transmit light information received from above (538).

Invertebrates

Moths and Flies

Truman and colleagues (585), have studied the mechanisms by which the daily eclosion rhythms of giant silkmoths (family *Saturniidae*) are regulated. The rhythms can be entrained by exposing the pupae to light cycles, and the critical area of exposure is the head region. Excising the brain still permits eclosion, but the daily rhythm is lost, suggesting that either the circadian pacemaker or the entrainment mechanism is in the brain. The timing of eclosion under an LD cycle is species characteristic, and this timing is preserved if the brain is excised and transplanted into the abdominal cavity. The claim that the brain contains both the photoreceptor and the entrainable oscillator is strengthened by the finding that cross-species transplants of the brain result in eclosion timing characteristic of the donor species rather than of the recipient in an LD cycle. The species-characteristic patterns of abdominal movements that precede eclosion are controlled by the abdominal ganglia and are therefore those of the recipient pupae. Normal moths emerge very close to dawn, but implanted animals emerge somewhat later in the morning. Transplanting an ocular anlage along with the brain into the abdomen restores more normal timing of emergence. Aparently the eyes contribute in some way to the timing of emergence, even though the principal photoreceptor is in the brain itself.

Studies on drosophilid flies have been the source of much of what we know about photic entrainment mechanisms, photoperiodism and the genetic basis of circadian organization (224, 440, 451), yet the neural mechanisms that control their rhythms are poorly studied. A number of genetic mutations that affect the expression of circadian rhythms have been mapped to parts of the brain, and the brain has been demonstrated to be the site of photoreception for behavioral entrainment (140, 224). Depletion of the pigment used by retinal photoreceptors and the resultant loss of visual sensitivity did not affect light sensitivity for entrainment (623).

The genetic analysis of the *Drosophila* pacemaker system has demonstrated that the organization is complex. Normal expression of the *per* locus is required for normal behavioral rhythmicity. The loss of overt circadian rhythmicity in genetically "arhythmic" individuals, which lack the *per* locus (*per-*) or carry the *per*[0] allele, may reflect the loss of appropriate coupling among higher frequency oscillators (136a). Restoration of copies of the *per* gene can restore circadian organization (32). The *per* gene is expressed in several tissues, and an analysis of mosaic flies has revealed that expression of *per*[S] (the allele causing short periods) in the head results in a short free-running circadian rhythm. By contrast, the expression of *per*[S] in the thorax is associated with shortening the period of the high-frequency courtship song (228, 311).

Roaches and Their Relatives
Moths and flies are holometabolous insects. The fact that primary photoreception for entrainment is in the brain in species from each of these groups suggests that the dramatic reorganization of more peripheral structures during metamorphosis may have selected for a brain site of photoreception because the brain is relatively conserved. Support for this hypothesis comes from studies of cockroaches, crickets, and grasshoppers. These hemimetabolous insects, which do not undergo dramatic morphological changes during development, use their main compound eyes as the exclusive route for photic entrainment information (511). Cockroaches have been studied for decades in attempts to localize both the pacemakers and the photoreceptors responsible for the generation and entrainment of their circadian rhythms. There is general agreement that blackening or excising the compound eyes renders roaches incapable of entrainment by photic cues, as does transecting the optic nerves that project from the eyes to the optic lobes. These lobes appear to contain the major light-entrainable pacemakers for activity rhythms, and each lobe appears to have a fully competent pacemaker (511).

Excising both optic lobes renders roaches arhythmic; transection of the neural connections from the optic lobes to the central portions of the brain also causes arhythmicity, indicating that a neural, not a humoral route, connects the pacemaker to lower order effector mechanisms. Lesion studies implicate a portion of the medulla and lobula of the optic lobes as being critical to rhythm generation (511).

Adult cockroaches have a remarkable ability to regenerate central neural connections that have been cut. About 40 days after a bilateral transection of the optic lobes or implantation of an optic lobe into a bilaterally lobectomized roach, central portions of the brain again show

neural responses to ocular illumination, and a circadian activity rhythm recovers (421). Studies in which donors' free-running periods were manipulated before surgery provide clear evidence that the implanted lobe contains the driving oscillator for the rhythm. If the donor roach had a markedly different free-running period from that expressed by the recipient when it was intact, the recipient expressed the period of the donor animal (Fig. 39) (420).

A native lobe retains an advantage over an implanted lobe in regulating rhythmicity. When a donor lobe was implanted

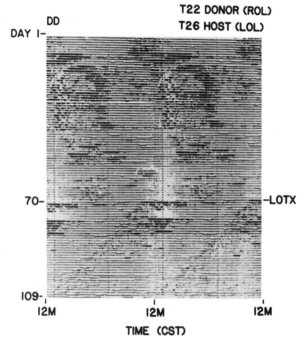

Figure 39. Activity record of a cockroach that was raised on a 26-h light-dark cycle (T26 Host). On the day prior to the first day of the record the right optic lobe of the animal was removed and an optic lobe from an animal that was raised in a 22-h light-dark cycle (T22 Donor) was transplanted in its place. The animal was maintained in constant darkness (DD) at constant temperature throughout. For the first 70 days of the record the animal exhibits a free-running rhythm of activity with a period characteristic of the host optic lobe. On day 70 the left optic tract was cut (LOTX) neurally isolating the host optic lobe from the midbrain. Beginning the day of surgery a free-running rhythm with a period characteristic of the donor's optic lobe was expressed. There was a large difference in phase between the pre- and postoperative rhythms (~10–12 h) (420).

in the presence of an intact native lobe on the contralateral side, only the native lobe controlled the free-running rhythm. When the native lobe was cut off from the brain but left *in situ*, the heterotypic free-running period, generated by the implanted lobe, appeared. Once the native lobe was allowed to regenerate its central connections, however, the roach might express two different free-running periods simultaneously (422).

There is other evidence from entrainment studies that the regenerated neural pathway from optic lobe to protocerebrum is somehow incomplete or inefficient relative to the normally connected lobe. If a lighting cycle is provided to a roach with one intact and one transplanted lobe but retinal input to the intact lobe is blocked, the roach free-runs as if in darkness. As soon as the intact lobe is separated from the brain, however, entrainment, mediated by the transplanted lobe and its retinal innervation, returns. Apparently the transplanted lobe is entrained by the lighting cycle but cannot control overt activity in competition with an intact lobe. In the absence of an intact lobe, however, the implanted tissue can entrain the locomotor activity system (422).

These results imply that the optic lobe pacemakers have their effects on activity by virtue of their neural connections to a mechanism in the central brain. This mechanism may itself be a secondary oscillator, located in or near the pars intercerebralis. The pattern of entrainment to temperature cycles in bilaterally lobectomized crickets (472) and cockroaches (423) is consistent with this conclusion.

Studies of other arthropod species have in general been consistent with findings in the well-studied insects (35, 419). The supraesophageal ganglion has been implicated as the site of rhythm generation, and neural efferents have been identified as the mode of communication to peripheral effectors in the scorpion (181),

crayfish (424), and *Limulus* (35). In addition, bilaterality of pacemakers and complexity of pacemaker structure are reported in other arthropods (307a).

Molluscs

The neural control of circadian rhythms has been extensively studied in two molluscs, *Aplysia* and *Bulla*, in which the eyes have been shown to contain self-sustaining circadian oscillators (50, 269, 272, 489a). The rhythms measured from the eyes *in vitro* are those of spontaneous compound action potentials (CAPs) propagated from the retina via the optic nerves. The frequency of these potentials increases during the day and decreases at night, and the rhythm persists in constant darkness. The rhythm is susceptible to phase shifting by photic cues and by a variety of drugs, notably inhibitors of protein synthesis, cyclic nucleotides, and those affecting ion transport (169, 170, 172). The origin of the CAP rhythm in *Bulla* has been traced to a small population of electrically coupled neurons near the base of the eye, called basal retinal neurons (BRNs). Photic effects on the CAP rhythm are mediated by light falling on the basal retina, not on the organized photoreceptor layer (49, 51). Excision of most of the retina is still compatible with the generation of a measurable rhythm, even when only a few BRNs remain.

Individual BRNs may be competent circadian oscillators (49). Intracellular recordings reveal a circadian rhythm of membrane potential in BRNs that oscillates in parallel to the CAP rhythm. The CAP rhythm is the product of a spontaneous rhythm of membrane potential in the population of electrically coupled BRNs, with CAPs being generated in largest numbers when the membrane is most depolarized (Fig. 40) (361). It is clear that protein synthesis is necessary for the operation of the ocular pacemaker (172, 617), but the role in cellular metab-

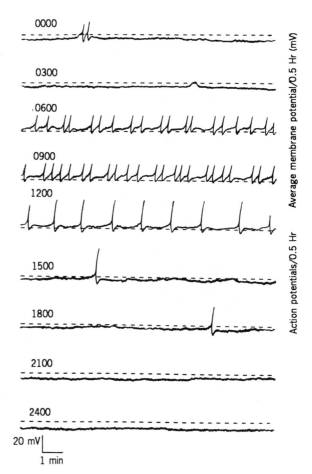

Figure 40. Long-term recording from a basal retinal neuron in a *Bulla* eye illustrating circadian rhythms in membrane potential and action potential frequency. Each line represents a 12-min segment from a continuous intracellular BRN recording. The time (EST) of each segment is shown on the left. This eye was taken from an animal entrained to an LD 12:12 light cycle with previous dawn at 0900. Dashed line in each trace represents the −60-mV level and serves to illustrate the circadian rhythm in membrane potential exhibited by the BRN. Action potential height is attenuated due to clipping (361).

olism of the protein(s) being synthesized is uncertain.

In general, events that depolarize the membrane of BRNs have the same effects on the rhythm as do light pulses (which are also depolarizing). Preventing depolarization during a light pulse can block the phase-shifting effect, suggesting that the depolarization is a critical mediator of photic effects (360). Drugs, such as serotonin, which hyperpolarize the membrane instead, can also phase shift the rhythm, but the sensitive phase is different from that for light pulses (104).

The two eyes of *Aplysia* are weakly and negatively coupled (265a) but in *Bulla* and in another species (*Bursatella*) the two eyes appear to be strongly positively coupled via a commissural pathway. Phase shifts induced in one eye of *Bulla* are transmitted to the other so that the two eyes eventually resynchronize to a common phase (484). The mechanism responsible for coupling the eyes to a common phase appears to involve the depolarizing (and therefore phase shifting) effects of CAPs arising in one eye and being transmitted to the BRNs of the contralateral eye (52). The differences between *Aplysia* and *Bulla* may reflect differences in behavior; the strongly nocturnal *Bulla* may rely on internal coupling mechanisms for the eyes to remain synchronized. In a diurnal species like *Aply-*

sia exposure to the external lighting cycle may be sufficient to maintain synchrony.

Biochemistry and Genetics: Molecular Models
J. W. Hastings

There have been extensive studies involving the analysis and modeling of endogenous oscillatory phenomena in the nervous system (83, 272, 584), and in higher organisms key aspects of the circadian mechanism may indeed involve the nervous system and cellular interactions (see the section on neural control of circadian rhythmicity). Nevertheless, it is well known that circadian expression and control occurs also in unicellular organisms, in the absence of a nervous system and other cellular interactions (236, 563). Thus, the components and organization of a fundamental cellular biological clock mechanism may be sought and modeled at the complexity level of the single cell.

The earliest of the cellular clock models, the "chronon" (150), made use of the discovery that clock function in the unicell *Gonyaulax* was blocked by inhibitors of both transcription and translation (286). This allowed the circadian system to be compared to a developmental system, in which differentiation recurs on a daily basis (200). In the chronon model, certain key segments of the DNA are transcribed sequentially once each day; with the synthesis of the relevant regulatory proteins, a loop is closed so as to repeat the sequence of genes transcribed, restarting at a point on the genomic loop corresponding to about 24 h earlier. A strong prediction of this model was the existence of time-of-day specific mRNAs. Such RNAs have now been discovered (62, 307, 340, 402, 427, 475a) but whether they are part of the actual clock mechanism, or simply controlled by the clock, remains to be established.

Based mostly on the effects of inhibitors of eukaryotic protein synthesis on circadian phase and period (175, 271, 286, 403, 417, 493, 572, 602), an important involvement of proteins in the clock mechanism is now firmly established. Several different inhibitors of 80S ribosome (eukaryotic) protein synthesis are similarly effective in shifting the phase of free-running circadian oscillations but chloramphenicol, an inhibitor specific for 70S ribosome (prokaryotic) protein synthesis, is not (286). Analogs of certain amino acids are also effective (240, 309), indicating that the clock function involves proteins as such, and not just their synthesis per se (137). Figure 41 illustrates the phase shifting effects of anisomycin, an inhibitor of 80S ribosome protein synthesis. Other inhibitors of eucaryotic protein synthesis act in a similar way; the magnitude and sign of phase shifting depends on the circadian time at which the drug is added, allowing the effects to be plotted as a drug-PRC (Fig. 42).

An important type of mathematical modeling is that of the limit cycle (432, 614). The oscillation may be described as an excursion in the form of a circle or closed loop located in one quadrant, with two "state" variables on the two axes, their amplitudes being given by the magnitude of the excursions (Fig. 43). In limit cycles, a series of quasiconcentric loops may be larger or smaller, depending on amplitudes, and somewhere near the center may be a point of metastable equilibrium between the two variables. At this "singularity" point the system is apparently arhythmic. In an oscillating system, critical pulses of some agent, such as light or a drug, may be able to move the trajectory to or near this singularity point. Experiments in several organisms have shown that a "critical" light pulse can indeed result in arhythmicity (151, 344, 432, 614). Remarkably, comparable light treatments on an individual human have resulted in not only phase shifting;

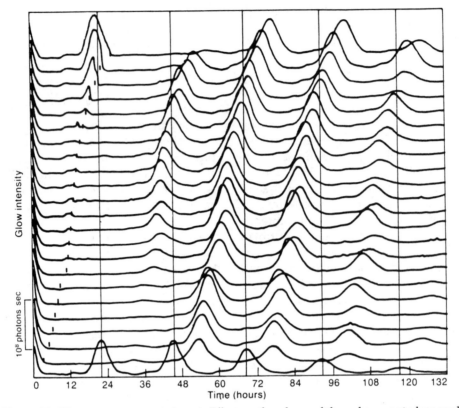

Figure 41. Phase response experiment. Effect on the phase of the subsequent glow peak with cultures of *Gonyaulax* initially in the same phase as the control. Pulses of 0.5 μM anisomycin applied for 1 h (black bars) at different times after entry into constant conditions, so as to span a full circadian cycle (573).

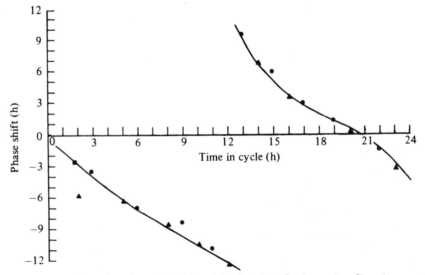

Figure 42. A PRC for 2-h pulses of 5 μM cycloheximide. Abscissa, circadian time modulo τ (23.2 h). Ordinate, phase shift measured after first (▲), second (●), and third (■) cycle (572).

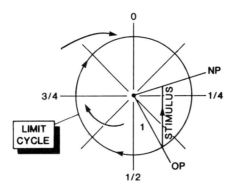

Figure 43. Limit cycle. Oscillation about a steady state, here placed at the origin, tends to unit amplitude, while angular velocity is strictly independent of amplitude. Isochronous radii are the equispaced radial lines. During stimulation the vertically portrayed variable is assumed to change, so that the change from one isochron to another is then simply described trigonometrically. NP new phase, OP old phase. This picture captures the essentials of phase resetting (614).

at certain times rhythmicity was attenuated (110, 111). Drug induced attenuation of rhythmicity has also been observed following the application of anisomycin to *Gonyaulax* in a critical dose (Fig. 44*a*) at a critical time of day (Fig. 44*b*; 570, 573).

The limit cycle description involves no particular limitations with regard to what the actual components of the circadian clock mechanism may be. Indeed, it could be compatible with many different models, with very different specific components. Several such models involve daily fluctuations of specific ions in cellular compartments and ion pump activity (91, 412, 560). Different cellular compartments might be involved; the concept of ion gradients and light-gated ion channels allows for the modeling of phase shifting and entrainment by light. The notion that the circadian period might somehow be related to membrane lipid fluidity, which is autoregulated in

many cells to a near constant value at different temperatures, was also advanced to account for temperature compensation of the circadian period (412). Both cell membranes and soluble proteins have been proposed to be involved directly in the clock mechanism (518).

A number of models have been centered on biochemical pathways associated with energy metabolism (109, 135, 203). With some exceptions, these models have not gained much support, but the more recent discovery that creatine has a large effect on the period of the clock in *Gonyaulax* (488) suggests that energy metabolism may indeed have a significant role in circadian timing.

The different models all involve proteins which, in one way or another, may vary in activity and/or cellular levels as related to time of day. Proteins and changes in their activities are also involved at the level of the expressed or overt rhythms; an important question is how the activities of such proteins might be controlled by the clock. Not many circadian-controlled proteins have been identified and analyzed, but two of those involved in bioluminescence of the unicell *Gonyaulax* have been. The enzyme luciferase and the substrate (luciferin) binding protein (LBP) constitute biochemical correlates of the rhythm, increasing and decreasing in activity in concert with the *in vivo* rhythm. Surprisingly, perhaps, these activity changes were found to be due to changes in the actual cellular concentration of the specific proteins (138, 277, 394) and, in the case studied, to involve a daily pulse of synthesis and a later degradative phase (Fig. 45). At the same time, the mRNA levels for this protein remain constant at different times of day, as does its *in vitro* translatability in a heterologous system. These facts indicate that the clock regulates the synthesis of this protein at the translational level, thus providing a clue as to how the clock

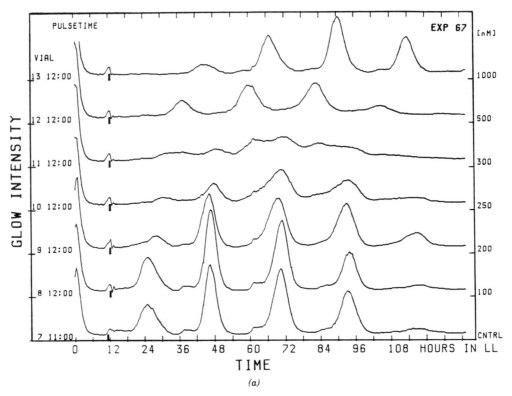

Figure 44a. Attenuation of rhythmicity by a critical dose of a drug (vial 11). Effect of anisomycin (concentration in nM at right) on the subsequent rhythm of bioluminescent glow in *Gonyaulax*. One hour pulses (black bars) applied 12 h after entry into constant dim light; a similar dose with pulses given at hours 11, 13, and 14 did not result in attenuation (573).

exerts its effect, but not necessarily one concerning how it "keeps time."

A model that we can suggest for the clock mechanism itself is that it involves a series of protein regulatory factors for mRNAs, each of which regulates the synthesis of the next regulatory protein at the translational level, eventually closing a loop such that the "last" protein regulates the synthesis of the first. These factors would thus constitute true "clock proteins." At the same time such proteins might control the synthesis of additional proteins outside the loop, which would be clock-controlled proteins. The regulatory clock proteins would thus serve dual roles as both clock components and as transducers to the driven overt rhythms. Each of the putative clock proteins would be encoded by clock genes whose transcripts are produced constitutively and are relatively stable (half-life, days). By contrast, the clock regulatory proteins themselves are unstable (half-life, hours) and synthesized in a similarly short burst, each one being regulated by a different and similarly short-lived protein.

A number of substances other than protein synthesis inhibitors have been reported to influence either phase or period of circadian rhythms in different organisms, but sites and mechanisms of action are largely unknown. Phase shifting effects of ion channel inhibitors and ion-

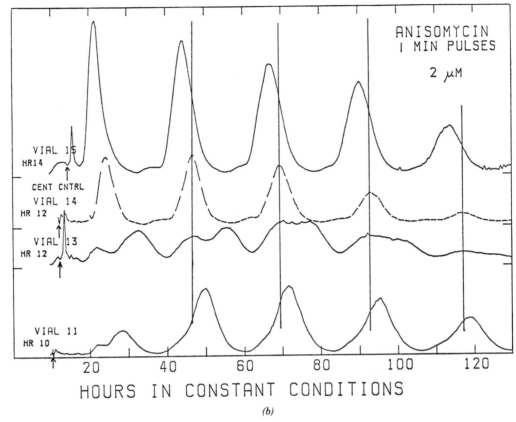

Figure 44b. Glow curves of cultures in vials pulsed with 2 μM anisomycin for 1 min at h-10, 12, and 14. There was an attenuation and alteration in the waveform of the culture treated at h-12 (vial 13). In the ones treated at h-10 and 14 (vials 11 and 15) the waveforms were smooth and amplitudes high, with delay and advance phase shifts, respectively. Centrifuged control, vial 14 (570).

ophores (514, 561) could relate to the participation of membranes in the rhythmic mechanism. Alcohols (75, 156) might also affect membranes, but chain length effects do not support this (571). In fact, the corresponding aldehydes are even more effective in *Gonyaulax* (569), and it is possible that aldehydes influence protein synthesis (600). Calcium, calmodulin, and cyclic AMP are implicated as potential clock elements in some species (170, 171, 616); the effects observed could relate in one way or another to protein phosphorylation. Still unexplained is the effect of heavy water which, in organisms ranging from unicells to invertebrates and

mammals, slows the biological clock (72a, 155, 358, 450). In all these cases it is important to appreciate that agents or mutations that appear to act directly on the pacemaker mechanism may in fact not be doing so (616).

The involvement of endogenous humoral agents in interactions between cells and tissues has been demonstrated in insect, avian, and mammalian systems (566, 586), but the identity of the specific agents remains unknown. With unicells and fungi, most studies have failed to reveal such effects (518, 555). An exception is the unicell *Gonyaulax*, where the activity peaks of two originally out of phase

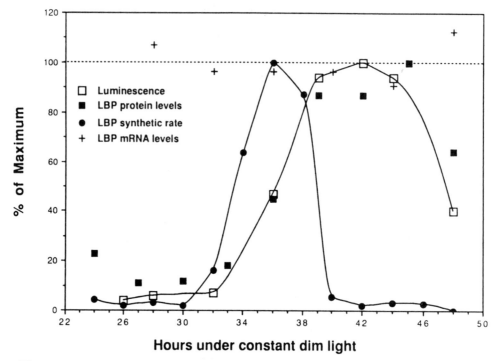

Figure 45. Densitometric scans of Western blots (■, LBP levels), Northern blots (+, LBP mRNA levels), and pulse labeled autoradiographs (●, LBP synthetic rates) plotted along with the *in vivo* levels of bioluminescence of cells kept in LL. Values were normalized relative to the band at 85 kD in Western blots and the band at 55 kD in pulse labeling. The rate of LBP synthesis is maximal prior to peaks in LBP protein level and luminescence, which rise concurrently to a maximum and remain high for ~6 h, while LBP mRNA remains constant (394).

cultures coalesced in mixed cultures, and this could be prevented by the replacement of the culture medium with fresh medium (Fig. 46) (67). Neither the mode of action (be it production, or consumption or both) nor the nature of the substance(s) involved is known, but the effect is related to cell density.

Creatine represents a naturally occurring substance that actually speeds up the clock in *Gonyaulax* (488) (Fig. 47). Creatine is well known in animal cells as a storage molecule for high energy phosphate, and it was recently found to function as a high energy shuttle molecule (581); creatine is actively taken up by animal cells via specific plasma membrane ports (339).

Whether or not it has similar or related effects on circadian systems of animals has not been determined. With 15-μM creatine in the medium, the period in *Gonyaulax* may be dramatically less, as low as 18 h. How this effect is mediated is not known; it also has an evident and persistent effect on motility, such that the cells are more active in its presence (489). However, creatine does not appear to be present in *Gonyaulax*, so it may be an analog for an endogenous molecule. Indeed, such a compound with period shortening effects has been isolated from *Gonyaulax* (Fig. 47).

Genetic studies have been carried out in several systems; mutants exhibiting al-

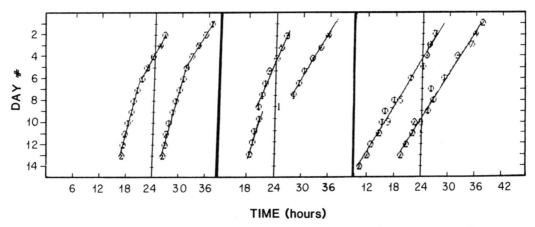

Figure 46. Free-running circadian rhythms of bioluminescent glow from two cultures that had been entrained to different phases presented in the form of plots of peak times on sequential days. The first panel at left shows the peaks of the two parent cultures maintained separately. The center panel shows the pattern of the peaks of luminescence from a vial containing a 1:1 mixture of the two cultures, while the right panel shows a similar mix in which the medium was replaced every 2 days during the first week (240).

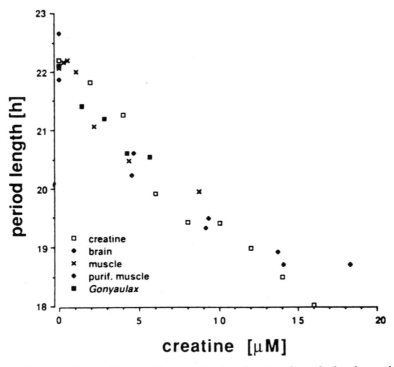

Figure 47. Creatine is a substance that accelerates the circadian clock when added to cultures of *Gonyaulax*. Its activity was originally detected in extracts of mammalian brain, and later in other tissues. Cultures were placed in constant dim white light at 19°C with creatine added to the different concentrations as indicated; periods of the circadian rhythm were measured as the average over the subsequent 12 days. A similar activity has been found in extracts of *Gonyaulax* but the substance responsible appears to be a different molecule (488).

tered clock properties can be extremely valuable in the analysis of the mechanism (176). Important results have been obtained with *Drosophila* (225, 618), *Neurospora* (177), *Chlamydomonas* (375a), and the hamster (460). In *Drosophila*, *per¹*, *perˢ*, and *per⁰* mutants were isolated as long, short, and aperiodic flies, respectively (310). These same alleles were later shown to similarly affect minute-long courtship rhythms in *Drosophila* (321). Thus, rhythms having very different frequencies can be affected by a single gene, but the way in which these effects relate to the mechanism itself remains to be discovered. The *per* gene has been cloned and characterized (31), and its introduction into mutant arhythmic flies has been shown to restore circadian rhythmicity (32, 620). The transcript of this locus, which is expressed rhythmically in the head (230a), appears to encode a proteoglycan-like molecule, which may act by modulating gap junction conductivity (33, 273, 463). *Per* homologs have been demonstrated in a number of other species, and it even possesses regions of homology with the frq gene in *Neurospora* (357). In *Drosophila*, a *per* homolog has been shown to be crucial to normal neurogenic development (108).

Evolution and Adaptive Significance

Z. Boulos

There is fossil evidence of daily rhythms dating back to the Middle Devonian (~375 million years) in corals (604) and to the Upper Ordovician (~420 million years) in nautiloids (285). These organisms, like many others (409), display daily growth layers that appear as fine ridges or growth lines on the shells of nautiloids and on coral skeletons. An interesting aspect of this research is the means it provides for estimating the changes in geophysical periodicities that may have taken place over the geological time scale. Thus, in addition to daily growth lines, corals show what appear to be annual growth rings; fossil specimens from the mid-Devonian show about 400 daily growth lines per annual ring (range 385–410), while specimens from the Pennsylvanian show some 385 lines and more recent specimens closer to 365 (604). Since the period of the earth's revolution around the sun is believed to have remained constant throughout that time, these results suggest a Devonian daylength of 21.9 h, in agreement with astronomical evidence that the earth's rotation around its axis has been decelerating by 2 s/100,000 years. The shells of modern nautiloid species show both daily growth lines and lunar monthly chamber walls, and these are visible in fossil material as well. Observations on the number of daily lines per chamber in nautiloid fossils suggest a lengthening of the lunar synodic month from 8 to 9 days, some 420 million years ago, to its present duration of 30 days (285).

But the ubiquity of circadian rhythms, at least among eukaryotes, points to an even earlier origin, and several hypotheses have been proposed as to when such rhythms may have first appeared and what original selection forces may have been involved in their evolution (147). Pittendrigh (439) speculates that the daily light cycle was the primary selective agent responsible for the emergence of circadian oscillations, serving as both proximate cause (in its role as entraining agent) and ultimate cause. Several cellular functions are known to be adversely affected by the absorption of visible light, and one way of avoiding these disruptive effects is to restrict such functions to the hours of darkness. A similar view (425) attributes the development of circadian rhythmicity to selection pressure from the daily light cycle and the increasing oxygen level. Circadian rhythms would thus have protected early eukaryotes from deleterious photooxidative effects caused by the increase in the level of free oxygen.

According to these hypotheses, circadian rhythms evolved in response to daily periodicity in the external environment. An alternative viewpoint is that the initial selective advantage of circadian rhythmicity was its provision of a reliable temporal framework for the internal coordination of cellular activities. Specifically, given the prevalent belief that circadian rhythms were unique to eukaryotes, and based on the endosymbiotic theory of the origin of eukaryotic cells, circadian rhythms were assumed to have allowed the necessary temporal coordination between host cells and their endosymbionts (304, 330).

An additional hypothesis, attributed to Bünning (439), suggests that historically the primary function of circadian rhythms lay in the mechanism it provided for photoperiodic time measurement. The absence of photoperiodism in many unicellular organisms, however, argues against this hypothesis.

Implicit in all these views is the assumption that circadian rhythms have a common historical origin, suggested by their unusual properties of precision and temperature compensation. However, several features of circadian systems could certainly have arisen independently in different organisms, any similarities between them being the result of convergent evolution (439).

Regardless of what their original function or functions may have been, there is general agreement that in later stages, circadian rhythms evolved primarily as adaptations to daily cycles in the environment (10, 154, 257). Initially, daily cycles in light, temperature, and humidity may have been the main agents of selection but later, the rhythmic activities of prey, predators, and competitors would have provided added selection pressure for circadian organization. Several examples of circadian rhythms showing apparently optimal phasing in relation to daily cycles in both the physical and biotic environment have already been described.

Endogenously generated oscillations entrained by external zeitgebers offer several advantages over exogenous, passively driven rhythms. One is that the timing of an animal's behavioral and physiological activities is less affected by irregular fluctuations in external conditions, such as may result from variations in weather, and by changes in light and other conditions experienced by an animal as a result of its own behavior. The latter is particularly important in nocturnal animals that spend most of the daylight hours in light-excluding shelters, and it is therefore of some interest that nocturnalism appears to have been adopted as a general strategy by several animal groups at crucial stages in their evolution. Thus, the more primitive terrestrial arthropods (98) and insects (303) are mostly nocturnal, as are the more primitive terrestrial vertebrates, the amphibians (115). Similarly, the earliest mammals are thought to have been nocturnal during the first 100 million years of their evolution (363).

Diurnal organisms, on the other hand, are normally exposed to light over all or most of the day, and the advantages of endogenous timing may therefore seem less obvious. There is, however, another major adaptive feature of circadian rhythms, applicable to both nocturnal and diurnal animals, namely, the ability they confer to anticipate and thus be prepared for cyclic environmental changes (385). Examples include the rise in body temperature and corticosteroid levels in humans before the end of the night (610), and the increase in hepatic and intestinal enzyme activity in rats immediately preceding a regularly scheduled daily feeding (59).

The temporal coordination of internal physiological and cellular biochemical events is often cited as an important benefit accruing from a circadian timing system, as "mutually interdependent events must not only occur at precise spatial locations but must also occur with appropriate timing. Similarly, incompatible processes, which may require different physicochemical conditions for their completion, can be separated just as effectively in time as in space" (386, p. 215). This is illustrated by recent data on oxygen-evolving photosynthesis and oxygen-labile nitrogen fixation in the unicellular cyanobacteria, *Synechococcus* (382). These two processes are incompatible and, in heterocystous cyanobacteria,

are made possible by spatial separation of the site of nitrogen fixation from that of photosynthesis. Nonheterocystous species, on the other hand, show endogenously generated daily rhythms in the two functions, with peak levels at opposite daily phases (Fig. 48). The results also show that the rhythm in nitrogen fixation (nitrogenase activity) is not a passive consequence of the rhythm in photosynthesis, as it persists in LL after complete inhibition of photosynthetic oxygen evolution. Thus, these organisms make use of circadian timing (564) as a mechanism for temporal segregation of incompatible functions.

A somewhat different example is the cycling of the luciferase molecule in the

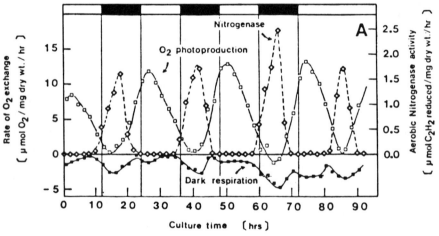

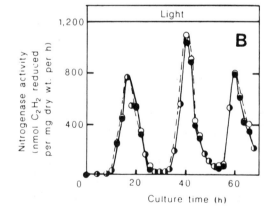

Figure 48. Circadian rhythms in synchronously growing cyanobacteria, *Synechococcus*. (*A*) *rhythms in photosynthetic oxygen* evolution (*open square*), nitrogenase activity (*open circle*), and respiratory oxygen uptake (*filled square*) during 3 days in LD 12:12 and 1 day in LL. Left-hand scale, oxygen exchange (μmol of O_2 per mg of dry weight per h); right-hand scale, nitrogenase activity (μmol of C_2H_2 reduced per mg of dry weight per h). (*B*) Rhythm in nitrogenase activity in LL after complete elimination of photosynthetic oxygen evolution by addition of DCMU to the culture medium (382).

bioluminescence rhythm of *Gonyaulax*. The protein is actually degraded to the level of amino acids and then synthesized anew every 24 h. As this organism lives in an environment limited by nitrogen but not energy, this protein turnover may represent a mechanism for conserving nitrogen by degrading enzymes whose function is temporarily unnecessary and reutilizing their amino acids in proteins having functions at a different time of day (277).

While there may be clear benefits to the use of circadian rhythmicity as a means of maintaining internal temporal order, it does not necessarily follow that such benefits alone are sufficient for the development of a circadian timing mechanism. In the cyanobacteria example described above, the rhythm of photosynthesis must have evolved in relation to the external light cycle, and it may be that only then did a rhythm in nitrogen fixation become advantageous.

One approach to this question is to examine the daily distribution of behavior and physiology in cave dwelling animals adapted to a permanently aperiodic environment (at least on a daily basis). In fact, most of the troglobitic species that have been studied, mainly arthropods and fishes, show either a complete absence of circadian rhythmicity under constant conditions, in some cases under daily LD cycles as well (158, 323, 364, 576), or highly variable activity patterns with only occasional periodicity in the circadian range (53, 275). Some individuals of the cave-dwelling fish *Anoptichthys jordani* showed strongly damped oscillations in constant conditions lasting 1–4 cycles, but these always had the same period as the preceding LD cycle, even when this was well outside the normal circadian range (159). The absence of circadian rhythms in energy metabolism in other species adapted to aperiodic environments was discussed in the section on circadian control of physiological functions.

These results would suggest that in the absence of daily environmental periodicity, the benefits of internal temporal coordination are insufficient for the maintenance of circadian timing. There is, however, one caveat to this argument: Some animals, including cats and guinea pigs, show diffuse behavioral patterns with little or no evidence of circadian rhythmicity in some situations, and yet cats and guinea pigs show clear circadian oscillations in CSF melatonin and vasopressin levels (476) and in SCN multiple unit activity (320), respectively. It remains possible, therefore, that cave-dwelling animals possess circadian rhythms in internal functions despite arhythmic activity patterns.

As several authors have pointed out, the advantages of circadian timing are by no means always evident (154, 257, 439). Nor is it always clear why daily rhythms should be endogenously generated in some cases and exogenously driven in others. For example, the horn shark, *Heterodontus francisci*, and the swell shark, *Cephaloscyllium ventriosum*, are both nocturnally active bottom-dwelling species with overlapping ranges, but while the activity rhythm of the swell shark anticipates the onset of darkness and free-runs in LL and DD, that of the horn shark appears to be entirely exogenous, as it always starts after lights-off and is promptly lost in constant conditions (406). Finally, given that most animals are rarely exposed to constant conditions for more than a few days, what is the selective advantage of self-sustained circadian oscillations capable of persisting indefinitely in the absence of periodic environmental input?

Although some of these questions may not be answered to everyone's satisfaction, enough is known about circadian timing, in its various manifestations, to justify the conclusion that circadian clocks represent a powerful and versatile physiological mechanism, one that has been put to a variety of different uses. In the words of Pittendrigh (443, p. 58), such clocks allow "the measurement of environmental time in two conceptually distinct ways: (1) in providing for a proper phasing of [a temporal] program to the cycle of environmental change, they, in effect, recognize local time; and (2) in assuring an appropriately stable temporal sequence in the program's successive events, they, in effect, measure the lapse of (sidereal) time."

Glossary of Abbreviations

LD	light-dark
LL	constant light
DD	constant darkness
τ	period of biological rhythm
T	period of zeitgeber
ψ	phase angle difference between zeitgeber and biological rhythm
φ	phase of biological rhythm
$\Delta\varphi$	phase shift of biological rhythm
CT	circadian time, where CT0 represents the beginning of an organism's subjective day and CT12 the beginning of subjective night. One hour in circadian time is equal to $\tau/24$ hours.
PRC	phase response curve
VRC	velocity response curve
HTP	high temperature pulse
LTP	low temperature pulse
Ta	ambient temperature
SCN	suprachiasmatic nucleus
RHT	retinohypothalamic tract
GHT	geniculohypothalamic tract

References

1. Abe, H. and S. Sugimoto. *Anim. Learn. Behav.* 15:353–359, 1987. Feeding anticipation in pigeons.

2. Abe, H. et al. *Physiol. Behav.* 45:397–401, 1989. Strain differences in responses to feeding schedules in mice.

3. Adler, K. In *Biochronometry* (M. Menaker, ed.). p. 342–350, Natl. Acad. Sci., Washington, DC, 1971. Pineal frontal organ and entrainment in frogs.

4. Albers, H. E. *Am. J. Physiol.* 250:R708–R711, 1986. PRCs for L-to-D and D-to-L step transitions.

5. Albers, H. E. and C. F. Ferris. *Neurosci. Lett.* 50:163–168, 1984. Phase shifts after NPY injections.

6. Allemand, R. and J. R. David. *Experientia* 32:1403–1405, 1976. Genetic latitudinal differences in circadian parameters in *Drosophila*.

7. Anderson, D. H. and B. A. Keafer. *Nature (London)* 325:616–617, 1987. An endogenous annual rhythm in a dinoflagellate.

8. Andreasson, S. and K. Müller. *Oikos* 20:171–174, 1969. Seasonal inversion of activity patterns in sculpins.

9. Aschoff, J. *Cold Spring Harbor Symp. Quant. Biol.* 25:11–28, 1960. Exogenous and endogenous components in circadian rhythms.

10. Aschoff, J. *Symp. Zool. Soc. London* 13:79–98, 1964. Survival value of diurnal rhythms.

11. Aschoff, J. *Rev. Suisse Zool.* 71:528–558, 1964. Seasonal effects on ψ-midpoint.

12. Aschoff, J. *Science* 148:1427–1432, 1965. Human circadian rhythms.

13. Aschoff, J. In *Circadian Clocks* (J. Aschoff, ed.), pp. 95–111. North-Holland Publ., Amsterdam, 1965. Phase response curves.

14. Aschoff, J. In *Circadian Clocks* (J. Aschoff, ed.), pp. 262–276. North-Holland Publ., Amsterdam, 1965. The phase angle of entrainment.

15. Aschoff, J. *Z. Tierpsychol.* 49:225–249, 1979. Effects of internal and external conditions on τ.

15a. Aschoff, J. In *Handbook of Behavioral Neurobiology*, Vol. 4 (J. Aschoff, ed.), pp. 81–93. Plenum, New York, 1981. Free-running and entrained circadian rhythms.

16. Aschoff, J. ed. *Handbook of Behavioral Neurobiology*, Vol. 4. Plenum, New York, 1981.

17. Aschoff, J. In *A Companion to Animal Physiology* (C. R. Taylor, K. Johansen, and L. Bolis, eds.), pp. 173–188. Cambridge Univ. Press, London, 1982. Body temperature rhythm as a function of body size.

18. Aschoff, J. *J. Therm. Biol.* 8:143–147, 1983. Circadian control of body temperature.

19. Aschoff, J. and A. Heise. In *Advances in Climatic Physiology* (S. Itoh, K. Ogata, and H. Yoshimura, eds.), pp. 334–348. Igaku Shoin, Tokyo; 1972. Daily rhythms in thermal conductance in man.

20. Aschoff, J. and H. Pohl. *Fed. Proc. Fed. Am. Soc. Exp. Biol.* 29:1541–1552, 1970. Rhythmic variations in energy metabolism.

21. Aschoff, J. and H. Pohl. *Naturwissenschaften* 65:80–84, 1978. Phase relation between a circadian rhythm and its zeitgeber.

22. Aschoff, J. and H. Tokura. *J. Biol. Rhythms* 1:91–100, 1986. Temperature entrainment of squirrel monkeys.

23. Aschoff, J. and C. von Goetz. *J. Biol. Rhythms* 1:267–276, 1986. Feeding cycle effects on squirrel monkeys.

24. Aschoff, J. and C. von Goetz. *J. Biol. Rhythms* 4:29–38, 1989. Masking in canaries by light and dark.

25. Aschoff, J. and U. von Saint Paul. *Physiol. Behav.* 10:529–533, 1973. Brain temperature and activity rhythms in chickens.

26. Aschoff, J. and R. Wever. *J. Ornithol.* 103:2–27, 1962. Onset and end of daily activity in free-living birds.

27. Aschoff, J. et al. *Oikos, Suppl.* 13:91–100, 1970. Seasonal changes in daily activity rhythms at the Arctic Circle.

28. Aschoff, J. et al. In *Biochronometry* (M. Menaker, ed.), pp. 3–29. Natl Acad Sci., Washington, DC, 1971. Properties of circadian activity rhythms in birds and man.

29. Aschoff, J., C. von Goetz, and K.-I. Honma. *Z. Tierpsychol.* 63:91–111, 1983. Food entrainment in rats.

30. Balzer, I. and R. Hardeland. *Int. J. Biometeorol.* 32:231–241, 1988. Influence of temperature on rhythms.

31. Bargiello, T. A. and M. W. Young. *Proc. Natl. Acad. Sci. U.S.A.* 81:2142–2146, 1988. Cloning of the per gene.

32. Bargiello, T. A., F. R. Jackson, and M. W. Young. *Nature (London)* 312:752–754, 1984. Restoration of rhythms by gene transfer in *Drosophila*.

33. Bargiello, T. A. et al. *Nature (London)* 328:686–691, 1987. Per gene product affects intercellular functional communication.

34. Barlow, G.W. *Ecology* 39:580–587, 1958. Daily movements and thermoregulation in desert pupfish.

35. Barlow, R. B., Jr. *J. Neurosci.* 3:856–870, 1983. Circadian system in *Limulus*.

36. Bartholomew, G. A. and T. J. Cade. *J. Mammal.* 38:60–72, 1957. Daily torpor in the little pocket mouse.

37. Bateman, M. A. *Aust. J. Zool.* 3:22–33, 1955. Entrainment by light and temperature cycles in fruit-flies.

38. Beck, S. D. *Bull. Entomol. Soc. Am.* 9:8–16, 1963. Circadian rhythms and photoperiodism in insects.

39. Beier, W. et al. *Ann. Epiphyt.* 19:133–144, 1968. Effects of external factors on the time sense of bees.

40. Beling, I. *Z. Vergl. Physiol.* 9:259–238, 1929. The time sense of bees.

41. Bennett, A. F. and H. John-Alder. *Copeia* 1986:57–64, 1986. Daily temperature preference rhythms in skinks.

42. Bennett, M. F. and M. Renner. *Biol. Bull. (Woods Hole, Mass.)* 125:416–430, 1963. Foraging activity rhythms in bees.

42a. Benoit, J. *Rec. Prog. Horm. Res.* 15:143–164, 1959. Light effects on avian reproductive function.

43. Berthold, P. *Naturwissenschaften* 65:546, 1978. Circannual rhythms in birds.

43a. Besharse, J. C. and P. M. Iuvone. *Nature* 305:133–135, 1983. Oscillator in the *Xenopus* retina.

44. Binkley, S. and K. Mosher. *Am. J. Physiol.* 251:R1156–R1162, 1986. After-effects of photoperiod in sparrows.

45. Binkley, S., J. B. Riebman, and K. B. Reilly. *Science* 202:1198–1199, 1978. Pineal NAT oscillation *in vitro*.

46. Binkley S., M. Hryshchyshyn, and K. Reilly. *Nature (London)* 281:479–481, 1979. NAT activity in the retina.

47. Blake, G. M. *Nature (London)* 183:126–127, 1959. Control of diapause in carpet beetles.

48. Bligh, J. and K. G. Johnson. *J. Appl. Physiol.* 35:941–961, 1973. Glossary of terms for thermal physiology.

49. Block, G. D. and D. G. McMahon. *J. Comp. Physiol.* 155:387–395, 1984. Pacemaker localization in *Bulla* eye.

50. Block G. D. and S. F. Wallace. *Science* 217:155–157, 1982. Circadian pacemaker in the eye of *Bulla*.

51. Block, G. D., D. G. McMahon, S. F. Wallace, and W. O. Friesen. *J. Comp. Physiol. A* 155A:365–378, 1984. Cellular analysis of the *Bulla* circadian clock.

52. Block, G. D., M. H. Roberts, and A. E. Lusska. *J. Biol. Rhythms* 1:199–217, 1986. Different impulses in ocular coupling in *Aplysia*.

53. Blume, J. et al. *Naturwissenschaften* 49:525, 1962. Activity pattern of cave-dwelling amphipods.

54. Bons, N. C. R. *C. R. Seances Soc. Biol. Ses Fil.* 181:274–281, 1987. Centrifugal projections to the rat retina.

55. Boulos, Z. and L. P. Morin. *J. Biol. Rhythms* 1:1–15, 1985. Entrainment of split rhythms in hamsters.

56. Boulos, Z. and B. Rusak. *J. Comp. Physiol.* 146:411–417, 1982. Dark pulse PRCs in hamsters.

57. Boulos, Z. and B. Rusak. In *Vertebrate Circadian Systems* (J. Aschoff, S. Daan, and G. Groos, eds.), pp. 215–223. Springer-Verlag; Berlin, 1982. Phase-shifting of split rhythms by dark pulses.

58. Boulos, Z. and M. Terman. *J. Comp. Physiol.* 134:75–83, 1979. Splitting of behavioral rhythms in rats.

59. Boulos, Z. and M. Terman. *Neurosci. Biobehav. Rev.* 4:119–131, 1980. Food availability and daily rhythms.

60. Boulos, Z., A. Rosenwasser, and M. Terman. *Behav. Brain Res.* 1:39–65, 1980. Feeding schedules and circadian rhythms.

61. Boulos, Z. et al. *Physiol. Behav.* 45:507–515, 1989. Feeding schedules and circadian rhythms in squirrel monkeys.

62. Brann, M. R. and L. V. Cohen. *Science* 235:585–587, 1987. mRNA for retinal transducin is rhythmic.

63. Brattström, B. H. *Ecology* 44:238–255, 1963. Thermal requirements of amphibians.

64. Brattström, B. H. *Am. Midl. Nat.* 73:376–422, 1965. Body temperatures of reptiles.

65. Brattström, B. H. In *Comparative Physiology of Thermoregulation* (G. C. Whittow, ed.) Vol. 1, pp. 135–166. Academic Press, New York, 1970. Thermoregulation in amphibians.

66. Brett, J. R. *Am. Zool.* 11:99–113, 1971. Role of thermoregulation in salmon physiology and behavior.

67. Broda, H. et al. *Cell Biophys.* 8:47–67, 1985. Circadian communication between unicells.

68. Broda, H. et al. *J. Biol. Rhythms* 1:251–263, 1986. Circadian bioluminescence rhythms in *Gonyaulax*.

69. Brown, F. A., Jr. *J. Interdiscip. Cycle Res.* 14:137–162, 1983. The exogenous timing hypothesis.

70. Brown, F. A., Jr. and M. Webb. *Physiol. Zool.* 21:371–381, 1954. Temperature independent period of rhythm in crab *Uca*.

71. Brown, F. A., Jr. et al. *Biol. Bull. (Woods Hole, Mass.)* 109:238–254, 1955. Rhythms in oxygen consumption in crabs and salamanders.

72. Bruce, V. G. and D. H. Minis. *Science* 163:583–585, 1969. Action spectrum for light pulses in moths.

72a. Bruce, V. C. and C. S. Pittendrigh (1960) *J. Cell Comp. Physiol.* 56:25031. Effect of D_2O on phase and period in *Euglena*.

73. Bubenik, G. A., G. M. Brown, and L. J. Grota. *Brain Res.* 118:417–427, 1976. Melatonin localization in SCN.

74. Bünning, E. *The Physiological Clock.* Springer-Verlag, Berlin, 1973.

75. Bünning, E. and I. Moser. *Proc. Natl. Acad. Sci. U.S.A.* 70:3387–3390, 1973. Effects of K+ and ethanol on leaf movement rhythm.

76. Bushinger, A. In *Effects of Temperature on Ectothermic Organisms* (W. Wieser, ed.), pp. 229–232. Springer-Verlag, New York; 1973. Daily temperature rhythms and brood development in ants.

77. Cabanac, M. et al. *J. Physiol. (London)* 257:275–291, 1976. Rhythmic thermoregulatory responses to thermal stimuli in man.

78. Cahill, G. M. and M. Menaker. *Brain Res.* 410:125–129, 1987. Blocking SCN responses to optic nerve stimulation.

79. Cahill, K. and S. Lustick. *Comp. Biochem. Physiol. A* 55A:355–357, 1976. Thermoregulation and oxygen consumption in bees.

80. Calhoun, J. B. *The Ecology and Sociology of the Norway Rat.* U.S. Public Health Service, Bethesda, MD, 1962.

81. Card, J. P. and R. Y. Moore. *J. Comp. Neurol.* 206:390–396, 1982. APP in the geniculate projection to SCN.

82. Cardé, R. T. et al. *Experientia* 31:46–48, 1975. Temperature effects on mating rhythms in moths.

83. Carew, T. and D. Kelley, eds. *Perspectives in Neural Systems.* Alan R. Liss, New York, 1989.

84. Cassone, V. M. and M. Menaker. *J. Exp. Zool.* 232:539–549, 1984. Structure of avian circadian system.

85. Cassone, V. M. and M. Menaker. *J. Comp. Neurol.* 266:171–182, 1987. Retinohypothalamic projection in the house sparrow.

86. Cassone, V. M., M. H. Roberts, and R. Y. Moore. *Neurosci. Lett.* 81:29–34, 1987. Melatonin effects on SCN metabolism.

87. Casterlin, M. E. and W. W. Reynolds. *J. Thermal Biol.* 3:143–145, 1978. Behavioral thermoregulation in tadpoles.

88. Cederwall, H. In *Cyclic Phenomena in Marine Plants and Animals* (E. Naylor and R. G. Hartnoll, eds.), pp. 309–316. Pergamon, New York, 1979. Oxygen consumption and activity rhythms in amphipods.

89. Chandrashekaran, M. K. *Z. Vergl. Physiol.* 50:137–150, 1965. Diurnal and tidal rhythms in crabs.

90. Chandrashekaran, M. K. *J. Interdiscip. Cycle Res.* 5:371–380, 1974. Phase shifting by temperature pulses in *Drosophila.*

91. Chay, T. R. *Proc. Natl. Acad. Sci. U.S.A.* 78:2204–2207, 1981. A model for biological oscillations.

92. Chiba, Y. et al. In *Chronobiology* (L. E. Scheving, F. Halberg, and J. E. Pauly, eds.), pp. 603–606. Igaku Shoin, Tokyo; 1974. Oxygen consumption in beetles.

93. Chong, G. et al. *Comp. Biochem. Physiol. A* 46A:827–829, 1973. Circadian rhythm of panting threshhold in lizards.

94. Clarke, P. B. S. et al. *J. Neurosci.* 5:1307–1315, 1985. Nicotinic binding sites in rat brain.

95. Clausen, R. G. *Ecology* 17:216–226, 1936. Oxygen consumption in fishes.

96. Claussen, D. L. *Comp. Biochem. Physiol. A* 44A:137–153, 1973. Daily temperature preference patterns in frogs.

97. Cloudsley-Thompson, J. L. *J. Exp. Biol.* 33:576–582, 1956. Humidity responses and nocturnal activity in woodlice.

98. Cloudsley-Thompson, J. L. *Annu. Rev. Entomol.* 7:199–222, 1962. Microclimates and the distribution of terrestrial arthropods.

99. Cloudsley-Thompson, J. L. In *Comparative Physiology of Thermoregulation* (G. C. Whittow, ed.) Vol. 1, pp. 15–77. Academic Press, New York, 1970. Thermoregulation in terrestrial invertebrates.

100. Coleman, G. J., S. Harper, J. D. Clarke, and S. Armstrong. *Physiol. Behav.* 29:107–115, 1982. 'Memory' for feeding times.

101. Comperatore, C. A. and F. K. Stephan. *J. Biol. Rhythms* 2:227–242, 1987. Entrainment of duodenal activity to periodic feeding.

102. Cooper, M. L., G. E. Pickard, and R. Silver. *Brain Res. Bull.* 10:715–718, 1983. Retinohypothalamic projection in the ring dove.

103. Cornélissen, G. et al. *Cell Biophys.* 8:69–85, 1985. Does *Gonyaulax* measure a week?

104. Corrent, G., D. G. McAdoo, and A. Eskin. *Science* 202:977–979, 1978. Serotonin phase shifts *Aplysia* eye.

105. Cowgell, J. and H. Underwood. *J. Exp. Zool.* 210:189–194, 1979. Circadian rhythm in behavioral thermoregulation in lizards.

106. Cragg, P. A. *J. Exp. Biol.* 77:33–56, 1978. Oxygen consumption rhythms in lizards.

107. Crawshaw, L. I. *J. Comp. Physiol.* 95:315–322, 1974. Temperature selection and activity rhythms in crayfish.

108. Crews, S. J. et al. *Cell (Cambridge, Mass.)* 52:143–151, 1988. A per homolog in *Drosophila* is crucial for normal neurogenic development.

109. Cummings, F. W. *J. Theor. Biol.* 55:455–470, 1975. cAMP model for circadian rhythmicity.

110. Czeisler, C. A. et al. *Sleep Res.* 17:367, 1988. Critical light pulse affects amplitude of human circadian rhythms.

111. Czeisler, C. A. et al. *Science* 244:1328–1333, 1989. Bright light resets human circadian pacemaker.

112. Daan, S. et al. *J. Biol. Rhythms* 4:267–283, 1989. Basal metabolic rate, temperature and body size in kestrels.

113. Daan, S. *Ibis* 118:223–236, 1976. Light intensity and the timing of daily activity of finches.

114. Daan, S. *Ann. N. Y. Acad. Sci.* 290:51–59, 1977. Tonic and phasic effects of light.

115. Daan, S. In *Handbook of Behavioral Neurobiology* (J. Aschoff, ed.), Vol. 4, pp. 275–298. Plenum, New York, 1981. Adaptive daily strategies in behavior.

116. Daan, S. and J. Aschoff. *Oecologia* 18:269–316, 1975. Seasonal and latitudinal variations in circadian rhythms.

117. Daan, S. and C. Berde. *J. Theor. Biol.* 10:297–313, 1978. Circadian pacemaker simulations based on dual-oscillator model.

118. Daan, S. and C. S. Pittendrigh. *J. Comp. Physiol.* 106:253–266, 1976. PRC variability in nocturnal rodents.

119. Daan, S. and C. S. Pittendrigh. *J. Comp. Physiol.* 106:267–290, 1976. Effects of LL on circadian period in nocturnal rodents.

120. Dark, J., G. E. Pickard, and I. Zucker. *Brain Res.* 332:201–207, 1985. Persistence of circannual rhythms in ground squirrels with SCN lesions.

121. Davis, F. C. In *Biological Rhythms* (J. Aschoff, ed.), pp. 257–274. Plenum, New York, 1981. Ontogeny of circadian rhythms.

121a. Davis, F. C. and R. A. Gorski. *J. Biol. Rhythms* 1:77–89, 1986. Prenatal entrainment of hamster circadian rhythms.

122. Davis, F. C. and M. Menaker. *Am. J. Physiol.* 239:R149–R155, 1980. Temporal structure of hamster rhythms.

123. Davis, F. C., S. Stice, and M. Menaker. *Physiol. Behav.* 40:583–590, 1987. Social effects on hamster behavior.

124. Dawson, W. R. *Ecol. Stud.* 12:443–473, 1975. Physiological significance of preferred body temperature in reptiles.

125. de Castro, J. M. *Physiol. Behav.* 21:883–886, 1978. Behavioral effects on body temperature in rats.

126. DeCoursey, P. J. Ph.D. Thesis, University of Wisconsin, Madison, 1959. Effects of light on circadian activity rhythms in flying squirrels.

127. DeCoursey, P. J. *Cold Spring Harbor Symp. Quant. Biol.* 25:49–55, 1960. Phase control of activity in a rodent.

128. DeCoursey, P. J. *J. Comp. Physiol.* 78:221–235, 1972. LD ratios and the entrainment of circadian activity in a nocturnal and a diurnal rodent.

129. DeCoursey, P. J. ed. *Biological Rhythms in the Marine Environment.* University of South Carolina Press, Columbia, 1976.

130. DeCoursey, P. J. *J. Comp. Physiol.* 159:161–169, 1986. Light-sampling and entrainment in flying squirrels.

131. DeCoursey, P. J. *Soc. Res. Biol. Rhythms Abstr.* 1:75, 1988. PRCs for monochromatic light pulses in flying squirrels.

132. Deguchi, T. *Proc. Natl. Acad. Sci. U.S.A.* 72:2814-2818, 1975. Ontogeny of NAT rhythm in fetal rat.

133. Demian, J. J. and D. H. Taylor. *J. Herpetol.* 11:131–139, 1977. Pineal role in newt circadian rhythms.

134. Demmelmeyer, H. and D. Haarhaus. *J. Comp. Physiol.* 78:25–29, 1972. Entrainment by daily color temperature cycles in zebra finches.

135. Deitzer, G. F. et al. *Planta* 117:29–41, 1974. Energy metabolism and clock mechanism.

136. Dowse, H. B. and J. D. Palmer. *Nature (London)* 222:564–566, 1969. Entrainment by electrostatic fields in mice.

136a. Dowse, H. B. and J. M. Ringo. *J. Biol. Rhythms* 2:65–76, 1987. *Drosophila* clocks as a population of ultradian oscillators.

137. Dunlap, J. C. and J. F. Feldman. *Proc. Natl. Acad. Sci. U.S.A.* 85:1096–1100, 1988. Role of protein synthesis in clock mechanism.

138. Dunlap, J. C. and J. W. Hastings. *J. Biol. Chem.* 256:10509–10518, 1981. Circadian changes in cellular amounts of luciferase.

139. Du Preez, H. H. et al. *Comp. Biochem. Physiol. A* 89A:283–294, 1988. Oxygen consumption in fishes.

140. Dushay, M. S., M. Rosbash, and J. C. Hall. *J. Biol. Rhythms* 4:1–27, 1989. Visual system mutations and *Drosophila* rhythms.

141. Earnest, D. J. and F. W. Turek. *J. Comp. Physiol.* 145:405–411, 1982. Effects of LL and DD on split rhythms.

142. Earnest, D. J. and F. W. Turek. *Naturwissenschaften* 73:565–567, 1986. Responses of split activity components in hamsters.

143. Earnest, D. J., S. M. DiGiorgio, and C. D. Sladek. *Soc. Neurosci. Abstr.* 14:385, 1988. K⁺ effects on rhythms in SCN.

144. Ebihara, S. and H. Kawamura. *J. Comp. Physiol.* 141:207–214, 1981. SCN lesions in Java sparrows.

145. Ebihara, S., K. Uchiyama, and I. Oshima. *J. Comp. Physiol. A* 154A: 59–69, 1984. Role of eyes and pineal in the pigeon.

146. Edmunds, L. N., Jr., ed. *Cell Cycle Clocks.* Dekker, New York, 1984.

147. Edmunds, L. N., Jr. *Cellular and Molecular Bases of Biological Clocks.* Springer-Verlag, Berlin and New York, 1988.

148. Edmunds, L. N., Jr. and K. J. Adams. *Science* 211:1002–1013, 1981. Clocked cell cycle clocks.

149. Ehret, C. F. *Cold Spring Harbor Symp. Quant. Biol.* 25:149–158, 1960. Action spectra and nucleic acid metabolism.

150. Ehret, C. F. and E. Trucco. *J. Theor. Biol.* 15:240–262, 1967. Molecular clock model.

151. Engelmann, W. *Physiol. Plant.* 43:68–76, 1978. Attenuation of rhythmicity in Kalanchoe by critical light pulses.

152. Enright, J. T. *Science* 147:864–867, 1965. Entrainment of a tidal rhythm.

153. Enright, J. T. *J. Theor. Biol.* 8:426–468, 1965. Search for rhythmicity in biological time series.

154. Enright, J. T. *Annu. Rev. Ecol. Syst.* 1:221–238, 1970. Ecological aspects of endogenous rhythmicity.

155. Enright, J. T. *Z. Physiol.* 72:1–16, 1971. Heavy water slows biological timing.

156. Enright, J. T. *Z. Physiol.* 75:332–346, 1971. The internal clock of drunken isopods.

157. Enright, J. T. *J. Biol. Rhythms* 4:295–304, 1989. Statistical testing of methods.

158. Erckens, W. and W. Martin. *Z. Naturforsch. C: Biosci.* 37C:1266–1273, 1982. Exogenous and endogenous control of swimming activity in cave-dwelling fishes.

159. Erckens, W. and F. Weber. *Experientia* 32:1297–1299, 1976. Time measurement in cavernicolous fishes.

160. Eriksson, L.-O. *Aquilo Ser. Zool.* 14:68–79, 1973. Seasonal inversion of activity pattern in fishes.

161. Eriksson, L.-O. (1978) In *Rhythmic Activity of Fishes* (J. E. Thorpe, ed.), pp.

69–89. Academic Press, New York. Seasonal inversions of daily activity patterns in fishes.

162. Eriksson, L.-O., M.-L. Källquist, and T. Mossing. *Oecologia* 48:64–70, 1981. Seasonal changes in daily activity patterns of reindeer.

163. Erkert, H. G. *Oecologia* 14:269–287, 1974. Effect of moonlight on activity rhythms of the night monkey.

164. Erkert, H. G. and E. Rothmund. *Comp. Biochem. Physiol. A* 68A:383–390, 1981. Temperature sensitivities of circadian systems in bats.

165. Erkert, S. *Z. Vergl. Physiol.* 67:243–272, 1970. L:D ratios and entrainment in bats.

166. Erkinaro, E. *Aquilo, Ser. Zool.* 8:1–29, 1969. Seasonal changes in activity pattern in voles.

167. Erkinaro, E. *Experientia* 25:649, 1969. Free-running circadian rhythms in natural daylight.

168. Eskin, A. In *Biochronometry* (M. Menaker, ed.), pp. 55–80. Natl. Acad. Sci., Washington, DC 1971. Some properties of the system controlling the circadian activity rhythm of sparrows.

169. Eskin, A. and G. Corrent. *J. Comp. Physiol.* 117:1–21, 1977. Chemical effects on *Aplysia* eye rhythms.

170. Eskin, A. and J. S. Takahashi. *Science* 220:82–84, 1983. Adenylate cyclase activation shifts *Aplysia* eye rhythm.

171. Eskin, A. et al. *Proc. Natl. Acad. Sci. U.S.A.* 79:660–664, 1982. Involvement of cAMP in phase shifting by serotonin.

172. Eskin, A., J. S. Takahashi, M. Zatz, and G. D. Block. *J. Neurosci.* 4:2466–2471, 1984. Cyclic GMP mimics light effects on *Aplysia* eye rhythms.

173. Evans, K. J. *Comp. Biochem. Physiol.* 19:91–103, 1966. Entrainment by light and temperature cycles in lizards.

174. Eyster, M. B. *Ecol. Monogr.* 24:1–28, 1954. Effects of photoperiod, temperature, and season on bird activity patterns.

175. Feldman, J. F. *Proc. Natl. Acad. Sci. U.S.A.* 57:1080–1087, 1967. Period of *Euglena* rhythm altered by cycloheximide.

176. Feldman, J. F. *Annu. Rev. Plant Physiol.* 33:583–608, 1982. Genetic approaches to circadian clocks.

177. Feldman, J. F. and J. Dunlap. *Photochem. Photobiol. Rev.* 7:319–368, 1983. *Neurospora* in the study of circadian rhythms.

178. Figala, J. and K. Müller. *Aquilo, Ser. Zool.* 13:13–20, 1972. Effects of season on activity pattern in a tropical fish.

179. Fingerman, M. and A. Lago. *Anat. Rec.* 125:616, 1956. Rhythms in metabolic rate and locomotor activity in crayfish.

180. Fitzgerald, K. M. and I. Zucker. *Proc. Natl. Acad. Sci. U.S.A.* 73:2923–2927, 1976. Circadian basis of estrous cyclces.

181. Fleissner, G. *Naturwissenschaften* 70:366–367, 1983. Efferent control of the scorpion circadian system.

182. Folk, G. E., Jr., et al. *Iowa Acad. Sci.* 68:594–602, 1961. Effects of locomotor activity on body temperature in hamsters.

183. Forel, A. H. *Das Sinnesleben der Insekten* (M. Semon, trans.), Munich, Ernst Reinhardt Verlag, 1910.

184. Foster, K. W. et al. *Nature (London)* 311:756–759, 1984. A rhodopsin function in phototaxis.

185. Foster, R. G. and B. K. Follett. *J. Comp. Physiol. A* 157A:519–528, 1985. Rhodopsin-like pigment in extraretinal photoreception.

185a. Francis, A. J. P. and G. J. Coleman. *Physiol. Behav.* 43:471–477, 1988. Effects of ambient temperature cycles in rats.

186. Frank, K. D. and W. F. Zimmerman. *Science* 163:685–689, 1969. Action spectra for phase shifts in *Drosophila*.

187. Frisch, B. and J. Aschoff. *Physiol. Entomol.* 12:41–49, 1987. Food entrainment in bees.

188. Fuller, C. A. et al. *Science* 199:794–796, 1978. Impaired thermoregulation in monkeys under LL.

189. Fuller, C. A. et al. *Physiol. Behav.* 34:543–546, 1985. Heat loss and heat production rhythms in squirrel monkeys.

190. Gander, P. H. *Int. J. Chronobiol.* 6:243–262, 1979. Temperature effects on activity rhythms in Orthopetera.

191. Gänshirt, G., S. Daan, and M. P. Gerkema. *J. Comp. Physiol. A* 154A:669–674, 1984. Effects of constant light on starling activity.

192. Gaston, S. In *Biochronometry* (M. Menaker, ed.), pp. 541–548. Natl. Acad. Sci., Washington, DC, 1971. The pineal and sparrow circadian rhythms.

193. Gaston, S. and M. Menaker. *Science* 160:1125–1127, 1968. Effects of pinealectomy in sparrows.

194. Geiser, F. *J. Comp. Physiol.* 158:25–37, 1988. Metabolism during hibernation and daily torpor in mammals and birds.

194a. Gern, W. A., D. W. Owens, and C. L. Ralph. *J. Exp. Zool.* 205:371–376, 1978. Melatonin rhythms in trout after pinealectomy and blinding.

195. Gibbs, F. P. *Am. J. Physiol.* 241:R17–R20, 1981. Temperature dependence of rat circadian pacemaker.

196. Gibbs, F. P. *Am. J. Physiol.* 244:R607–R610, 1983. Temperature dependence of hamster circadian pacemaker.

197. Gilbert, F. S. *Oikos* 42:249–255, 1984. Thermoregulation and the structure of swarms in hoverflies.

198. Gillette, M. U. *Soc. Neurosci. Abstr.* 13:51, 1987. K^+ phase shifts SCN firing rate rhythm.

199. Glotzbach, S. F, C. M. Cornett, and H. C. Heller. *Brain Res.* 419:279–286, 1987. Firing rate changes in single SCN cells during sleep and waking.

200. Goodwin, B. *Temporal Organization in Cells.* Academic Press, New York, 1963.

201. Gordon, S. A. and G. A. Brown. In *Biochronometry* (M. Menaker, ed.), pp. 363–371. Natl. Acad. Sci, Washington, DC, 1971. Spectral sensitivity of circadian entrainment in pocket mice.

202. Goss, R. J. *J. Exp. Zool.* 170:311–324, 1969. Photoperiodic control of annual cycles in deer.

203. Goto, K. et al. *Science* 228:1284–1288, 1985. Biochemical modeling of circadian clock in *Euglena*.

204. Gould, J. L. *Anim. Behav.* 35:1579–1581, 1987. Learned foraging behavior according to time of day in bees.

205. Graf, R. In *Symposium on Temperature Regulation in Birds*, pp. 331–335. Deutsche Ornithologen-Gesellschaft, Berlin, 1980. Diurnal rhythms of thermoregulation in pigeons.

206. Graf, R. *Pfluegers Arch.* 386:181–185, 1980. Thermoregulatory responses to thermal stimuli in pigeons.

207. Graf, R., S. Krishna, and H. C. Heller. *Am. J. Physiol.* 256:R733–R738, 1989. Regulated nocturnal hypothermia induced in pigeons by food deprivation.

208. Graham, T. E. and V. H. Hutchison. *Comp. Biochem. Physiol. A* 63A:299–305, 1979. Temperature affects daily activity pattern in turtles.

209. Gratz, R. K. and V. H. Hutchison. *Physiol. Zool.* 50:99–114, 1977. O_2 consumption and activity in water snakes.

210. Grobbelaar, N. et al. *FEMS Microbiol. Lett.* 37:173–178, 1986. An endogenous rhythm in a procaryote.

211. Guimond, R. W. and V. H. Hutchison. *Comp. Biochem. Physiol.* 27:177–195, 1968. Effects of temperature and photoperiod on oxygen consumption in frogs.

212. Güldner, F.-H. *Exp. Brain Res.* 50:373–376, 1983. Numbers of neurons in the suprachiasmatic nuclei.

213. Gundlach, H. *Z. Tierpsychol.* 25:955–1005, 1968. Activity patterns in wild boar.

214. Gwinner, E. *Experientia* 22:765–766, 1966. Entrainment by song cycles in birds.

215. Gwinner, E. *J. Ornithol.* 109:70–95, 1968. Circannual rhythm of migratory behavior in birds.

216. Gwinner, E. *Science* 185:72–74, 1974. Testosterone induces splitting of circadian rhythms in starlings.

217. Gwinner, E. *J. Comp. Physiol.* 103:315–328, 1975. Effects of season and testosterone on τ in starlings.

218. Gwinner, E. *J. Comp. Physiol.* 126:123–129, 1978. Effects of pinealectomy in starlings.

219. Gwinner, E. *Circannual Rhythms.* Springer-Verlag, Berlin, 1986.

220. Gwinner, E. and I. Benzinger. *J. Comp. Physiol. A* 127A:209–213, 1978. Melatonin entrainment of starlings.

221. Gwinner, E., R. Subbaraj, C. Bluhm, and M. Gerkema. *J. Biol. Rhythms* 2:109–120, 1987. Pinealectomy effects on feeding and activity.

222. Halberg, F. et al. In *Photoperiodism and Related Phenomena in Plants and Animals* (R. B. Withrow, ed.), Publ. No. 55, pp. 803–878. Am. Assoc. Adv. Sci., Washington, DC, 1959. An early review and introduction of term circadian.

224. Hall, J. C. and M. Rosbash. *J. Biol. Rhythms* 2:153–178, 1987. Genetic analysis of biological rhythms.

225. Hamblen, M. et al. *J. Neurogenet.* 3:249–291, 1986. Germ line transformation of circadian properties.

226. Hamner, K. C. et al. *Nature (London)* 195:476–480, 1962. Circadian rhythms measured near the South Pole.

227. Handelman, G., R. Ravizza, and W. J. Ray. *Horm. Behav.* 14:107–115, 1980. Social effects on hamster estrous cycles.

228. Handler, A. M. and R. J. Konopka. *Nature (London)* 279:236–238, 1979. Pacemaker transplantation in *Drosophila*.

229. Hanegan, J. L. and J. E. Heath. *J. Exp. Biol.* 53:611–627, 1970. Activity patterns and energetics in moths.

230. Hardeland, R. and I. Balzer. *Int. J. Biometerol.* 32:149–162, 1988. Geophysical periodicity and the circadian oscillator.

230a. Hardin, P. E., J. C. Hall, and M. Rosbash. *Nature* 343:536–540, 1990. *Per* gene mRNA levels undergo circadian oscillations in *Drosophila* head.

231. Harrington, M. E. and B. Rusak. *J. Biol. Rhythms* 1:309–325, 1986. PRC changes after intergeniculate leaflet lesions.

232. Harrington, M. E. and B. Rusak. *Physiol. Behav.* 42:183–189, 1988. Intergeniculate leaflet lesions alter responses to constant light.

233. Harrington, M. E. and B. Rusak. *Vis. Neurosci.* 2:367–375, 1989. Photic responses of geniculohypothalamic tract cells.

234. Harrington, M. E., D. M. Nance, and B. Rusak. *Brain Res. Bull.* 15:465–472, 1987. NPY projection to SCN from geniculate.

235. Harrington, M. E., D. M. Nance, and B. Rusak. *Brain Res.* 410:275–282, 1987. Double-labelling study of geniculohypothalamic tract.

236. Hastings, J. W. *Annu. Rev. Microbiol.* 13:297–312, 1959. Unicellular clocks.

237. Hastings, J. W. *Cold Spring Harbor Symp. Quant. Biol.* 25:131–144, 1960. Biochemical aspects.

238. Hastings, J. W. and B. M. Sweeney. *Proc. Natl. Acad. Sci. U.S.A.* 53:804–811, 1957. Q.10 for period of rhythm less than 1.

238a. Hastings, J. W. and B. M. Sweeney. *Biol. Bull.* 115:440–458, 1958. Light pulse PRC in *Gonyaulax*.

239. Hastings, J. W. and B. M. Sweeney. *J. Gen. Physiol.* 43:697–706, 1960. Action spectrum for phase shifts in *Gonyaulax*.

240. Hastings, J. W., H. Broda, and C. H. Johnson. In *Temporal Order* (L. Rensing and N. I. Jaeger, eds.), pp. 213–221. Springer-Verlag, Berlin, 1985. Amino acid analogs phase shift the *Gonyaulax* clock.

241. Hastings, J. W. et al. *Photochem. Photbiol.* 46:86S, 1987. Action spectrum for phase shifting in *Chlamydomonas*.

242. Hastings, M. H., J. Herbert, N. D. Martensz, and A. C. Roberts. *Ann. N. Y. Acad. Sci.* 453:182–204, 1985. Synchronization of annual rhythms by light.

243. Hayden, P. and R. G. Lindberg. *Science* 164:1288–1289, 1969. Entrainment by cyclic pressure changes in pocket mice.

244. Heath, J. E. *Science* 138:891–892, 1962. Temperature-independent morning emergence in lizards.

245. Hebrant, F. *J. Insect Physiol.* 16:1229–1235, 1970. Oxygen consumption rhythms in termite colonies.

246. Heckrotte, C. *Anim. Behav.* 10:193–207, 1962. Temperature effects on phase and pattern of activity rhythm in snakes.

247. Heckrotte, C. *J. Interdiscip. Res.* 6:279–290, 1975. Temperature affects daily activity pattern in garter snakes.

248. Heusner, A. *C. R. Seances Soc. Biol.* 150:1246–1249, 1956. Body temperature rhythms independent of activity.

249. Heusner, A. and J. T. Enright. *Science* 154:532–533, 1966. Long term activity recordings in aquatic animals.

250. Heusner, A. and T. Stussi. *Insectes Soc.* 11:239–265, 1964. Energy metabolism in bees.

251. Hicks, J. W. and M. L. Riedesel. *J. Comp. Physiol.* 149:503–510, 1983. Oxygen consumption rhythms in garter snakes.

252. Hildebrandt, G. In *Chronobiology* (L. E. Scheving, F. Halberg, and J. E. Pauly, eds.), pp. 234–240. Igaku Shoin, Tokyo, 1974. Circadian variations of thermoregulatory response in man.

253. Himmer, A. *Biol. Rev. Cambridge Philos. Soc.* 7:224–253, 1932. Thermoregulation in social insects.

254. Hiroshige, T., K.-I. Honma, and K. Watanabe. *J. Physiol.* (*London*) 325:507–519, 1982. Maternal effects on rhythms in infant rats.

255. Hoffmann, K. *Oecologia* 3:184–206, 1969. Entrainment by temperature cycles.

256. Hoffmann, K. In *Biochronometry* (M. Menaker, ed.), pp. 134–148. Natl. Acad. Sci., Washington, DC, 1971. Splitting of circadian rhythms in tree shrews.

257. Hoffmann, K. In *The Molecular Basis of Circadian Rhythms* (J. W. Hastings and H. G. Schweiger, eds.), pp. 63–75. Abakon Verlagsgesellschaft, Berlin, 1976. Adaptive significance of biological rhythms corresponding to geophysical cycles.

258. Homma, K. and J. W. Hastings. *J. Biol. Rhythms* 3:49–58, 1988. Quantized generation times in *Gonyaulax*.

259. Honma, K.-I. and T. Hiroshige. *Jpn. J. Physiol.* 28:159–169, 1978. Circadian rhythms of locomotor activity and body temperature in the rat.

260. Honma, K.-I., S. Honma, and T. Hiroshige. *Physiol. Behav.* 40:767–774, 1987. Methamphetamine effects on circadian rhythms.

261. Honma, K.-I. et al. *Physiol. Behav.* 30:905–913, 1983. Effects of restricted daily feeding in rats.

262. Honma, S. et al. *Physiol. Behav.* 39:211–215, 1987. Persistence of anticipatory rhythms in VMH-ablated rats.

262a. Horton, T. H. *Biol. Reprod.* 31:499–504, 1984. Prenatal photoperiod affects growth and reproductive development in voles.

263. Horton, T. H. *Biol. Reprod.* 33:934–939, 1985. Prenatal photoperiod effects in voles.

264. Hudson, J. W. In *Strategies in Cold: Natural Torpidy and Thermogenesis* (L. C. H. Wang and J. W. Hudson, eds.), pp. 67–108. Academic Press, New York, 1978. Daily torpor as a thermoregulatory adaptation.

265. Hudson, J. W. and S. L. Kimzey. *Comp. Biochem. Physiol.* 17:203–217, 1966. After-effects of photoperiod duration in sparrows.

265a. Hudson, D. J. and M. E. Lickey. *Brain Res.* 183:481–485, 1980. Desynchronization between two eyes of *Aplysia*.

266. Hutchison, V. H. and M. A. Kohl. *Z. Vergl. Physiol.* 75:367–382, 1971. Effect of photoperiod on daily oxygen consumption rhythms in toads.

267. Hutchison, V. H. and J. D. Maness. *Am. Zool.* 19:367–384, 1979. Behavioral thermoregulation in ectotherms.

268. Hutchison, V. H. and K. K. Spriestersbach. *Copeia* 1986:612–618, 1986. Temperature preference rhythms in salamanders.

268a. Inouye, S. T. and H. Kawamura. *Proc. Natl. Acad. Sci. U.S.A.* 76:5962–5966, 1979. Circadian neural rhythm in a hypothalamic island.

269. Jacklet, J. W. *Science* 164:562–564, 1969. Circadian rhythm in isolated *Aplysia* eye.

270. Jacklet, J. W. *J. Comp. Physiol.* 90:33–45, 1974. Effects of light on isolated *Aplysia* eye.

271. Jacklet, J. W. *Biol. Bull.* (*Woods Hole, Mass.*) 160:199–227, 1981. Cellular mechanisms of circadian rhythms.

272. Jacklet, J. W., ed. *Cellular and Neuronal Oscillators.* Dekker, Amsterdam, 1989.

273. Jackson, F. R. et al. *Nature (London)* 320:185–188, 1986. Product of *per* locus shares homology with proteoglycans.

274. Jameson, E. W. et al. *Comp. Biochem. Physiol. A* 56A:73–79, 1977. Daily oxygen consumption patterns in lizards.

275. Jegla, T. C. and T. L. Poulson. *J. Exp. Zool.* 168:273–282, 1968. Daily patterns of activity and oxygen consumption in cave-dwelling crayfish.

276. Jilge, B., H. Hörnicke, and H. Stähle. *Am. J. Physiol.* 253:R46–R54, 1987. Food entrainment in rabbits.

277. Johnson, C. H., J. Roeber, and J. W. Hastings. *Science* 223:1428–1430, 1984. Cellular changes in luciferase concentration during circadian cycle.

278. Johnson, C. H., I. Miwa, T. Kondo, and J. W. Hastings. *J. Biol. Rhythms* 4:405–416, 1989. Light responsiveness of circadian system in *Paramecium.*

279. Johnson, R. F., L. P. Morin, and R. Y. Moore. *Brain Res.* 462:301–312, 1988. Retinal projections to rodent hypothalamus.

280. Johnson, R. F. et al. *Proc. Natl. Acad. Sci. U.S.A.* 85:5301–5304, 1988. Geniculate lesions block phase-shifts to a benzodiazepine.

281. Johnston, P. G. and I. Zucker. *Am. J. Physiol.* 244:R338–R346, 1983. Lability of circadian rhythms in cotton rats.

282. Joshi, D. and M. K. Chandrashekaran. *J. Exp. Zool.* 230:325–328, 1984. Flashes of 0.5 msec reset the circadian clock of a bat.

283. Joshi, D. and M. K. Chandrashekaran. *Naturwissenshaften* 72:548–549, 1985. PRCs for white light pulses of different spectral composition.

284. Joshi, D. and M. K. Chandrashekaran. *J. Comp. Physiol.* 156:189–198, 1985. Spectral sensitivity of photoreceptors in phase shifting in bats.

285. Kahn, P. G. K. and S. M. Pompea. *Nature (London)* 275:606–611, 1978. Nautiloid growth rhythms and dynamical evolution of the earth-moon system.

286. Karakashian, M. W. and J. W. Hastings. *Proc. Natl. Acad. Sci. U.S.A.* 48:2130–2137, 1962. Clock effects of macromolecular biosynthesis inhibitors.

287. Kasbohm, P. *Helgol. Wiss. Meersuntersuch.* 16:157–178, 1967. Oxygen consumption rhythms in frogs.

288. Kavaliers, M. *Can. J. Zool.* 56:2591–2596, 1978. Seasonal changes in the circadian period of the lake chub.

289. Kavaliers, M. *Rev. Can. Biol.* 38:281–292, 1979. Role of the pineal in fish.

290. Kavaliers, M. *Behav. Neural Biol.* 30:56–57, 1980. Retinal and extraretinal entrainment action spectra in the lake chub.

291. Kavaliers, M. *J. Comp. Physiol.* 136:215–218, 1980. Effects of season and pinealectomy on circadian period in the burbot.

292. Kavaliers, M. *Can. J. Zool.* 58:1399–1403, 1980. Social effects on fish activity rhythms.

293. Kavaliers, M. *Comp. Biochem. Physiol. A* 68A:127–129, 1981. Effects of pinealectomy in white suckers.

294. Kavaliers, M. *J. Comp. Physiol.* 146:235–243, 1982. Temperature preference rhythms in fishes.

295. Kavaliers, M and D. M. Ross. *Can. J. Zool.* 59:1326–1334, 1981. Seasonal changes in entrained and free-running rhythms in the lake chub.

296. Kayser, C. *Arch. Sci. Physiol.* 19:369–413, 1965. Energy metabolism in hibernating animals.

297. Kayser, C. and A. A. Heusner. *J. Physiol. (Paris)* 59:3–116, 1967. Circadian rhythms of thermoregulation and energy metabolism.

297a. Kayser, C. and J.-P. Schieber. *Arch. Sci. Physiol.* 23:365–382, 1969. Oxygen consumption rhythms in frogs.

298. Kayser, C. et al. *C. R. Seances Soc. Biol.* 161:918–921, 1967. Oxygen consumption in hibernating marmots.

299. Kenagy, G. J. *Ecology* 54:1201–1219, 1973. Daily activity patterns in kangaroo rats.

300. Kenagy, G. J. *Oecologia* 24:105–140, 1976. Seasonal changes in daily activity rhythms of kangaroo rats.

301. Kenagy, G. J. *J. Comp. Physiol.* 128:21–36, 1978. Seasonality of circadian parameters in rodents.

302. Kenagy, G. J. and D. Vleck. In *Vertebrate Circadian Systems* (J. Aschoff, S. Daan, and G. Groos, eds.), pp. 322–328. Springer-Verlag, Berlin, 1982. Daily metabolic patterns and ecological characteristics of small mammals.

303. Kennedy, C. H. *Ecology* 9:367–379, 1928. Evolutionary level in relation to geographic, seasonal and diurnal distribution of insects.

304. Kippert, F. *J. Interdiscip. Cycle Res.* 16:77–84, 1985. Endosymbiosis and the origin of circadian rhythms.

305. Klein S. E. B., S. Binkley, and K. Mosher. *Photochem. Photobiol.* 41:453–457, 1985. PRCs for light and dark pulses in sparrows.

306. Klemm, E. and H. Ninnemann. *Photochem. Photobiol.* 24:369–371, 1976. Action spectrum for phase delays in *Drosophila*.

307. Kloppstech, K. *Planta* 165:502–506, 1985. Circadian mRNA in plants.

307a. Koehler, W. K. and G. Fleissner. *Nature* 274:708–710, 1978. Desynchronization of retinal oscillators.

308. Koltermann, R. In *Experimental Analysis of Insect Behaviour* (L. Barton Browne, ed.), pp. 218–227. Springer-Verlag, New York, 1974. Periodicity and learning performance of the honeybee.

309. Kondo, T. *Plant Physiol.* 88:953–958, 1988. Phase shift of circadian rhythm in *Lemma* by amino acid analogs.

310. Konopka, R. J. and S. Benzer. *Proc. Natl. Acad. Sci. U.S.A.* 68:2112–2116, 1971. Isolation of clock gene mutants *Drosophila*.

311. Konopka, R. J., S. Wells, and T. Lee. *Mol. Gen. Genet.* 190:284–288, 1983. Mosaic analysis of *Drosophila* mutants.

312. Kosh, R. J. and V. H. Hutchison. *Copeia* 1968:244–246, 1968. Daily rhythm of temperature tolerance in turtles.

313. Kow, L. and D. W. Pfaff. *Brain Res.* 297:275–286, 1984. Neurotransmitter effects on SCN neurons *in vitro*: Effects of estrogen.

314. Kramm, K. R. *Am. Nat.* 116:452–453, 1980. Relation between the value and precision of τ.

315. Krüll, F. *Oecologia* 24:141–148, 1976. Sun position as possible Zeitgeber in the arctic.

316. Krüll, F. *Oecologia* 24:149–157, 1976. Zeitgebers in the arctic midsummer.

317. Krüll, F. *Oecologia* 25:301–308, 1976. Effects of L:D intensity ratio in birds.

318. Krüll, F., H. Demmelmeyer, and H. Remmert. *Naturwissenschaften* 72:197–203, 1985. Circadian rhythms at high polar latitudes.

319. Kureck, A. *Oecologia* 40:311–323, 1979. Temperature affects eclosion time in midges.

320. Kurumiya, S. and H. Kawamura. *J. Comp. Physiol.* 162:301–308, 1988. Circadian rhythm of multiple unit activity in guinea pig SCN.

321. Kyriacou, C. P. and J. C. Hall. *Proc. Natl. Acad. Sci. U.S.A.* 77:6729–6733, 1980. Circadian rhythm mutations affect high frequency rhythms.

322. Laitman, R. S. and F. W. Turek. *J. Comp. Physiol.* 134:339–343, 1979. Pinealectomy and entrainment in sparrows.

323. Lamprecht, G. and F. Weber. *Int. J. Speleol.* 10:351–379, 1978. Activity patterns of cave-dwelling beetles.

324. Lang, J. W. *Science* 191:575–577, 1976. Daily rhythms in amphibious behavior of alligators.

325. Lankinen, P. *J. Comp. Physiol.* 159:123–142, 1986. Latitudinal variation in circadian eclosion rhythm and photoperiodic diapause in *Drosophila*.

326. Lee, T. M., M. S. Carmichael, and I. Zucker. *Am. J. Physiol.* 259:R831–R836, 1986. Circannual variations in circadian rhythms in ground squirrels.

327. Lees, J. G. et al. *J. Comp. Physiol.* 153:123–132, 1983. Phase shifting of split components by dark pulses.

328. Lehman, M. N. et al. *J. Neurosci.* 7:1626–1638, 1987. Circadian rhythm restoration with SCN transplants.

329. Leinweber, F. *Z. Bot.* 44:337–364, 1956. Effect of temperature on period of *Phaseolus* rhythm.

330. Levandowsky, M. *Ann. N. Y. Acad. Sci.* 361:369–374, 1981. Endosymbiosis and the origin of circadian rhythms.

331. Licht, P. and A. G. Brown. *Ecology* 48:598–611, 1967. Daily temperature preference patterns in newts.

332. Lindberg, R. G. and P. Hayden. *Chronobiologia* 1:356–361, 1974. Entrainment by temperature cycles in the little pocket mouse.

333. Liou, S. Y. et al. *Brain Res. Bull.* 16:527–531, 1986. Release of amino acids into SCN after optic nerve stimulation.

334. Lloyd, D. *Biochem. J.* 242:313–321, 1987. Biochemistry of the cell cycle.

335. Lloyd, D., S. W. Edwards, and J. C. Fry. *Proc. Natl. Acad. Sci. U.S.A.* 79:3785–3788, 1982. Temperature compensated ultradian oscillations in *Acanthamoeba castrelanii*.

336. Lloyd, D., R. K. Poole, and S. W. Edwards. *The Cell Division Cycle.* Academic Press, London, 1982. Different length periods.

337. Loher, W. *Behav. Ecol. Sociobiol.* 5:383–390, 1979. Influence of male song on female locomotion in crickets.

338. Lohmann, M. and J. T. Enright. *Comp. Biochem. Physiol.* 22:289–296, 1967. Entrainment by sound cycles in finches.

339. Loike, J. D. et al. *Proc. Natl. Acad. Sci. U.S.A.* 84:807–813, 1988. Creatine ports and cellular uptake.

340. Loros, J. J. et al. *Science* 243:385–388, 1989. Circadian messages in *Neurospora.*

341. MacBride, S. E. Ph.D. Thesis, University of Pittsburgh, Pittsburgh, PA, 1973. Pineal rhythms in chickens.

342. Mackay, R. S. *Nature (London)* 204:355–358, 1964. Telemetric recording of body temperature in tortoise and marine iguana.

343. Mackay, R. S. *Copeia* 1968:252–259, 1968. Diurnal rhythms in peristaltic activity and body temperature in reptiles.

344. Malinowski, J. et al. *J. Comp. Physiol. B* 155:257–267, 1985. Singular light pulses attenuate amplitude of division rhythm in *Euglena.*

345. Marais, J. F. K. *Mar. Biol. (Berlin)* 50:9–16, 1978. Oxygen consumption rhythms in fishes.

346. Marimuthu, G. *Oecologia* 61:352–357, 1984. Effects of season on circadian rhythms in tropical bats.

347. Marimuthu, G., S. Rajan, and M. K. Chandrashekaran. *Behav. Ecol. Sociobiol.* 8:147–150, 1981. Social entrainment in bats.

348. Mason, R. *Physiol. (London)* 377:1–13, 1986. Serotonin sensitivity rhythms in SCN and LGN.

349. Mason, R. and A. Brooks. *Neurosci. Lett.* 95:296–301, 1988. Melatonin sensitivity rhythms in rat SCN cells *in vitro.*

350. Mason, R., M. E. Harrington, and B. Rusak. *Neurosci. Lett.* 80:173–179, 1987. NPY effects on hamster SCN neurons *in vitro.*

351. Mattern, D. L., L. R. Forman, and S. Brody. *Proc. Natl. Acad. Sci. U.S.A.* 79:825–829, 1982. A mutation affecting temperature compensation.

352. Mautz, W. J. *Copeia* 1979:577–584, 1979. The metabolism of reclusive lizards.

353. May, M. L. *Ecology* 58:787–798, 1977. Thermoregulation and reproductive activity in tropical dagonflies.

354. May, M. L. *Annu. Rev. Entomol.* 24:313–349, 1979. Insect thermoregulation.

355. McClintock, M. K. *Nature (London)* 229:244, 1971. Menstrual synchronization in humans.

356. McClintock, M. K. *Horm. Behav.* 10:264–276, 1978. Estrous synchrony mediated by pheromones.

357. McClung, C. R., B. Fox, and J. Dunlap. *Nature (London)* 339:558–562, 1989. *Neurospora* clock gene shares sequences with *Drosophila* period gene.

358. McDaniel, M., F. Sulzman, and J. W. Hastings. *Proc. Natl. Acad. Sci. U.S.A.* 71:4389–4391, 1974. D_2O slows the clock in *Gonyaulax.*

359. McGuire, R. A. et al. *Science* 181:956–957, 1973. Action spectrum for entrainment in rats.

360. McMahon, D. G. and G. D. Block. *J. Comp. Physiol. A* 161A:335–346, 1987. Membrane potential effects on *Bulla* eye rhythms.

361. McMahon, D. G., S. F. Wallace and G. D. Block. *J. Comp. Physiol. A* 155A:379–385, 1984. Neurophysiological basis of rhythmicity in *Bulla* eye.

362. McMillan, J. P., S. A. Gauthreaux, Jr., and C. W. Helms. *BioScience* 29:1259–1260, 1970. Role of circadian rhythms in migratory restlessness in birds.

363. McNab, B. K. *Am. Nat.* 12:1–12, 1978. The evolution of endothermy in the phylogeny of mammals.

364. Mead, M. and J.-C. Gilhodes. *J. Comp. Physiol.* 90:47–52, 1974. Locomotor activity patterns in cave-dwelling millipedes.

365. Meijer, J. H. and W. J. Rietveld. *Physiol. Rev.* 69:671–707, 1989. Neurophysiology of the SCN in rodents.

366. Meijer, J. H., G. A. Groos, and B. Rusak. *Brain Res.* 382:109–118, 1986. Luminance coding in rat and hamster SCN neurons.

367. Meijer, J. H., E. van der Zee, and M. Dietz. *J. Biol. Rhythms* 4:333–348, 1988. Carbachol injections effects on hamster rhythms.

368. Meijer, J. H., E. van der Zee, and M. Dietz. *Neurosci. Lett.* 86:177–183, 1988. Glutamate effects on rhythms in hamsters.

369. Meijer, J. H., B. Rusak, and M. E. Harrington. *Brain Res.* 501:315–323, 1989. Photic responses of SCN cells in a ground squirrel.

370. Menaker, M. *Nature (London)* 184:1251–1252, 1959. Endogenous rhythms of body temperature in hibernating bats.

371. Menaker, M. *J. Cell Comp. Physiol.* 57:81–86, 1961. Seasonal variation in τ at low body temperature in bats.

372. Menaker, M. *Proc. Natl. Acad. Sci. U.S.A.* 59:414–421, 1968. Extraretinal entrainment in sparrows.

373. Menaker, M. In *Biochronometry* (M. Menaker, ed.), pp. 315–332. Natl. Acad. Sci., Washington, DC, 1971. Extraretinal photoreceptors in sparrows.

374. Menaker, M. and A. Eskin. *Science* 154:1579–1581, 1966. Sound entrainment in sparrows.

375. Menaker, M. and S. Wisner. *Proc. Natl.* *Acad. Sci. U.S.A.* 80:6119–6121, 1983. Melatonin rhythm in isolated *Anolis* pineal.

375a. Mergenhagen, D. *Eur. J. Cell Biol.* 33:13–18, 1984. Genetic characterization of a *Chlamydomonas* clock mutant.

376. Meyer, A. *Naturwissenschaften* 55:234–235, 1968. Entrainment by sound cycles in hamsters.

377. Meyer, D. C. and W. B. Quay. *Endocrinology (Baltimore)* 98:1160–1165, 1976. Daily rhythms in SCN uptake of serotonin.

378. Michel, U. and R. Hardeland. *J. Interdiscip. Cycle Res.* 16:17–23, 1985. Temperature effects on an ultradian oscillation in *Tetrahymena*.

379. Miller, J. D., D. M. Murakami, and C. A. Fuller. *J. Neurosci.* 7:978–986, 1987. Light and nicotinic responses of SCN cells.

380. Miller, M. M., J. Silver, and R. B. Billiar. *Brain Res.* 290:67–75, 1984. Steroid effects on alpha-bungarotoxin binding in SCN.

381. Mistlberger, R. E. and B. Rusak. *Neurosci. Lett.* 72:357–362, 1986. Carbachol effects on rhythms in ovariectomized rats.

382. Mitsui, A. et al. *Nature (London)* 323:720–722, 1986. Rhythms in photosynthesis and nitrogen fixation in cyanobacteria.

383. Moore, R. Y. *Brain Res.* 49:403–409, 1973. Retinohypothalamic projections in mammals.

383a. Moore, R. Y. In *The Neurosciences: Third Study Program* (F. O. Schmitt and F. G. Worden, eds.), pp. 537–542. MIT Press, Cambridge, MA, 1974. Neural regulation of circadian rhythms.

384. Moore, R. Y. and D. C. Klein. *Brain Res.* 71:17–33, 1974. SCN control of pineal gland.

385. Moore-Ede, M. C. *Am. J. Physiol.* 250:R735–R752, 1986. Circadian physiology: predictive versus reactive homeostasis.

386. Moore-Ede, M. C. and F. M. Sulzman. In *Handbook of Behavioral Neurobiology* (J. Aschoff, ed.), Vol. 4, pp. 215–241. Plenum, New York, 1981. Internal temporal order.

387. Moore-Ede, M. et al. *The Clocks That Time Us.* Harvard Univ. Press, Cambridge, MA, 1982.

388. Morgan, K. R. *J. Exp. Biol.* 128:107–122, 1987. Temperature regulation, energy metabolism and mate-searching in rain beetles.

389. Morgan, K. R. and B. Heinrich. *J. Exp. Biol.* 133:59–71, 1987. Temperature regulation in syrphid flies.

390. Mori, S. *Cold Spring Harbor Symp. Quant. Biol.* 25:333–344, 1960. Circadian rhythms in the sea pen.

391. Morin, L. P. *J. Biol. Rhythms* 3:237–248, 1988. Effects of age on hamster rhythms.

392. Morin, L. P. and L. A. Cummings. *Physiol. Behav.* 29:665–675, 1982. Effects of steroid hormones on splitting.

393. Morrison, P. In *Comparative Physiology of Temperature Regulation* (J. P. Hannon and E. Viereck, eds.), pp. 389–419. Arctic Aeromedical Laboratory, Fort Wainswright, 1962. Thermoregulation in mammals from the tropics and from high altitudes.

394. Morse, D. et al. *Proc. Natl. Acad. Sci. U.S.A.* 86:172–176, 1989. Circadian change in luciferin binding protein involves translational control.

395. Mote, M. I. and K. R. Black. *Photochem. Photobiol.* 34:257–265, 1981. Action spectrum for entrainment in cockroaches.

396. Mrosovsky, N. *J. Comp. Physiol.* 136:355–360, 1980. Effects of temperature on circannual cycles.

397. Mrosovsky, N. *J. Comp. Physiol. A* 162A:35–46, 1988. Social entrainment in hamsters.

398. Müller, K. *Naturwissenschaften* 55:140, 1968. Free-running rhythm in natural daylight in the arctic.

399. Müller, K. *Oikos,* Suppl. 13:108–121, 1970. Daily and annual rhythms in sculpins at high latitudes.

400. Müller, K. *Aquilo, Ser. Zool.* 14:1–18, 1973. Circadian rhythms of aquatic organisms in the subarctic summer.

401. Müller, K. In *Rhythmic Activity of Fishes* (J. E. Thorpe, ed.), pp. 91–104. Academic Press, New York, 1978. The flexibility of the circadian system of fishes at different latitudes.

402. Nagy, F. et al. *Genes Dev.* 2:376—388, 1988. Circadian control of message production.

403. Nakashima, H. et al. *Science* 212:361–362, 1981. Genetic evidence tht protein synthesis is required for clock function.

404. Navaneethakannan, K. and M. K. Chandrashekaran. *Exp. Biol.* 45:267–273, 1986. PRCs for light and dark pulses in palm squirrels.

405. Naylor, E. *Symp. Soc. Exp. Biol.* 39:63–93, 1985. Tidal rhythms.

406. Nelson, D. R. and R. H. Johnson. *Copeia* 1970:732:739, 1970. Diurnal activity rhythms in sharks.

407. Nelson, R. J. and I. Zucker, *Comp. Biochem. Physiol. A* 69A:145–148, 1981. Absence of extraretinal photoreceptors in mammals.

408. Neumann, D. In *Handbook of Behavioral Neurobiology* (J. Aschoff, ed.), Vol. 4, pp. 351–389. Plenum, New York, 1981. Tidal and lunar rhythms.

409. Neville, A. C. *Biol. Rev. Cambridge Philos. Soc.* 42:421–441, 1967. Daily growth layers in animals and plants.

410. Newman, G. C. and F. E. Hospod. *Brain Res.* 381:345–350, 1986. *In vitro* rhythm of 2-deoxyglucose uptake.

411. Nishino, H., K. Koizumi, and C. McC. Brooks. *Brain Res.* 112:45–59, 1976. SCN stimulation effects on pineal function.

412. Njus, D. et al. *Nature (London)* 248:116–120, 1974. Membrane model of the circadian clock.

413. Njus, D. et al. *J. Comp. Physiol.* 117:335–344, 1977. Conditionality of circadian rhythmicity.

414. Nordtug, T. and T. B. Melø. *Holarctic Ecol.* 11:202–209, 1988. Diurnal variations in natural light conditions in arctic and subarctic areas.

415. Norgren, R. B. and R. Silver. *Brain, Behav. Evol.* 34:73–83, 1989. Retinohypothalamic projections in birds.

416. Nyholm, E. S. *Ann. Zool. Fenn.* 2:77–123,

1965. Seasonal inversion of activity pattern in bats.

417. Olesiak, W. et al. *J. Biol. Rhythms* 2:121–138, 1987. Correlation between protein synthesis inhibition and phase shifting.

418. O'Reilly, H, S. M. Armstrong, and G. J. Coleman. *Physiol. Behav.* 38:471–476, 1986. Food anticipation in a marsupial.

419. Page, T. L. In *Handbook of Behavioral Neurobiology* (J. Aschoff, ed.), Vol. 4, pp. 145–172. Plenum, New York, 1981. Neural and endocrine control in invertebrates.

420. Page, T. L. *Science* 216:73–75, 1982. Transplantation of the pacemaker in cockroaches.

421. Page, T. L. *J. Comp. Physiol.* 152:231–240, 1983. Regeneration of rhythms in lobectomized roaches.

422. Page, T. L. *J. Comp. Physiol.* 153:353–363, 1983. Optic tract regeneration and coupling in roaches.

423. Page, T. L. *J. Insect Physiol.* 31:235–242, 1985. Temperature effects on lobectomized cockroaches.

424. Page, T. L. and J. L. Larimer. *J. Comp. Physiol.* 97:59–80, 1975. Neural control of rhythms in crayfish.

425. Paietta, J. *J. Theor. Biol.* 97:77–82, 1982. Photooxidation and the evolution of circadian rhythmicity.

426. Palmer, J. D. *Mar. Behav. Physiol.* 13:201–219, 1988. A circalunidian rhythm.

426a. Palmer, J. D. *Mar. Behav. Physiol.* 14:231–243, 1989. Translocation experiment with circadian rhythms.

427. Paulsen, H. and L. Bogorad. *Plant Physiol.* 88:1104–1109, 1988. Circadian mRNA for chl a/b binding protein in tobacco.

428. Peiponen, V. A. *Arch. Soc. Vanamo* 17:171–178, 1962. Activity patterns of voles in the arctic summer.

429. Pengelley, E. T. et al. *Comp. Biochem. Physiol. A* 53A:273–277, 1976. Circannual rhythmicity in the ground squirrel.

430. Peterson, E. L. *Nature (London)* 280:677–679, 1979. Circadian oscillators in constant bright light.

431. Peterson, E. L. *Biol. Cybernet.* 40:171–179, 1980. Circadian pacemaker recovery from extended light exposure.

432. Peterson, E. L. and D. S. Saunders. *J. Theor. Biol.* 86:256–277, 1980. Limit cycle interpretation of circadian oscillator.

433. Pickard, G. E. *J. Comp. Neurol.* 211:65–83, 1982. HRP tracing of retinal projections to the SCN.

434. Pickard, G. E., R. Kahn, and R. Silver. *Physiol. Behav.* 32:763–766, 1984. Splitting of temperature rhythm in hamsters.

435. Pickard, G. E., M. R. Ralph, and M. Menaker. *J. Biol. Rhythms* 2:35–56, 1987. Role of intergeniculate leaflet in light effects on rhythms.

436. Pittendrigh, C. S. *Evolution (Lawrence, Kans.)* 4:43–63, 1950. Activity patterns of mosquitoes.

437. Pittendrigh, C. S. *Proc. Natl. Acad. Sci. U.S.A.* 40:1018–1029, 1954. Temperature independence in the *Drosophila* clock.

438. Pittendrigh, C. S. In *Perspectives in Marine Biology* (A. A. Buzzati-Traverso, ed.), pp. 239–268. Univ. of California Press, Berkeley, 1958. Perspectives in the study of biological clocks.

439. Pittendrigh, C. S. In *Science in the Sixties* (D. L. Arm, ed.), pp. 96–111. Univ. of New Mexico, Office of Publications, Albuquerque, 1965. Biological clocks.

440. Pittendrigh, C. S. *Z. Pflanzenphysiol.* 54:275–307, 1966. Light pulse PRCs and entrainment in *Drosophila*.

441. Pittendrigh, C. S. In *The Neurosciences: Third Study Program* (F. O. Schmitt and F. G. Worden, eds.), pp. 437–458. MIT Press, Cambridge, MA, 1974. The circadian organization of multicellular systems.

442. Pittendrigh, C. S. In *Handbook of Behavioral Neurobiology* (J. Aschoff, ed.), Vol. 4, pp. 95–124. Plenum, New York, 1981. Circadian systems.

443. Pittendrigh, C. S. In *Handbook of Behavioral Neurobiology* (J. Aschoff, ed.), Vol. 4, pp. 57–80. Plenum, New York, 1981. Circadian systems: Entrainment.

444. Pittendrigh, C. S. and V. G. Bruce. In *Photoperiodism and Related Phenomena in Plants and Animals* (R. B. Withrow, ed.), Publ. No. 55, pp. 475–505. Am. Assoc. Adv. Sci., Washington, DC, 1959. Daily rhythms as coupled oscillator systems.

445. Pittendrigh, C. S. and S. Daan. *J. Comp. Physiol.* 106:231–252, 1976. Stability and lability of circadian period in nocturnal rodents.

446. Pittendrigh, C. S. and S. Daan. *J. Comp. Physiol.* 106:291–331, 1976. Circadian entrainment in nocturnal rodents.

447. Pittendrigh, C. S. and S. Daan. *J. Comp. Physiol.* 106:333–355, 1976. Circadian pacemakers in nocturnal rodents.

448. Pittendrigh, C. S. and D. H. Minis. *Am. Nat.* 98:261–294, 1964. Entrainment by skeleton photoperiods in *Drosophila*.

449. Pittendrigh, C. S., V. Bruce, and P. Kaus. *Proc. Natl. Acad. Sci. U.S.A.* 44:965–973, 1958. Coupled oscillator interpretation of transient cycles in *Drosophila*.

450. Pittendrigh, C. S. et al. *Proc. Natl. Acad. Sci. U.S.A.* 70:2037–2041, 1973. Are D$_2$O effects compensated in circadian systems?

451. Pittendrigh, C. S., J. Elliot, and T. Takamura. *Ciba Found. Symp.* 104:26–47, 1984. The circadian component in photoperiodic induction.

452. Pohl, H. *Z. Vergl. Physiol.* 45:109–153, 1961. Thermoregulatory and metabolic rhythms during hibernation.

453. Pohl, H. *Z. Vergl. Physiol.* 58:364–380, 1968. Effects of temperature on circadian period.

454. Pohl, H. *Z. Vergl. Physiol.* 58:381–394, 1968. Effects of temperature on LD-entrained rhythms.

455. Pohl, H. *Comp. Biochem. Physiol. A* 56A:145–153, 1977. Circadian rhythms of metabolism in finches as a function of light intensity and season.

455a. Pohl, H. *Comp. Bioch. Physiol. B* 76B:723–729, 1983. Circadian activity patterns of diurnal ground squirrels.

456. Prosser, R. A. and M. U. Gillette. *J. Neurosci.* 9:1073–1081, 1989. SCN firing rhythm *in vitro* and cAMP effects.

457. Protomastro, J. J. *Physis (Buenos Aires)* 32C:123–128, 1973. Temperature preference and brood transport rhythms in ants.

458. Raab, J. L. and M. S. Brady. *Nature (London)* 260:38–39, 1976. Do nocturnal rodents run more efficiently at night?

459. Ralph, C. L. *Physiol. Zool.* 50:41–55, 1957. Rhythms of activity and oxygen consumption in the earthworm.

460. Ralph, M. R. and M. Menaker. *Science* 241:1225–1227, 1988. Hamster rhythm mutant.

461. Ralph, M. R. and M. Menaker. *J. Neurosci.* (in press), 1988. GABA regulation of circadian responses to light.

462. Ralph, M. R., F. C. Davis, and M. Menaker. *Soc. Neurosci. Abstr.* 14:462, 1988. SCN transplants convey donor period to the recipient.

462a. Rapp, P. E. *J. Exp. Biol.* 81:281–306, 1979. An atlas of cellular oscillations.

463. Reddy, P. et al. *Cell (Cambridge, Mass.)* 46:53–61, 1986. *Per* locus codes for a proteoglycan.

464. Redman, J., S. Armstrong, and K. T. Ng. *Science* 219:305–328, 1983. Entrainment of rats by melatonin.

465. Reebs, S. G. *Ethology* 80:172–181, 1989. Acoustical entrainment in sparrows.

466. Reebs, S. G. and N. Mrosovsky. *J. Biol. Rhythms* 4:39–48, 1989. Phase shifts caused by induced running.

467. Reebs, S. G., R. J. Lavery, and N. Mrosovsky. *J. Comp. Physiol. A* 165:811–818, 1989. Running mediates dark pulse effects on hamsters.

468. Regal, P. J. *Copeia* 1966:588–590, 1966. Thermophilic reponse following feeding in reptiles.

469. Regal, P. J. In *Chronobiology* (L. E. Scheving, F. Halberg, and J. E. Pauly, eds.), pp. 709–711. Igaku Shoin, Tokyo, 1974. Circadian rhythms in the temperature preference of a lizard.

470. Remmert, H. *Biol. Zentralbl.* 79:577–584,

1960. Daily temperature preference patterns in insects.

471. Remmert, H. *Arctic Animal Ecology* Springer-Verlag, Berlin, 1980.

472. Rence, B. and W. Loher. *Science* 190:385–387, 1975. Entrainment by temperature in bilobectomized crickets.

473. Rensing, L. *Z. Vergl. Physiol.* 53:62–83, 1966. Rhythms in energy metabolism in *Drosophila*.

474. Reppert, S. M. and W. J. Schwartz. *Science* 220:969–971, 1983. Maternal coordination of prenatal circadian rhythms.

475. Reppert, S. M. and W. J. Schwartz. *J. Neurosci.* 4:1677–1682, 1984. Circadian metabolic rhythms in fetal rat SCN.

475a. Reppert, S. M. and G. R. Uhl. *Endocr.* 120:2483–2487, 1987. Circadian mRNA for vasopressin.

476. Reppert, S. M. et al. *Am. J. Physiol.* 243:E489–E498, 1982. Vasopressin and melatonin rhythms in cat CSF.

477. Reynolds, W. W. and M. W. Casterlin. *Comp. Biochem. Physiol. A* 64A:25–28, 1979. Behavioral thermoregulation and activity in lobsters.

478. Reynolds, W. W. and M. W. Casterlin. *Hydrobiologia* 67:179–182, 1979. Thermoregulatory behavior of the pink shrimp.

479. Reynolds, W. W. and M. E. Casterlin. In *Environmental Physiology of Fishes* (M. A. Ali, ed.), pp. 497–518. Plenum, New York, 1979. Thermoregulatory rhythms in fishes.

480. Richter, C. P. *Biological Clocks in Medicine and Psychiatry*. Thomas, Springfield, IL, 1965.

481. Rismiller, P. D. and G. Heldmaier. *Physiol. Zool.* 58:71–79, 1985. Thermal behavior as a function of time of day in lizards.

482. Rismiller, P. D. and G. Heldmaier. *Oecologia* 75:125–131, 1988. Effects of photoperiod on temperature selection in lizards.

483. Roberts, L. A. *Ecology* 49:809–819, 1968. Oxygen consumption in lizards.

484. Roberts, M. H. and G. D. Block. *J. Biol. Rhythms* 1:55–75, 1986. Mutual coupling between *Bulla* eyes.

485. Roberts, S. K. de F. *J. Cell. Comp. Physiol.* 58:175–186, 1962. Phase shifting and entrainment by temperature in cockroaches.

486. Robertson, L. M. and J. S. Takahashi. *J. Neurosci.* 8:12–21, 1988. Circadian rhythm in pineal cell cultures.

487. Roenneberg, T. and J. W. Hastings. *Naturwissenschaften* 75:206–207, 1988. Two photoreceptors control the circadian clock in a unicell.

488. Roenneberg, T., H. Nakamura, and J. W. Hastings. *Nature (London)* 334:432–434, 1988. Creatine accelerates the circadian clock in *Gonyaulax*.

489. Roenneberg, T. et al. *J. Biol. Rhythms* 4:201–216, 1989. The motility rhythm in *Gonyaulax*.

489a. Rosenwasser, A. M. *Prog. Psychobiol. Physiol. Psychol.* 13:155–226, 1988. Neurobiology of circadian pacemakers.

490. Rosenwasser, A. M. et al. *Physiol. Behav.* 32:25–30, 1984. Memory for daily feeding time in rats.

491. Rosenwasser, A. M., G. Trubowitsch, and N. T. Adler. *Neurosci. Lett.* 58:183–187, 1985. Deoxyglucose uptake rhythms in different brain regions.

492. Ross, L. G. and R. W. McKinney. *Comp. Biochem. Physiol. A* 89A:637–643, 1988. Respiratory cycles in fishes.

493. Rothman, B. and F. Strumwasser. *J. Gen. Physiol.* 68:359–384, 1976. Phase shifting *Aplysia* rhythm by protein in synthesis inhibitors.

494. Ruby, N. F. et al. *Am. J. Physiol.* 257:R210–R215, 1989. Role of the SCN in daily torpor and temperature rhythms in Siberian hamsters.

495. Rusak, B. *J. Comp. Physiol.* 118:145–164, 1977. Suprachiasmatic nuclei and circadian rhythms in hamsters.

496. Rusak, B. In *Handbook of Behavioral Neurobiology* (J. Aschoff, ed.), Vol. 4, pp. 183–213. Plenum, New York, 1981. Vertebrate behavioral rhythms.

497. Rusak, B. In *The Pineal Gland* (R. J. Reiter, ed), Vol. 3, pp. 27–51. CRC Press, Boca Raton, FL, 1982. Pineal role in circadian system of birds and mammals.

498. Rusak, B. *J. Biol. Rhythms* 4:121–134, 1989. Modelling the mammalian circadian system.

499. Rusak, B. and G. Bina. *Annu. Rev. Neurosci.* 13:387–401, 1990. Neurotransmitters in the circadian system.

500. Rusak, B. and G. A. Groos. *Science* 215:1407–1409, 1982. SCN electrical stimulation shifts rhythms.

501. Rusak, B. and R. Mason. *Soc. Neurosci. Abstr.* 14:1298, 1988. Melatonin responses in hamster SCN.

502. Rusak, B. and I. Zucker. *Physiol. Rev.* 59:449–526, 1979. Neural regulation of circadian rhythms.

503. Rusak, B. et al. *J. Comp. Physiol. A* 164A:165–171, 1988. Entrainment of hamsters to hoarding opportunities.

504. Rusak, B., J. H. Meijer, and M. E. Harrington. *Brain Res.* 493:283–291, 1989. Stimulation of intergeniculate leaflet phase shifts hamster rhythms.

505. Saint-Girons, H. and M. C. Saint-Girons. *Vie Milieu* 7:133–226, 1956. Activity rhythms and thermoregulation in reptiles.

506. Saint-Paul, U. *Comp. Biochem. Physiol. A* 89A:675–682, 1988. Diurnal oxygen consumption rhythms in fishes.

507. Sandeen, M. I. et al. *Physiol. Zool.* 27:350–356, 1954. Daily and tidal rhythms of oxygen consumption in marine snails.

508. Sanders, C. J. and G. S. Lucuik. *Can. Entomol.* 104:1751–1762, 1972. Temperature effects on calling rhythm in budworms.

509. Sato, T. and H. Kawamura. *Neurosci. Res.* 1:67–72, 1984. Suprachiasmatic nucleus rhythms in a diurnal squirrel.

510. Saunders, D. S. *Science* 181:358–360, 1973. Thermoperiodism in wasps.

511. Saunders, D. S. *Insect Clocks*. Pergamon, Oxford, 1976.

512. Sawaki, Y., I. Nihonmatsu, and H. Kawamura. *Neurosci. Res.* 1:67–72, 1984. Neonatal SCN transplants restore circadian rhythms.

513. Scheck, S. H. and E. D. Fleharty. *Physiol. Zool.* 52:390–397, 1979. Daily energy budget and activity patterns in ground squirrels.

514. Schmidt, H. P. and W. Engelmann. *J. Interdiscip. Cycle Res.* 16:324, 1986. Electrolyte and ion channel inhibitor effects.

515. Schmidt-Nielsen, K. et al. *Am. J. Physiol.* 188:103–112, 1957. Body temperature of the camel and its relation to water economy.

516. Schwartz, W. J., L. Davidsen, and C. Smith. *J. Comp. Neurol.* 189:157–167, 1980. Deoxyglucose uptake rhythms in rat suprachiasmatic nucleus neurons.

517. Schwartz, W. J., R. J. Gross, and M. T. Morton. *Proc. Natl. Acad. Sci. U.S.A.* 84:1694–1698, 1987. Tetrodotoxin effects on SCN function.

518. Schweiger, H.-G. and M. Schweiger. *Int. Rev. Cytol.* 51:315–342, 1977. Molecular mechanism of cellular circadian rhythms.

518a. Shibata, S. and R. Y. Moore. *J. Biol. Rhythms* 3:265–276, 1988. NPY and vasopressin effects on rat SCN cells in vitro.

519. Shibata, S., S. Y. Liou, and S. Ueki. *Neurosci. Lett.* 39:187–192, 1983. Neurotransmitter effects on SCN neurons *in vitro*.

520. Shibata, S., S. Y. Liou, and S. Ueki. *Neuropharmacology* 28:403–409, 1986. Excitatory amino acids and optic nerve transmission to SCN.

521. Shibuya, C. A. et al. *Naturwissenschaften* 67:45–47, 1980. Simultaneous splitting of drinking and locomotor activity rhythms.

522. Shido, O. *J. Therm. Biol.* 12:273–279, 1987. Day-night variation of thermoregulatory responses to intraperitoneal heating.

523. Silver, R. et al. *Cell Tissue Res.* 253:189–198, 1988. Opsin immunoreactivity in avian brain.

524. Silyn-Roberts, H. et al. *J. Interdiscip. Cycle Res.* 17:181–187, 1986, Temperature dependence of *Thalassomyxa australis* rhythmicity.

525. Simpson, J. *Science* 133:1327–1333, 1961. Nest climate regulation in honey bee colonies.

526. Simpson, S. M. and B. K. Follett. *J. Comp. Physiol.* 144:381–389, 1981. Pinealectomy and hypothalamic lesions in quail.

527. Smith, E. N. *Am. Zool.* 19:239–247, 1979. Behavioral and physiological thermoregulation of crocodilians.

528. Smits, A. W. *Copeia* 1984:689–696, 1984. Activity patterns and thermal biology of toads.

529. Solem, J. O. *Aquilo, Ser. Zool.* 14:80–83, 1973. Daily activity patterns in Ephemeroptera at high latitudes.

530. Sollars, P. J. and D. P. Kimble. *Soc. Neurosci. Abstr.* 14:49, 1988. Cross-species transplants of fetal hypothalamus.

531. Songdahl, J. H. and V. H. Hutchison. *Herpetologica* 28:148–156, 1972. Oxygen consumption rhythms in lizards.

532. Southwick, E. E. and J. N. Mugaas. *Comp. Biochem. Physiol. A* 40A:935–944, 1971. Homeothermy of honeybee hives.

533. Sower, L. L., H. H. Shorey, and L. K. Gaston. *Ann. Entomol. Soc. Am.* 64:488–492, 1971. Temperature affects calling rhythm in moths.

534. Spangler, H. G. *Ann. Entomol. Soc. Am.* 65:1073–1076, 1972. Daily activity rhythms of bees.

535. Spangler, H. G. *Ann. Entomol. Soc. Am.* 66:449–451, 1973. Effects of light on circadian rhythms in bees.

536. Spellerberg, I. F. and K. Hoffmann. *Naturwissenschaften* 59:517–518, 1972. Circadian rhythm in lizard critical minimum temperature.

537. Springer, A. D. and J. S. Gaffney. *J. Comp. Neurol.* 203:401–424, 1981. Retinal projections to hypothalamus of goldfish.

538. Springer, A. D. and A. S. Mednick. *Brain Res.* 323:293–296, 1984. Retinal neurons innervating the goldfish SCN.

539. Stebbins, R. C. and R. E. Barwick. *Copeia* 1968:541–547, 1968. Radiotelemetric study of thermoregulation in a lace monitor.

540. Stephan, F. K. *J. Comp. Physiol.* 143:401–410, 1981. Entrainment limits for food-entrainment in rats.

541. Stephan, F. K. *Physiol. Behav.* 32:663–671, 1984. Phase shifts of food-entrained rhythms.

542. Stephan, F. K. *Physiol. Behav.* 38:537–544, 1986. Food entrainment of dominant pacemaker in rats.

543. Stephan, F. K., J. M. Swann, and C. L. Sisk. *Behav. Neural Biol.* 25:346–363, 1979. Food entrainment after SCN lesions.

544. Stephenson, L. A. et al. *Am. J. Physiol.* 246:R321–R324, 1984. Circadian rhythm in sweating and cutaneous blood flow.

545. Stetson, M. H., J. A. Elliott, and B. D. Goldman. *Biol. Reprod.* 34:664–669, 1986. Prenatal photoperiod effects on reproduction in Djungarian hamsters.

546. Stokkan, K.-A., A. Mortensen, and A. S. Blix. *Am. J. Physiol.* 251:R264–R267, 1986. Seasonal changes in feeding rhythms of ptarmigans at high latitudes.

547. Stussi, T. *Arch. Sci. Physiol.* 26:131–159, 1972. Heterothermy in bees.

548. Stussi, T. and A. Heusner. *C. R. Seances Soc. Biol.* 157:1509–1512, 1963. Oxygen consumption rhythms in insects.

549. Stussi, T. and A. Heusner. *C. R. Seances Soc. Biol. Ses Fil.* 157:2304–2307, 1963. Oxygen consumption and muscular activity in isolated bees.

550. Subbaraj, R. and M. K. Chandrashekaran. *Oecologia* 29:341–348, 1977. "Rigid" internal timing of bat flight activity.

551. Subbaraj, R. and M. K. Chandrashekaran. *J. Comp. Physiol.* 127:239–243, 1978. PRCs for dark pulses in bats.

552. Sulzman, F. M. and S. A. Sickles. *Physiologist* 25, Supl.: 165–166, 1982. Daily rhythms of activity and temperature of *Macaca nemestrina*.

553. Sulzman, F. M., C. A. Fuller, and M. C.

Moore-Ede. *Physiol. Behav.* 18:775–779, 1977. Entrainment of monkeys by food cycles.

554. Sulzman, F. M. et al. *J. Appl. Physiol.* 43:795–800, 1977. Environmental synchronizers of squirrel monkey circadian rhythms.

555. Sulzman, F. M. et al. *Cell Biophys.* 4:97–103, 1982. No detectable circadian interaction in mixed out of phase cultures.

556. Sulzman, F. M. et al. *Science* 225:232–234, 1984. *Neurospora* circadian rhythm in an earth orbiting space vehicle.

557. Swade, R. H. and C. S. Pittendrigh. *Am. Nat.* 101:431–466, 1967. Circadian locomotor rhythms of rodents in the arctic.

558. Swade, R. H. *J. Theor. Biol.* 227–239, 1969. Tonic effects of light and parametric entrainment.

559. Sweeney, B. M. *Cold Spring Harbor Symp. Quant. Biol.* 25:145–148. Photosynthesis rhythm in isolated single cells.

560. Sweeney, B. M. *Int. J. Chronobiol.* 2:25–33, 1974. A physiological model for circadian rhythms.

561. Sweeney, B. M. *Plant Physiol.* 53:337–342, 1974. Phase shifting by valinomycin and ethanol.

562. Sweeney, B. M. *Plant Physiol.* 64:314–344, 1979. Bright light does not immediately stop clock.

563. Sweeney, B. M. *Rhythmic Phenomena in Plants.* Academic Press, San Diego, CA, 1987.

563a. Sweeney, B. M. and J. W. Hastings. *Cold Spring Harbor Symp. Quant. Biol.* 25:87–104. Review of temperature effects in different species.

564. Sweeney, B. M. and M. B. Borgese. *J. Phycol.* 25:183–186, 1989. Evidence for a circadian mechanism in a prokaryote.

565. Takahashi, J. S. and M. Menaker. *J. Neurosci.* 2:815–828, 1982. The SCN and sparrow circadian rhythms.

566. Takahashi, J. S. and M. Menaker. *J. Comp. Physiol. A* 154A:435–440, 1984. Multiple circadian oscillators in isolated pineal fragments.

567. Takahashi, J. S., H. Hamm, and M.

Menaker. *Proc. Natl. Acad. Sci. U.S.A.* 77:2319–2322, 1980. Melatonin rhythm from pineal *in vitro*.

568. Takahashi, J. S. et al. *Nature (London)* 308:186–188, 1984. Spectral sensitivity of mammalian entrainment system.

569. Taylor, W. R. and J. W. Hastings. *J. Comp. Physiol.* 130:359–362, 1979. Aldehydes phase shift the *Gonyaulax* clock.

570. Taylor, W. R., and J. W. Hastings. *Naturwissenschaften* 69:94–96, 1982. Minute long drug pulses cause phase shifts of hours.

571. Taylor, W. R., V. Gooch, and J. W. Hastings. *J. Comp. Physiol.* 130:355–358, 1979. Ethanol shortens the period and shifts phase in *Gonyaulax*.

572. Taylor, W. R., J. C. Dunlap, and J. W. Hastings. *J. Exp. Biol.* 97:121–136, 1982. Protein synthesis inhibitors phase shift.

573. Taylor, W. R. et al. *J. Comp. Physiol.* 148:11–25, 1982. Drug induced singularity.

574. Templeton, J. R. In *Comparative Physiology of Thermoregulation* (G. C. Whittow, ed.), Vol. 1, pp. 167–221. Academic Press, New York, 1970. Thermoregulation in reptiles.

575. Terman, M. and J. Terman. *Ann. N.Y. Acad. Sci.* 453:147–161, 1985. A pacemaker for visual sensitivity rhythms.

576. Thinès, G. et al. *Ann. Soc. R. Belg.* 96:61–116, 1965. Activity patterns of a cave-dwelling fish.

577. Thomson, A. M., D. C. West, and I. G. Vlachonikolis. *Neurosci. Lett.* 52:329–334, 1984. Firing patterns of SCN neurons.

578. Timbal, J. et al. *Pfluegers Arch.* 335:97–108, 1972. Thermal balance in man during 24 h in controlled environment.

579. Tokura, H. and J. Aschoff. *Am. J. Physiol.* 245:R800–R804, 1983. Effects of temperature on the circadian rhythm of pig-tailed macaques.

580. Tokura, H. and T. Oishi. *Comp. Biochem. Physiol. A* 81A:271–275, 1985. Effect of temperature cycles in Djungarian hamsters.

581. Tombes, R. M. and B. M. Shapiro. *Cell (Cambridge, Mass.)* 41:325–344, 1985. Creatine as an energy shuttle molecule.

582. Tomioka, K. and Y. Chiba. *J. Comp. Physiol.* 147:299–304, 1982. Rhythm reversal during post-embryonic development in crickets.

583. Toutain, P. L. and Y. Ruckebusch. *Experientia* 31:312–314, 1975. Cyclic arousal in the hedgehog.

584. Traub, R. D. et al. *Science* 243:1319–1325, 1989. Model of the origin of rhythmic oscillations in the hippocampal slice.

585. Truman, J. W. In *Biochronometry* (M. Menaker, ed.), pp. 483–501. Natl. Acad. Sci., Washington, DC, 1971. Role of the brain in ecdysis in silkmoths.

586. Truman, J. W. *J. Comp. Physiol.* 81:91–114, 1972. The silkmoth brain as the location of the biological clock controlling eclosion.

587. Truman, J. W. *Science* 182:727–729, 1973. Temperature affects phase of activity in silkmoths.

588. Tucker, V. A. *Ecology* 47:245–252, 1966. Diurnal torpor in pocket mice.

589. Turek, F. W. et al. *Recent Prog. Horm. Res.* 40:143–183, 1984. Role of the circadian system in reproductive phenomena.

590. Tweedy, D. G. and W. P. Stephen. *Experientia* 26:377–379, 1970. Temperature pulses synchronize emergence rhythm in bees.

591. Uhl, G. R. and S. M. Reppert. *Science* 232:390–393, 1986. Vasopressin mRNA rhythm in rat SCN.

592. Underwood, H. *Science* 195:587–589, 1977. Pinealectomy effects in lizards.

593. Underwood, H. *J. Comp. Physiol.* 141:537–547, 1981. Effects of pinealectomy and blinding in lizards.

594. Underwood, H. In *The Pineal Gland* (R. J. Reiter, ed.), Vol. 3, pp. 1–25. CRC Press, Boca Raton, FL, 1982. Pineal role in lower vertebrates.

595. Underwood, H. and T. Siopes. *J. Exp. Zool.* 232:557–566, 1984. Role of eyes and pineal in quail.

596. Van den Pol, A. N. *Am. J. Physiol.* 240:R16–R22, 1981. Rhythm of hypothalamic protein synthesis.

597. Van den Pol, A. N. and K. L. Tsujimoto. *Neuroscience* 15:1049–1086, 1985. Neurotransmitters of the suprachiasmatic nucleus.

598. Vaněček, J., A. Pavlik, and H. Illnerová. *Brain Res.* 435:359–362, 1987. Melatonin receptors in hypothalamus.

599. Viswanathan, N. and M. K. Chandrashekaran. *Nature (London)* 317:530–531, 1985. Maternal entrainment of mouse pups.

600. Volknandt, W. and R. Hardeland. *Comp. Biochem. Physiol. C* 78C:51–54, 1984. Effects of aldehydes and alcohols on protein synthesis.

601. Walker, E. P. *Mammals of the World.* Johns Hopkins Univ. Press, Baltimore, MD, 1975.

602. Walz, B. and B. M. Sweeney. *Proc. Natl. Acad. Sci. U.S.A.* 76:6443–6447, 1979. Kinetics of cycloheximide induced phase shifting.

603. Watts, A. G., L. W. Swanson, and G. Sanchez-Watts. *J. Comp. Neurol.* 258:204–229, 1987. Efferent projections of the suprachiasmatic nuclei.

604. Wells, J. W. *Nature (London)* 197:948–950, 1963. Coral growth and geochronometry.

605. Welsh, D. K. et al. *J. Comp. Physiol.* 158:827–834, 1986. Precision of circadian timing in the mouse.

606. Wenger, C. B. et al. *J. Appl. Physiol.* 41:15–19, 1976. Nocturnal lowering of thresholds for sweating and vasodilation.

607. West, G. C. *Ecology* 49:1035–1045, 1968. Bioenergetics of ptarmigans under natural conditions.

608. West, G. C. and H. Pohl. *J. Comp. Physiol.* 83:289–302, 1973. Effects of L:D intensity and duration ratios in chaffinches.

609. Wever, R. Z. *Vergl. Physiol.* 55:255–277, 1967. Effects of twilight on circadian rhythms.

610. Wever, R. *The Circadian System of Man: Experiments Under Temporal Isolation.* Springer-Verlag, New York, 1979.

611. Wheal, H. V. and A. M. Thomson. *Neuroscience* 13:97–104, 1984. Electrical properties of SCN neurons.

612. Whitten, W. K., F. H. Bronson, and J. A. Greenstein. *Science* 161:584–585, 1968. Estrous cycle modification by male odors.

613. Winfree, A. T. *The Geometry of Biological Time.* Springer-Verlag, New York, 1980.

614. Winfree, A. T. *When Time Breaks Down.* Princeton University Press, Princeton, NJ, 1987.

615. Wood, S. C. et al. *J. Comp. Physiol.* 127:331–336, 1978. Aerobic metabolism in lizards.

616. Woolum, J. C. and F. Strumwasser. *J. Comp. Physiol.* 151:253–259, 1983. Is the period of the circadian oscillator homeostatically regulated in *Aplysia*?

617. Yeung, S. J. and A. Eskin. *J. Biol. Rhythms* 3:225–236, 1988. Protein synthesis inhibitors and *Aplysia* eye rhythms.

618. Young, M. W. et al. *Adv. Biosci.* 73:43–53, 1988. The molecular genetic approach to studies of rhythms in *Drosphila*.

619. Zatz, M. et al. *Brain Res.* 438:199–215, 1988. Photoendocrine trandsuction in cultured chick pineal cells.

620. Zehring, W. A. et al. *Cell (Cambridge, Mass.)* 39:369–376, 1984. *Per* gene in *Drosophila*.

621. Zielinski, W. J. *Physiol. Behav.* 38:613–620, 1986. Anticipation of feeding in small carnivores.

622. Zimmerman, N. H. and M. Menaker. *Proc. Natl. Acad. Sci. U.S.A.* 76:999–1003, 1979. Pineal transplants in sparrows.

623. Zimmerman, W. F. and T. H. Goldsmith. *Science* 171:1167–1168, 1971. *Drosophila* photosensitivity after carotenoid depletion.

624. Zimmerman, W. F., C. S. Pittendrigh, and P. Pavlidis. *J. Insect Physiol.* 14:669–684, 1968. Phase shifting and entrainment by temperature in *Drosophila*.

625. Zucker, I., M. Boshes, and J. Dark. *Am. J. Physiol.* 244:R472–R480, 1983. Circannual and circadian rhythms in ground squirrels.

626. Prosser, C. L. and J. E. Heath. In *Environmental and Metabolic Animal Physiology* (C. Ladd Prosser, ed.). pp. 109–165. Wiley-Liss, Inc., New York, 1991.

Chapter 9 | *Central Nervous Systems*

Section A
Introduction

Fred Delcomyn and C. Ladd Prosser

The principal goal of neurobiology is to explain behavior in terms of the action of individual nerve cells. Comparative neurophysiology has contributed significantly to elucidation of the nature of *nerve impulses* and *synaptic transmission* and to the analysis of *sensory reception* and *muscle contraction*. Furthermore, understanding of *central nervous integration* of sensory input and patterned motor output, as well as the role of identified interneurons, has advanced enormously in recent years.

Methods of Study

Neurobiologists use different methods to study different problems of neural function. The following is a list of techniques that may provide a contemporary picture of neural function in different animals.

1. Behavioral studies include observations of animals under natural situa-tions and measurements under controlled laboratory conditions. It is increasingly evident that restraint or subtle manipulations of animals bias behavioral observations. Telemetering methods provide new possibilities of observations under natural conditions.

2. Lesions and pharmacological treatments, which bring about behavioral deficits (and reinforcements), provide a first gross step toward localization of neural function. It is difficult to separate primary from secondary effects of such treatments.

3. Physiological measurements of input–output relations, or reflex function, in anesthetized and unanesthetized animals are useful in analyzing movements such as types of locomotion, feeding, and escape responses.

4. To understand function in nervous systems requires detailed knowledge of structure and neuroanatomy at all levels—gross connections, microscopic structure, and ultrastructure.

5. Comparison of the neurochemistry of different regions of a nervous system

permits some understanding of synaptic transmitters and of metabolism. Neuropharmacology holds promise of chemical manipulation of neural function.

6. Recording of electrical activity and measurements of membrane constants are powerful techniques. Two approaches are useful in the analysis of central nervous integration: (a) Records can be obtained from single neurons by either extracellular or intracellular electrodes. Unit recordings are useful for all classes of neurons: afferent, efferent, and interneurons. It is now even possible to record intracellularly from relatively small interneurons in the neuropile. (b) In nervous systems with a cortical or layered arrangement, as in the cerebellum and cerebrum of vertebrates, it is possible to record field potentials that represent the algebraic sum of electrical activities of a population of neurons, particularly summed synaptic currents. Profiles of these currents give information concerning the activation and interaction of masses of neurons. Activity in fields of neurons can be measured optically by electrosensitive dyes.

7. Models of neural circuits, particularly computer models of networks, have suggested hypotheses and measurements to be made with living systems, and methods for simulation of neural functions.

General Properties of Neural Integration

Most of the functional properties of nervous systems that can be measured are found in nearly all metazoans, but some kinds of animals are especially favorable for study of certain properties.

Distributing Systems: Polarity and Delay

The simplest of nervous coordinating systems is a switchboardlike distributing system. The output is some function of the input. This may be a one-to-one transfer, either with or without summation of converging inputs. A single input may be insufficient to trigger an output, and multiple inputs may be additive (*summation*) or may be multiplicative (*facilitation*). One input may reduce the response to another; that is, it may be *inhibitory*. A single interneuron may activate several motorneurons (*divergence*) or several interneurons may *converge* on a single motorneuron.

In complex networks, information is coded for selective activation of specific neurons and for retention of memory patterns. Much central selectivity is based on specific interneuronal connections. Because of their complexity, these connections (wiring patterns) are often difficult to diagram. Gradation and patterning of signals may be by voltage amplitude where transmission is electrotonic, or it may be by frequency or pattern of all-or-none spikes. Coding may also be chemical, by specific membrane proteins, each determining a given pattern and amount of ion permeability.

A few definitions applicable to fixed responses or reflexes are needed before discussion of central levels of neuronal organization in different groups of animals can be meaningful. Nonpolarized electrical junctions occur in septate giant fibers and in a few central systems where there is close synchrony between neurons. Electrical synapses (polarized) provide for speed and synchrony but permit relatively little integration. Synaptic polarization (either chemical or electrical) guarantees one-way traffic, since it prevents antidromic impulses from entering a neural pathway. However, many neurons, particularly those with axons *leaving* a center, have recurrent branches, which *feed back* via interneurons (Fig. A-1). Usually this feedback is inhibitory and prevents repetitive discharge of the efferent neuron. Examples are the Renshaw cells of the mammalian spinal cord and recur-

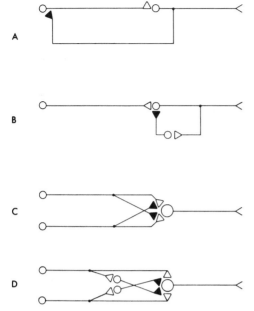

Figure A-1. Diagrams of feedback and feedforward inhibitory circuits. Open triangles, excitatory synapses; closed triangles, inhibitory synapses. (*A*) Direct feedback inhibition. (*B*) Feedback by interneuron. (*C*) Crossed presynaptic feedforward inhibition. (*D*) Crossed postsynaptic feedforward inhibition via interneurons.

rents from Purkinje cells in the cerebellum and from Mauthner neurons in fish. Occasionally, recurrent collaterals are excitatory, causing repetitive firing or circus conduction, as in nerve nets. A different type of control is by neurons *entering* a center and giving off inhibitory collaterals, which end presynaptically on a target neuron that is later excited by a parallel pathway or by a second impulse in the primary pathway; thus, the target neuron is inhibited or has depressed excitability before the excitation arrives. This is known as *feedforward* inhibition. It prevents the interneuron from firing in response to a trivial input. An example is found in *Aplysia* ganglia and in parallel fibers of the cerebellum, which activate basket cells that are inhibitory on Purkinje

cells (which are themselves excited by the parallel fibers). Feedback control is shown by exiting neurons; feedforward control is shown by entering neurons or by a branch from one interneuron in parallel with another.

Within certain neural centers, some communication may be between dendrites, as in pacemaker neurons of crustacean cardiac ganglia. In many multipolar neurons of invertebrate central nervous systems (CNSs) there is no distinction between dendrites and axons, and conduction in one process need not be one-way. Another mechanism of coding in branching neurons is brought about by regions of different fiber diameter and the resulting differences in safety factor and limiting frequency at branch points. In summary, the microcircuitry of nerve centers provides for patterning by means of polarized connections between neurons.

Another type of local control is that due to external fields created by synchronized impulses and junction potentials. For example, slow potentials spread out from the spinal cord in dorsal and ventral roots. In a frog, antidromic stimulation of some ventral root axons depolarizes other nearby motorneurons, and this interaction is enhanced by orthodromic (sensory) input. Hence, there is electrical interaction between overlapping dendrites of adjacent motorneurons. Antidromic impulses in ventral root axons also initiate potentials that can be recorded externally in dorsal roots, and these, unlike the motorneuron effects, are abolished by synaptic blocking agents; hence, they are due to synaptic connections of recurrent fibers of the motorneurons. Presumably the antidromic impulses are equivalent to activity in the recurrent collaterals of the orthodromic pathway (12). There is a correlation between field potentials and probability of firing of cortical neurons.

Summation and Facilitation

Sherrington pioneered the analysis of behavior in terms of the actions of individual neurons. Studying reflex output as a function of inputs of different strengths allows one to infer neuronal interactions. Early studies showed that two or more inputs may be required for a response to occur. Furthermore, if a single input elicits one response, multiple inputs will cause larger responses. The requirement for two inputs may be spatial (inputs via different converging fibers) or temporal (two inputs from the same fiber within a short time). In this context, *summation* means simple addition of responses; *facilitation* means that the two inputs give a greater effect than the algebraic sum of the individual responses. Thus, for two converging fibers, *spatial facilitation* ($a + b$) $> a + b$; and for closely spaced inputs in one fiber, *temporal facilitation* ($2a$) $> a + a$.

As investigation of summation and facilitaton at the level of synaptic potentials progressed, a more modern use of these terms has developed. These modern definitions are based on the postsynaptic response to one or more inputs. *Summation* means the interaction of postsynaptic potentials such that the amplitude of the final potential is the algebraic sum of the individual potentials that generate it. *Spatial summation* results from two or more inputs from different fibers; *temporal summation* results from successive inputs from the same fiber. *Facilitation* refers to a short-term increase in the amplitude of a postsynaptic response to repetitive stimulation of a single fiber. Facilitation has been most thoroughly studied at neuromuscular junctions, but can occur at neuron-to-neuron synapses as well. It is thought to be the result of the release of increased amounts of transmitter substance in successive activations of the synapse, due to an accumulation of calcium, or a calcium–protein complex, in the presynaptic terminal (5). An increase in the effectiveness of individual calcium ions in inducing release of transmitter substance has also been suggested (17). In addition to this short-term synaptic effect, several much longer lasting phenomena also occur, variously termed *long-term facilitation, potentiation, posttentanic potentiation* (PTP), or *long-term potentiation* (LTP). They are discussed later in relation to synpatic plasticity and learning.

Other types of short-term synaptic changes have been described. *Antifacilitation* (also called *disfacilitation* or *depression*) means the decrease in amplitude of a postsynaptic response with repetitive stimulation. It is due to a decrease in the amount of transmitter substance released by successive action potentials in a neuron (5). *Heterosynaptic facilitation* is facilitation across more than one synapse. The best known example is in *Aplysia*, in which pairing a weak stimulus (test input) with a strong one in another pathway (priming input) can cause a postsynaptic response in a neuron to increase in amplitude as a result of excitation by the test input alone. The phenomenon is thought to be due to presynaptic facilitation (13).

Inhibition

In most integrative centers, inhibition is as important a synaptic function as excitation. Neural output represents a balance between the two processes. Inhibition may be *presynaptic*, reducing the liberation of excitatory transmitter and preventing postneuron discharge even when no potential change occurs in the membrane of the inhibited neuron. Inhibition may also be *postsynaptic*, typically causing a hyperpolarization of the postsynaptic membrane and thus reducing the response to excitatory transmitter. Presynaptic inhibition can occur in the spinal cord. For example, terminals of afferent fibers from muscle spindles receive presynaptic inhibition from afferents arising in tendons during a twitch, so the muscle afferents are prevented from liberating transmitter to motorneurons. Postsynaptic inhibition occurs as well. For example, collateral branches from motorneuron axons excite Renshaw interneurons, and these in turn inhibit the motorneuron and prevent repetitive firing after an initial excitation (1).

Generally, postsynaptic inhibition is mediated via interneurons, sometimes with considerable latency. Inhibition modulates spontaneous activity and sensory input. The output of neurons in a center normally depends on the algebraic sum of excitation and inhibition, and a given neuron has receptor membrane patches for both excitatory and inhibitory transmitters. One method of central coding may be by variation of the relative size of the patches.

Repetitive Responses and Spontaneous Activity

Many nerve fibers discharge repetitively at frequencies higher than refractory period limitation during prolonged depolarizing stimulation. In some neural centers, certain neurons show *after-discharge* or repetitive firing as part of a reflex response. Many central neurons are repetitively active in the absence of sensory or other input; that is, they fire spontaneously. Incoming signals may modulate the ongoing discharge to increase or decrease its rate; this permits more gradation of information transmission than if a neuron were silent and the only effect of an input were to start it firing. Spontaneous activity is general in integrative nervous systems; it is less usual in distributive centers such as sympathetic ganglia.

Two types of central nervous *spontaneity* are recognized. Individual cells are spontaneously active in many invertebrate ganglia and in most integrative regions of the brain of vertebrates. The frequency varies from cell to cell and there is little synchronization. One cell may fire regularly or may fire in bursts interrupted by silent periods, as in crustacean central and cardiac ganglia. At any one time only a few cells are spontaneously active. Spontaneity is temperature sensitive and requires energy from oxidative metabolism, but the coupling between metabolism and rhythmic conductance changes is unknown. A second pattern of spontaneity is found mainly in cortical structures such as cerebral cortex, cerebellum, and optic tectum of vertebrates and optic lobes of cephalopods. Very many cells are synchronized to give large rhythmic waves not necessarily of impulses but rather of slow changes in membrane potential. The synchronization is due partly to reverberating circuits and partly to electrotonic coupling of neurons.

In hermit crabs and lobsters, the motor rhythm of ventilation is controlled by a pair of neurons, the membrane potentials of which oscillate without spiking but elicit spikes in motorneurons serving the muscles of ventilation (15). In *Aplysia*, certain neurons fire uniformly during a burst, others fire at decreasing frequency, and in still others the frequency increases and then decreases (these are parabolic bursters). If spiking is prevented by use of TTX in a Ca^{2+}-free medium, rhythmic slow waves corresponding to the burst frequency remain. These slow waves are abolished by ouabain or in Na^+-free medium, and they may represent a rhythmic sodium pump (18). It is probable that several different mechanisms are involved in oscillating membrane potentials and repetitive spiking (2, 4), and that spontaneous rhythmicity is characteristic of integrative neural centers.

Numerous examples will be given later for the behavioral meaning of spontaneous activity and for programmed sequences in nerve centers. Observations on deafferented animals show that many rhythmic or repetitive movements do not require any sensory feedback for their expression (6, 10). Instead, rhythmic motor patterns underlying behaviors such as walking, swimming, flying, chewing, and sound production in vertebrate and invertebrate animals can be generated in the complete absence of sensory feedback. Nevertheless, sensory feedback may be essential for generation of the properly timed, normal behavior. The influence of sensory signals on the timing of components of a motor rhythm can be relatively minor, as in the case of escape swimming in the mollusc *Tritonia* (9), or major, as in the case of walking in insects (7) or swimming in dogfish sharks (11).

Adaptations for Speed of Conduction

Conduction of electrical signals in nerves was compared with conduction in rows of epithelial cells in Chapter 12, and conduction in nerve fibers is compared with that in a multijunctional nerve net in Section 20B. Fast conduction in axons and reflex pathways may give an animal an advantage in escaping from predators, in prey capture, and in response to stress. Slow reactions may be advantageous in having a low energy cost and in slow types of behavior.

Fast conducting nerve fibers have shorter excitation times than slow fibers. The threshold for exciting a slowly rising impulse is higher than for a rapidly rising pulse. A faster rate of rise of a conducted action potential results in larger local currents and faster excitation of adjacent regions of an axon. Speed is inversely related to membrane electrical capacity because the time needed to charge a large capacity is long. Speed of conduction is inversely related to intracellular specific resistance; low intracellular resistance means greater longitudinal current for a given voltage.

Long neuronal processes favor speed in a nervous system because interposed synapses slow conduction. Speed of conduction is related to fiber diameter by a proportionality constant. In large myelinated fibers of vertebrates, velocity in meters per second numerically equals the diameter in micrometers times a factor of 6–8; in small nonmyelinated vertebrate fibers velocity is directly proportional to fiber diameter. In giant axons of squid, velocity is an exponential function ($V = DP^{0.61}$). A myelin sheath confers fast conduction by lowering C_m, increasing R_m, and increasing the space constant. The fastest vertebrate fibers are of medium size (4–15 μm in diameter), but have thick sheaths (30–50% of diameter). The fastest fibers of invertebrates are large (15 to several hundred μm) and have thin sheaths (<1% of diameter). A 4-μm fiber of a cat saphenous nerve at 38°C conducts at about the same velocity (25 m/s) as a 650-μm squid giant axon at 20°C. Conduction in vertebrate myelinated axons is saltatory; an impulse jumps from node to node. The local circuit for conduction involves inflowing and outflowing current only at the nodes; internodes have high resistance and low capacitance. External current flows outside the myelin sheath. Branch points in nonmyelinated giant axons of crustaceans (shrimp) make for saltation.

A giant fiber is an *axon* whose diameter is significantly larger than the diameters of other axons in the nervous system of the same animal. It should not be confused with a giant neuron, a term reserved for cells with very large somata. Giant fiber systems are examples of simple rapid distribution systems. They are efficient, since one impulse in a giant fiber may activate muscles over a wide body area. They are of several types and occur in a series of increasing complexity (from through-conducting fibers, to septate neurons with electrical septal junctions, to giant fiber chains connected through chemical synapses), and include some of the most complex single neurons known. Giant fiber systems are chiefly associated with rapid startle, escape, and grasping reactions that are initiated abruptly and are not usually graded (3, 8). Giant fibers are distributed widely throughout the animal kingdom, including vertebrates. They may be single large neurons (often polyploid), or may (in invertebrates) result from fusion of several smaller fibers. Fibers of both morphological types seem to have evolved several times. They should be considered specialized rather than primitive.

The best known example is the famous squid giant axon used by Hodgkin and Huxley to work out the ionic basis of the action potential. This axon is the third-order neuron in the escape circuit that acts to excite the muscles of the mantle. The first-order giants originate in the posterior palloisceral ganglion of the brain; they synapse in the visceral part of the brain onto the second-order giants. These in turn synapse in the stellate ganglion onto the giant motorneurons used by Hodgkin and Huxley (14, 19). The giant motorneurons are formed by fusion of many (300–1500) smaller cells. They may reach 1 mm in diameter, and conduct at 5–25 m/s. Contraction of the mantle muscle by stimulation via the motor giant forces water out of the mantle cavity through the funnel, sending the animal rapidly forward or back according to the funnel angle (16). Other examples are discussed below in association with the groups in which they occur.

References

1. Andersen, P. and J. C. Eccles. *Nature (London)* **196**:645–647, 1963. Inhibitory phasing of neuronal discharge.

2. Arvanitaki, A. and N. Chalazonitis. *Bull. Inst. Oceanogr.* **1164**:1–83, 1960; **1224**:1–15, 1961. Rhythmic synaptic potentials, *Aplysia* neurons.

3. Bullock, T. H. In *Neural Mechanisms of Star-* tle Behavior (R. C. Eaton, Ed.), pp. 1–13. Plenum, New York, 1984. Comparative neuroethology of escape and giant fiber systems.

4. Callec, J. J. and J. Boistel. *C. R. Seances Soc. Biol. Ses Fil.* **160**:1943–1947; 2418–2424, 1966. Rhythmic synaptic potentials, abdominal ganglia, *Periplaneta*.

5. Charlton, M. P., S. J. Smith, and R. S. Zucker. *J. Physiol. (London)* **323**:173–193, 1982. Calcium and facilitation, depression in squid.

6. Delcomyn, F. *Science* **210**:492–498, 1980. Neural basis of rhythmic behavior in animals.

7. Delcomyn, F. In *Comprehensive Insect Physiology Biochemistry and Pharmacology* (G. A. Kerkut and L. I. Gilbert, eds), Vol. 5, pp. 439–466. Pergamon, Oxford, 1985. Walking and running.

8. Dorsett, D. A. *Trends Neurosci.* **3**:205–208, 1980. Design and function of giant fibre systems.

9. Getting, P. A. *Symp. Soc. Exp. Biol.* **37**:89–128, 1983. Neural control of swimming in *Tritonia*.

10. Grillner, S. *Science* **228**:143–149, 1985. Neurobiological bases of rhythmic activity in vertebrates.

11. Grillner, S. and P. Wallen. *J. Exp. Biol.* **98**:1–22, 1982. Peripheral factors and the central pattern generator for swimming in the dogfish.

12. Grinnel, A. D. *J. Physiol. (London)* **182**:612–648, 1966. Motorneurons in frog spinal cord.

13. Kandel, E. and L. Tauc. *J. Physiol. (London)* **181**:1–27, 28–47, 1965. Heterosynaptic facilitation in neurons of abdominal ganglion of *Aplysia*.

14. Martin, R. and R. Miledi. *Philos. Trans. R. Soc. London, Ser. B* **312**:355–377, 1986. The morphology of the squid giant synapse.

15. Mendelson, M. *Science* **171**:1170–1173, 1971. Oscillator neurons in crustacean ganglia.

16. Packard, A. *Nature (London)* **221**:875–877, 1969. Jet propulsion and giant fiber responses of *Loligo*.

17. Stanley, E. F. *J. Neurosci.* **6**:782–789, 1986.

The basis of facilitation at the squid giant synapse.

18. Strumwasser, F. J. *J. Psychiatr. Res.* **8**:237–257, 1971. Rhythmic activity in molluscan neurons in culture.

19. Young, J. Z. *Q. J. Microsc. Sci.* **78**:367–386, 1936; *Philos. Trans. R. Soc. London, Ser.* B. **229**:465–501, 1939. Giant fiber systems of cephalopods.

Section B
A Protoneuronal System in Sponges*; Cnidarian Nervous Systems

Andrew Spencer

Although there has been considerable debate (33, 68) recent ultrastructural evidence confirms that sponges do not have a true nervous system (51). There is evidence that the glass sponges (Hexactinellida) use a syncytial trabecular tissue, a specialized type of epithelium, as a substratum for conducting signals throughout the body. This unique conducting system was discovered by monitoring the velocity of the exhalant stream of water emitted from the osculum using a thermistor flowmeter (53). Like other sponges the genus *Rhabdocalyptus* pumps a considerable volume of water through the internal passages and chambers using flagellae. If mechanically disturbed, pumping is inhibited and may stop completely. This is probably a response to detritus being dislodged and then carried by the inhalant stream into the sponge where it can block passages and interfere with feeding. By electrically stimulating a slab of body wall and recording the in-

*Within the phylum Porifera there are three classically recognized classes: the Hexactinellida, Calcarea, and Demospongiae. However, because of several important ultrastructural and electrophysiological differences between the Hexactinellida and the other two classes it may be necessary to erect two new subphyla: the Symplasma (present Hexactinellida) and the Cellularia (Calcarea and Demospongiae).

creasing delay of the arrest response at increasing distances from the point of stimulation Mackie and collaborators have shown that this inhibitory signal is conducted at ~0.26 cm/s (Fig. B-1). The exact nature of the conducted signals is not known, but it is believed they are conventional action potentials that are propagated through the diffuse pathway of the trabecular network by direct current flow. Since the conduction velocity in this system is very low a novel conducting mechanism cannot be discounted.

In many sponges the exhalant osculum contracts when mechanically stimulated. Myocytes that contain both thick and thin filaments are the contractile elements. Contractions are local, graded, and require Ca^{2+} together with a univalent cation that can be Na^+, K^+, or Li^+. Apparently, membrane depolarization is not required for contraction that is not conducted (69).

The Origin of Neurons and Nerve Nets

It is sometimes assumed that neurons evolved only once, probably in the Cnidaria. Nevertheless, it is quite possible that neurons arose on several occasions during evolution. Whenever there was selection pressure for epithelial cells to specialize for transduction (i.e., to act as sensory cells), integration (to act as interneurons), or for transmission of electrical impulses, cells that we would describe as neurons could have evolved. The properties of epithelial cells in extant animals suggest that these cells were preadapted for a role as neurons. In hydrozoan jellyfish and polyps, ctenophores, molluscs, tunicates, and amphibian embryos there are examples of epithelal tissues that propagate action potentials (5) (Chapter 1). In most cases these eithelial conducting systems are used to mediate escape or defensive behaviors. It is apparent that many of the channel proteins nec-

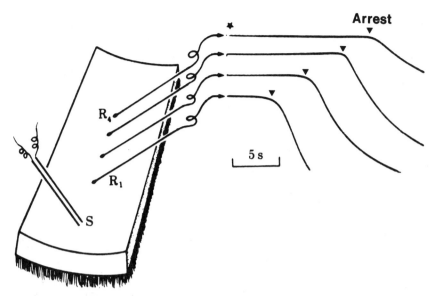

Figure B-1. Conduction of flagellar arrests in a slab of body wall in the sponge *Rhabdo-calyptus dawsoni*. A thermistor flowmeter (R) was used to monitor the decrease in the velocity of the water current produced by the flagellae when pumping was arrested by electrical shocks delivered to the slab through electrode *S*. As the flowmeter was moved further from the point of stimulaton (R_1 to R_4) the delay to initiation of the arrest response increased. The asterisk marks the time when the shock was given. The mean conduction velocity was 0.26 cm/s. [From Mackie et al., (53).]

essary for electrical excitability may have been present in epithelial cells. One scenario for the origin of neurons has been drawn by Horridge (32), which shows neurons forming from insinking epithelial cells that are at first electrically coupled to the external layer but which later are connected by true neurons with chemical synapses. This evolutionary scenario proposes that electrical synapses preceded the appearance of chemical synapses. This is only one theory that, although popular, is not universally accepted.

Many biologists use the term "nerve-net" to describe the organization of primitive nervous systems. The nervous system of *Hydra* is often used as the archetypal nerve-net. However, a nervous system organized as a diffuse network may be more a reflection of the constraints set by the symmetry of the animal and the required properties of the nervous system, than of the *primitiveness* of that animal. Diffuse networks of neurons, often appearing as two-dimensional sheets, can be found in the peripheral nervous systems of most animals. For exmaple, in molluscs there are frequently motor networks associated with peripheral ganglia that form local reflex pathways. In some instances they may be used to coordinate quite complex behaviors such as locomotion. In every phylum, including the chordates, nerve-nets innervate various organs such as the gut or gonads, which are either surrounded by sheets of smooth muscle or contain secretory cells. In some cases the nerve-nets may have a sensory function, while in others they may be used to modulate the intrinsic activity of an organ by releasing neurotransmitters or peptide modulators. In such groups as the Cnidaria, Ctenophora, Echinodermata, and Hemichordata a large proportion of the CNS, in addition to the peripheral nervous system, is in the form of nerve-nets. Condensation of nervous tissue, when it does occur, appears as nerve-rings or trunks rather than discrete ganglia. This degree of condensation may be the maximum that the radiate phyla (coelenterates and echinoderms) can tolerate before integration centers become too distant from sensory structures.

Cnidaria*

The cnidarian nervous system consists of one or several, diffuse, two-dimensional nerve-nets. Action potentials are presumed to conduct in all directions and may or may not require facilitation for through conduction. These are the major features of "the nerve-net concept," which is the culmination of many elegant physiological studies on medusae, hydroid polyps, and sea anemones during the period up to 1960 when electrophysiological recording techniques were introduced. Romanes is considered the father of this school with his carefully designed experiments involving complex cuts to the bells of medusae demonstrating that conduction pathways in the nerve-net are diffuse (Fig. B-2) (73, 74). Other early workers, such as Loeb, Parker, and Horridge used anemones (class Anthozoa) and true jellyfish (class Scyphozoa) almost exclusively for their experiments (28, 44, 64). In some ways this was unfortunate as it is now apparent that the physiology of the nervous system of these two classes is quite different from

*The phylum Cnidaria is represented by three classes: Anthozoa (sea anemones and corals), Scyphozoa (true jellyfish), and Hydrozoa (hydroids and their medusae) and is characterized by the presence of cnidae (nematocysts). The comb jellies are in a separate phylum, the Ctenophora, which generally lack nematocysts.

that of the third cnidarian class, the Hydrozoa (hydroids and their medusae). The nerve-net concept does not accommodate many properties of hydrozoan systems. Although the general morphological features of the anthozoan, scyphozoan, and hydrozoan nervous systems are similar, their functioning is very different. In all classes there are normally several overlapping nerve-nets with condensation into nerve rings and simple ganglia, particularly in medusae. Gap junctions, which mediate electrical current flow between coupled cells, have never been seen in any anthozoan or scyphozoan tissues whereas in the Hydrozoa gap junctions are ubiquitous, coupling both epithelial cells and neurons (26, 52). In hydrozoan nerve-nets, action potentials propagate through electrical synapses joining neighboring neurons, whereas in the Scyphozoa and probably the Anthozoa, the synapses between adjacent neurons in a nerve-net are chemical (6, 85). Chemical synapses are also present in hydrozoan nervous systems but in this case they appear to be used for communication between different nerve-nets or for exciting effectors (86, 88).

Electrical coupling also allows hydrozoans to use epithelial sheets as conduction pathways (5). This conduction path was first described by Mackie in several species of siphonophores (*Hippopodius*, *Forskalia*, *Chelophyes*, and *Abylopsis*), where the pathways are used to control bioluminescent flashes and radial muscle contraction in the margin (14, 45). Similar epithelial conducting systems are found in hydroid polyps and hydromedusae where they mediate protective behaviors such as column contraction and crumpling. Analogous epithelial systems have not been definitively described in either anthozoans or scyphozoans.

Epithelial and neuronal conducting systems can interact. All the external ectodermal epithelium in many hydrome-

Figure B-2. An example of the complex geometrical cuts used by Romanes to demonstrate that the pathway for conduction of the swimming contraction in the scyphomedusa *Aurelia aurita* is diffuse. The dark tubular structures are part of the gastrovascular system and not the nerve-net of the muscle used during swimming. [From Bullock and Horridge (15a), after Romanes (74a).]

dusae is mechanically sensitive and can propagate the resulting action potentials (54). These action potentials can invade the CNS at the margin and thereby initiate contraction of radial muscles in the tentacles and radii, resulting in a defensive behavior known as crumpling (37). Since crumpling makes it impossible for the jellyfish to swim, the epithelial action potentials concurrently inhibit activity in the swim motor neuron network by producing large, long duration hyperpolarizations (85). As this effect is not blocked by Mg^{2+} it is assumed that it is a direct field effect resulting from the proximity of the ectodermal epithelium to the motor neurons. Chemically mediated inhibitory synapses have not yet been described in any cnidarian.

Anthozoan Nervous Systems

This class consists of exclusively polypoid individuals, which may be solitary, or colonial and polymorphic, as in the corals. In addition to controlling relatively simple behaviors such as column contraction or mouth gaping, the nervous system is responsible for coordinating colonial activities such as polyp retraction in corals and complex behaviors such as shell climbing in *Calliactis* and swimming in *Boloceroides* (30, 43, 59). In many cases the exact cellular substrates for a particular system, whether neuronal or epithelial, are unknown. A well-described system in anemones is the "through-conducting nerve-net" (TCNN), which is found in the mesenteries, oral disk, and tentacles (Fig. B-3). Spikes in this system of mostly bipolar neurons conduct at up to 100 cm/s without decrement and are responsible for both tentacle and mesenteric contractions (56). Fast muscle contractions showing facilitation are seen in response to spike trains with frequencies greater than ~1 Hz. Slow contractions, which show strong facilitation, are produced by low-frequency spike trains (<0.2 Hz in *Metridium*) with each muscle field commencing contraction at a different frequency (62, 72). It is possible that slow contractions are produced indirectly by activation of a local nerve-net rather than directly by the TCNN (60). Many anthozoans exhibit spontaneous expansion and contraction rhythms, which are presumably controlled by spontaneous activity in nerve-nets.

In coral colonies a nerve-net connecting individual polyps is responsible for conducting events that lead to polyp retraction that spreads out in waves from the point of stimulation. In the coral *Tubipora* the first stimulus leads to colony-wide polyp withdrawal, and in *Heteroxenia* each successive stimulus (at 1 Hz) spreads with increasing increments while in *Goniopora* the spread is with decreasing increments. Horridge explained these patterns as being due to different requirements for interneural facilitation in the nerve-net (30). Although interneural facilitation has not been clearly demonstrated in anthozoans it is assumed that it involves priming of a synapse by an action potential, which then allows succeeding action potentials to be transmitted across the synapse. However, it now appears that in most cases [*Goniopora* and *Lophelia* being exceptions (3, 83)] the colonial nervous system is through conducting and thus is analogous to the

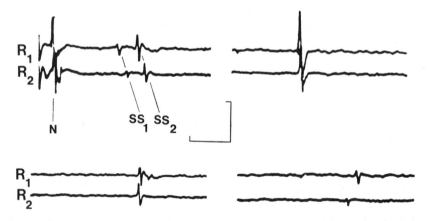

Figure B-3. Electrical activity recorded from the tentacles of *Calliactis parasitica* by suction electrodes. Two electrodes, R_1 and R_2, 2 cm apart were attached to different tentacles. Upper left, response to a single, high intensity shock to base of the column showing that all three pulse types are evoked: N, nerve net pulse; SS_1, pulse from slow system 1; SS_2, pulse from slow system 2. Upper right, spontaneous nerve-net pulse. Lower left, spontaneous SS_1 pulse. Lower right, spontaneous SS_2 pulse. Scale bars 20 μV and 500 ms. [From McFarlane (58a).]

TCNN of solitary anthozoans. Thus the patterns observed are more likely a property of the labile connections between the colonial nervous system and that of individual polyps (79). In the pennatulids (sea pens and pansies), and perhaps other anthozoans, it has been observed that the conduction velocities of action potentials in the colonial nerve-net increase with the first few pulses (81). Thus distant polyps receive spike trains at a higher frequency. Such spike trains are more likely to produce polyp retraction because of their facilitatory effect, resulting in a behavioral expression of through conduction. In other species, such as the hard coral *Porites* and the gorgonian *Muricea*, nerve-net spikes are conducted at decreasing velocities during repetitive stimulation and thus distant polyps experience a lower frequency spike train (Fig. B-4) (76, 82). If the interspike interval is too short to produce neuromuscular facilitation at distant polyps, no retraction is seen in these polyps despite the spikes being through conducted to all polyps. Neuromuscular facilitation had been pre-

viously described by Pantin in solitary anemones (62, 63). The ability of polyps to respond to activity in the colonial nerve-net can also be limited by the properties of the junctions betwen this net and the local polyp nerve-net. These junctions may require a specific "gating" signal (76).

Colonial anthozoans also possess a slowly conducting system that controls synchronous polyp expansion (80,82). The cellular basis for this conducting system is unknown. Similar conducting systems have been described in solitary anthozoans (Fig. B-3). An ectodermal system, the slow system 1 or SS1, conducts at between 5 and 12 cm/s in *Calliactis*, and is responsible for oral-disk expansion, column extension, and pedal disk detachment (56–58, 60). In *Stomphia*, SS1 spikes initiate swimming, while in *Tealia felina* they are involved in feeding chemoreception, which leads to relaxation of ectodermal muscles (41, 42). In the endoderm there is another slow system, the SS2, conducting at 3–6 cm/s, which is involved with mouth opening and pharynx

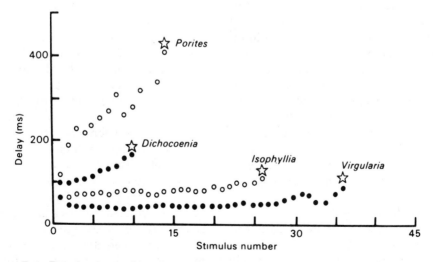

Figure B-4. Development of increasing delay (i.e., decreasing conduction velocity) with increasing stimulus number in four species of corals following repetitive stimulation at 1 Hz. Recordings by suction electrodes of the colonial impulse were made 2 cm from the point of stimulation. [From Shelton (82).]

protrusion (56). Although it has been suggested that both the SS1 and SS2 are epithelial conduction systems, evidence such as blockade by Mg^{2+} and the absence of gap junctions indicate that these systems are in fact neuronal (51, 56). There must also be transmesogleal pathways that enable ectodermal sensory signals to reach endodermal effectors.

Anemones are capable of surprisingly complex behaviors. An example of such behaviors is shell climbing in *Calliactis*. The preferred attachment site for *Calliactis parasitica* is a shell occupied by hermit crabs. An anemone that is attached to some other surface, if allowed to contact a shell with its tentacles, will go through the following repertoire: (i) tentacles attach, (ii) the pedal disk detaches, (iii) the column bends, swinging the pedal disk to the shell, (iv) the pedal disk attaches to the shell, and (v) tentacles are released. In this way the anemone transfers itself to the shell. This response can be artificially initiated by electrical stimulation of the SS1 and SS2 systems at 0.2 Hz (59). Under natural conditions, SS1 and SS2 pulses are evoked when tentacles contact the shell. The SS1 spikes cause detachment of the pedal disk and SS2 activity inhibits pacemakers, which cause column contraction. Thus SS1 activity then initiates peristaltic circular muscle contraction, which causes column elongation. This elongation reflexly excites the parietal muscles of one side, via the sensory nerve-net, causing column bending. Bending triggers a burst of through-conducting NN pulses, which leads to pedal disk expansion and attachment (60).

Scyphozoan Nervous Systems
The nervous systems of anthozoans and scyphozoans share many features such as multiple conducting systems, chemically mediated transmission between neurons of a nerve-net, frequency dependent facilitation at neuromuscular synapses, and spontaneous activity in certain nerve-nets.

Much of the experimental work with jellyfish has concentrated on the control of swimming as this is their most obvious behavior. The swimming muscle sheets are formed by the subumbrellar ectoderm, which also contains at least two nerve-nets, the motor nerve-net (or GFNN) made up of large, mostly bipolar neurons, which innervate the swimming muscle and a network of smaller often multipolar neurons, the diffuse nerve-net (9, 27). In each of the eight sensory rhopalia, which are located in the bell margin, is a ganglion like condensation of nervous tissue, which are the pacemaker sites for the swimming rhythm (65, 73). The pacemakers form a dominance hierarchy with the fastest pacemaker resetting the others through the motor net (31). Modulation of the swimming rhythm comes from local sensory structures such as ocelli and statocysts and from the diffuse nerve-net. Activity in the diffuse nerve-net also causes a wave of tentacle contraction and can facilitate the swimming musculature so that subsequent motor nerve-net spikes are more effective (29, 65).

Intracellular microelectrode recordings taken from neurons of the motor nerve-net of cubozoan (*Carybdea*) and scyphozoan (*Cyanea*) jellyfish show that their excitability is conventional (10, 77). Short duration (2–6 ms) action potentials arise from a resting potential of about -60 mV, and are through conducted in the motor nerve-net. Using the whole-cell voltage clamp technique on isolated neurons of the motor network Anderson has shown that the upstroke of the action potential is produced by a voltage-gated Na^+ current that is insensitive to the usual pharmacological blocking agents for such currents, for example, tetrodotoxin (TTX), saxitoxin (STX), and conotoxin (Fig. B-5) (7). In some respects this Na^+

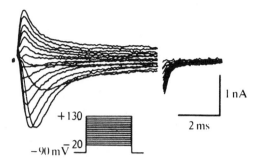

Figure B-5. A family of currents recorded by the whole-cell, patch clamp technique from a neuron of the motor nerve-net of the jellyfish *Cyanea capillata*. The voltage clamp protocol used to elicit these currents is shown in the inset and consists of a series of steps incrementing by 10 mV from a holding potential of −90 mV. All potassium current is pharmocologically blocked by 4-aminopyridine in the external and electrode solutions. Additionally, TEA was added to the electrode solution and all K^+ was replaced by Cs^+. Note that the fast, transient component is a Na^+ current, which is insensitive to TTX. A slowly inactivating Ca^{2+} current is also present in these recordings. [From Anderson (7).]

current has many features of more primitive Na^+ currents. The repolarizing currents appear to be conventional K^+ currents which are blocked by 4-aminopyridine and tetraethylammonium (TEA).

Propagation of action potentials in the motor nerve-net (at ∼ 50 cm/s) depends on synaptic transmission at each of the chemical synapses that are formed where neuronal processes cross *en passant*. These synapses have unusual properties, which allow them to transmit in either direction, and they have therefore been called bidirectional synapses (Fig. B-6) (6). Electron micrographs show synaptic vesicles in both neurons facing each other across the synaptic cleft (8). Bidirectional synapses have a delay of less than 1 ms between the presynaptic spike and the large (up to 70 mV) EPSP. Recordings often show an "echo" EPSP on the repolarizing

phase of the presynaptic action potential owing to release of transmitter from the postsynaptic neuron (6). The postsynaptic neuron must spike for it to cause transmitter release onto the neuron, which was originally presynaptic. Each action potential in the motor nerve-net produces an action potential in the overlying striated swimming muscle resulting in a pulsation of the bell. The first 10 or so spikes in a burst give graded, facilitating junctional potentials (Fig. B-7) (75).

Hydrozoan Nervous Systems
Alternation of generation between polypoid and medusoid phases predominates in this class and it is therefore convenient to treat these two forms separately. Of the 15 or so species of hydrozoan polyps that have been examined electrophysiologically two genera of hydroid polyps, *Hydra* and *Tubularia*, have been studied most extensively. It is worthwhile remembering that *Hydra* may not be an ideal model for other hydropolyps as it is undoubtedly specialized for specific features of its freshwater habitat. The ectodermal nerve-net of *Hydra* is diffuse with the concentration of neurons being greatest in the hypostome, peduncle, pedal disk, and tentacles (19, 38, 91). Scattered neurons are found in the endoderm. In some species there is a thin ectodermal nerve ring in the hypostome running between the bases of each tentacle (22). Neurons arise in the mid-column and migrate towards the "head" and "foot," where they are sloughed off together with epithelial cells. They are believed to differentiate into various neuronal types during this migration (39). Although histologists describe a single nerve-net, behavioral and physiological studies indicate that there are several overlapping nerve-nets.

One physiologically identified system is the contraction pulse (CP) system, which causes column contraction (34, 67). Contraction pulses can arise sponta-

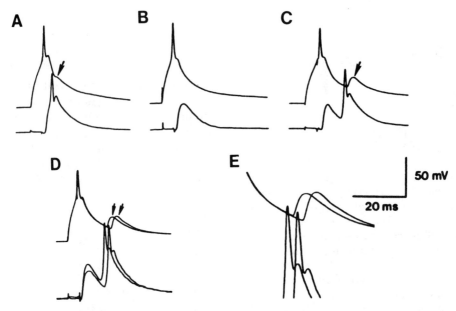

Figure B-6. Transmission at an excitatory, chemical, bidirectional synapse in the motor nerve-net of *Cyanea capillata*. All recordings are intracellular from the same pair of neurons and all stimuli were delvered to the neuron in the upper trace. (*A*) An action potential is elicited in the presynaptic cell (upper trace) by current injection, which produced a large EPSP and action potential in the postsynaptic cell (lower trace); note a notch (arrow) on the repolarizing phase of the "presynaptic" spike due to release of transmitter by the "postsynaptic" cell onto the presynaptic cell. This potential has been called an echo potential. (*B*) After several trials the synapse fatigues and the postsynaptic cell (lower trace) no longer produces an action potential owing to the reduced EPSP amplitude. In this case a notch is not seen in the presynaptic spike. (C) A delayed EPSP and a resultant action potential (produced by a different presynaptic cell) elicits an echo potential of increased amplitude owing to the increased input resistance of the cell as the potassium conductance begins to decrease. (*D* and *E*) The delay between the peak of the postsynaptic action potential and the echo potential (arrows) was constant irrespective of the timing of the postsynaptic action potential. [From Anderson (6).]

neously with the frequency being modulated by light, mechanical stimulation, or food. They are through conducted at 3–8 cm/s. Since *Hydra* in which all the neurons have been removed by treatment with colchicine are still capable of conducting contraction pulses, albeit at a slower velocity, it is apparent that action potentials are propagated in the ectodermal epithelium while the ectodermal nerve-net is responsible for their initiation (16). A second type of electrical impulse has been recorded from *Hydra* using suction electrodes, the rhythmical potential (RP) (66). These spontaneous events occur repetitively and are thought to be through conducted (at ~5 cm/s) in the endoderm (36). These potentials cause contraction of the column circular muscle thus acting as antagonists to the longitudinal muscle activated by contraction pulses. A third pulse type, the tentacle contraction pulse (TCP) is associated with tentacle contractions. Feeding behavior in *Hydra* has been well described but the conduction systems responsible are not known.

There is far more complexity in the

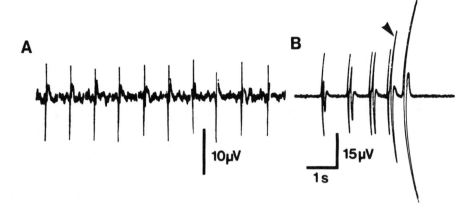

Figure B-7. Suction electrode recordings from the swimming system of the cubomedusa. *Carybdea rastoni.* (*A*) Spontaneous neuronal impulses recorded from the pacemaker area of a rhopalium. Each impulse propagates into the motor nerve-net of the swimming muscle and results in a swimming contraction. (*B*) Spontaneous motor neuron and muscle impulses recorded over the nerve ring showing that the muscle response (one example marked with an arrowhead) facilitates. [From Satterlie (75).]

conducting systems of colonial species of the marine hydroid *Tubularia* (35). In each polyp there are two pacemaker systems: the neck pulse system produces NPs, which cause neck contractions, and the hydranth pulse system produces HPs, often as bursts, causing synchronous oral flexion of the tentacles. Within the stalks there are three conducting systems: a labile slow system (SS) of unknown function conducting at ~6 cm/s, a faster (16 cm/s1) distal opener system (DOS), which causes aboral bending of the distal tentacles and inhibition of the NP and HP pacemakers, and the triggering system (TS), which conducts impulses at 17 cm/s through the stolons to trigger NPs in neighboring polyps. Homologies of the NP and HP systems are present in related hydroids such as *Corymorpha palma* and the chondrophores *Porpita* and *Velella* (13, 21). It is also likely that the CP system of *Hydra* and the HP system of *Tubularia* are homologous (35). The cellular substrates for these various hydroid conducting systems can be either neuronal or epithelial or a combination of the two.

Much of our knowledge of the cellular physiology of cnidarian nervous systems comes from studies of hydromedusae and siphonophores. These jellyfish have far more complex nervous systems than their polyps since there are often multicellular sensory structures such as ocelli and statocysts, which send afferent fibers to rings of condensed nervous tissue at the bell margin that form the CNS. In addition to the CNS there are peripheral nerve-nets in the tentacles, manubrium, and gonads.

Several neuronal systems in hydromedusae and siphonophores, particularly those controlling swimming and protective behaviors, have neurons that are sufficiently large to be penetrated by intracellular microelectrodes (47, 78). Neuronal and epithelal action potentials arise from K^+-dependent resting potentials of -40 to -85 mV and are normally overshooting with amplitudes up to ~130 mV (17, 79). The depolarizing phase of the action potential is mostly sodium dependent though there is often an associated inward Ca^{2+} current (4, 89). The motor giant axon of the trachymedusa

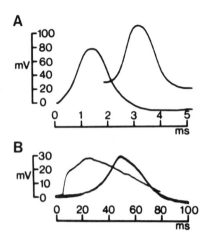

Figure B-8. Dual action potentials in the giant motor axons of the trachymedusan jellyfish *Aglantha digitalie*. (A) Intracellular recordings from two sites, 2.5 mm apart, along a motor axon of a rapid, Na^+-dependent action potential, which elicited escape swimming. The calculated conduction velocity was 1.4 m/s. (B) Recording at two sites, 6.9 mm apart, of a low-amplitude, Ca^{2+}-dependent action potential from the same axon during a slow swimming contraction. The calculated conduction velocity was 0.3 m/s. [From Mackie and Meech (49).]

Aglantha is capable of independently propagating sodium and calcium spikes, which are used to control fast and slow swimming, respectively (Fig. B-8) (49).

Like their polyps, hydromedusae have multiple conducting systems. Besides an epithelial conduction system, which is used to mediate defensive contraction of the tentacular and radial muscles (37, 50), the anthomedusae such as *Sarsia*, *Stomatoca*, and *Polyorchis* have at least four neuronal systems (46, 84). One of these, the swimming motor neuron (SMN) system consists of a condensed network of large, electrically coupled neurons lying in the inner nerve ring and radial nerves. These neurons innervate the striated, swimming muscle on the inner surface of the bell. This network has distributed pacemaker properties so that action potentials

can arise anywhere in the nerve ring. In *Polyorchis* these motor spikes invade all neurons of the network at velocities that can vary from ~100 cm/s to many meters per second depending on whether distant parts of the network are close to threshold or not (85). Electrical coupling between neurons of this network is very strong, which is presumably a specialization for increasing conduction velocity that aids in synchronizing the swimming muscle. An interesting feature of action potentials in this network is that the duration of action potentials decreases from ~25 ms at the initiation site to 8 ms as it propagates to the most distant sites (Fig. B-9) (86). Voltage clamp experiments on isolated motor neurons have shown that the loss of the plateau is due to the spike arising from a more hyperpolarized resting potential, which removes inactivation of a fast, transient K^+ current, which is the major repolarizing current (70, 89). These changes in the duration of motor spikes are functionally significant. Long duration spikes close to the initiation site produce small EJPs with long delays of the muscle action potential while short motor spikes at distant sites result in large EJPs with short delays (Fig. B-9) (86). This decreased delay automatically compensates for the conduction time of the motor spike through the network. In the smaller medusa *Aglantha*, giant ring and radial motor neurons are sufficiently large to ensure that motor spikes are rapidly conducted resulting in synchronous contraction of the swimming muscle (71). *Aglantha* and other medusae use giant fibers to produce the rapid conduction velocities required for escape behaviors (48, 61).

Iontophoretic injection of dyes, such as lucifer-yellow, permits identification, both morphologically and physiologically of two other central nerve-nets in *Polyorchis* (88). Bursts of spikes in one of these systems, the burster or B system, cause synchronous tentacle contractions and produce facilitating EPSPs in the SMN.

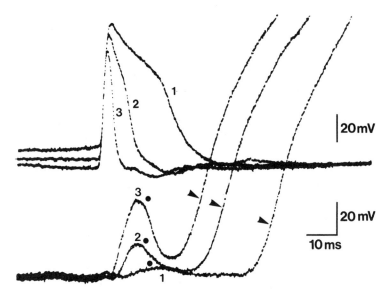

Figure B-9. Transmission at a neuromuscular, chemical synapse in the hydrozoan jellyfish *Polyorchis penicillatus*. The upper trace is an intracellular recording from a motor neuron controlling swimming, while the lower trace simultaneously monitors the membrane potential in a postsynaptic myo-epithelial cell. Three spontaneous motor neuron action potentials of varying duration elicit junctional potentials (solid circles) of varying amplitude. The longest duration action potential produced the smallest junctional potential and the shortest action potential the largest junctional potential. Also the larger junctional potentials are associated with shorter delays to the resulting muscle action potentials (arrowheads). [Adapted from Spencer et al. (89).]

Another system, the oscillator or O system is formed from photoreceptive neurons in the ocelli, which project into the outer nerve ring and synapse onto both the SMNs and B neurons. The O system neurons do not produce spikes but display oscillatory depolarizations, which are modulated by light (11, 12). Like the SMN system, the B and O systems are formed from networks of identical, electrically coupled neurons. The synapses between different networks are always chemical.

Cnidarian Neurotransmitters
Despite considerable efforts we do not yet have conclusive evidence for the identity of any transmitter substances in the Cnidaria (55). This is unfortunate as it was presumably in this group that chemical neurotransmission first evolved. That

chemical synapses exist in this phylum is not in question. Besides ultrastructural studies showing the expected specializations, such as synaptic vesicles; apposed, parallel, electron dense membranes; and a synaptic cleft, there is convincing physiological evidence for chemical neurotransmission (86, 90). For example, excess magnesium causes synaptic blockade and the synaptic delay in the appearance of conventional postsynaptic potentials is constant (1 to ~8 ms depending on the preparation) (6, 86). Of the numerous putative transmitters tested in various preparations the best evidence suggests that catecholamines and neuropeptides may be transmitters.

Immunohistochemical studies of a wide range of cnidarians from all classes have shown that discrete populations of neurons containing peptides with the carboxy terminus, Arg-Phe-amide are ubiquitous (24). This family of peptides is related to molluscan FMRFamide (Phe-Met-Arg-Phe-amide), which is known to have transmitter functions (23). Preliminary electrophysiological data show that pepetides with the RFamide (Arg-Phe-amide) terminus produce long duration excitation of motor neurons in *Polyorchis*, which is known to contain three RFamide neuropeptides (25, 87). In the anthozoan *Renilla*, RFamide peptides cause rachidial contractions and potentiate activity in the TCNN (1). Catecholamines, especially dopamine, have been found in the nervous systems of both anthozoans and hydrozoans by immunohistochemistry, high-pressure liquid chromatography, and gas chromatography–mass spectrometry (18, 20, 40). Epinephrine acts as a transmitter for control of luminescence and modulation of rachidial contraction in *Renilla* (2).

Pacemakers and Spontaneous Rhythmic Behavior

Many behaviors in cnidarians are rhythmical; swimming in medusae, peristaltic pumping in hydranths, burrowing in anemones, and expansion–contraction rhythms in anemones. The idea that such activity could arise endogenously was novel when it was proposed by Romanes and Eimer in the 1870s to explain pulsations in scyphomedusae (74). Many believed that rhythmical motor activities are produced by reflexes. In all cases where a cellular analysis is possible it has been shown that such activity arises endogenously in pacemaker neurons. This is most clearly seen in hydromedusae where swimming is controlled by a network of motor neurons that exhibit pacemaker potentials analogous to those seen in molluscs. The oscillating membrane potential is most probably produced by a calcium activated K^+ current, which leads to membrane hyperpolarization following a burst. The following burst arises when the membrane spontaneously depolarizes as calcium is sequestered and the K^+ current is reduced allowing expression of an inward Na^+ current (4). This pacemaker property is distrubuted throughout the swim motor network of the inner nerve ring (85). This contrasts with the situation in scyphomedusae where pacemaker neurons are located in each of the eight rhopalia. The frequency of spikes resulting from the pacemaker potentials can be modulated by sensory input. For example, in *Polyorchis* a reduction in light intensity produces a transient depolarizaton of the SMNs that is superimposed on the underlying pacemaker potentials and results in an increased swimming frequency (11). Many anthozoan polyps undergo spontaneous postural changes, often involving column contraction and expansion (15). It is likely that these movements aid in irrigating the gastrovascular cavity for respiratory purposes. The cellular substrate for this pacemaker activity is still in question. Although the TCNN is capable of spontaneous activity it is also possible that the initial source for pacemaker activity is the SS_2 system, which is active between TCNN system bursts. A reduction in SS_2 activity could release the TCNN system from inhibition, resulting in retractor muscle contraction (60).

Nerve Nets as Central Nervous Systems

The term CNS usually brings to mind either the vertebrate brain and spinal cord or the centralized ganglia of arthropods and molluscs. The degree of condensation of nervous tissue and the position of the brain or cerebral ganglia are most likely a result of the selection pressures associated with bilateral symmetry and evolution from ancestors, which crawled on a hard surface and moved anterior end first. Cephalization (or the tendency to develop a head region) was the result of selection for major sensory structures to be located at the anterior end. Consequently, it was advantageous to position the major integrating center, the brain, close to these sense organs. A longitudinal nerve cord, either ventral or dorsal, has been selected as the structure most suited for transmitting motor signals to effectors distributed bilaterally along an elongated body. In contrast, the selection pressures acting on radially symmetrical animals are very different. The two major phyla having radially symmetrical adults are the Cnidaria and Echinodermata. For these animals, whether sessile or planktonic, there is little polarity in the direction from which sensory information is most likely to come. Thus a distribution of sensory structures throughout 360° becomes advantageous and is often associated with a ring-shaped CNS that integrates this information. It has been shown that a "brain" in the shape of a ring is necessary for distinguishing between local and general sensory input in hydrozoan jellyfish (88). Nerve rings in hydromedusae appear to have evolved by the collapse and condensation of nerve-nets. The ring nerves together with radial nerves are preadapted for conducting motor signals to radially arranged effectors.

In conclusion, we should be aware that it is sometimes inappropriate to use the term "primitive" to describe the netlike nervous system of radially symmetrical animals. The organization of their nervous systems into condensed nerve rings as the CNS and peripheral nerve-nets is undoubtedly the best evolutionary solution for radially symmetrical animals.

Studies of cnidarian nervous systems are particularly valuable because of their evolutionary position and the relative simplicity of their neuronal organization. Now that techniques are available for culturing cnidarian neurons (70) it is likely that we will learn a great deal about the evolution of excitability in early nervous systems.

References

1. Anctil, M. *J. Comp. Physiol. B* **157**:31–38, 1987. Bioactivity of FMRF amide and related peptides on a contractile system of the coelenterate *Renilla köllikeri*.

2. Anctil, M., D. Boulay, and L. LaRiviere. *J. Exp. Zool.* **223**:11–24, 1982. Monoaminergic mechanisms associated with control of luminescence and contractile activities in the coelenterate, *Renilla köllikeri*.

3. Anderson, P. A. V. *J. Exp. Biol.* **65**:381–393, 1976. An electrophysiological study of mechanisms controlling polyp retraction in colonies of the scleractinian coral *Goniopora lobata*.

4. Anderson, P. A. V. *J. Exp. Biol.* **78**:299–302, 1979. Ionic basis of action potentials and bursting activity in the hydromedusan jellyfish *Polyorchis penicillatus*.

5. Anderson, P. A. V. *Prog. Neurobiol.* **15**:161–203, 1980. Epithelial conduction: its properties and functions.

6. Anderson, P. A. V. *J. Neurophysiol.* **53**:821–835, 1985. Physiology of a bidirectional, excitatory, chemical synapse.

7. Anderson, P. A. V. *J. Exp. Biol.* **133**:231–248, 1987. Properties and pharmacology of a TTX-insensitive Na$^+$ current in neurones of the jellyfish *Cyanea capillata*.

8. Anderson, P. A. V. and U. Grunert. *Synapse* **2**:606–613, 1988. Three dimensional structure of bidirectional, excitatory, chemical synapses in the jellyfish.

9. Anderson, P. A. V. and W. E. Schwab. *J. Morphol.* **170**:383–399, 1981. The organization and structure of nerve and muscle in the jellyfish *Cyanea capillata* (Coelenterata: Scyphozoa).

10. Anderson, P. A. V. and W. E. Schwab. *J. Neurophysiol.* **50**:671–683, 1983. Action potential in neurons of the motor nerve net of *Cyanea* (Coelenterata).

11. Arkett, S. A., and A. N. Spencer. *J. Comp. Physiol. A* **159**:201–213, 1986. Neuronal mechanisms of a hydromedusan

shadow reflex. I. Identified reflex components and sequence of events.

12. Arkett, S. A. and A. N. Spencer. *J. Comp. Physiol. A* **159**:215–225, 1986. Neuronal mechanisms of a hydromedusan shadow reflex. II. Graded response of relex components, possible mechanisms of photic integration, and functional significance.

13. Ball, E. E. *Biol. Bull. (Woods Hole, Mass.)* **145**:223–242, 1973. Electrical activity and behavior in the solitary hydroid *Corymorpha palma*. I. Spontaneous activity in whole animals and in isolated parts.

14. Bassot, J. M., A. Bilbaut, G. O. Mackie, L. M. Passano, and M. Pavans de Ceccatty. *Biol. Bull (Woods Hole, Mass.)* **155**:473–479, 1978. Bioluminescence and other responses spread by epithelial conduction in the siphonophore *Hippopodius*.

15. Batham, E. J. and C. F. A. Pantin. *J. Exp. Biol.* **27**:290–301, 1950, Inherent activity in the sea anemone *Metridium senile* (L.).

15a. Bullock, T. H. and G. A. Horridge. *Structure and Function in the Nervous System of Invertebrates,* p. 463. Freeman, San Francisco, CA, 1965.

16. Campbell, R. D., R. K. Josephson, W. E. Schwab, and N. B. Rushforth. *Nature (London)* **262**:388–390, 1976. Excitability of nerve-free hydra.

17. Chain, B. M., Q. Bone, and P. A. V. Anderson. *J. Comp. Physiol.* **143**:329–-338, 1981. Electrophysiology of a myoid epithelium in Chelophyes (Coelenterata: Siphonophora).

18. Chung, J. M., A. N. Spencer, and K. H. Gahm. *J. Comp. Physiol. B* **159**:(in press), 1989. Dopamine in tissues of the hydrozoan jellyfish *Polyorchis penicillatus* as revealed by HPLC and GC/MS.

19. Davis, L. E., A. L. Burnett, and J. F. Haynes. *J. Exp. Zool.* **167**:295–332, 1968. Histological and ultrastructural study of the muscular and nervous system of *Hydra*. II. Nervous system.

20. DeWaele, J. P. and M. Anctil. *Can. J. Zool.* **65**:2458–2465. Biogenic catecholamines in the cnidarian *Renilla köllikeri*: Radioenzymatic and chromatographic detection.

21. Fields, G. W. and G. O. Mackie. *J. Fish. Res. Board Can.* **28**:1595–1602, 1971. Evolution of the Chondrophora: evidence from behavioral studies on *Velella*.

22. Grimmelikhuijzen, C. J. P. *Cell Tissue Res.* **241**:171–182, 1985. Antisera to the sequence Arg-Phe-amide visualize neuronal centralization in hydroid polyps.

23. Grimmelikhuijzen, C. J. P., D. Graff, A. Groeger, and I. D. McFarlane. *NATO ASI Ser., Ser. A* **141**:105–132, 1987. Neuropeptides in invertebrates.

24. Grimmelikhuijzen, C. J. P., D. Graff, and A. N. Spencer. pp. 199–218 In *Invertebrate Peptides and Amines* (M. C. Thorndyke and G. Goldsworthy, Eds.). Cambridge Univ. Press, London and New York. Structure, locaton and possible actions of Arg-Phe-amide peptides in coelenterates.

25. Grimmelikhuijzen, C. J. P., M. Hahn, K. L. Rinehart, and A. N. Spencer. *Brain Res.* **475**:198–203, 1988. Isolation of <Glu-Leu-Leu-Gly-Gly-Arg-Phe-NH_2, a novel neuropeptide from hydromedusae.

26. Hand, A. R. and S. Gobel. *J. Cell Biol.* **52**:397–408, 1972. The structural organization of the septate and gap junctions of *Hydra*.

27. Horridge, G. A. *Q. J. Microsc. Sci.* **95**:85–92, 1954. Observations on the nerve fibers of *Aurellia aurita*.

28. Horridge, G. A. *J. Exp. Biol.* **31**:594–600, 1954. The nerves and muscles of medusae. I. Conduction in the nervous system of *Aurellia aurita* Lamarck.

29. Horridge, G. A. *J. Exp. Biol.* **33**:366–383, 1956. The nerves and muscles of medusae. V. Double innervation in scyphozoa.

30. Horridge, G. A. *Philos. Trans. R. Soc. London* **240**:495–529, 1957. The coordination of the protective retraction of coral polyps.

31. Horridge, G. A. *J. Exp. Biol.* **36**:72–91, 1959. The nerves and muscles of medusa. VI. The rhythm.

32. Horridge, G. A. *Interneurons.* Freeman, San Francisco, CA, 1968.

33. Jones, W. C. *Biol. Rev. Cambridge Philos.*

Soc. **37**:1–50, 1962. Is there a nervous system in sponges?

34. Josephson, R. K. *J. Exp. Biol.* **47**:179–190, 1967. Conduction and contraction in the column of hydra.

35. Josephson, R. K. and G. O. Mackie. *J. Exp. Biol.* **43**:293–332, 1965. Multiple pacemakers and the behaviour of the hydroid *Tubularia.*

36. Kass-Simon, G. and L. M. Passano. *Am. Zool.* **9**:113, 1969. *Hydra* conduction pathways.

37. King, M. G. and A. N. Spencer. *J. Exp. Biol.* **94**:203–218, 1981. The involvement of nerves in the epithelial control of crumpling behaviour in a hydrozoan jellyfish.

38. Kinnamon, J. C. and J. A. Westfall. *J. Morphol.* **168**:321–329, 1981. A three-dimensional reconstruction of neuronal distributions in the hypostome of a hydra.

39. Koizumi, O. and H. R. Bode. *Dev. Biol.* **116**:407–421, 1986. Plasticity in the nervous system of adult *Hydra.* I. The position-dependent expression of FMRF amide-like immunoreactivity.

40. Kolberg, K. J. S. and V. J. Martin. *Dev. Biol.* **103**:249–258, 1988. Morphological, cytochemical and neuropharmacological evidence for the presence of catecholamines in hydrozoan planulae.

41. Lawn, I. D. In *Coelenterate Ecology and Behavior* (G. O. Mackie, ed.), pp. 581–590. Plenum, New York, 1976. Chemoreception and conduction systems in sea anemones.

42. Lawn, I. D. *Nature (London)* **262**:708–709, 1976. Swimming in the sea anemone *Stomphia coccinea* triggered by a slow conduction system.

43. Lawn, I. D. and D. M. Ross. *Proc. R. Soc. London, B Ser.* **216**:315–334, 1982. The behavioural physiology of the swimming sea anemone *Boloceroides mcmurrichi.*

44. Loeb, J. *Pfluegers Arch. Gesamte Physiol. Menschen Tiere* **59**:415–420, 1895. Zur Physiologie und Psychologie der Actinien.

45. Mackie, G. O. *Am. Zoo.* **5**:439–453, 1965.

Conduction in the nerve-free epithelia of siphonophores.

46. Mackie, G. O. *J. Neurobiol.* **6**:357–378, 1975. Neurobiology of *Stomotoca.* II. Pacemakers and conduction pathways.

47. Mackie, G. O. *Mar. Behav. Physiol.* **5**:325–346, 1978. Coordination in physonectid siphonophores.

48. Mackie, G. O. In *Neural Mechanisms of Startle Behavior* (R. C. Eaton, ed.), pp. 15–42. Plenum, New York. 1984. Fast pathways and escape in Cnidaria.

49. Mackie, G. O. and R. W. Meech. *Nature (London)* **313**:791–793, 1985. Separate sodium and calcium spikes in the same axon.

50. Mackie, G. O. and L. M. Passano. *J. Gen. Physiol.* **52**:600–621, 1968. Epithelial conduction in hydromedusae.

51. Mackie, G. O. and C. L. Singla. *Philos. Trans. R. Soc. London, Ser. B* **301**:365–400, 1983. Studies on hexactinellid sponges. I. Histology of *Rhabdocalyptus dawsoni* (Lambe, 1873).

52. Mackie, G. O., P. A. V. Anderson, and C. L. Singla. *Biol. Bull. (Woods Hole, Mass.)* **167**:120–123, 1984. Apparent absence of gap junctions in two classes of Cnidaria.

53. Mackie, G. O., I. D. Lawn, and M. Pavans de Ceccatty. *Philos. Trans. R. Soc. London, Ser. B* **301**:401–418, 1983. Studies on hexactinellid sponges. II. Excitability, conduction and coordination of responses in *Rhabdocalyptus dawsoni* (Lambe, 1873).

54. Mackie, G. O., L. M. Passano, and M. Pavans de Ceccatty. *C. R. Hebd. Séances Acad. Sci.* **264**:466–469, 1967. Physiologie du comportement de l'Hydroméduse Sarsia tubulosa Sars. Les systèmes à conduction aneurale.

55. Martin, S. M. and A. N. Spencer. *Comp. Biochem. Physiol. C.* **74C**:1–14, 1983. Neurotransmitters in coelenterates.

56. McFarlane, I. D. *J. Exp. Biol.* **51**:377–385, 1969. Two slow conduction systems in the sea anemone *Calliactis parasitica.*

57. McFarlane, I. D. *J. Exp. Biol.* **51**:387–396, 1969. Coordination of pedal-disc detach-

ment in the sea anemone *Calliactis parasitica*.

58. McFarlane, I. D. *J. Exp. Biol.* **53**:211–220, 1970. Control of preparatory feeding behaviour in the sea anemone *Tealia felina*.

58a. McFarlane, I. D. J. *J. Exp. Biol.* **58**:77–90, 1973. Spontaneous electrical activity in sea anemone Calliactis.

59. McFarlane, I. D. *J. Exp. Biol.* **64**:431–446, 1976. Two slow conducting systems coordinate shell-climbing behaviour in the sea anemone *Calliactis parasitica*.

60. McFarlane, I. D. pp. 243–265. In *Electrical Conduction and Behaviour in 'Simple' Invertebrates* (G. A. B. Shelton, ed.), pp. 243–265. Oxford Univ. Press (Clarendon), London and New York, 1982. *Calliactis parasitica*.

61. Mills, C. E., G. O. Mackie, and C. L. Singla. *Can. J. Zool.* **63**:2221–2224, 1985. Giant nerve axons and escape swimming in *Amphigona apicata*.

62. Pantin, C. F. A. *J. Exp. Biol.* **12**:119–138, 1935. The nerve net of the Actinozoa. I. Facilitation.

63. Pantin, C. F. A. *J. Exp. Biol.* **12**:389–396, 1935. The nerve-net of the Actinozoa. IV. Facilitaton and the 'staircase.'

64. Parker, G. H. *J. Exp. Zool.* **22**:87–94, 1917. Nervous transmission in actinians.

65. Passano, L. M. *Am. Zool.* **5**:465–481, 1965. Pacemakers and activity patterns in medusae: homage to Romanes.

66. Passano, L. M. and C. B. McCullough. *Proc. Natl. Acad. Sci. U.S.A.* **48**:1376–1382, 1962. The light response and the rhythmic potentials of hydra.

67. Passano, L. M. and C. B. McCullough. *Nature (London)* 119:1174–1175, 1963. Pacemaker hierarchies controlling the behavior of hydra.

68. Pavans de Ceccatty, M. *Perspect. Biol. Med.* **17**:379–390, 1974. The origin of the integrative systems: A change in view derived from research on coelenterates and sponges.

69. Prosser, C. L. *Z. Vergl. Physiol.* 54:109–120, 1967. Ionic analyses and effects of ions on contractions of sponge tissues.

70. Przysiezniak, J. and A. N. Spencer, *J. Exp. Biol.* **142**:97–113, 1989. Primary culture of identified neurones from a cnidarian.

71. Roberts, A. and G. O. Mackie. *J. Exp. Biol.* **84**:303–318, 1980. The giant axon escape system of a hydrozoan medusa, *Aglantha digitale*.

72. Robson, E. A. and R. K. Josephons. *J. Exp. Biol.* **50**:151–168, 1969. Neuromuscular properties of mesenteries from the sea anemone *Metridium*.

73. Romanes, G. J. *Philos. Trans. R. Soc. London* **166**:269–313, 1877. Preliminary observations on the locomotor system of medusae.

74. Romanes, G. J. *Philos. Trans. R. Soc. London* **166**:659–752, 1878. Further observations on the locomotor system of medusae.

74a. Romanes, G. J. *Jellyfish, Starfish,* and *Sea Urchins, Being a Research on the Primitive Nervous Systems.* International Science Series, Appleton, New York. 1985.

75. Satterlie, R. A. *J. Comp. Physiol.* **133**:357–367, 1979. Central control of swimming in the cubomedusan jellyfish *Carybdea rastonii*.

76. Satterlie, R. A. and J. F. Case. *J. Exp. Biol.* **79**:191–204, 1979. Neurobiology of the gorgonian coelenterates, *Muricea californica* and *Lophogorgia chilensis*.

77. Satterlie, R. A. and A. N. Spencer. *Nature (London)* **281**:141–142, 1979. Swimming control in a cubomedusan jellyfish.

78. Satterlie, R. A. and A. N. Spencer. *J. Comp. Physiol.* **150**:195–207, 1983. Neuronal control of locomotion in hydrozoan medusae: A comparative study.

79. Satterlie, R. A. and A. N. Spencer. *NATO ASI Ser., Ser. A* **141**:213–264, 1987. Organization of conducting systems in 'simple' invertebrates: Porifera, Cnidaria and Ctenophora.

80. Satterlie, R. A., P. A. V. Anderson, and J. Case. *Mar. Behav. Physiol.* **7**:25–46, 1980. Colonial coordination in anthozoans: Pennatulacea.

81. Shelton, G. A. B. *J. Exp. Biol.* **62**:571–578,

1975. Colonial conduction systems in the Anthozoa: Octocorallia.

82. Shelton, G. A. B. *Proc. R. Soc. London, Ser. B* **190**:139–256, 1975. Colonial behaviour and electrical activity in the Hexacorallia.

83. Shelton, G. A. B. *J. Mar. Biol. Assoc. U.K.* **60**:517–528, 1980. *Lophelia pertusa* (L.): Electrical conduction and behaviour in a deep-water coral.

84. Spencer, A. N. *J. Neurobiol.* **9**:143–157, 1978. Neurobiology of *Polyorchis*. I. Function of effector systems.

85. Spencer, A. N. *J. Exp. Biol.* **93**:33–50, 1981. The parameters and properties of a group of electrically coupled neurones in the central nervous system of a hydrozoan jellyfish.

86. Spencer, A. N. *J. Comp. Physiol.* **148**:353–363, 1982. The physiology of a coelenterate neuromuscular synapse.

87. Spencer, A. N. *Can. J. Zool.* **66**:639–645, 1988. Effects of Arg-Phe-amide peptides on identified motor neurons in the hydromedusa *Polyorchis penicillatus.*

88. Spencer, A. N. and S. A. Arkett. *J. Exp. Biol.* **110**:69–90, 1984. Radial symmetry and the organization of central neurones in a hydrozoan jellyfish.

89. Spencer, A. N., J. Przysiezniak, J. Acosta-Urquidi, and T. A. Basarsky. *Nature (London),* **340**:636–638, 1989. Presynaptic spike broadening reduces junctional potential amplitude.

90. Westfall, J. A. *NATO ASI Ser., Ser. A* **141**:3–28, 1987. Ultrastructure of invertebrate synapses.

91. Westfall, J. A., D. R. Argast, and J. C. Kinnamon. *J. Morphol.* **178**:95–103, 1983. Numbers, distribution, types of neurons in the pedal disk of *Hydra* based on a serial reconstruction from transmission electron micrographs.

Section C
Nonsegmented Worms

Fred Delcomyn

The flatworms, the pseudocoelomate and the minor coelomate phyla represent an evolutionary transition between the nerve-net organization of coelenterates and the fully developed CNS of higher invertebrates and vertebrates. Two main trends are apparent: cephalization and the concentration of neurons into a CNS.

Flatworms represent an intermediate stage in the transition. Most flatworms have a small but clearly defined brain and a ladderlike pair of nerve trunks extending the length of the body, as well as an extensive nerve net anastomosing throughout the periphery of the body (Fig. C-1*A* and *B*) (12). Neurons have morphologies typical of those from higher groups: monopolar, bipolar and multipolar types have all been identified (Fig. C-1*C*) (6). Most neurons are short; flatworms seem to lack the long-axon through-conducting pathways found in arthropods. Even in these primitive animals, individual neurons in the brain can be identified (characterized morphologically and physiologically), and found in different individuals of the same species (6, 9). Functionally, the nerve-net exerts local control, including reflex control, over the muscles. The brain acts as a decision making center (12). Animals can continue to carry out some acts, like feeding, in the absence of a brain, using local reflexes to coordinate their actions; however, control over behavior such as ceasing to eat when the stomach is full, is lost. Full rapid recovery follows transplantation of the brain of one individual into the brainless body of another, even if the new brain is inverted or otherwise disoriented. Nerve stumps fuse with the wrong peripheral nerves, but after some time, processes in the fused roots can be seen to turn 180° back toward the correct side (3).

The physiology of the flatworm nervous system is as complex as that in the more complex invertebrates. Action potentials in some neurons have multiple ionic mechanisms: two sodium, one cal-

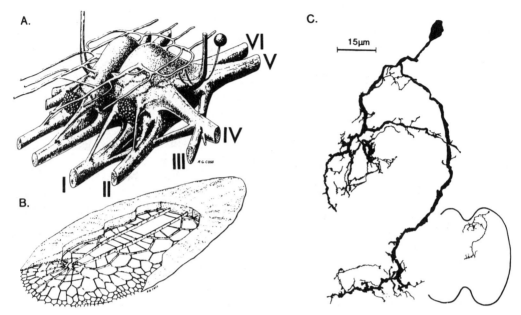

Figure C-1. The nervous system and a representative neuron of *Notoplana acticola*. (*A*) Drawing of the whole brain, based on a model built up from 15-μm serial sections. Anterior is to the left. The grapelike clusters are cell bodies that lie outside the main neural mass, but send their axons into the neuropile. (*B*) Cut-away view of the nervous system. Note the ribbonlike shape of the main nerve trunks. (*C*) Drawing of a monopolar neuron stained with cobalt choloride and drawn. The small inset shows the location of the neuron relative to a profile of the brain. [From Keenan et al. (6).]

cium, and two potassium currents have been shown to contribute (9). Sodium channels may be TTX sensitive (7). Inhibitory mechanisms have been demonstrated, as have simple synaptic interactions like facilitation (13). Facilitation can also be demonstrated in the simpler nervous system of the parasitic cestode *Gyrocotyle fimbriata* (11). The identity of any chemical neurotransmitter in a flatworm nervous system is not known, but the CNS of both parasitic (*Gyrocotyle fimbriata*) and free-living (*Notoplana acticola*) flatworms are affected by amino acids or GABA (γ-aminobutyrate). Electrical activity in the cord of *Notoplana* is depressed by GABA and glycine (10), an action that may be mediated by a chloride channel. In *Gyrocotyle*, the longitudinal nerve cords are excited when exposed to glutamate and aspartate at concentrations of $10^{-5} M$ and lower (8).

The nematodes, even though they have their own specializations, represent another stage in CNS development. These animals have dispensed with the peripheral nerve-net and concentrated all their neurons in central nerve cords. The cords are not ganglionic; cell bodies are distributed along their entire lengths. There is generally a nerve ring in the anterior end that serves as a brain (17). Defecation and egg laying can proceed in the absence of the anterior end of *Caenorhabditis elegans* (2). Morphologically, neurons in the nematode CNS are simpler than those in flatworms, consisting mainly of bipolar types (14). Motorneurons in *Ascaris* can be either excitatory or inhibitory on muscle, and can also interact directly with other motorneurons. There are only seven morphological types of motorneurons in *Ascaris* (Fig. C-2) (16). Electrical activity propagates along the nerve cord

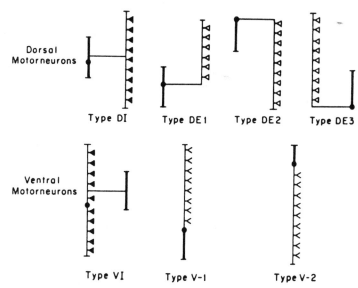

Figure C-2. Drawings of the seven types of motorneurons in *Ascaris*. The small triangles and open Ys show the regions of synaptic contact with muscle: open triangles represent excitatory output, filled triangles represent inhibitory output, and the open Ys represent synapses whose sign has not yet been determined. The thickened lines represent the dendritic input area of each neuron. The thin horizontal lines represent a branch through a commissure that connects one of the two main nerve cords to the other. [From Walrond et al. (16).]

at ~16 cm/s. This is much faster than the wave of contraction that occurs during swimming, so the CNS must actively regulate muscle activity (18). Serotonin and octopamine can alter the behavior of *Caenorhabditis*; the former stimulates egg laying and pharyngeal pumping, whereas the latter depresses egg laying (but not pharyngeal pumping) (4). γ-aminobutyrate immunoreactivity has been reported in the motorneurons of *Ascaris* (5).

Using only the criterion of large axon diameter relative to other neurons in the same animal, giant fibers are present in many of these minor phyla. They have been described in Platyhelminths (12), Nermatines, Echiuroids, Onychophora, and Phoronidia (1). Except for the latter, nothing is known of their physiology. In Phoronida, conduction velocities of fibers ~25–35 μm in diameter have been measured at 5–6 m/s (19). Other species have smaller (4 μm) or larger (80 μm) diameter axons that may occupy from 0.3 to 10% of the entire body diameter (15).

References

1. Bullock, T. H. and G. A. Horridge. *Structure and Function in the Nervous Systems of Invertebrates*. Freeman, San Francisco, CA, 1965.

2. Croll, N. A. and J. M. Smith. *J. Zool.* **184**:507–517, 1978. Feeding behaviour in *Caenorhabditis*.

3. Davies, L., L. Keenan, and H. Koopowitz. *J. Exp. Zool.* **235**:157–173, 1985. Behavioral recovery following brain transplantation in a flatworm.

4. Horvitz, H. R., M. Chalfie, C. Trent, J. E. Sulston, and P. D. Evans. *Science* **216**:1012–1014, 1982. Serotonin and octopamine in the nematode *Caenorhabditis elegans*.

5. Johnson, C. D. and A. O. W. Stretton. *J. Neurosci.* **7**:223–235, 1987. GABA-immunoreactivity in motor neurons of the nematode *Ascaris*.

6. Keenan, C. L., R. Coss, and H. Koopow-itz. *J. Comp. Neurol.* **195**:697–716, 1981. Cytoarchitecture of flatworms brains.

7. Keenan, L. and H. Koopowitz. *J. Exp. Zool.* **215**:209–213, 1981. TTX-sensitive action potentials in the brain of the flatworm, *Notoplana acticola*.

8. Keenan, L. and H. Koopowitz. *J. Neurobiol.* **13**:9–21, 1982. Putative aminergic neurotransmitters in *Gyrocotyle fimbriata*, a parasitic flatworm.

9. Keenan, L. and H. Koopowitz. *J. Comp. Physiol. A* **155**:197–208, 1984. Ionic bases of action potentials in identified flatworm neurones.

10. Keenan, L., H. Koopowitz, and K. Bernardo. *J. Neurobiol.* **10**:397–407, 1979. Action of aminergic drugs and blocking agents on activity in the nerve cord of the flatworm *Notoplana acticola*.

11. Keenan, L., L. Koopowitz, and M. H. Solon. *J. Parasitol.* **70**(1):131–138, 1984. Electrical activity in the nerve cords of the parasitic flatworm, *Gyrocotyle fimbriata*.

12. Koopowitz, H. In *Electrical Conduction and Behaviour in 'Simple' Invertebrates* (G. A. B. Shelton, ed.), pp. 359–392. Oxford Univ. Press (Clarendon), London and New York, 1982. Free-living Platyhelminthes.

13. Koopowitz, H., K. Bernardo, and L. Keenan, *J. Neurobiol.* **10**:367–381, 1979. Primitive nervous systems: electrical activity in ventral nerve cords of the flatworm, *Notoplana acticola*; Koopowitz, H., L. Keenan, and K. Bernardo, *J. Neurobiol.* **10**:383–395, 1979. Primitive nervous systems: electrophysiology of inhibitory events in flatworm nerve cords.

14. Stretton, A. O. W., R. E. Davis, J. D. Angstadt, J. E. Donmoyer, and C. D. Johnson, *Trends Neurosci.* **8**:294–300, 1985. Neural control of behaviour in *Ascaris*.

15. Thorpe, J. P. In *Electrical Conduction and Behaviour in 'Simple' Invertebrates* (G. A. B. Shelton, ed.), pp. 393–439. Oxford Univ. Press (Clarendon), London and New York, 1982.

16. Walrond, J. P., I. S. Kass, A. O. W. Stretton, and J. E. Donmoyer, *J. Neurosci.* **5**:1–8, 1985. Identification of excitatory and inhibitory motoneurons in the nematode *Ascaris* by electrophysiological techniques; Walrond, J. P. and A. O. W. Stretton, *J. Neurosci.* **5**:9–15, 1985. Reciprocal inhibition in the motor nervous system of the nematode *Ascaris*: direct control of ventral inhibitory motoneurons by dorsal excitatory motoneurons.

17. Ware, R. W., D. Clark, K. Crossland, and R. L. Russell. *J. Comp. Neurol.* **162**:71–110, 1975. Sensory input and motor output in the nematode *Caenorhabditis elegans*.

18. Weisblat, D. A. and R. L. Russell. *J. Comp. Physiol.* **107**:239–307, 1976. Electrical activity in nerve cord and muscle of *Ascaris lumbricoides*.

19. Wilson, D. M. and T. H. Bullock. *Anat. Rec.* **132**:518–519, 1958. Electrical recording from giant fiber and muscle in phoronids.

Section D
The Molluscan Nervous System

Rhanor Gillette

Studies of the molluscan nervous system in the last 40 years have been numerous and the body of knowledge is extensive. This section aims at an interpretive overview of function and structure in the molluscan nervous system.

The molluscs arose in the pre-Cambrian era and are anciently allied with arthropod and annelid lines. Presently, they comprise more than 80,000 species, widely diversified in marine, aquatic, and terrestrial environments. An abbreviated lineage is shown in Figure D-1 for the reader's convenience in referring to the relationships of the genera mentioned here.

The Character of the Molluscan Nervous System

The Peripheral and Central Nervous Systems in Soft-Bodied Animals

An extensive peripheral nervous system (PNS) is a salient character of the phylum Mollusca, which must be accounted in all

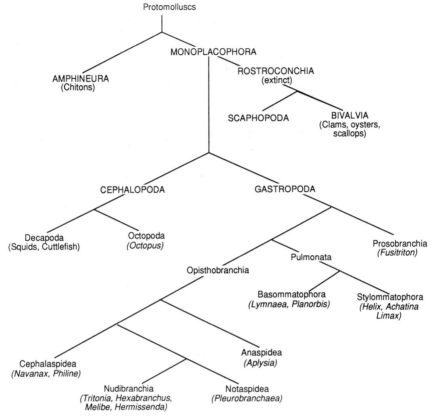

Figure D-1. The descent of molluscs. An abbreviated cladogram of a plausible molluscan lineage, with particular reference to genera discussed in this section.

neurophysiological analyses of molluscan behavior. The PNS is capable of mediating many simple behavioral reflexes autonomously. The PNS comprises both a subepidermal nerve plexus as well as small, distributed peripheral ganglia serving reflex control or primary sensory processing of special organs (e.g., suckers, palps, rhinophores, and gills). Many gastropods with the CNS removed show continuous forward locomotion for hours, mediated by the PNS. The potential for complex coordinative capacity in the cephalopod PNS was demonstrated amply by Rowell (165) and others (Fig. D-2).

The mutual contributions of PNS and CNS to molluscan behavior are well documented (152). The gill–siphon withdrawal reflex of the sea slug *Aplysia*, triggered by tactile stimuli, has both central and peripheral components; both components support the relatively higher integrative functions of habituation and sensitization of the withdrawal reflex (35, 41, 94, 122, 156, 157). In the sea slug *Pleurobranchaea*, reflex withdrawal of the skin and oral veil to mechanical stimuli is also mediated both centrally and peripherally (117). Chemosensory reflex function is also shared: amino acids such as betaine (trimethyglycine) and glycine, which initiate feeding in the intact animal, when applied to food chemosensory areas (oral veil and rhinophores) of a deganglionated preparation cause appetitive flaring of chemosensory papillae and movements of lateral tentacles. A specifically aversive amino acid, taurine, applied anywhere on the body at concentrations as low as 10^{-6}

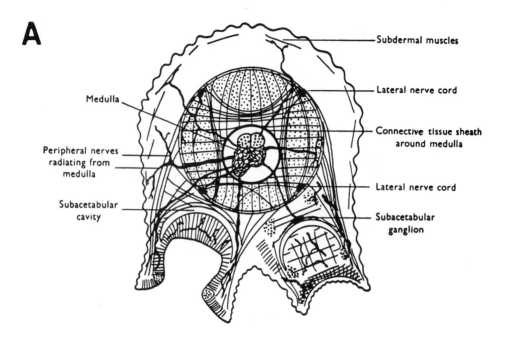

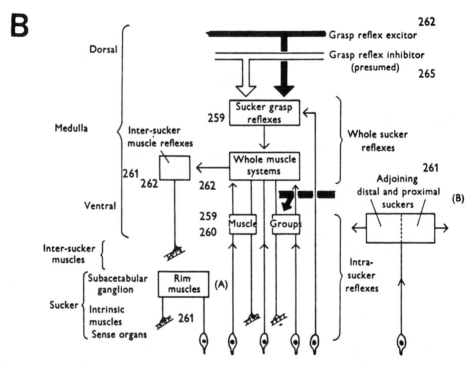

Figure D-2. Computations performed in the peripheral ganglia integrate and direct much of the sensory-motor traffic between the CNS and the environment. (*A*) Transverse section of an octopus arm showing the main muscles, nerves, and ganglia. (*B*) The nervous organization in the medulla affecting the muscles of one sucker of the arm. [From Rowell (165).]

M causes local withdrawal and active aversive locomotion (100); local withdrawal to taurine remains in preparations without a CNS.

Thus, the PNS commands significant function in molluscs having either very reduced and simplified CNS, as in the opisthobranch gastropod, or as in the cephalopods having very highly developed and complex brains. It is likely that the extensiveness of the molluscan PNS is a specific adaptation to the soft body and hydrostatic skeleton of the animals. Musculature of the skin is typically not discretely organized into well-ordered bundles. The absence of a peripheral plexus, demands a great precision of direct central innervation of the muscle. The local control of motor function and reflexes by the PNS, interposed between CNS and periphery, acts to reduce the precision necessary for CNS innervation of the body wall and thereby reduces the number of central neurons necessary to innervate large areas of periphery.

This interpretation of CNS–PNS interaction is supported by consideration of the exception: The gastropod and cephalopod buccal mass, which is the feeding apparatus for biting, swallowing, and rejection movements. The buccal mass has a skeletal structure provided by chitin and tough connective tissue, discrete muscle bundles, and is centrally innervated with a high precision directly from central motorneurons. Even the sensory, stretch receptor neurons of the buccal musculature originate centrally. The lack of PNS in this nonsoft, quasiarticulate structure is consistent with the role of the PNS as a specific adaptation to soft-body coordination.

For soft-bodied molluscs, the lack of joints precludes a precise positional sense; without the positional information available from stretch receptor innervation of discrete, articulated joints, the blinded octopus does not know the position of its arms. Nor can an intact octopus estimate the weight of an object, only size and texture. Wells (190) used the analogy of a human riding a camel to describe CNS control of a soft body. Central motor commands can be elaborated to adapt to local conditions by the peripheral system, with very little feedback about the details, save through the driver's direct exteroceptors. The inability of octopus and other molluscs to learn to manipulate levers or learn mazes is laid to the nature of their motor control and proprioceptive sense in the periphery (190).

The elaboration of PNS in the phylum may also favor growth of larger body sizes. Elaboration of the PNS for control of larger body wall areas might entail lesser demand for increased innervation of the periphery by the CNS. Some species of molluscs, with their soft bodies and extensively developed PNS, are the largest invertebrates living or extinct, including the abyssal squid *Architeuthis* (18 m in length), the *Octopus dofleini* (10 m in diameter), some opisthobranch snails of the genera *Pleurobranchus* (0.3 m in length), *Pleurobranchaea* (3.1 kg), *Hexabranchus* (0.3 m in length), and *Aplysia* (3.6 kg), pulmonate snails such as *Achatina* (1 kg), tridacnid clams (270 kg), and many prosobranch snails of the abalones, strombids, cassids, and whelks. Extinct cephalopod ammonites (the spirally coiled shell of *Pachydiscus seppenradensis* reached 2.5 m in diameter) and the clamlike rudists of the upper Cretaceous also attained great bulk.

The Molluscan Neuron

MORPHOLOGY. The nervous system has been most completely studied in the gastropods, followed by work on cephalopods. The studies on gastropods have for the most part been motivated by the relative simplicity of behavior and CNS, which suggested that neural mechanisms were more accessible to investigation. In contrast, work on cephalopods has been motivated by the complexity of their intelligence, and lured by the presence of great neural complexity in a line evolved independent of that leading to vertebrates.

The ganglia of the gastropods in general exhibit a simplicity characteristic of small, primitive ganglia of most animals, such as the peripheral sympathetic ganglia of vertebrates. They are structured on a simple plan where the somata of neurons lie on the outside of the ganglia. Somata send axons centrally to form the neuropil where synapses are made between fine axonal processes (dendrites) (Fig. D-3). Morphologically well-defined and clear focal synaptic contacts are rare in electron micrographs (47, 93), and the role of other types of contacts in synaptic transactions remains for better definition. Conventional "gap" junctions mediate direct electrical coupling between neurons (17, 93, 107). Neuron somata generally lack synaptic contacts (47, 183). Axons may terminate in neuropil or exit in nerve trunks to conduct spikes to or from peripheral structures or other ganglia. Such simplicity of structure arises from an equally simple developmental plan. This structure is common to ganglia of all phyla where small mass and fewness of neurons do not require a higher order of structure (Fig. D-4). The superimposed orderliness of plexiform and lamellar structure of larger and more complex ganglia (like the brains of cephalopods, insects, and vertebrates) seems to be associated with a need to facilitate the making of specific synaptic connections among large numbers of different neuron types in development.

The molluscan neuron typically has a much larger surface area than would be initially estimated in the light microscope

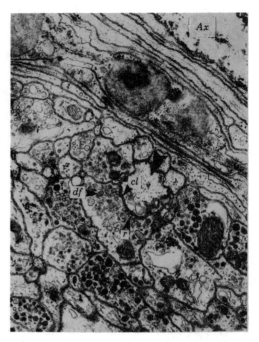

Figure D-3. A small process (large arrow) of the L10 interneuron of the *Aplysia* abdominal ganglion receives a presynaptic contact from another neuron. c1, synaptic density; df, diffuse vesicle; r, round clear vesicle. Part of the large axon of L10 is visible (Ax). [From Gillette and Pomeranz (93).]

(up to 8×), due to extensive invagination of the membrane (Fig. D-5). The reduction in input resistance that this large surface might cause is compensated by a very high specific membrane resistance (98). A consequence of the infolding is that axons can be severely stretched (several times their length) without significant change in conduction velocity (144). In fact, the ganglia and nerves are protected by only a muscular sheath and are normally subject to much mechanical distortion by movements of daily life. Snail neurons are much more tolerant of mechanical stress during dissection than those of arthropods and vertebrates. Thus, the infolding of the neurons lends elasticity and resilience, which may

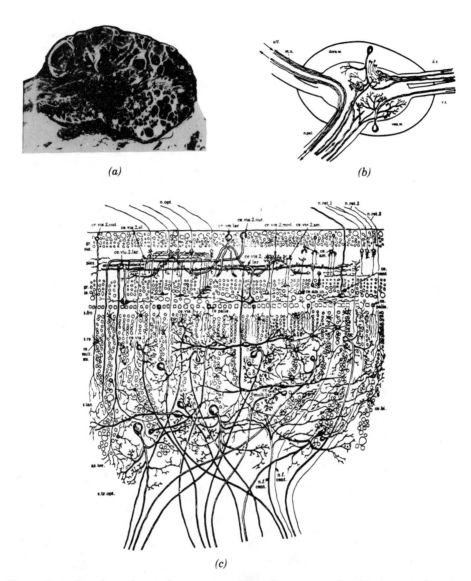

(a)

(b)

(c)

Figure D-4. Simple and complex structure in molluscan ganglia. (*a*) A vertical section through the right parietal ganglion of *Achatina* sp., showing typical simple ganglion structure where neuron cell bodies are arranged peripherally to surround a core of neuropil. Giant cell somata (gc) and smaller somata send axons into the neuropil (npt). Some axons form bundles exiting in the right pallial nerve (rpn). [From Nisbet (154).] (*b*) Similar structure is depicted in a drawing of the simple stellate ganglion of *Octopus* (195). (*c*) The structural complexity of the optic lobe of the squid *Loligo* contrasts with (*a*) and (*b*), reflecting its functional and integrative complexity (197, plate 38).

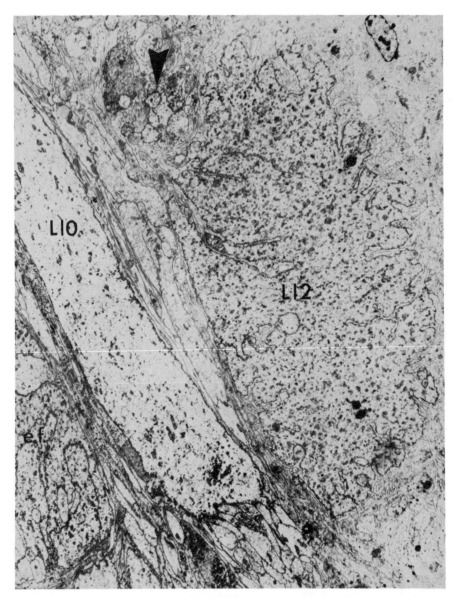

Figure D-5. The surface area of molluscan neurons is disproportionately large because of extensive invaginations of the membrane. A low-power electron micrograph shows a cross section of the axon of the neuron L12, within the neuropil of the abdominal ganglion of *Aplysia*, illustrating the infolding of the axon surface. A primary dendrite of a presynaptic neuron, L10, courses nearby and synaptic contact is made at finer processes (arrowhead). Both neurons are intracellularly stained with osmiophilic polymer. [From Gillette and Pomeranz (93).]

be seen as an important neural adaptation to a soft body.

PHYSIOLOGY. Gastropod neurons are particularly well studied with respect to their ion conductances (142); yet, new types of ion current continue to emerge as methods of study are refined. Most studies are conducted by intracellular recording and voltage clamping of the neuron somata, which are excitable. The most prominent species of ion conductance are similar to those described in arthropod, annelid, and vertebrate neurons. Individual differences in representation of the conductances among identified neurons are pronounced and reflect the neurons' functional roles. Typically, in identified neurons of the cerebral and buccal ganglia of *Pleurobranchaea*, motorneurons and interneurons with no significant presynaptic role in their ganglia of origin have negligible gCa component to their somata action potentials. The spikes are supported by gNa and are of relatively short duration. However, sensory neurons and interneurons with demonstrable local synaptic outputs may show a considerable (slower) Ca^{2+} component in their somata action potentials, consistent with a role in Ca^{2+}-dependent transmitter output in the electrically nearby dendrites— for this reason it is often presumed that dendritic membrane physiology of the molluscan neuron is similar to that of the soma.• Regardless of soma physiology, rapid action potentials in peripheral axons and interganglionic connectives are primarily suited for conduction, their inward current being largely Na^+ dependent with rapid kinetics (108).

The K^+ currents contribute greatly to shaping the activity of molluscan neurons. The phyletically widespread K^+ currents, the delayed rectifier current, transient "A" current, and Ca^{2+}-activated K^+ current, were all first characterized electrophysiologically in neurons of ce-

phalopod and opisthobranch (see Chapter 12).

IDENTIFIED NEURONS. In the pulmonate and opisthobranch nervous systems many individual neurons are identifiable on the basis of ganglion position, hue, axon path, synaptic connectivity, and other characters that are associated with specific function (79). Prominent in these subclasses is the reduced number and relative large size of the neurons in the CNS. In the higher prosobranch gastropods and cephalopods, molluscs with larger and more complex behavioral repertories, neuron numbers are greater and cells are generally smaller; the concept of identifiability is more usefully exchanged for the less exact notion of addressability of populations of similar neurons (29).

NEURON GIANTISM AND ITS SIGNIFICANCE. A prominent feature of the opisthobranch and pulmonate nervous systems is the large size of many identifiable neurons, up to a millimeter in diameter for the somata of certain giant neurons. This condition has enabled many studies of neural circuitry and neuron biophysics. The size of specific neurons, in soma diameter, axon diameter, and dendritic field, increases with the size of the animal during growth (47, 48, 79). For most cell types, increasing cell size is generally accompanied by increase in the actual mass of the DNA of the genome and cell RNA (145; cf. 43); this is effected either through polyploidy or polyteny. In molluscan neurons, increasing polyploidy occurs with increasing neuron size. The nuclei of the largest neurons of mature *Aplysia* contain >0.2 μg of DNA, >200,000 times the haploid amount (125). Neurons of *Achatina* with soma diameters of >9 μm (nuclear diameter >7 μm) were found to be polyploid (45). The frequency distribution of the DNA content in *Achatina* (45) and *Planorbis* (132) neurons indicated that the endoreplicaton during growth probably represented selective gene amplification, rather than simple sequential doubling. However, sequential doubling may occur in growth of *Aplysia* (48, 125). While the condition of neuron giantism has been cheerfully accepted by molluscan neurophysiologists, the actual significance is obscure. The following paragraphs offer a possible explanation.

Neuron giants may be understood first as neurons that innervate larger postsynaptic target areas than nongiants, and it proceeds that their condition probably functions as an adaptation allowing increased animal size without proportional increase in central neuron number. Giant cells in most systems are *in toto* more metabolically active than smaller cells and are frequently associated with transport and secretory processes. A familiar example is the giant polytene cells of dipteran salivary glands, malpighian tubules, and gut, all of which are notably active in ion and peptide transport and exocytotic secretion. Thus, elaboration of DNA, RNA, and protein in many giant cells is indicative of enhanced synthetic capacity, presumably to serve the needs of increased cell activity. In giant neurons, these needs would likely be connected with increased axon transport and secretion processes at synaptic terminals.

The evidence that neuron size is directly related to the extent of postsynaptic innervation comes from consideration of the functions of the better known giants. The largest neurons of opisthobranch and pulmonate ganglia act as effectors, which innervate large areas of the periphery. Two of the largest neurons known, the neurons R2 and LP11 of *Aplysia*, innervate large areas of the skin; their electrical activity stimulates mucus secretion (46, 102, 161). A bilateral pair of serotonergic neurons identified across many snail species are commonly the largest neuron somata of the cerebral ganglion; they innervate large areas of the buccal mass, the

esophagus and lip region, as well as having an interneuronal function in the buccal ganglion.

The well-studied buccal ganglia provide more examples. The largest neurons of opisthobranch buccal ganglia are typically motorneurons; sensory neurons are on the average much smaller (33, 71, 83, 171, 174). The largest known buccal cells may be those of the buccal ganglion of *Navanax*, which innervate the musculature of the large pharynx and drive its expansion during prey capture (173). Interneurons with only central synaptic outputs tend to be smaller than interneurons of dual function, having both CNS output as well as peripheral axons innervating muscle. For instance, both the identified VWC and B3I neurons of *Pleurobranchaea* are capable of driving intense cyclic motor output in the buccal oscillator network; however, the VWC also innervates the muscular esophagus and its soma is nearly three times the B3I soma diameter (90). Identified neurons with purely central outputs also may differ in size according to the extent of their postsynaptic output. The paired SO interneurons of the buccal ganglion of *Lymnaea* have a large postsynaptic field and their somata are three times the size of the interneurons of the N1, N2, and N3 populations, which have collectively rather similar function as oscillator elements but have smaller postsynaptic fields in the ganglion (74, 75). This also exemplifies the trade-off of neuron size for neuron number.

The relation between the extent of the field of postsynaptic innervation and neuron soma size has been recognized for years in arthropod neurobiology (143). The soma diameter of serially homologous motor neurons in the segmental nervous system of crayfish is roughly proportional to the area of the serially homologous muscle they innervate (146). Amputation of the specialized snapping claw of the snapping shrimp *Alpheus* causes the contralateral claw and its musculature to enlarge into a snapping claw with subsequent molts; the soma of the claw opener motorneuron enlarges with the size of its target organ (143). Similarly, the largest neurons of the opisthobranch CNS are peripheral effectors; their size increases with the growth of their target organs. It follows that the sensory neurons and interneurons themselves must increase in size in order to effectively innervate the enlarged peripheral effectors, a form of load matching. In summary, giantism in opisthobranch neurons may be seen as an adaptation for control of a relatively large body by a nervous system operating with a relatively small number of nerve cells.

The Nervous System and Behavior

Complexity of Brain and Behavior

GASTROPODS. The ancestors of the opisthobranchs and pulmonates may have split from the prosobranch snails as lines of burrowing, mud-dwelling, and possibly scavenging snails (23). Reduction and loss of the shell in the opisthobranch species correlates generally with loss of burrowing behavior and is consonant with acquisition of the notable chemical defenses of many free-ranging species (acid and toxic allomone secretions (182)). Pulmonate divergence from the opisthobranchs into land and fresh waters presented less stable and more complex environments to which some species responded evolutionarily by development of somewhat more complex brains and behavior. Radiation of the more ancient prosobranchs into diverse niches of the marine environment matches the success of the crustaceans, and is reflected in diversity of CNS structure (26, 70).

Central nervous system development is directly associated with sensory and behavioral ability. The simplicity of the structure of the opisthobranch nervous system and its sense organs is accompanied by an endearing stupidity. The behavior of opisthobranchs, like their nervous systems, lacks the complexity of that of many prosobranch snails; the number of behavioral subroutines they utilize in daily living is obviously smaller than those of animals living in more complex ecological niches.

Large, more complex brains, with large numbers of small neurons, are associated with development of sense organs for high-resolution analysis of the environment and greater complexity of behavior. In the predatory prosobranch whelks the many tiny neurons, relatively large ganglia, and eyes mediate similarly complex behaviors. The whelk *Fusitriton oregonesis* devotes considerable behavioral strategy to reproduction. Mating pairs form seasonally and endure up to 4 months. Subsequently, a parent attaches its clutch of eggs to a rock surface and patrols them against predators (72). Potential predators, perhaps sensed in part by the animal's well-developed eyes, are attacked with twisting movements of the whelk's shell to dislodge them; failing that, the whelk may directionally squirt an aversive acid secretion. No opisthobranch snails, with their rudimentary-at-best vision and small numbers of CNS neurons, come anywhere near such complexity of behavior.

CEPHALOPODS. Complexity of cephalopod behavior and brain has been the subject of numerous studies (22, 191, 196), anecdotes (124), and media documentaries. Squids and octopi live a highly motile life, largely three dimensional, and actively predatory. Cephalopods have developed equilibrium receptor systems (statocysts) paralleling in structure and function those of vertebrates (27), and even more highly differentiated (Fig. D-6). The highly developed rheoreceptor system of squid demonstrated by Budelmann and Bleckmann (25), analogous to the lateral line of

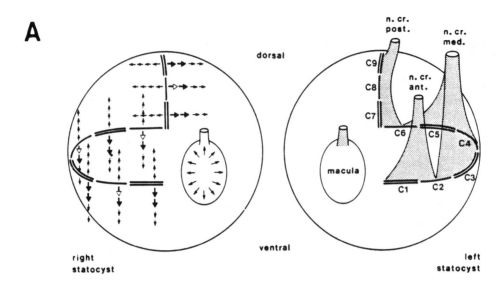

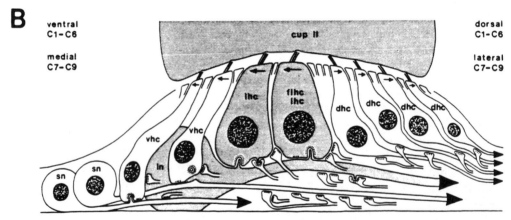

Figure D-6. Cephalopod equilibrium sense. (*A*) Organization of octopus statocysts. Left: Polarization of hair cells in the nine cristae. Arrows indicate patches of hair cells and their polarization. Right: Innervation of cristae by anterior, medial, and posterior crista nerves, carrying both afferent and efferent fibers. The macula is composed of a calcareous otolith lying on a patch of hair cells; it acts as a gravity receptor affecting rotatory eye reflexes. (*B*)Cross section of a crista. Angular acceleration causes movement of endolymph, which in turn deflects the cupula. This shears the kinocilia of the underlying hair cells, exciting or inhibiting them according to their direction of polarization (191). Two size classes of crista–cupula may serve octopus in slow crawling versus fast swimming locomotion. The drawing shows a large type II cupula (cup II), small and large 1° afferent neurons (sn, 1n), primary hair cells (dhc), and large and small secondary hair cells (1hc, vhc). The extraordinary degree of efferent innervation is also depicted. [From Budelmann *et al.* (26).]

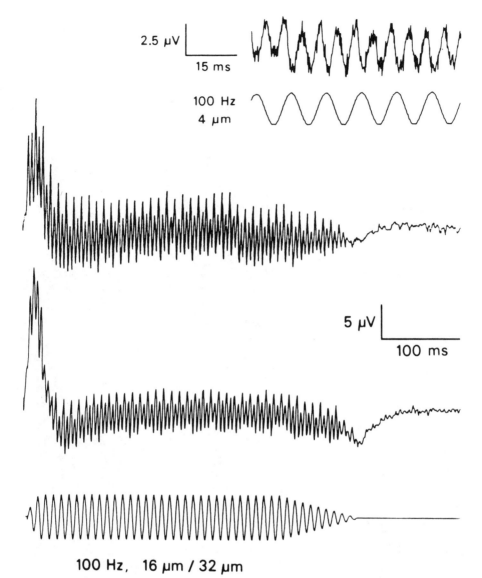

2.5 μV

15 ms

100 Hz
4 μm

5 μV

100 ms

100 Hz,   16 μm / 32 μm

Figure D-7. Microphonic potentials recorded from dorsal epidermal head lines of a squid, *Lolliguncula* (upper record), and the cuttlefish *Sepia* (middle) responding to a 100-Hz vibrational stimulus (lower). The response occurs at twice the stimulus frequency (inset upper right). [From Budelmann and Bleckmann (25).]

pelagic fishes, is a substrate of a vibrational sense usable in predator and prey detection, and in schooling (Fig. D-7). Some cephalopods show visual abilities comparable to those of mammals; they have an interesting and complex central visual physiology (cf. 98). The considerable learning abilities of octopi and squids extend to a variety of tactile and visual discriminations (6, 190, 196). The localization of function and memory processes has been examined by behavioral and lesioning techniques (190, 196). The animal has shown itself to have some potential as an electrophysiological preparation; however, while intracellular recordings are easily obtained, the resting potentials are very low (126). Little progress has been made in the investigation of CNS physiology. Effort in this direction should yield rich results.

The posterior buccal and subfrontal lobes of *Octopus* are necessary brain structures for learning of a tactile discrimination between textured objects; preliminary results of electron microscopy of the neuropil of the lobes suggest that extension of filopodial processes by neurons

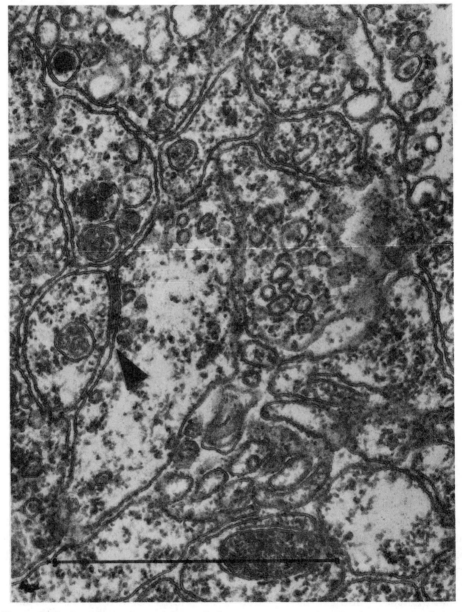

Figure D-8. A synapse (arrow) is made between two neurites in the posterior buccal lobe of *Octopus vulgaris*. The synapsing processes lie within a small nest of filopodia. Micrograph courtesy of J. D. Robertson and P. Lee. Calibration, 10 μm.

are *sequelae* of the process of memory formation, and may participate in synapse formation (162; Fig. D-8).

The complexity and orderliness of the cephalopod CNS parallels that of vertebrates and higher arthropods in numbers of neurons, cytoarchitecture, and laminar organization (Fig. D-4B, optic lobe). Unlike opisthobranchs and arthropods, the electroencephalogram of octopus brain indicates that many neurons are functionally synchronized to generate low frequency field potentials; these are comparable to those of vertebrate brain (30). Octopus brain may have more than 10^8 cells. The PNS is also elaborate: Each sucker on the arms has its own peripheral ganglion interfacing with the CNS. The ganglia handle local chemosensory and mechanosensory integration, and motor control of the suckers (165; cf. Fig. D-2). Thus, in molluscs as in other animals the size of the CNS is closely associated with both body size and motility of life-style. The numbers of neurons of the CNS seem generally to increase with body size and complexity of behavior.

Control of Behavior

The CNS and PNS act in parallel and serial pathways. Complex behaviors, however, tend to originate centrally even if they interface to some extent with the PNS in sensory initiation and control of musculature. As in other animals, complex behaviors in molluscs depend, to varying extents, on sensory inputs. Examples are discussed in the context of central pattern generation.

CENTRAL PATTERN GENERATION. The patterning of behavior in its timing and intensity, and in the squences of its elements, is determined to varying degree by the properties of the CNS and the sensory inputs from the animal's periphery and the environment. That is, behaviors differ in the relative contribution of central and peripheral components to the final expression. Thus, complex behaviors may be arranged over a spectrum of dependence on central versus peripheral control of pattern generation. At one extreme end of this spectrum lie the *fixed action patterns*, episodic behaviors which, once triggered by sensory releasors, automatically carry through to an end. They are centrally generated and insensitive to sensory influences (87, 135). At the other extreme lie behaviors heavily dependent on continuous sensory inputs for their expression; visual tracking is an example.

The central component of behavioral patterning is recognized in the more rigid and stereotypic types of behavior, in particular in rhythmic behaviors whose oscillatory nature is central in origin. For such behaviors, the central components are ofen demonstrable in their expression when the CNS is isolated from sensory inputs by dissection. These observations have given rise to the concept of the *central pattern generator* (66).

Neural Mechanisms of Molluscan Fixed Action Patterns

Fixed action patterns are definable on the basis of six characters:

1. Triggered by specific sensory releasors.
2. Episodic (all or none) in expression.
3. Rigid and stereotypic in expression.
4. Low sensitivity to sensory inputs.
5. Instinctual, or inborn, hard-wired behaviors.
6. Refractory for a period following expression.

The genesis of these features lies in the neural command systems that drive the behaviors.

ESCAPE SWIMMING IN TRITONIA. One of the best examples of a fixed action pattern whose neural circuitry is rather well understood is the swimming episode of the sea slug *Tritonia*. An episode of escape swimming in this marine slug is released by noxious stimuli; in particular, contact with the mucus of carnivorous starfish or of the predatory sea slug *Pleurobranchaea*. The swim consists of a stereotyped series of alternating dorsal and ventral flexions of the body that lift the animal off the substrate into the current, and away from predation (Fig. D-9*A*). The fixed action pattern is driven by a small central pattern generator that provides the timing for the rhythmic behavior (85, 179).

The central pattern generator is composed of four groups of premotor interneurons linked through chemical and electrical synaptic connections. The pattern generated by this system during swim motor output is shown in Figure D-9*B*. The pattern generator can be considered an elaboration on a simple two-phase, half-cell oscillator, where antagonist cells are mutually inhibitory (Fig. D-9*C*). The interneuron groups are known as the Dorsal Swim Interneurons (DSIs), the Ventral Swim Interneurons (VSIs; a heterogeneous group divisible into populations *a* and *b*), Cerebral Cells 2 (C2s), and the hypothetical Inhibitor neurons (I) (84, 85). Triggering and maintenance of the swim pattern is through sensory initiation of a long-lasting depolarization in one group of neurons, the DSIs.

A high threshold for the behavior is maintained by a potent negative feedback relation between DSI and I, DSI activity excited by subthreshold stimuli is damped out by inhibitory feedback from I. The behavioral episode is triggered by stronger sensory excitation of C2, which both inhibits I and excites DSI. The cyclic pattern of activity emerges primarily from the reciprocal inhibitory connections between DSI and VSI, the skeleton of a half-cell oscillator. The slowly decaying depolarization, or "ramp," of the DSIs re-

sults from self-reexcitation, from positive feedback from the C2s, and from other unknown sources. The excitatory drive of the oscillator thus originates largely in DSI and is distributed via connections with C2. The amplitude and decay rate of the DSI ramp therefore determine fixed action pattern duration and intensity.

Insensitivity to sensory inputs during expression of the fixed action pattern arises in part from polysynaptic inhibition from the command system onto neurons not directly involved in the swim behavior (179). The network is refractory for a period following a swim episode through mechanisms not yet known. The entire basis for insensitivity of the fixed action pattern to sensory inputs is not yet clear in this system.

GILL VENTILATORY BEHAVIOR IN APLYSIA. Episodes of respiratory pumping in *Aplysia* consist of a stereotyped, recurrent motor sequence of contraction of gill, siphon mantle shelf, and parapodia with concomitant inhibition of heartbeat and vasomotor tone (Fig. D-10*A*). This motor sequence circulates fresh seawater through the mantle cavity and pumps freshly gas-exchanged blood through the quiescent heart from the gill. Spontaneous episodes of respiratory pumping last for 1 to several minutes and consist of a variable number of cycles (73). The adequate sensory stimuli for eliciting the behavior have been variously reported as mechanical stimuli to the rhinophores and siphon (110, 121) or lowered pH and increased p_{CO_2} (51); O_2 consumption and pumping rate are not correlated (131). These observations suggest that the behavior may deal with CO_2 buildup and acidification, as occurs intermittently in tidepools and the muddy coastal environments where the animals congregate. With the addition of fecal matter expulsion (73), the behavior appears as a useful

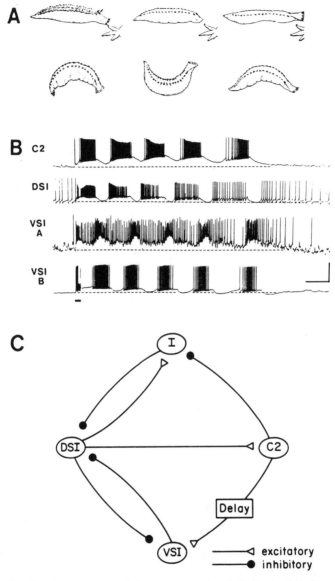

Figure D-9. Fixed action pattern of the escape swim in *Tritonia*. (*A*) A swim triggered by contact with a starfish. From Willows, 1971. (*B*)Neural activity underlying the swim in the pattern generator neurons C2, a dorsal swim interneuron (DSI), and ventral swim interneurons (VSI A and B). [From Getting and Dekin (84).] (*C*) Circuit structure of the central pattern generator. Activity is initiated by simultaneous sensory excitation of C2 and DSI, which causes inhibition of a hypothetical inhibitory interneuron (I) and initiates a long-lasting depolarization in DSI.

multifunctional action pattern, rather like a sneeze.

The known structure of the small command system of the abdominal ganglion underlying the behavioral episode is shown in Fig. D-10*B*. The pattern generator (Fig. D-8*C*) consists of two interneuron populations: L25 and R25, found on the left and right sides of the ganglion, respectively. The 20–30 neurons of the network have extensive excitatory interconnections, at least partly through direct electrical synapses (31, 115, 116). The L25–R25 system drives the variety of motorneurons involved in the behavior and inhibits other major interneurons, L24 and L10, involved in conflicting behaviors (Fig. D-10*C*).

Burst production results from a combination of cellular and network properties (31,115). Slow depolarizing potentials, triggered by depolarization support spiking and reverberatory excitation within the population. The threshold properties of the network confer the all-or-none nature of the behavior. Burst and fixed action pattern termination are produced by endogenous hyperpolarizing afterpotentials common to all network members as well as slow IPSPs produced by some network members. Observations on sensory insensitivity of fixed action pattern and command system during activity are as yet scant.

Habituation of the response may be mediated in part by synaptic depression in sensory pathways. However, refractoriness of the response may also be partly based on the same activity-dependent mechanisms that terminate the bursts, as found in other bursting systems.

In both of these rigid and stereotypic behaviors the patterning of the behavior is certainly almost entirely central in origin. The central pattern generators produce motor output that is unmodified by sensory inputs. The genesis of fixed action patterns has been associated with *command systems,* specialized neural loci that generate the defining characters of the behaviors out of their intrinsic physiology (87). A difference between escape swimming in *Tritonia* and gill ventilation in *Aplysia* is that in the former case the command system may be defined as being the small central pattern generator that produces the motor output, while in gill ventilation the command system comprises the L25/R25 neuron populations, segregated from the central pattern generator.

Neural Mechanisms of Nonfixed Behavior

PATTERN GENERATION IN THE BUCCAL GANGLIA. Feeding behavior in the gastropods is a good example of complex behavior comprising many elements. It is subject to regulation by both external stimuli and internal state, and it is multimodal in its motor output. This class of behavior has received attention in various species from laboratories worldwide.

In general, feeding behavior is conveniently divisible into appetitive and consummatory stages, characterized by food-seeking and ingestive actions, respectively. The final phase of successful appetitive behavior and the main portion of ingestive behavior commonly involve rhythmic rasping or grasping with the radula and a phase-locked swallowing sequence where food items are passed into the esophagus. The rhythmic part of feeding behavior is the product of a central pattern generator whose rhythmic motor output is frequently demonstrable in the isolated CNS. The pattern generator is distributed between the buccal and cerebral ganglia (61, 133); however, it is most robust in the buccal ganglia that drive the muscles of the radula and buccal

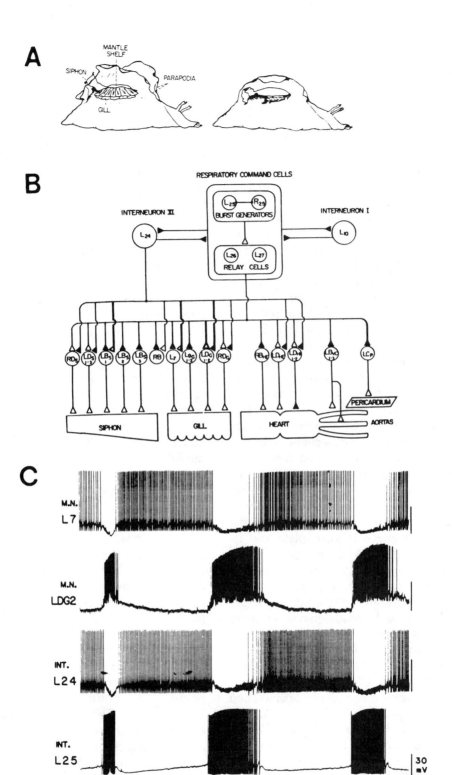

Figure D-10. (*A*) Gill ventilatory behavior in *Aplysia*. (*B*) The pattern generator and its relation to other interneurons and motorneurons. Modified from Byrne and Koester (34). (*C*) Activity in L25 and its followers during a fictive ventilatory episode in an isolated abdominal ganglion. [From Byrne (31).]

mass (64, 123, 193). In *Aplysia*, biting is lost when the buccal ganglion is removed, but appetitive search and orienting behavior remain intact (120).

The circuitry of the feeding network is conserved to an appreciable extent; homologous individual neurons are recognizable across species and even subclasses. These include paired giant cerebral serotonergic neurons of opisthobranchs and pulmonates (18, 89, 97, 169, 188), gill motorneurons of notaspid, nudibranch and anaspid opisthobranchs (67–69), neurosecretory cells of pulmonates and opisthobranchs (44), and cerebrobuccal coordinating interneurons of opisthobranchs and pulmonates (92, 141).

Rhythmic activity is frequently expressed in isolated buccal ganglia and termed "fictive feeding." However, this assignment of behavioral value is usually tentative. The pattern generator of the buccal ganglion can sustain more than one type of rhythmic pattern (53, 138, 139, 176). In the absence of the chemosensory and proprioceptive inputs to the pattern generator of the isolated ganglion, the pattern generator is incomplete and its output suspect. While isolated ganglia must be used for circuitry studies, the ways that rudimentary pattern-generating properties are influenced by the periphery to yield useful behaviors require careful analysis.

Much partial information is available on the circuitry of the pattern generators of the buccal ganglion of different species. Some of the most complete work has been done on *Lymnaea* (74, 75, 95, 163, 164, 165). The feeding pattern in this pulmonate pond snail is a three-phase rhythm generated in interneuron populations N1, N2, and N3; their sequential activity corresponds to radular movements of protraction, rasp (retraction 1) and swallow (retraction 2) (Fig. D-11). Data from many simultaneous intracellular recordings re-

sult in a model based on simple recurrent inhibition and endogenous burstiness (Fig. 11*B*). Oscillation in this central pattern generator, as in most others known, is due to both endogenously bursty membrane properties and synaptic connections. During a fictive feeding cycle, a burst in the N1 group inhibits N3 and slowly depolarizes N2 to threshold. Activity in N2 activates endogenous plateau properties, suppresses N1 and further inhibits N3; spike frequency adaptation in N2 releases N3 to rebound from inhibition to activity which causes depolarizing IPSPs in N1. Adaptation in N3 and consequent recovery of N1 ends the cycle. An unpaired large "Slow Oscillator" (SO) interneuron is another element of the oscillator with abundant outputs, making reciprocal excitatory connections with N1, inhibiting N2, and receiving inhibition from N2 and N3 (75). The SO is a modulatory neuron, which by itself can drive the fictive feeding rhythm. Arshavsky et al. (7) found that synaptic relations in the buccal ganglion of another basommatophoran pulmonate, *Planorbis*, are similar to those of *Lymnaea*, with the exception that N3 analogs were not found. The buccal rhythm in *Planorbis* also emerges from endogenously bursting properties of the N1 and N2 analogs (8, 9).

Inspection of intracellular recordings from the motor networks of the buccal ganglia of other pulmonate and opisthobranch snails show many similarities in the phases of the motor cycle (7, 27, 74, 111, 158, 172, 193). The origin of pulmonate and opisthobranch lines from a common ancestor suggests that comparative work can describe how variation on an ancestral oscillator structure has produced the feeding adaptations found among these lines. It may be that the structure of the central pattern generator of the buccal ganglion is highly conserved; the most

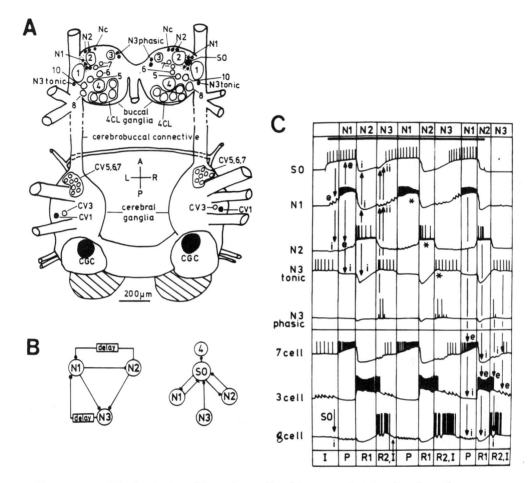

Figure D-11. (A) Cerebral and buccal ganglia of *Lymnaea* showing location of neurons of the feeding network. (B) Circuitry relations of the pattern generating neurons. Filled circles are inhibitory, and barred endings are excitatory synapses. (C) Activity in the pattern generating neurons and motorneurons and its relation to behavioral phases of protraction (P), retraction (R1, R2) and a rest interval (I). [From Benjamin and Elliot (16).]

major adaptive changes in evolution for feeding on different types of food, changing from herbivory to carnivory and from grazing to raptorial bolting are through peripheral changes in musculature and structure of the feeding apparatus and the physiology of the gut. Changes from continuous grazing behavior to episodic predation might also reside largely in control of the pattern generator by sensory and command inputs, rather than in the intrinsic structure of the pattern generator.

Such conservation of detailed neural circuitry would parallel that observed across widely differing taxa in arthropods (146) and annelids (130).

The nudibranch *Melibe leonina* has no buccal mass, but retains the buccal ganglion to innervate the muscular esophagus. A planktonic feeder on captured copepods, little is known about neural control of feeding. The animal uses the smooth musculature of the buccal cavity–esophagus both to swallow and to reject objects through its mouth, apparently peristaltically (personal observation). Knowledge of central control by the buccal ganglion might indicate how far the structure of a central pattern generator may be retained when radical peripheral changes occur.

The investigation of central pattern generation in the buccal ganglion of *Octopus* has had a promising beginning in the recording of spontaneous cyclic movements and electrical activity in isolated ganglion–buccal mass preparations (21).

THE ORGANIZATION OF FEEDING BEHAVIOR: REGULATION BY INTERNAL AND EXTERNAL FACTORS, AND ITS INTERACTIONS WITH OTHER BEHAVIORS. Within the organization of gastropod behavior, feeding occupies a central position, and it is useful to approach the animals' behavioral ecology in terms of feeding regulation. The behavior generated by motor activity in buccal ganglia of most snails thus far examined is similar: It functions to transfer food into the buccal cavity and then swallow it to the esophagus. This is so in grazers and raspers, such as *Lymnaea, Helisoma, Planorbis, Helix, Limax, Aplysia,* and *Tritonia,* as well as for episodic predators like *Pleurobranchaea* and *Philine.* However, the notaspid *Pleurobranchaea* and particularly the cephalaspid *Navanax* have evolved ballistic bite–strike behaviors, which function to seize invertebrate prey preparatory to swallowing (62, 194).

Feeding behavior is in major part regulated by an organism's state of satiety, that is, its motivation to feed. Reproductive state, external stimuli and learning experience also affect the willingness to feed over performance of other behaviors. The regulation of feeding behavior by these factors has been studied in several snails (63, 120, 175).

The structural organization of the animal's behavioral repertoire (sometimes called a heirarchy) is strongly affected by the motivation to feed. In the hungry *Pleurobranchaea,* feeding behavior causes suppression of other behaviors, such as righting in an upsidedown slug or reflex withdrawal from a mechanical stimulus to the oral veil (62). In the satiated animal, ingestive feeding behavior is suppressed and other behavior may be released (see later). Suppression of behavior incompatible with feeding may be mediated by active inhibition of other neural circuitry by the feeding network. Kovac and Davis (117) identified neurons in the buccal ganglion activated tonically by rhythmic activity in the pattern generator; these neurons suppressed excitation of withdrawal motorneurons by tactile stimuli. Additional evidence for *mutual* inhibition between circuitry for feeding and withdrawal delineate an attractive general principle for organization of antagonistic behaviors (117, 118).

Food stimuli initiate or enhance a state of arousal of the feeding behavior, underlain by activity in the underlying neural circuitry and musculature. Weak dilutions of a food stimulus applied to the oral veil and rhinophores of *Pleurobranchaea* and *Aplysia* stimulate orienting search and locomotion; increasingly stronger food stimulation causes first proboscis extension and then biting (62, 120, 128, 129). Strong stimulation of the mouth region suppresses locomotion and stimulates biting activity. These relations have obvious relation to prey catching. Arousal of the feeding network appears to occur through sensory induction of greater excitability in network members and buccal musculature by neuromodulatory pathways (86, 112, 187).

Initiation and maintenance of rhythmic feeding behavior appear to be sustained by the overall state of excitation of the network oscillator. While a number of different neurons have been identified that drive feeding behavior when stimulated and are active during feeding, most do not appear to act as classical command neurons responsible for driving the behavior. Rather, they are elements of potent positive feedback pathways within the network oscillator; regulation of activity within these pathways by learning- and motivation-dependent processes influences the expression of feeding behavior (60, 61). Initiation of feeding behavior appears to result from distributed sensory excitation of the feeding neural network (134). Substances drawn into the mouth of a typical gastropod are effectively tested for palatability; if they are unaccompanied by chemical food stimulus or accompanied by a noxious stimulus, they are rejected by rhythmic movements of the buccal mass. The cyclic motor pattern of rejection is generated by elements of the same buccal network that also generate the feeding rhythm. It is not yet clear whether mechanisms of motor program switching between feeding and rejection involve triggering of a fixed action pattern type of central motor program, driven by a command locus, or simply reorganization and recoordination of the feeding oscillator by specific sensory pathways (52, 86, 87, 140). Intensive effort has been directed at characterizing neural elements responsible for directing "feeding" versus "rejection" rhythm patterns in isolated preparations (53–55, 140). This approach has been so far unable accurately to correlate predicted behavioral roles of specific neural elements (the ventral white cells; see below) with the activity of those elements either during actual behavior or in response to appropriate sensory releasors (86).

Smaller animals swallowed whole by *Pleurobranchaea* are quickly subdued by H_2SO_4 copiously emptied into the buccal cavity from an acid gland (180). Food stimulation of the buccal cavity further excites the feeding oscillator. In particular, food stimuli in the buccal cavity induce minutes-long endogenous bursts in feeding command neurons by neuromodulatory action. The command neurons, the ventral white cells of the buccal ganglion, accelerate both frequency and intensity of feeding cycles and thus facilitate the rapid bolting of large quantities of food when it is available (86). The ventral white cells drive episodes of feeding behavior corresponding in their character to classic fixed action patterns (87,88) (Fig. D-12). Triggered by a specific stimulus class, they are episodic in nature, relatively insensitive to sensory stimuli once in motion, and refractory for an interval following an episode. The features of the fixed action pattern correspond in each detail to aspects of the physiology of the command neurons, largely based on their inducible capacity for prolonged endogenous burst activity (87). The ventral white cells thus confer a significant aspect of the opportunistic gluttony typical of active predation.

Satiety in gastropods is likely mediated partially by bulk stretch of the esophagus (177); synaptic mechanisms are undoubtedly of interest, but presently unknown. Humoral mechanisms of satiety may also operate (106). Satiety, measured by the elevated threshold strength of a chemosensory stimulus necessary for stimulating feeding behavior, decays with digestion (63).

AVOIDANCE BEHAVIOR AS AN ALTERNATIVE TO FEEDING. Sensory stimuli may elicit different behaviors, depending on the context in which they occur. In the satiated *Pleurobranchaea*, or one trained in a food-avoidance conditioning paradigm, food stimuli frequently cause avoidance behavior, consisting of a turn away from the stimulus source and rapid locomotion; conversely, in the hungry animal normally aversive chemical stimuli elicit transient components of feeding behavior before avoidance behavior occurs (100).

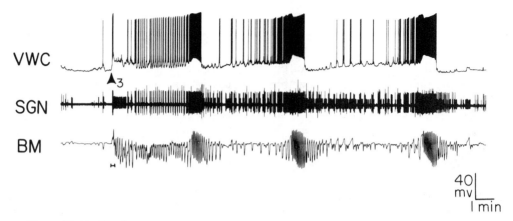

Figure D-12. Fixed action pattern activity in the feeding network of the buccal ganglion of *Pleurobranchaea* is initiated by delivery of a food stimulus into the animal's buccal cavity (arrowhead). The VWC neuron is driven passively by the network until it enters a long endogenous burst episode, which drives rapid and intense biting and swallowing action of the buccal mass (BM, force transducer record; SGN, stomatogastric nerve). [From Gillette and Gillette (86).]

Since locomotion is suppressed during ingestive feeding behavior, a neural model postulates inhibition of locomotor circuits by rhythmic activity of the feeding network, with food and aversive sensory pathways having access to both networks. Whether either behavior is expressed in *Pleurobranchaea* may be determined by the activity state of the feeding network, itself influenced by the motivational factor of satiety. Similar data exist for *Helix* (137). Active avoidance to various stimuli is seen in many gastropods.

Unwillingness to feed in *Pleurobranchaea*, due to satiety or learning, is not due to overall inhibition of food sensory pathways or to inhibition of the feeding oscillator. Rather, behavioral feeding suppression is attended by suppression of activity in protractor neurons and hyperactivity in retractor neurons of the oscillator, thus locking the feeding oscillator in a bias toward retraction (134). Potentially, the tonic retractor activity may either excite or disinhibit avoidance locomotion.

The significance of environment chemosensory signals is probably critical for understanding the neuroethology of any opisthobranch, since chemosensation conveys more information than all other senses in their transactions with the environment. Feeding behavior is stimulated notably by betaine, glycine, cysteine, and aspartate in the carnivore *Pleurobranchaea* (19,20,100), by glutamate and aspartate in herbivorous *Aplysia* (103), and also by phosphatidylcholine and glycerolipid in *Aplysia* (168). Specifically aversive amino acids have also been identified for these animals, taurine for *Pleurobranchaea* (100) and cysteine for *Aplysia* (103), which are present in appreciable amounts in the animals' normal food and may also act as feeding deterrents, perhaps aiding in regulating the size of a meal. Their role in the physiological ecology of the organisms is not adequately understood. Other known natural aversive stimuli for *Pleurobranchaea* are dilute acid (pH 1–3) and the mucus of other members of the species (91). *Pleurobranchaea* and its relatives secrete acid when annoyed (180,181). The animal's own acid secretion may potentiate its avoidance behavior (91). Intraspecific aversiveness is likely defensive, since cannibalism is the only well-documented form of predation on *Pleurobranchaea*. Aversiveness to acid, mucus, and taurine habituates with repeated presentation; such habituation may be important for mating.

Understanding of aversive behavior and its interactions with feeding depends on adequate description of the neural bases of locomotion, directional orienting, and how directional orienting processes can become reversed to effect directional avoidance to the same stimulus. Some information is available on the neural bases of locomotion. Locomotion in *Tritonia*, similar to that of many snails, occurs by the sweep of myriad cilia on the foot through secreted mucus (105). The ciliary beat is regulated by neurons in the pedal ganglia (10). The neurons are stimulated by aversive stimuli; contact of the animal with food suppresses the ciliary beat (11). *Aplysia* locomotes via a front-to-back pedal wave of contraction; the wave rhythm is generated by an oscillator network in the pedal ganglia. Cerebral neurons excited by food and noxious stimuli distribute activating excitation to pedal oscillator neurons, which are also sensitive to proprioceptive feedback (80). Thus, although fine details are not yet known, the general structure of the network is evident. Most gastropods are incapable of reverse locomotion; rather, they rely on the considerable turning ability of their soft bodies and forward crawling to get out of tight situations. Those species carrying large shells employ a variety of twisting and dodging maneuvers to carry them by obstacles.

Our present understanding of feeding behavior in several snails, although incomplete, serves well for synthesis of a testable general model for the organization of complex behavior (Fig. D-13). Orienting and aversive pathways must be served by separate crossed and uncrossed sensory paths converging on turning motor paths (to unilateral, longitudinally shortening muscles). The model provided by the above descriptions interprets the mandated expression of behavior in terms of reactive, shifting forms of coordination in the nervous system integrating the influences of environment and internal state on neural circuitry with varied pattern-generating capacity.

Plasticity in Behavior and Neural Activity

Much behavior is relatively hard wired, as in the fixed action pattern. Such behavior is advantageous in fitting the organism with a ready-made adaptive response to a potentially critical situation. Extreme types of fixed action patterns need no or little learning experience for their expression, and are commonly associated with escape, feeding, and reproductive behaviors where it can be important to get it right the first time. On the other hand, the need to adapt certain behaviors to uneveness and changes in the environment is reflected in modifiability in the underlying neural circuitry. In molluscs this has been a particularly strong focus of investigation, and presently there is an interesting variety of documentation of circuitry and cellular mechanisms of behavioral plasticity. These include modification of simple reflexes, taxes, and complex behaviors.

Simple Reflexes

HABITUATION. For many opisthobranchs, repeated tactile stimulation results in decrement of local withdrawal responses

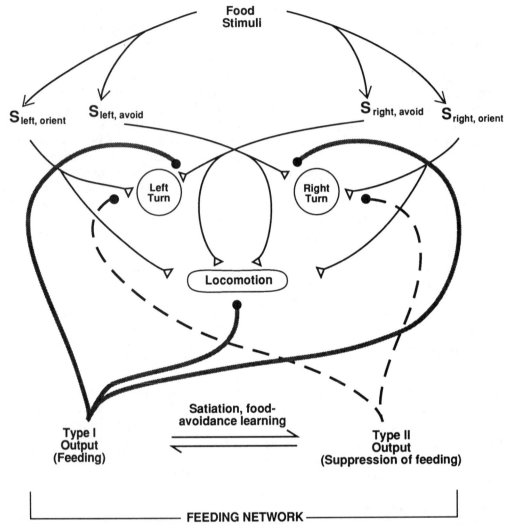

Figure D-13. A model of the minimal functional connections necessary to organize interactions of feeding and food-avoidance behavior in gastropods. The model is based on output of two different types of activity from the feeding network. The type of feeding network output modulates input pathways to left and right "turn" motorneurons (longitudinal shortening), and regulates activity in locomotory effectors. Food stimuli excite chemosensory (S) pathways conveying excitation to locomotory neurons, activating pedal cilia (*Pleurobranchaea, Tritonia*) or pedal oscillator neurons (*Aplysia*). Directional orientation to food is performed by ipsilateral stimulation of left- and right-turn motorneurons via sensory pathways from left ($S_{L,O}$), and right ($S_{R,O}$) sides. Crossed sensory pathways ($S_{L,A}$, $S_{R,A}$) stimulating contralateral (aversive) turns are inhibited by feeding (type I) output from the motor network. Stronger type I output inhibits locomotory neurons, overcoming food-induced excitation. In *Pleurobranchaea*, type I output may correspond to tonic activity in identified feeding interneurons known to suppress other behavior (117). Avoidance behavior is stimulated by type II output, which inhibits ipsilateral sensory stimulation of turn neurons, and disinhibits locomotion and contralateral stimulation of turn neurons. Type II output is seen in tonic retractor feeding neuron activity in learned or satiation-induced feeding suppression in *Pleurobranchaea* (134).

(59,159). For the gill–siphon withdrawal reflex of *Aplysia*, habituation is mediated in parallel by both the peripheral and the CNSs (35, 159, 160). The central circuitry mediating both the reflex and its habituation includes both simple sensory–motor synapses and interneuronal intermediaries. Use-dependent decrement of transmitter release from sensory neuron terminals underlies central habituation (41, 42), and results in part from accumulating inactivation of a presynaptic Ca^{2+}-current component of the action potential (114). Depletion of synaptic vesicles adjacent to the presynaptic active zone is also correlated with short-term habituation (14).

Long-term habituation of the gill–siphon reflex in *Aplysia*, lasting for days, is achieved by repeated training sessions (37) and may serve naturally as a behavioral compensation for animals living in a turbulent environment.

SENSITIZATION. The gill withdrawal reflex and the defensive tail withdrawal reflex may be both dishabituated or sensitized by noxious stimuli to the head or tail, accompanied by increased amplitude of the EPSPs from the sensory neurons (40, 184). Multiple mechanisms underlie sensitization, both pre- and postsynaptic and involving different ion channels, but all may be mediated by cyclic AMP (81). One mechanism occurs through a well-studied process of heterosynaptic facilitation, wherein the activated heterosynaptic pathways induce medium-term (minutes to hours) changes in ion channel activity through neuromodulatory mechanisms. The neuromodulatory transmitters used by the identified facilitator neurons may be serotonin or one or more small peptides (2, 24). At least one facilitatory mechanism in the presynaptic neurons involves receptor-mediated increase in cyclic AMP levels, which cause a decrease in K^+ conductance ("S" current) (113, 114, 170). Reduction of the K^+ current causes slight prolongation of the action potential, permits greater Ca^{2+} influx during the spike, and thus causes more transmitter release. Direct effects of cyclic AMP on Ca^{2+} entry remain possible. Long-term sensitization over days occurs (160).

ASSOCIATIVE LEARNING. Learning mechanisms permit the integration of neural pathways in adaptive and frequently novel ways. In *Aplysia*, previously ineffectual light tactile stimuli come to elicit reflex gill–siphon withdrawal after repeated pairing with more effective mechanical or electrical stimuli (36). The general mechanisms underlying the association appear to be elaborated from those underlying sensitization, and have been termed activity-dependent modulation (186). Concurrent activation of the sensory pathway with the heterosynaptic facilitating path produces long-lasting greater and more long-lasting synaptic facilitation than unpaired activation (99,185). The enhancing effect of activity in the sensory pathway is due to Ca^{2+} influx (99,185). The increased levels of Ca^{2+} in the presynaptic cell lead to higher levels of cyclic AMP formed in response to the neuromodulator released by the heterosynaptic pathway, perhaps due to synergistic effects of Ca^{2+} on a Ca^{2+}–calmodulin-dependent adenylate cyclase activated by the neuromodulator (155). Thus, cyclic AMP mediates both the short-term effects of associative learning in the form of regulation of ion channels by phosphorylation mechanisms, and it must initiate longer term changes in the neuron, which lead to enhanced efficacy in the pathways. Enhanced long-term synaptic efficacy in the gill–siphon sensory neurons is correlated with an increased number of presumptively synaptic varicosities (15).

Taxis and Associative Learning

The nudibranch *Hermissenda crassicornis* is normally positively phototaxic, tending to move up a gradient of light. However, when in the laboratory flashes of light are paired with periods of rotation for 150 times over 3 days the animals suppress phototaxis specifically due to paired stimulation (57). The suppression of phototaxis is due to increased latency in light-induced initiation of locomotion (58, 76). Moreover, conditioned animals encountering a shadow edge are more apt to enter the shadow and spend time than unconditioned animals, which tend to reverse direction at the shadow edge (127).

The explanation for this learning must arise from interaction of the statocysts and the five photoreceptors of the eyes. Photoreceptors are light-sensitive neurons and are classified as types A and B. Conditioning causes an initial increase in the dark-adapted spontaneous firing rate of type B photoreceptors and an increase in intrinsic input resistance, leading to an enhanced and long-lasting light response. The cumulative depolarization of associative conditioning is mimicked by pairing depolarization of a statocyst hair cell, or an identified optic ganglion neuron, and light depolarization of photoreceptor (77). It is supposed that the enhanced response decreases excitation of the locomotory network. However, when the photoresponse is measured under conditions of constant illumination like those of the conditioning trials, the response is actually found to be diminished (56). Thus, it is possible that suppression of phototaxis is due to less light-induced excitation of locomotion. Conditioning produces sustained decreases in the transient I_A (4) and may also

decrease a Ca^{2+}-activated K^+ current (5) while increasing I_{Ca} (78). Such various changes are tied to phosphorylation mechanisms (153) stimulated by Ca^{2+} through Ca^{2+}-calmodulin-dependent kinase or protein kinase C. While not yet wholly resolved, the temporal specificity of light-rotation pairing may be due to the augmentation of Ca^{2+} levels in the B cells.

Complex Behavior and Associative Learning

Associatively learned behavior is response modification due to temporally contiguous pairing of distinct sensory stimuli. The nature of the response modification, the plasticity itself, can be considered hard wired; not all behavior can be altered. For example, the resistance of the escape swim of *Tritonia* to associative sensitization was shown by Abraham and Willows (1). The capacity for associative learning is widely spread in phylogeny; the ability allows the animal to effectively anticipate environmental flux with adaptive behavior shaped by experience. A particular advantage over the hard wiring of the fixed action pattern response to specific releasers is the adaptability of motor behavior to changes in sign or modality of sensory cues.

Various gastropods have shown capacities for associative learning similar in character to those defined in the laboratory for vertebrates. First-order conditioning, association of conditioned and unconditioned stimuli (CS and US, respectively) has been shown for *Pleurobranchaea* (150, 151), *Aplysia* (39, 136), *Limax* (82), *Helix* (137), and *Lymnaea* (3, 12). Second-order conditioning—an association of one CS, formerly paired on a US—with another CS has been demonstrated for *Limax* (167). Sahley et al. (167) also showed "blocking" in *Limax*; prior training of CS_1–US prevents subsequent successful training on another CS presented with the first pairing; the association CS_2–US is blocked.

Foraging snails can learn to recognize potentially harmful prey as well as to preferentially focus on the most available food items (cf. 104). The ability to learn discrimination of different foods is clearly important to foraging strategy. Thus, *Pleurobranchaea* and *Limax* rapidly learn to avoid food after pairing with a noxious UCS (82, 101, 151), and both animals can discriminate different foods in avoidance conditioning paradigms (65, 99, 148, 149, 166). *Aplysia* learns to avoid food stimuli that are inaccessible for consumption (178), and *Lymnaea* rapidly learns to respond with feeding behavior to a previously neutral amyl acetate stimulus when it has been paired with a food stimulus (12).

The circuitry and mechanisms underlying learning of food avoidance and discrimination in *Pleurobranchaea* and *Limax* could well turn out to be similar, since so much circuitry in the feeding motor networks appears to have been conserved during the evolution of opisthobranchs and pulmonates. In *Pleurobranchaea* associative training with food and shock causes hyperexcited responses to food stimuli in potent interneurons (Int-2s) that suppress protraction and lock the oscillatory feeding network in the retraction state (134). Learning mechanisms are also distributed; food avoidance training also weakens positive feedback among protractor interneurons (119), in part through reducing postsynaptic cholinergic responsiveness (147).

Elaboration of Behavioral Complexity through Learning Mechanisms

In both *Aplysia* and *Hermissenda*, where cellular mechanisms have been examined to some degree, short- and long-term manifestations of enhanced synaptic connectivity are caused by temporal association of conditioned and unconditioned stimuli. Intracellularly, this pairing leads to stimulation of one or two intracellular messengers. In the short term, effects may be mediated rather directly by ion channel regulation by either one of the messengers. However, the temporal association of the two stimuli provides an augmented intracellular signal activating a more complex response leading to long-lasting changes in ion channel availability or synaptic contacts, or both. Conceptually, this is simple, and is able to account in principle for the simple modifications of circuitry that can support the range of learning observed in molluscs. The possibility of more than one type of associative learning mechanism at the cellular level is consistent with the redundancy observed in many other aspects of nervous system structure and function.

Opisthobranch and pulmonate behavior is not complex in the sense that the number of behavioral subroutines in the repertory is limited by the snail's simplified CNS. However, complexity must also be reckoned in terms of the number of sensory discriminations an animal can make. The number of different sensory pairings, somatotopic, chemo-, photo-, or mechanospecific in which these animals might be associatively conditioned is potentially large. *Pleurobranchaea, Limax,* and *Aplysia* are capable of various chemosensory discriminations (ibid.; 50), and *Aplysia* shows context-specific learning (49). Thus, within the limitations of their motor repertory, the motile, foraging gastropods must be capable of an appreciable range of learned behavioral adaptation to the environment.

Squids and octopi readily learn to recognize different shapes visually. While not able to discriminate shapes, weights, and sizes by touch, they can sensitively recognize texture. They can also recognize individual flavors labeled as sweet, sour and bitter (189), and no doubt a large variety of other chemostimulants. Quite probably the number of discriminations the octopus can make is much larger than the capacity of any opisthobranch, and is supported by the elaborate sensory apparatus and neural processing system which contain large numbers of neurons. In the opisthobranchs, known mechanisms of associative memory are based on modifications in neurons having specific access to labeled sensory pathways. In the more complex cephalopods, associative learning mechanisms link much enhanced sensory processing and a greater motor repertory to support complex behavior and discriminative ability rivaling vertebrates. Do learning mechanisms at circuitry and cellular levels differ greatly between simple and complex mollusc ganglia? The question is an important one in comparative neuroscience, and the answer is likely to have broad significance.

The similarities of molluscan and mammalian learning may suggest similar underlying mechanisms of learning. However, a more certain lesson is that the adaptive values and selective pressures that shape behavior have produced the same functions in different phyla. Apparently, this has occurred independently in phylogeny, as for the multiple origins of many complex traits (e.g., focusing eyes) in phylogeny. What mechanistic similarities do occur at the cellular level are likely to have arisen from the availability of similar, and homologous, ion conductances, and cellular physiologies. Withal, the opportunities offered by the molluscs for comparative analysis of neural principles in simpler and more complex animals, related within the same phylum, are extremely rich.

Other excellent reviews should be consulted for more and specialized detail on the neurobiology and behavior of the mollusca (13, 21, 28, 32, 38, 70, 96, 109, 190, 192, 196).

References

1. Abraham, F. D. and A. O. D. Willows. *Commun. Behav. Biol.* **6**:271–280, 1971. Plasticity of a fixed action pattern in the sea slug *Tritonia diomedia*.

2. Abrams, T. W., T. J. Carew, R. D. Hawkins, and E. R. Kandel. *Proc. Natl. Acad. Sci. U.S.A.* **81**:7956–7960, 1984. Two endogenous neuropeptides modulate the gill and siphon withdrawal reflex in *Aplysia* by presynaptic facilitation involving cAMP-dependent closure of a serotonin-sensitive potassium channel.

3. Alexander, J. E., Jr., T. E. Audesirk, and G. J. Audesirk. *J. Neurobiol.* **15**:67–72, 1984. One trial reward learning in the snail *Lymnaea stagnalis*.

4. Alkon, D. L., I. Lederhendler, and J. Shoukimas. *Science* **215**:693–695, 1982. Primary changes of membrane currents during retention of associative learning.

5. Alkon, D. L., M. Sakakibara, R. Forman, J. Harrigan, I. Lederhendler, and J. Farley. *Behav. Neural Biol.* **44**:278–300, 1985. Reduction of two voltage-dependent K⁺ currents mediates retention of a learned association.

6. Allen, A., J. Michels, and J. Z. Young. *Mar. Behav. Physiol.* **11**:271–282, 1985. Memory and visual discrimination by squids.

7. Arshavsky, Yu. I., T. G. Deliagina, E. S. Meizerov, G. N. Orlovsky, and Yu. V. Panchin. *Exp. Brain Res.* **70**:310–322, 1988. Control of feeding movements in the freshwater snail *Planorbis corneus*. I. Rhythmical neurons of buccal ganglia.

8. Arshavsky, Yu. I., T. G. Deliagina, G. N. Orlovsky, and Yu. V. Panchin. *Exp. Brain Res.* **70**:310–322, 1988. Control of feeding movements in the freshwater snail *Planorbis corneus*. II. Activity of isolated neurons of buccal ganglia.

9. Arshavsky, Yu. I., T. G. Deliagina, G. N. Orlovsky, and Yu. V. Panchin. *Exp. Brain*

Res. **70**:310–322, 1988. Control of feeding movements in the freshwater snail *Planorbis corneus*. III. Organizaton of the feeding rhythm generator.

10. Audesirk, G. *Nature (London)* **272**:541–543, 1978. Central neuronal control of cilia in *Tritonia diomedea*.

11. Audesirk, G. *J. Comp. Physiol.* **128**:259–267, 1978. Properties of central motor neurons exciting locomotory cilia in *Tritonia diomedea*.

12. Audesirk, T. E., J. E. Alexander, Jr., G. J. Audesirk, and C. M. Moyer. *Behav. Neural Biol.* **36**,379–390, 1982. Rapid, non-aversive conditioning in a frehwater gastropod. I. Effects of age and motivation.

13. Audesirk, T. E. and G. Audesirk. In *The Mollusca* (A. O. D. Willows, ed.), Vol. 8, Part 1, pp. 2–94. Academic Press, Orlando, FL, 1985. Behavior of gastropod molluscs.

14. Bailey, C. H. and M. Chen. *J. Neurosci.* **8**:2452–2459, 1988. Morphological basis of short-term habituation in *Aplysia*.

15. Bailey, C. H. and M. Chen. *Proc. Natl. Acad Sci. U.S.A.* **85**:2373–2377, 1988. Long-term memory in *Aplysia* modulates the total number of varicosities of single identified sensory neurons.

16. Benjamin, P. R. and C. J. H. Elliott. In *Neuronal and Cellular Oscillators* (J. Jacklet, ed.), pp. 173–214. Dekker, New York, 1989. Snail feeding oscillator: The central pattern generator and its control by modulatory interneurons.

17. Berdan, R. C., R. R. Shivers, and A. G. M. Bulloch. *Synapse* **1**:304–323, 1987. Chemical synapses, particle arrays, pseudo-gap junctions and gap junctions of neurons and glia in the buccal ganglion of *Helisoma*.

18. Berry, M. S. and V. W. Pentreath. *J. Exp. Biol.* **80**:119–135, 1976. Properties of a symmetric pair of serotonin-containing neurones in the cerebral ganglia of *Planorbis*.

19. Bicker, G., W. J. Davis, E. M. Matera, M. P. Kovac, and D. J. Stormo-Gipson. *J.*

Comp. Physiol. **149**:221–234, 1982. Chemoreception and mechanoreception in the gastropod mollusc *Pleurobranchaea californica*. I. Extracellular analysis of afferent pathways.

20. Bicker, G., W. J. Davis, and E. M. Matera. *J. Comp. Physiol.* **149**:235–250, 1982. Chemoreception and mechanoreception in the gastropod mollusc *Pleurobranchaea californica*. II. Neuroanatomical and intracellular analysis of afferent pathways.

21. Boyle, P. R., K. Mangold, and D. Froesch. *J. Zool.* **188**:53–67, 1979. The mandibular movements of *Octopus vulgaris*.

22. Boyle, P. R. In *The Mollusca* (A. O. D. Willows, ed.), Vol. 9, Part 2, pp. 1–99. Academic Press, Orlando, FL, 1986. Neural control of cephalopod behavior.

23. Brace, R. C. *Philos. Trans. R. Soc. London, Ser. B* **300**:463–491, 1983. Observations on the morphology and behaviour of *Chilina fluctuosa* Gray (Chilinidae), with a discussion on the early evolution of pulmonate gastropods.

24. Brunelli, M., V. Castellucci, and E. R. Kandel. *Science* **194**:1178–1181, 1976. Synaptic facilitation and behavioral sensitization in *Aplysia*: Possible role of serotonin and cyclic AMP.

25. Budelmann, B. U. and H. Bleckmann. *J. Comp. Physiol.* **164**:1–5, 1988. A lateral line analogue in cephalopods.

26. Budelmann, B. U., M. Sachse, and M. Staudigl. *Philos. Trans. R. Soc. London, Ser. B* **315**:305–343, 1987. The angular acceleration receptor system of the statocyst of *Octopus vulgaris*: Morphometry, ultrastructure, and neuronal and synaptic organization.

27. Bulloch, A. G. M. and D. A. Dorsett. *J. Exp. Biol.* **79**:7–22, 1979. The functional morphology and motor innervation of the buccal mass of *Tritonia hombergi*.

28. Bullock, T. H. In *Structure and Function in the Nervous Systems of Invertebrates*, (T. H. Bullock and G. A. Horridge, eds.), Vol. 2, pp. 1273–1515. Freeman, San Francisco, CA, 1965. The Mollusca.

29. Bullock, T. H. In *Information Processing in the Nervous System* (H. M. Pinsker and W. D. Willis, Jr., eds.), pp. 199–240. Raven Press, New York, 1980. Reassessment of neural connectivity and its specification.

30. Bullock, T. H. and E. Basar. *Electroencephalogr. Clin. Neurophysiol.* **57**:473–483, 1984. Ongoing compound field potentials from octopus brain are labile and vertebrate-like.

31. Byrne, J. H. *J. Neurophysiol.* **49**:491–508, 1983. Identification and initial characterization of a cluster of command and pattern-generating neurons underlying respiratory pumping in *Aplysia californica*.

32. Byrne, J. H. *Physiol. Rev.* **67**:329–439, 1987. Cellular analysis of associative learning.

33. Byrne, J. H., V. F. Castellucci, and E. R. Kandel. *J. Neurophysiol.* **37**:1041–1064, 1974. Receptive fields and response properties of mechanoreceptor neurons innervating siphon skin and mantle shelf in *Aplysia*.

34. Byrne, J. H. and J. Koester. *Brain Res.* **143**:87–105, 1978. Respiratory pumping:neuronal control of a centrally commanded behavior in *Aplysia*.

35. Carew, T. J., V. F. Castellucci, J. H. Byrne, and E. R. Kandel. *J. Neurophysiol.* **42**:497–509, 1979. A quantitative analysis of the relative contribution of central and peripheral neurons to the gill withdrawal reflex in *Aplysia*.

36. Carew, T. J., R. D. Hawkins, and E. R. Kandel. *Science* **182**:1158–1160, 1983. Differential classical conditioning of a defensive withdrawal reflex in *Aplysia californica*.

37. Carew, T. J. and E. R. Kandel. *Science* **182**:1158–1160, 1973. Acquisition and retention of long-term habituation in *Aplysia*: Correlation of behavioral and cellular processes.

38. Carew, T. J. and C. L. Sahley. *Annu. Rev. Neurosci.* **9**:435–437, 1986. Invertebrate learning and memory: From behavior to molecules.

39. Carew, T. J., E. T. Walters, and E. R.

Kandel. *J. Neurosci.* **1**:1426–1437, 1981. Classical conditioning in a simple withdrawal reflex in *Aplysia californica*.

40. Castellucci, V. F. and E. R. Kandel. *Science* **194**:1176–1178, 1976. Presynaptic facilitation as a mechanism for behavioral sensitization in *Aplysia*.

41. Castellucci, V. F., H. M. Pinsker, I. Kupfermann, and E. R. Kandel. *Science* **167**:1745–1748, 1970. Neural mechanisms of habituation and dishabituation of the gill withdrawal reflex in *Aplysia*.

42. Castellucci, V. F., T. J. Carew, and E. R. Kandel. *Science* **202**:1306–1308, 1978. Cellular analysis of long-term habituation of the gill withdrawal reflex of *Aplysia californica*.

43. Cavalier-Smith, T. *J. Cell Sci.* **34**:247–279, 1978. Nuclear volume control by nucleoskeletal DNA, selection for cell volume and cell growth rate, and the solution of the DNA C-value paradox.

44. Chase, R. and H. E. Goodman. *Cell Tissue Res.* **176**:109–120, 1977. Homologous neurosecretory cell groups in the land snail *Achatina fulica* and the sea slug *Aplysia californica*.

45. Chase, R. and B. Tolloczko. *J. Neurobiol.* **18**:395–406, 1987. Evidence for differential DNA endoreplication during the development of a molluscan brain.

46. Cobbs, J. S. and H. M. Pinsker. *J. Neurobiol.* **9**:121–141, 1979. *In vivo* responses of paired giant mechanoreceptor neurons in *Aplysia* abdominal ganglion.

47. Coggeshall, R. E. *J. Neurophysiol.* **30**:1263–1287, 1967. A light and electron microscope study of the abdominal ganglion of *Aplysia californica*.

48. Coggeshall, R. E., B. A. Yaksta, and F. J. Swartz. *Chromosoma* **32**:205–212, 1970. A cytophotometric analysis of the DNA in the nucleus of the giant cell, R-2, in *Aplysia*.

49. Colwill, R. M., R. A. Absher, and M. L. Roberts. *J. Neurosci.* **8**:4434–4439, 1989. Context-US learning in *Aplysia californica*.

50. Colwill, R. M., R. A. Absher, and M. L. Roberts. *J. Neurosci.* **8**:4440–4444, 1989.

Conditional discrimination learning in *Aplysia californica*.

51. Croll, R. P. *J. Exp. Biol.*, **117**:15–27, 1985. Sensory control of respiratory pumping in *Aplysia californica*.

52. Croll, R. P. and W. J. Davis. In *Higher Brain Functions: Recent Explorations of the Brain's Emergent Properties* (S. P. Wise, ed.), pp. 157–179. Wiley, New York, 1987. Neural mechanisms of motor program switching in *Pleurobranchaea*.

53. Croll, R. P., W. J. Davis, and M. P. Kovac. *J. Neurosci.* **5**:48–55, 1984. Neural mechanisms of motor program switching in the mollusc *Pleurobranchaea*. I. Central motor programs underlying ingestion, egestion, and the "neutral" rhythm(s).

54. Croll, R. P., W. J. Davis, and M. P. Kovac. *J. Neurosci.* **5**:56–63, 1984. Neural mechanisms of motor program switching in the mollusc *Pleurobranchaea*. II. Role of the ventral white cell, anterior ventral, and B3 buccal neurons.

55. Croll, R. P., W. J. Davis, and M. P. Kovac. *J. Neurosci.* **5**:64–71, 1984. Neural mechanisms of motor program switching in the mollusc *Pleurobranchaea*. III. Role of the paracerebral neurons and other identified brain neurons.

56. Crow, T. J. *J. Neurosci.* **5**:209–214, 1985. Conditioned modification of phototactic behavior in *Hermissenda*. I. Analysis of light intensity.

57. Crow, T. J. and D. L. Alkon. *Science* **201**:1239–1241, 1978. Retention of an associative behavioral change in Hermissenda.

58. Crow, T. J. and N. Offenbach. *Brain Res.* **271**:301–310, 1983. Modificaton of the initiation of locomotion in *Hermissenda*: Behavioral analysis.

59. Crozier, W. J. and L. B. Arey. *J. Exp. Zool.* **29**:261–310, 1919. Sensory reactions of *Chromodoris zebra*.

60. Davis, W. J. and R. Gillette. *Science* **199**:801–804, 1978. Neural correlate of behavioral plasticity in command neurons of *Pleurobranchaea*.

61. Davis, W. J., R. Gillette, M. P. Kovac, R. P. Croll, and E. M. Matera. *J. Neuro-*

physiol. **49**:1557–1572, 1983. Organization of synaptic inputs to paracerebral feeding command interneurons of *Pleurobranchaea californica*. III. Modifications induced by experience.

62. Davis, W. J. and G. J. Mpitsos. *Z. Vergl. Physiol.* **75**:207–232, 1971. Behavioral choice and habituation in the marine mollusk *Pleurobranchaea californica*.

63. Davis, W. J., G. J. Mpitsos, J. M. Pinneo, and J. L. Ram. *J. Comp. Physiol.* **117**:99–125, 1977. Modification of the behavioral hierarchy of *Pleurobranchaea*. I. Satiation and feeding motivation.

64. Davis, W. J., M. V. S. Siegler, and G. J. Mpitsos. *J. Neurophysiol.* **36**:258–274, 1973. Distributed neuronal oscillators and efference copy in the feeding system of *Pleurobranchaea*.

65. Davis, W. J., J. Villet, D. Lee, M. Rigler, R. Gillette, and E. Prince. *J. Comp. Physiol.* **138**:157–165, 1980. Selective and differential avoidance learning in the feeding and withdrawal behavior of *Pleurobranchaea californica*.

66. Delcomyn, F. *Science* **210**:492–498, 1980. Neural basis of rhythmic behavior in animals.

67. Dickinson, P. S. *J. Comp. Physiol.* **131**:277–283, 1979. Homologous neurons control movements of diverse gill types in nudibranch molluscs.

68. Dickinson, P. S. *J. Comp. Physiol.* **139**:11–16, 1989. Gill control in the notaspidean *Pleurobranchaea* and possible homologies with nudibranchs.

69. Dickinson, P. S. *J. Comp. Physiol.* **139**:17–23, 1980. Neuronal control of gills in diverse *Aplysia* species: Conservative evolution.

70. Dorsett, D. A. In *The Mollusca*, (A. O. D. Willows, ed.), Vol. 9, Part 2, pp. 101–187. Academic Press, New York, 1986. Brains to cells: The neuroanatomy of selected gastropod species.

71. Doresett, D. A. and J. N. Sigger. *J. Exp. Biol.* **94**:77–93, 1981. Sensory fields and properties of the oesophageal proprioceptors in the mollusc, *Philine*.

72. Eaton, C. M. M. S. Thesis, University of

Washington, Seattle, 1970. The reproductive and feeding biology of the prosobranch gastropod *Fusitriton oregonensis* (Redfield) (Fam. Cymatiidae).

73. Eberly, L., J. Kanz, C. Taylor, and H. Pinsker. *Behav. Neural Biol.* **32:**21–34, 1981. Environmental modulation of a central pattern generator in freely behaving *Aplysia*.

74. Elliot, C. J. H. and P. R. Benjamin. *J. Neurophysiol.* **54:**1396–1411, 1985. Interactions of pattern-generating interneurons controlling feeding in *Lymnaea stagnalis*.

75. Elliot, C. J. H. and P. R. Benjamin. *J. Neurophysiol.* **54:**1412–1421, 1985. Interactions of the slow oscillator interneuron with feeding pattern-generating interneurons in *Lymnaea stagnalis*.

76. Farley, J. and D. L. Alkon. *J. Neurophysiol.* **48:**785–807, 1982. Associative neural and behavioral change in *Hermissenda:* Consequences of nervous system orientation for light and pairing specificity.

77. Farley, J. and D. L. Alkon. *J. Neurophysiol.* **57:**1639–1688, 1988. In vitro associative conditioning of *Hermissenda:* Cumulative depolarization of Type B photoreceptors and short-term associative behavioral changes.

78. Farley, J., M. Sakakibara, and D. L. Alkon. *Soc. Neurosci. Abstr.* **10:**270, 1984. Associative-training correlated changes in I_C and I_{Ca} in *Hermissenda* Type B photoreceptor.

79. Frazier, W. T., E. R. Kandel, I. Kupfermann, R. Waziri, and R. E. Coggeshall. *J. Neurophysiol.* **30:**1288–1351, 1967. Morphological and functional properties of identified neurons in the abdominal ganglion of *Aplysia californica*.

80. Fredman, S. M. and B. Jahan-Parwar. *J. Neurophysiol.* **49:**1092–1117, 1983. Command neurons for locomotion in *Aplysia*.

81. Frost, W. N., G. A. Clark, and E. R. Kandel. *J. Neurobiol.* **19:**297–334, 1988. Parallel processing of short-term memory for sensitization in *Aplysia*.

82. Gelperin, A. *Science* **189:**567–570, 1975. Rapid food-aversion learning by a terrestrial mollusk.

83. Getting, P. A. *J. Comp. Physiol.* **110:**271–286, 1977. Afferent neurons mediating escape swimming of the marine mollusc *Tritonia*.

84. Getting, P. A. and M. S. Dekin. *J. Neurophysiol.* **53:**466–480, 1985. Mechanisms of pattern generation underlying swimming in *Tritonia*. IV. Gating of central pattern generator.

85. Getting, P. A. and M. S. Dekin. In *Model Neural Networks and Behavior* (A. I. Selverston, ed.), pp. 1–20. Plenum, New York, 1985. *Tritonia* swimming: A model system for integration within rhythmic motor systems.

86. Gillette, M. U. and R. Gillette. *J. Neurosci.* **3:**1791–1806, 1983. Bursting neurons command consummatory feeding behavior and coordinated visceral receptivity in the predatory mollusk *Pleurobranchaea*.

87. Gillette, R. In *Aims and Methods in Neuroethology* (D. M. Guthrie, ed.), pp. 46–79. Manchester Univ. Press, 1987. The role of neural command in fixed action patterns of behaviour.

88. Gillette, R. In *Pulsatility in Neuroendocrine Systems* (G. Leng, ed.), pp. 205–220. CRC Press, Boca Raton, FL, 1988. Second messengers as elements of the endogenous neuronal oscillator.

89. Gillette, R. and W. J. Davis. *J. Comp. Physiol.* **116:**129–159, 1977. The role of the metacerebral giant neurone in the feeding behavior of *Pleurobranchaea*.

90. Gillette, R., M. U. Gillette, and W. J. Davis. *J. Neurophysiol.* **43:**669–685, 1980. Action potential broadening and endogenously sustained bursting are substrates of command ability in a feeding neuron of *Pleurobranchaea*.

91. Gillette, R. and R.-C. Huang. *Soc. Neurosci. Abstr.* **13:**175.14, 1987. Self-loathing and evasiveness in notaspid snails: Possible regulation of avoidance behavior by defensive acid secretion.

92. Gillette, R., M. P. Kovac, and W. J. Davis. *J. Neurophysiol.* **47:**885–908, 1982. Control of feeding motor output by paracerebral neurons in the brain of *Pleurobranchaea californica*.

93. Gillette, R. and B. Pomeranz. *J. Neurobiol.*

6:463–474, 1975. Ultrastructural correlates of onterneuronal function in the abdominal ganglion of *Aplysia californica*.

94. Goldberg, J. I. and K. Lukowiak. *Can. J. Physiol. Pharmacol.* **61**:749–755, 1983. Transfer of habituation between stimulation sites of the siphon withdrawal reflex in *Aplysia californica*.

95. Goldschmeding, J. T. *Proc. K. Ned. Akad. Wet., Ser. C* **80**:97–115, 1977. Motor control of eating cycles in the freshwater snail *Lymnaea stagnalis*.

96. Gorman, A. L. F. and M. Mirolli. *J. Physiol. (London)* **227**:35–49, 1972. The passive electrical properties of the membrane of a molluscan neurone.

97. Granzow, B. and S. B. Kater. *Neuroscience* **2**:1049–1063, 1977. Identified higher-order neurones controlling the feeding motor program of *Helisoma*.

98. Hartline, P. H. and G. D. Lange. In *Comparative Physiology of Sensory Systems* (L. Bolis, R. D. Keynes, and S. H. P. Maddrell, eds.), pp. 335–355. Cambridge Univ. Press, London and New York, 1984. Visual systems of cephalopods.

99. Hawkins, R. D., T. W. Abrams, T. J. Carew, and E. R. Kandel. *Science* **219**:400–405, 1983. A cellular mechanism of classical conditioning in *Aplysia*: Activity-dependent amplification of presynaptic facilitation.

100. Huang, R.-C. and Gillette, R. *Soc. Neurosci. Abstr.* **11**:78.2, 1985. Mixed signals in chemosensory regulation of feeding behavior: Motivation tips the balance in *Pleurobranchaea*.

101. Huang, R.-C. and R. Gillette. *Soc. Neurosci. Abstr.* **12**:237.10, 1986. Prey avoidance learning in *Pleurobranchaea californica*.

102. Hughes, G. M. and L. Tauc. *J. Exp. Biol.* **40**:469–486, 1963. An electrophysiological study of the anatomical relations of two giant nerve cells in *Aplysia depilans*.

103. Jahan-Parwar, B. *Am. Zool.* **12**:525–537, 1972. Behavioral and electrophysiological studies on chemoreception in *Aplysia*.

104. Jensen, K. R. *J. Molluscan Stud.* **55**:79–88, 1989. Learning as a factor in diet selection by *Elysia viridis* (Montagu) (Opisthobranchia).

105. Jones, H. D. In *Pulmonates* (V. Fretter and J. Peake, eds.), Vol. 1. Academic Press, New York, 1975. Locomotion.

106. Jones, P. G., S. J. Rosser, and A. G. M. Bulloch. *Brain Res.* **437**:56–68, 1987. Glutamate suppression of feeding and the underlying output of effector neurons in *Helisoma*.

107. Kaczmarek, L. K., M. Finbow, J. P. Revel, and F. Strumwasser. *J. Neurobiol.* **10**:535–550, 1979. The morphology and coupling of *Aplysia* bag cells within the abdominal ganglion and in cell culture.

108. Kado, R. T. *Science* **182**:843–845, 1973. *Aplysia* giant cell: Soma-axon voltage clamp current differences.

109. Kandel, E. R. *Behavioral Biology of Aplysia*. Freeman, San Francisco, CA, 1979.

110. Kanz, J. E., L. B. Eberly, J. S. Cobbs, and H. M. Pinsker. *J. Neurophysiol.* **42**:1538–1556, 1979. Neuronal correlates of siphon withdrawal in freely-behaving *Aplysia*.

111. Kater, S. B. *Am. Zool.* **14**:1017–1036, 1974. Feeding in *Helisoma trivolvis*: The morphological and physiological bases of a fixed action pattern.

112. Kirk, M. D. and R. H. Scheller. *Proc. Natl. Acad. Sci. U.S.A.* **83**:3017–3021, 1986. Egg-laying hormone of *Aplysia* induces a voltage-dependent slow inward current carried by Na^+ in an identified motoneuron.

113. Klein, M., E. Shapiro, and E. R. Kandel. *J. Exp. Biol.* **89**:117–157, 1980. Synaptic plasticity and the modulation of the Ca^{2+} current.

114. Klein, M., J. Camardo, and E. R. Kandel. *Proc. Natl. Acad. Sci. U.S.A.* **79**:5713–5717, 1982. Serotonin modulates a specific potassium current in the sensory neurons that show presynaptic facilitation in *Aplysia*.

115. Koester, J. *Soc. Neurosci. Abstr.* **9**:158.13, 1983. Respiratory pumping in *Aplysia* is mediated by two coupled clusters of interneurons.

116. Koester, J., E. Mayeri, G. Liebeswar, and E. R. Kandel. *J. Neurophysiol.* **37**:476–496,

1974. Neural control of circulation in *Aplysia*. II. Interneurons.

117. Kovac, M. P. and W. J. Davis. *J. Comp. Physiol.* **139**:77–86, 1980. Reciprocal inhibition between feeding and withdrawal behaviors in *Pleurobranchaea*.

118. Kovac, M. P. and W. J. Davis. *J. Neurophysiol.* **43**:469–487, 1980. Neural mechanism underlying behavioral choice in *Pleurobranchaea*.

119. Kovac, M. P., W. J. Davis, E. M. Matera, A. Morielli, and R. P. Croll. *Brain Res.* **331**:275–284, 1985. Learning: Neural analysis in the isolated brain of a previously trained mollusc, *Pleurobranchaea californica.*

120. Kupfermann, I. *Behav. Biol.* **10**:89–97, 1974. Dissociation of the appetitive and consumatory phases of feeding behavior in *Aplysia:* A lesion study.

121. Kupfermann, I. *Behav. Biol.* **10**:1–26, 1974. Feeding behavior in *Aplysia:* A simple system for the study of motivation.

122. Kupfermann, I. and E. R. Kandel. *Science* **164**:847–850, 1969. Neuronal controls of a behavioral response mediated by the abdominal ganglion of *Aplysia*.

123. Kyriakides, M. A. and C. R. McCrohan. *J. Exp. Biol.* **136**:103–123, 1988. Central coordination of buccal and pedal neuronal activity in the pond snail *Lymnaea stagnalis.*

124. Lane, F. W. *Kingdom of the Octopus. The Life History of the Cephalopoda.* Jarrolds, London, 1957.

125. Lasek, R. J. and W. J. Dower. *Science* **172**:278–280, 1971. *Aplysia californica:* Analysis of nuclear DNA in individual nuclei of giant neurons.

126. Laverack, M. S. *Mar. Behav. Physiol.* **7**:155–169, 1980. Electrophysiology of the isolated central nervous system of the northern octopus *Eledone cirrhosa.*

127. Lederhendler, I. and Alkon, D. L. *Behav. Neural Biol.* **47**:227–249, 1987. Associatively reduced withdrawal from shadows in *Hermissenda:* A direct behavioral analog of photoreceptor responses to brief light steps.

128. Lee, R. M. and R. J. Liegeois. *J. Neurobiol.*

5:545–564, 1974. Motor and sensory mechanism of feeding in *Pleurobranchaea*.

129. Lee, R. M. and R. A. Palovcik. *Behav. Biol.* **16**:251–266, 1976. Behavioral states and feeding in the gastropod *Pleurobranchaea.*

130. Lent, C. M., M. H. Dickinson, and C. G. Marshall. *Am. Zool.* (in press), 1989. Serotonin and leech feeding behavior: Obligatory neuromodulation.

131. Levy, M., Y. Achituv, and A. J. Susswein. *J. Exp. Biol.* **141**:389–405, 1989. Relationship between respiratory pumping and oxygen consumption in *Aplysia depilans* and *Aplysia fasciata*.

132. Lombardo, F., O. Sonetti, and E. Baraldi. *Nucleus* **23**:30–36, 1980. Differential staining and fluorescence of chromatin in populations of neuronal nuclei from *Planorbis*.

133. London, J. A. and R. Gillette. *J. Exp. Biol.* **108**:471–475, 1984. Rhythmic and bilaterally coordinated motor activity in the isolated brain of *Pleurobranchaea*.

134. London, J. A. and R. Gillette. *Proc. Natl. Acad. Sci. U.S.A.* **83**:4058–4062, 1986. Mechanism for food avoidance learning in the central pattern generator of feeding behavior of *Pleurobranchaea californica*.

135. Lorenz, K. and N. Tinbergen. *Z. Tierpsychol.* **2**:1–29, 1938. Taxis und Instinkthandlung in der Eirollbewegung der Graugans. I.

136. Lukowiak, K. and C. Sahley. *Science* **212**:1516–1518, 1981. The *in vitro* classical conditioning of the gill withdrawal reflex in *Aplysia californica.*

137. Maximova, O. A. and P. M. Balaban. *Brain Res.* **292**:139–149, 1984. Neuronal correlates of aversive learning in command neurons for avoidance behavior of *Helix lucorum L.*

138. McClellan, A. D. *J. Exp. Biol.* **98**:195–211, 1982. Movements and motor patterns of the buccal mass of *Pleurobranchaea* during feeding, regurgitation and rejection.

139. McClellan, A. D. *J. Exp. Biol.* **98**:195–211, 1982. Re-examination of presumed feeding motor activity in the isolated nervous system of *Pleurobranchaea*.

140. McClellan, A. D. *J. Neurophysiol.* **50:**658–670, 1983. Higher order neurons in buccal ganglia of *Pleurobranchaea* elicit vomiting motor activity.

141. McCrohan, C. R. *J. Exp. Biol.* **108:**257–272, 1984. Properties of ventral cerebral neurones involved in the feeding system of the snail *Lymnaea stagnalis.*

142. Meech, R. W. In *The Mollusca* (A. O. D. Willows, ed.), Vol. 9, Part 2, pp. 189–277. Academic Press, Orlando, FL, 1986. Membranes, gates and channels.

143. Mellon, DeF., Jr., J. A. Wilson, and C. E. Phillips. *Brain Res.* **223:**134–140, 1981. Modification of motor neuron size and position in the central nervous system of adult snapping shrimps.

144. Mirolli, M. and S. R. Talbott. *J. Physiol. (London)* **227:**35–49, 1972. The geometrical factors determining the electrotonic properties of a molluscan neurone.

145. Mirsky, A. E. and S. Osawa. In *The Cell* (J. Brachet and A. E. Mirsky, eds.), pp. 677–770. Academic Press, New York, 1961. The interphase nucleus.

146. Mittenthal, J. E. and J. J. Wine. *J. Comp. Neurol.* **177:**311–334, 1978. Segmental homology and variation in flexor motoneurons of the crayfish abdomen.

147. Morielli, A. D., E. M. Matera, M. P. Kovac, R. G. Shrum, K. J. McCormack, and W. J. Davis. *Proc. Natl. Acad. Sci. U.S.A.* **83:**4556–4560, 1986. Cholinergic suppression: A postsynaptic mechanism of long-term associative learning.

148. Mpitsos, G. J. and C. S. Cohan. *J. Neurobiol.* **17:**469–486, 1986. Discriminative behavior and pavlovian conditioning in the mollusc *Pleurobranchaea.*

149. Mpitsos, G. J. and C. S. Cohan. *J. Neurobiol.* **17:**487–497, 1986. Differential pavlovian conditioning in the mollusc *Pleurobranchaea.*

150. Mpitsos, G. J. and W. J. Davis. *Science* **180:**317–320, 1973. Learning: Classical and avoidance conditioning in the mollusk *Pleurobranchaea.*

151. Mpitsos, G. J. and S. D. Collins. *Science* **188:**954–957, 1975. Learning: Rapid aversion conditioning in the gastropod mollusk *Pleurobranchaea californica.*

152. Mpitsos, G. J. and K. Lukowiak. In *The Mollusca* (A. O. D. Willows, ed.), Vol. 9, Part 2, pp. 95–267. Academic Press, Orlando, FL, 1986. Learning in gastropod molluscs.

153. Neary, J. T., T. Crow, and D. L. Alkon, *Nature (London)* **293:**658–660, 1981. Changes in a specific phosphoprotein following associative learning in *Hermissenda.*

154. Nisbet, R. H. *Proc. R. Soc. London, Ser. B* **154:**267–287, 1961. Some aspects of the structure and function of the nervous system of *Archachatina (Calachatina) marginata* (Swainson).

155. Ocorr, K. A., E. T. Walters, and J. H. Byrne. *Proc. Natl. Acad. Sci. U.S.A.* **82:**2548–2552, 1985. Associative conditioning analog selectively increases cAMP level of tail sensory neurons in *Aplysia.*

156. Peretz, B. *Science* **169:**379–381, 1970. Habituation and dishabituation in the absence of a central nervous system.

157. Peretz, B., J. W. Jacklet, and K. Lukowiak. *Science* **191:**396–399, 1976. Habituation of reflexes in *Aplysia.* Contribution of the peripheral and central nervous systems.

158. Peters, M. and U. Altrup. *J. Neurophysiol.* **52:**389–408, 1984. Motor organization in pharynx of *Helix pomatia.*

159. Pinsker, H. M., I. Kupfermann, V. F. Castelucci, and E. R. Kandel. *Science* **167:**1740–1742, 1970. Habituation and disgabituation of the gill withdrawal reflex in *Aplysia.*

160. Pinsker, H. M., W. A. Hening, T. J. Carew, and E. R. Kandel. Long-term sensitization of a defensive withdrawal reflex in *Aplysia californica.*

161. Rayport, S. G., R. T. Ambron, and J. Babiarz. *J. Neurophysiol.* **49:**864–876, 1983. Identified cholinergic neurons R2 and LP1$_1$ control mucus release in *Aplysia.*

162. Robertson, J. D., P. Lee, and J. Z. Young. *Soc. Neurosci. Abstr.* **14:**316.5, 1988. Filopodia are present in synaptic glomerulae

in the tactile learning neuropils of the posterior buccal lobes of *Octopus vulgaris*.

163. Rose, R. M. and P. R. Benjamin. *J. Exp. Biol.* **92**:187–201, 1981. Interneuronal control of feeding in the pond snail, *Lymnaea stagnalis*. I. Initiation of feeding cycles by a single buccal interneurone.

164. Rose, R. M. and P. R. Benjamin. *J. Exp. Biol.* **92**:187–201, 1981. Interneuronal control of feeding in the pond snail, *Lymnaea stagnalis*. II. The interneuronal mechanism generating feeding cycles.

165. Rowell, C. H. F. *J. Exp. Biol.* **40**:257–270, 1963. Excitatory and inhibitory pathways in the arm of *Octopus*.

166. Sahley, C., A. Gelperin, and J. W. Rudy. *Proc. Natl. Acad. Sci. U.S.A.* **78**:640–642, 1981. One-trial associative learning modifies food odor preference in a terrestrial mollusc.

167. Sahley, C., J. W. Rudy, and A. Gelperin. *J. Comp. Physiol.* **144**:1–8, 1981. An analysis of associative learning in a terrestrial mollusc. I. Higher-order conditioning, blocking and a transient US pre-exposure effect.

168. Sakata, K. M., M. Tsuge, Y. Kamiya, and K. Ina. *Agric. Biol. Chem.* **49**:1905–1907, 1985. Isolation of a glycerolipid (DGTH) as a phagostimulant for a seahare, *Aplysia juliana*, from a green alga, *Ulva pertusa*.

169. Senseman, D. and A. Gelperin. *Malacol. Rev.* **7**:51–52, 1973. Comparative aspects of the morphology and physiology of a single identifiable neurone in *Helix aspersa*, *Limax maximus*, and *Ariolimax californica*.

170. Siegelbaum, S. A., J. S. Camardo, and E. R. Kandel. *Nature (London)* **299**:413–417, 1982. Serotonin and cyclic AMP close single K^+ channels in *Aplysia* sensory neurons.

171. Siegler, M. V. S. *J. Exp. Biol.* **71**:27–48, 1977. Motor neurone coordination and sensory modulation in the feeding system of the mollusc *Pleurobranchaea californica*.

172. Siegler, M. V. S., G. J. Mpitsos, and W. J. Davis. *J. Neurophysiol.* **37**:1173–1196, 1974. Motor organization and generation of rhythmic feeding output in buccal ganglion of *Pleurobranchaea*.

173. Spira, M. E. and M. V. L. Bennett. *Brain Res.* **37**:294–300, 1972. Synaptic control of electrotonic coupling between neurons.

174. Spray, D. C., M. E. Spira, and M. V. L. Bennett. *Brain Res.* **182**:253–270, 1980. Peripheral fields and branching patterns of buccal mechanosensory neurons in the opisthobranch mollusc, *Navanax inermis*.

175. Susswein, A. M. and M. V. L. Bennett. *J. Neurobiol.* **10**:521–534, 1979. Plasticity of feeding behavior in the opisthobranch mollusc *Navanax*.

176. Susswein, A. J. and J. H. Byrne. *J. Neurosci.* **8**:2049–2061, 1988. Identification and characterization of neurons initiating patterned neural activity in the buccal ganglia of *Aplysia*.

177. Susswein, A. J. and I. Kupfermann. *Behav. Neural Biol.* **13**:203–209, 1975. Bulk as a stimulus for satiation in *Aplysia*.

178. Susswein, A. J. and M. Schwartz. *Behav. Neural Biol.* **39**:1–6, 1983. A learned change of response to inedible food in *Aplysia*.

179. Taghert, P. H. and A. O. D. Willows. *J. Comp. Physiol.* **123**:253–259, 1978. Control of a fixed action pattern by single, central neurones in the marine mollusk, *Tritonia diomedea*.

180. Thompson, T. E. *J. Mar. Biol. Assoc. U.K.* **39**:115–122, 1960. Defensive acid secretion in marine gastropods.

181. Thompson, T. E. *Comp. Biochem. Physiol. A* **74A**:615–621, 1983. Detection of epithelial acid secretions in marine molluscs: Review of techniques, and new analytical methods.

182. Thompson, T. E. *J. Molluscan Stud.* **50**:66–67, 1984. Histology of acid glands in Pleurobranchomorpha.

183. Tremblay, J. P., M. Colonnier, and H. McLennan. *J. Comp. Neurol.* **188**:367–389, 1979. An electron microscope study of synaptic contacts in the abdominal ganglion of *Aplysia californica*.

184. Walters, E. T., J. H. Byrne, T. J. Carew, and E. R. Kandel. *J. Neurophysiol.*

50:1543–1559, 1983. Mechanoafferent neurons innervating the tail of *Aplysia*. II. Modulation by sensitizing stimulation.

185. Walters, E. T. and J. H. Byrne. *Science* **219**:405–408, 1983. Associative conditioning of single sensory neurons suggests a cellular mechanism for learning.

186. Walters, E. T. and J. H. Byrne. *J. Neurosci.* **5**:662–672, 1985. Long-term enhancement produced by activity-dependent modulation of *Aplysia* sensory neurons.

187. Weiss, K. R., J. Cohen, and I. Kupfermann. *Brain Res.* **99**:381–386, 1975. Potentiation of muscle contraction: A possible modulatory function of an identified serotonergic cell in *Aplysia*.

188. Weiss, K. R. and I. Kupfermann. *Brain Res.* **117**:33–49, 1976. Homology of the giant serotonergic neurones (metacerebral cells) in *Aplysia* and pulmonate molluscs.

189. Wells, M. J. *J. Exp. Biol.* **40**:187–193, 1963. Taste by touch: Some experiments with *Octopus*.

190. Wells, M. J. *Octopus: Physiology and Behaviour of an Advanced Invertebrate*. Chapman & Hall, London, 1978.

191. Williamson, R. and B. U. Budelmann. *J. Comp. Physiol.* **156**:403–412, 1985. An angular acceleration receptor system of dual sensitivity in the statocyst of *Octopus vulgaris*.

192. Willows, A. O. D. In *Invertebrate Learning* (W. C. Corning, J. A. Dyal, and A. O. D. Willows, eds.), Vol. 2, pp. 187–274. Plenum, New York, 1973. Learning in gastropod mollusks.

193. Willows, A. O. D. *J. Neurophysiol.* **44**:849–861, 1980. Physiological basis of feeding behavior in *Tritonia diomedea*. II. Neuronal mechanisms.

194. Woolacott, M. H. *J. Comp. Physiol.* **94**:69–84, 1974. Patterned neural activity associated with prey capture in *Navanax* (Gastropoda, Aplysiacea).

195. Young, J. Z. *Philos. Trans. R. Soc. London, Ser. B* **263**:409–429, 1963. The organization of a cephalopod ganglion.

196. Young, J. Z. *A Model of the Brain*. Oxford Univ. Press (Clarendon), London and New York, 1964.

197. Young, J. Z. *Philos. Trans. R. Soc. London, Ser. B* **267**:263–302, 1974. The central nervous system of *Loligo*. I. The optic lobe.

Section E
Annelids

Fred Delcomyn

The annelids were the first major invertebrate group with a significantly condensed CNS. The CNS consists of a ladderlike chain of ventral ganglia and connectives. The ganglia are fused in some species. Nerve cell bodies tend to be concentrated in the ganglia in leeches (25), with ~400 neurons per ganglion (22). In other orders, cell bodies are also located along the length of the cord. A well-developed stomatogastric nervous system exists in many annelids (8). Neurons are typically monopolar in form, and in those annelids with well-defined ganglia, the cell somata form a rind around the central neuropile region (25). Earthworms, at least, have an extensive network of fine nerves below the epidermis. It has been suggested that there is a direct connection between sensory and motor nerves in this network, but if such connections exist, their role in a behavior like peristaltic creeping seems minimal (14, 29).

Several of the common chemical transmitter substances have been found to be present in annelids (14, 33, 34). Serotonin has been shown to have specific behavioral effects in leeches, turning on feeding (21) and inducing swimming (19). It may therefore act as a neuromodulator. Furthermore, it has been found in interneurons whose activity intitiates the swim cycle in leeches (19).

The functional organization of the annelid nervous system has been studied mainly with respect to its role in three different behaviors: escape behavior involving the giant fibers, regulation of the heartbeat, and locomotion (swimming and crawling).

Giant Fibers

Giant axons (also called giant fibers) are widespread throughout most annelid groups, where they mediate powerful escape withdrawal responses. The arrangement is variable in oligochaetes and polychaetes. There may be more than a dozen giant interneurons, typically three to five in *Lumbricus*, a single one, as in *Myxicola*, or even none at all in *Aphrodite* (4). The nerve cords of some species have giant motorneurons in addition to giant interneurons.

Hirudinea

Most authors do not cite the Hirudinea among the annelids that have giant axons. However, the CNS of leeches do possess one fiber (Rohde's fiber, ~3 μm in diameter) or a bundle of fibers (Faivre's nerve) that are large relative to other axons in these animals (10). These fibers are segmental, but are electrotonically coupled from one ganglion to the next. The "fast conducting system" so formed [conduction velocity ~1.0 m/s (20)] has been shown to receive input from various tactile and other sensory receptors (11). Behaviorally, leeches show a rapid withdrawal response similar to that exhibited by other annelids, contracting to 70% of the relaxed body length in response to a strong mechanical stimulus on the head (30). It has not been conclusively demonstrated that the fast conducting system is responsible for this rapid shortening.

Polychaetes

A giant axon in a polychaete can be formed from a single intersegmental neuron, from the serial fusion of several intersegmental neurons, or from the fusion of several neurons in a single segment (9, 10). It may range in size from an enormous 1.7 mm in the sabellid *Myxicola* (8), down to a more typical 15–200 μm in most polychaetes (4). Conduction velocities range from 21 m/s in *Myxicola* (8), to 2.5–4.5 m/s in *Nereis* (18). The giant interneurons are excited by a variety of sensory inputs, including both tactile and chemical stimuli from the body wall. In worms that have more than one giant, excitation of a giant fiber on one side of the body frequently causes excitation of the contralateral partner as well. In *Sabella* and *Nereis*, this cross excitation is due to electrotonic connections between the paired neurons (9). In *Nereis*, which has medial, lateral and paramedial giants, activation of different combinations of giant fibers causes different behavioral responses. The lateral giants respond to strong touch stimuli anywhere and tend to use a general body shortening. The medial and paramedial groups respond selectively to touch on the anterior and posterior parts, respectively (18), and primarily use shortening of the appropriate body part. They also cause parapodial movement directed away from the stimulus (8). These responses are brought about by synapses between the giants and the muscles (directly in *Myxicola*), or between the giants and (usually giant) motorneurons.

Most worms show behavioral habituation of the escape response to repeated stimuli. Habituation to tactile stimuli in *Nereis* is caused by loss of transmission at the junction between the lateral giant and the giant motorneurons with which it synapses (18). There is no evidence for sensory habituation. In *Nereis*, after habituation has been established, recruitment of the medial giant by raising the stimulus strength not only causes a full response in the muscles, but also restoration of the full response to the next stimulation of the lateral giant alone. The behavioral response of a worm to stimulation also depends on the animal's ongoing activity. Stimuli that in a restrained preparation always elicit 1:1 transmission between the giant interneuron and the motorneuron frequently fail to elicit a response when the worm is crawling.

Oligochaetes

Nearly all oligochaetes have a single large median (MGF) and two smaller lateral (LGF) giant fibers in the dorsal region of the nerve cord (10). They are typically formed by end-to-end fusion of serially arranged interneurons. During development in earthworms, some of each septum between pairs of cells remains, while the rest disappears (17). Electrical transmission occurs at gap junctions on the septa (2). In addition to the LGF and MGF giants, oligochaetes also have giant *motorneurons*, other giant interneurons (that feed sensory information to the MGFs and LGFs) and in some cases, a more ventral system of giant axons of unknown function (10). Diameters of the through-conducting axons range from ~60 μm down to only 1–4 μm in small aquatic oligochaetes; conduction velocities range upwards from 80 to 90 cm/s (37). Conduction velocities for the large fibers in oligochaetes are high (nearly 30–38 m/s in an earthworm) due to a myelinlike outer sheath (10).

Oligochaetes withdraw anteriorly if touched on the posterior end, posteriorly if touched on the anterior end. These different behavioral responses are mediated by the different giant fiber systems: anterior touch excites the medial giant fiber, whereas touch on the rear evokes activity in the pair of lateral giants. The switch from anterior withdrawal to posterior withdrawal occurs about two thirds of the distance back from the head in earthworms (16). Touch-sensitive neurons in the body wall generate postsynaptic potentials (psps) in the giants that are proportional to the position of the mechanoreceptor; that is, anterior units yield large psps, mid-body units yield small psps and posterior units yield no psps in the MGF. The converse is true for mechanoreceptors and LGF excitation (31). Coupling between the giant fibers and the giant motorneurons may be via electrical synapses (15). In aquatic oligochaetes, the tail withdrawal response is easier to elicit and less susceptible to habituation, perhaps a reflection of different evolutionary pressures on these worms, which often extend their tails above their burrow openings (37). The reaction time of these worms is only 7 ms (38). The extremely rapid escape is made possible by an extraordinary asymmetry of the LGFs, the left fiber being 40–90 μm in diameter, the right 20–40 μm (39). This asymmetry allows the smaller fiber to be used as a center for spike initiation, the larger exclusively for rapid conduction.

Heartbeat

The neural basis of the beating of the tubular hearts in leeches has received attention as an example of neural control of a rhythmic behavior. The two tubular hearts of a leech lie laterally; they are innervated by segmental heart excitor (HE) and heart accessory (HA) motorneurons, whose activity in turn is regulated by interneurons (HN) in the CNS. Each heart has two modes of coordination. One is a peristaltic mode, in which the heart muscles contract in sequence rear to front, forcing the blood forward. The other is a synchronous mode, in which all the muscles contract and relax more or less together, excluding blood from one heart altogether during the contraction phase. The hearts do not contract in the same mode at the same time, but switch from one mode to the other approximately every 50 beats (7).

Surgically isolated hearts have an intrinsic rhythmicity, but this is slower than that shown by hearts whose neural innervation is intact, and does not exhibit the two modes of coordination (7). The normal rhythm and coordination is thus regulated by the network of HN neurons, which in turn controls the activities of the HA and HE neurons. Timing of the heartbeat is entirely determined by reciprocal inhibitory interactions between four sets of HN cells located in the third and fourth abdominal ganglia (HN1, HN2, HN3, and HN4 cells) and the spontaneous activity of these cells (27). The coordination of the beat along the length of the heart is also established by interactions between these four cells, and the drive they put on the motorneurons (Fig. E-1) (5, 23, 28). A variety of sensory inputs, some of which work via the HA neurons, can influence the heartbeat. Activity of the HA neurons can change the intrinsic properties of the heart beat so that the heart muscle responds differently to the activity of the HE cells (6), thereby adjusting heart action to the behavioral needs of the animal.

Locomotion

Annelids move by creeping or swimming. Creeping in oligochaetes and polychaetes is thought to be under the control of a

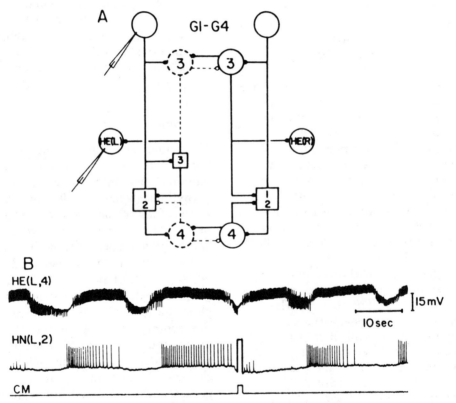

Figure E-1. (A) Diagram of the connections between the HN interneurons and some HE motorneurons in the leech heart. The dotted circles and lines represent cells whose cell somata (circles) or processes (lines) have been photoinactivated in an experiment to determine the connections between other cells. (B) Records of activity in HE4 and HN2 cells. A pulse of current injected into cell HN2 causes cessation of the strong ipsps (inhibitory postsynaptic potentials) in HE4 that represent input from HN3. Therefore, HN2 cells must inhibit HN3 neurons. [From Peterson (28).]

central network of neurons, with contributions from peripherally located sensory receptors (4, 8, 14).

Swimming has been extensively investigated in leeches. The main mechanosensory neurons of the body wall are the nociceptive (N), pressure (P), and touch (T) cells. These, when stimulated in an intact leech, induce undulatory swimming (1). The N and P cells in the head and first body segment make monosynaptic connections with trigger (Tr1) interneurons in the subesophageal ganglion. Brief bursts of activity in a trigger cell initiate longer plateaus of depolarization in several swim-initiating interneurons (also called *gating* neurons), principally cell numbers 204 and 205. These in turn excite cell 208, a member of the network of neurons that generates the swim rhythm (Fig. E-2). The rhythm-generating neurons then drive the motorneurons that produce the undulatory movement (3, 3a, 24, 32, 36).

According to early descriptions, the "oscillator" network of neurons that generate the swim rhythm was a distributed one, consisting of four cells, 27, 28, 33,

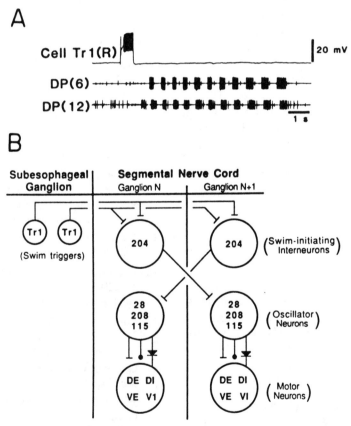

Figure E-2. Trigger cells (Tr1) can initiate swimming in leeches. (*A*) Records showing the initiation of the swimming motor pattern when one Tr1 cell is stimulated. The second and third traces show the motor output of single motorneurons in different body segments (the numbers in parentheses). (*B*) Model circuit by which activity of Tr1 cells can initiate activity in the swim oscillator network in the CNS. Connections between individual oscillator cells are not shown. Motorneurons are dorsal (D) or ventral (V) excitors (E) or inhibitors (I). *T*-junctions are excitatory synapses, filled circles inhibitory, solid triangles with a *T*-junction rectifying electrical junctions. [From Brodfuehrer and Friesen (3).]

and 123, in each of several ganglia (32). Modeling of interactions of these showed that they were capable of generating an appropriately timed, patterned output (13), but only if several sets of cells, representing several ganglia, were included. Later work, however, showed that a single ganglion was capable of generating appropriate motor output (35), and furthermore, that none of the original four oscillator cells received any synaptic input from the swim-initiating neurons that presumably were driving them (36). The recently discovered oscillator cell 208 does receive such input; it and other oscillator neurons that have been discovered (12) must be incorporated into an operational model of the system. Intersegmental coordination, but not production of the rhythmic motor output within a single segment, has been modeled (26). The exact role of the trigger neurons is not known. The trigger cells in the head excite the swim-initiator neurons, but the isolated (headless) nervous system is still fully capable of generating the swimming rhythm; therefore, there must be another pathway that bypasses the trigger neurons.

References

1. Blackshaw, S. E. In *Neurobiology of the Leech* (K. J. Muller, J. G. Nicholls, and G. S. Stent, eds.), pp. 51–78. Cold Spring Harbor Lab., Cold Spring Harbor, New York, 1981. Sensory cells and motor neurons.

2. Brink, P. and L. Barr. *J. Gen. Physiol.* **69**:517–536, 1977; Jaslove, S. W. and P. Brink. In *Cell Interactions and Gap Junctions* (N. Sperelakis and W. C. Cole, eds.), Vol. 1, pp. 203–238. CRC Press, Boca Raton, FL, 1989. Electrical resistance of septal membranes in earthworm giant fibers.

3. Brodfuehrer, P. D. and W. O. Friesen. *J. Comp. Physiol. A* **159**:489–502; 503–510; 511–519, 1986. Swimming and trigger neurons for leech swimming.

3a. Brodfuehrer, P. D. and W. O. Friesen. *Science* **234**:1002–1004, 1986. Swimming and trigger neurons for leech swimming.

4. Bullock, T. H. and G. A. Horridge. *Structure and Function in the Nervous Systems of Invertebrates.* Freeman, San Francisco, CA, 1965.

5. Calabrese, R. L. *Am. Zool.* **19**:87–102, 1979. Mechanisms of coordination of heartbeat in the leech.

6. Calabrese, R. L. and E. A. Arbas. In *Model Neural Networks and Behavior* (A. I. Selverston, ed.), pp. 69–85. Plenum, New York, 1985. Modulation of rhythmicity in the leech heartbeat.

7. Calabrese, R. L. and E. Peterson. *Symp. Soc. Exp. Biol.* **37**:195–221, 1983. Neural control of heartbeat in the leech.

8. Dorsett, D. A. In *Physiology of Annelids* (P. J. Mill, ed.), pp. 115–160. Academic Press, London, 1978. Organization of the nerve cord.

9. Dorsett, D. A. *Trends Neurosci.* **3**:205–208, 1980. Design and function of giant fibre systems.

10. Drewes, C. D., In *Neural Mechanisms of Startle Behavior* (R. C. Eaton, ed.), pp. 43–91. Plenum, New York, 1984. Escape reflexes in earthworms and other annelids.

11. Friesen, W. O. *J. Exp. Biol.* **92**:255–275, 1981. Water motion detection in the leech.

12. Friesen, W. O. *J. Comp. Physiol. A* **156**:231–242, 1985. Neuronal control of leech swimming movements.

13. Friesen, W. O., M. Poon, and G. S. Stent. *J. Exp. Biol.* **75**:25–43, 1978. Neuronal control of swimming in the leech.

14. Gardner, C. R. *Biol. Rev. Cambridge Philos. Soc.* **51**:25–52, 1976. Neuronal control of locomotion in the earthworm.

15. Günther, J. *Comp. Biochem. Physiol.* **42**:967–974, 1972. Giant motor neurons in the earthworm.

16. Günther, J. *Naturwissenschaften* **11**:521–522, 1973. Overlapping sensory fields of the giant fiber systems in the earthworm.

17. Günther, J. *J. Neurocytol.* **4**:55–62, 1975.

Neuronal syncytia in giant fibres of earthworms.

18. Horridge, G. A. *Proc. R. Soc. London Ser. B* **150**:245–262, 1959. Analysis of rapid responses in *Nereis* and *Harmothoe*.

19. Kristan, W. B., Jr. and M. P. Nusbaum. *J. Physiol. (Paris)* **78**:743–747, 1983. The dual role of serotonin in leech swimming.

20. Laverack, M. S. *J. Exp. Biol.* **50**:129–140, 1969. Mechanoreceptors, photoreceptors, and rapid conduction pathways in the leech.

21. Lent, C. M. *Brain Res. Bull.* **14**:643–655, 1985. Serotonergic modulation of the feeding behavior of the leech.

22. Macagno, E. R. *J. Comp. Neurol.* **190**:283–302, 1980. Number and distribution of neurons in leech segmental ganglia.

23. Maranto, A. R. and R. L. Calabrese. *J. Comp. Physiol. A* **154**:381–391, 1984. Neural control of the hearts in the leech.

24. Nusbaum, M. P., W. O. Friesen, W. B. Kristan, Jr., and R. A. Pearce. *J. Comp. Physiol. A* **161**:355–366, 1987. Excitation of swim oscillator neurons by swim initiator neurons in leeches.

25. Payton, B. In *Neurobiology of the Leech* (K. J. Muller, J. G. Nicholls, and G. S. Stent, eds.), pp. 35–50. Cold Spring Harbor Lab., Cold Spring Harbor, New York. 1981. Structure of the leech nervous system.

26. Pearce, R. A. and W. O. Friesen. *Biol. Cybernet.* **58**:301–311, 1988. Model for intersegmental coordination in the leech.

27. Peterson, E. L. *J. Neurophysiol.* **49**:611–626, 1983. Oscillation of heart neurons in isolated leech ganglia.

28. Peterson, E. L. *J. Neurophysiol.* **49**:627–638, 1983. Intersegmental coordination in leech heartbeat.

29. Prosser, C. L. *J. Exp. Biol.* **12**:95–104, 1935. Impulses in earthworm segmental nerves.

30. Sawyer, R. T. In *Neurobiology of the Leech* (K. J. Muller, J. G. Nicholls, and G. S. Stent, eds.), pp. 7–26. Cold Spring Harbor Lab., Cold Spring Harbor, New York. 1981. Leech biology and behavior.

31. Smith, P. H. and J. E. Mittenthal. *J. Comp. Physiol.* **140**:351–353, 1980. Variation in pathways to giant interneurons of the earthworm.

32. Stent, G. S., W. B. Kristan, Jr., W. O. Friesen, C. A. Ort, M. Poon, and R. L. Calabrese. *Science* **200**:1348–1357, 1978. Neuronal generation of the leech swimming movement.

33. Tashiro, N. and H. Kuriyama. In *Physiology of Annelids* (P. J. Mill, ed.), pp. 207–242. Academic Press, New York, 1978. Neurosecretion and pharmacology of the nervous system.

34. Wallace, B. G. In *Neurobiology of the Leech* (K. J. Muller, J. G. Nicholls, and G. S. Stent, eds.), pp. 147–172. Cold Spring Harbor Lab., Cold Spring Harbor, New York, 1981. Neurotransmitter chemistry.

35. Weeks, J. C. *J. Neurophysiol.* **45**:698–723, 1981. Neuronal basis of leech swimming.

36. Weeks, J. C. *J. Comp. Physiol.* **148**:253–263; 265–279, 1982. Synaptic basis of swim initiation in the leech.

37. Zoran, M. J. and C. D. Drewes. *J. Comp. Physiol. A* **161**:729–738, 1987. Rapid escape reflexes in aquatic oligochaetes.

38. Zoran, M. J. and C. D. Drewes. *J. Exp. Biol.* **137**, 487–500, 1988. Rapid tail withdrawal in a tubificid worm.

39. Zoran, M. J., C. D. Drewes, C. R. Fourtner, and A. J. Siegel. *J. Comp. Neurol.* **275**:76–86, 1988. Structural and functional asymmetry in giant fibers.

Section F
Arthropods

Fred Delcomyn

Morphology

Arthropod nervous systems resemble those of annelids. Primitively, they consist of a chain of ventrally located ganglia joined by paired connectives. In many species the ganglia in the head, metathoracic segment, and terminal abdominal segment are formed by the fusion of ganglia from several segments. Less primitive arthropods such as crabs and some of the insects (e.g., flies), may have only a single fused ganglionic mass in the thorax, plus the head ganglia. Typical ganglia may contain several hundred neurons. Crayfish have ~600 in the thoracic ganglion, half of which are motorneurons (121). Crayfish abdominal ganglia have ~500 neurons, whereas the subesophageal ganglion has ~6000 and the "brain" (the superesophageal ganglion) has ~10,000 (130). The ganglia of insects contain comparable numbers.

Interganglionic connectives consist of the axons of interneurons and primary sensory fibers. The circumesophageal connectives in various decapod crustacea contain anywhere from 1500 to 9600 axons (fibers), indicating the extreme variability that can exist between species (112). In the crayfish *Orconectes*, connectives in the thorax have ~9500 fibers, whereas thoracic–abdominal connectives contain ~5300 (119).

Insect connectives also contain many axons. There are ~2000 in the stick insect cord (63), 2500 in the locust thoracic–abdominal connective (101), 5000 in the cricket neck connective (3), 8000 in the locust thoracic connective (101), and over 11000 in the thoracic connectives of a cockroach (72). Well over 50% of the fibers in every case are less than 1 μm in diameter. It is thus possible that some of the counts of fibers that yielded relatively small numbers missed some of these very fine axons. The relatively large fibers from which recordings can routinely be taken usually comprise less than 10% of all the fibers in a connective.

Neurons in the CNS are monopolar in type. The cell bodies are arranged around the periphery of the ganglia, which contain the dendrites and some axons in the center. Synaptic contacts are all between neurites. Sensory neurons are typically bipolar, with the cell body in the periphery near the sense organ, although there are exceptions to both cell type and cell body location.

Cellular Properties

Arthropod interneurons and motorneurons generally have electrically inexcitable cell somata that do not receive synapses, although spikes spreading passively from the axon can be recorded (1, 83). Probably most central neurons are capable of responding to excitatory input by generating an action potential, but some are not. These cells, referred to as nonspiking neurons, are generally small and without axons, and seem to leak transmitter substance continually. Excitatory or inhibitory input to these cells causes a few millivolts of depolarization or hyperpolarization, causing a graded increase or decrease of transmitter substance release, but does not evoke an action potential. Nonspiking cells seem especially to be used in local circuits controlling static (postural) behavior, or in biasing other networks on the basis of incoming sensory signals (93).

Arthropods show chemical excitatory and inhibitory synapses, as well as presynaptic inhibition. Electrical synapses seem quite rare (1, 15). γ-aminobutyrate and glutamate are typical transmitters at neuromuscular junctions, ACh (acetylcholine) and others at central synapses (1 , 15). In addition, a number of important substances have recently been discovered to have nontraditional roles, acting as neuromodulators rather than as typical neurotransmitters. Included among these substances are octopamine, serotonin, and proctolin.

Neural Circuits and Behavior

Research on arthropod nervous systems supports the concept that each individual behavior of an animal can be understood in terms of the actions of a set of neurons that themselves are recognizable (identifiable) electrically and by staining from individual to individual in a species. Accordingly, the thrust of much current research is to locate and characterize the neurons involved in a particular behavior, and to analyze the circuit of neurons that they form and within which they interact. Several systems are understood well enough that the behavior they control can be explained almost entirely by what we know about the properties and interactions of individual, identifiable neurons. Other systems are not understood in such detail, but point the way to additional organizational principles.

Single cell studies have shown the widespread occurrence and importance of nonspiking interneurons in arthropods (10, 106, 110, 134). These neurons have been reported to serve three different functions in arthropods: (1) sensory processing or integration [crab stretch receptor (13); crayfish mechanosensory processing (86); fly compound eye (104)], (2) postural control [locusts (10)], and (3) rhythmic motor control [ventilation in crabs (111); pyloric and gastric movements in crayfish (41); swimmeret movements in crayfish (49); walking in cockroaches (78)]. The role of nonspiking neurons in the control of rhythmic motor activity is discussed below in the context of the general control of rhythmic behavior.

Postural control has been studied extensively in locusts. Nonspiking neurons are confined to the half-ganglion within which the cell body is located. They are therefore referred to as *local interneurons*, in contrast to neurons that crossover into other parts of the ganglion, or interganglionic neurons. In locusts, nonspiking neurons form only inhibitory synapses with each other, but either inhibitory or excitatory connections with motorneurons (Fig. F-1). Each interneuron may have influence over many different motorneurons; some nonspiking cells are

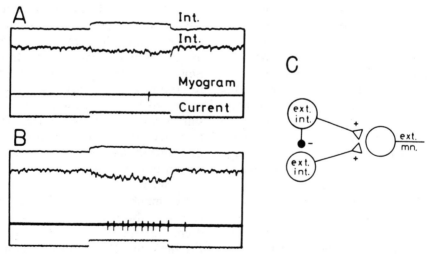

Figure F-1. Graded inhibitory interactions between two nonspiking local interneurons, both of which excite a single motorneuron. Stimulation of one interneuron (*A*) inhibits the second and causes little activity in the motorneuron. Increasing the stimulus current (*B*) causes greater inhibition of the interneuron, and excitation of the motorneuron as well. Stimulation of the other interneuron (not shown) has a similar effect. (*C*) Diagram of the proposed interactions between these neurons. [From Burrows (8).]

able to adjust a limb to a particular position by appropriate control of specific sets of motorneurons (9). Nonspiking neurons may also exert a biasing effect on motorneurons. They tend to slightly depolarize or hyperpolarize populations of leg extensor or flexor motorneurons depending not only on the position of the leg at any given moment, but also on where the leg was before it reached that position (105).

Many of these effects depend on significant integration of sensory input. In crustacea, first-order processing of sensory input may involve nonspiking neurons (86). Nonspiking neurons receive direct input from some exteroceptors in insects as well (62), but sensory processing may also be done by small, local *spiking* interneurons that receive considerable direct sensory input from leg receptors (12, 107). Some of the sensory input that they receive also goes directly to the leg motorneurons (11). The receptive fields of the spiking interneurons are functionally organized so as to assist the locust in making appropriate behavioral responses to specific patterns of sensory stimuli; for example, touch to a particular part of the leg excites a set of local spiking cells that in turn activates certain motorneurons so as to move the leg away from the stimulus. Touch to another part will bring about movement in a different direction (108).

Work on rhythmic motor activities such as walking, swimming, and flying has shown that networks of neurons in the CNS, termed central pattern generators (CPGs), can generate rhythmic output in the absence of any rhythmic sensory input (29). It is clear that many neurons contribute to the generation of a rhythmic motor pattern, but that not all of them are absolutely essential to the production of the basic rhythmicity (68). The role of sensory feedback from the body parts whose movements are being regulated is important, even in systems previously thought to operate largely independent of such feedback (77).

Walking in arthropods is strongly influenced by sensory feedback. There is some evidence for the presence of a CPG for walking in insects (78), but most studies have concentrated on the basic leg movements or effects of sensory feedback, not on discovering the properties of the CPG (4). Nonspiking interneurons seem to be involved in the control of walking, at least in insects (78). In both insects and crustacea, sensory feedback signals play an important role in "fine tuning" the movements of individual legs, adjusting them to the requirements of the walking surface (20, 30). Sensory effects can be shown if an animal is experimentally manipulated so that it receives incorrect sensory input, as can be done by surgically repositioning sense organ receptor strands. When a chordotonal organ of a stick insect is so manipulated, the insect walks with the operated leg held stiffly up off the ground (40). Interlimb coordination also is strongly influenced by sensory feedback. In insects and crustaceans, experiments involving the amputation or deafferentation of one or more legs have dramatic effects on the timing of the remaining legs (20, 30, 109). In both cases, this effect has been interpreted to be due to the loss of sensory signals from the amputated legs. The current view of walking in all arthropods is that sensory feedback is essential for the proper timing of leg movements in an intact animal.

Swimmerets are paired appendages located on the ventral surface of the abdomen in many crustacea. They beat rhythmically. The pattern-generating interneurons controlling the movements of each swimmeret are situated in the ganglion containing the motorneurons for the swimmerets. The local networks are coordinated in their actions by several intersegmental coordinating interneurons (75). These interneurons serve to keep the swimmerets in different segments properly phased relative to one another. Nonspiking interneurons are known to be part of the pattern generating network. Nonspiking neurons can excite return stroke or power stroke motorneurons and also reset ongoing rhythmic motor output in crayfish (49). Furthermore, nonspiking neurons whose activity can initiate the entire swimmeret rhythm in all ganglia of the nervous system have also been found; at least one motorneuron is electrically coupled to a nonspiking neuron that has strong influence over the swimmeret rhythm (74). Sensory feedback from the moving swimmerets can be modulated by output from the pattern-generating network that generates the movements (73).

Swimming has been studied at the cellular level mainly in crayfish, as an expression of the animal's rapid escape response. This response consists of powerful flexions of the abdomen (tail flips), mediated by the action of the giant interneurons. The behavior is relatively well understood in terms of the action of individual neurons. There are actually two separate systems for escape. One is a slow, nongiant mediated system that has a latency of 50–500 ms from stimulus to the first flexor muscle activity, and is evoked by a relatively slow buildup of excitatory postsynaptic potentials (epsps) in members of a small population of interneurons and in the fast flexor motorneurons (58). These epsps begin to excite the flexor motorneurons only after the initial, fast-axon-mediated tail flip is over. The stimulated interneurons excite specific subpopulations of the fast flexors, so that the precise behavioral form of the tail flip is adjusted according to the stimulus that evoked it (58) (Fig. F-2).

The other escape system is a faster, giant-mediated system that has a latency of only 3–10 ms from stimulus to response (87). It consists of four groups of neurons with giant axons: the paired lateral (LG) and medial (MG) giant interneurons, the segmental giants and the motor giants (moG) (59, 136). These fibers range in size from 70 to 250 μm in diameter, and conduct at ~10–20 m/s (7). The lateral giants are serially segmental, each separate neuron in the chain being electrically coupled to the next. The left and right LG chains are also cross-connected electrically, ensuring that they fire together (135). Each medial giant is a single neuron that runs the length of the nerve cord and has its soma in the brain. The two synapse with each other in the brain. All the segmental and moG axons arise from individual cells (135).

Giant-mediated escape is complex (Fig. F-3). If the triggering stimulus is rostral (on the head, legs, or thorax), the crayfish generates a powerful backward-directed tailflip. The stimulus excites the medial giant. The MG in turn synapses with all the segmental and motor giants, causing rapid contraction of all the abdominal flexor muscles. There is only weak, if any, direct synaptic contact between the MG and the fast flexor motorneurons (which are independent of the moG neurons). If the triggering stimulus is caudal (on the tail or abdomen), the crayfish contracts only the anteriormost flexor muscles, causing the animal to flip upward rather than back, since the tail does not curl completely under the body. A caudal stimulus excites the *lateral* giants. The LGs then excite the anterior segmental and motor giants. The LG to segmental synapses (at least) are electrotonic. Other neurons involved in the circuit ensure that each abdominal flexion is followed by an extension, that the initial escape blends smoothly into rapid swimming that does not require the involvement of the LG and MG neurons, and that the movements that are produced are suitably coordinated. Current research is focusing on

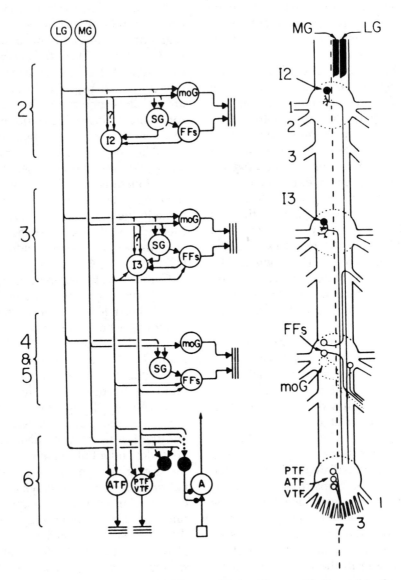

Figure F-2. The motor circuitry for tailflip production diagramatically and in the context of the ventral nerve cord. Numbers at the left are segment numbers. In the left-hand diagram, arrowheads represent excitatory synapses, filled circles inhibitory ones. In the right-hand diagram, the presence of the fast flexors (FF) and motor giants (moG) are represented for clarity in one ganglion only. Note that the lateral giant (LG) does not synapse with the moG in the caudal segments. I2 and I3 are interneurons involved in the production of nongiant tailflips. Other symbols: A—Another identified interneuron; ATF, PTF, VTF—motorneurons controlling muscles in the tailfan; MG—medial giant; SG—segmental giant. [From Krasne and Wine (59).]

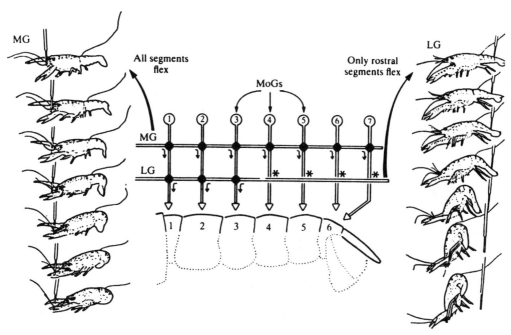

Figure F-3. Connectivity between the giant axons and motor giants determines the escape behavior of the crayfish. The medial giants (MG) respond to tactile input to the rostral end of the animal, and stimulate all the motor giants (MoGs), causing the animal to flex all the abdominal segments and swim backwards. The lateral giant (LG) responds to caudal tactile input and stimulates only the anterior motor giants, causing the animal to move up rather than back. [From Dumon and Wine (34).]

several issues: the role of the segmental giants (47), differences between the giant neuron connections in different abdominal ganglia (34, 69), the neurons involved in initiation of the escape response, the circuitry of the nongiant escape system, and the plastic properties (habituation and dishabituation) of the system.

Giant fiber systems are also found in other crustacea. The giants in hermit crabs, for example, control a rapid withdrawal of the animal into its shell (19). The response is mediated by several sets of giant fibers: A single pair of intersegmental giants, thought to be homologous to the crayfish MG (116), segmental giants (48), and motor giants (117). Synapses between the main giants and the ipsilateral segmental and motor giants are apparently electrotonic (48). Evidence for a chemical synapse between the main giants and contralateral motor giants (117) has been disputed (48). Other crustacea, such as the stomatopod mantis shrimp, may possess giant fiber systems as well (7).

Flying in locusts was one of the first rhythmic behaviors for which the involvement of a CPG was clearly established (133). The meso- and metathoracic ganglia contain a number of spiking interneurons that are active in phase with the motor activity of flight (Fig. F-4) (94). The physiology and morphology of some of these neurons have been described (95, 96). They show extensive bilateral branching in both ganglia, with predominantly inhibitory interactions, some graded. At one time it was thought that the flight pattern was produced by these neurons

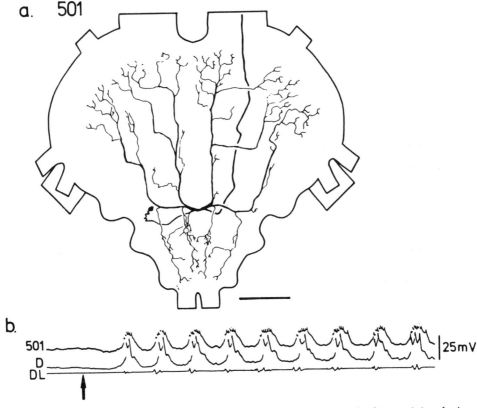

Figure F-4. A flight interneuron (501) in a locust (*a*) and a record of its activity during flight (*b*). This neuron is one of several with similar patterns of branching in the meta-thoracic ganglion, but with the cell soma located in regions with different embryological origins. This cell shows strong phases of depolarization during flight, in synchrony with the depressor muscles (DL, depressor longitudinal neuron). [From Robertson and Pearson (94).]

acting as a single network distributed over the mesothoracic, metathoracic, and first three fused abdominal ganglia (96). However, cutting the meso- or metathoracic ganglion in half does not significantly impair the ability of a locust to fly, nor does it destroy the rhythmic production of motor output of the ganglia in deafferented preparations (100). Since a distributed neural network would be destroyed by these procedures, as yet undiscovered mechanisms must be at work.

What is the role of sensory feedback in this system? The alternating activity of elevator and depressor motorneurons and the relatively constant delay between el-

evator activity in the meso- and metathoracic segments are two of the main temporal features of the locust flight system. These are retained even after the CNS is deafferented by surgical isolation from sensory feedback from the moving wings (133). Nevertheless, sensory feedback plays an essential role in normal flight, since deafferentation reduces the flight frequency by as much as 50%. Stimulation of the nerves from wing stretch receptors in a deafferented locust can reset an ongoing flight rhythm, and (if continued), elevate the flight frequency significantly (82, 88). Stretch receptors show extensive connections with the pattern generating interneurons in the CNS (96), and there are subtle but important differences between the intracellularly recorded activity of elevator motorneurons in animals with and without stretch receptor feedback (137). The powerful role of sensory feedback has therefore been emphasized (77, 138). Since other recent work suggests that under the influence of the neuromodulatory substance octopamine an isolated locust CNS can produce a motor pattern that shows normal timing (118), the issue is, however, not yet entirely settled.

Stomatogastric System

Control of the gastric and pyloric chewing rhythms in the stomach of crustacea is probably the best understood rhythmic control system (103). There are two regions of the stomach. In crayfish the anterior stomach has a gastric mill con-

taining teeth moved by 25 muscles driven by 9 motorneurons. The pyloric stomach is driven by 14 motorneurons. The stomatogastric ganglion of the lobster *Panulirus* contains 33 neurons, 12 neurons in the gastric mill, and 14 in the pyloric region, activating a total of 30 muscles (Fig. F-5). The rhythmic activity of the gastric mill consists of (1) opening and closing of two lateral teeth and (2) power and return strokes of a single medial tooth (68, 101a, 102). Movement of the pylorus consists of alternating dilation and constriction. The neurons for each region are coupled by electrical and chemical synapses and generate separate interconnected rhythms. Pyloric bursts occur at 0.5 s intervals, gastric bursts at 3 s intervals. The functions of neurons of the stomatogastric ganglion have been ascertained by (1) simultaneous recordings from each of several neurons, (2) photodestruction of single cells by spot illumination after

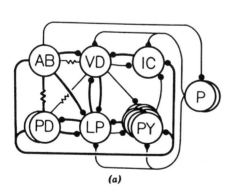

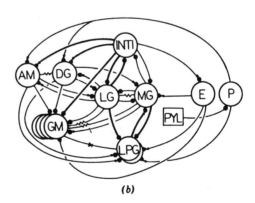

(a)   (b)

Figure F-5. Abbreviated circuit diagrams of stomatogastric ganglia of lobster. (*a*) Pyloric and (*b*) gastric pattern generators. Inhibitory synapses indicated by solid circles, excitatory synapses by triangles, electrical junctions by resistors. Only strong synapses shown; synergistic neurons grouped. a, Pyloric generator: AB and P endogenous bursters, inhibitory to other pyloric neurons, PD pyloric dilators, VD ventricular dilator, IC inferior cardiac, LP and PY pyloric neurons. P in commissural ganglion (of CNS) is general phasic excitor of network. b, Gastric pattern generator: E and P in commissural ganglion are phasic excitors, INT1 inhibits E, LG lateral gastric and MG median gastric close medial tooth, LPG lateral posterior gastric opens lateral teeth, DG dorsal gastric and AM anterior median, GM and AM are triggered bursters tonically active when E cells are active. P indicated as tonically acting on both circuits. [From Prosser, *Adaptational Biology*, p. 481. Original from Selverston et al. (102a).]

staining with lucifer yellow, and (3) reversible blocking of cells by hyperpolarizing them.

Two different mechanisms generate stomatogastric rhythms. (1) The pyloric rhythm is generated by cellular bursting and synaptic interaction (Fig. F-5a). The AB (anterior burster) cell is an endogenous burster, but the other cells can be made to burst when supplied with input from extrinsic ganglia. Reciprocal inhibition, especially between the PD (pyloric dilator) and LP (lateral pyloric) cells also contributes to alternating bursting. The synapses are arranged so that when PD and AB fire, all of the other cells are inhibited. On release from inhibition, LP and IC (inferior cardiac) fire first, then cells PY (pyloric) and VD (ventricular dilator) (68, 101a, 102). Thus all connections between the cells are inhibitory; discharge normally occurs after stepwise removal of inhibition.

Muscles of the stomach are activated by neurons in succession. The basic gastric mill rhythm appears to be due to reciprocal inhibition between the LG–MG (lateral gastric-medial gastric) pair (which close the lateral teeth) bursting between the two components. When the LG and MG fire, they inhibit the LGPs (lateral posterior gastrics), which open the lateral teeth. When interneuron one fires, it excites the DG–AM (dorsal gastric-anterior median), which resets the medial tooth and inhibits the GM (gastric mill) cells, which pull the medial tooth forward. The GMs are also inhibited by the DG and AM so that these two groups fire out of phase with each other. The gastric mill network has been simulated by a computer program that fires in a pattern like the ganglion.

The mechanism of endogenous bursting in pyloric pacemakers in crustaceans consists of voltage-sensitive slow waves, much like those in bursting molluscan neurons. If the cells (PD) are hyperpolarized, spikes are stopped and sinusoidal membrane oscillations occur. Depolarizing current is carried by Na^+ and Ca^{2+} and outward currents are carried by K^+.

The gastric and pyloric rhythms are endogenous but are modulated by input from other neural centers. For example, Int-1 cells inhibit E (excitatory) cells of the commissural ganglion, which in turn excite gastric mill neurons. Coupling between pyloric gastric mill systems results in synchrony of bursts. Neurons in the CNSs excite PD neurons, and food in the stomach probably reflexly activates PD (Fig. F-5b, cells E, P). No single mechanism is sufficient, but one can substitute for another, resulting in redundancy in the total stomach system. Neurons have been found that can influence both the gastric and pyloric CPGs (32).

Escape Responses

The best known nonrhythmic behavior to have been studied is the escape response of the American cockroach, *Periplaneta americana*, which is mediated by giant fibers (84). These insects show a rapid escape in response to air disturbances (98). The escape consists of an initial turn away from the air stimulus, followed by a straight run (16, 17). The nervous system has seven pairs of giants, arranged in a dorsal group of four and a ventral group of three. Each of the giants has a single cell body in the last abdominal ganglion (Fig. F-6A), and a long axon that probably terminates in the subesophageal ganglion in the head (28). They range from 25 to 60 μm in diameter and have conduction velocities of ~4–7 m/s in the posterior end of the abdomen (115). The fibers narrow as they course toward the head.

The dorsal and ventral groups of giant interneurons are different functionally as well as morphologically (89, 90). One group, the three large ventral giants, are inhibited while the animal walks spontaneously (27). At the same time, the

A.

B.

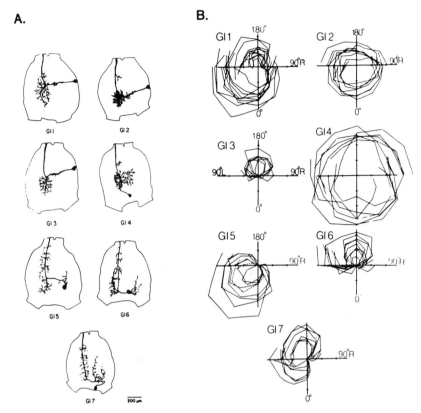

Figure F-6. Morphology (*A*) and pattern of response to wind from different directions (*B*) of the seven giant axons of the American cockroach. Each axon can be identified on the basis of its pattern of branching in the terminal (sixth) abdominal ganglion. Most also have distinctive responses to wind from different angles. Panel (*B*) shows polar plots of the number of spikes with which each neuron responded when wind from different angles were directed at the cerci. (*A*) from Daley et al. (28). (*B*) from Westin et al. (127).

group of smaller, dorsally located giants is excited (31). It has been suggested that only the ventral group is actually responsible for initiating an escape turn and run, while the dorsal group acts to keep the escape system responsive during spontaneous walking (18).

It is not clear exactly what role either the ventral or the dorsal group giants actually plays in escape. Some of the giants respond to puffs of wind from some directions only (Fig. F-6*B*) (127), due to differential synaptic connections received from morphologically polarized filiform hairs on the cerci (25, 26). It is hypothesized that the information about wind direction that is thereby encoded is used in controlling the direction of the escape turn. This idea is supported by the predictable disruption of the normal turn caused by selective destruction of one or several giants (21, 128). However, how the giants actually control muscle activity during the turn is not understood. Stimulation of a giant that is sensitive to wind from one side causes excitation of motorneurons to extensor leg muscles on that same side, and motorneurons of flexor muscles on the other. This is necessary for the animal to turn away from the source of the wind stimulus (89). Stimulation of individual giants does not excite leg motorneurons strongly enough to drive the escape turn, so several giants must act together (89). Interneurons may be interposed between the giant interneurons and the motorneurons (90). Only giants of the ventral group can excite the thoracic interneurons interposed between them and leg motorneurons (129). Selective excitation of the thoracic interneurons by the giants confers to the thoracic neurons directional sensitivity to wind input (91, 129). Destruction of all of the giant fibers in the ventral nerve cord does not prevent escape behavior (although it does slow down the reaction time to wind stimuli); this suggests that nongiant pathways may play a significant role even in normal escape (22).

Several insect groups besides cockroaches have been shown to have giant interneurons. The most primitive is the firebrat *Thermobia*. This insect has several paired abdominal giant fibers (ranging in size from ~10 to 25 μm in diameter) as well as paired abdominal cerci and wind sensitive mechanoreceptors on the cerci (35). Air puffs directed at these cerci elicit electrical activity in the giant fibers, and escape behavior in intact animals. Silverfish, primitive wingless relatives of *Thermobia*, seem to have a similar escape system (35). Many orthopteroid insects besides cockroaches also have giant fibers in their abdominal nerve cords. Crickets have large medial and lateral giants that are thought to control defensive kicking against predators (39). Similar fibers have been described in mantides (6), locusts (23), and grasshoppers (126). Based on the morphological and physiological similarities between systems of giant fibers in different primitive insects, it has been suggested that the use of giant fibers in the abdominal cord for predator evasion was an early evolutionary feature of the group (35). Later evolutionary developments, such as sucking–piercing mouthparts and holometaboly, often involved a

less mobile life style for at least part of the life cycle. They thus made other modes of escape necessary, and the ancient escape system was abandoned.

Most orthopteroid insects use running as the primary means of escape, presumably because most are relatively weak fliers. Some cockroaches can fly, however, and locusts and grasshoppers are strong fliers. In these insects, the giants can trigger escape via the air. In cockroaches, the dorsal giants can elicit flight if the tarsi are without contact (92). In locusts, the four pair of abdominal giants can also elicit flight (5). In strong fliers, other means are also used to escape via flight. In locusts, the large, fast conducting DCMD (descending contralateral movement detector) neurons respond to visual input to trigger the jump that precedes flight; other pathways respond to sound or other stimuli (80). In flies, a similar, prominent pair of neurons has been described in several species. These giant interneurons receive input from the head (including visual and mechanical) and terminate in the fused thoracic ganglion (2). The neurons seem to initiate the flies' take off jump in preparation for flight, which serves as escape for these insects (120).

Another example of a neural circuit is that which controls jumping and kicking in locusts. To execute either behavior, a locust first flexes its hind tibiae, then strongly cocontracts both extensor and flexor muscles of the tibia, and finally, inhibits the flexor muscle activity, leaving the powerful extensor muscle unopposed in rapidly extending the tibia (46). The main difference between jumping and kicking seems to be the angle of the femur of each of the hind legs. A pair of identified interneurons (C neurons) is at least partly responsible for the cocontraction of extensor and flexor muscles (81). Another pair, the M neurons, are multimodal, receiving strong input from a variety of sensory modalities, including the visual

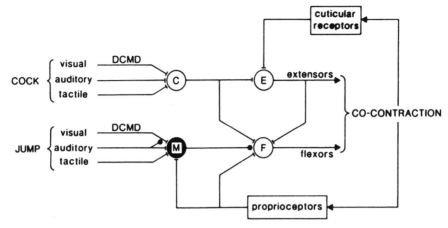

Figure F-7. Neural circuit for jumping in the locust. C neurons strongly excite both extensor (E) and flexor (F) motorneurons in the hindlegs. Then, strong input from the proprioceptors in the legs are required to fire the M neuron that in turn inhibits the flexor motorneurons, allowing the leg to extend due to the action of the extensors. Additional input from elsewhere in the CNS is important in triggering the jump by exciting the M neuron. [From Pearson (76).]

system (via DCMD), the auditory system, and tactile mechanoreceptors on the body (79). The M neurons supply powerful inhibition to the flexor motorneurons, and therefore have been called the trigger neurons for the jump (76). The M neurons are inhibited during the cocontraction phase, and the trigger for the jump is probably generated by several higher order neurons that turn off this inhibition (Fig. F-7) (42). However, the system is complex. It had been postulated that development of suitable levels of tension in the extensor tibialis muscles of the legs was a prerequisite to jumping–kicking. However, kicks can be induced even if tension is experimentally reduced (45).

The neural control of any behavior is complex (33). The flight control system, for example, involves dozens of neurons in several ganglionic areas. Many of the interneurons involved in producing the flight pattern have their cell somata in the parts of the metathoracic ganglion that represent the embryologically fused abdominal ganglia 1, 2, and 3. These neurons show strong serial homology (94). It

has been suggested that this distribution of neurons reflects an origin of wings from pleural appendages carried by both thoracic and abdominal segments in primitive insects, rather than from paranotal lobes, which were confined to the thorax (97). The networks of neurons that are present in the animals that serve as experimental subjects today are the result of considerable evolutionary modification, sometimes away from their original purpose. Therefore, we should not think that a neural circuit is always constructed in the most functionally efficient way (33).

Higher Order Control

The neural circuits described in the previous section are mostly confined to one or a few ganglia. Each circuit acts to generate the appropriate motor output for the expression of a coordinated sequence of behavior. As a rule, circuits operate almost exclusively at a local ganglionic level, except for interganglionic coordination pathways. Local circuits are controlled by higher centers. All behavior is ultimately under the control of the animal's brain [the main ganglionic mass(es) in the head]. Individual "command" fibers were once believed to control behavior (61, 131), but now control is thought to be exerted via groups of axons. In locusts, such a group of what has been termed "recommendation" fibers regulates walking (determining speed, direction, turning, etc.) (56). Control of abdominal posture in crayfish is effected by a similar means, although in this system the fibers are referred to as "command elements" (55).

Selection, initiation, maintenance, vigor, and duration are all aspects of behavior that are controlled by the "brain." If an orientational component is also possible, as in walking or flying, that also is regulated by the brain. If one considers the brain the main neural mass in the head, then arthropods actually have a two part brain, consisting of the superesophageal ganglion and the subesophageal ganglion. Functionally, this description is more sensible than the traditional terminology, which reserves the word brain for the superesophageal ganglion only. Both head ganglia play important roles in regulating behavior in insects (70). The same is presumably true in other arthropods. Very little detail is available about the organization or function of the head ganglia. In general, the superesophageal ganglion in insects has an inhibitory effect on the subesophageal ganglion, which in turn has an excitatory effect on many behaviors. This can be shown by the loss of spontaneous behavior when the head of an insect is removed, and the hyperexcitability of the animal when only the circumesophageal connectives are cut (30, 99). A number of identified descending interneurons have been described in locusts whose activity can initiate, influence, or terminate particular behavioral acts, but little is known about how they are able to perform these functions (57, 85, 132).

Among the complex behavior shown by arthropods is learning. Most studies of arthropod learning have been strictly behavioral (24), but Horridge (50) developed a preparation that holds the potential for cellular analysis (14). The idea is to train the animal to maintain a particular leg position by using an aversive (electrical shock or noise) or other (food or heat) stimulus, and study the CNS effects that accompany the behavioral change. Horridge's original experiments were done on cockroaches. Since then other insects [locust (37, 52) and weta (54)] and crustacea [ghost crab (51)] have been shown to be trainable as well. Individual animals can be trained to maintain a flexed or extended position of a leg within only a few minutes under some circumstances. In locusts, the pacemaker activity of a leg-muscle motorneuron can be influenced by the training procedure. An electrical shock to the leg nerve when the frequency of firing of the motorneuron was above a preset maximum resulted in long-term reduction in pacemaker activity. Delivering shock when the frequency fell below a minimum resulted in an increase in pacemaker activity (139).

Hormonal and Neuromodulatory Effects

Some synaptic transmitter substances can have effects on the nervous system that go far beyond the traditional short-term, local electrical actions normally associated with chemical transmitters. For example, injection of octopamine or serontonin into the abdomen of an unrestrained lobster produces a characteristic static posture; octopamine produces hyperextention of the abdomen and the extremities, whereas serotonin produces a strong flexion of the abdomen and the limbs (66). The two substances are normally released directly into the hemolymph by different neurons (67). Octopamine and serotonin have direct peripheral (muscle) effects, but there the effects are both excitatory rather than antagonistic to each other (36); peripheral effects can therefore enhance but do not cause the behavioral reactions.

Application of octopamine to an isolated CNS with attached abdominal muscles results in a pattern of activity in motorneurons that would lead to abdominal extension in an intact lobster. Serotonin tends to produce patterns associated with flexion, although this effect is more variable (44). In the presence of octopamine or serotonin, stimulation in an abdominal nerve cord of interneurons that cause patterns of abdominal extension or flexion yields greatly augmented responses in the abdominal muscles; octopamine enhances extension activity and serotonin enhances flexion activity (43). In addition, octopamine increases the responsiveness of the crayfish lateral giant fiber to sensory stimuli, whereas serotonin depresses it (38). These observations suggest that serotonin and octopamine act to bias the behavior of the animal toward specific responses when the animal encounters certain sensory stimuli (60).

Even more dramatic effects result from

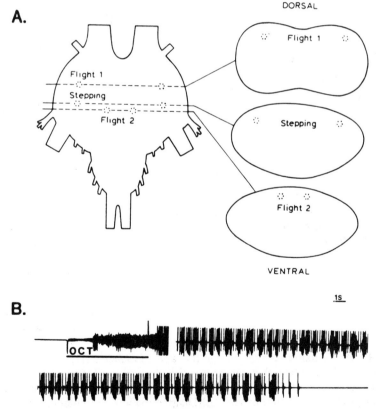

Figure F-8. (*A*) Sites in the metathoracic ganglion of a locust at which "stepping" or "flight" motor patterns can be evoked following iontophoresis of octopamine. (*B*) Example of bout of rhythmic flexion activity evoked by release of octopamine at sites shown above. Records are extracellular from the flexor tibiae muscle. [From Sombati and Hoyle (113).]

the focal application of octopamine in an insect. Microelectrode injection into small regions of the neuropile of the metathoracic ganglion of a locust evoke unilateral stepping movements or flight, depending on precisely where the injection is made (113). The areas in which injection is effective are sharply defined; movement of the microelectrode by as little as 5 μm can cause a loss of the behavioral response (Fig. F-8). Octopamine also has a strong dishabituating effect on conditioned reflexes (114). The interneuronal dorsal unpaired median (DUM) neurons are octopaminergic, branch extensively to specific ganglionic regions, and often fire just before a specific behavior. For these

reasons, it has been suggested that locusts select and organize (orchestrate) their behavior via action of these DUM neurons, using octopamine as a neuromodulator to turn on specific neural circuits (53).

Neuromodulators are released by what appear to be typical neurons, yet act much like hormones. But even classical hormones may have strong effects on the nervous system. In insects, this has been shown clearly in the mechanisms by which molting behavior is controlled (124). In the holometabolous insects, the animal not only changes bodily form from a larval stage through a pupa to the adult, it also changes its behavior. Caterpillars

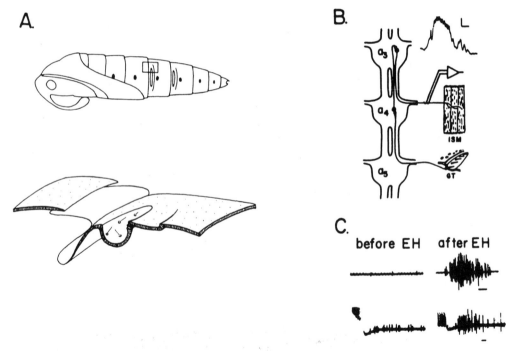

Figure F-9. (*A*) diagram of the gin trap in the larva of *Manduca sexta*. The lower diagram shows the pit in the cuticle that constitutes the gin trap itself. Stimulation of the hairs in the pit causes the cuticle on the left to slide over the pit, trapping or crushing the object that produced the stimulation. (*B*) Diagrammatic representation of the gin trap circuitry in the pupa CNS, and the muscle response that is evoked by stimulation of the gin trap hairs. (*C*) the effect of ecdysone on the motorneurons in the gin trap reflex in a caterpillar just ready to molt to a pupa. The reflex is not yet active at this time (left), but after addition of EH, stimulation of the sensory hairs evokes a strong response. (*A*) Courtesy of J. H. Cocatre-Zilgien. (*B* and *C*) From Levine and Truman (65).

can crawl, moths can fly, and pupae can do neither. When a pupa is physically ready to molt, it undergoes eclosion, the shedding of its old cuticle, under the influence of eclosion hormone. Eclosion hormone acts directly on the pupal CNS of a silkmoth (*Hyalophora cecropia*) or tobacco hawkmoth (*Manduca sexta*) to turn on eclosion behavior, which is a stereotyped series of rotational and peristaltic movements (123, 125). The hormone also activates other behavior. In Cecropia, the ability of the insect to express adult behavior is turned on and the ability to express pupal behavior is turned off by the presence of eclosion hormone even for quite short (10–15 min) periods of time (122). The effect is mediated by cyclic GMP, which acts as a second messenger to phosphorylate two identified proteins. The proteins, in turn, appear in neurons sensitive to eclosion hormone after they are exposed to ecdysone (molting hormone) (71).

Much of the behavior of an insect is stage specific. In *Manduca*, pupae exhibit a specific behavior, the gin-trap reflex. It is evoked by stimulation of hairs in a pit-like depression in the body wall between two segments (Fig. F-9A). The response is a strong movement of the segments together, which crushes any small animal in the pit. Larvae do not have gin traps. However, the hairs that invest the gin trap are present on caterpillars; they undergo some modification during metamorphosis, as do the sensory neurons that innervate them (64). Stimulation of the sensory hairs in caterpillars causes a weak, diffuse withdrawal response, not the strong, localized response of the gin trap reflex. The neural circuitry that allows behavioral responses to touch of the sensory hairs must therefore be modified during metamorphosis. This modification of the neural circuit, along with other pupal morphological modifications, must be complete before the molt to the pupal

stage takes place. At this time, the gin trap reflex is not active. Only after the CNS has been exposed to eclosion hormone is the reflex operational (Fig. F-9B and C) (65). It is hypothesized that the activation of the reflex by eclosion hormone is the result of *in*activation of an inhibitory interneuron that after physical metamorphosis but before molting, suppresses contraction of the muscles that close the gin trap.

References

1. Atwood, H. L. In *The Biology of Crustacea* (H. L. Atwood and D. C. Sandeman, eds.), vol. 3, pp. 105–150. Academic Press, New York, 1982. Synapses and neurotransmitters.

2. Bacon, J. P. and N. J. Strausfeld. *J. Comp. Physiol. A* **158**:529–548, 1986. The dipteran 'Giant fibre' pathway.

3. Bentley, D. *J. Comp. Physiol.* **116**:19–38, 1977. Interneurons and cricket song.

4. Bowerman, R. F. *Comp. Biochem. Physiol.* **56A**:231–247, 1977. The control of arthropod walking.

5. Boyan, G. S., S. Ashman, and E. E. Ball. *Naturwissenschaften* **73**:272–274, 1986. Flight initiation and modulation by a giant interneuron in locusts.

6. Boyan, G. S. and E. E. Ball. *J. Comp. Physiol. A* **159**:773–789, 1986. Wind-sensitive interneurones in the terminal ganglion of praying mantids.

7. Bullock, T. H. and G. A. Horridge. *Structure and Function in the Nervous Systems of Invertebrates.* Freeman, San Francisco, CA, 1965.

8. Burrows, M. *J. Neurophysiol.* **42**:1108–1123, 1979. Graded interactions between nonspiking interneurons in locusts.

9. Burrows, M. *J. Physiol. (London)* **298**:213–233, 1980. Local interneurons and motorneurons in the locust.

10. Burrows, M. In *Model Neural Networks and Behavior* (A. I. Selverston, ed.), pp. 109–125. Plenum, New York, 1985. Lo-

cust nonspiking and spiking local interneurons.

11. Burrows, M. *J. Neurosci.* **7**:1064–1080, 1987. Processing of proprioceptive signals by spiking local interneurons in the locust.

12. Burrows, M. *J. Comp. Physiol. A* **164**:207–217, 1988. Input from the femoral chordotonal organ to spiking local interneurons in locusts.

13. Bush, B. M. H. In *Neurones Without Impulses* (A. Roberts and B. M. H. Bush, eds.), pp. 147–176. Cambridge Univ. Press, London and New York, 1981. Non-impulsive stretch receptors in crustaceans.

14. Byrne, J. H. *Physiol. Rev.* **67**:329–439, 1987. Cellular analysis of learning.

15. Callec, J. J. In *Comprehensive Insect Physiology Biochemistry and Pharmacology* (G. A. Kerkut and L. I. Gilbert, eds.), Vol. 5, pp. 139–180. Pergamon, Oxford, 1985. Synaptic transmission in CNS.

16. Camhi, J. M. and A. Levy. *J. Comp. Physiol. A* **163**:317–328, 1988. Escape behavior of the cockroach.

17. Camhi, J. M. and W. Tom. *J. Comp. Physiol.* **128**:193–201, 1978. Escape behavior of the cockroach.

18. Camhi, J. M., W. Tom, and S. Volman. *J. Comp. Physiol.* **128**:203–212, 1978. Detection of predators by the cockroach.

19. Chapple, W. D. *J. Exp. Biol.* **45**:65–80, 1966. Motor system asymmetry in the hermit crab.

20. Clarac, F. In *Feedback and Motor Control in Invertebrates and Vertebrates* (W. J. P. Barnes and M. H. Gladden, eds.), pp. 379–400. Croom Helm, London, 1985. Reflexes and walking in crustacea.

21. Comer, C. M. and J. P. Dowd. *J. Comp. Physiol. A* **160**:571–583, 1987. Effect of CNS lesions on turning behavior of the cockroach.

22. Comer, C. M., J. P. Dowd, and G. T. Stubblefield. *Brain Res.* **445**:370–375, 1988. Effect of giant fiber elimination on escape in cockroaches.

23. Cook, P. M. *Q. J. Microsc. Sci.* **92**:297–305, 1951. Giant fibres in locusts.

24. Corning, W. C., J. A. Dyal, and A. O. D. Willows, eds. *Invertebrate Learning*, Vol. 2. Plenum, New York, 1973.

25. Daley, D. L. *Brain Res.* **238**:211–216, 1982. Cercal hairs and wind directionality of giant interneurons in cockroaches.

26. Daley, D. L. and J. M. Camhi. *J. Neurophysiol.* **60**:1350–1368, 1988. Connectivity between cercal input and giant fibers in cockroach.

27. Daley, D. L. and F. Delcomyn. *J. Comp. Physiol.* **138**:231–239, 1980. Modulation of the excitability of cockroach giant interneurons during walking.

28. Daley, D. L., N. Vardi, B. Appignani, and J. M. Camhi. *J. Comp. Neurol.* **196**:41–52, 1981. Morphology of cockroach giant interneurons.

29. Delcomyn, F. *Science* **210**:492–498, 1980. Neural basis of rhythmic behavior in animals.

30. Delcomyn, F. In *Comprehensive Insect Physiology Biochemistry and Pharmacology* (G. A. Kerkut and L. I. Gilbert, eds.), Vol. 5, pp. 439–466. Pergamon, Oxford, 1985. Walking and running.

31. Delcomyn, F. and D. L. Daley. *J. Comp. Physiol.* **130**:39–48, 1979. Central excitation of cockroach giant interneurons during walking.

32. Dickinson, P. S., F. Nagy, and M. Moulins. *J. Exp. Biol.* **136**:53–87, 1988. Control of two stomatogastric CPGs by a single neuron.

33. Dumont, J. P. C. and M. Robertson. *Science* **233**:849–853, 1986. Neuronal circuits: an evolutionary perspective.

34. Dumont, J. P. C. and J. J. Wine. *J. Exp. Biol.* **127**:249–277, 279–294, 295–311, 1987. The telson flexor neuromuscular system of the crayfish.

35. Edwards, J. S. and G. R. Reddy. *J. Comp. Neurol.* **243**:535–546, 1986. Giant interneurons in the firebrat.

36. Fischer, L. and E. Florey. *J. Exp. Biol.* **102**:187–198, 1983. Effect of serotonin and octopamine on crayfish muscle.

37. Forman, R. R. *J. Neurobiol.* **15**:127–140,

1984. Leg position learning by an insect using heat.

38. Glanzman, D. L. and F. B. Krasne. *J. Neurosci.* **3**:2263–2269, 1983. Effect of serotonin and octopamine on the crayfish escape reaction.

39. Gnatzy, W. and R. Heusslein. *Naturwissenschaften* **73**:212–214, 1986. Digger wasp antipredator strategies against crickets.

40. Graham, D. and U. Bässler. *J. Exp. Biol.* **91**:179–193, 1981. Effects of afference sign reversal on motor activity in walking stick insects.

41. Graubard, K. *J. Neurophysiol.* **41**:1014–1025, 1978. Input-output properties of a non-spiking presynaptic neuron.

42. Gynther, I. C. and K. G. Pearson. *J. Exp. Biol.* **122**:323–343, 1986. Neural mechanisms of jump control in the locust.

43. Harris-Warrick, R. M. *J. Comp. Physiol. A* **156**:875–884, 1985. Amine modulation of motor activity in the lobster abdomen.

44. Harris-Warrick, R. M. and E. A. Kravitz. *J. Neurosci.* **4**:1976–1993, 1984. Modulation of posture by serotonin and octopamine in the lobster.

45. Heitler, W. J. and P. Bräunig. *J. Exp. Biol.* **136**:289–309, 1988. Manipulation of motor activity in the extensor tibialis muscle during kicking in locusts.

46. Heitler, W. J. and M. Burrows. *J. Exp. Biol.* **66**:203–219, 1977. The motor programme for the locust jump.

47. Heitler, W. J. and S. Darrig. *J. Exp. Biol.* **121**:55–75, 1986. The segmental giant neurone of the crayfish.

48. Heitler, W. J. and K. Fraser. *J. Exp. Biol.* **125**:245–269, 1986; **133**:353–370, 1987. The segmental giant neuron of the hermit crab.

49. Heitler, W. J. and K. G. Pearson. *Brain Res.* **187**:206–211, 1980. Nonspiking interneurones in the crayfish swimmeret system.

50. Horridge, G. A. *Proc. R. Soc. London, Ser. B* **157**:33–52, 1962. Learning leg position in cockroaches.

51. Hoyle, G. *Behav. Biol.* **18**:147–163, 1976. Leg position learning in the ghost crab.

52. Hoyle, G. *J. Neurobiol.* **11**:323–354, 1980. Leg position conditioning in the locust.

53. Hoyle, G. In *Comprehensive Insect Physiology Biochemistry and Pharmacology* (G. A. Kerkut and L. I. Gilbert, eds.), Vol. 5, pp. 607–622. Pergamon, Oxford, 1985. Role of octopamine in generation of motor activity in locusts.

54. Hoyle, G. and L. H. Field. *J. Neurobiol.* **14**:285–298, 1983. Leg position learning in the weta.

55. Jellies, J. and J. L. Larimer. *J. Comp. Physiol. A* **156**:861–873, 1985. Motor control of abdominal posture in crayfish.

56. Kien, J. *Proc. R. Soc. London, Ser. B* **219**:137–174, 1983. An alternative to the command concept in walking locusts.

57. Kien, J. and J. S. Altman. *J. Insect Physiol.* **30**:59–72, 1984. Control of locust behavior by descending interneurones.

58. Kramer, A. P. and F. B. Krasne. *J. Neurophysiol.* **52**:189–211, 1984. Production of tailflips without giant fiber activity in crayfish.

59. Krasne, F. B. and J. J. Wine. In *Neural Mechanisms of Startle Behavior* (R. C. Eaton, ed.), pp. 179–211. Plenum, New York, 1984. The production of crayfish tailflip escape responses.

60. Kravitz, E. A. *Science* **241**:1775–1780, 1988. Hormonal control of behavior in lobsters.

61. Kupfermann, I. and K. R. Weiss. *Behav. Brain Sci.* **1**:3–39, 1978. The command neuron concept.

62. Laurent, G. J. and M. Burrows. *J. Comp. Physiol. A* **162**:563–572, 1988. Excitation of nonspiking neurons by hair or campaniform sensillar input in locusts.

63. Leslie, R. A. *Z. Zellforsch. Mikrosk. Anat.* **145**:299–309, 1973. Fine structure of interganglionic connectives in the newly hatched and adult stick insect.

64. Levine, R. B., C. Pak, and D. Linn. *J. Comp. Physiol. A* **157**:1–13, 1985. The metamorphic reorganization of sensory neurons in *Manduca sexta* larvae.

65. Levine, R. B. and J. W. Truman. *Brain Res.* **279**:335–338, 1983. Peptide activation of a simple neural circuit.

66. Livingstone, M. S., R. M. Harris-Warrick, and E. A. Kravitz. *Science* **208**:76–79, 1980. Serotonin and octopamine produce opposite postures in lobsters.

67. Livingstone, M. S., S. F. Schaeffer, and E. A. Kravitz. *J. Neurobiol.* **12**:27–54, 1981. Serotonergic nerve endings in the lobster.

68. Miller, J. P. and A. I. Selverston. In *Model Neural Networks and Behavior* (A. I. Selverston, ed.), pp. 37–48. Plenum, New York. 1985. Neural mechanisms of the lobster pyloric motor pattern.

69. Miller, L. A., G. Hagiwara, and J. J. Wine. *J. Neurophysiol.* **53**:252–265, 1985. Segmental differences in pathways connecting crayfish giant axons and flexor motorneurons.

70. Mobbs, P. G. In *Comprehensive Insect Physiology Biochemistry and Pharmacology* (G. A. Kerkut and L. I. Gilbert, eds.), Vol. 5, pp. 299–370. Pergamon, Oxford, 1985. Brain structure.

71. Morton, D. B. and J. W. Truman. *J. Neurosci.* **8**:1338–1345, 1988; *Nature* (*London*) **323**:264–267, 1986; *J. Comp. Physiol. A* **157**:423–432, 1985. cGMP-mediated activation of neurons.

72. Nunnemacher, R. F., W. J. Fiske, and R. G. Sherman. *J. Insect Physiol.* **20**:2123–2134, 1974. Axon counts in connectives of the cockroach *Blaberus craniifer*.

73. Paul, D. H. *J. Exp. Biol.* **141**:257–264, 1989. CPG output to a sense organ during a rhythmic behavior.

74. Paul, D. H. and B. Mulloney. *J. Neurophysiol.* **54**:28–39, 1985. Nonspiking local interneurons controlling the crayfish swimmeret system.

75. Paul, D. H. and B. Mulloney. *J. Comp. Physiol. A* **158**:215–224, 1986. Intersegmental coordination of swimmeret rhythms in isolated nerve cords of crayfish.

76. Pearson, K. G. *J. Physiol.* (*Paris*) **78**:765–771, 1983. Neural circuits for jumping in the locust.

77. Pearson, K. G. In *Feedback and Motor Control in Invertebrates and Vertebrates* (W. J. P. Barnes and M. H. Gladden, eds.), pp. 307–315. Croom Helm, London, 1985. Are there central pattern generators for walking and flight in insects?

78. Pearson, K. G. and C. R. Fourtner. *J. Neurophysiol.* **38**: 33–52, 1975; Fourtner, C. R. In *Neural Control of Locomotion* (R. M. Herman, S. Grillner, P. S. G. Stein, and D. G. Stuart, eds.), pp. 401–418. Plenum, New York, 1976. Nonspiking interneurons and CNS control of cockroach walking.

79. Pearson, K. G., W. J. Heitler, and J. D. Steeves. *J. Neurophysiol.* **43**:257–278, 1980. Triggering of locust jump by multimodal inhibitory interneurons.

80. Pearson, K. G. and M. O'Shea. In *Neural Mechanisms of Startle Behavior* (R. C. Eaton, ed.), pp. 163–178. Plenum, New York, 1984. The locust jump and its initiation by visual stimuli.

81. Pearson, K. G. and R. M. Robertson. *J. Comp. Physiol.* **144**:391–400, 1981. Interneurons coactivating hindleg flexor and extensor motoneurons in the locust.

82. Pearson, K. G., D. N. Reye, and R. M. Robertson. *J. Neurophysiol.* **49**:1168–1181, 1983. Sensory feedback effects on flying in locusts.

83. Pichon, Y. and F. M. Ashcroft. In *Comprehensive Insect Physiology Biochemistry and Pharmacology* (G. A. Kerkut and L. I. Gilbert, eds.), Vol. 5, pp. 49–84. Pergamon, Oxford, 1985. Nerve and muscle: Electrical activity.

84. Pumphrey, R. J. and A. F. Rawdon-Smith. *Proc. R. Soc. London* **122**:106–118, 1937. Synaptic transmission of nerve impulses through the last abdominal ganglion of the cockroach.

85. Ramirez, J.-M. *J. Comp. Physiol. A* **162**:669–685, 1988. Flight-initiating interneurons in locusts.

86. Reichert, H., M. R. Plummer, and J. J. Wine. *J. Comp. Physiol.* **151**:261–276, 1983. Nonspiking interneurons and lat-

eral inhibition of crayfish mechanosensory interneurons.

87. Reichert, H. and J. J. Wine. *J. Comp. Physiol.* **153**:3–15, 1983. Coordination of lateral giant and non-giant systems in crayfish escape behavior.

88. Reye, D. N. and K. G. Pearson. *J. Comp. Physiol. A* **162**:77–89, 1988. Entrainment of the flight CPG by stretch receptor activity in locusts.

89. Ritzmann, R. E. and A. J. Pollack. *J. Comp. Physiol.* **143**:61–70, 1981. Motor responses to paired stimulation of giant interneurons in the cockroach.

90. Ritzmann, R. E. and A. J. Pollack. *J. Comp. Physiol. A* **159**:639–654, 1986. Thoracic interneurons that mediate giant interneuron-to-motor pathways in the cockroach.

91. Ritzmann, R. E. and A. J. Pollack. *J. Neurobiol.* **19**:589–611, 1988. Wind-activated thoracic interneurons in cockroaches.

92. Ritzmann, R. E., A. J. Pollack, and M. L. Tobias. *J. Comp. Physiol.* **147**:313–322, 1982. Flight evoked by giant fiber stimulation in cockroaches.

93. Roberts, A. and B. M. H. Bush, eds. *Neurones Without Impulses.* Cambridge Univ. Press, London and New York, 1981.

94. Robertson, R. M. and K. G. Pearson. *J. Comp. Neurol.* **215**:33–50, 1983. Properties of interneurons in the flight system of the locust.

95. Robertson, R. M. and K. G. Pearson. In *Model Neural Networks and Behavior* (A. I. Selverston, ed.), pp. 21–35. Plenum, New York, 1985. Neural networks controlling locomotion in locusts.

96. Robertson, R. M. and K. G. Pearson. *J. Neurophysiol.* **53**:110–128, 1985. Neural circuits in the flight system of the locust.

97. Robertson, R. M., K. G. Pearson, and H. Reichert. *Science* **145**:177–179, 1982. Flight interneurons in the locust and the origin of insect wings.

98. Roeder, K. D. *J. Exp. Zool.* **108**:243–261, 1948. The giant fiber system in the American cockroach.

99. Roeder, K. D. *Nerve Cells and Insect Behavior.* Harvard Univ. Press, Cambridge, MA, 1967.

100. Ronacher, B., H. Wolf, and H. Reichert. *J. Comp. Physiol. A* **163**:749–759, 761–769, 1988. Effect of ganglionic hemisection on synaptic input to flight motorneurons and on flight in locusts.

101. Rowell, C. H. F. and A. E. Dorey. *Z. Zellforsch. Mikrosk. Anat.* **83**:288–294, 1967. The number and size of axons in the thoracic connectives of the locust.

101a. Selverston, A. I. and B. Mulloney, *J. Comp. Physiol.* **91**:33–51, 1974. Organization of the stomatogastric ganglion of the spiny lobster. II. Neurons driving the medial tooth.

102. Selverston, A. I. In *Identified Neurons of Arthropods* (G. Hoyle, ed.) Chap. 12, pp. 209–225. Plenum, New York, 1977; in *Simpler Networks and Behavior* (J. Fentress, ed.) Chap. 6, pp. 82–98. Sinauer, Sunderland, MA, 1979; J. L. Ayers, Jr. and A. I. Selverston *J. Comp. Physiol.* **129**:5–17, 1979; A. I. Selverston, and J. P. Miller *J. Neurophysiol.* **44**:1102–1121, 1980; M. Gola and A. I. Selverston *J. Comp. Physiol.* **145**:191–207, 1981; J. P. Miller and A. I. Selverston *J. Neurophysiol.* **48**:1416–1432, 1982. Network analysis of patterned discharge in stomatogastric ganglion of *Panulirus*.

102a. Selverston, A. I., J. P. Miller, and M. Wadepuhl, *Symp. Soc. Exp. Biol.* **37**:55–87, 1983. Cooperative mechanisms for the production of rhythmic movements.

103. Selverston, A. I. and M. Moulins, eds. *The Crustacean Stomatogastric System.* Springer-Verlag, Berlin and New York, 1987.

104. Shaw, S. R. In *Neurones Without Impulses* (A. Roberts and B. M. H. Bush, eds.), pp. 61–116. Cambridge Univ. Press, London and New York, 1981. Anatomy and physiology of non-spiking cells in the eyes of Diptera.

105. Siegler, M. V. S. *J. Neurophysiol.* **46**:310–323, 296–309, 1981. Nonspiking neurons and postural adjustments in locusts.

106. Siegler, M. V. S. *J. Exp. Biol.* **112**:253–281, 1984. Local interneurones and local interactions in arthropods.

107. Siegler, M. V. S. and M. Burrows. *J. Neurophysiol.* **50**:1281–1295, 1983. Spiking local interneurons and processing of mechanosensory information in the locust.

108. Siegler, M. V. S. and M. Burrows. *J. Neurosci.* **6**:507–513, 1986. Receptive fields of motor neurons in the locust.

109. Sillar, K. T., F. Clarac, and B. M. H. Bush. *J. Exp. Biol.* **131**:245–264, 1987. Interneuronal activity during walking in crayfish.

110. Simmers, A. J. In *Neurones Without Impulses* (A. Roberts and B. M. H. Bush, eds.), pp. 177–198. Cambridge Univ. Press, London and New York, 1981. Nonspiking interactions in crustacean rhythmic motor systems.

111. Simmers, A. J. and B. M. H. Bush. *Brain Res.* **197**:247–252, 1980. Nonspiking neurones controlling ventilation in crabs.

112. Skobe, Z. and R. F. Nunnemacher. *J. Comp. Neurol.* **139**:81–92, 1970. Axon count in the circumesophageal nerve in decapod crustaceans.

113. Sombati, S. and G. Hoyle. *J. Neurobiol.* **15**:481–506, 1984. Generation of specific behaviors in a locust by stimulation via octopamine.

114. Sombati, S. and G. Hoyle. *J. Neurobiol.* **15**:455–480, 1984. Sensitization and dishabituation in an insect by the neuromodulator octopamine.

115. Spira, M. E., I. Parnas, and F. Bergmann. *J. Exp. Biol.* **50**:615–627, 1969. Organization of the giant axons of the cockroach.

116. Stephens, P. J. *J. Neurobiol.* **16**:361–372, 1985. Morphology and physiology of the giant interneuron of the hermit crab.

117. Stephens, P. J. *J. Exp. Biol.* **123**:217–228, 1986. Synapses between giant interneurons and giant flexor motor neurones of the hermit crab.

118. Stevenson, P. A. and W. Kutsch. *J. Comp. Physiol. A* **161**:115–129, 1987. Reconsideration of the central pattern generator concept for locust flight.

119. Sutherland, R. M. and R. F. Nunnemacher. *J. Comp. Neurol.* **132**:499–512, 1968. Microanatomy of crayfish thoracic cord and roots.

120. Tanouye, M. A. and D. G. King. *J. Exp. Biol.* **105**:241–251, 1983. Giant fibre activation of direct flight muscles in *Drosophila*.

121. Titova, V. A. *J. Evol. Biochem. Physiol.* **11**:378–380, 1976. Fiber count in the cord of the crayfish.

122. Truman, J. W. *J. Comp. Physiol.* **107**:39–48, 1976. Development and hormonal release of adult behavior patterns in silkmoths.

123. Truman, J. W. *J. Exp. Biol.* **74**:151–173, 1978. Hormonal release of motor programmes the isolated cecropia nervous system.

124. Truman, J. W. and L. M. Riddiford. *Adv. Insect Physiol.* **10**:297–352, 1973. Hormonal mechanisms underlying insect behaviour.

125. Truman, J. W. and J. C. Weeks. In *Model Neural Networks and Behavior* (A. I. Selverston, ed.), pp. 381–399. Plenum, New York, 1985. Activation of neuronal circuits by circulating hormones in insects.

126. Weidler, D. J., P. J. Gardner, A. M. Earle, and G. G. Myers. *Comp. Biochem. Physiol.* **44A**:807–812, 1973. Anatomy and ionic content of *Melanoplus differentialis* nerve cord.

127. Westin, J., J. J. Langberg, and J. M. Camhi. *J. Comp. Physiol.* **121**:307–324, 1977. The effects of wind puffs on giant interneurons of the cockroach.

128. Westin, J. and R. E. Ritzmann. *J. Neurobiol.* **13**:127–139, 1982. The effect of single giant interneuron lesions on wind-evoked motor responses in the cockroach, *Periplaneta americana*.

129. Westin, J., R. E. Ritzmann, and D. J. Goddard. *J. Neurobiol.* **19**:573–588, 1988. Response of thoracic interneurons to wind puffs in cockroaches.

130. Wiersma, C. A. G. In *Physiology of Crustaceans* (T. Waterman, ed.), Vol. 2, pp.

241–279. Academic Press, New York, 1961. Central nervous systems, crustacea.

131. Wiersma, C. A. G. and K. Ikeda. *Comp. Biochem. Physiol.* **12:**509–525, 1964. Interneurons commanding swimmeret movements in the crayfish, *Procambarus clarkii* (Girard).

132. Williamson, R. and M. D. Burns. *J. Comp. Physiol.* **147:**379–388, 1982. Large neurones in locust neck connectives.

133. Wilson, D. M. In *Neural Theory and Modeling* (R. R. Reiss, ed.), pp. 331–345. Stanford Univ. Press, Stanford, CA, 1964. The origin of the flight-motor command in grasshoppers.

134. Wilson, J. A. and C. E. Phillips. *Prog. Neurobiol.* **20:**89–107, 1983. Premotor non-spiking interneurons.

135. Wine, J. J. and F. B. Krasne. *J. Exp. Biol.* **56:**1–18, 1972. The organization of escape behaviour in the crayfish.

136. Wine, J. J. and F. B. Krasne. In *The Biology of Crustacea* (D. C. Sandeman and H. L. Atwood, eds.), Vol. 4, pp. 241–292. Academic Press, New York, 1982. The cellular organization of crayfish escape behavior.

137. Wolf, H. and K. G. Pearson. *J. Comp. Physiol. A* **160:**269–279, 1987. Motor patterns in the intact and deafferented flight system of the locust.

138. Wolf, H. and K. G. Pearson. *J. Comp. Physiol. A* **161:**103–114, 1987. Flight motor patterns in isolated nerve cord of *Locusta migratoria*.

139. Woollacott, M. and G. Hoyle. *Proc. R. Soc. London, Ser. B* **195:**395–415, 1977. Neural events underlying learning in insects.

Section G
Vertebrate Nervous Systems

C. Ladd Prosser and
William T. Greenough

Introduction; Anatomical Overview

The basic organization of the nervous system is similar in all vertebrates. The nervous system originates as a neural tube from which, during development, outpouchings become sensory structures or lobes of the brain. Vertebrates are bilaterally symmetrical, but in some vertebrates, one side of the brain becomes dominant over the opposite side for certain functions. In general, sensory nerves are dorsal, motor nerves are ventral. Related to bilaterality is decussation of both sensory and motor pathways. Spinal pathways mostly cross above the spinal cord. In some vertebrates the visual pathway from retina is partly crossed, in others it is completely crossed. Auditory pathways have both crossed and uncrossed fibers and crossings occur at all levels in the ascending pathway. In contrast, olfactory and gustatory paths are not crossed. Motor pathways, particularly the pyramidal neurons from motor cortex, are crossed.

A morphological difference between vertebrate and most invertebrate nervous systems is the arrangement of white and gray matter fiber tracts and somata. In vertebrates, the fiber tracts are mostly peripheral to the cell bodies; in many invertebrate ganglia the nerve cells are around the outside, the fibers inside (neuropil).

Cephalization, formation of dominant anterior regions, increases in the series from cyclostomes to mammals; sense organs become concentrated at the anterior end and integrative functions become established in rostral portions of the nervous system. By this arrangement conduction time from receptors to integrative centers is minimized.

The relative sizes of homologous regions of brains of various animals are shown in Figure G-1 (drawn to different scales). In all vertebrates, the spinal cord mediates locomotor reflexes; a single segment or group of segments is capable of some autonomy. Segments isolated from the spinal cord of cyclostomes, turtles (and probably other vertebrates), are ca-

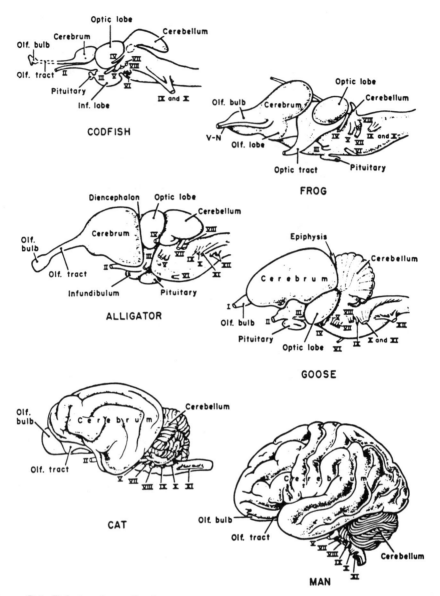

Figure G-1. Relative sizes of brain regions in several vertebrates. Brains drawn at different scales. [From Strong and Elwyn (145a).]

pable of rhythmic and coordinated loco-motor functions (fictive rhythms). Spinal cats are capable of walking-type movements, sex-receptive spinal rats show lordosis. These behaviors indicate longitudinal and crossed coordination.

The caudal part of the brain, metencephalon, has several centers for auto-nomic (vegetative) functions that differ little from class to class. These centers serve cardiac regulation, respiratory movement, regulation of digestive motility, modulation of blood pressure. The hindbrain also contains clusters of motorneurons that receive motor input from more rostral regions of the brain. The medulla has synaptic relays for sensory input via several cranial afferent nerves. One notable specialization in hindbrain is the very large vagal lobes of teleosts, centers for taste input. Electric fish have a prominent motor region for activation of electric organs.

The mesencephalon, or midbrain, has associative functions to varying degrees in different classes of vertebrates. The tectum or dorsal portion mediates visual responses in fishes and amphibians; the optic tectum has several layers of cells and synapses. In mammals, the tectum (superior colliculus) serves oculomotor responses and is a visual pathway in parallel with the retino-diencephalic path; the inferior colliculus serves as an auditory relay in mammals. The ventral midbrain includes the tegmentum, an important motor center.

Lying above the midbrain and medulla and connected to midbrain by axonal peduncles is the cerebellum, a structure with highly conserved circuitry in all vertebrate classes. The cerebellum sits as a modulator of sensorimotor traffic up and down the brainstem. Recent evidence indicates an important function of the cerebellum in some types of associatively conditioned responses. In addition, the cerebellum functions in an array of learned motor tasks, ranging from learned limb positions to vestibulo-ocular reflexes that can be modified by experience.

The diencephalon contains in its ventral part the hypothalamus, which governs several hormones of reproduction, growth, and water balance. Rostral to the hormone synthesizing and controlling regions are cells that function in reproductive behavior; anterior to this region is the thermoregulating center—behavioral in aquatic and terrestrial poikilotherms, metabolic-circulatory in homeotherms. The dorsal part of the diencephalon is the thalamus, which interacts with optic tectum in fishes and amphibians; in birds and mammals the thalamus contains integrative relays for vision (lateral geniculate and pulvinar) and audition (medial geniculate).

The telencephalon, or forebrain, has changed more than other regions of the brain during vertebrate evolution. In cyclostomes, elasmobranchs, and primitive teleosts the olfactory function occupies most of the forebrain (109, 110). The functions of nonolfactory portions of the forebrain of fishes are closely related to the diencephalon (136). In reptiles the pallium or outer layer becomes infolded into the ventricle and forms a dorsal ridge that receives auditory and visual projections. The pallial cortex of reptiles consists of a single layer of neurons (23). In birds, the infolded pallium is massive, the dorsal ventricular ridge (ectostriatum) receives auditory fibers from medial geniculate, and the dorsomedial cortex (Wulst) sends motor fibers to the spinal cord (125).

In mammals the pallium consists of (1) isocortex that contains (a) specific sensory areas that receive projections from thalamus and (b) motor areas with pyramidal neurons, (2) pyriform cortex, mainly olfactory, and (3) medial cortex or hippocampus. The amygdala of mammalian forebrain may be the homolog of the reptilian and avian paleostriatum. The limbic, nonolfactory part of the archipallium in reptiles and birds becomes in mammals the hippocampus and amygdaloid complex. In monotremes and marsupials the neocortex is in layers, differently organized from those in "higher" mammals; the mammalian neocortex contains six recognized layers, which in primates may be as much as 100 cells thick (23). Association areas are regions that neither receive direct sensory input nor elicit specific movements on stimulation; they contain many cortico-cortical connections. Each of the sensory areas—visual, auditory, and somesthetic—consists of several functionally differentiated regions. For example, the visual cortex of cat has 13 identifiable areas of visual function. Each sensory area has neurons arranged in columns, all cells in each column serve a similar function (24).

General Principles of Organization

Structural and Functional Properties

In addition to the anatomical principles listed in the introduction, other general organizational characteristics of vertebrate nervous systems are noteworthy.

1. Hierarchial Organization Is a Consequence of Cephalization of Sensory and Motor Functions. Integrative capacity has increased in the rostral direction during evolution from agnathans to mammals. In agnathans and cartilaginous fishes, the spinal cord can mediate near-normal swimming and visceral functions. By contrast, primates with high spinal transections lack coordination of limbs and are deficient in visceral control. In fishes and amphibians, the processing of visual patterns is primarily mesencephalic. In birds and mammals, diencephalon and forebrain are necessary for visual function. An example of hierarchial control is temperature regulation in birds and mammals; spinomedullary circuits carry out local control, preoptic-hypothalamic centers combine peripheral and preoptic temperature sense; the forebrain receives from multiple inputs and modulates the controls in lower centers.

2. Redundancy and Parallel Pathways Are Shown by Persistance of a Function after Several Fibers or a Brain Region Have Been Damaged. Several parallel neural routes transmit signals from retina to forebrain, each carrying slightly different information. There are several cortical projection areas for a given sense, a primary area and two or more secondary areas. These regions are not strictly equivalent but one can serve in the absence of others. Impairment of multisensory tasks such as maze learning by rats is proportionate to the amount of association cortex removed. Divergence and convergence usually characterize parallel pathways. In the retina the number of rods and cones exceeds the number of ganglion cells; there are more ganglion cells than neurons in lateral geniculate, but the number of visual cortical neurons greatly exceeds the LGN neurons; thus in the visual system convergence is succeeded by divergence. Parameters of a sensory message, for example from the lateral line system of fishes, are altered as a result of central convergence and divergence.

3. Feedback, Feedforward, and Gain Control Occur in All Vertebrate Nervous Systems. Feedback control is usually presynaptic, gain control by recurrent collaterals postsynaptic, usually with intervening interneurons. Neurons in a hierarchy exert feedback control from rostral to caudal levels, loops and circles are established in sensory pathways. For several modalities there is feedforward modulation by efferents to the receptors, for example, central control of hair cells in cochlea and lateral line. The efferent fibers from olivary nucleus to cochlea are crossed. An example of gain control by a collateral that acts via one or more interneurons is in Mauthner neurons (Fig. G8). Gain control via an identified interneuron is in motor regulation via the Renshaw cells of the spinal cord. An example of a feedback loop is the gamma loop of muscle spindles back to motorneurons.

4. Localization of Function in Integrative Centers Is More or Less Precise. In cerebral cortex of mammals there is localization by lobes—vision in occipital lobes, hearing in temporal lobes, somatosensory in postcentral gyrus of parietal lobes, motor

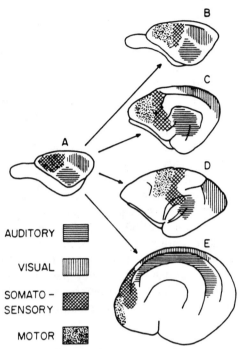

AUDITORY

VISUAL

SOMATO – SENSORY

MOTOR

Figure G-2. Schema showing sensory and motor areas projected onto the brain surface in a variety of mammals. (*A*) Hypothetical common ancestor; (*B*) rabbit, (*C*) cat; (*D*) rhesus monkey; (*E*) dolphin. In dolphin note displacement of visual and auditory areas onto dorsolateral convexity and lack of intervening cortex. [From Labygina et al. (78b).]

function in precentral gyrus of frontal lobes facing the somatosensory area (Fig. G-2). Within each lobe there is very precise topography of sensory projection areas. In addition to primary sensory areas there are multiple sensory areas in other lobes for each modality, each area different from others in sensitivity and processing of information. In subcortical sensory nuclei there is point-for-point projection from sense organs. In fishes and amphibians each point on the optic tectum receives from a specific region of the retina. Recordings from single neurons show localization to be very precise. In a frog, specific components of the species call are recognized in single midbrain cells. In the inferior temporal cortex of a monkey, objects such as a face can evoke

responses in single neurons. Stimulation of local regions of motor areas elicits movements not only of limbs and joints but of single muscles. Association areas may show localization of set patterns, as in bird song.

5. *The Most Specialized Integrative Regions of Vertebrate Nervous Systems Are Organized in Horizontal Layers.* Visual input to optic tectum is largely to outer layers, less to deep layers. Cell types vary in layer distribution (Fig. G-3); in tectum of frogs and fishes the most abundant cells are the pyriform neurons close to the central ventricle. In lateral geniculate of mammals the visual neurons are in laminae, each projecting to the visual cortex. The number of layers identified in mammalian cerebral cortex varies according to the investigator as well as to the region (69). In general there are six layers: (1) outer molecular layer of small axons, and dendrites some neurons, (2) small pyramidal neurons, (3) medium and large pyramidials, (4) stellate cells, (5 and 6) deep layers of large

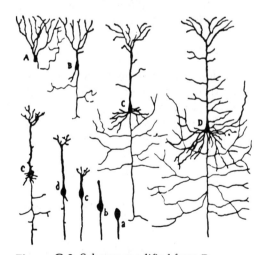

Figure G-3. Schema modified from Ramon y Cajal of cortical neurons in (*A*) frog; (*B*) lizard; (*C*) mouse; (*D*) human. Phylogenetic trend paralleled by ontogeny of pyramidal neurons of mouse, *a* to *e* (at lower left). [From Ramon y Cajal (127a).]

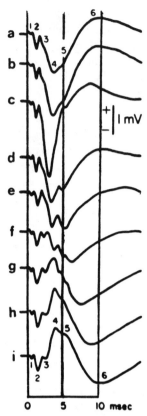

Figure G-4. Field potentials elicited by optic nerve stimulation and recorded at depths in optic tectum indicated from *a* to *i* along radial electrode penetration. [From Vanegas (152a).]

ity of many cells and (1) may be evoked by specific stimuli, (2) may be graded according to strength of stimulus, (3) reverse in sign as an electrode penetrates through an active layer (Fig. G-4), (4) may have a fixed latency for a given stimulus such as a tone, and (5) may be spontaneous as in brain waves. Field potentials are measured as near-field or as far-field potentials according to their proximity to the point of generation. Scalp-recorded responses in humans to a tone are a short (100 ms) latency negativity (N-100) a positive response at 300 ms (P-300) modifiable by cognition, a short latency positive response (P-200) related to short-term memory and a long-latency negative response (N-400) that may be related to stimulus interpretation. A dolphin shows

pyramidal cells. The neurons of different layers project to different regions in the cortex and midbrain.

Within each sensory area, the neurons are organized in vertical columns that traverse the layers. Columns provide for mapping projections and for interaction of cells of similar functions. Columns have been identified in visual, somatosensory, auditory, and motor cortex.

6. Electrical Activity of CNSs Is of Several Types. Two general categories are field potentials and unitary spikes. Field potentials arise from the synchronous activ-

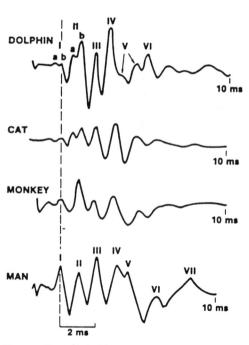

Figure G-5. Dolphin averaged brainstem response to brief click stimulation compared to those of cat, monkey, and human. Dashed line is through wave I for each species. [From Ridgway et al. (127b).]

responses to orienting tones P-25, small N-200, broad P-550 (160) (Fig. G-5). In optic tectum of frogs and fishes, there are two levels of polarity reversal corresponding to entry of visual afferent bundles.

Rhythmic field potentials are called "brain waves." In humans, the alpha waves (8–13 Hz) arise from frontal cortex and are interrupted by beta waves during mental activity such as solving an arithmetic problem. Beta waves are of low amplitude, frequency 13–30 Hz. Delta waves, 0.5–4 Hz, and theta waves, 4–7 Hz, are prominent in slow-wave sleep. Theta waves arise in the hippocampus, probably in basket cells, that are synchronously active (10). Electrical stimulation of hippocampus elicits maximum spike responses when delivered at the theta frequency (10, 79). Cat visual cortex is organized in orientation columns that oscillate at ~40 Hz. Neuronal firing correlates with the phase and amplitude of the oscillating field potentials (47). Unitary spikes carry specific messages, simultaneous recordings from multiple units indicate cooperativity (45).

In summary, evoked potentials and rhythmic field potentials result from synchronized activity, synaptic potentials and spikes, of many cells and correlate with neuronal excitability. Evoked potentials have been much used for mapping sensory fields in integrative structures in the vertebrate brain. Unit responses may be graded in short neurons, as in the retina. They are all-or-none spikes in most central neurons. Unit potentials are precise in the coding of signals; they are evoked by appropriate inputs in sensory interneurons, in internuncial neurons, and in motorneurons. Unit responses are very specific for information processing. Short nonspiking neurons carry signals electrotonically for short distances, for example, in intraretinal cells.

7. A Given Structure May Serve Several Functions. Also, functions of the same or homologous structures may have changed during evolution, changes that correspond to changes in life style. An example of transfer of function is in temperature regulation. Table 7, Chapter 3 lists mechanisms in reptiles in which thermoregulatory function in heat or cold makes use of the same effectors used for different functions (105). In lizards, snakes, and turtles, postural behaviors (burrowing, basking) use locomotor circuits, panting uses breathing mechanisms, skin vasodilation and constriction use vascular control, color changes in warming reflect concealment and aggression, muscle contraction as a heat source derives from locomotion in body warming and in brooding of eggs.

An example of coupling of two functions with a common mechanism is the coordination of locomotion with respiration. In many fishes, locomotor movements of fins and body synchronize with breathing; both kinds of movement increase in frequency as the fish accelerates. In some midwater and pelagic fishes respiratory rates of swimming breathing becomes continuous in that the mouth remains open—ram ventilation—and the work of gill ventilation is transferred to swimming (129). In cats, running at increasing speeds on a treadmill, frequency of spinal motor units increases in parallel with increase in respiratory rate and blood pressure. A common neural command center serves both locomotion and breathing (153).

8. Different Neurotransmitters Function at Synapses in Different Regions of Vertebrate Nervous Systems and a General Principle is That the Same Transmitters Occur in Comparable Regions in Animals of All Classes. In the spinal cord substance P is abundant in dorsal roots, acetylcholine in ventral roots. Approximately 30% of neurons in the cerebral cortex are sensitive to acetylcholine. Cortical cholinergic synapses are muscarinic, spinal cholinergic synapses are nicotinic. Important excitatory agents in the spinal cord are glutamate and aspartate and these are excitatory to the fictive rhythms in the isolated spinal cord of lampreys. γ-Aminobutyrate is an important inhibitory transmitter in all vertebrate classes, GABA and glycine in the spinal cord. Catecholamines, dopamine, and norepinephrine are excitatory at many higher centers, especially in midbrain and diencephalic nuclei. Norepinephrine, in addition to activating postsynaptic potentials, may modulate cell responses to different inputs. The forebrain contains several neuropeptides, for example, neuropeptide Y, CCK, bioopiates, and somatostatin (Chapter 1);

these are predominantly modulators rather than transmitters.

BRAIN SIZE. Brain size in vertebrates is a function of body weight. The ratio of brain weight to body weight (cephalization coefficient) follows an exponential relation, brain size increasing as an exponent (0.57–0.85) of body mass (88). Relative brain size as a function of body weight is as follows (162):

Relative Brain Weight	$= g_{(brain)} \times kg^{-0.67}_{(body\ wt.)}$
Humans	114
Hominoids	26
Other mammals	12
Songbirds	23
Other birds	4.3
Lizards	1.2
Frogs	0.9

The slopes of plots of forebrain size as a function of body size for present day animals are similar to those for fossil forms (142b, 143). The position of the curve for mammals is higher than for birds and much above those for reptiles and fishes of comparable size (Fig. G-6). The value of the brain allometric exponent in mammals varies from 0.57 in insectivores to 0.79 in primates and within orders (e.g., rodents) there is much variation. Also, the allometric constant for different brain

regions varies: 0.48 for lateral geniculate and 0.67 for neocortex in primates. Folivores have lower cephalic indexes than frugivores and insectivores (88). Among elasmobranch fishes the squalomorphs have low ratios and carcharhinoform sharks have higher brain–body ratios (111). It appears that locomotor activity influences brain size more than taxonomy.

Data for brain volume of primates are (5):

Primate	Brain Volume (mL)	Body weight (g)	Cephalic Index $\left(\dfrac{\text{brain weight}}{\text{body weight}}\right)$ $(\times 10^2)$
Human	1330	65,000	2
Chimpanzee	420	46,000	0.2
Gorilla	465	165,000	0.2

Cortical volume as a function of brain volume is shown in Fig. G-7.

The nervous systems of a rotifer or nematode has some 300 neurons, an octopus 30×10^6, and whales and elephants

10×10^9–10×10^{12}. The human brain has fewer neurons than the elephant. In mammals, the brain uses 18% of its blood O_2 but the brain is only ~2% body mass (159). A conservative estimate places the

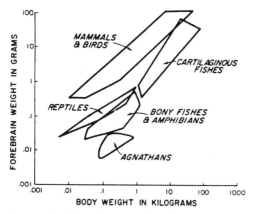

Figure G-6. Forebrain and body weights for major vertebrate groups expressed as minimum convex polygons. [From Northcutt (111).]

number of neurons in a human brain at 10^{11}, each with 10^3 synapses for a total of 10^{14} synapses; at a firing rate of 100 Hz there could be 10^{16} interconnects/s.

The CNS of vertebrates is generally organized with a greater abundance of neurons than in any invertebrate other than cephalopods. As in invertebrate animals, small neurons (50–20 μm) and small diameter fibers far outnumber large neurons. In a lumbar segment of a dog one dorsal root has 12,000 fibers, most of them afferent; a ventral root has 6000 fibers. There may be 375,000 neurons in one segment of the spinal cord, many small, few large (59). Sensory cell bodies are in dorsal ganglia outside the cord and spinal neurons are interneurons and motorneurons, each of the latter with many synaptic boutons. There is potential for much integration in a single segment of spinal cord. In goldfish, the optic nerve contains 68,000 unmyelinated fibers, 90% of which are 1 μm or less in diameter (68).

Small neurons in cerebrum and other regions, especially cerebellum, are significantly more numerous in primates than in other mammals. The percentage of pyramidal cells and fusiform cells in the visual and motor cortex of the monkey and cat is similar but stellate cells are twice as abundant in the monkey as in the cat

(100). In cerebellum, the ratio of granule cells to large Purkinje neurons is in frog 100, monkeys 950, and in human 2500 (100). It is of interest that in the octopus, the most intelligent of invertebrates, the highest integrative region of the brain, the verticalis complex, has predominantly small neurons. Dolphins have a large fissured brain with surface area greater than that of humans; however the cortex thickness in humans is twice that in the dolphin (101). Satellite cells (various forms of glia) are abundant, in some parts of the brain many more than the number of neurons.

In fishes and amphibians, the forebrain is nonconvoluted and has largely, but not exclusively, olfactory input. The forebrain of reptiles has a cortical structure but is not convoluted. In mammals the convoluted cerebral cortex provides a large surface area for synaptic interaction; the surface areas of several species are: mouse 4 cm², human 2000 cm², chimpanzee 800 cm², and dolphin 3000 cm². For a series of mammals, the slope for brain surface area as a function of brain volume is 0.91, mean cortex thickness 0.08, and cortex

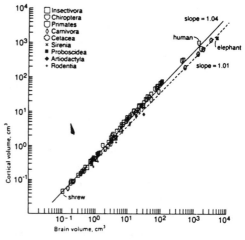

Figure G-7. Cortical volume as a function of brain volume. The slope of the regression line is 1.04 ± 0.01. [From Prothero and Sundsten (126).]

volume 1.04 (126). The ratio of neocortex to cortical volume relative to an insectivore taken as unity is: Homo 156, chimpanzee 58, gorilla 32, lemur 20 (142c, 143).

The size of different areas of the cortex correlates well with behavior, for example, the olfactory cortex of the dog, vibrissal area in rats, face and mouth in carnivores, auditory cortex in nocturnal animals, tail in spider monkey, and the hand and fingers in humans.

Brain contains more kinds of proteins than other organs (99, 147). It is estimated that the rat brain has expressed DNA sequences sufficent for coding 240,000 different proteins, the liver has sequences for 57,000 and the kidney 37,000 proteins. Of the total genome only 20% of genes are expressed in kidney, nearly 100% in brain (17, 18). Many nonpolyadenylated mRNAs are formed during postnatal development (18). Drosophilia has some 5000–10,000 genes and most of them are expressed in the nervous system. Two thirds of the mRNAs of brain are specific to that organ.

MAUTHNER NEURONS. The only truly giant identifiable neurons in vertebrates include the Mauthner and Muller cells of fishes and amphibians (32). Comparative morphological studies of teleost fish show differences in Mauthner cell (M cell) size; somata can be categorized as either large or small. M cells are not found in certain fish; most M-cell studies have been conducted on cyprinids many of which have M cells (26, 27, 75, 166).

The Mauthner neurons of the goldfish have two major dendrites, one that extends caudally and laterally and the other ventrally and rostrally (Fig. G-8). The axon initial segment arises from a prominent axon hillock and is surrounded by an "axon cap" containing a powerful crossed inhibitory input from the contralateral M neuron. This cap is composed of an outer and inner zone and this struc-

ture creates a high electrical resistivity critical to the functioning of the cell. The axon decussates and then turns caudally, synapsing on inter- and motorneurons as it travels the length of the spinal cord. The conduction velocity of the axon in the medullary region is ~82 m/s at 19–22.5°C (40).

The Mauthner cell receives multiple inputs including afferents from the visual and octavolateralis systems. The best characterized input is from saccular afferents to the distal lateral dendrite. Single afferents are capable of mediating both electrotonic and chemical epsps (Fig. G-8).

M-cell activity is modulated by a group of inhibitory interneurons, which are activated by afferents as well as by a M-cell recurrent collateral pathway. A single interneuron can inhibit the M cell by (1) an early "field effect" involving no synapse. As the action potentials of these interneurons enter the outer portion of the

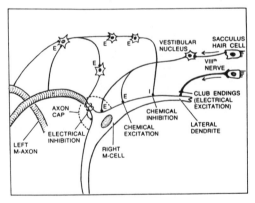

Figure G-8. Diagram of Mauthner neuron showing electrical and chemical synapses. Lateral dendrites receive electrical and chemical excitatory synapses from ipsilateral input and chemical inhibition from contralateral input via interneurons. Ventral dendrite is shown below cell body. Axon crosses to the opposite side. Electrical inhibition is at the axon cap. [From Kuffler and Nicholls (78a). Figure modified from Furukawa (41a).]

axon cap, propagation fails, creating an external positivity (extrinsic hyperpolarizing potential, EHP) not recorded intracellularly and thus the transmembrane potential is hyperpolarizing. (2) A later Cl^--dependent chemical inhibition at synaptic contact on dendrites of the M cell (32a).

There is also a crossed inhibitory network in the spinal cord. If an action potential in an ipsilateral M axon leads that in the contralateral one by 0.15 ms or longer, muscular activity on the contralateral side is inhibited. If the interval between action potentials is less, then muscular activity on both sides is inhibited (4). One group of axons from the M cell activate cranial relay neurons that activate inhibitory neurons that in turn produce ipsps in M cells and also inhibit auditory neurons. Inhibition of auditory responses prevents excitation during a tail flip evoked by M cells. Similar negative feedback occurs with lateral line (82).

Recordings of M cell activity in unrestrained goldfish have implicated the M cell in sound and visually evoked startle responses (167). M-cell activity initiates the first phase or C-bend component of a startle response, which can propel a fish forward at ~0.5–1.5 body lengths within 100 ms (27). There is a parallel system that can mediate startle responses in the absence of M cells (26). The non-M-cell initiated C bends have a longer latency after stimulation but are otherwise identical to M cell initiated responses (27).

Comparative physiological studies indicate that the medullary inhibitory networks are a common feature of M cells independent of their size but that the inputs to the cell may vary between fish species (166).

Endogenous Functions of the Spinal Cord and Lower Brainstem

The principles of reflex behavior were first established by observations of Sherrington on functioning of the spinal cord in laboratory animals, mostly cats (140).

Input to the spinal centers of mammals include the following: (1) Muscle spindles respond to stretch and spindle afferents (group Ia), are excitatory monosynaptically to extensor motorneurons and are inhibitory to motorneurons of antagonistic muscles. (2) Slow afferent fibers from muscle spindles (group II afferents) act via interneurons to inhibit extensor motorneurons and to excite flexors. (3) Tendon organs of Golgi respond to tension and their fibers (group Ib afferents) inhibit ipsilateral alpha motorneurons, especially to extensors. (4) Cutaneous receptors provide local information, which may evoke scratch reflexes. (5) Central descending pathways are from pyramidal cells of cerebral cortex, also rubrospinal, vestibulospinal, and reticulospinal tracts; each of these triggers and modulates spinal cord motor responses.

Output from spinal centers are (1) Large alpha efferents elicit maximum tension from leg muscles. (2) Small alpha efferents activate small extrafusal muscle fibers that produce low tension. (3) Gamma efferents elicit contraction of intrafusal muscles that stretch muscle spindle sense organs; this initiates the so-called gamma loop, which functions in preparation for and in smooth initiation of movement and in coordination of movements and body positions. (4) Ascending tracts are in dorsal and lateral columns, for example, spinothalamic and spinocerebellar tracts.

In addition to sensory and motor fibers, the spinal cord has many interneurons that coordinate both ipsilateral and contralateral activity. These also constitute the networks of central program generators; many interneurons are spontaneously active. The best known are the Renshaw cells that are activated by collaterals from motorneurons; this postsynaptic feedback blocks continuing discharge of the same or adjacent motorneurons. Renshaw neurons give high frequency bursts in response to single antidromic impulses in motorneurons.

When the spinal cord of a mammal (cat or dog) is transected in the upper thoracic region, motor reflexes are at first depressed by spinal shock. Flexor reflexes return before extensor reflexes. A "spinal" mammal cannot stand but when suspended it may make treading motions. Local stimulation of skin can evoke a scratch reflex. Autonomic reactions of digestion, excretion, and reproduction persist or gradually recover function. The neural network for rhythmic leg movements is modulated by sensory inputs but after transection of dorsal roots, patterns of alternate flexor and extensor movements persist, that is, there is no causal dependence on afferent inputs for stepping to occur.

Autonomy of spinal function is greater in amphibians and fishes than in mammals. Amphibians and reptiles do not have differentiated intrafusal and extrafusal muscle fibers and there is nothing analogous to the gamma loop of mammals. Fish lack muscle spindles but the skin is so tightly applied to musculature that cutaneous receptors respond to body movements. A spinal frog can jump in a coordinated fashion when stimulated. The normal pattern for walking movement of a toad is: right hind (RH), right front (RF), LH, LF (58); deafferentiation results in abnormal rhythms. A frog shows a "wiping" reflex in response to local skin stimulation, even after sensory input from the leg is eliminated. Spinal tadpoles show S shaped bending orientation away from a site of touch. Each side of the spinal cord can generate its own rhythm if the cord is bisected (127c, 128). Intracellular recordings from motorneurons of embyros of *Xenopus* show periodic tonic depolarization and on the depolar-

ized potential, epsps give rise to spikes; after-hyperpolarizing ipsps due to Cl⁻ conductance follow the spikes (Fig. G-9) (128).

Turtles show coordinated leg movements after upper spinal transection; localized stimulation of the plastron or of the spinal cord elicits a persistent rhythm—fictive locomotion that can be recorded by electrodes on ventral roots. Movement of each limb is out of phase with the ipsilateral limb; protraction of one limb is accompanied by retraction of its antagonist. Each limb is driven by its own spinal center and interlimb coordination is brought about by spinal network (145a).

Fishes show much species variation in autonomy of the spinal cord. In spinal teleosts and elasmobranchs fin reflexes can be elicited by local stimulation of the skin. In an eel, high section of the cord interferes very little with rhythmic undulatory movements; the rhythmic wave can pass down the cord through a region from which skin and muscles have been removed (48). A spinal dogfish swims much like a normal fish and shows body motions at 40 waves/min. Rhythmic movements in dogfish are abolished by complete deafferentation but a rhythmic pattern can pass down the cord through some 12 denervated segments (154). Neurons on each side of the cord of a dogfish fire alternately, but the locomotor wave is disrupted in the absence of sensory excitation.

Fictive locomotion has been examined in cyclostomes (lampreys) by recording from ventral roots or directly from spinal motorneurons (130, 131). Locomotor rhythms can be started in isolated cords by stimulating the rostral end with amino acid agonists, for example, glutamate or aspartate acting on NMDA receptors. Each hemisegment contains some 500 neurons that constitute a patterned network (11, 50–52). During each fictive

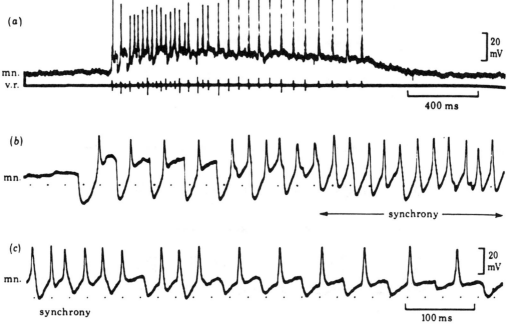

Figure G-9. Intracellular recordings from the spinal cord of larval Xenopus during fictive swimming. (*a*) motorneuron record (mn) and ventral root record (vr). A short spontaneous episode of about 20 cycles of asychronous discharge (19–28 ms), then swimming alternating pattern (56–78 ms); (*b* and *c*) faster time base shows transition from swimming to synchrony and vice versa. At transitions the cycle is halved or doubled. [From Roberts and Kahn (127c).]

cycle, a motorneuron shows an excitatory phase followed by an inhibitory phase. A model circuit includes ipsilateral excitatory and inhibitory neurons and contralateral inhibitory neurons [Fig. G-10; (51)]. The oscillator is TTX resistant and 5-HT potentiates by depressing postsynaptic hyperpolarization. The fictive rhythms can be obtained from 2 to 3 segments of lamprey spinal cord. The two sides alternate and anterior–posterior segments lag by one-half a cycle. Each motorneuron receives excitation from interneurons (52).

It is concluded that in all vertebrates, locomotor rhythms can be centrally programmed. Once initiated, central program generators may continue in the absence of rhythmic sensory inputs. A phase lag occurs between hind and fore leg programs and reciprocal coupling exists between ipsilateral longitudinal centers and between the program of the two sides.

Visual Pathways*
Properties of that portion of the brain that forms the retina were described in Chapter 16. At levels of bipolar, horizontal, amacrine, and ganglion cells much processing of visual information occurs. Axons of the ganglion cells form the optic nerve. In fishes and amphibians all optic nerve fibers cross to the opposite side but in reptiles, birds, and mammals some portion of the fibers do not cross (37). The

*Joseph Malpeli contributed to this section. Central pathways for audition, olfaction, and taste are given in sensory chapters.

A D-GLUTAMATE, 5 min

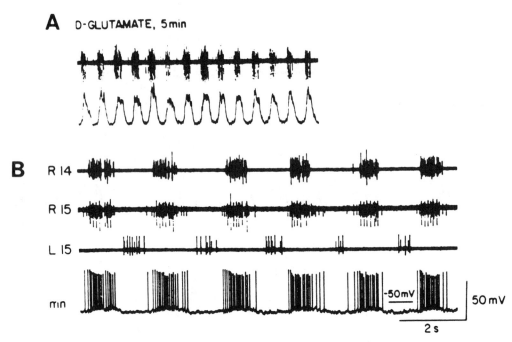

Figure G-10. Rhythmic activity recorded from isolated pieces of spinal cord of lamprey. (A) Fictive swimming induced by adding D-glutamate (0.4 mM) to the bath. Upper record from ventral root by suction electrode. Lower record recorded from motorneuron. (B) Rhythmic activity of isolated spinal cord induced by D-glutamate. Upper three records from spinal motor roots of segments 14–16. Bottom record, intracellular recording from motorneurons. [From Russel and Wallen (131a).]

proportion of crossed fibers is determined by the overlaps of visual fields; where there is no binocular overlap all fibers are crossed, whereas in binocular regions the fibers are uncrossed. In primates, fibers from the nasal one half of the retina cross in the optic chiasma while those from the temporal retina go in the ipsilateral tract (Fig. G-11). In turtles all retinal fibers cross, in mammals, the visual path is bilateral (27a). The general pattern of visual pathways is similar in different classes of vertebrates. All vertebrates have retinothalamic, retinohypothalamic, retinotectal, retinopretectal, and retinotegemental tracts in various proportions but in all vertebrates there are two independent pathways (37). The paths in mammals are (1) retina → lateral geniculate → striate cortex and (2) retina → superior colliculus →

lateral posterior nucleus of thalamus → visual areas of cortex. In birds the paths are (1) retina-optic thalamus-forebrain Wulst, and (2) retina-optic tectum of thalamus → ectostriatum of forebrain (27a).

In teleosts and amphibians, the optic fibers are mainly distributed to the optic tectum (OT) (mesencephalon); smaller tracts pass to thalamus and hypothalamus (Fig. G-12). The OT contains variable numbers of layers—six in catfish (Fig. G-13); some optic fibers terminate in the superficial layer, others pass to deep layers (92, 137). The tectum contains several cell types—pyramidal cells, fusiform cells, horizontal cells and small pyriform cells in the paraventricular layer, which comprise 90% of tectal neurons (Fig. G-14). Output from the tectum goes mainly to the cerebellum and spinal cord, some to

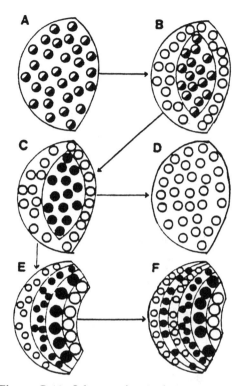

Figure G-11. Schema of retinal projection to thalamus. Open circles indicate contralateral, solid circles ipsilateral. The ancestral retinal projections to the dorsal thalamus are thought to have been bilateral, with overlap from the two eyes on the same cells [(*A*) half-filled circles]. This condition is best seen in some amphibians. (*B*) In reptiles many neurons appear to have a monocular input from the contralateral eye. (*C*) In primitive mammals (e.g., hedgehogs) only monocular layers are found. (*D*) Birds have only an input from the contralateral eye. (*E*) Some primates have four monocular layers in the lateral geniculate. (*F*) Adult rhesus monkeys have six monocular layers. [From Ebbesson (27a).]

diencephalon and forebrain. Axons from pyriform cells cross to the opposite tectum.

Local stimulation of the retina shows that projection to the tectum (superior colliculus of mammals) is topographically ordered: nasal quadrants project posteriorly, temporal anteriorly, lower retina

medially, and upper retina laterally. In animals with binocular visual fields, such as frogs, the central part of the field is projected to each tectum, the binocular field subtending 100° along the vertical meridian (44, 110).

After removal of the OT from fishes, visual behavior is somewhat impaired. However, nurse sharks from which the entire OT has been removed can learn to discriminate black from white objects, horizontal from vertical stripes. Similarly, a goldfish that lacks tectum can detect light but is less sensitive and less precise in visual responses than normally (64, 65, 163).

The field potential profile of goldfish optic tectum evoked by optic nerve stimulation reverses in polarity just below the outer plexiform (fiber and synaptic) layer (Fig. G-4). Current density was recorded as maximum just below the reversal point, that is, there is vertical current flow. Some neuronal units in optic tectum respond to both visual and auditory inputs (148).

Recordings from the optic tectum of goldfish have been made from terminal axons of retinal ganglion cells and from tectal neurons. Responses in axons of ganglion cells to stimulation by light are (1) tonic units, 2–12 spikes per burst and interburst intervals 20–300 ms, (2) phasic units with concentric or flanking ON–OFF fields, response frequency depends on a balance between excitation and inhibition; (3) directional units for stimuli moving in the nasotemporal direction; (4) fibers in mid-tectum with very small visual fields; (5) color-opponent cells, especially in superficial layers. Tectal neuron responses have larger visual fields than the retinal axons, show spontaneity and bursts of spikes, have long latency, fail to follow above stimuli at 60 Hz, and show plasticity of response. Responses of some tectal cells in upper layers habituate, others do not. Pyriform cells in periventri-

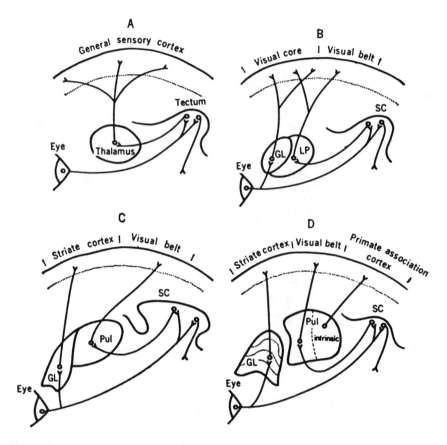

Figure G-12. Diagram of four postulated steps in evolution of visual projections in thalamus and cortical visual areas. (*A*) Reptilian ancestor; (*B*) early mammal represented by hedgehog; (*C*) tree shrew striate cortex; (*D*) monkey—subdivision of pulvinar; SC, superior colliculus. [From Diamond (23).]

cular layer are very abundant, give small endogenously generated spikes in bursts at intervals of several seconds and respond to visual stimulation by interruption of the spontaneous firing (114).

In the bass Micropterus, neurons of the dorsal median thalamus respond to optic nerve stimulation after latencies of ∼7 ms, neurons of nucleus rotundus after 15.7 ms; thus there is a pathway to diencephalon in parallel with that to tectum. Diencephalic neurons respond to visual stimulation with small receptive fields and may show some pattern discrimination. The two pathways—tectal and diencephalic—show some crossover, that is, are not completely without interaction.

Stimulation of telencephalic neurons elicits tectal responses via a direct telencephalo-tectal path. The apparent lack of visual deficit after lesions to telencephalon in fish may be due to inadequate testing (38).

In goldfish, some ganglion cells show antagonism between excitation of two sets of cones—color opponent responses. Two sets of cones are sensitive to either 467 or 620 nm and excitation by one color is accompanied by inhibition by the other (68, 114).

In frogs several types of tectal neurons have been identified. Convergence of visual, tactile and auditory input, as well as from contralateral tectum has been ob-

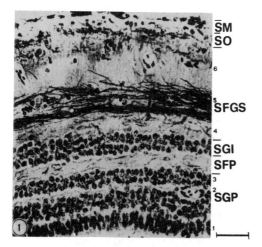

SM
SO

6

SFGS

4

SGI
SFP

3

SGP

Figure G-13. Cross section of optic tectum of fish *Calamoichthys calabaricus*. SM, marginal stratum; SO, optic stratum; SFGS, fibrous stratum; SGI, interneurons; SFP, deep fibrous stratum; SGP, periventricular layer. [From Mazzi (92).]

served. Tectal units of frog include: (1) sustained edge detectors with or without inhibitory surround, (2) convex edge detectors, (3) ON–OFF units, (4) dimming detectors in deep layers; tectal units habituate on repetitive stimulation (54, 55). In superficial layers of frog tectum, units respond to slow movement of small objects, are inhibited by large objects. In deeper layers, units respond to any movement, to dark edges or dimming (64). In frogs, the thalamic and tectal visual systems are independent, motion is detected by O.T. and fixed edges are detected by thalamus (64, 65). In the frog some visual signals go directly to the thalamus without a tectal relay (64). In the dorsal tegmentum of frog the nucleus isthmi receives output from the tectum and after removal of the nucleus isthmi, the frog no longer responds to food prey (53). In the toad, the caudal thalamus (pretectum) has units that are (1) spontaneous, (2) respond to tactile stimuli, (3) respond to small (15–30°) fields, (4) respond to large fields, (5) are fast habituating, (6) respond to dark objects

moving toward the eye, (7) respond to dimming light, (8) are spontaneous units with excitatory or inhibitory fields, and (9) have prolonged responses (30, 31). Prey-catching and avoidance behavior in a toad is mediated by optic tectum. Electrical stimulation of the tectum elicits the feeding sequence; stimulation of pretectum and caudal thalamus leads to avoidance (29, 30). Small moving objects are treated as food, large objects as enemies. The medial reticulum of the medulla receives inputs from tectum and caudal thalamus.

In the turtle *Pseudemys* two routes from retina to forebrain are (1) via diencephalic geniculate and (2) via tectum, then nucleus rotundus of thalamus. After removal of the OT the turtle can distinguish vertical from horizontal stripes but shows reduced visual discrimination (7).

In birds the visual system is highly developed with some 2.4 million retinal ganglion cells (pigeon). A large fraction of optic fibers cross to the opposite tectum; output from tectum goes to nucleus rotundus of thalamus and from there to ectostriatum of forebrain. Some optic fibers run directly to thalamus and then to the forebrain. In pigeon, individual tectal cells have visual fields much larger than those of retinal ganglion cells; most cells are directionally selective, some are sensitive to concentric and contrasting fields. Direction and movement are also registered by thalamic neurons. The importance of the striatum (central forebrain) in birds was shown by a deficit in discrimination of brightness and pattern (but not of color) after a lesion in hyperstriatum (125). The bird pretectum is essential for horizontal optokinetic nystagmus, accessory optic nucleus for nystagmus in other directions (94).

In mammals, the mesencephalic relay stations are in superior colliculi (SC) that are homologous with optic tecta of frogs and fishes (Fig. G-12). Units of the superior colliculus of rat and cat show pro-

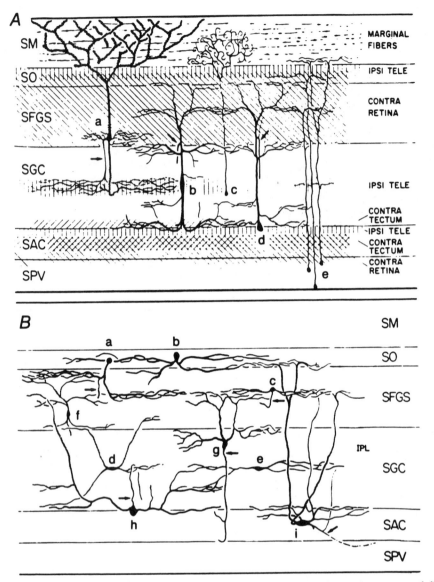

Figure G-14. (*A*) Principal types of vertically oriented neurons in optic tectum of fish *Eugerres*; a, pyramidal; b, fusiform; c, small pyriform; d, large pyriform; e, periventricular neurons. (*B*) Other neuron types g, h, and i are multipolar. [From Vanegas (152a, p. 45).]

jection from contralateral retina. Most collicular units give responses to binocular stimulation and each cell has an optimum rate and direction of stimuli.

In mammals, the diencephalic centers are the laminated lateral geniculate (LGN) that receives fibers directly from retina and the pulvinar that receives input from the superior colliculus, and in some species from the retina. Both lateral geniculate and pulvinar project to striate visual cortex (area 17) and prestriate cortex (areas 18, 19) although there is much species diversity as to projection pattern (98, 122). In the monkey the LGN projects only to area 17; in the cat, projection is to areas 17–19 and sparse projection to so-called Claire-Bishop region. The LGN of diencephalon is generally divided into separate layers, with each layer receiving inputs from one eye. In addition, layers differ in the types of retinal ganglion cell afferents they receive. For example, in the monkey, the dorsal (parvocellular) layers receive inputs from color-opponent cells whereas the ventral (magnocellular) layers receive inputs from cells that are not strongly color selective (60, 61).

The visual cortical areas have six layers in cat that are interconnected (especially from striate area 17 to prestriate areas 18, 19), as well as to visual association cortex (117a). Twelve areas of cat cortex respond to visual input. The SC sends fibers to nuclei controlling eye movements; in addition to the parallel pathways to cortex there is much divergence. It is estimated that in a monkey $\sim 10^6$ optic nerve fibers terminate in one LGN, that each LGN has 2×10^6 cells, that visual cortex has 150,000 neurons/mm², amounting to some 200×10^6 cells in the visual cortex on one side. Thus there are more than 200 cortical cells for each LGN neuron and twice that number for each ganglion cell (122). In addition to the parallel ascending pathways and divergence there is considerable feedback from the primary and secondary visual cortex to LGN. In the absence of the primary visual cortex, the SC can mediate gross localization of a visually perceived object. Also collicular receptive fields are modulated from the cortex. In monkeys, destruction of the visual cortex causes an initial apparent blindness, but after extensive training visual responses are obtained, mediated by subcortical regions (63).

In the visual cortex (cat and monkey) responses of single neurons have been classified as simple, complex, and hypercomplex (Fig. G-15) (60, 61). In LGN, excitatory and inhibitory areas are concentric but in the visual cortex these cells are arranged along the sides of a straight boundary. The inclination of this boundary determines the orientation of a luminosity edge in one direction; some cells give equal discharges in response to

motion in either direction at the "preferred" orientation, whereas others are direction specific. Neurons of LGN are much less sensitive to orientation of stimuli than cortical neurons (34). The next level of complexity (complex cells) includes cells that respond tonically and at high frequency during movement in a given direction and at a certain angle anywhere in the receptive field. Hypercomplex cells respond to moving oriented boundaries of a particular length, excitatory areas are flanked by inhibitory regions. Most cells in the visual cortex respond binocularly with more units responding to contralateral than to ipsilateral stimuli.

Columns in cat and monkey reflect ocular dominance (right alternating with left) and orientation tuning. Ocular dominance columns are revealed by autoradiography after aminoacid injection into one eye (Fig. G-16) (60).

Single units of the dorsolateral frontal cortex of awake monkeys show highly specific responses in delayed-response tasks. These tasks require memory for stimulus locations over a several-seconds period; they are impaired by dorsolateral frontal cortex lesions. Some units are active only during an imposed delay between stimulus (patterned light) and pushing a lever for reward as if holding the memory for the response. Some units respond 200 ms before muscle activity of the response; some are differentially active for right or left presentations, some units depend on stimulus location, others correlate with direction (107).

Visual cortical areas of primitive mammals—hedgehogs and tree shrews—have been compared with those in monkeys. In the primitive species, the deeper layers of the superior colliculus are needed for guided eye and head movement; lesions of the geniculo-striate systems and area 17 have less behavioral effect than lesions of the tectopulvinar system.

Retinal cells are diverse in anatomy

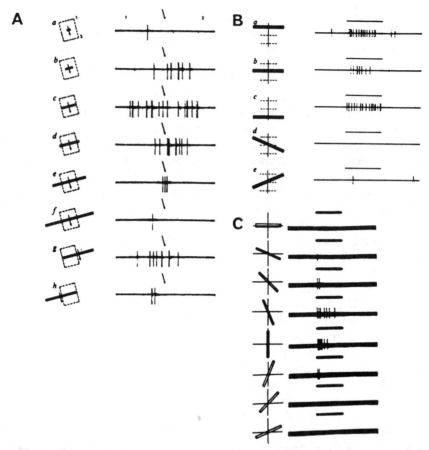

Figure G-15. Characterization of receptor neurons in visual cortex of cat; responses to slit stimuli. (A) Responses to a rectangular slit of light arranged with center slit superimposed on the center of the receptive field in various orientations; receptive field with axis vertically oriented. (B) Complex cell responding best to a black, horizontally oriented rectangle. Response when the rectangle is horizontal, not when it is tilted. (C) Hypercomplex cell responding to black bar at an angle and pointed downward. [From Hubel (60).]

and physiology in all groups of vertebrates (Chapter 5). In some mammals, three distinct categories of ganglion cells have been described (Fig. G-17). Recording from individual cells followed by staining in the cat retina gave a functional correlation with cell type. The Y cells (histologically) have large somata, thick dendrites, sparse branching; they respond quickly, some have "on" center, some have "off" center responses, have rapidly conducting axons. X cells are of medium size, have bushy dendrites, give sustained responses to flashes, conduct more slowly than Y cells. W cells are small, have thin dendrites, give weakly persistent responses with center-surround fields; the W cells may be the most primitive but they are poorly known because their small size makes penetration difficult (132, 142b).

The three cell types maintain independent distribution in the LGN and visual cortex. A diagram of the distribution of

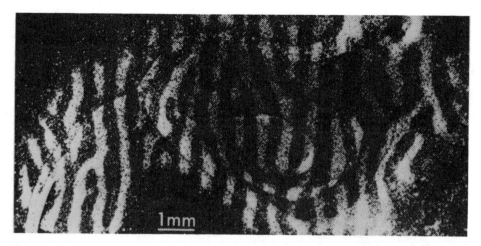

Figure G-16. Autoradiograph of striate cortex of adult macaque monkey in which contralateral eye had been injected with tritiated proline–fucose. Microscopic sections tangential to surface of occipital lobe. Columnar organization inferred by recordings of similarly sensitive cells in vertical columns corresponding to labelled neurons. [From Hubel (60).]

the three types is shown in Figure G-18. The ratio of X:Y cells in cat retina is 10:1, in geniculate 1:1 and this distribution continues to the cortex. Each retinogeniculate and each geniculocortical Y axon innervates more neurons than each X-cell axon. Thus a minority of retinal ganglion cells (Y cells) come to exert a much greater influence in visual cortex than expected from their numbers in the retina. The superior colliculus receives only Y and W axons (62, 139b). Somewhat different morphologies and responses have been described for rabbit and monkey; corre-

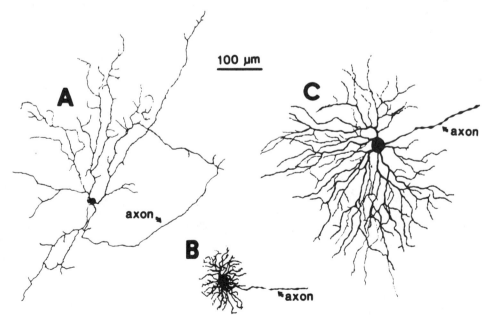

Figure G-17. Drawings of ganglion cells of retina of cat. (*A*) W cell; (*B*) X cell; and (*C*) Y cell. [From Stanford (142a). Fig. 9, p. 255.]

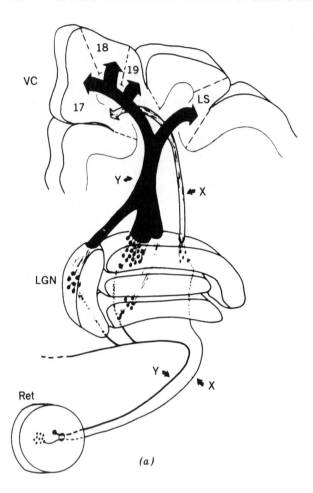

(a)

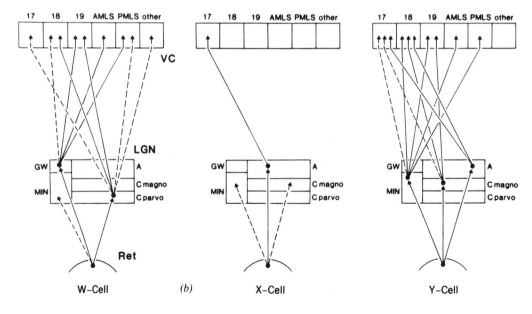

(b)

Figure G-18. (*a*) Diagram of retin-ogeniculocortical X- and Y-cell pathways. VC, visual cortex; LGN, lateral geniculate nucleus; RET, retina. (*b*) Diagram of W-, X-, and Y-cell pathways from retina through lateral geniculate to numbered areas of cortex. [From Sherman (139a).]

sponding cell types have not been described in nonmammals.

In summary, vision in vertebrates is by two parallel pathways, the function of each varies in different classes. In general, the tectal system is most concerned with visual orientation and the thalmao-cortical with complex pattern recognition. However, the two systems are not independent and connections at levels of diencephalon and tectum provide some interaction. In fishes and amphibians the tectum is the dominant visual integrator although there is some thalamic visual and telencephalic projection. In tectal integration, visuomotor activity is direct and stereotyped, for example, frogs snapping at small moving objects and avoiding large objects. In mammals, and to a lesser degree in birds, the tectum is more specific for control of eye movement; in primates the tectum coordinates saccadic eye movement. Visual perception in mammals is increasingly by the geniculate (thalamic) relays to forebrain, and multiple integrative areas occur in visual cortex where single neurons can respond to very specific visual patterns. There is a phylogenetic tendency for pattern perception to occur in the visual cortex while movement perception continues in the tectum (e.g., ground squirrels).

SOMATOSENSORY PATHWAYS. In nonmammalian vertebrates, projections of sensory input from tactile and proprioceptive receptors are to several regions of the brain. In all vertebrates, much integration of input from mechanoreceptors occurs in spinal cord. Fishes show responses in cerebellum and tectum to fin and body stimulation. Facial lobes of goldfish show somatotopic projections from fins and from the skin of the head and trunk (121a). In a salamander, the contralateral body surface is projected topographically in the tectum. Thalamic nuclei of the frog and turtle give somatotopic responses.

The telencephalon of reptiles (caiman) has a dorso-ventricular ridge, which is the target for ascending somatosensory paths (124).

The cerebral cortex of a monotreme, *Echidna*, is folded into sulci that have little resemblance to the gyral organization of eutherians; a large area in the posterior half of the cortex receives projections from tactile stimulation of the body. Insectivores, such as the hedgehog, have an unspecialized forebrain with two dorsal regions of visual projection, a large rostral somatosensory projection and a postero-ventral auditory region (71). The sensorimotor cortex of a sloth, *Bradypus*, has a primary area with large representation of forelimbs. Marsupials have an extensive rostral somatosensory region. In both sloths and marsupials (opossum and wallaby) there is coincident overlap of sensory and motor representation of body regions (135).

In mammals of higher orders, the somatosensory pathways follow a common pattern (Fig. G-19). In the spinal cord, fibers of the dorsal columns carry messages from touch, pressure, and motion to the medulla where the trigeminal nerve enters with similar information from the face and head. From the medulla, tracts go to the thalamus where the mapped projection is more closely proportional to sensory innervation density than to body geometry (102a). From thalamic nuclei in humans there is projection to at least three somatosensory areas of the cerebral cortex. The cat has five somatosensory areas. The tactile areas correspond in size to importance of sensory input; hand and face areas are large in monkeys, mouthparts in rabbits, claws and forelimbs in cats, face in dogs, vibrissae in rabbits and rats, mouth in sheep, and forepaws in raccoon and monkeys. In new world monkeys, the size of the area representing the tail corresponds to its prehensile use (96, 97).

Figure G-19*A* is a diagram of the or-

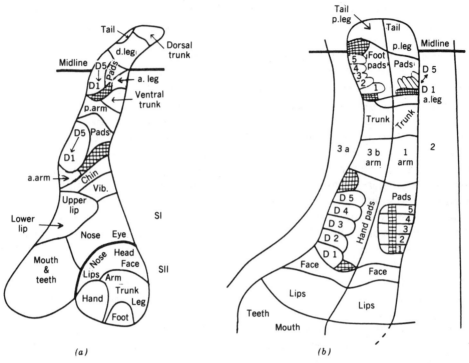

Figure G-19. (*a*) Organization of somatosensory cortex in primitive mammal, the tree shrew. Body surface representation in SI and SII (primary and secondary) areas. Map of representation of body areas. (*b*) Somatosensory cortex of advanced primate, owl monkey; body surface representations. [From Kaas (71).]

ganization of S1 in tree shrews (insectivores). The head areas of tree shrews and opossums emphasize the glabrous nose. In the S1 of eutherians, the head is represented with dorsal surface posterior, ventral surface anterior. Proximal surfaces of hands and feet are posterior and digit tips anterior. In bats the positions of digits are the reverse of that in other mammals, that is, digit representation is caudal (12). In general, SII is arranged so that the top of the head joins the corresponding area in SI. Figure G-19*B* is a diagrammatic representation of somatosensory cortex in an owl monkey that has much in common with apes and humans. For two areas corresponding to the same body region, individual cortical neurons distinguish tonic from phasic inputs, tac-

tile from proprioceptive, and pressure from light touch.

In eutherian mammals, the primary somatosensory cortex is in the postcentral gyrus. Directly opposite on the precentral gyrus is the primary motor cortex. The arrangement of motor areas is such that stimulation of the upper part of the gyrus evokes movements of lower extremities and stimulation of the lower part of the gyrus evokes movement of upper limbs, body, and head. In humans, the areas for movement of eyes and neck are forward and separate from the primary motor area. The orientation on the postcentral gyrus is the same as the motor orientation of the precentral gyrus. The primary motor output is via pyramidal neurons that descend to motorneurons and interneurons in the brainstem and spinal cord. The motor cortex is modulated by premotor cortex and, to some degree, by parietal cortex. Pyramidal neurons begin their discharge prior to an actual movement. Central motor control is hierarchial: spinal cord, brainstem, motor cortex, premotor, and supplementary motor cortex. Cortical control is evolutionarily recent. Motor activity is primary spinal and by brainstem in fishes and amphibians. Modulation by cerebellum probably occurs in all vertebrates.

In animals with vibrissae, layer IV of SI has a barrel organization resembling the columns in the visual cortex (142, 161). Each barrel receives input primarily from one vibrissa and if a lesion is made in a vibrissal base at birth, its barrel does not develop. Five rows of moustache vibrissae, posterior to anterior, are mirrored in five rows of barrels.

There is some plasticity in somatosensory specificity. Topographic maps may be correlated with temporally related inputs (19, 97). Experimentally connecting the skin of two fingers of a young monkey increases the correlation of inputs from finger skin. When a sensory nerve to part of a hand is cut, neurons in the deprived cortex become responsive to stimuli on the remaining adjacent innervated parts of the hand (71).

Temperature and pain signals travel through antero-lateral spinal pathways. These pathways project diffusely and connect extensively in reticular formation and only weakly to thalamic centers. Cortical projection is not to specific loci.

CEREBELLUM. The cerebellum is an integrative portion of the brain that consists of circuitry that is similar in all classes of vertebrates. It receives inputs via two cell types from many sensory nuclei and sends outputs via a single cell type to deep cerebellar nuclei or to a variety of motor centers (66, 67). This prominent cell type onto which all other cells converge is the Purkinje cells (PC), neurons with extensively branched dendritic tree and single efferent axons (Fig. G-20). The total number of PCs is estimated in human cerebellum as $14-15 \times 10^6$, in rat $2.7-5.5 \times 10^5$ (66, 67). Two types of input to PCs are (1) climbing fibers that end directly on main dendrites of PCs and which in mammals arise in inferior olive and in other animals from various sensory nuclei (33); and (2) mossy fibers that terminate on granule cells the axons of which form parallel fibers running at right angles to and converging on the fan of

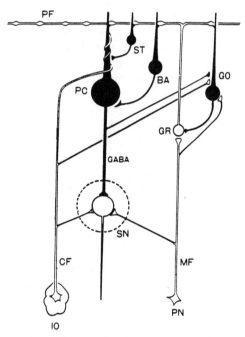

Figure G-20. Basic neuronal circuitry in mammal cerebellum. Principal cerebellar neuron, Purkinjke cell (PC); Intrinsic neurons, Golgi cell (GO), basket cell (BA), stellate cell (ST), granule cell (GR), parallel fiber (PF); Input paths, mossy fiber (MF), climbing fiber (CF), precerebellar neuron (PN) from which mossy fibers arise and inferior olive (IO) from which climbing fibers arise. Output from PC goes via cerebellar deep nucleus (SN). (Modified from several sources.)

spiny PC dendrites. Climbing fibers arising in inferior olive and probably use aspartate as transmitter. Granule cells are the most abundant cell type in the cerebellum. In humans the total number of granule cells is 10^{10}–10^{11}; the ratio of granule cells to PCs is in humans 1600–4000 to 1 (66). Granule cells appear to use glutamic acid as a transmitter from parallel fibers to PCs.

In mammals the inputs are (1) climbing fibers that arise from the inferior olive on which many sensory fibers converge and (2) mossy fibers that have inputs from vestibular, proprioceptive, somatic sensory, and cerebral tracts. In cat, stimulation of various leg nerves, rapidly adapting skin receptors, and stimulation of muscle afferents can activate climbing fibers in the inferior olive. Purkinje cells are spontaneously active, largely because of continual bombardment by mossy fiber and granule cell impulses.

When a stimulus is applied to a cerebellar peduncle (afferent) three excitatory responses can be distinguished in PCs: (1) antidromic spikes, (2) short latency responses to climbing fiber (CF) stimulation—complex epsps with superimposed multiple spikes—(8), (3) single or multiple long-latency epsps that are responses to the mossy fiber–granule cell parallel–fiber complex.

In addition to the excitatory responses of PCs there are three types of inhibitory neurons, variable in abundance in different classes. In mammals, parallel fibers activate Golgi cells, providing feedforward inhibition on PCs. Golgi cells probably use GABA as an inhibitory transmitter. Stellate cells and basket cells, excited by parallel fibers, inhibit PCs, axons of stellate cells end on PC dendrites, axons of basket cells end on the soma or axon hillock of PCs. A basket cell may elicit an ipsp lasting 100 ms or longer. In mammals, collaterals of climbing fibers also excite each type of inhibitory neuron.

After a burst of spikes in a PC activated by climbing fibers the activity of the PC is interrupted in its excitation as tested by antidromic stimulation; the interruption is due to action from the inhibitory neurons (66). Stellate cell inhibition has been demonstrated in dogfish, salmon, frog, goldfish, lizard, and alligator.

A climbing fiber innervates 1–10 PCs and makes many contacts on main dendrites, in cat some 2000, in frog 300 contacts. Parallel fibers make synaptic contacts on spiny branches near the ends of PC dendrites. In cat a single PC has some 200,000 synaptic spines, most of which represent synapses with parallel fibers.

All of the output from cerebellar cortex is by axons of Purkinje cells. In mammals most of the PCs project to deep cerebellar nuclei (a few go directly to vestibular nuclei). The output from cerebellum is uniformly inhibitory on motor systems. The output to oculomotor neurons provides for error correction in the rapid and fine adjustments of eye position and movement. Inputs converge on cerebellar flocculus from labyrinth, retina (via pretectum and tegmentum), eye proprioceptors, and neck afferents; output is to vestibular nucleus and oculomotor nucleus; the net effect is correction of eye position with regard to head and body movement (66).

In lampreys the cerebellum is a simple bridge spanning the medulla. The lamprey cerebellum receives vestibulo-lateral line, also somatosensory afferents.

In fishes and amphibians the cerebellar circuitry is similar to that in mammals. However, in fishes the deep cerebellar nuclei may be lacking and 80% of Purkinje axons exit directly. A schematic diagram for the cell types and typical neuronal responses in goldfish is given in Figure G-21. Identified inputs in goldfish are via mossy fibers that evoke simple spike responses via parallel fibers (76). Climbing

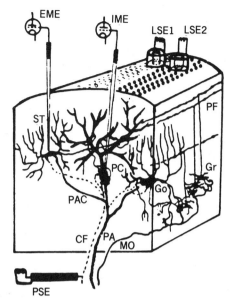

Figure G-21. Diagram of neuronal circuitry of goldfish cerebellum and position of extracellular electrodes (EME) for recording from stellate cells (ST); IME for recording from Purkinje cells. LSE1 and LSE2 for stimulating parallel fibers; climbing fibers (CF), Golgi cell (Go), mossy fiber (Mo), Purkinje cell axon (PA), Purkinje cell axon collateral (PAC), peduncle stimulating electrode (PSE). [From Kotchabhakdi (76).]

fibers enter via the inferior olive and give characteristic complex spike responses (Fig. G-22); CFs are activated from vestibular, spinal, and fore- and midbrain inputs. There is much covergence of inputs via each input fiber type and a PC can respond to several simultaneous inputs. Seventy-five percent of the spontaneous activity in PCs is due to continued input via mossy fibers, 25% is endogenous. The PC output from goldfish cerebellum, is to tegmentum and ventral thalamus, also to reticulum and oculomotor and vestibular nuclei (36).

In the frog, stimulation of a vestibular nerve or physiological stimulation of saccule or utricle activates climbing fibers; semicircular canal stimulation activates mossy fibers. Responses to auditory and visual inputs vary with class; in bony

fishes visual stimulation is via a relay in the optic tectum. Mammals have input from the forebrain via a corticocerebellar tract and pontine nucleus. A general function of cerebellum is to modulate motor response, especially in oculomotor and vestibular centers.

Lesions in the cerebellum produce varying degrees of disturbance in posture and movement. In a dogfish, lesioning fails to interfere with swimming although after hemicerebellectomy the fish frequently circles toward the side of the lesion. In carp and goldfish there is muscle hypotonia, side-to-side swaying and impairment of equilibrium. Small lesions of cerebellum in birds and mammals impair posture and equilibrium.

Spiny branchlets of Purkinje cell dendrites, thought to be sites of efficacy change during motor learning, degenerate during aging. Loss of spiny branchlets was less in rats housed in an enriched environment than in these housed singly or in standard cages (49).

Recent evidence implicates the cerebellum in conditioned motor responses. It was proposed by Marr (91) that when PCs are activated by climbing fibers in temporal conjunction with parallel fibers, the efficacy of parallel fiber synapses on PCs may be altered so that granule cells alone can evoke responses not seen prior to conditioning by CFs. An alternative hypothesis (1) is that learning diminishes the efficacy of parallel fiber input. These hypotheses were tested on monkeys by recording by implanted electrodes from PCs while motor response of wrist muscles were being learned. Purkinje cells gave trains of simple spikes in response to parallel fibers and complex spikes in response to climbing fibers. When the motor response was conditioned, the frequency of simple spikes was decreased, that is, parallel fiber effects on PCs were reduced (46). In other experiments, the eyelid response (nictitating membrane ex-

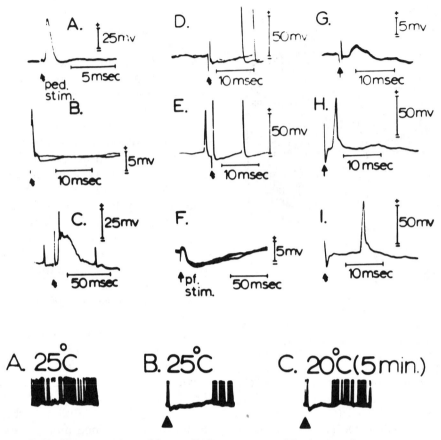

Figure G-22. Upper portion of figure: Different types of Purkinje cell responses in cerebellum of goldfish. (*A*) Antidromic response to stimulation of Purkinje axon. (*B*) Short latency IPSP produced by stimulation of cerebellar peduncle, probably mediated via recurrent collaterals of PC axons. (*C*) Monosynaptic climbing fiber response. (*D* and *E*) Disynaptic excitatory responses due to stimulation of mossy fiber via granule cell–parallel fiber system (*F*) IPSP produced by off-beam parallel fiber stimulation, a disynaptic system of parallel fiber–stellate cell synapses. (*G*) Monosynaptic EPSP produced by on-beam parallel fiber stimulation. (*H*) Spike elicited by parallel fiber stimulation. (*I*) Disynaptic response elicited by mossy fiber stimulation. Lower portion of figure: Ongoing spikes in a Purkinje cell. (*A*) control activity. (*B* and *C*) Inhibition (interruption) of firing due to stimulation of mossy fibers in cerebellar peduncle probably mediated via Golgi cells acting on granule cells. [From Friedlander et al. (38a).]

tension over eyeball) to corneal stimulation was conditioned (150). Responses of the deep nucleus (nucleus interpositus) developed during the conditioning. Lesions of this nucleus abolished establishment of the conditioned reflex (93). Others report that cerebellar cortical ablation also abolishes conditioned eyeblink responses (164).

HIPPOCAMPUS. While the hippocampal formation takes many forms across species, it has been of particular interest because of its apparent involvement in the formation of long-term memory in humans. Studies of a neurological patient (138) first drew widespread attention to the importance of the temporal lobe and underlying limbic regions in memory. The patient who underwent surgical resection of the anterior temporal lobe, amygdala, and much of the hippocampus for relief of intractable epilepsy in the mid-1950s is still incapable of forming memories of facts and experiences he has subsequently encountered. Recent findings with other patients, some with precise hippocampal lesions, specifically implicate the hippocampus in these memory disorders. Significantly, the patient is capable of learning new perceptual and motor skills, a result that suggests that the brain, at least in humans and some other primates, has two or more separate memory systems, one for facts and experiences, another for skills (144). Analagous findings have been described in nonhuman primates.

In rodents, the hippocampus is less involved in memory and seems to play roles in spatial or "place" working memory or behavioral context (78). For example, damage to the hippocampal formation or disconnection by fornix transection impairs spatial orientation. Rats with such damage cannot learn to swim directly to a platform submerged in opaque liquid or to retrieve without error food pellets placed at the ends of a set of arms radiating in all directions from a central starting location. Single neurons in the hippocampus often are activated when the rat enters specific places in the environment and are virtually quiescent otherwise (79). Different neurons encode different places, such that a "map" of space in the hippocampus has been proposed (10). Spatial maps cannot account for all hippocampal functions (102). Some nonspatial tasks, particularly those requiring rats to take note of changes in the contingencies between behavior and reward, are also impaired. These findings seem quite different from those from human patients. With some olfactory tasks, memory failure is seen in rats with ventral hippocampal lesions (144a).

The principal connections of the hip-

pocampus are relatively simple. However, the widespread connectivity of the hippocampal formation, its afferent structures, and its target structures may make it one of the most integrative structures of the forebrain (158). The hippocampal formation in prototypical mammals is often described in terms of the "tri-synaptic circuit" of the lamellae illustrated in Figure G-23. The entorhinal region of the cerebral cortex, which receives input from a number of neocortical regions, projects to the dentate gyrus in a "many-to-many" fashion, in which each axon passes through the dendrites of many granule cells. Axons of the granule cells, the mossy fibers, connect with CA3 pyramids at thorny excrescence glomerular synapses near the soma. The CA3 pyramids project, via Schaffer collateral axons, to CA1 pyramid dendrites, which, in turn, project to the subiculum, the primary source of hippocampal output through the fimbria-fornix pathway to the basal forebrain. Less commonly studied are an array of ipsilateral and contralateral and longitudinally organized intrahippocampal connections (84a).

This simple circuit, which is repeated throughout the length of the hippocampal formation, provides very good experimental control and has led to use of the hippocampus as a model system. A feature first discovered in the hippocampal formation and since demonstrated in many other CNS and PNS regions is long-term potentiation (LTP). Following brief bursts of high frequency activation of any of the trisynaptic connections, the amplitude of the compound or "population" epsp, and usually of the population spike, is increased. This alteration in the strength of a synaptic pathway persists for days or weeks, possibly indefinitely, given proper inducing conditions (144a). It has been considered a possible model for memory in the CNS. Proposed mechanisms range from increases in the diameter of the stem of the postsynaptic

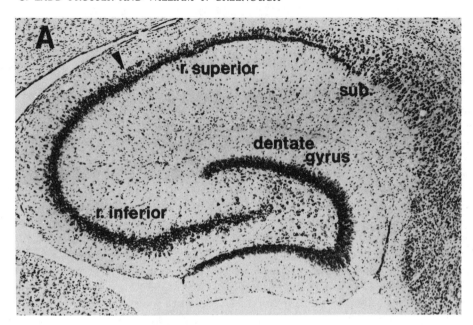

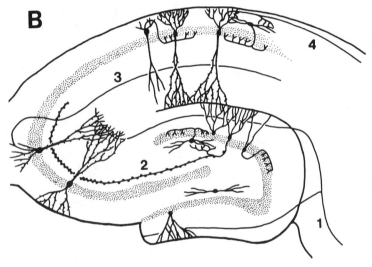

Figure G-23. (*A*) Nissl-stained section transverse to the dorsomedial-ventrolateral axis of the hippocampal formation, revealing the major subdivisions. Arrow marks transition between regio inferior and regio superior, which, together are also referred to as *Cornu Ammon* (CA) or Ammon's horn. Sub = subiculum. (*B*) Schematic of the primary intrinsic connections in the transverse plane: (1) entorhinal cortex axons project to dendrites of the granule cells of the dentate gyrus and to outer dendrites of regio inferior pyramidal neurons; (2) granule cell axons project from dentate gyrus to regio inferior (CA3 and CA4) pyramidal cells; (3) regio inferior pyramidal cells project, via Schaffer collateral axons, to regio superior (CA1) pyramidal cell dendrites; (4) regio superior axons project to the subiculum, a source of many of the major hippocampal formation efferent projections. Collateral axons to inhibitory interneurons are depicted in the dentate gyrus and regio superior. [From Rose (129a).]

spine, reducing resistance to synaptic current, to changes in neurotransmitter release or receptor density, to the formation of new synapses (25).

Sleep

Sleep is a reversible state in which animals are behaviorally quiescent, assume a characteristic posture, and are not readily aroused (13). In mammals, birds and reptiles, brain waves (EEG) show a characteristic rhythmic sequence. Behavioral sleep has been observed in fishes, insects (cockroach, bee), and molluscs (*Aplysia*). Sleep is periodic, usually circadian but in the presence of constant illumination sleep periods may drift while body temperature cycle changes less. Times of onset and termination of sleep can be altered by temperature, light, nutrition, and psychological state. Some animals sleep by day, some at night. In rats, the suprachiasmatic nucleus is an important part of the circadian clock, and lesions to this nucleus interrupt sleep rhythms (Chapter 8). Why nervous systems require sleep as a state of periodic interruption of their activity is not known (13, 72).

During a sleep period (8–12 h) changes in electroencephalographic (EEG) patterns occur periodically (human and cat) (156). Four stages in EEG activity are recognized in human slow-wave sleep plus one stage in paradoxical or rapid eye movement (REM) sleep. As a person enters sleep, the EEG record becomes nearly flat (stage 1), then gradually through two stages low-frequency, synchronized, high-amplitude brain waves appear (stage 4). After ~1 h of sleep the brain waves become desynchronized (low voltage), heart rate and blood pressure increase, respiration is accelerated, muscle tone declines but rapid eye movements occur. Dreams occur during the REM period; as sleep goes on the REM periods lengthen. There is a reciprocal relation between REM and stage 4 (high-voltage slow waves) (156). Like mammals, birds

alternate slow-wave and paradoxical sleep. Slow-wave sleep has been observed in a tortoise.

Several areas of the brain may interact to induce sleep—hippocampus, raphe, locus coeruleus, suprachiasmatic nucleus, and midbrain tegmentum. Some neurons in these regions are serotonergic; blocking 5-HT synthesis temporarily leads to wakefulness. Adrenergic neurons in pontine tegmentum and locus coeruleus appear necessary for deep REM sleep. A population of cholinergic neurons in tegmentum are active during REM sleep. A sleep promoting peptidoglycan with a muramic acid residue, molecular weight less than 500, derived from cerebral-spinal fluid of sleep-deprived goats, when injected into other animals increases slow-wave sleep but not REM sleep (118).

Many animals—frogs and beetles—show tonic immobilization or death feigning in response to threatening stimuli. The neural mechanisms are unknown.

Hibernating mammals sleep for long periods in winter and interrupt sleep periods by occasionally awaking (Chapter 3). Electroencephalographic recordings from hibernating mammals show slow wave but essentially no REM sleep. Activity in the hippocampus is the last to go out during entry into hibernation. The turnover of 5-HT in brain increases very much during hibernation and 5-HT applied to hippocampal slices modulates spike activity at temperatures below those at which slow waves are altered.

The functions of sleep remain largely unknown. That it plays important roles is indicated by studies of mood and memory disorders associated with sleep deprivation or dysfunction.

Development of Nervous Systems; Embryonic Motility

Developmental neurobiology is concerned with two general problems: (1)

growth and differentiation of neural structures, and (2) development of behavior in relation to sensory, central, and effector organs. Neural development is largely similar in all animals, there are differences in details (127).

Specific features of nervous system development are coded in the genome and are expressed sequentially. Motorneuron connectivity, neuronal death and synapse death are activity-dependent. The pattern of neuronal development is determined (1) genetically, (2) by cellular environment, and (3) by experience. The actions of the three determinants provide for location of nerve cells and for pathfinding by processes of neurons. In neural development two patterns are recognized: (1) cell lineage, determined early in development, stereotypy; (2) precursor cells that can give rise to any of several types of neurons, induction of differentiation. Development by lineages has been more studied in invertebrate animals and common precursors are commonly identified in vertebrates. Lineages have been identified by marking specific neurons early in development and following their progeny. The nematode *Caenorhabditis* has a fixed number (302) of neurons; all of these arise in accord with a stereotyped sequence of proliferation and cell death. Specified cells are derived from identified progenitor neuroblasts (14). Nerve cell lineage has been followed in leeches and in insects (grasshoppers); these animals are characterized by mosaicism in eggs, rapid differentiation, and limited ability of embryos to replace missing cell lines after ablation. Recent evidence from following labeled neuroblasts in mammalian embryos indicates that neurons of cerebral cortex and some other regions of the brain are formed by cell lineages (85).

A general property of neural development in vertebrates is induction or the altered differentiation of one tissue by contact or proximity with another. Several mechanisms of induction have been proposed. Experiments with vertebrates (transplantation, induction, and making of chimeras) show wide ranges of possible fates of neuroblasts. In an early (neurula) stage, a dorsal neural plate differentiates from ectoderm and becomes grooved; this forms a neural tube with the central canal that ultimately becomes the brain and spinal cord. Early in this sequence, mesoderm migrates beneath neuroectoderm. Ectoderm does not form a neural crest and tube in the absence of induction by underlying mesoderm (152).

Extension of neural processes involves terminal swellings called growth cones (Fig. G-24). Filopodia extending from growth cones draw them along, some filopodia and growth cones succeeding while others withdraw. Filopodial extension involves polymerization of tubulin, actin, and other cytoskeletal elements. Extension and withdrawal are made evident by labeling experiments (e.g., by HRP, horseradish peroxidase) and are influenced by the immediate cellular environment. How decisions are made regarding persistence, withdrawal, and establishment of ordered arrays of connections is a fundamental problem of neural development.

A well-known example of specificity of connections is the distribution of retinal ganglion cell fibers over the optic tectum in fishes, frogs, and salamanders. Optic nerve fibers enter the optic tectum and receive information that guides the fibers to appropriate locations: dorsal retina to ventral tectum and ventral retina to dorsal tectum. If the optic nerve is crushed in mature frogs, retinal axons labeled with HRP migrate by sprouting and branches spread randomly over the tectum while other branches are retracted or atrophied (142a). In early stages (prior to stage 29

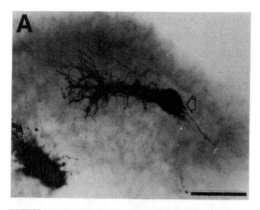

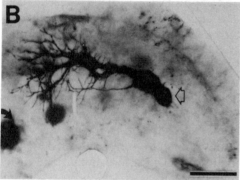

Figure G-24. Illustrations of growth cones. Photographs of growing neuron from *in vivo* preparation of develop in insect embryo. [From Caudy and Bentley (13a).]

in frog embryos) there is no such specificity and retinal fibers do not go to specific targets. If the optic nerve is crushed in adult frog or late embryo and the eye is inverted and axons allowed to regrow, electrical recordings from the tectum show an inversion of the normal visual field projection pattern. In *Xenopus,* surgical rotation of the eye at stage 28 results in normal retinal projection, but after rotation at stage 30, orientation of the projection is inverted in the nasotemporal axis but not in the dorsoventral axis; at stage 31 and later, the projection is rotated in both axes (68). In chickens, no defects in visual responses occur after the eye has been rotated and reimplanted before 70 h of incubation.

Specification of neural development is shown by exchange of patches of skin between the belly and back of tadpoles and frogs. In early tadpoles, stimulation of transplanted skin patches elicits reflexes normal to that body region but in late stage tadpoles, the specification of sensory neurons has occurred and the reflex responses are misdirected. It is concluded that sensory neurons instruct central connections and that sensory neurons become specified according to their body location during late stages of development (68). Experience may also affect the specification of connections, particularly in mammals (142a).

In kittens, closure of the lid of one eye for a few days during the period from 4 to 6 weeks after birth results in loss of pattern discrimination as recorded from cells in the striate cortex (139c). In other experiments, kittens were reared, surrounded by either vertical or horizontal stripes or were reared with goggles to present vertical stripes to one eye, horizontal to the other eye; after several months, binocularly sensitive neurons of the visual cortex showed preference for the experienced orientation. A large proportion of cells of the visual cortex had one orientation specificity; it is uncertain whether this enhancement is due to death of some cells or transformation of developing neurons (139c). A general conclusion is that certain developmental periods are critical for neuronal connections and that neuronal specificity increases with development.

Hypotheses accounting for the differentiation and guidance of growing axons toward their targets are (127)

1. Field effects: The brain has polarities unspecified in nature, that guide growth of processes to specific locations.

2. Chemotropism: Some chemical compound or trophic factor of the target

provides guidance. Such a substance is nerve growth factor (NGF) a protein of 130,000 MW, consisting of three subunits in the ratio $2\alpha:\beta:2\gamma$. Nerve growth factor acts primarily on sympathetic ganglia and sensory ganglia of neural crest origin (79a). Hormones that influence general growth may also have regional effects. Aggressive behavior of mice causes massive release of NGF (3, 79a). Other hormones, such as thyroid hormone, modulate neural development (106).

3. Mechanical support, stereotropism: Growing axons may be guided by other axons and by glia. In tissue cultures, axons follow supporting cells. In the cerebellum, developing granule cells migrate along supporting glia. In cerebral cortex radial glia guide neuronal migrations from subependymal proliferative zones into position. Glial bridges appear instrumental in specifying axon trajectories in the corpus callosum.

4. Galvanotropism: Electrical gradients occur in many polarized structures, for example, between regions of nervous systems. In Cnidarians, an electrical gradient can be measured between head and base of polyps. When a field of reverse polarity is applied, regeneration can be reversed. In neuronal tissue cultures, electrical fields alter growth of processes (70).

5. Cell adhesion molecules (CAMs): These are glycoproteins of MW 180,000–250,000, varying in sugar composition with brain region. These substances aid cell–cell attachment. Different CAMs have been isolated from neurons, glia, and liver (28).

6. Transmitters and receptors: Reciprocal inductive effects occur between presynaptic terminals and postsynaptic receptors. Sympathetic neurons normally liberate norepinephrine (Chapter 1), but when cultured with parasympathetically innervated cells such as heart, the sympathetic neurons are caused to produce acetylcholine and some cells produce both transmitters (41). An example of postsynaptic effect is regulation of myosin production in muscle by motor nerves; when nerves are crossed between "fast" and "slow" muscles, the myosin type can be reversed and physiological properties of the muscles reversed (Chapter 2). It is uncertain whether this action is caused by differences in discharge patterns of motor nerve firing or by some unidentified trophic factor. Ganglia of leeches contain large serotonergic neurons, Retzius cells (Rz), that normal innervate skin but in segments 5–6 they innervate gonads. Retzius cells of most segments are activated from pressure receptors of the skin, but not by the skin pressure in segments 5–6. Removal of gonads early in development causes Rz cells of those segments to develop so as to respond to skin stimulation, hence the functional identity of these neurons is specified by the target they contact during development (83).

7. Competition between neurons and death of neurons are events of normal development. In the silkmoth *Manduca*, one half of the neurons in abdominal ganglia die in the transition between pupa and adult. Cell death is regulated by the hormone ecdysone. In anurans, the Mauthner cells of tadpoles disappear during metamorphosis. In chick embryos, some 40% of spinal cord motorneurons die after 5–9 days incubation; in *Xenopus* embryos 75% of motor neurons die at metamorphosis. Neuron population is adjusted proportionally to the size of a projection field and misdirected neurons are eliminated. In rat cerebellum added thyroxin increases granule cell death (22).

A preganglionic column in chick embryos has 9300 neurons on day 8; 6900 on day 10. Injection of NGF produced an increase in volume of caudal thoracicolumbar sympathetic ganglia and in the number of motorneurons in corresponding preganglionic cell column; NGF also decreased the number of cell deaths (117). Chronic treatment of chick embryos with curare or bungarotoxin reduces death (116, 117) hence death of motorneurons involves competition for targets.

8. Genetically programmed growth; in early development of the CNSs, connections can be made in the absence of neural input. When salamander larvae were anesthetized for many hours, neural development proceeded and on emergence from anesthesia, reflexes were the same as if the animal had been performing the usual intermediate stages of behavior. Explants of rat spinal cord in culture can be anesthetized by xylocaine, which blocks the normal electrical activity and yet neuronal development continues (22a).

9. Sensory deprivation results in developmental deficits both centrally and in sense organs. Sensory dendrites fail to develop in the absence of axons that normally terminate on them. Visual deprivation reduces concentrations of RNA, AChEs (acetylcholinesterase) in retina. In kittens, effects on cortical neurons of alterations in visual patterns were described previously.

10. Patterned sensory influences modulate development of ordered arrays of peripheral input connections. In general, asymmetric deprivation has more profound effects than symmetric deprivation; for example, monocular deprivations in cats and monkeys. Central processing involves competition; in the visual cortex, overlapping axonal projections of each eye normally shrink back to equal size columns for each eye. After monocular deprivation that eye's columns shrink excessively while columns corresponding to the open eye become expanded. The competition is activity based. Impulse blockage by TTX in one retina is like monocular

deprivation. Electrical stimulation of an optic nerve results in control of cortex neurons (144a). Data for other sensory systems suggest similar affects of sensory activity on brain development:

In rats, each olfactory bulb receives its entire afferent input from the ipsilateral olfactory mucosa. Olfactory deprivation can be produced by lesions in newborn pups, by application of a poisonous solution to nasal membranes, or by obstructing one of the nares. Unilateral olfactory deprivation in young rats results in reduced size of the corresponding olfactory bulb. The anosmic bulb is also reduced in RNA and DNA content and in Na^+-K^+ ATPase. Deprivation results in reduction in numbers of mitral and tufted cells in olfactory glomeruli of the bulb (95). In olfactory deprived bulbs, a catecholamine-synthesizing enzyme in the glomeruli was reduced but not a GABA synthesizing enzyme (95).

Auditory input is essential for development of normal electrical responses to sound in inferior colliculi of rats. Monaural deprivation reduces collicular responses of adults if the deprivation is in a critical period between 10 and 60 days after birth (20).

Development of Behavior

Some kinds of embryos can move before neural control is established. In the earthworm *Eisenia*, following a ciliated gastrula stage, the embryo shows spontaneous contractions around the stomodeum, then in response to mechanical stimulation both local and conducted contractions occur. After ganglionic organization, the anterior end turns away from the point of stimulation. Peristalsis spreads backward as the nervous system develops (125a).

In embryos of a dogfish *Sciliorhinus*, rhythmic contractions of myotomes occur on each side independently before neural connections to them are present and also after the nerve cord is removed; myogenic contractions are replaced by neurally elicited ones (154).

Embryos of a salamander, *Ambystoma*, show the following sequence: A nonmotile stage of responses to stimulation but no spontaneity, a stage of simple flexure and spontaneous bending, an S stage, and finally locomotor waves of contraction. The first responses are of gross regions, local reflexes are individuated later (21).

In chick embryos at the limb-bud stage, random movements occur, first in the neck, then in the trunk, and finally in the limbs. Recordings from the spinal cord show spontaneous electrical activity correlated with the random movements. Deafferentation was performed by removal from 2-day embryos of the dorsal half of the lumbar cord and transection of thoracic cord; periodic leg movements were similar in control and deafferented embryos. The adult pattern of coordinated reflexes occurs by day 17. It is concluded that in birds, also probably in mammals, motor coordination can occur before peripheral sensory input modulates motor neurons. Apparently, spinal neurons show endogenous activity very early and organized motor activity is established before the nervous system receives sensory information.

Learning and Memory

The ability to acquire from experience is one of the most essential adaptations of organisms with large forebrains, particularly birds and mammals. The ability to know one's own environment, as opposed to reacting to each situation as if it were novel, and employing stereotyped behavioral patterns, would appear to confer enormous survival advantages, as suggested by the amount of total metabolic capacity devoted to the nervous system in mammals (as much as 20% of basal oxygen consumption in humans). The nervous system is capable of storing essentially any arbitrary piece of information (73).

Learning may be roughly categorized as follows:

1. Very simple nonassociative learning—changes in the state of responsiveness of a fixed system. Habituation and sensitization are examples of a nonassociative learning. Habituation is a decrease in the response to a regularly repeated stimulus, such as the mild tactile sensation produced by a water jet applied to the siphon of the mollusc *Aplysia californica*. Sensitization is a generalized increase in responsiveness to relatively neutral stimuli following presentation of a highly salient stimulus, such as one that induces pain. These processes are discussed elsewhere in this chapter.

2. Associative learning, of which Pavlovian conditioning is the prototypical example. Presentation of a relatively neutral stimulus in some consistent temporal or otherwise predictive relationship with a comparatively salient stimulus brings about the learning of preparative responses to the neutral stimulus. A dog presented with a bell followed by food powder in the mouth comes to salivate upon presentation of the bell alone.

3. Manipulative, or operant learning, in which the organism performs specific behaviors that bring about positive consequences or mitigate negative consequences. This learning demonstrates knowledge of a relationship between behavior and its consequences.

This classification scheme, while viable for simpler organisms, falls short of the range of complexity encountered in the most elegant vertebrate nervous systems. Attempts to further classify learned behaviors with terms such as "insight," "problem solving," and the vastly more complicated behaviors implied by cognitive psychologists' terms "procedural, declarative, semantic, and episodic memory," "knowledge representation," and "quantitative reasoning" have highlighted what are probably important aspects of the capabilities of complex brains. In a related arena, the ability of higher mammalian brains to regain functional capacities following brain or sensory-motor system damage probably involves many similar phenomena at cellular and circuitry levels. Learning may be short-term, often by single trials, or long-term, associations persisting for long periods. Auditory conditioning at several brain levels occurs very rapidly (157).

Of interest is whether the *brain systems* underlying these types of learning and the *cellular mechanisms* underlying their storage are naturally or experimentally dissociable. Different types of learning may be associated with different kinds of plastic changes in neurons, and their characteristics may also emerge from properties of different neuronal networks. By "brain system" we mean the complete pattern of circuitry underlying a behavior and the functional changes that occur within the pattern when learning takes place. By "cellular mechanism" we mean the changes within individual cells that give rise to changes in circuitry.

One example of an associatively learned response that is reasonably well defined at a system level is eyeblink or nictitating membrane conditioning. This has been studied in numerous mammalian species, including extensive study in humans. In eyeblink conditioning, a neutral conditioned stimulus (CS), such as a tone, is paired with an unconditioned stimulus (US) that elicits an eyeblink unconditioned response (UR), such as a puff of air to the eye; after a few tens or more paired presentations, the CS comes to elicit an eyeblink conditioned response (CR). Speed of learning depends on a number of variables: time between CS and US onset (optimally around 0.5 s), whether the CS terminates (trace conditioning), or overlaps with the US (delay conditioning), and so on. Neural substrates of eyeblink conditioning in the rabbit have been studied by Thompson and others (86, 150). In rabbits, the eyeblink involves retraction of the eyeball by the retractor bulbi muscle, which allows an inner eyelid, the nictitating membrane, to slide across the eye. A combination of lesion, stimulation, and unit recording studies, established that (1) the forebrain is unnecessary for simple delay conditioning, (2) a circuit involving the cerebellum is required for the appearance of the CR, (3) noncerebellar brainstem structures are sufficient to mediate the UR, and (4) spike activity in both cerebellar cortical Purkinje neurons and neurons in the deep nuclei to which the PC cells project is altered preceding the conditioned response following presentation of the CS. As is evident in the circuit diagram (Fig. G-25), the CR pathway could course either through the cerebellar cortex or the interpositius nucleus (or both), a point that is currently disputed (164). When electrical stimulation of the dorsolateral pontine nucleus is substituted for the tone CS, cerebellar cortical lesions abolish CRs, indicating mandatory cortical involvement. This procedure (49) demonstrated that training alters the size of the spiny branchlets on Purkinje cells, the site of parallel fiber termination (119). This indicates possible involvement of new synapse formation in the conditioning process, a putative memory mechanism that has found additional support from other associative learning studies (9, 35). For example, Tsukahara et al. (151) proposed changes in synaptic connections between the somatosensory cortex and red nucleus-mediated fore-

Sectioned Folia HVI of the Rabbit Cerebellum

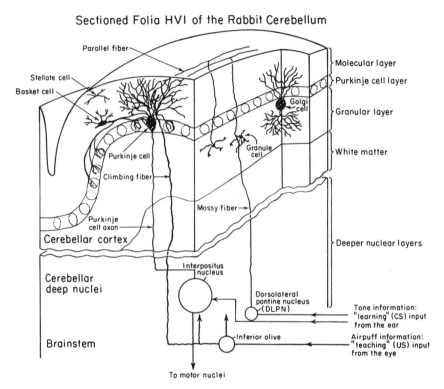

Figure G-25. Proposed circuit underlying the conditioning of the nictitating membrane response. The CS tone information passes through the dorsolateral pontine nucleus (DLPN) and is projected via the mossy fiber input to Purkinje cell dendrites and to inhibitory interneurons. The airpuff information from the eye is projected through the inferior olive to both the cerebellar cortex, via climbing fibers, and to the interpositus nucleus. The Purkinje cells, as the only output neurons of the cerebellar cortex, integrate the parallel and climbing fiber input to modify the nictitating membrane response through projections to the interpositus nucleus. Morphological changes were detected in the spiny branchlets of Purkinje cells, the sites of parallel fiber synapses, which are associated with the conditioning process. [Figure adapted from Ghez and Fahn (45a) and Thompson and Donegan (150a).]

limb conditioning. Sprouting of cortical afferents to red nucleus neurons, which increased the aggregate cortical connectivity, occurs in cats conditioned with electrical stimulation of the cerebral peduncle as the CS and forelimb shock as the US.

While cerebellar and brainstem structures have been found to be essential to the performance of several conditioned responses, forebrain structures also show involvement in associative learning processes. Neuronal activity in hippocampal

formation matches the CR in much the same manner that cerebellar neurons do, and failure of the neural response corresponds well to instances of failure of the CR. However, hippocampal damage does not impair delay conditioning, suggesting that the hippocampus may monitor the conditioning process without being critically involved in the response. Similarly, during auditory conditioning (involving an electrical shock US), altered neuronal responsiveness to the CS is evident in auditory afferent structures such as the medial geniculate nucleus and primary auditory cortex (15, 16, 42, 43). There is also evidence for alteration of response characteristics of motor cortex neurons during conditioning of very short-latency eyeblink response in the cat (160a). The motor cortex neurons increase membrane resistance as a result of an extensive training process.

Manipulative learning systems have helped to elucidate the organization of neural systems underlying behavioral choice. Mishkin (99a, b) has elaborated a detailed model of temporal cortical, limbic, and thalamic structures involved in two categories of memory in the monkey. The first, *recognition memory*, is a rapidly acquired memory for information such as the characteristics of objects. The prototypical task is delayed nonmatching to sample (DNMS), in which the monkey sees an object and shortly thereafter must select a novel object in preference to the previously seen object (for food reward). Based primarily upon studies of monkeys with damage to various limbic and temporal lobe areas, Mishkin proposed that this object recognition learning depends on circuits involving temporal areas of the cerebral cortex, the hippocampus, the amygdala, and the limbic (anterior and medial dorsal) thalamus. The second type of memory, *habit*, requires repeated practice with a restricted set of pairs of objects, such that the monkey learns that a particular object of a pair is *always* the correct choice. This habit memory is largely insensitive to damage to the structures involved in recognition memory and may be more specifically associated with primary sensory cortex regions or with non-neocortical regions. This model has many characteristics in common with the model of human memory derived from hippocampal damage studies that is discussed elsewhere in this chapter.

Recognition memory and habit may interact in hippocampal and other limbic cortical areas (42, 43). Rabbits avoid a shock by stepping in response to a tone CS. Neurons in the limbic thalamic nuclei increase their firing in response to the tone CS but not in response to a different one that does not predict shock. Lesions in these nuclei block conditioning. However, lesions of the hippocampus and limbic cortex increase the firing of limbic thalamic neurons and behavioral hyperresponsiveness to the CS. These finding indicate that neurons in the limbic thalamic nuclei participate in a type of habit, that is, cue-driven behavioral CR. The limbic cortical memory system inhibits the thalamic excitatory system when something unusual occurs in the learning environment. Thus, the limbic cortical memory system has veto power over habit system-mediated behavioral performance. It uses data from the neocortex to decide when and when not to allow the behavior to occur (104).

Most of the preceding accounts of learning deal with observations on mammals. Conditioning is known for other classes of vertebrates but neural mechanisms are less well established. Birds, for example, pigeons, readily learn to select grain of different colors. Fishes, especially goldfish, have been conditioned to visual and auditory CS for a variety of responses, bradycardia, respiratory arrest, feeding, color selection, and operation of levers for food. In addition, focal stimulation in telencephalon in fish can serve as a CS. Examples of conditioning in molluscs, insects, and cnidarians were given in early sections of this chapter. A general property of nervous systems is plasticity, more in integrative than in relay or reflex systems. In the superior temporal sulcus of awake monkeys, some units respond to visual stimulation by a face—human or monkey, some cells are sensitive to orientation, full face or profile, many respond to gaze, some to rotation of face (121). Neurons that respond to faces do not respond to bars and edges; a cell may be selective for one view of a head. Short-term memory may be by different mechanisms than long-term memory (56).

Measurements on mammals indicate that different types of memories are associated with different functional systems. In contrast, some evidence suggests that multiple mechanisms may work in parallel at the cellular level in the memory process. Moreover, similar mechanisms appear to occur in organisms as diverse as molluscs and mammals. In the marine mollusc *Hermissenda*, Alkon (2) and colleagues reported that associative conditioning of a defensive response to US using a photic CS, involved a reduced conductance of K^+ channels in the photoreceptor cell. In parallel, (25) similar potassium conductance reduction underlies increased excitability of hippocampal neurons involved in the eyeblink conditioning paradigm described above. Similarly, the formation of new synapses appears to be intimately tied to the formation of long-term memory in both molluscs and mammals (6). An extensive body of data has linked learning to the formation of new synapses in mammals. For example, rats learning maze tasks show increased branching of dendrites in visual cortex (49). This anatomical effect is restricted to one hemisphere if the eye on the opposite side is blocked and the corpus callosum that connects the hemispheres is severed (15, 16). Brightness discrimination learning, but not the physical activity of performing in the training apparatus, increases synapse numbers in the hippocampus of rats (158). Similarly, motor skill learning is associated with synapse formation in the rat cerebellar cortex that does not occur in animals performing physical tasks that do not involve significant new learning (9). In chickens, learning to avoid an aversive substance in feeding increases axodendritic synapse numbers in the intermediate medial hy-

perstriatum ventrale, a region thought to be involved in this learning (119). Paralleling this at the molluscan level, sensitization of the gill withdrawal reflex in *Aplysia* is associated with an increase in synapse numbers on the sensory neuron axonal arboration, while habituation is associated with a decrease. Thus, while the relationship of these changes to behavioral changes has yet to be conclusively established, available data suggest that, in addition to synapse formation, a variety of cellular mechanisms, many of which are common to nervous systems at diverse phyletic levels, are involved in the process of learning and memory.

Section H
Behavioral Overview

C. Ladd Prosser

Comparative cognitive science is a description of complex behavior in terms of the CNS in a variety of animals and in animals at different stages of development. Many descriptions of animal behavior are couched in anthropomorphic terms. Comparative cognitive science starts with a classification of behavior. Eight levels are recognizable (125b):

Direct Response to Environmental Change or Stimulation
This is the simplest behavior. The response may be positive or negative; it is genetically programmed and stereotyped. Several kinds of direct responses are recognized: thermotaxis, phototaxis, geotaxis, galvanotaxis, and thigmotaxis. In animals without nervous systems taxic responses are mediated by cell membranes. Paramecia respond to touch—thigmotaxis—by reversing the beat of cilia, triggered by an action potential. Mutant paramecia such as "pawn" lack Ca^{2+} channels and fail to respond to mechan-

ical stimulation. Photoactic flagellates orient toward a light. When presented with two lights, flagellates go to the brighter of two lights, some go between two according to relative intensity. Animals with nerve nets or simple ganglionic nervous systems—larvae of invertebrates, adult crustaceans and insects—make direct taxic responses. More complex animals respond directly by chains of reflexes often in a heirarchy of levels. A response to one stimulus may be potentiated or antagonized by another stimulus. Direct responses are summed into complex behavior.

A Direct Response May Be Modified by Experience

Memory may be short or long term. Modification may be by habituation or by conditioning with association of conditioned stimuli and unconditioned stimuli. For neurons of a single ganglion in gastropod molluscs conditioning by paired touch and shock stimuli, delivered to the nervous system by different afferent nerves, has been described. This is associative or learning at the cellular level.

An example is conditioning of feeding behavior in *Pleurobranchea* that occurs in a central pattern generator consisting of an interneuron that receives input from chemoreceptors and sets up a cycle of alternating excitation of protractor and retractor neurons. After food-avoidance conditioning, the interneuron fires tonically to cause persistent excitation of retractor neurons and inhibition of protractor neurons, hence feeding is suppressed, a state that persists for hours or days (84).

Reactions to a Complex Environment Result in Selection of Relevant Elements Followed by Decision Making

In a snail *Pleurobranchaea*, escape reactions have the lowest threshold, other behaviors in decreasing dominance are egg laying, feeding, gill, and siphon withdrawal; righting and mating are equally less in strength than feeding. A foraging bee makes decisions as to which flower it will visit according to color, abundance, nectar, and recency of visitation. Many animals devote more time and effort to locating food and to feeding than to any other behavior. A caterpillar may feed on only one kind of leaf. A carnivore must balance the energy used in a hunt against potential nutritional value to be derived. At some seasons feeding is less important than mating, reproduction, and care of young; this applies especially to some birds and mammals. A decision whether to feed or to seek a mate is made on the basis of direct responses modulated hormonally or neurally. Gulls and crows on a rocky seashore feed on snails that they carry to 5 m and drop to break open, then eat; from a population of snails the birds select the largest (165).

One useful breakdown of memory proposes two kinds: working memory is short term and recent actions; reference memory is long term and refers to information content, codes, and uses. Both short- and long-term memories are shown by rats running in mazes. Working memory recalls turns and areas recently experienced, reference memory is retention of a spatial map of the maze. Working memory contains up to ~12 items (115). Marsh tits hide food—nuts and insects— under leaves or bark and recover the stored food a few hours later. If one eye is covered during the hiding and then exposed while the other eye is covered, the birds fail to remember where the food was hidden (141). Retrieval of stored food depends on spatial memory and damage of hippocampus in food-storing birds disrupts memory of sites (77).

Communication between Individual Animals of the Same or Other Species

Communication uses several kinds of sensory modalities—chemical pheromones, sound vocalizations, electric discharges, and luminescent flashes. Communication can be genetically determined, learned, or modified by experience during development.

Chemical signals serve many functions in insect communication—species recognition, attraction to the oppposite sex, aggregation, and colony formation. An example of a sex attractant is bombykol given off by a female silkmoth that can attract a male silkmoth in extremely low concentrations—a few hundred molecules per milliliter of air. Several insect pheromones are produced by different glands in various insects. Most pheromones are long-chain alcohols or aldehydes. Specialized chemoreceptors are present in recipient insects. Insect pheromones are used in reproductive behavior (Lepidoptera), in establishing trails (ants), in signaling aggregation (pine bark beetles). Chemical communication is used by mammals as sex attractants, in establishment of territorial boundaries, and in species and individual recognition. The evolution of chemical signals, specific receptors and the complex behavior resulting in attraction or repellence presumably occurred in coordinated fashion. An example of chemical signals to other species is the production of a distasteful or toxic substance that is repellent to predators, for example, substances in monarch butterflies that repel birds. In lizard colonies, aggressive behavior leads to social hierarchies; lesion to forebrain (amygdala) results in demotion of a dominant individual (149).

Electric fishes communicate by electrical discharges especially in turbid water. Frequency of discharge changes with posture, locomotor attacks, dominance, and sexual actions. Individual fish have a "preferred" frequency range within which discharge is maintained with very little variation. When an electric fish encounters another fish that is discharging close to its own frequency, the two fish avoid jamming by changing discharge rate up or down in the jamming avoidance reaction (Chapter 6).

Communication by sound is used by insects, fishes, amphibians, birds, and mammals. Male crickets sing three kinds of song—calling, courtship, and aggression, each delimited in frequency and pattern (Chapter 5).

Frogs use patterned sounds for communication. A male bullfrog gives at least five distinct calls—mating, release, distress, warning, and rain. Mating calls are species specific. The auditory systems of frogs and toads are adapted to receive cells of specific frequency and pattern (Chapter 5).

Sound communication, innate and learned, is highly developed in birds. Some birds such as mockingbirds and myhnas are imitators and an individual can give many kinds of calls. Some birds, such as robins, sing only a few songs, species specific, probably inherited and improved by practice. In chickens, the territorial dominance crowing of a rooster is unlike the cackling of a hen after egg laying, clucking to chicks, or alarm cry on sighting a hawk. Songbirds derive song patterns from innate templates, from hearing other birds of their species and from hearing their own song. A male song sparrow's innate template is sufficient for development of its specific song if the bird can hear itself; a white-crowned sparrow can sing its song only if it has heard other white-crowned sparrows. A zebra finch memorizes its song components during the first 14–15 days after hatching, well before it begins to sing (74, 89, 90, 108). An indigo bunting in isolation develops abnormal songs. Local dialects are recognized in several bird species, learning is necessary for proper song development. Bird songs are used to establish and maintain territories and for mate attraction. Singing by canaries is lateralized to one side of the brain as is language in humans (112–113a).

Male canaries sing only during the breeding season and one bird may sing 25 patterns of syllables grouped in phrases. The repertoire is redeveloped each spring and new songs are added. If a singing canary is deafened, its old patterns remain but new ones do not develop. If the left hypoglossal nerve is transected, the song becomes disturbed; cutting the right nerve has no such effect. However, if the left side is denervated in a young canary, singing can be learned with the right side (4, 112). Singing can be induced in female canaries by giving them testosterone, the cooing by ring doves is activated by the presence of testosterone in centers that control vocalization.

In canaries, three forebrain nuclei must be present for normal singing—hyperstriatum ventrale, pars centrale (HVc), the nucleus robustus archistriatalis (RA), and

a poorly defined area (X) (Fig. H-1). These regions are larger in male canaries than in the female and lesioning results in deficits in proportion to the amount of tissue removed. The two nuclei (HVc and RA) are normally 99 and 76% larger in the spring when new songs are learned than in the fall. Neurons show increased dendritic branching in the spring, also some hyperplasia. Hyperstriatum ventrale is also involved in hearing (4, 113, 113a).

Male fiddler crabs and some other semi-terrestrial crabs lure females into their burrows by species-specific signaling. Some species wave the large chelipeds by day; some send vibrations in the sand by night, either by tapping with walking legs or by vibrating legs just above the sand (57, 133, 134). Vibration receptors on the legs respond either phasically or tonically with species optima of frequency. Interneurons in thoracic ganglia receive input from several legs and project to cells in the tritocerebrum of the brain (57). Spe-

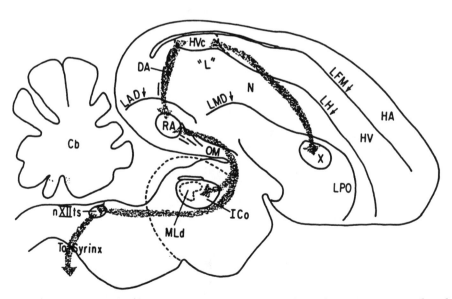

Figure H-1. Schematic view of forebrain of canary to indicate brain structure and pathways implicated in song control. Unilateral lesions to hyperstriatum ventrale, pars caudale (HVc) result in song disruptions. Lesions to robustus archistriatalis (RA) also disrupt song and pathways connect HVc to RA and area X. A direct pathway from RA goes to the motor nucleus (NXIIts) innervating the syrinx. [From Nottebohm et al (113a).]

cies that use visual signals have large optic lobes in the brain (57).

Fish make color displays by expansion of chromatophores. Many male birds are more brightly colored than females and some species display, for example, peacocks and turkeys. Baboons and mandrills display brightly colored buttocks.

Symbolic Communication

In symbolic communication signs or symbols are substituted for actions or objects. The distinction between symbolic communication and direct visual or auditory signals is not a sharp one.

Scout bees returning to the hive from foraging, communicate by waggle dance patterns that are signals for direction, distance of food source, and by nectar samples they are able to communicate the kind of available flowers. The dance converts a map to gravity (on a vertical comb) or visual patterns (on a horizontal comb), which are symbols for the map. Vervet monkeys emit different alarm calls meaning leopard, eagle, snake, or gorilla. The call is a symbol for seeing the predator and elicits appropriate evasive behavior by other vervets in the colony (139a).

Dolphins learn to execute particular behaviors when they hear certain sounds (commands); dolphins also respond with actions to signs made by their trainer. The most extensive repertoire of symbols associated with objects (usually food or toys) and actions (limb movements) is in trained chimpanzees. They interpret a trainer's sign gestures, colored blocks, or keyboard symbols as symbols for objects or actions. They distinguish foods and generalize between fruit and vegetable.

Pigeons have been trained to associate a color with pecking at an object. Other birds discriminate shapes; for example, they distinguish shapes of oak from shapes of other leaves.

The Capacity to Put Symbols Together in Meaningful Sequence

The ability to make sentences of words is the next most complex level of behavior (87). A few chimpanzees that have been trained to associate symbols with objects can assemble blocks or keyboard signs to ask or answer queries. There are individual chimpanzees who have been taught 100–200 symbols after many months of training and are able to make appropriate responses to questions or to ask questions. Several animals have originated "sentences" independently. Prolonged training in associating colored plastic blocks with objects led to enhancement of ability to match objects by size or shape and to fill in incomplete designs; however, training made no improvements in locating items in space or drawing inferences (123). Common characteristics of training with symbols are the long time required, the association with familiar objects and the trainer. It is difficult to exclude cues from trainer and surroundings, cues truly different from the inferred interpretation, the so-called "clever Hans effect."

One parrot was trained during several years to name shapes, colors, and numerical quantities (120).

Human Language Has a Cognitive Component

Language as used by humans is more than assembling symbols as sentences, more than communication by sounds and gestures (87). It has a cognitive component by which sound and words take on meaning and can be used in thought processes without speech. Human language has a grammar, arrangement of words into subject and predicate with a sentence structure. One view is that thought preceded speech. Deaf and mute persons can communicate by signing with fingers, arms, body, and facial expressions and can process visual and tactile inputs into complex reasoning.

What neurological features are so specialized as to permit human language; how and when did these features evolve? To what extent are these features specializations of properties found in higher apes? One anatomical specialization is the degree of lateralization of the cerebral cortex (155). The corpus callosum is extremely large in the human brain and continues to develop postnatally. Lesion studies show that the left side of the brain serves for spoken language; this is to some extent independent of handedness. The right hemisphere is usually nonlinguistic but recognizes and remembers sound stimuli. Lesions to the right hemisphere impair tone discrimination, the recognition of music, facial recognition, and spatial reconstruction abilities. The right side mediates emotions more than does the left. In apes, some parts of the two sides of the brain differ but lateralization is less than in humans. Lateral specialization is also seen in other mammalian groups, for example, spatial cognitive abilities in the left hemisphere of rodents (155).

A very important difference between humans and higher apes is the mechanism of vocalization. The human larynx is so constructed that vowels and consonants can be pronounced. Study of brain casts and reconstructions of the laryngeal region in fossil humans indicates that Neanderthal man could not have made the sounds of the vowels a, i, and u or the step consonants k and g. However, the presence of tools, weapons, burial artifacts, and art indicate that Neanderthal men communicated. A variety of evidence puts the origin of human language between 30,000 and 100,000 years ago (80, 81).

The most distinctive features of the human brain are the speech areas, identified by effects of lesions and by electrical stimulation. Wernicke's area is in the temporal lobe and is necessary for meaningful recognition of words and sentences. Broca's area in the left lateral frontal lobe anterior to the Sylvian fissure is necessary for controlling vocal mechanisms of speech. The planum temporale is near Wernicke's area and supplements the function of that area. No cortical areas strictly comparable to Wernicke's and Broca's speech areas are found in apes and monkeys. However, stimulation of the singular area of cortex in squirrel monkey evokes vocalization (103).

The Ability to Reason and to Think in Abstract Terms Is Primarily a Human Character

A degree of "reasoning" is found in the ability of apes to match objects and to use "tools." It is unlikely that the vocal larynx, brain areas of speech, and reception of meaningful sound evolved suddenly. It is even more unlikely that grammar, syntax, and abstract reasoning appeared without precursors. Possibly the function of the left cortex for "thought" preceded its functioning in language. Elements of language are present in higher apes, which lack the structure for vocalization, the degree of lateralization and the speech areas of the cortex found in humans.

The preceding eight categories of behavior are not sharply delimited but there is much intergradation between levels; complex behavior includes elements of the simpler grades. Examples of behavior that embraces several categories in varying amounts are territoriality, aggression, mate acquisition, mating, care of young, food location and consumption, predation, avoidance of predators, and social association.

Modern cognitive science as a discipline uses computer models that simulate some functions of human behavior. Black-box models are behaviorally heuristic and are not concerned with neural mechanisms. Neural network models represent interactive neural elements. These models have threshold functions, on–off responses, excitatory–inhibitory antagonisms, synaptic connections of different and adjustable weights, patterned organizations of elements, and modifiability of response (learning). Several real neural systems have been modeled—synaptic interactions of retinal neurons, feedforward and feedback interactions between different levels in the nervous system, open and closed loops of locomotor behavior, endogenous activity, and feature extraction in visual systems. Some nerve network models simulate neuronal systems; they can provide a quantitative and condensed description and often suggest additional measurements to be made. Unfortunately, much of cognitive science does not take account of biological plausibility and validity and gives little insight into biological processes. Much modeling lacks a neural basis (139).

References for Sections G and H

1. Albus, J. S., *Math. Biosci.* 10:25–61, 1971. Theory of learning by cerebellum.

2. Alkon, D. L. In *Memory Systems of the Brain: Animal and Human Cognitive Processes* (N. M. Weinberger, J. L. McGaugh, and G. Lynch, eds.), pp. 9–26. Guilford Press, New York, 1985. Conditioning-induced changes of ion channels in *Hermissenda* channels: Relevance to mammalian brain function.

3. Aloe, L. and R. Levi-Montalcini. *Proc. Natl. Acad. Sci. U.S.A.* 83:6184–6187, 1986. Aggressive behavior induced by isolation causes release of nerve growth factor.

4. Alvarez-Buylla, A. and F. Nottebohm. *Proc. Natl. Acad. Sci. U.S.A.* 85:8722–8726, 1988. Birth of neurons in vocal center of canary forebrain during and after song learning.

5. Armstrong, E. and D. Falk. *Primate Brain Evolution.* Plenum, New York, 1982.

6. Bailey, C. H. and M. Chen. *Proc. Natl. Acad. Sci. U.S.A.* 85:2373–2377, 1988. Long-term memory in *Aplysia* modulates the total number of varicosities of single identified sensory neurons.

7. Bass, A. H., M. B. Pritz, and R. Northcutt. *Brain Res.* 55:455–460, 1973. Telencephalic and tectal ablation effects on visual behavior of turtle Podocnemis.

8. Bell, C. C. and T. Kawasaki, *J. Neurophysiol.* 35:155–169, 1972. Responses of Purkinje cells to climbing fiber input.

9. Black, J. E., A. L. Jones, B. J. Armstrong, K. R. Isaacs, A. A. Alcantara, and W. T. Greenough. *Soc. Neurosci. Abstr.* 13:1596, 1987. Cerebellar plasticity: Learning, rather than repetitive motor exercise alters cerebellar cortex thickness in middle-aged rats.

10. Bland, B. H., P. Anderson, T. Ganes, and O. Sween. *Exp. Brain Res.* 38:205–219, 1980. Rhythms in neurons of rabbit hippocampus.

11. Brodin, L. and S. Grillner. *Brain Res.* 360:139–148, 1985. Initiation of locomotor rhythms in lamprey spinal cord.

12. Calford, M. B., M. Graydon, M. Hardon, J. Kaas, and J. D. Pettigrew. *Nature* 313:477–479, 1985. Somatotopic map in a bat.

13. Campbell, S. and J. Tobler. *Neurosci. Behav. Rev.* 8:269–300, 1984. Animal sleep duration across phylogeny.

13a. Candy, M. and D. Bentley, *J. Neurosci.* 6:364–379, 1986. Nerve growth cones in grasshopper embryo legs.

14. Chalfie, M. and M. Au. *Science* 243:1027–1033, 1989. Genetic control of neurons in Caenorhabditis.

15. Chang, F.-L. and W. T. Greenough. *Brain Res.* 232:283–292, 1982. Lateralized effects of monocular training on dendritic branching in adult split-brain rats.

16. Chang, F.-L. and W. T. Greenough. *Brain Res.* 309:35–46, 1984. Trainsient and enduring morphological correlates of synaptic activity and efficacy in rat hippocampus slices.

17. Chaudhari, N. and W. Hahn. *Science* 200:924–928, 1983; 220:824–828, 1983. Genetic expression of mRNAs in brain.

18. Chikaraishi, D. M. *Biochemistry* 18:3249–3256, 1976. Poly-A RNA in rat brain.

19. Clark, S., T. Allard, W. Jenkins, and M. Merzenich. *Nature (London)* 332:444–445, 1988. Plasticity of somatosensory map.

20. Clopton, B. M. and M. S. Silverman. *J. Neurophysiol.* 40:1275–1280, 1977. Plasticity of binaural interaction.

21. Coghill, G. *Anatomy and the Problem of Behavior.* Macmillan, New York, 1929.

21a. Constantine-Paton, M., T. Cline, and E. Debski. *Ann. Rev. Neurosci.* 13:129–152, 1990. Patterned activity, synaptic convergence and NMDA receptors in developing visual pathways.

22. Cowan, W., J. Fawcett, D. O'Leary, and B. Stanfeld. *Science* 225:1258–1265, 1984. Regressive events in neurogenesis.

22a. Crain, S. M. et al. *Brain Res.* 8:363–372, 1968. Tissue culture of spinal neurons.

23. Diamond, I. T. and W. Hall. *Science* 164:251–262, 1969. Evolution of neocortex.

24. Diamond, I. T. *Fed. Proc., Fed. Am. Soc. Exp. Biol.* 35:60–67, 1976. Organization of visual cortex.

25. Disterhoft, J. F., D. A. Coulter, and D. L. Alkon. *Proc. Natl. Acad. Sci. U.S.A.* 83:2733–2737, 1986. Conditioning-specific membrane changes of rabbit hippocampal neurons measured *in vitro.*

26. Eaton, R. C. and J. T. Hachett. In *Neuronal Mechanisms of Startle Behavior* (R. C. Eaton, ed.), pp. 213–266. Plenum, New York, 1984. Role of Mauthner neurons in fast startle in fishes.

27. Eaton, R. C. et al. *J. Comp. Physiol. A* 145:485–496, 1982. Startle responses in goldfish following Mauthner cell lesions.

27a. Ebbesson, S. *Behav. Brain Sci.* 7:321–366, 1984. Evolution and ontogeny of brain circuits.

28. Edelman, G. M. *Sci. Am.* 250:119–129, 1984. Cell-adhesion molecules.

29. Ewert, J. P. *Z. Vergl. Physiol.* 74:81–102, 1971. Responses of toad thalamic neurons to visual objects; Ewert, J. P. et al. *Naturwissenschaften* 71:590–591, 1984. Motor patterns in medulla of toads.

30. Ewert, J. P. *Brain, Behav. Evol.* 3:36–56, 1970. Neural mechanisms of prey capture and avoidance behavior in Bufo.

31. Ewert, J. P. and A. Wietersheim. *J. Comp. Physiol. A* 126:35–42, 1978. Toad visual system.

32. Faber, D. S. and H. Korn. *Neurobiology of Mauthner Cell.* Raven Press, New York, 1978.

32a. Faber, D. S. and H. Korn. *J. Neurophysiol.* 48:654–678, 1982. Inhibition of M cells.

33. Feng, A. S. *Brain Res.* 364:167–171, 1986. Afferent and efferent innervation patterns of superior olivary nucleus of frog.

34. Fester, D. and C. Koch. *Trends Neurosci.* 10:487–492, 1987. Neuronal connections for orientation selectivity in cat visual cortex.

35. Fifkova, E. and C. L. Anderson. *Exp. Neurol.* 74:621–627, 1981. Stimulation-induced changes in dimensions of stalks of dendritic spines in the dentate molecular layer of cortex.

36. Finger, T. In *Fish Neurobiology* (G. Northcutt and R. Davis, eds.), pp. 261–284, 285–310. Univ. of Michigan Press, Ann Arbor, 1986. Organization of cerebellum; gustatory system in teleost fishes.

37. Fite, K. V. *Brain, Behav. Evol.* 26:71–90, 1985. Pretectal and accessory optic visual nuclei of fish, reptiles, and amphibians.

38. Friedlander, M. In *Fish Neurobiology* (N. Northcutt and R. Davis, eds.), pp. 91–111. Univ. of Michigan Press, Ann Arbor, 1983. Visual prosencephalon of teleosts.

38a. Friedlander, M., N. Kochabakhdi, and

C. L. Prosser. *J. Comp. Physiol. A* 112:19–45, 1976. Temperature effects on goldfish cerebellum.

39. Fujisawam, H. et al. *Dev. Biol.* 90:43–57, 1982. Branching of regenerating retinal axons and selection of appropriate branches in newt.

40. Funch, P. G. and D. Faber. *J. Neurophysiol.* 47:1214–1231, 1982. Propagation in Mauthner neurons.

41. Furshpan, E. J. *Proc. Natl. Acad. Sci. U.S.A.* 73:4225–4229, 1976. Chemical transmission between rat sympathetic neurons and cardiac myocytes in culture.

41a. Furukawa, T. *Prog. Brain Res.* 21A:44–70, 1966. Synaptic interactions at Mauthner cell of goldfish.

42. Gabriel, M., S. E. Saltwick, and J. D. Miller. *Science* 189:1108–1109, 1975. Conditioning and reversal of short-latency multiple-unit responses in the rabbit medial geniculate.

43. Gabriel, M., S. Sparenborg, and N. Stolar. In *The Hippocampus* (R. L. Isaacson and K. H. Pribram, eds.), Vol. 4, pp 1–31. Plenum, New York, 1986.

44. Gaze, R. M. and M. Jacobson. *Proc. R. Soc. London, Ser. B* 157:430–448. Retinotectal projection in frog; *J. Physiol. (London)* 169:1P–3P, 1963. Convexity detectors in frog.

45. Gerstein, G., D. Perkel, and J. Dayhoff. *J. Neurosci.* 5:881–889, 1985. Computer modelling of cooperative firing in simultaneously recorded neurons.

45a. Ghez, C. and S. Fahn. In *Principles of Neural Science* (E. R. Kandel and J. H. Schwartz, eds.), p. 507. Elsevier, Amsterdam, 1985. The cerebellum.

46. Gilbert, P. and E. W. Thatch. *Brain Res.* 128:309–328, 1977. Purkinje cell activity in motor learning.

47. Gray, C. M. and W. Singer. *Proc. Natl. Acad. Sci. U.S.A.* 86:1698–1702, 1989. Stimulus-specific neuronal oscillations in orientation columns of cat visual cortex.

48. Gray, J. *J. Exp. Biol.* 13:181–191, 200–218, 1936. Locomotor reflexes in fishes.

49. Greenough, W. T., J. M. Juraska, and F. R. Volkmar. *Behav. Neural Biol.* 26:287–297, 1979. Maze training effects on dendritic branching in occipital cortex of adult rats.

50. Grillner, S. et al. *Acta Physiol. Scand.* 113:549–551, 1984; 120:393–405, 1975; *Brain Res.* 88:367–371, 1975; *Exp. Brain Res.* 34:241–261, 1979. NMDA receptors and fictive activity in spinal cord of lamprey.

51. Grillner, S. *Science* 228:143–149, 1985; 236:313. Rhythmic activity in lamprey spinal cord.

52. Grillner, S. and P. Wallen, *Annu. Rev. Neurosci.* 8:233–261, 1985; *Trends Neurosci.* 10:34–41, 1987. Pattern generators for locomotion in lampreys.

53. Gruberg, E. R. Nucleus isthmi in tegmentum of frog. Personal communication.

54. Grusser-Cornehls, V. and O. J. Grusser. In *Neurophysiologie und Psychophysiologie des visuellen System*, (R. Jung and H. Kornhuber, eds.), pp. 275–286, 1960.

55. Grüsser-Cornehls, U., O. Grüsser, and T. H. Bullock. *Science* 141:820–822, 1963. Visual responses of frog tectum.

56. Gustafsson, B. and H. Wigström. *Trends Neurosci.* 11:156–162, 1988. Physiological mechanisms underlying long-term potentiation.

57. Hall, J. C. *J. Comp. Physiol. A* 157:91–104, 105–113, 1985. Processing of vibration patterns in interneurons of Uca.

58. Harcombe, E. S. and R. Wyman. *J. Exp. Biol.* 53:255–263, 1970. Diagonal locomotion in deafferented toads.

59. Henneman, E. et al. *J. Neurophysiol.* 28:560–580, 1965. Significance of cell size in spinal motorneurons.

60. Hubel, D. G. *Nature (London)* 299:515–524, 1982. The primary visual cortex.

61. Hubel, D. G. and T. Wiesel. *J. Physiol. (London)* 195:215–243, 1968; *Proc. R. Soc. London, Ser. B* 108:1059, 1977; *J. Comp. Neurol.* 158:267–293; 177:361–380, 1978. Organization of monkey striate cortex.

62. Humphrey, A. L., M. Sur, D. Uhlrich, and S. Sherman. *J. Comp. Neurol.*

233:159–189, 1985. Projection patterns of X and Y axons from lateral geniculate to area 17 of cortex.

63. Humphrey, N. *Brain, Behav. Evol.* 3:324–337, 1970. What frog's eye tells monkey's brain.

64. Ingle, D. *Science* 180:422–424, 1973. Habituation of neurons in frog tectum.

65. Ingle, D. *Science* 181:1053–1055, 1973. Effects of removal of one optic tectum, frog.

66. Ito, M. *The Cerebellum and Neural Control.* Raven Press, New York, 1984.

67. Ito, M. *Annu. Rev. Neurosci.* 5:275–296, 1983. Cerebellar control of vestibulo-ocular reflex.

68. Jacobson, M. and R. M. Gaze. *J. Exp. Physiol. Cogn. Med. Sci.* 49:199–209, 1964. Responses of units in optic tectum of goldfish; *ibid.*, 384–393. Spectral sensitivity of tectal units and responses to light.

69. Jacobson, M. In *Golgi Centennial Volume* (M. Santini, ed.), pp. 147–151. Raven Press, New York, 1975. Cell types in mammalian brain.

70. Jaffe, L. and M. Poo. *J. Exp. Zool.* 209:115–125, 1979. Neurites grow faster toward cathode than anode in constant electric field.

71. Kaas, J. H. In *Encyclopedia of Neuroscience* (G. Adelman, ed.), Vol. II, pp. 1113–1117. Birkhaueser, Boston, MA, 1987. Somatosensory cortex.

72. Kelly, D. D., In *Principles of Neural Science* (E. Kandel and J. Schwartz, eds.), pp. 648–658. Am. Elsevier, New York, 1985. Sleep and dreaming.

73. Kolb, B. and I. Whishaw. *Fundamentals of Neuropsychology.* Freeman, New York, 1985.

74. Konishi, M. *Annu. Rev. Neurosci.* 8:125–170, 1985; *Proc. Nat. Acad. Sci. U.S.A.* 82:5997–6000, 1985. Bird song.

75. Korn, H. In *Encyclopedia of Neuroscience* (G. Adelman, ed.), Vol. II, pp. 617–619. Birkhaueser, Boston, MA, 1987. The Mauthner cell.

76. Kotchabhakdi, N. *J. Comp. Physiol. A* 112A:47–93, 1976. Structure and function of goldfish cerebellum.

77. Krebs, J. et al. *Proc. Natl. Acad. Sci. U.S.A.* 86:1388–1392, 1989. Hippocampal specialization of food-storing birds.

78. Kubie, J. L. and J. B. Ranck, Jr. In *Neurobiology of the Hippocampus* (W. Seifert, ed.), pp. 433–447. Academic Press, New York, 1983. Sensory behavioral correlates in individual hippocampal neurons in three situations: Space and context.

78a. Kuffler, S. W. and J. G. Nicholls. *From Neuron to Brain,* p. 342, Sinauer, Sunderland, MA, 1975.

78b. Labygina, T. F., A. M. Mass, and R. Supin. *Zh. Zyssh. Napv., Deiab.* 28:1047–1054, 1978. Multiple sensory projections in dolphin cerebral cortex.

79. Larson, J., G. Lynch, and D. Wong. *Brain Res.* 368:347–350, 1986. Patterned stimulation of theta frequency is optimal for hippocampal potentiation.

79a. Levi-Montalcini, R. *Prog. Brain Res.* 45:235–258, 1976. Nerve growth factor; Levi-Montalcini, R. and L. Aloe. In *Autonomic Ganglia* (L. Elfin, ed.), pp. 401–426. Wiley, New York, 1983. Effects of nerve growth factor on ganglion cells.

80. Lieberman, P. *The Biology and Evolution of Language.* Harvard Univ. Press, Cambridge, MA, 1984.

81. Lieberman, P. *On the Origins of Language.* Macmillan, New York, 1975.

82. Lin, J. and D. Faber. *J. Neurosci.* 8:1302–1312, 1988. Synaptic transmission by single club endings on Mauthner cell.

83. Loeb, C. M. and W. R. Kristian. *Science* 244:64–66, 1989. Central synaptic inputs to identified leech neurons determined by peripheral target.

84. London, J. and R. Gillette. *Proc. Natl. Acad. Sci. U.S.A.* 83:4058–4062, 1986. Food avoidance learning in neurons of Pleurbranchea.

84a. Lorente de No, R. *J. Psychol. Neurol.* 46:113–177. Studies on the structure of the cerebral cortex. II. Continuation of the study of the ammonic system.

85. Luskin, M., A. Pearlman, and J. Shanes. *Neuron* 1:635–647, 1988. Cell lineage in cerebral cortex of mouse studied *in vivo* and *in vitro*.

86. Lynch, G. and M. Baudry. *Science* 224:1057–1063, 1984. The biochemistry of memory: A new and specific hypothesis.

87. Macphail, E. M. *Brain and Intelligence in Vertebrates*, Chapter 8. Oxford Univ. Press (Clarendon), London and New York, 1982.

88. Mann, M. D., G. E. Glickman, and A. L. Tomer. *Brain, Behav. Evol.* 31:111–124, 1988. Brain/body relations among myomorph rodents.

89. Margoliash, D. and M. Konishi. *Proc. Natl. Acad. Sci. U.S.A.* 82:5997–6000, 1985. Auditory representation of song in white-crowned sparrow.

90. Margoliash, D. *J. Neurosci.* 6:1643–1661, 1986. Preferences for autogenous song by auditory neurons in white-crowned sparrows.

91. Marr, D. *J. Physiol. (London)* 202:437–470, 1969. Theory of cerebellar cortex function.

92. Mazzi, V., A. Fasolo, and M. Franzoni. *Cell Tissue Res.* 182:491–503, 1971. Optic tectum of polypteriform fish *Calamolihthys.*

93. McCormick, D. A. and R. Thompson. *Science* 223:296–299, 1984; *J. Neurosci.* 4:2811–2822, 1984. Role of cerebellum in conditioned eyelid response.

94. McKenna, O. and J. Wallman. *Brain, Behav. Evol.* 26:91–116, 1985. Optic tectum and pretectum of birds.

95. Meisami, E. *Prog. Brain Res.* 48:211–229, 1978. Development of olfactory bulb.

96. Merzenich, M., J. Kaas, J. Wall, R. J. Nelson, M. Sur, and D. Telleman. *Neuroscience* 8:33–55, 1983. Topographic reorganization of somatosensory areas in monkey.

97. Merzenich, M. and J. Kaas. *Trends Neurosci.* 5:434–436, 1982. Reorganization of somatosensory cortex after peripheral nerve injury.

98. Michael, C. R. *Proc. Natl. Acad. Sci. U.S.A.* 85:4914–4918, 1988. Retinal projections to lateral geniculate of monkey.

99. Milner, R. G. et al. *Curr. Top. Dev. Biol.* 21:117–150, 1987. Brain specific genes.

99a. Mishkin, M. *Philos. Trans. R. Soc. London, Ser. B* 298:85–95, 1982. A memory system in monkeys.

99b. Mishkin, M., B. Malamut, and J. Bachevalier. In *Neurobiology of Learning and Memory* (G. Lynch, J. L. McGaugh, and N. M. Weingberger, eds.), pp. 65–77. Guilford Press, New York, 1984. Memories and habits: Two neural systems.

100. Mitra, M. L. *J. Anat.* 89:467–483, 1955. Cell types in mammalian cortex.

101. Morgane, P. J. et al. In *Dolphin Cognition and Behavior* (R. Schustman et al., eds.), Chapter 1. Erlbaum, Hillsdale, NJ, 1986. Evolutionary morphology of the dolphin brain.

102. Morris, R. G. M. In *Neurobiology of the Hippocampus* (W. Seifert, ed.), pp. 406–432. Academic Press, New York, 1983. An attempt to dissociate ''spatial-mapping'' and ''working memory'' theories of hippocampal function.

102a. Mountcastle, V. *J. Neurophysiol.* 20:408–434, 1957. Column organization in somatosensory cortex.

103. Muller-Preuss, P. and U. Jurgens. *Brain Res.* 103:29–43, 1976. Projections from cingular vocalization area in squirrel monkey.

104. Murakami, F., S. Higashi, H. Katsumaru, and Y. Oda. *Brain Res.* 437:379–382, 1987. Formation of new corticorubral synapses as a mechanism for classical conditioning in the cat.

105. Nelson, D., J. E. Heath, and C. L. Prosser. *Am. Zool.* 24:791–804, 1984. Evolution of temperature regulation in nervous system.

106. Nicholson, J. L. and J. Altman. *Science* 176:530–532, 1972. Synaptogenesis in the rat cerebellum: Effects of early hypo- and hyperthyroidism.

107. Niki, H. *Brain Res.* 68:185–301, 1974; 70:346–349, 1974. Prefrontal unit responses during right and left delayed responses in monkeys.

108. Nordeen, K. and E. Nordeen. *Nature (London)* 334:149–151, 1988. Projection neurons in vocal pathway during learning in zebra finches.

109. Northcutt, R. G. *Am. Zool.* 17:411–429, 1977. Organization of elasmobranch nervous system.

110. Northcutt, R. G. In *Fish Neurobiology* (R. Davis and G. Northcutt, eds.), Vol. 2, pp. 1–42. Univ. of Michigan Press, Ann Arbor, 1983. Evolution of optic tectum in fin-rayed fishes.

111. Northcutt, R. G. *Annu. Rev. Neurosci.* 4:301–350, 1981. Evolution of telencephalon in nonmammals.

112. Nottebohm, F. and M. Nottebohm. *J. Comp. Physiol.* 108:171–192, 1976. Left hypoglossal dominance in control of canary and white crowned sparrow song.

113. Nottebohm, F. *J. Comp. Neurol.* 165:457–486, 1976. Central control of bird song.

113a. Nottebohm. F., T. Stores, and C. Leonard. *J. Comp. Neurol.* 165:457–486, 1976. Central control of song in the canary, *Serinus carius*.

114. O'Benar, J. *Brain Res. Bull.* 1:529–541, 1976. Electrophysiology of neural units in goldfish optic tectum.

115. Olton, D. S. *Am. Psychol.* 34:583–596, 1979. Mazes, maps and memory.

116. Oppenheim, R. et al. *Neurosciences* 1:141–151, 1981; *J. Comp. Neurol.* 210:174–189, 1982; Cell death of motorneurons in chick embryo spinal cord.

117. Oppenheim, R. W. *J. Neurosci.* 1:141–151, 1981. Cell death in chick embryo spinal cord.

117a. Orban, G. A. *Neuronal Operations in Visual Cortex.* Springer-Verlag, Berlin, 1984.

118. Pappenheimer, J. R. *Sci. Am.* 235:155–376, 1976. The sleep factor.

119. Patel, S. J. and M. G. Stewart. *Brain Res.* 449:36–46, 1988. Changes in the number and structure of dendritic spines 25 hours after passive avoidance training in the domestic chick, *Gallus domesticus*.

120. Pepperberg, I. *Anim. Learn. Behav.* 11:179–185, 1983. Cognition in an African parrot.

121. Perrett, D. S. et al. *Neurosci. Newsl., Suppl.* 3:5358, 1979; *Exp. Brain Res.* 47:329–342, 1982; *Proc. R. Soc. London, Ser. B* 223:293–317, 1985. Visual cells in cortex sensitive to face and gaze direction.

121a. Peterson, R. A. *Copeia,* 816–819, 1972. Tactile response of facial lobe of goldfish.

122. Poggio, G. In *Medical Physiology* (V. B. Mountcastle, ed.), 13th ed., pp. 497–535. Mosby, St. Louis, MO, 1974. Divergence in visual pathways in brain.

123. Premack, D. *Behav. Brain Sci.* 6:125–167, 1983. The codes of man and beasts.

124. Pritz, M. B. and R. G. Northcutt. *Exp. Brain Res.* 40:342–345, 1980. Ascending somatosensory pathway to telencephalon in crocodiles.

125. Pritz, M. B. et al. *J. Comp. Neurol.* 140:81–100, 1970. Effects of Wulst ablation in pigeons.

125a. Prosser, C. L. *J. Comp. Neurol.* 38:603–630, 1934. Development of behavior in earthworm embryos.

125b. Prosser, C. L. *Adaptational Biology.* Wiley, New York, 1986.

126. Prothero, J. and Sundsten. *Brain, Behav. Evol.* 24:152–167, 1984. Folding of cerebral cortex in mammals.

127. Purves, D. and J. Lichtman. *Principles of Neural Development.* Sinauer Associates, Sunderland, MA, 1985.

127a. Ramon y Cajal. *Histologie du Systéme Nerveux de l'Homme at des Vertébrés.* Transl. L. Azoulay. Maloine, Paris 1909.

127b. Ridgway, S. H. et al. Proc. Nat. Acad. Sci. 78:943–947, 1981. Auditory brainstem responses in dolphin.

127c. Roberts, A. and J. A. Kahn. *Philos. Trans. R. Soc. London, Ser. B* 296:213–228, 1983. Fictive swimming in amphibian larvae.

128. Roberts, A., S. R. Soffe et al. *Symp. Soc. Exp. Biol.* 37:261–284, 1983; *Philos. Trans. R. Soc. London, Ser. B* 296:213–228. Control of swimming in amphibian embryos.

129. Roberts, J. and D. Rowell. *Can. J. Zool.*

66:182–190, 1985. Periodic respiration in gill-breathing fish.

129a. Rose, G. In *Neurobiology of the Hippocampus* (W. Seifert, ed.), p. 450. Academic Press, New York, 1983. Physiological and behavioral characteristics of dentate granule cells.

130. Rovainen, C. and K. L. Birnberger. *J. Neurophysiol.* 34:974–982, 1971. Motorneuron functions in sea lampreys.

131. Rovainen, C. *Physiol. Rev.* 59:1007-1077, 1979. Neurobiology of lampreys.

131a. Russell, D. and P. Wallen, *Acta Physiol. Scand.* 117:161-170, Fictive swimming in lamprey spinal cord.

132. Saito, H. *J. Comp. Neurol.* 221:279–288, 1983. Morphology of X, Y and W type retinal ganglion cells of cat.

133. Salmon, M. and G. Hyatt. In *The Biology of Crustacea* (D. E. Bliss, ed.), Vol. 7, pp. 1–10. Academic Press, New York, 1983. Acoustic calling by fiddler crabs.

134. Salmon, M. In *Studies in Adaptation* (S. Rebach and D. Dunham, eds.), pp. 143–169. Wiley, New York, 1983. Behavioral adaptations of fiddler crabs.

135. Sarawa, E. S. and B. M. Castor. *Brain Res.* 90:181–194, 1975. Sensory of cortex of sloth.

136. Savage, G. E. In *Comparative Physiology of Telencephalon* (S. Ebbeson, ed.), pp. 129–167. Plenum, New York, 1980. The fish telencephalon.

137. Schroeder, D. *J. Comp. Neurol.* 175:287–299, 1977. Cytoarchitecture of tectum in fishes.

138. Scoville, W. B. and B. Milner. *J. Neurol., Neurosurg. Psychiatry* 20:11–21, 1957. Loss of recent memory after bilateral hippocampal lesions.

139. Sejnowski, T. et al. *Science* 241:1299–1306, 1988. Computational neuroscience.

139a. Seyfarph, R. M., D. L. Cheney, and P. Marler. *Science* 210:801–803, 1980. Monkey responses to different alarm calls: predator classification and semantic communication.

139b. Sherman, S. M. *Prog. Psychobiol. Physiol. Psychol.* 11:233–314, 1985. Functional organization of the W-, Y-, and X-cell pathways in cat.

139c. Sherman, S. M. and P. D. Spear. *Physiol. Rev.* 63:788–855, 1982. Organization of visual pathways in visually deprived cats.

140. Sherrington, C. S. *The Intergrative Action of the Nervous System.* Scribner's, New York, 1906.

141. Sherry, D., J. Krebs, and R. Cowie. *Anim. Behav.* 29:1260–1266, 1981. Memory for location of stored food in marsh tits.

142. Simons, D., J. T. Woolsey, and D. Durham. *Somatosens. Res.* 1:207–245, 1984. Organization of somatosensory barrel cortex.

142a. Sperry, R. W. In *The Biopsychology of Development.* (L. Tobard, R. Aronson, and E. Shaw, eds.). Academic, New York, 1971. How the developing brain becomes properly wired.

142b. Stanford, L. *Brain Res.* 297:381–386, 1984. Retinal ganglion cells of cat.

142c. Stephan, H. et al. *Comp. Primate Biol.* 4:1–38, 1988. Comparative size of brains and brain components.

143. Stephan, H. and O. J. Andy. *Am. Zool.* 4:59–74, 1969; *Ann. N. Y. Acad. Sci.* 167:370–387, 1969. Quantitative studies on primate brain structure.

144. Squire, L. R. *Memory and Brain.* Oxford Univ. Press, New York, 1978.

144a. Staubli, U., D. Fraser, M. Kessler, and G. Lynch. *Behav. Neural. Biol.* 46:432–444, 1986. Retrograde and anterograde amnesia of olfactory memory after denervation of hippocampus.

145. Stein, P. *J. Comp. Physiol.* 124:203–210, 1979; 146:401–409, 1982. Fictive movements in spinal turtles.

145a. Strong, O. S. and A. Elwyn, *Neuroanatomy,* 5th ed. Williams & Wilkins, Baltimore, MD, 1964.

146. Stryker, M. P. et al. *J. Neurophysiol.* 41:896–709, 1978. Effect of restricting early experience on cat visual cortex.

147. Sutcliffe, J. G. *Annu. Rev. Neurosci.* 11:157–198, 1988. mRNA in mammalian brain.

148. Sutterlin, A. and C. L. Prosser. *J. Neurophysiol.* 33:36–45, 1970. Distribution of current fields in optic tectum of goldfish.

149. Tarr, R. *Physiol. Behav.* 18:1153–1158, 1977. Role of amygdala in aggressive behavior of western fence lizard.

150. Thompson, R. F. *Trends Neurosci.* 11:152, 1988. Neural basis for learning; cerebellum, hippocampus.

150a. Thompson, R. F. and N. H. Donegan. In *Learning and Memory: A Biological View* (R. P. Kesner and J. L. Martinez, eds.), p. 40. Academic Press, Orlando, FL, 1986. The search for the engram.

151. Tsukahara, N., Y. Oda, and T. Notsu. *J. Neurosci.* 1:72–79, 1981. Classical conditioning mediated by the red nucleus in the cat.

152. Ulinski, P. S. In *Behavior and Neurology of Lizards* (N. Greenberg and P. MacLean, eds.), pp. 121–132. U.S. Department of Public Health, Natl. Inst. Mental Health, Bethesda, MD, 1978. Organization of dorso-ventral ridge of reptiles and birds.

152a. Vanegas, H. In *Fish, Neurobiology* (G. Northcutt and R. Davis, eds.), Vol. 2, p. 59. Univ. of Michigan Press, Ann Arbor, 1983. Tectal anatomy.

153. Viala, D. et al. *Neurosci. Lett.* 74:49–52, 1987. Relation between phrenic discharge and hindlimb extension in fictive locomotion.

154. von Holst, E. *Z. Vergl. Physiol.* 20:582–599, 1934; 26:481–528, 1939; *Pfluegers Arch.* 236:149–158, 1935. Reflexes and locomotion in fishes.

155. Walker, S. F. *Br. J. Psychol.* 71:329–367, 1980. Lateralization of function in vertebrate brain. A review.

156. Webb, W. B. nd M. G. Dube. *Handb. Behavior and Neurobiology* (J. Aschoff, ed.), Vol. 4, pp. 499–522. Plenum, New York, 1981. Temporal characterization of sleep.

157. Weinberger, N. M. and D. M. Diamond. *Prog. Neurobiol.* 29:1–55, 1987. Physiological plasticity in auditory system: Rapid induction by learning; many levels in brain of cat.

158. Wenzel, J. and H. Matthies. In *Memory Systems of the Brain: Animal and Human Cognitive Processes* (N. M. Weinberger, J. L. McGaugh, and G. Lynch, eds.), pp. 150–170. Guilford Press, New York, 1985. Morphological changes in the hippocampal formation accompanying memory formation and long-term potentiation.

159. Williams, R. W. and K. Herrup. *Annu. Rev. Neurosci.* 11:433–453, 1988. Control of neuron number.

160. Woods, D. L. et al. In *Dolphin Cognition*, R. Schastman, ed., pp. 61–77. Erlbaum, Hillsdale, N.J., 1986. Middle and long latency auditory responses in cortex.

160a. Woody, C. D. *Memory, Learning, and Higher Function.* Springer-Verlag, New York, 1982.

161. Woolsey, T. and H. Van der Loos. *Brain Res.* 17:205–242, 1970; *J. Comp. Neurol.* 164:79–94, 1975. Sensory barrels in somatosensory cortex.

162. Wyles, J., J. Kunkel, and A. C. Wilson. *Proc. Natl. Acad. Sci. U.S.A.* 80:4394–4397, 1983. Birds, behavior and anatomical evolution.

163. Yager, D., S. Sharma, and B. Grover. *Brain Res.* 137:267–275, 1977. Visual function in goldfish: Tectal ablation.

164. Yeo, C. H., M. J. Hardiman, and M. Glickstein. *Exp. Brain Res.* 60:99–113, 1985. Cerebellar lesions abolish conditioned eyeball responses.

165. Zach, R. *Behaviour* 67:134–148; 68:106–117, 1979. Selection and dropping of whelks by northwestern crows.

166. Zottoli, S. et al. *J. Comp. Neurol.* 219:100–111, 1983. Functioning of Mauthner cells in fish brain.

167. Zottoli, S. and D. S. Faber. *Neuroscience* 5:1287–1302, 1980. Interneurons projecting to Mauthner cells.

168. Prosser, C. L. and J. E. Heath. In *Environmental and Metabolic Animal Physiology* (C. Ladd Prosser, ed.). pp. 109–165. Wiley-Liss, Inc., New York, 1991.

Chapter 10 | Endocrines

Aubrey Gorbman and Kenneth Davey

Hormones are chemicals that carry messages from one more or less well defined part of an organism to the whole organism or to another tissue or organ. The source of the hormone may be a single cell or a small group of cells, as in certain neuroendocrine cells of the brain, or it may be a more organized tissue or gland, as in the thyroid gland. The target of the chemical message may be another endocrine organ, a tissue, or a specific organ. Typically, the identification of an endocrine function has involved the following steps: (1) histological evidence of secretory activity in an organ or tissue is related to some physiological or developmental process in the organism under investigation; (2) surgical removal or destruction of the suspected endocrine tissue halts or alters the process; (3) the process is restored by reimplanting the extirpated tissue or by injecting extracts of the suspected endocrine tissue; (4) hormonal activity is detectable in the blood or body fluids of the organism at the appropriate times; (5) ultimately, the activity is extractable from the tissues or body fluid,

the extract ought to be purified, and the chemical identity of the hormone established. Chemicals that may act as endocrine or hormonal substances in one function may possess other nonendocrine functions, for example, as neurotransmitters (see below), or as paracrine substances, which act locally by diffusion to neighboring cells.

Endocrine mechanisms play an important role in the coordination and control of most of the major physiological processes of organisms. There is rich variety in the precise mechanisms involved, and it is impossible within the scope of a single chapter to do justice to more than a few. Those endocrine mechanisms that are involved with the major physiological systems, such as digestion, excretion, osmoregulation, and circulation are more properly considered as part of the physiology of those systems. This chapter will concentrate on the hormonal control of growth, development, and reproduction. Since much of our knowledge of hormones rests on what has been discovered in higher organisms such as the vertebrates and insects, the subject is considered in reverse evolutionary order.

Among the sources of hormones in many animals are specialized nerve cells, which are referred to as neurosecretory cells (76). Neurosecretory cells are usually recognized by a cluster of anatomical characteristics as revealed by light and electron microscopy. Thus, they are recognized in the light microscope by their affinity for various stains, usually, but not invariably, associated with a high content of sulfur-containing proteins. In the electron microscope, neurosecretory cells are recognized by their prominent content of membrane-bound secretory vesicles, and by the presence of abundant granular endoplasmic reticulum and Golgi (199). There are many hormones that originate in nonnervous tissue: The thyroid gland and pancreas are two such examples.

Not all neurosecretory cells are clearly endocrine in function. In order to qualify as neuroendocrine cells (Fig. 1) there must be clear evidence that the secretion is released into the blood or other body fluids, or at least into a space of greater dimensions than those characteristic of a synapse or myo-neural junction. Secretion by exocytosis usually occurs at sites at the terminals of the secretory neurones, and concentrations of such sites are usually referred to as "neurohemal organs," as in the pituitary of mammals or the corpus cardiacum of insects (198, 199). Many neurosecretory cells, particularly in invertebrates, may have all of the characteristics of classical neurosecretory cells. Indeed, they may contain hormonal substances, and still behave as normal neurons in which the neurosecretory peptide acts, not as a hormone, but as a neurotransmitter (121).

Neurosecretory cells, as defined by their structural characteristics, may produce a wide variety of secretory materials. Neuroendocrine cells, on the other hand, typically secrete peptide hormones such as oxytocin or vasopressin (Fig. 2).

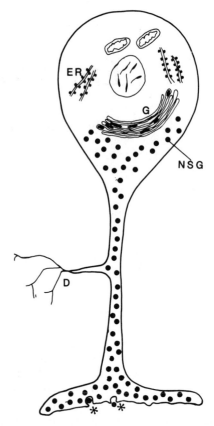

Figure 1. Diagram of a neurosecretory neuron, which is the source of neurohormones. The neuropeptides are synthesized as part of a larger protein in the endoplasmic reticulum, ER, and packaged into membrane bound granules in the Golgi, G. The neurosecretory granules, NSG, travel along the axon to neurohemal sites where they are released (*) into the circulation. The protein in the granules is further processed during transport along the axon so as to produce the neurohormonal peptide. Neurosecretory cells usually have prominent dendritic trees through which connections to the nervous system are established.

These peptides are usually synthesized as part of a larger protein, which is cleaved or processed in secretory granules. In at least some cases in vertebrates and insects, a peptide hormone, which may be rather small (fewer than a dozen amino acid residues), is associated with a larger

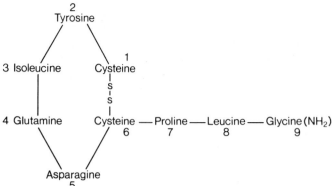

2
Tyrosine

3 Isoleucine   Cysteine 1

          S
          |
          S

4 Glutamine   Cysteine — Proline — Leucine — Glycine (NH₂)
                    6   7   8   9

          Asparagine
              5

Figure 2. Structural formula for oxytocin, an octapeptide from the neurohypophysis. The cysteines at 1 and 6 are joined by a disulfide bond, a frequently occurring feature of peptide neurohormones. The amino acids at positions 1, 5, and 9 are in the form of amides. These two characteristics are shared by all of the known neurohypophysial octapeptides. [From (102).]

protein, a neurophysin, which is thought to protect the hormone from degradation in the granule, and to prevent the leakage out of the secretory granule of the much smaller peptide hormone (24, 84).

There are two broad modes of hormone action at the cellular and molecular levels, both of which involve proteins that bind the hormones in a specific way. These receptor proteins, which may be altered during the process of binding, represent the first step in the transduction of the message carried by the hormone into some alteration in the physiology or biochemistry of the target cell. The first broad model of hormone action (Fig. 3) is represented by the "second-messenger" concept and involves receptor proteins in the cell membrane. The hormone molecule, typically a peptide, arrives at the cell surface and binds to a receptor protein. The receptor–hormone complex in turn activates the membrane-bound enzyme adenylate cyclase, which converts intracellular ATP into one of the cyclic nucleotides, cAMP or cGMP. The cyclic nucleotide then acts as a second messenger in the cell, usually by interacting with protein kinases leading ultimately to the phosphorylation and activation of other enzymes. The cyclic nucleotide is inactivated by a phosphatase. In many systems calcium ion is important in the action of hormones, and in some cases, Ca^{2+} may itself play the role of a second messenger (13, 42).

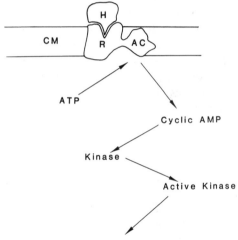

Enzyme activation by phosphorylation

Figure 3. Diagram illustrating the action of hormones involving the "second-messenger" principle. This mode of action is characteristic of neuropeptides and many other hormones. The hormone, H, arrives at the surface of the cell membrane, CM, and there binds to a receptor protein, R, in the membrane. Part of the receptor complex is the enzyme adenylate cyclase, AC, which becomes activated as a result of the binding of the hormone to the receptor. Adenylate cyclase acts with ATP to increase the amount of cyclic AMP (cAMP) in the cell. The second messenger, cAMP, initiates a cascade of events that leads to phosphorylation of various enzymes. The second messenger may also be cyclic GMP (cGMP). Ca^{2+} may also become involved through *calmodulin*, a protein that regulates Ca^{2+}-dependent adenylate cyclase and other enzymes important to the second-messenger system. Calmodulin is a cytoplasmic receptor for Ca^{2+}.

The second general model for hormone action refers to a different class of chemicals, the steroids (Figs. 12 and 13), which alter the development of the cell. In this model (Fig. 4), it has thus far not been necessary to invoke receptors at the cell surface, and the hormones are thought to enter the cell by diffusion. Inside the cell, they bind to receptor proteins in the cytosol. During binding, the receptor undergoes an allosteric alteration, which renders it susceptible to uptake by the nucleus. Inside the nucleus, the hormone–receptor complex in some way, which is far from being understood, stimulates transcription at specific loci leading to messenger RNA production, and ultimately, the production of specific proteins. This classical model, based primarily on the action of progesterone

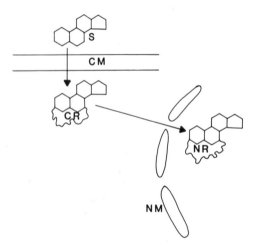

Figure 4. Diagram illustrating the action of steroid hormones. The steroid molecule, S, enters the cell by diffusion across the cell membrane, CM, and binds to a cytoplasmic receptor protein, CR. The steroid–receptor complex crosses the nuclear membrane, NM, where the steroid binds to a nuclear receptor protein, NR. This steroid–receptor complex interacts with the genome, activating various genes. In some cases, the NR protein is the only receptor present in the cell, and a separate cytoplasmic receptor appears not to be essential.

on chick oviduct (203), may require modification in the case of specific steroids. In the case of ecdysteroid receptors in insects it is not necessary to invoke a receptor in the cytosol: The steroid enters the cell by diffusion and binds to receptors resident in the nucleus (165).

Vertebrates

Endocrine Glands

In vertebrate organisms, there is typically a specialized endocrine tissue source for each of the various hormones. It is important to recognize, however, that some hormones may be produced by, or at least occur in, several tissues. For example, several of the hormones thought to be characteristic of the wall of the digestive tract (insulin, gastrin, and cholecystokinin), and known to take part in the integration of various digestive and metabolic functions, are also found in the brain. Conversely, one of the typical brain hormones, TRH (thyrotropin releasing hormone), the normal function of which is to release the thyrotropic hormone from the pituitary gland, is found in large amounts in the skin of amphibians, as well as in other organs. The functional significance of these distributions is far from clear (102).

Nevertheless, there are tissues or glands that act as the primary source of the released hormone. These classical endocrine glands (Fig. 5) release their product as a result of some signal, which in itself may be hormonal in nature. Since vertebrate endocrine glands release their products into the blood, one of their common structural features is an intimate and efficient relationship to the blood. The three most common patterns of cellular organization of vertebrate endocrine glands are illustrated in (Fig. 6). The most general of these is the cell-cord–sinusoid arrangement in which interweaving

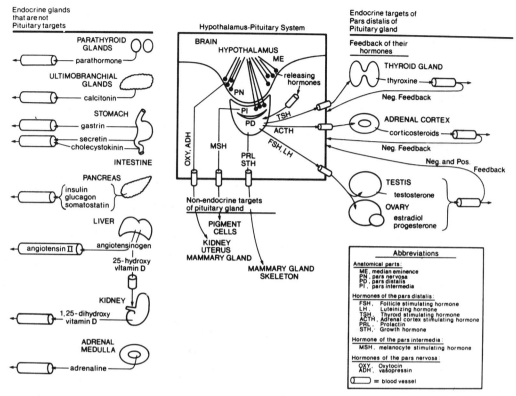

Figure 5. The principal endocrine organs of vertebrates and the hormones that they secrete. The central box contains the pituitary gland and shows that the pars nervosa, PN, and median eminence, ME, of the pituitary are actually extensions of the brain and contain the endings of neurosecretory cells. Neurosecretions (releasing hormones) from the ME are conducted by small vessels to the pars distalis, PD, of the pituitary, where they regulate secretion of six hormones (TSH, ACTH, FSH, LH, PRL, and STH). As shown, four of these regulate secretion of hormones by three other endocrine centers (thyroid, adrenal cortex, and gonads). Two of the PD hormones (PRL and STH) control still other functions. On the left are shown seven organs and tissues that secrete hormones and that are relatively independent of brain–pituitary regulation. These comprise the majority of the "classical" vertebrate endocrine organs. However, this figure does not by any means illustrate all of the known hormone sources of vertebrates. Not included are the peptide growth factors found in many tissues and the hormones involved in hemopoiesis and the immune response. [From (102).]

strings of secretory cells are separated by blood sinusoids (102).

Hormonal Control of Growth

There are numerous genes that code for particular peptide and protein regulators of growth. If body proportions are regulated by the programmed expression of many such genes, how is their transcription turned on or off in a timely fashion so as to produce an organism of the appropriate size and shape? These are important questions in developmental biology.

Although growth is a pervasive phenomenon, it is important to be clear about how it can be measured. Increase in mass

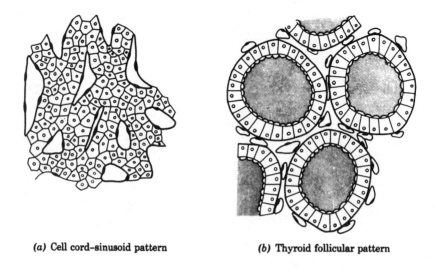

(a) Cell cord–sinusoid pattern   (b) Thyroid follicular pattern

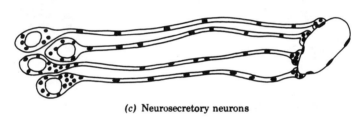

(c) Neurosecretory neurons

Figure 6. The three main types of cellular patterns in vertebrate endocrine tissues. The commonest is the cell-cord–sinusoid type (a), in which cords of cells one to three cells thick are separated by blood spaces (sinusoids). This is seen in the pars distalis and pars intermedia of the pituitary, in the adrenal gland, the parathyroids, the insulin secreting cell islet of the pancreas and in the liver. The follicular pattern (b) is less common and is typical of the thyroid gland. The neurosecretory arrangement (c) is seen in the pars nervosa and median eminence of the pituitary. In each case, all secretory elements are in direct contact with, or very close to, a blood vessel. [From (102).]

alone may be misleading because of large changes in water content or in metabolically inactive stores such as fat or calcium. Increase in skeletal length has been a useful parameter, and forms the basis of an assay for growth hormone. Incorporation of amino acids into protein is frequently used as a short-term measurement of growth, but this can be confounded in tissues in which there is a large synthesis of secretory proteins.

The two principal elements in tissue growth are hypertrophy or cell enlarge-ment and hyperplasia or increase in cell number by mitotic proliferation. Hyper-trophy, if it occurs alone, is a limited phase of tissue growth in vertebrates, and continued growth eventually requires cel-lular mutiplication.

Pituitary Growth Hormone
When young vertebrates are surgically deprived of their pituitaries, they stop growing, and the defect can be at least partially remedied by the administration of extracts of pituitary glands. Pituitary

growth hormone has been extracted and purified from all major vertebrate groups except the Agnatha, where the control of growth by the pituitary has been little investigated (100). In many species the hormone has been sequenced, but the physiology of growth hormone secretion, secretory control, and action have been thoroughly studied only in mammals. Pituitary growth hormones are proteins with a molecular weight of 21,000–24,000 daltons, and containing ~200 amino acid residues. The gene coding for human growth hormone has been cloned in bacteria and translated, yielding a product that can be administered to children whose growth is stunted because of a lack of the hormone.

In hypophysectomized (pituitary surgically removed) animals, thyroid hormone augments the action of administered growth hormone. It is considered that this is a result of a "permissive" action of the thyroid hormone, which is regarded as providing metabolic support for the action of growth hormone by remedying some of the metabolic defects in hypophysectomized animals, which growth hormone does not affect.

The principal basic effects of pituitary growth hormone treatment of vertebrates are increased skeletal and soft tissue growth, increased glucose levels in the blood and release of insulin from the pancreas. At the cellular level, the important effect is the increased transport of amino acids into responsive cells. This is accompanied by a decreased destruction of intracellular amino acids (e.g., by gluconeogenesis), and an increase in protein synthesis.

The control of growth hormone secretion appears to be of two types: metabolic or neuroendocrine, or a combination of the two. Growth hormone secretion is quickly responsive to changes in glucose levels in the blood of mammals, being depressed by high levels and stimulated by low levels. At least two neurosecretory brain hormones, secreted into short blood vessels in the hypothalamus and carried directly to the pars distalis of the pituitary, regulate growth hormone release: somatostatin (SRIF) is an inhibitor, and growth hormone releasing hormone (GRH) is a stimulant. In mammals there are diurnal cycles of growth hormone levels in the blood such that the level increases during sleep. Normally, control of growth hormone release involves a balance of the two types of neuroendocrine control (76).

Peptide Growth Factors

An unexpected finding by Daughaday was that growth hormone, which is a strong stimulant of cartilage cell multiplication *in vivo*, lacks this action *in vitro* unless a factor extractable from normal blood serum is added (48). This observation led to the eventual isolation of two peptides, the somatomedins (110), now more commonly referred to as insulin-like growth factors (IGFs) (258). These growth factors are formed by a variety of tissues, but in large part by the liver (39). Their secretion is stimulated by growth hormone and they are the principal agents through which the effects of growth hormone are mediated. Indeed, there are probably few, if any, effects of growth hormone that are direct or exclusive effects of growth hormone action. Many tissues have receptors for both growth hormone and IGFs (180). Growth hormone may thus be simply a regulator of IGF formation and action, particularly where growth stimulation and tissue DNA and RNA synthesis are involved (167).

There is a close structural similarity between IGFs and insulin, and this explains why, in some instances, receptors for insulin and IGFs will recognize both kinds of molecules (180). Insulin is itself a growth hormone in that it (like IGF) stimulates amino acid transport into cells, inhibits gluconeogenesis, and stimulates protein synthesis. It is clear that insulin and the IGFs must have shared an evolutionary history. Indeed, since immunoreactive insulin-like molecules are known to occur in many invertebrate phyla, in protozoa, and even in plants (142), the structure appears to have been conserved throughout evolution. The functional significance of this conservation of the insulin-like molecule is not at all clear, but it is important to remember that the molecule may not always function as a hormone or as a growth factor (85).

The insulin–IGF family includes other molecular relatives that are less closely similar in structure to insulin or IGF (11, 28, 67). These include the hormone relaxin, and nerve growth factor (NGF). The nerve growth factor is a promoter of growth in nerve cells (22, 71). Relaxin is released from the ovary and acts on the fibrous and cartilaginous tissues of the female tract in order to prepare the way for the birth of the fully developed infant (73).

Table 1 summarizes information about a series of peptide growth factors in addition to the IGFs. These are the best known of the peptides extractable from a variety of tissues, but numerous others, less well characterized, have been claimed. All the growth factors in the list are stimulants of tissue growth except transforming growth factor (TGF). Although TGF has some stimulatory actions on fibrous tissues, it generally antagonizes the growth promoting actions of other factors, and receptors for TGF may be an important negative control for the stimulatory peptides. Peptide growth factors, especially the IGFs, can be found free or bound to carrier proteins in the blood. However, the concentrations in the plasma are low, and much lower than in the tissues that produce them. Most of their actions are believed to be localized to their tissues of origin, and these same tissues contain receptors for the growth factors on their cell membranes. Thus, while peptide growth factors may have an endocrine-like action via the blood (e.g., IGFs), they are believed to be distributed regionally by diffusion (paracrine action) and their actions are therefore localized (58). Indeed, it is possible that the action of some of these peptide growth factors may be autocrine, that is, the action is confined to the producing cell (218).

It is a frequent finding that the membrane receptors for hormones or parahormones that stimulate cell multiplication are linked to a tyrosine kinase enzyme (23, 47). Activation of the receptor and its associated tyrosine kinase results in tyrosine phosphorylation, and this appears to be an initiator in the chain of events that evokes mitosis. Most of the growth stimulatory hormones listed in Table 1, and insulin as well, are tyrosine kinase activators through their receptors.

Almost all of the peptide growth factor research has been confined to humans and several rodent species (77, 151). Insulin-like growth factors have been detected in the blood of fish (trout), toads, turtles, and chickens in addition to several species of mammals. Snake (*Vipera russeli*) venom has served well as a rich source of NGF. Furthermore, NGF has actions on the nerve cells of fish and amphibians (22, 223). FGF is strongly angiogenic for the chorioallantoic membrane of the chick embryo, and this forms the basis for a bioassay for FGF (82, 94, 103). Comparative information remains fragmentary at best, and additional research is needed.

Oncogenes and Growth Factors

Oncogenes are virus-carried genetic factors that can cause abnormal cell transformation and growth after entry of the

TABLE 1. Peptide Growth Factors[a]

Peptide Names (Acronyms)	Molecule (daltons) Number of a.a. Residues	Tissue Sources	Actions
Insulin-like growth, factor-I, IGF-I Somatomedin C, SM-C	7500 D (insulin-like)	Blood, liver placenta, adipose cells, chick fibroblasts, fetal myoblasts, ovary. In blood of mammals, chickens, turtles, toads, rainbow trout	Stimulates skeletal growth, collagen synthesis, fibroblast multiplication, antler growth, muscle development; suppresses growth hormone secretion. Stimulates ovarian granulosa cell differentiation and progesterone synthesis (synergy with pituitary FSH); stimulates testicular steroid hormone synthesis and glucose metabolism
Insulin-like growth factor-II, IGF-II Multiplication stimulating factor (MSF)	7500 D (insulin-like)	Liver, fetal rat myoblasts Found in blood of mammals, chickens, turtles	Actions more like insulin's than IGF-I. Stimulates fibroblasts, carbohydrate metabolism, muscle cell differentiation and growth
Nerve growth factor (NGF)	13,250 D. Normally the molecule is a dimer of two similar units 118-Amino acid β chain, best known, is insulin-like (18% homology between mouse NGF and rate proinsulin) Chicken β-NGF, 118 residue, is 85% homologous to mouse NGF	Mouse salivary gland, snake venom, guinea pig prostate and cultures of various embryonic and adult cell types	Typical action is stimulation of growth of nerve fibers from cultures of chick nervous ganglia. Stimulates growth and function of cells from sympathetic and sensory ganglia and cholinergic cells in the brain. *In vivo* action on ganglion cells in fish and tadpole has been shown experimentally
Epidermal growth factor (EGF)	53 Residues. Surprisingly large prohormone has 1200 residues, EGF molecule is similar to urogastrone	Mouse salivary gland, anterior pituitary, urine, Brunner's gland (intestine). Measurable in blood	Proliferation and keratinization of skin epidermis; stimulates eye opening and tooth eruption in newborn rodents. Mitogenic for epithelium of many organs including kidney, entire digestive tract, and lung, as well as rat uterus and mammary gland. EGF receptor is related to the product of the *erb-B* oncogene

TABLE 1. (*Continued*)

Peptide Names (Acronyms)	Molecule (daltons) Number of a.a. Residues	Tissue Sources	Actions
Transforming growth factor α (TGFα)	Human TGFα: 50 residues; rat TGFα differs slightly Structure related to EGF No structural relationship to TGFβ	Found mostly in embryonic or tumorous tissues; not free in blood	Has stimulatory EGF-like actions on epidermis of skin. Will bind to EGF receptors. Probably limited to autocrine or paracrine actions. Synthesis by mammary tumor cells stimulated by estrogen
Transforming growth factor β (TGFβ)	Homo-dimer of identical units, each 112 residues, 12,500D. Some homology to ovarian inhibin, and to Mullerian inhibiting factor	Blood platelets, bone, kidney, placenta, tumorous tissues. Found in blood serum	Stimulates growth and differentiation of fibroblasts but is inhibitory to growth of many cell types like endothelium, epidermis, lymphocytes, liver. TGFβ-producing cells also have receptors for TGFβ. TGFβ blocks activation of some oncogenes by EGF and PDGF
Fibroblast growth factor (FGF)	Subunits of prohormone: acidic b-FGF-1, 140 residues basic b-FGF-2, 146 residues 55% homology to each other	Bovine pituitary, bovine brain, bovine adrenal	Potent mitogen for cultured fibroblasts, chondrocytes, endothelial cells. Angiogenic for corpus luteum and chick chorio-allantoic membrane. Stimulates endothelial cell multiplication
Corpus luteum angiogenic factor	Single chain, 15000 D.	Corpus luteum	Stimulates ovarian granulosa and adrenal cortex
Chondrocyte growth factor	Single chains, 15000 and 16000 D		Stimulates growth of cartilage. Human interleukin-1 has 30% sequence with FGF-1. Slight homology with neuromedin C, bombesin, substance P
Platelet-derived growth factor (PDGF)	28,000–35,000 D Heterodimeric molecule in man; homodimer in pigs Oncogene v-*sis* of simian sarcoma virus codes a protein closely related to a major part of PDGF. Cellular oncogene c-*sis* may do the same	Blood platelets. Little or no free PDGF in normal blood. Smaller amounts found in monocytes, macrophages, endothelium, some smooth muscle	Potent mitogen for cultured fibroblasts, preadipose cells, pig aortic smooth muscle cells. Synergistic with IGF-I in stimulating growth of human fibroblasts. PDGF may cause local release of IGF-I and interferon (INF). TGF may activate fibroblast *sis* gene to form PDGF to stimulate growth

TABLE 1. (*Continued*)

Peptide Names (Acronyms)	Molecule (daltons) Number of a.a. Residues	Tissue Sources	Actions
Bone-derived growth factor (BDGF) (SGF)	70,000 D (purified, not sequenced)	Adult mineralized bone, extract Similar factor released by chick embryo bones in culture	Local action on bone cells. Mitogenic and osteoinductive. Also called bone morphogenetic protein, BMP

[a]Data in this table are from references 22, 39, 58, 71, 72, 82, 85, 88, 94, 159, 187, 213, 219, 221, and 250.

virus into a cell. For a small number of known oncogenes, normal closely related counterparts have been identified in the genome of cells, and these have been located on particular chromosomes.

Two oncogenes, v-erb (avian erythroblastosis virus) and v-neu, code for a protein that is very similar to the receptor for epidermal growth factor, EGF (25, 64). The translated product of v-erb is a truncated version of the EGF receptor, in which the external part of the receptor is at least partially lacking, leaving the internal part, which contains the tyrosine kinase (47). One interpretation of this is that it leaves the enzyme permanently in the "on" position. The c-fms protooncogene codes for the receptor of M-CSF (macrophage colony stimulating factor) (231).

The oncogene, v-sis from the monkey (simian) sarcoma virus, codes for a protein that is a homodimer of the PDGF (platelet derived growth factor) B-chain molecule. The normal counterpart of v-sis, c-sis, codes for the B chain of PDGF. There is evidence that v-sis infection leads to the production of large amounts of PDGF-like product, and that this acts via the normal cell surface receptor for PDGF leading to tyrosine kinase activation (120, 187, 250).

These demonstrated relationships be-

tween cellular genetic elements that regulate growth and cancer-causing viral agents has given rise to interesting speculation. It is believed that the viral agents are derived by altering normal cellular DNA that has been incorporated by the virus. It would be adaptive for a virus to exploit a normal cellular mechanism for host cell multiplication because this process produces more host cells in which the virus can grow. In the above instances, the virus utilizes one of two strategies: it leads to increased production of a peptide growth factor, or it leads to production of a receptor for the growth factor. The latter mechanism could lead, by mutation (truncation in the case of the receptor for EGF), to the production of a form of the receptor that produces the effect on growth without the intervention of the hormone (25, 150, 231).

Hormones in Hemopoiesis and the Immune Response

About one half of the blood volume of vertebrates consists of blood cells of a variety of types, which are distinguishable by their cytoplasmic contents. Those that contain hemoglobin, the erythrocytes, function in respiration. The remaining types function principally in the protection of the organism against invasive and foreign materials, cells, viruses, and so on. The primary protective device against disease-causing organisms and toxic foreign substances is the immune response, which will be described below (2).

Hemopoiesis

The vertebrate blood cells are ultimately all derived from the same primordial embryonic cell type, the hemocytoblast. The offspring of the hemocytoblast differentiate in several lines. Hemopoiesis is a continuous process in the adult since the blood cells, especially the erythrocytes, have a limited life span. The most frequent sites of hemopoiesis are the bone marrow, liver, spleen, lymph nodes, and thymus. The brain membranes of some fish, and the wall of the entire intestine of hag fish, are hemopoietic. Generally, each separate blood cell type forms within separate clusters called colonies, and colonies of completely different cell types may be adjacent to one another in the same hemopoietic tissue, such as bone marrow. The multiplication of the separate blood cell types is under the regulatory control of parahormones and hormones acting only on those cellular colonies that carry the appropriate receptor for that hormone (157, 188). The principal regulator for erythrocyte formation is erythropoietin, a peptide hormone of ~130 amino acid residues that in mammals is made in the kidney (81). Erythropoietin-like activity has been found in the blood of most vertebrate classes, but only the mammalian hormone has been isolated. At least four other mammalian hemopoietic colony stimulating hormones (or parahormones) (CSFs) have been isolated and purified and some have been sequenced. These have varying specificities: some stimulate only one type of cell; one, however, multipotential colony stimulating factor, multi-CSF or interleukin-3 (see Fig. 8), is mitogenic for all blood cell types (136, 205).

It should be realized that the colony stimulating factors are peptide growth factors, with their actions directed more or less specifically towards the blood cells. Only one CSF, in addition to erythropoietin, is readily demonstrable in blood (granulocyte colony stimulating factor, g-CSF). Thus, a majority of CSFs are parahormones released in the site of their cellular action (157).

Immune Response

The immune response (Figs. 7 and 8) is in fact a succession or cascade of cellular interactions from the ingestion of a foreign antigen by a blood cell to the production of an antibody or of a natural "killer" or cytotoxic cell type. The immune cellular interactions, differentiation, and multiplication are regulated by a series of hormones and parahormones that are known as lymphokines, interleukins, interferons, and CSFs. A full description of the immune response is not possible here. More detailed descriptions can be found in textbooks of cell biology and immunology (75). The main cellular events in the immune response are summarized in Figure 7. Figure 8 represents the same phenomenon but emphasizes the hormones that are involved.

The principal blood cells that conduct the immune response are the lymphocytes and macrophages, which together comprise the so-called mononuclear leucocytes. These cells are able to leave the small blood capillaries, enter the intercellular spaces and the lymphatic vessels, and, through the latter, to reenter the blood. This continual movement has been characterized as a kind of surveillance or patrolling of the body fluids and intercellular spaces (191).

The lymphocytes are structurally indistinguishable, but they are biochemically differentiated into two types, T and B lymphocytes. Whether a given lymphoblast (immature lymphocyte) becomes

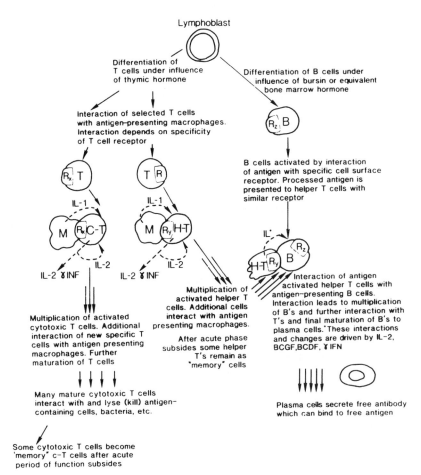

Figure 7. Only a few lymphocytes (T or B) carry plasma membrane receptors (R_x, R_y, etc.) appropriate for interaction with a particular antigen. The immune response is for this reason a means for multiplying the numbers of cells that can interact with, or manufacture the antibody for that antigen. The T cells may interact with the antigen only after it has been processed by macrophages, M. At the time of this interaction between macrophages and the appropriate T cells, interleukin-1, IL-1, secreted by the macrophage stimulates IL-2 and interferon secretion by the T lymphocyte. One action of the IL-2 is to stimulate multiplication and further differentiation of the T cells as helper T, suppressor T, or cytotoxic T cells. The multiplied T cells are capable of further interaction with antigen presenting and IL-1 secreting macrophages. This augments by rapid logarithmic multiplication the number of T cells capable of responding to the particular antigen. The B lymphocytes are capable of direct interaction by means of a particular cell membrane receptor for that antigen. However, they cannot multiply until they contact appropriate T lymphocytes that have been stimulated by the same antigen. At this contact the helper T cells secrete a number of factors (IL-2, B-cell growth factor, BCGF, B-cell differentiation factor, BCDF, and interferon) that stimulate B-cell multiplication and differentiation. The differentiated B cells manufacture and secrete free antibody to interact with and neutralize soluble free antigens.

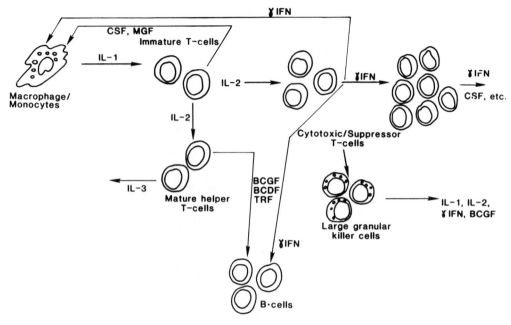

Figure 8. The lymphokine regulatory network. The emphasis in this figure is on the hormones and parahormones that are formed by macrophages and lymphocytes in different phases of the immune response. The cellular interactions and differentiations are better shown in Figure 7. The macrophages are the source of interleukin 1, IL-1; this stimulates T cells at various stages of maturation and differentiation to secrete T-cell growth factor, IL-2. More differentiated T cells (helper, suppressor, cytotoxic, and killer), as shown, variously secrete interferon, IFN, various blood colony stimulating factors, CSF, including IL-3, which is multi-CSF (stimulates formation of all blood cell types). The helper T cells secrete substances that stimulate B lymphocyte differentiation, multiplication, and activity (BCGF, BCDF, and TRF). Furthermore, the several types of T cells, through secreted IFN, CSF, and MGF (macrophage granulocyte stimulating factor) act back on the macrophage system, creating a positive feedback circuit. The only regulatory (inhibitory) actions in this system come from suppressor T cell action and from the steroid hormones shown in Figure 9.

one or the other depends (in mammals) on whether it has been acted upon by inducing substance from bone marrow (B) or thymus (T) (148). The inductive substance(s) from the thymus have been isolated and identified as "thymosine" or "thymopoietin" (111). The bone marrow factor has not been identified, but seems to be the same as "bursin," a factor isolated from a lymphoid organ in birds, the bursa of Fabricius (which induces B cells in birds) (5, 200). It is not clear whether these inductive substances are exclusively parahormonal, acting within the thymus

or bone marrow, or whether they can also act via the blood as true hormones.

Once differentiated, T and B cells interact with antigenic substances in different ways. The B cells bear specific receptors on their cell membranes, which interact with specific antigens (3, 75). The antigen–receptor complexes are internalized, partly broken up and ultimately presented once more on the B-cell surface in conjunction with a protein of the major histocompatibility complex (MHC). On the other hand, the T cells do not interact directly with unaltered antigen. They en-

counter the antigen through macrophages that have nonspecifically ingested
it, partly digested it, and then presented
it on the macrophage cell surface in conjunction with a MHC protein. The MHC
proteins are considered to function in order to permit the recognition of foreign
protein versus "self"-proteins. Like the B
cells, the T cells bear a specific surface
receptor that recognizes the specific antigen. There must be a great variety of
surface receptors to recognize sterically,
and interact with the almost infinite variety of possible antigens. This is made
possible by a special system of variable
combinations of DNA segments that together code for the antigen-interacting T-
and B-cell receptors (148). There can be
only a few lymphocytes available that can
fortuitously recognize a particular antigen. A first problem to be solved in the
immune response is to multiply the reactive lymphocytes to a degree that can
successfully cope with a massive invasion
of the organism by an antigenic component.

As shown in Figure 7, when the anti-
gen-presenting macrophage contacts the
appropriate T lymphocyte, it secretes parahormonal interleukin (28, 77). The antigen reaction plus IL-1 stimulate mitoses
in the T cell, as well as secretion of IL-2
and γ-interferon (γ-INF). The IL-2 is a potent mitogen for the activated T cells, producing in a short time many more T cells
of the type that can interact with the an-
tigen-presenting macrophages. Depending on the type of MHC protein on their
surface, the multiplying T cells further
differentiate in either of two directions. In
one they become cytotoxic T (killer) cells
that can attach to and lyse foreign anti-
gen-containing cells that have entered the
organism (3). In the other, they become
helper T, or suppressor T cells (75, 148).

The immune response interacts with a
complex of hormones. Thus, substances
such as the interleukins and hepatocyte

stimulating factor produced by responding macrophages act via the pituitary–adrenal axis to induce the release of
glucocorticosteroids from the adrenal cortex, thereby slowing interleukin production (2, 217) (Fig. 9).

Aside from the bursin finding, there is
relatively little information available from

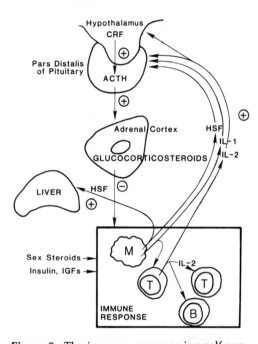

Figure 9. The immune response is a self-aug-
menting system, since the interleukins and
interferon produced by responding macrophages, M, and T lymphocytes stimulate their
own multiplication and action on B lymphocytes, B. Here is shown how IL-1 and IL-2
stimulate the production of ACTH by the pituitary, in part by stimulating CRF production
in the hypothalamus. ACTH then stimulates
glucocorticosteroid production by the adrenal
cortex. This hormone has a strong immuno-
suppressive action. Glucocorticosteroids not
only reduce IL production, but lymphocyte
multiplication in general, even at the thymic
level. Also shown is the action of hepatocyte
stimulating factor, HSF, (IL-b), produced by
macrophages. It not only stimulates liver cells,
but also ACTH production. The immune response is also affected by the sex hormones
and is augmented by insulin and IGFs.

lower vertebrates. The demonstration that cyclostomes and fishes reject homoplastic skin or scale transplants (and second transplants are rejected more quickly than the first) indicates a cytotoxic T-cell like response. Immune response-generated antibodies, free in the blood, have been found in all vertebrate groups, including cyclostomes. However, to what extent the system that forms them resembles mammalian B cells and plasma cells requires study. Such specific humoral antibodies in cyclostomes and elasmobranchs require a long time (40–60 days) to appear and they attain only low concentrations. Many substances that are antigenic in mammals and birds evoke no immune response in the fish groups. In frogs both helper T-like and supressor T-like functions have been found. Such facts merely indicate functional similarities, but the cellular and hormonal basis for the immune responses in cold blooded vertebrates needs clarification (190).

With the cellular immune systems so poorly defined in nonmammalian vertebrates, it is difficult to demonstrate lymphokines or interleukins. Nevertheless, Rijkers has described a fish interferon, and IL-2-like activity has been claimed in supernatants of cultures of mitogen-stimulated fish lymphocytes (30).

Another way to show similarity or difference between the immune systems of lower and higher vertebrates is to expose them to supranormal levels of glucocorticosteroid hormones. In such experiments, inhibitory or destructive changes in the thymus of fishes, amphibia, and reptiles have been found. Thymic involution in a small fish, the Japanese medaka, follows desoxycorticosterone, but not cortisone, treatment, arguing that the steroid hormone receptors of the thymic cells of this fish are highly specific. Hydrocortisone has been found to lower the number of lymphocytes in the blood of a lizard by about one half. Comparative im-

munoendocrinology is a virtually unexplored field, but it holds much interest and promise to those who wish to understand the evolution of the immune system (2, 91, 92, 190, 194).

Hormones in Reproduction

Sexual reproduction in its different phases is the most conservative and constant of all functions, and yet it is at the same time also the most variously adapted to special conditions. Sexual reproduction always involves the union of the gametes, the ovum, and the spermatozoon. In order to attain maximum reproductive success in a variety of habitats and specialized life cycles, the conditions under which the gametes are brought together and the disposition and care of the zygote must be extremely varied (232).

Species vary enormously in the number of eggs fertilized per reproductive cycle, in the amount of yolk they deposit in the egg for early embryonic nutrition, and in the degree to which they protect the early development of the zygotes, embryos, and still vulnerable young. Patterns of reproductive behavior are the most varied of all, and probably are unique for each species. In at least the female of every vertebrate, reproduction is cyclic, and often linked to season, to assure that the young are born, or first exposed to the environment, at a favorable time that promises optimal survival. Reproductive patterns are so characteristic and differentiated that in a textbook of comparative physiology one could endlessly cite specific patterns and discuss their individual variations. We will avoid losing ourselves in a sea of interesting detail by emphasizing common features of reproductive control and mentioning interesting examples where they are pertinent.

The central organizing scheme for a

large part of the vertebrate endocrine system is an anatomical link between the brain and the pituitary gland (78, 100, 102). This linkage is of two kinds. The simplest is the ending of a series of neurosecretory neurones in the pars nervosa of the pituitary (Fig. 10*a* and *b*). Here the neurones secrete the octapeptides, oxytocin, and vasopressin, which are carried by the blood to peripheral responsive organs. The second brain–pituitary relationship involves secretory neurones that terminate on blood vessels in the median eminence of the pituitary. Short portal vessels carry the secreted neuropeptides (releasing hormones) to the pars distalis of the pituitary gland (Fig. 10*a* and *c*).

The brain–pituitary relationship makes possible the linkage of environmental events detected by various sense organs to the reproductive organs via the endocrine system. For example, suckling by nursing young mammals yields milk at the nipple within <1 min. In this "neuroendocrine reflex arc" (Fig. 10*b*), the tactile stimulation of the nipple sends nerve impulses to the spinal cord and ascending impulses to the hypothalamic part of the brain. Synaptic activation of the appropriate neurones in the preoptic nucleus leads to activation of the neurosecretory cells in the pars nervosa of the pituitary, and the release into the blood of the octapeptide oxytocin. The oxytocin, carried by the blood to the mammary gland, stimulates the muscles surrounding the storage elements in the alveoli of the gland, causing expulsion of the milk. This entire process requires only a few seconds. Similarly, an afferent stimulus from the distension of the lowest part of the oviduct by the descent of the completed egg in the hen causes release of arginine vasotocin (AVT), which stimulates egg laying. However, if the egg arrives at the end of the oviduct during the evening or night, egg laying is delayed until early morning.

In the hen, there is a circadian clock that regulates the responsiveness of the hypothalamus to nervous signals from the oviduct (102, 232).

Certain amphibians and fishes appear to make use of behaviorally evoked release of neurohormones from the pars nervosa during mating, although this is highly variable among individual species. In several amphibian species, neurohypophysial hormones have a role in reproductive behavior, and their release may be triggered by a preceding behavioral event or contact. The same has been claimed for mating in fish (*Fundulus*, the killifish) (169, 232).

A second type of neuroendocrine control is illustrated by the ovulation response, which occurs in rabbits ~10 h after copulation (Fig. 10*c*). The inciting stimulus originates in tactile receptors in the wall of the vagina, and the afferent impulse is routed via spinal nerves to the spinal cord, and up the cord into the hypothalamus, stimulating neurosecretory cells in the supraoptic nucleus. These neurones terminate in the median eminence of the neurohypophysis, where they release a decapeptide GnRH (gonadotropin releasing hormone) into the portal blood vessels, which conduct the GnRH to the nearby pars distalis, where gonadotropic cells release one or both of the gonadotropic hormones. These, via the systemic circulation, reach the ovary, and set in motion the changes that culminate in ovulation. The ovulated egg is moved from the ovary into the oviduct by ciliary action, and by that time the spermatozoa have reached the oviduct as well, and fertilization takes place (232).

The rabbit ovulation neuroendocrine reflex can be taken as the prototype for many gonadal regulatory events in which an environmental stimulus, detected by a sense organ, eventually has an effect upon as complex a phenomenon as reproduction (see also Fig. 11). The most

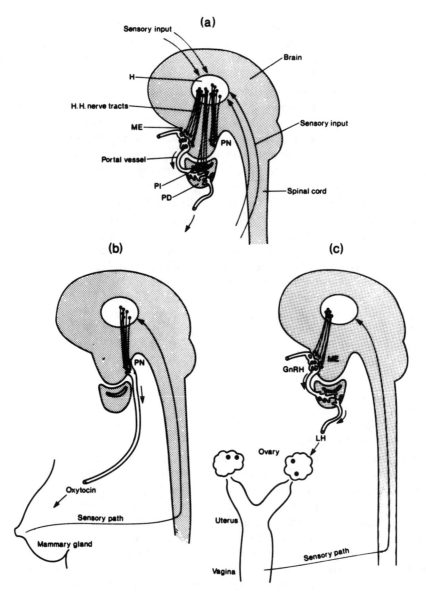

Figure 10. Three representations of the vertebrate central nervous system, CNS. In (*a*) the essential structural relationships to the pituitary gland are shown. The oval area in the center of the brain represents the hypothalamus, H, or ventral diencephalon that contains the cell bodies of many neurosecretory neurones. The lobe on the ventral side of the brain is the neurohypophysis. In (*a*) it can be seen that there are three kinds of relationships. In one, the neurones from H end in the pars nervosa, PN, of the neuro-hypophysis. A second type (*b*) of neurone from H terminates within the pars intermedia, PI, of the adenohypophysis. In the third type (*c*), the secretory neurones end in the median eminence, ME. From the ME, portal vessels carry the neurosecretion to the pars distalis, PD, of the adenohypophysis. The ways in which these structures are used in the neurosecretory reflex are illustrated in (*b*) and (*c*). In (*b*), the elements of the suckling reflex are shown. Stimulating of the nipple of the mammary gland initiates sensory

common feature of vertebrate reproduction, seasonal cyclicity, is of this type. Here the usual stimulus is a prolonged one, the gradual increase in the amount of daylight per day (photoperiod) in the spring. The photoreceptor is most often the eye, but in birds, there appear to be photoreceptors in the brain itself. The mechanism for measuring the photoperiod remains uncertain (232).

There are many slight variations, but in all cases the adaptive advantage of seasonality involves producing young at the most favorable time of the year for their survival. In environments that are equally favorable at all times of the year, seasonal cyclicity tends to be lost or reduced. In equatorial regions, where annual temperature and photoperiod fluctuations are small, annual reproductive periodicity may be keyed to other recurrent seasonal clues, like rainfall and availability of food, or annual periodicity may be lost completely. In some species the reproductive cycle is keyed to short days, instead of long days. In sheep, for example, which have a gestation period of ~5 months, ripe gametes and mating must occur in the short days of winter in order for birth to take place in the spring (6, 27). In hagfish, which live deep in the ocean, where there is no annual cycle of photoperiod or temperature, reproductive activity is continuous, the only cycle depending on how long it requires a female to produce a new batch of eggs.

When the reproductive cycle is repeated at intervals that are <1 year, external environmental clues no longer operate. In the female rat, for example, the cycle is repeated every 4 or 5 days unless pregnancy intervenes. In women it is 28 days, in sheep and guinea pigs 16 to 17 days, and even the female elephant has a cycle that is <1 month. Since such individual female cycles are not necessarily synchronized, the males must necessarily be fertile continuously, without cycles, always prepared to provide spermatozoa when a female has mature eggs (145). The hormonal regulation of short cycles is complex and not yet completely explained.

In many short cycle females the repetition of the cycles is "automatic" or spontaneous, a new batch of eggs being produced each time. If these are not fertilized, they degenerate (i.e., they become atretic) and the ovary reenters a phase of maturation of another group of eggs from a limited stock of oocytes. A variation on this pattern is seen in those mammalian females that are "induced ovulators." In such animals (cats, ferrets, rabbits, and camels) the cycle is halted when the ovary contains mature eggs, ready for fertilization. A nervous signal, provided by copulation, is required for ovulation. Copulation is followed by a surge in secretion of gonadotropic hormone, which, in turn, stimulates ovulation (Fig. 10c). The essential difference between sponta-

---

impulses that rise in the CNS to H. Eventually, secretory neurones are stimulated in H, and these end in the PN, where oxytocin is secreted into the blood. The oxytocin stimulates contraction of cells in the mammary gland, squeezing milk down to the nipple. A neurosecretory reflex that utilizes neurones ending in ME is shown in (c), which refers to ovulation in the mated rabbit. Stimulation of the vaginal wall during mating results in sensory impulses that spread to neurosecretory cells in H, which end in the ME, where the decapeptide hormone GnRH is released into the portal blood vessels. The GnRH is carried to the PD, where it stimulates cells to release a gonadotropic hormone, LH, which acts on the ripe ovarian follicles, causing them to release their eggs for fertilization in the oviducts, and implantation in the uterus. [From (102).]

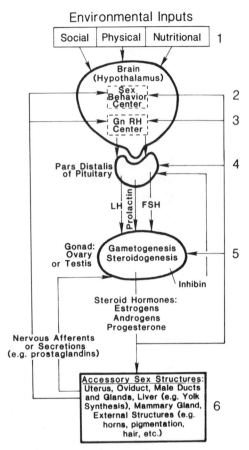

Figure 11. Hierarchies in the regulation of reproduction in the vertebrates and feedback on the regulators. In this diagram, the brain (2, 3), pituitary (4), gonads (5), and sex accessories (6) are shown as a descending hierarchy. Superimposed on this series are environmental influences (1), which often are triggers of modifiers for reproductive events. Downward arrows in the middle show the stimulatory primary hypothalamo–pituitary–gonadal axis. At either side are upwards pointing arrows indicating feedback from lower to higher elements in the axis. These are generally, but not always, negative or inhibitory.

neous and induced ovulators lies in whether the timed release of an ovulation inducing quantity of pituitary gonadotropin is programmed by an external or internal signal. This difference, one step further back in the mechanism, depends

on the stimulus to which the GnRH center is programmed to respond (232).

Both spontaneous and induced ovulators can be seasonal breeders as well: In certain seasons they cease short cycling altogether. This implies that a seasonal clock can be superimposed on the hypothalamic regulatory mechanisms for short (less than annual) cycles as in animals that breed only once yearly. The advantage of this annual shutdown of reproduction is that young are born only at favorable times of the year. Furthermore, even a circadian clock can be superimposed on the short cycle. An example is the rat, in which the ovulation inducing surge of gonadotropin is delayed in the hypothalamus so as to occur shortly after midnight. Thus, ovulation occurs in the early morning hours and coincides with female mating behavior readiness. It will be recalled that a similar circadian clock is superimposed on the vasotocin releasing system in the hen. There is a similar circadian restraint on egg laying (vasotocin induced) in some fishes (e.g., *Brachydanio*) in which eggs are deposited always just after dawn (232).

To summarize, reproduction in vertebrates, though extremely varied in pattern, is regulated by the same general nervous and endocrine factors in all vertebrates (Fig. 11). However, these factors may have different actions in different vertebrates, or interact with each other in different ways, or over different time periods. One of the most common features of vertebrate reproduction is cyclicity and this feature is regulated by a combination of nervous and endocrine factors, the so-called hypothalamo–hypophysial–gonadal axis. Cyclicity is most commonly seasonal, coordinated with external environmental cues to produce young at the most favorable time to insure maximum survival. However, many species, particularly those in minimally cyclic environments such as the tropics, and also domesticated species, may have shorter than annual cycles in addition to, or instead of, annual cycles. In the following section some of the specific variation in adapted reproductive systems will be examined in the order indicated in Fig. 11.

Environmental Influences on Reproduction

In the most primitive vertebrates, the cyclostomes (hagfishes and lampreys), gonadal cycling is extremely different in the two groups (101). In most hagfishes, which inhabit deep-sea noncycling environments, there appears to be no synchronized gonadal cycling; females captured at any one time are in all phases of egg production (257). However, in the Japanese hagfish, *Eptatretus burgeri*, there is a reported annual migration between deeper and colder waters (in summer) and warmer shallower waters (in winter), and a correlated synchronization of gonadal development among females. Such simultaneity of reproduction correlated with season indicates a neuroendocrine mechanism, but none has been described thus far.

In lampreys, on the other hand, all species breed only once per lifetime, and the final development of the gonad and breeding are seasonally defined. However, there has been no experimental analysis in lampreys, which links gonadal cycling to particular environmental events. There is some correlative evidence that water temperature determines the final spawning migration and gonadal maturation (210, 211).

In the elasmobranch fishes that have been examined (primarily temperate zone species), breeding is seasonal, but there is little information concerning the nature of the environmental regulators for this cyclicity. Some elasmobranch species have been claimed to breed throughout the year (60).

In teleosts there is considerable evidence that photoperiod is a major determinant for the sex cycle in a number of temperate zone species (169). However, temperature also plays a role, primarily in setting a "threshold" (e.g., in goldfish and rainbow trout). Nevertheless, whether a threshold control or not, temperature can obviously affect or limit the hypothalamic programming of gonadotropic control. Other environmental influences on teleostean reproduction include food supply, salinity, tidal rhythms, olfaction, and visual stimuli. The olfactory and visual stimuli are probably more important as evokers of sex behavior: Their role in gonadal development is not known.

In temperate zone amphibians, photoperiod, temperature, humidity, and food supply all affect reproduction, but to different degrees in different species (124). Olfaction and audition (through mating calls) also play roles, especially in evoking behavior. There is some evidence that chorusing by males may precipitate ovulation in tree frogs (*Hyla*). Olfaction, presumably through following a pheromone, an odorant chemical released by the females, leads male toward female salamanders (*Taricha*).

In seasonally breeding lizards (e.g., the American chameleon, *Anolis carolinensis*), photoperiod apparently supplies the primary seasonal clue to the endogenous hypothalamic clock, but, as mentioned for fishes, temperature is an important modifier of the photoperiodic response. Below a particular temperature, manipulation of the photoperiod is ineffective. Temperature not only regulates secretion of pituitary gonadotropins (presumably through GnRH stimulation), but it also affects the sensitivity of the gonads to the pituitary hormones. With respect to breeding behavior itself, a variety of external cues, including tactile (direct contact), auditory (vocal), and olfactory (pheromones) stimuli have been described as important regulators for lizards and snakes. Some of the pheromones that evoke male courting behavior in snakes have been extracted from the female skin or internal tissues. Visual cues in the sometimes complex courtship displays and behaviors are of obvious value in guiding courtship behavior, especially in some lizards (144).

Reproduction in birds can follow extremely varied patterns. This is to be expected in view of the wide variety of habitats, social behaviors, diets, migrations, climates, and latitudes experienced by individuals in a given species. Although annual breeding cycles are the most common among birds, especially in temperate latitudes, some species are continuous breeders, some have regular cycles of <1 year, and some are opportunistic breeders.

The effect of photoperiod varies among bird species, and even among races (e.g., northern versus southern). Exposure of male sparrows to long days can result in increases in testicular weight of several thousand percent over a short time. The photoreceptor involved in the detection of photoperiod in birds is not the eye, but the hypothalamic region of the brain itself. No specialized photoreceptor has been found in this region, but surgically implanted light sources in this region (but not others) evoke gonadotropic and gonadal responses.

Temperature influences are minor, as might be expected in a homeothermic animal. Similarly, olfactory influences play a lesser role in birds than in other vertebrates, although deprivation of olfactory function in ducks affects mating behavior. Auditory stimuli (song and other vocal patterns) are stimuli for gonadal maturation and growth in budgerigars, ring doves, and canaries, and probably other species as well. Visual stimuli, particularly in combination with other sensory modalities that are part of courtship be-

havior undoubtedly are influential, but have received less study, particularly as to how they affect the neuroendocrine gonadotropic system. If they were not important, it would be difficult to explain the evolution of elaborate courtship behaviors and their accompaniments (territoriality, sex limited anatomical structures and pigmentation, etc.). Removal of some or all of the eggs from the nest of the American robin results in ovarian growth and the laying of additional eggs. The stimuli here may be visual and possibly tactile. Other demonstrated environmental regulators include the availability of a nest of appropriate dimensions, presence or absence of particular foods, and nutritional status in general (232).

In some mammals, increasing photoperiod has been shown to be stimulatory (horses, voles, raccoons, hamsters, the hare *Lepus*, and ferrets). In a few mammals, decreasing photoperiod is stimulatory (sheep, goats, white-tailed deer, and silver fox). The eye is the principal photoreceptor for this response, but may not be the only one (78).

Diet can influence gonadal development. In voles (*Microtus* sp.), for example, eating of young plants (grasses, alfalfa, and wheat) is correlated with reproductive success and fertility. Ambient temperature changes would not be expected to figure largely in species that maintain a constant body temperature. However, extremes of temperature within the tolerance range will reduce reproductive efficiency (reduced litter size, fertility, gametogenesis, and lactation). These effects may constitute a response to stress, leading to adrenal corticosteroid secretion and a negative effect on reproduction.

Olfaction plays an important role in the reproduction of many mammalian species. Reproductive behavior in a number of species includes sniffing and licking by the male of pheromone-containing secretions near the vaginal opening of the female. The appearance of such secretions, which may be of vaginal or urinary origin, is often the result of stimulation by sex steroids (usually estrogens), which are relatively abundant when the ovary contains eggs ready for ovulation (estrus). Sex hormones are thus often signals and stimulants for sexual activity by males when females have eggs ready for fertilization. Such signals are more important in species with short female cycles, or with cycles that are not synchronized among individuals. In highly seasonal, annual breeding females, pheromonal signals are less important since all females provide mature eggs at the same time. Behavioral sex pheromones can be effective over considerable distances: In dogs or pigs, for example, the females can exert strong attraction to the male from great distances. Some of these attractant compounds have been isolated: They are relatively simple organic compounds.

Behavioral sex pheromones can also be produced by the male (145). In pigs, the boar releases a urinary pheromone that is a metabolite of testosterone. When detected by an estrous sow, this substance evokes mating readiness, a kind of stiff-legged stance that invites male copulatory behavior. In mice, male urinary pheromones have been demonstrated in several different ways. Grouped female mice eventually cycle irregularly. However, placing a male or even a small amount of male urine in the group will cause the females to resume normal cycling (Lee–Boot effect). Furthermore, in the continued presence of a male or male urine, the cycles of the caged females become synchronized (Whitten effect). If the paternal male is replaced by a strange male in the cage, or if urine from a strange male is introduced, pregnancy in the female in the cage is interrupted (Bruce effect).

It should be emphasized that the environmentally evoked reproductive phenomena are of two kinds, behavioral and endocrine. The environmental influences on sex behavior can be direct (Fig. 11). However, the sex behavior center in the hypothalamus is itself the target of action by sex hormones and other influences. The preoptic sex behavior centers have receptors for the sex steroids and their thresholds for arousal are understandably regulated by the sex steroids. The environmental actions on endocrine secretion must, as a rule, be channeled through the GnRH center, and these actions can be stimulatory and inhibitory. In the Whitten and Lee–Boot effects, GnRH and gonadotropin secretion are increased. In the Bruce effect the opposite occurs and abortion results.

The hypothalamic gonadotropin releasing factor, GnRH, is a decapeptide found in all vertebrates except the myxinoid (hagfish) cyclostomes (78). In some species (e.g., chickens and lampreys) more than one type of GnRH can be found, differing by one or two amino acids from the common mammalian type. One of the lamprey GnRHs differs by 5 amino acids from the 10 residue mammalian type, and it is not surprising to find that it has no effect in mammals (21). The functional significance of multiple GnRHs is still not clear. There are two pituitary gonadotropins, FSH and LH, and it might be thought that each was released by a specific GnRH. However, if a particular GnRH has any action, both gonadotropins are released.

Cyclostomes have no portal blood vessels between the neurohypophysis and the adenohypophysis (101). In dogfish sharks (only one species has been studied extensively) >90% of gonadotropic activity is in the ventral lobe of the adenohypophysis, which has no direct blood connection to the neurohypophysis (60). How then does hypothalamic GnRH in lampreys and dogfish reach the gonadotropic cells of the adenohypophysis? In lampreys there is indirect evidence that

GnRH can diffuse in sufficient concentration from the adjacent neurohypophysis. In dogfish, large quantities of GnRH would have to be released into the systemic blood, and high concentrations of GnRH have been found in the peripheral blood in the dogfish. In the part of the dogfish adenohypophysis that has a well-developed portal vessel communication to the neurohypophysis (a true median eminence), there is little, if any, gonadotropin. These facts raise some puzzling questions concerning evolution of the brain–pituitary relationship in this group of vertebrates (100).

Obviously, full clarification of mechanisms controlling reproductive cycling, and responses to environmental influences, must include an understanding of the controls over GnRH release in individual species. These controls are of three types: (1) direct afferent nervous synaptic regulation, primarily those controls excited by various pertinent sensory modalities; (2) nervous afferents modulated by circadian or annual endogenous clocks; and (3) hormonal and other feedback controls (illustrated in Fig. 11).

Steroid hormone feedback on hypothalamic GnRH release is most generally negative or inhibitory, an expected relationship that permits the setting of a balance (homeostasis) between the stimulatory (pituitary) hormones and the targets of their stimulation, the gonads (35). However, in cycling female mammals, birds, and even teleost fish, gonadotropin surges occur at appropriate times (e.g., at ovulation), implying that the balance must be changed or reversed to allow surging. At these times in female mammals, there is a positive feedback of estrogenic steroids on hypothalamic GnRH centers. In this sense, the hypothalamus is sexually differentiated, no positive feedback being seen in male mammals. Exploration of lower vertebrates for sexually differentiated positive feedback GnRH centers has not yet been done as systematically as in mammals, and no generalizations can yet be made about evolution of this mechanism.

The *pituitary gonadotropic hormones* in all vertebrates but the fishes, classified on the basis of their action on the ovary, are of two types: follicle stimulating hormone (FSH) and luteinizing hormone (LH) (6). There is no good evidence that teleostean hormones are differentiated in this way and gonadotropin molecules (if any) of elasmobranchs and cyclostomes are not well enough characterized for conclusions to be drawn. The FSH and LH, like pituitary thyroid stimulating hormone, TSH, are glycoproteins. Their molecules are made up of two easily separable subunits, α and β, each of which contains ~100 amino acid residues. Molecular similarity between the two subunits suggests that they evolved, probably by gene duplication, from a common parent molecule. The α subunits are almost identical among FSH, LH, and TSH. By themselves, the subunits have little or no biological activity. The specificity of the action of FSH, LH, and TSH has been shown to reside in the β subunit. Thus, the binding of any of these three pituitary hormones to cells in the gonads is determined by receptor affinity for the β subunit, but no binding occurs without the α subunit.

The teleosts (e.g., salmon) may have more than one molecular form of gonadotropin, but these forms are not differentiated as FSHs or LHs (169). If their gonadotropic action differs at all it may be with respect to yolk formation in the female during ovogenesis. Biochemists have speculated that the first step in the evolution of pituitary hormones was differentiation of separate β subunits for gonadotropin (an FSH-like molecule) and TSH. Further evolution (in amphibians?) yielded separate LH-like and TSH subunits. It is important to determine whether gonadotropins or TSH exists in lampreys. Hagfish are not known to have gonadotropic hormones, but a closer examination of *Eptatretus burgeri*, the annual cycling hagfish, might be worthwhile.

In general, the important biological actions of FSHs are on gamete production, one of the two primary functions of the testis and ovary; the primary actions of LHs are on steroidogenesis, the other important gonadal function. However, in species in which adequate studies have been carried out, certain gonadal gametogenic and steroidal cell units have receptors for both gonadotropins. The Sertoli cells of the testis, for example, which have roles in both spermatogenesis and steroidogenesis, have receptors for both FSH and LH (145). The same is true of the follicle cells in the mammalian ovary.

Although GnRH releases both FSH and LH, titers of FSH and LH in the blood do not vary together in the cycling female. This may be the result of a differential sensitivity of the secreting cells to negative feedback from gonadal steroids. As is shown in Fig. 11, steroid feedback action is exerted at both the hypothalamic and pituitary level. Another selective feedback on gonadotropic secretion is by the peptide hormone inhibin. Inhibin is secreted by the testicular Sertoli cells and by the ovarian follicle. It selectively inhibits FSH secretion in mammals by a direct action on the pituitary (78).

Both of the primary gonadal functions, gametogenesis and steroidogenesis, are relatively conservative. That is, they are quite comparable in all vertebrate organisms. Spermatogenesis always produces haploid spermatozoa of comparable size and structure. Ovogenesis always produces ova in which the final meiotic division is arrested until fertilization, although the ova may be of different sizes depending on the requirement of the embryo for yolk.

Similarly, gonadal steroidogenesis is comparable in most vertebrates (35). The predominant testicular steroid hormone is testosterone or a similar molecule (Fig. 12), and the typical ovarian steroids are estradiol and progesterone (Fig. 13). All of these steroids are four-ring compounds that can be formed from precursor cholesterol. The ability of a particular cell to form one or another of these steroids depends on the array of enzymes that has been programmed genetically for that cell (see Fig. 12). Steroidogenesis in both the ovary and testis involves side chain splitting of cholesterol (removal of most of the long chain attached to the 5-carbon ring) and the production of testosterone and/ or androstenedione. The important molecular change in producing estrogens from androgens involves unsaturation (aromatizing) of the first 6-carbon ring and removal of one of the angular methyl groups (Fig. 13). The occurrence of the steroids is not strictly sex specific: Androgens can be found in the gonads and blood of some female vertebrates, and estrogens can be found in males, sometimes in large quantities (e.g., horses). A number of peripheral organs that possess the appropriate enzymes can take part in the metabolism of steroid hormones. Aromatase enzymes in certain centers in the brain, for example, can aromatize androgens, converting them to estrogens. The testis and some target organs of the male hormones (seminal vesicle, brain) contain 5α-reductase, which converts testosterone to a more active androgen, 5α-dihydrotestosterone (see Fig. 12).

The actions of the sex hormones are many and diverse since the strategies for sexual reproduction are so specialized and adapted (143). One generally applicable action involves the feedback on the hypothalamo–hypophyseal system, thereby achieving a balance between that system and the gonad. A second class of general actions is gametogenesis. There

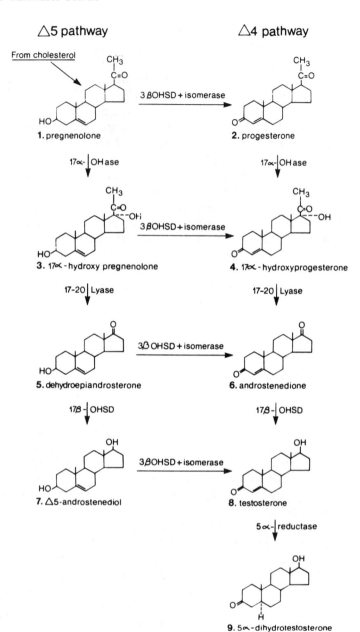

Figure 12. Two alternate pathways for stepwise production of androgenic steroids from cholesterol and pregnenolone. The arrows indicate the possible directions of the metabolic conversions, with the enzymes involved indicated at the arrows. In different species and different tissues different pathways may be used, and the conversions may stop at one of the intermediate positions instead of proceeding to testosterone or 5α-dihydrotestosterone. [From (102).]

is good evidence that androgenic steroids are essential for the final maturational changes during spermatogenesis, whereby the spermatids become fully functional spermatozoa. A third class of actions is on sex behavior. Sex hormones act on the brain to evoke appropriate mating behavior at the precise time when ripe

gametes are available. A fourth very general action of sex hormones is on the secondary sex organs (the ducts through which gametes and developing products of fertilization must pass) and accessory sex structures (brain, mammary gland, and external structures). Sex hormones may have profound metabolic actions as

Ovarian Steroidogenesis

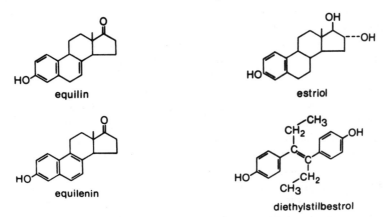

Other Estrogens

Figure 13. Incomplete representation of biosynthetic pathways for production of estrogens from androgens. The two principal estrogens, estrone and estradiol, are produced from androstenedione or testosterone, respectively, principally through the action of the enzyme aromatase. Other estrogenic steroids are shown below. Equilin and equilenin are found in large quantities in horses. Diethylstilbestrol, found in coal tar distillates, does not occur naturally in vertebrates. It is strongly estrogenic because it binds to estradiol receptors and because there are no natural enzymes for its conversion to inactive forms. [From (102).]

well. For vitellogenesis to proceed, the production of yolk proteins by the liver is stimulated by estrogens, and androgens have a strong anabolic action on muscle growth in mammals, an action well known to athletes.

In *cyclostomes*, relatively little is known about the actions of the sex steroids. Neither the lampreys or hagfish have sex ducts from the single testis or ovary. Eggs and sperm are shed into the body cavity and they exit through a small aperture in the cloacal wall. The cloacal wall contains glands that are larger in males and which increase in size at the time of breeding. These glands may secrete a pheromone that brings breeding animals together, but the physiology of the glands is unknown. Estradiol stimulates vitellogenesis in the hagfish, so that this estrogenic function has already evolved in these primitive vertebrates (229, 257). Under normal conditions there is little measurable sex hormone in hagfish blood, nor is there a high affinity steroid binding protein in the blood plasma, as occurs in other vertebrates, where it acts as a carrier for relatively insoluble sex steroids. Lampreys exhibit variations in sex steroid levels in blood that parallel the reproductive cycle, with the highest levels being found just before spawning when the gonad is maturing (236). Both estradiol and testosterone stimulate intestinal degeneration, a normal developmental event in migrating spawning sea lampreys (210, 211).

In the *elasmobranchs* plasma concentrations of androgens and estrogens are comparable to those in other vertebrates, or even higher. Most elasmobranchs produce at one time rather few large yolky eggs that are enclosed by a leathery shell secreted in the oviduct. However, the ovary of a very large viviparous shark, *Cetorhinus*, produces several million relatively small (50 mm) eggs at one time. Fertilization is internal as a rule, sperm being introduced into the female by a modified pelvic fin in the male. The male ducts have associated glands in their walls and a broadened sperm storage structure near the terminus of the male ducts. A majority of elasmobranch species is viviparous, retaining the eggs in the oviduct during development (256). In such species the shell is exceedingly thin or absent. Wourms has shown that the weight of the developing embryo increases markedly during development in the mother. It is thus clear that the embryos have received nourishment from the mother and that development is viviparous. A correlation has been found between raised levels of progesterone in the blood and "pregnancy" in viviparous species. In oviparous species, there is no increase in progesterone in the blood at egg laying or thereafter. The source of the progesterone appears to be an ovarian structure, the corpus luteum, which develops from follicular tissues after ovulation (255).

The many species of *teleosts* are found in a much larger variety of habitats than the elasmobranchs: fresh waters of different characteristics (temperature, oxygen tension, rate of flow, and presence of vegetation) as well as saline waters of various different depths, temperatures and relation to shore features. In addition, some species, in different phases of the life cycle, migrate between fresh and salt waters. With these different life styles, teleosts have evolved many different scenarios for reproduction (60). Some deposit thousands of eggs with virtually no care or protection, depending on the statistical survival of a small fraction. Other species, like the stickleback, deposit fewer eggs in sometimes elaborately constructed nests and then protect those eggs by parental behavior. A few teleosts are "oral incubators," keeping the eggs in the mouth of the parent, which is behaviorally bound not to eat them. The female bitterling develops a long hormonally regulated ovipositor, a tube that can be inserted into a mussel shell where the eggs may develop in relative safety. Some species are viviparous and may construct placenta-like temporary organs for the nutritive and respiratory needs of the embryos. The commonest method of fertilization is external, but some species have evolved in the males a modified anal fin structure for insertion of spermatozoa directly into the female.

The development, timing, sequence, and intensity of these various structural, metabolic, and behavioral items in the reproductive processes of teleosts utilize the same hypothalamo–hypophysio–gonadal axis for their regulation and linkage to environmental cues, as in most other vertebrates. The majority of structural and behavioral postovulatory items in the more specialized teleostean reproductive repertoires are regulated by steroid hormones. This is a logical relationship since maximal sex steroid secretion can be related in time to maximal gonadal development. A steroid that is typically involved in the final phases of maturation of ova in fishes is 17, 20-dihydroxyprogesterone, a steroid that is not usually found in other vertebrates. Similarly, the androgenic steroid typical for fishes is 11-ketotestosterone; it is rarely if ever seen in other vertebrates (169, 170).

In most *Amphibia*, the pattern of reproduction is fairly similar, although there are a few species with highly adapted and specialized modes of reproduction. In the species that have been appropriately studied, gamete maturation and steroidogenesis are under the control of separate FSH and LH gonadotropins from the pituitary (143, 206). Most commonly, amphibians engage in amplexus, a behavior in which the male grasps the female, thereby stimulating expulsion of ovulated eggs. Release of spermatozoa is coordinated with egg deposition. Mating generally takes place in the water, and many semiterrestrial amphibians migrate to water to breed. In Anurans, vocalization by males, or "mate calling" appears to guide the females to the water and to the males. There has been some investigation of the endocrine control of this phenomenon in frogs and of amplexus in the salamander *Taricha*. Even large doses of testosterone will not induce mate calling in frogs. On the other hand, testosterone accompanied by one of the neurohypophysial octapeptides is very effective.

In a few species there is direct transfer of spermatozoa to the female by a fairly simple cloacal intromittent organ. In a number of urodele (tailed) amphibians, masses of spermatozoa are bundled into packets with secretions from the male ducts, and the male deposits these spermatophores on the bottom during mating. The female follows and picks up the spermatophore with the cloacal lips. In one frog, *Nectophrynoides*, and a salamander, the young develop in the uterus in a viviparous manner, but there appears to be little structural accommodation for viviparity. In two frog species, the young develop in unusual internal locations, the vocal sacs and the stomach, and are born through the mouth. In the marsupial toad, the female places fertilized eggs in pockets of skin on her own back, where they develop. The possible endocrine basis for these modifications of behavior and structure has had very little study.

The *reptilian* ovary produces a relatively small number of large yolky eggs, and, depending on species, from one to a few eggs are ovulated at any one time (144). Gametogenesis in both sexes is under the control of both FSH and LH from the pituitary. Fertilization in reptiles is internal, as might be expected in a group that is primarily terrestrial. Most reptiles are oviparous, the egg being covered by a leathery shell that is secreted during the descent through the oviduct. However, several species are viviparous, and produce a corpus luteum, formed in the ovary from the follicular tissue of each ovulated egg, which secretes progesterone. In the viviparous snake, *Nerodia*, plasma progesterone levels rise sharply after ovulation. Like oxytocin in mammals, arginine vasotocin (AVT) appears to be the stimulant for the uterine contractions that accompany birth. Progesterone inhibits, and estradiol potentiates, the action of AVT on the contractions of the uterus.

Hypothalamic control over the reptilian reproductive axis is responsive in its turn to a number of factors, and temperature appears to be a dominant influence. In temperate zones, reproduction is most commonly annual and seasonal, with photoperiod a relatively more important environmental key. Rainfall, and accompanying abundance of prey, is an important factor for some species, particularly those in desert environments.

A common reptilian feature is long-term viability of spermatozoa. The mature sperm may be stored in either the male or the female so that the periods of active spermatogenesis need not coincide with periods of ovogenesis and with mating behavior. In the American chameleon, *Anolis*, reproductive activity in both sexes is suspended during the winter months (September–January). Males become continually spermatogenic and able to engage in reproductive activity during the rest of the year, and they defend territories against other males. Females ovulate a single egg at intervals of ~2 weeks, and are sexually receptive as an egg becomes mature. Fertilization must occur in the upper oviduct before the shell is deposited.

Hormonal control of reproduction in reptiles has not been studied exhaustively. Viviparity, for example, is influenced by progesterone from a corpus luteum, but generally is not completely dependent on it. Ovariectomy during pregnancy, particularly after it is well established, has relatively little effect. Whether this is due to a placental supply of the hormone is not known. In viviparous reptiles, a simple placenta formed either from the yolk sac or from chorioallantoic membranes nourishes the embryo. Expulsion of the eggs in oviparous species or embryos from viviparous species can be induced by injection of pars nervosa hormones, oxytocin, or arginine vasotocin.

Mating and territorial aggressive behavior of the male is dependent on testosterone or dihydrotestosterone (145). Estrogens stimulate female mating behavior (receptivity) in *Anolis*, particularly if followed by progesterone. Female attractiveness to males decreases after ovariectomy, and is restored by estrogen treatment, suggesting that in female *Anolis* formation of a pheromonal sex attractant is under estrogenic control in individuals with a fully matured egg. There is evidence that such a pheromone is made by the liver, which also responds to estrogen by synthesis of egg yolk proteins.

Reproduction in *birds* and *mammals* has received much more intensive study than it has in any of the other vertebrate groups, and more species of birds have been studied than of mammals. Most birds fly, and seasonally build nests in which hard-shelled eggs are laid. The eggs are usually few in number and are incubated by the parents; there is specialized parental behavior associated with the care and feeding of the young. There are, of course, many variants on this general theme, since the latitudes, temperatures, and physical and biotic features of the environments for different avian species vary so widely. Additional variants are created by species behavioral differences, social organization, and seasonal migration.

Studies of environmental influences on

reproduction in birds have emphasized photoperiodic control. Indeed, the discovery of the relationship between photoperiod and gonadal growth was made through Rowan's observations on starlings. Since then the avian photoperiodic response has been analyzed in fine detail using artificial illumination with other factors kept constant. In birds (at least in ducks and sparrows) there is a photosensitive clock in the hypothalamic and/or the thalamic brain tissue itself that appears to measure the daily ration of light. Particular wavelengths are more effective than others. Even low intensities of light that reach the sensitive region through bone and feathers are effective. Painting the top of the head with light-impervious material blocks the response to artificial long days; blinding does not.

Other external factors affect reproduction through hypothalamic routes. Bird song and other vocal signals, the sight of other birds, the presence or absence of a mate, the character and amount of food eaten, are all influential. Temperature has some influence, but as to be expected in a warm-blooded animal it is less important than in fishes or amphibians.

The male (Wolffian) ducts of birds are relatively simple and lack prominent glandular structures. The male Japanese quail has cloacal foam glands that produce a medium in which spermatozoa are transferred to the female. In all birds, fertilization is internal, but as a rule, there is no penis for the transfer of sperm. In some larger birds (geese, ostrich) the wall of the male cloaca is extended as a false penis.

The large yolked eggs are laid in clutches. In temperate or cold latitudes there may be only one or two ovulations of clutches per year, while in tropical or domesticated species (e.g., chickens) breeding may be continuous, with intervals between ovulations determined by the time required for the maturation of a new clutch of eggs or a single egg. In a continuous breeder like the chicken, a single large egg is ovulated each time, and, as in some reptilian ovaries, there is a hierarchy of maturing eggs that succeed each other.

Avian FSH and LH have been purified and are similar in molecular character to mammals. The hypothalamic releaser for these gonadotropins is GnRH and, as discussed earlier in this chapter, two GnRHs have been isolated from chicken brains. The follicle stimulating hormone appears to have an important role in early phases of ovogenesis while LH appears to be important in the later stages and in ovulation. The follicle cells that surround the growing eggs secrete progesterone, estradiol, and prostaglandin.

As the avian egg moves through the oviduct, it is invested with layers of albumin by glands in the wall of this organ. Secretory activity by these glands is stimulated by estrogens and progesterone, acting sequentially (203). The shell is secreted by a glandular structure that surrounds the lower end of the oviduct. Expulsion of the egg is stimulated by the action of prostaglandin and AVT. Arginine vasotocin (AVT) is considered to be released in response to afferent nervous signals resulting from the presence of an egg in the lower oviduct. Prostaglandins are believed to be released at the time that the egg breaks out of the follicle. As mentioned earlier, however, the neuroendocrine secretion of AVT can be delayed by a hypothalamic diurnal clock to occur at first daylight.

Male aggressiveness and copulatory behavior are primarily stimulated by testosterone and 5α-dihydrotestosterone acting on hypothalamic nervous centers (145). By implanting pellets containing testosterone in different parts of the brain it can be shown that aggressiveness and copulatory behavior can be selectively and separately stimulated. Estradiol, which has no action on male behavior in male fowl, stimulates male behavior (crowing) in Japanese quail. Estradiol is important for female behavior in courting ring doves; estrogen values rise dramatically at this time.

Other hormone-influenced behaviors include incubation, which seems to be under the organizing influence of prolactin and, in the dove, progesterone in both sexes. Prolactin also stimulates mitotic proliferation in, and the shedding of, the epithelial cells of the crop in pigeons and doves. This material is regurgitated as "crop milk" and fed to nestling young.

Singing in canaries is stimulated by androgen administration, but not in some other species (Zebra finches). In a flock of domestic fowl a hierarchical social relationship is eventually established, known as a "peck order." Androgen injection tends to raise, and gonadectomy to lower, the position of a bird in the hierarchy. In some birds there is a correlation between plasma testosterone levels and dominance–submissiveness of two individuals in a social (aggressive) encounter. In Japanese quail this correlation is not distinct and corticosteroids play a role as well.

The *mammals* include species with a surprisingly wide array of reproductive patterns. The most primitive mammals, the monotremes (platypus and echidna) are oviparous, giving birth to partly developed embryos in a large yolky egg. After birth, the eggs are incubated in a nest until the young hatch and begin feeding by licking milk from the mother's skin, there being no nipples for the mammary glands. The marsupials (e.g., kangaroos and opossum) give birth to incompletely developed fetuses, which must complete their development in a skin pouch, the marsupium, while attached to a nipple of the mammary gland. In all other mammals young are retained in the uterus for completion of a much longer proportion of development, being nourished through a placenta. The young of higher mammals are born in various degrees of development in terms of relative size and mobility and length of dependence on parental care. Thus, one of the evolutionary trends in mammalian reproduction is for decreasing dependence on egg yolk and lactation for support of early embryonic development, and an increasing reliance on intrauterine placental support of early development.

Fertilization in all mammals is internal, occurring in the upper end of the oviduct not far from the ovary. Temperate zone mammals generally have a single annual reproductive cycle that is commonly cued to the season by photoperiodic responsiveness of the hypothalamo–hypophysial–gonadal system. However, many mammals, and especially tropical and domesticated species, have shorter cycles that are repeated spontaneously until pregnancy occurs. Such autonomous female cycles cannot depend on external environmental cues, but are integrated by an internal regulatory neuroendocrine mechanism. A variant of the short (less than annual) cycling pattern is seen in the cats, rabbit, ferret, and camel, which are known as induced ovulators. Here the ovarian cycle halts at the estrous stage when a mature egg is available, needing only a burst of pituitary LH for ovulation to occur. The stimulus for this surge of LH is the copulatory act in which tactile receptors in the vaginal wall originate afferent nervous impulses that ascend the spinal cord. Here they evoke hypothalamic GnRH release to the pituitary and consequently LH secretion. Induced ovulators do not remain permanently in estrus. If mating and ovulation do not occur, the egg in its follicle degenerates (undergoes atresia) after a variable period, depending on the species, and a new cycle of egg maturation begins.

Many of the peptide and steroid hormones of mammals have been purified and characterized, making possible the development of radioimmunoassays for the determination of hormone titres throughout the various cycles. The details of the hormonal cycles are well known in the human female and in a number of other domesticated and laboratory mammals. As a rule males do not cycle other than annually, if at all. During the period when females are cycling, the males remain continually able to provide spermatozoa.

Although there are differences in detail in spontaneous hormonal patterns in short cycle female mammals, there are several important constant features (6). Ovulation of a ripe ovarian egg is the result of a surge of pituitary LH secretion. In induced ovulaters, as described above, release of the egg from the ovary is the eventual result of a neuroendocrine GnRH response triggered by copulation. In spontaneous ovulaters (e.g., sheep, humans, and rodents), the trigger is the high level of estrogen secreted by the large ovarian follicles at this time. This is coupled with the positive feedback action of the estrogen on the hypothalamic GnRH center. That is, the increasing levels of ovarian estrogen stimulate more GnRH and gonadotropin secretion. These, in turn, further augment the estrogen secretion, resulting in an explosive peak of estrogen that just precedes the peak of LH. This, of course, is an example of positive feedback, and, if unchecked, such systems will become unstable. In this case the system is turned off as a result of secretion of progesterone by the corpus luteum, formed by the follicle cells of the egg that has been ovulated. Progesterone in rodents has the effect of reducing the number of steroid receptors in the hypothalamic GnRH centers, in this way ending the positive feedback phase of the cycle. Thus, cycling in spontaneous short cycle female mammals is due to an interplay between the changing levels of steroids from the sequential succession of ovarian structures and the hypothalamo–pituitary system. In some species termination of the period of secretion of progesterone by the corpus luteum is due to the action of prostaglandin from the progesterone-stimulated uterus. This is, in effect, negative feedback. The prostaglandin-induced luteolysis is one of the events in the cycle that must be blocked if pregnancy is to occur and the corpus luteum is to persist.

Both the ovary and testis manufacture inhibin, a glycoprotein, that has a negative feedback action on pituitary FSH secretion. Its precise role in the female cycle is difficult to understand except that it makes possible a period of predominant LH secretion, as at the time of the preovulatory LH surge. In the rodent male the role of inhibin is even more difficult to define. Normally the male rat pituitary secretes more FSH than LH.

The roles of the steroid hormones in postovulatory events and in establishment and maintenance of pregnancy have been worked out in some detail (108). Hormones affect the transport of the egg and spermatozoa to the site of fertilization, and the subsequent movement of the zygote to the uterus. Steroid hormones are needed prior to implantation of the embryo in the uterine wall. It is of interest that some hormones needed at the time of implantation are made by the young embryo (blastocyst) itself, including a gonadotropic hormone. In different species the placentas secrete an array of steroid and peptide hormones. For some of these there is a clear role in the maintenance of pregnancy; for others, like CCK, a hormone that typically has a role in digestion, it is difficult to conceive what their role might be in the placenta.

The principal hormonal basis for male behavior is one or more of the androgenic steroids testosterone, androstenedione, or 5α-dihydrotestosterone. For females, estrogen (usually estradiol) stimulates sexual receptiveness or other facets of female sexual behavior (e.g., lordosis). In females, testosterone may evoke reproductive behaviors, but this steroid is aromatized in the brain to estrogen. Progesterone, in the luteal phase of the cycle, is generally inhibitory for sex behavior, except in the rat. In the rat, there is a pulse of progesterone just before ovulation, but after the LH surge, which evokes female sex behavior. This illustrates how easily a new or different use of a hormone may evolve if it serves an adaptive purpose in the reproductive scenario of a particular species.

In addition, there are other hypothalamus-mediated modifiers of hormone-regulated sex behaviors. For example, deprivation of olfactory function (anosmia) makes it impossible for the male rat to mate despite an adequate endocrine status. Female monkeys secrete a vaginal pheromone under estrogenic stimulation that attracts the sexual attentions of males. Descriptions of sex behaviors of various mammals include accounts of sniffing or licking, usually by the male, that precedes mounting and intromission. This implies that pheromones play an important enabling influence or key role in the chain of events that comprise sex behavior. Generally these pheromones are sex hormone evoked, since they, like other facets of reproduction, must be timed to play their role at a particular time when sex hormone levels are high and ripe gametes are available. Furthermore, since males are more or less continually fertile, it is the female that must supply the pheromonal signal to the male. Nevertheless mammalian male sex attractants are known, as in the male pig, whose "boar odor" evokes female responsiveness. Also, the male pheromones of mice affect pregnancy and cycling in females. In the vole (*Microtus*

ochrogaster), male urine contains a pheromone that will in a few days evoke estrus in an anestrus female by causing GnRH-stimulated LH secretion. Furthermore, removal of the male of a breeding pair will interrupt pregnancy in the early stages. Numerous examples have been documented of similar phenomena in other species.

Insects

The Endocrine System

It is possible to discern a general pattern of organization in the anterior or "retrocerebral" endocrine system of insects (Fig. 14). There are two major groups of neurosecretory cells in the protocerebral lobes of the brain. The median cells of the pars intercerebralis, near the median furrow delineating the two protocerebral hemispheres, are also sometimes referred to as the A cells. These send their axons in a posteroventral direction to emerge from the brain as the paired nervi corpora cadiacii I (NCC I). Within the brain most of these axons in most species decussate, so that the NCC I of one side serves perikarya in the contralateral hemisphere. The other major group of secretory neurons is comprised of the lateral cells, sometimes called B cells, situated in the pars lateralis of the protocerebrum. Their axons do not decussate, but emerge from the brain as the NCC II on the ipsilateral side. There are also less prominent groups of cells in the deutocerebrum and in the tritocerebrum; the axons of the latter usually emerge from the brain as the NCC III (152, 174, 245).

The axons of the NCC I, II, and III enter the CC, and many of them terminate there. The CC is a prominent neuroendocrine structure situated posterior to the brain and ventral to the dorsal aorta, with which it is intimately associated. Two regions are usually recognized. The storage

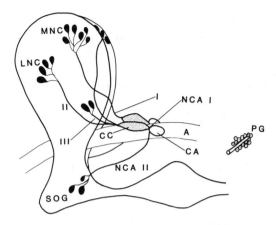

Figure 14. Diagram illustrating the insect neuroendocrine system in the head. Median neurosecretory cells, MNC, and lateral neurosecretory cells, LNC, located in the pars intercerebralis of the brain contribute their axons to the nerves NCC I (I) and NCC II (II), respectively. The axons of a third group of cells at the posterior of the brain emerge as NCC III (III). These nerves enter corpus cardiacum, CC, which acts as a storage organ and release site for many of the neurohormones associated with the neurosecretory cells of the brain. The CC is intimately associated with the dorsal aorta, A. At least some of the fibers in NCC II pass through the CC and emerge as NCA I, which connects the nonnervous corpus allatum, CA, to the CC. The CA also receives innervation from neurosecretory cells in the suboesophageal ganglion, SOG, via NCA II. Also shown are the prothoracic glands (PG), usually associated with trachea in the thoracic segments. They are of nonnervous origin.

portion of the CC consists of the expanded axons of neurosecretory cells originating in the brain: The neurosecretory products may be stored in this region before being released at typical release sites in the CC. The CC also contains additional endocrine cells, the so-called intrinsic cells of the CC. These cells have many of the characteristics of neurosecretory cells that lack axons, including a prominent content of membrane bound granules (198, 208, 245, 246).

Not all of the axons of the NCC terminate in the CC. Some fibers of NCC I and II pass to the corpus allatum (see below), where they may form synaptoid structures or typical neurohemal structures (175). In some insects, the CC contains rather few terminations of the NCC, which instead form release sites on the wall of the aorta (230).

Attached to the posterior end of the CC

are the corpora allata (CA), generally paired nonnervous endocrine structures originating from ectodermal invaginations in the maxillary segment (62). These structures receive an innervation via the nervi corpora allati (NCA). The NCA I connect the CA to the CC and contain fibers from the NCC I and II. The CA contain both synaptoid terminals containing typical neurosecretory granules in close association with the cells of the CA (31), as well as more superficially situated neurosecretory terminals, which may release their products into the blood (204). The CA are also innervated by the NCA II from the subesophageal ganglion (176, 245).

Functionally associated with the retrocerebral complex, but anatomically distinct from it, are the prothoracic glands, paired endocrine structures consisting of diffuse strings of cells arranged along the trachea in the thoracic segments. They are ectodermal derivatives of the labial segment (63). Their innervation is far from clear (175), although there is no doubt that neurosecretory terminals and neuroglandular junctions occur in the glands (163, 186). There are many other neuroendocrine cells in insects: For example, each ganglion of the ventral chain contains several prominent neurosecretory cells.

Control of Growth and Development

Growth in all insects is punctuated by a number of molts, a process in which: the epidermal cells withdraw from the existing cuticle (apolysis), increase in number by mitosis, secrete some layers of the new cuticle, and secrete the molting fluid that digests the inner layers of the existing cuticle, the resulting products being resorbed by the epidermal cells; behavioral changes occur by which the remaining old cuticle is ruptured and the next developmental stage wriggles free in the act of ecdysis; the newly ecdyzed insect then inflates itself before the hardening and darkening of the new cuticle occurs to establish its enlarged exoskeleton. Cuticle formation often proceeds between bouts of molting. Each of these phases—cuticle formation, ecdysis, and hardening—is under the control of the endocrine system.

Superimposed on this pattern of growth by molting is the phenomenon of metamorphosis. In insects, there are two major developmental patterns. In the Endopterygota (Holometabola), a series of larval stages culminates in the production of a pupa, an immobile stage during which many of the larval tissues undergo autolysis to be replaced by adult tissues such as wings, antennae, and legs, growing up from anlagen or developmental disks in the larva. Within the pupal cuticle, the adult develops to emerge as the final developmental stage. Thus, although growth, as defined by increase in weight, is continuous in insects (or at least occurs between molts), development, in terms of the appearance of the various stages, is discontinuous or saltatory. Metamorphosis and molting are closely associated in that metamorphosis, whether more gradual, as in the case of the Exopterygota, or more abrupt, as in the Endopterygota, can become manifest only when the insect molts.

Early "classical" studies involving surgical approaches indicated that the brain was essential for cuticle formation. Thus Kopec (134) showed that caterpillars of the gypsy moth were unable to form pupae when a ligature was tied around the neck. Since severing the nerve cord did not prevent molting to the next stage, he concluded that the effect was humoral. But it was the work of Wigglesworth using the blood sucking hemipteran *Rhodnius prolixus* that provided the best early evidence for control of molting by a brain hormone. *Rhodnius* takes a single large blood meal, which serves as a signal to initiate the molting process. Wigglesworth brought forward three lines of evidence demonstrating that one or more hormones from the head are essential for molting. If a *Rhodnius* larva is allowed to feed to repletion, and is decapitated within a few hours of feeding, it will not initiate the molting process; if, on the other hand, decapitation is delayed for a few days, a new cuticle is formed, although the larva, lacking its brain, is unable to ecdyse fully. Some factor in the head is required for a period after feeding: That period is known as the "head critical period." If a larva decapitated immediately after feeding, and hence destined not to form a new cuticle, is joined surgically to a fed larva with its head intact in a procedure known as parabiosis, both members of the parabiotic pair form a new cuticle at the same time. Finally, fragments of cuticle and underlying epidermis from one larva implanted into the hemocoel of another larva will form a new cuticle in unison with the tissues of the host (238, 239). By implanting various parts of the brain from recently fed insects into the hemocoel of fed and decapitated larvae, Wigglesworth (241) was later able to show that the hormonal influence in the head emanated from the pars intercerebralis, where many of the neurosecretory cells of the brain are to be found. The brain as a source of a "brain hormone" required for molting in insects has now been demonstrated for a wide variety of insects (178).

Early evidence for a thoracic center controlling development was derived from experiments in which lepidopterous pupae were ligatured at various levels. Ligatures posterior to the thorax prevented further development, while those anterior to the thorax did not (15, 107). By a series of ligatures and transplants in the larva of the commercial silkworm *Bombyx mori*, Fukuda (86, 87) was able to demonstrate that the source of the thoracic factor was the prothoracic gland, a paired structure in the thorax, consisting of a diffuse string of cells associated with tracheae; a thoracic gland transplanted into the posterior half of a larva ligatured posterior to the thorax induced the formation of a new cuticle.

The link between the brain and the prothoracic gland was established by C. M. Williams, working on the diapausing pupal stage of the giant silkmoth, *Hyalophora cecropia*, and its relatives. In these insects, diapause (see below) interrupts development during the pupal stage, and a period of chilling at 5°C is required to reinitiate development leading to the production of the adult moth. Relying on the classical surgical techniques of extirpation and implantation, Williams demonstrated that a chilled brain implanted into an isolated abdomen would not initiate

development, whereas adult cuticle was produced if the brain were implanted together with a prothoracic gland (16, 242, 247, 248).

It is now well established that the insect brain secretes a hormone that activates the prothoracic gland to secrete a second hormone. It is this second hormone that acts directly on the epidermal cells to initiate and sustain the complex series of events leading to the production of a new cuticle. The brain hormone is required for only a relatively brief period, the "head-critical period," at the beginning of the moulting cycle. After that time, the prothoracic glands are able to continue to function in the absence of the brain hormone (215).

The product of the prothoracic glands is ecdysone, a steroid isolated from extracts of 500 kg of pupae of the silkworm by Butenandt and Karlson (26), although the complete structure was not known until 10 years later (128). A second active material, ecdysterone (β-ecdysone, 20-OH ecdysone) was isolated from the same extracts and identified somewhat later (114). It is clear that prothoracic glands *in vitro* secrete ecdysone, but not 20-OH ecdysone (21, 36, 131). Ecdysone is converted by many tissues to 20-OH ecdysone (130), and receptor proteins in target cells appear to have an affinity for 20-OH ecdysone, which is three orders of magnitude greater than for ecdysone (166). These facts have led to the widely accepted view that ecdysone, as secreted by the prothoracic glands, is a prohormone, which becomes physiologically active after conversion to 20-OH ecdysone.

The precise chemical nature of the brain hormone is less clear. Several laboratories have found extracts of insect heads that have the property of inducing prothoracic glands *in vitro* to make ecdysone, thus confirming the action of the brain hormone, and giving it the name "prothoracicotropic hormone," or PTTH.

The active material in the extracts prepared from *Manduca* is proteinaceous, and two peptides with PTTH activity have been isolated: a "small PTTH" of ~8 kD and a "big PTTH" of ~25 kD (18, 132). Prothoracicotropic hormone is produced by neurosecretory cells in the brain, the precise location of which varies from species to species; in the tobacco hornworm, *Manduca sexta*, PTTH activity is confined to two lateral neurosecretory cells. Release of the hormone occurs at neurosecretory terminals located in the corpus cardiacum or corpus allatum (19).

Metamorphosis

While the process of molting is under the control of the brain–prothoracic gland axis, the progress of metamorphosis is controlled by the corpus allatum. Simple surgical experiments of early workers demonstrated that metamorphosis was under hormonal control. In *Rhodnius*, decapitation of a fourth stage larva just after the critical period had passed resulted in some insects exhibiting partially adult characters rather than the larval characters of the normal fifth stage (238, 239). Conversely, implanting the corpus allatum from a fourth stage larva into a fed fifth stage larva of *Rhodnius* resulted at the next molt, not in a normal adult, but in a giant sixth stage larva. Moreover, adult *Rhodnius* could be induced to form a new cuticle by parabiosis with a fifth stage larva; if such molting adults were also supplied with active corpora allata, the new cuticle was partially larval (240, 241). Similar experiments performed on a vari-

ety of insects have confirmed the existence of a hormonal control over metamorphosis emanating from the corpus allatum (171–173). Because the hormone apparently encouraged the appearance of larval or juvenile characters, the hormone was designated juvenile hormone or JH.

Juvenile hormone has been isolated, identified and synthesized (Fig. 15). Five closely related substances with JH activity have been isolated from various insects. All are methyl esters of farnesenic acid, and all have an epoxide ring in the 10–11 position. They differ only in the side chains at C-3, C-7, and C-11. Three of the compounds have been found in larval insects. In JH III, the most commonly occurring form, all three side chains are methyl groups. Juvenile hormone II is identical, but the side chain at 11 bears an ethyl group, and in JH I, C-7, and C-11 bear ethyl groups. Juvenile hormone 0, in which all three side chains are ethyl groups, and iso-JH 0 or 4-methyl JH I, have been isolated only from embryos. The isomeric configuration of the molecule is an important determinant of biological activity (129, 185, 243).

The relationship between ecdysone and JH in their cellular actions is far from clear. They are closely related, for JH cannot exert its effect on metamorphosis unless new cuticle is formed. 20-OH Ecdysone has a myriad of cellular effects, ranging from a stimulation of cell division in the epidermis through the secretion of cuticle to the synthesis of specific enzymes such as dopa decarboxylase (133, 182). While it is generally agreed that the action of 20-OH ecdysone conforms to the general model for steroidal

Figure 15. Structure of insect juvenile hormone I. The functionally important characteristics are the epoxide bond and the ester group. Without both of these, the hormonal activity of the molecule is sharply reduced.

hormone action, whereby steroid–receptor complexes influence the activities of various genes (165), definitive proof of such action is still lacking. The complexity of the cascade of events that is controlled by the hormone is glimpsed from experiments in which sequences of "puffing" at specific sites on polytene chromosomes were induced by 20-OH ecdysone. Puffing is a visible manifestation of RNA synthesis, and late appearing puffs are dependant on earlier puffs or on the protein products of earlier puffs (4, 40). Given that 20-OH ecdysone is in some way, probably via receptor–hormone complexes, interacting with the DNA, causing various genes to be activated in an appropriate sequence, how does JH intervene to determine whether the ensuing molt will result in the appearance of a larva, a pupa, or an adult? There are no clear answers to this intriguing question. Juvenile hormone seems to be essential for the complete action of 20-OH ecdysone in terms of activating all of the DNA normally associated with the production of larval cuticle. In studies on the larval–pupal transformation in the cuticle of *Manduca*, it has been shown that JH first disappears from the hemolymph, to be followed by two peaks of ecdysteroid of which the second initiates the larval pupal molt. The first secretion of ecdysteroid is considered to be essential to "reprogram" the epidermis so that it will cease making larval cuticle. Precisely what is involved in reprogramming remains a mystery, but there is some evidence that JH might influence the chromatin (185, 251).

The timing of the release of PTTH is the result both of internal cues, such as nutritional state, or stretching of the abdomen, and of external cues, principally photoperiod. In larval moults, a single release of PTTH, corresponding to the "head-critical period" (see above) appears to initiate the molting cycle, while

metamorphic moults involve two successive releases of PTTH, corresponding to the two peaks of release of 20-OH ecdysone characteristic of these molts (216). In the tobacco hornworm, *Manduca sexta*, it is clear that the release of the first of the two surges of PTTH, leading to the early or reprogramming peak of ecdysteroid depends partly on the decline in titre of JH, which is essential for the metamorphic molt (20). The release of PTTH at this first metamorphic moult is inhibited by the presence in the hemolymph of JH. The decline in titer of JH is initiated as soon as the insect reaches a weight of 5 g, suggesting that proprioceptive information, depending perhaps on stretch, is integrated by the brain, causing the CA to switch off. This results in a decline in the hemolymph of the titre of JH. There is evidence that the release of PTTH occurs at a particular time on the day following the disappearance of JH from the hemolymph, adding a circadian influence to the control of the release (216).

In the exopterygote insect *Rhodnius prolixus*, two peaks of ecdysteroid release are also seen at the metamorphic molt. In this case, there is good evidence that the second, and larger, surge of ecdysteroid is responsible for shutting down the release of PTTH, which initiated the surge, and that this action is independent of JH (216).

Other hormones intervene in the molting process. A hormone controlling the complex act of eclosion (emergence) of the adult of giant silk moths from the pupa was first demonstrated by Truman (227). Eclosion hormone (EH) is a peptide released from the CC, and has its primary action on the nervous system of the insect, switching the behavior from one that is primarily pupal, in which stimulation results in abdominal twitching, to a behavior that results in eclosion together with behavior that is characteristic of an adult moth. Some of these changes in behavior have been traced to alterations in individual neurons (228).

Sclerotization, or the hardening of the cuticle that follows closely upon eclosion, is also controlled by a neurohormone, bursicon. First described in the pupariation of blowflies (46, 83), it is now known to occur in a wide variety of insects (181).

Diapause

In many insects, development is interrupted at a specific point in the life cycle. This interruption in development, or diapause, characterized by a lowered metabolic rate and an absence of the normal correlates of development, such as cell division, enables the insect to survive seasonal periods of hostile environmental conditions, such as drought or winter. Diapause can occur in any of the developmental stages—egg, larva, pupa, or adult—and usually occurs as a result of an environmental signal, such as day length (140, 196). The environmental signal has its effect on development via the endocrine system.

The endocrinology of diapause was first worked out in lepidopterous pupae, and that system is perhaps best known. It is clear that diapause in the pupae of giant silkworms results from a failure in the release of PTTH: Indeed, the electrical activity of the entire brain is relatively low (202). This failure in the secretion of PTTH in giant silkmoths appears to result from the action of photoperiod directly upon the brain (37, 249). In the giant silkmoths, pupal diapause appears to be exclusively dependent on PTTH, but in other lepidopteran pupae, the control of diapause appears to reside in the prothoracic glands. In such cases, the brain may be competent to secrete PTTH, but the prothoracic glands of diapausing pupae are incapable of responding to PTTH by se-

creting ecdysone, an incapacity that may result from the effects of low temperature directly upon the prothoracic gland itself. The injection of ecdysteroids terminates diapause and initiates development in most pupae. The role of JH in pupal diapause is far from clear. In the giant silkmoths, it appears to play no role, while in the pupal diapause of the fleshfly *Sarcophaga*, a pulse of JH secretion may signal the termination of diapause (59).

In larval diapause, by contrast, the brain remains competent to secrete PTTH, but is restrained in some way from doing so; simply transplanting the brains appears to activate release of PTTH and lead to the termination of diapause (41, 220). The CA of diapausing larvae exhibit signs of continued secretion during diapause. In addition, the titer of JH is relatively high during diapause. JH is suspected of preventing the release of PTTH in the last larval stage (see above), and it is possible that high titers of JH seen during larval diapause may function to suppress the release of PTTH (59).

Diapause may intervene in eggs at various stages of development, and in some species, diapause may occur before the endocrine organs have differentiated, although eggs may incorporate both ecdysteroids and JH while they are in the ovary of their mother, and before development has been initiated (12, 115). Only in the commercial silkworm, *Bombyx mori*, is the initiation of diapause understood. If the embryos and young larvae are exposed to a combination of long day length and relatively elevated temperature, the female moths that develop from such larvae will be programmed to lay eggs that will enter diapause. This effect is mediated by a diapause hormone, a peptide that is released from the suboesophageal ganglion of the affected female during the pupal stage (59).

Diapause also occurs in adult insects,

where it is characterized by a cessation of reproduction, particularly egg production.

Reproduction

Hormones intervene in the reproductive process of insects at several levels. In only one species is the determination of sex and secondary sexual characters known to be influenced by hormones. In the glow-worm, *Lampyris noctiluca*, Naisse (160, 161), by a series of elegant surgical maneuvers, demonstrated a very complex set of hormonal controls over sex determination. During early larval development, the gonads exhibit no obvious differences between the sexes, but at the end of the fourth larval stage, the male develops at the apex of the gonad some secretory tissue of mesodermal origin that disappears at the end of the next larval stage. This apical tissue, it was shown by a series of transplants, releases an androgenic hormone that leads to the further differentiation of the testis and to the development of the male secondary characters in this highly dimorphic species. In the absence of the apical tissue, the gonadal tissues become ovarian in character. Moreover, the development of the apical tissue is itself dependant on a neurosecretory hormone from the pars intercerebralis of the brain of males. Presumptive ovaries develop apical tissue and become testes if transplanted into castrated males, and male brains transplanted into female abdomens induce presumptive ovaries to develop apical tissue and become male in character.

Hormones also regulate gametogenesis. In the males of most insects, mature sperms are present before eclosion, and in some insects, spermatogenesis continues in the adult male. The proliferative stages of spermatogenesis are under hormonal control in at least some insects. In *Rhodnius*, the mitotic divisions of the spermatogonia occur at a low basal rate, which can be accelerated by ecdysone. The ecdysone-stimulated rate, but not the basal rate, is inhibited by JH. As a result, there is a burst of spermatogenesis that occurs in the last larval instar, when JH is absent (66, 67).

In many male insects, the timing of the appearance of mature spermatozoa is regulated by autolysis of the most differentiated elements of the testis until the stage immediately preceding the appearance of the adult, when autolysis ceases, permitting the spermatozoa to complete their development (66). In the pupa of the moth *Samia cynthia*, ecdysone appears to play a permissive role, rendering the connective tissue sheath surrounding the spermatocytes permeable to a "macromolecular factor" from the hemolymph, which in turn permits development of the spermatozoa (126).

The accessory glands of the male, which are involved in the formation of the spermatophore, are controlled both by JH and a neurosecretory peptide in *Rhodnius*. Both factors appear to act directly on the gland, and both are essential for the normal production of the secretory product of the glands (7, 8).

The control of the proliferative processes in the ovary is by no means clear, and the process of egg production in many insects is complicated by the fact that some oocyte sister cells become nurse cells, contributing their substance to the oocyte. In most insects, meiosis is delayed until after fertilization. In *Rhodnius*, allatectomy inhibits proliferation of the nurse cells, but this is not considered to be a direct effect (177).

Insect eggs contain abundant yolk, of which the principal component is a protein, vitellin. In the great majority of species, most of the vitellin has its origin as vitellogenin, which is synthesized in the fat body, and released into the hemolymph. The vitellogenin finds its way to the oocyte surface, and the oocyte takes it up by endocytosis, altering the vitellogenin so as to form vitellin. This complex of processes by which the oocyte accumulates yolk is encompassed by the term vitellogenesis.

Vitellogenesis has been known to be under the control of the CA since the pioneering surgical experiments of Wigglesworth (239) on *Rhodnius*. For most of the major groups of insects, evidence is now available which demonstrates that removal of the CA inhibits or halts vitellogenesis, while reimplantation of the CA, or hormone replacement therapy with JH, restores vitellogenesis. In most insects (see below for exceptions), JH acts on the fat body to initiate and sustain the synthesis and release of vitellogenin (74). While definitive data are not available, it is assumed that this action of JH conforms to the model described for steroids that act on the genome.

In *Rhodnius*, and probably in many other insects, JH also acts directly on the ovary to govern the access of the vitellogenin from the hemolymph to the oocyte surface. It acts on the follicular epithelium, a single layer of cells that surrounds the oocyte, causing them to shrink, so that they lose ~50% of their volume. This opens up large spaces between the cells, giving the vitellogenin access to the oocyte surface. Juvenile hormone binds to the cell membranes of the follicle cells, causing the activation of an existing JH-sensitive Na^+-K^+ ATPase, which in some way leads to a loss of fluid from the cell. Juvenile hormone also acts on the follicle cells earlier in development, rendering them sensitive to JH by stimulating the production of binding sites for JH and of the special JH sensitive Na^+-K^+ ATPase in the membrane (1, 51, 119).

The action of JH on the follicle cells is subject to the influence of an antigonadotropin, a peptide released from segmental abdominal neurosecretory organs. Originally described in *Rhodnius* (57), there is evidence for an antigonadotropic factor in several other species (117).

Ovulation is under hormonal control in a number of species, but the phenomenon has been most extensively investigated in *Rhodnius*. In this species, ovulation is effected as the result of contractions of muscles in the ovary that expel the completed eggs into the lateral oviducts. A myotropic neuropeptide, emanating from 10 large neurosecretory cells in the pars intercerebralis of the brain, and released from the CC, increases the force of these contractions, and is essential for normal ovulation (135). A peak of myotropic activity appears in the hemolymph at the time of ovulation, and this peak is absent from virgin females, which do not ovulate (56).

Two inputs are necessary for the release of the myotropin, and thus for ovulation. First, the insect must have mated, and this signal is provided by a hormone emanating from the spermathecae of mated females, the spermathecal factor. This is not a sufficient condition for the release of the hormone as measured by electrical activity in the CC: A second input is represented by an ecdysteroid (probably ecdysterone) from the ovary, which signals that mature eggs are present in the ovary. The ecdysteroid does not act directly on the neuroendocrine cells, but via an aminergic pathway (168, 192, 193).

The control of the CA in the female insect is complex and varies in detail from species to species. In most species, severing the NCA I leads to maximal activation of the CA, an increased rate of synthesis of JH (225) and, it is generally agreed that the CA in such species is under inhibitory control from the brain via

NCC I and NCA I (216). An exception is the locust, *Schistocerca gregaria*, in which severing NCA I has the opposite effect (226).

Perhaps the most detailed examination of the control of the CA has been carried out on the ovoviviparous cockroaches, *Diploptera punctata* and *Leucophaea maderae*, which incubate their eggs until they hatch, and thus larviposit. In these species, the control of the CA is precise, so as to ensure that eggs are produced at the appropriate time. Egg development is inhibited while the female is incubating eggs, as well as in virgin females. The CA undergoes a cycle of activity that is correlated with vitellogenesis. Severing the NCA I of virgin females leads to activation of the CA, but the denervated CA undergoes the same cycle of activity as an intact CA in a mated female: it increases, then decreases as the eggs in the ovary mature, suggesting that hormonal factors are also involved. Two factors from the ovary appear to operate. The previtellogenic ovary produces an allatotropic factor that causes the CA to increase JH synthesis, while the mature ovary inhibits synthesis; the latter effect is probably due to ecdysterone secreted by the ovary (214). There is also an allatotropin emanating from the pars intercerebralis, which operates only when the brain is free of the inhibitory influence of the NCA I. Finally, the titer of JH in the hemolymph is among the factors that acts via the NCA I to affect the capacity of the CA to synthesize JH (224). These basic elements, direct nervous inhibition, an allatotropin from the brain, and influences from the ovary appear to be a feature of the control of the CA in many species, but it is not yet clear how the environmental information, such as feeding, mating and various seasonal influences, is integrated to affect egg development via the titer of JH.

In some insects, diapause occurs in the adult female, and is characterized by a cessation of egg development or gonotrophic dissociation. This results from a suppression of the activity of the CA as the result of some environmental signal. In the best studied example, the potato beetle, *Leptinotarsa decemlineata*, the effect of short days on the CA appears to be exerted via neuroendocrine centers in the brain (59).

In mosquitoes, and possibly some other Diptera, ecdysterone from the ovary is the hormone that stimulates the synthesis and release of the vitellogenin from the fat body (109). The production of ecdysterone by the ovary is under the control of egg development neurosecretory hormone (EDNH) (59, 139). Juvenile hormone is also essential to egg production: it promotes the growth of the egg follicles up to the immediately previtellogenic stage, and it is essential in order to render the fat body sensitive to the effects of ecdysterone (106). Ecdysterone is also essential for the early growth of the second generation of follicles (101).

Crustacea

Endocrine Organs

The endocrine system of the crustacea is closely associated with the nervous system and is dominated by neurohormones. The neurosecretory cells are organized into three major systems, characterized by the location and nature of the neurohemal organs involved.

The *X organ–sinus gland* complex consists of neurosecretory cells located in the ganglia of the head together with their neurohemal and storage organ, the sinus gland. In those crustaceans with eyestalks, the X organ and sinus gland are situated in the eyestalks, a location that has made it possible to extirpate them easily by simply removing the eyestalks. The sinus gland may receive neurosecretory innervation from neurosecretory cells located outside the X organ, but the great majority originate in the X organ (29, 44).

The *pericardial organ* consists of the neurohemal terminations of neurosecretory cells located in the thoracic ganglia as well as in the pericardial organs themselves. The organs are situated in the sinus that surrounds the heart. The pericardial organs appear to be the release sites for various cardioexcitatory peptides and amines (44).

The *post-commissural organs* are neurosecretory terminations associated with a commissure that joins the two circumoesophageal connectives. The cells supplying the terminations are thought to reside in the brain, but there is little known about the organs (29, 44).

In addition to these neurosecretory endocrine organs, two other structures of nonnervous origin have been implicated as endocrine sources. The Y organ is a pair of glands in the thoracic region, situated in the maxillary or antennary segment; it is known to produce ecdysone. A similar nonnervous gland, the mandibular gland, has also been postulated to have an endocrine function.

Control of Molting

Development in Crustacea is characterized by a series of molts demarcating the various larval stages. After metamorphosis to the adult form, molting cycles continue, and the crustacean may continue to grow as an adult. In many forms, the metamorphic molt to the familiar adultoid form may not be the molt that leads to sexual maturity: Additional molts may be required to produce the copulatory appendages. Some species do not molt again after reaching sexual maturity, while others alternate molts with bouts of reproductive activity.

Molting in crustaceans has been divided into five major stages, A–E. Stage A, which begins immediately after ec-

dysis, the shedding of the cuticle from the previous molt, normally lasts only a few days. Stage B is characterized by the beginning of the formation of the new endocuticle beneath the epicuticle and exocuticle, which were formed at the end of the previous cycle, just before ecdysis. During stage C, endocuticle formation is completed, and the animal enters a period of anecdysis, or absence of molting activity, a stage of variable length, which can last for as long as a year. During stage D, or proecdysis, a number of changes are initiated. Lost limbs are regenerated, much of the old cuticle is digested, the calcium in the cuticle being stored in gastroliths, and the new epicuticle and exocuticle are synthesized. At stage E, ecdysis occurs and the cycle begins again (207).

It has been known for many years that removal of the eyestalks in crustaceans results in an acceleration of molting. If eyestalks are removed at the beginning of anecdysis, for example, the animal rapidly enters proecdysis. Implanting eyestalks or injecting extracts of the sinus gland leads to reversal of this effect and prevents or inhibits molting or many of the phenomena associated with it (207). Such experiments, performed on a wide variety of crustaceans, have formed the basis for the conclusion that the eyestalk contains and releases a molt inhibiting hormone (MIH), which is contained in the X organ–sinus gland complex.

Ecdysone is also implicated in the control of moulting in Crustacea. Removal of Y organs results in a cessation of molting, and treatment with 20-OH ecdysone restores molting in such animals. Injection of 20-OH ecdysone into animals in anecdysis stimulates early molting. The steroid is synthesized from cholesterol by the Y organs, released into the hemolymph, and is converted to 20-OH ecdysone by various tissues, especially the mid-gut (33). Radioimmunoassay of ecdysteroids shows that the titer in the hemolymph of the crab rises sharply during proecdysis (164). Cytoplasmic and nuclear binding proteins have been demonstrated in the epidermal cells of the crayfish (146). These facts indicate that 20-OH ecdysone is a hormone controlling molting in Crustacea.

What is the relationship between the Y organ and MIH? Ablation of the eyestalks in the crab *Pachygrapsus crassipes* results in a rise in the ecdysteroid titer. Injection of eyestalk extracts prevents such a rise, and halts the rise once it has begun. Extracts of sinus gland inhibit the *in vitro* production of ecdysone by isolated Y organs (164). Clearly, the MIH exerts its inhibitory effect on molting by suppressing the synthesis of ecdysone by the Y organ. In a pattern that is similar to that in insects, ecdysteroids feed back onto the sinus gland, inhibiting the release of MIH activity, but not the synthesis (153).

How does the MIH act on the Y organ? It has been hypothesized that MIH, which is a peptide (212), activates adenylate cyclase in the membrane of the cells of the Y organ. The resultant rise in cAMP inside the cells decreases the synthesis of ecdysone by means which are not yet understood. The cAMP-modulated suppression of ecdysone synthesis is itself antagonized by Ca^{2+}, which activates a Ca^{2+}-calmodulin-sensitive cAMP phosphodiesterase, resulting in reduction of intracellular cAMP, and the stimulation of ecdysone synthesis (153). Removal of Y organs does not completely eliminate ecdysone from the hemolymph, suggesting that there is an alternate source for the hormone, possibly the ovaries (137).

Other endocrine factors are involved in the control of molting in crustaceans. Limb growth inhibiting factor is a small peptide from the X organ–sinus gland complex, which inhibits the regeneration of lost limbs. It is considered to be differ-

ent from MIH. There have also been suggestions that an ecdysial hormone, analogous to that described in insects, may occur in Crustacea (207).

Metamorphosis

The control of metamorphosis is poorly understood in the Crustacea. Insect JH, or analogs of the hormones, have been demonstrated to retard or prevent metamorphosis in a variety of Crustacea. In addition, materials with JH activity have been extracted from Crustacea, and the mandibular gland has been shown to synthesize methylfarnesoate, a precursor of JH in insects (99, 113, 138, 201).

Reproduction

While some crustaceans cease molting with the onset of reproductive activity, many continue to molt, and periods of reproductive activity alternate with molting: There is a coordination between the events involved in the molting process and those in reproduction, particularly in the female. In such species, the females develop various secondary sexual characters associated with the incubation of the eggs at a precopulatory molt, and lose them again at the ensuing molt, the timing of which must be correlated with the cycle of egg development.

In males, the androgenic gland, a group of endocrine cells closely associated with the distal end of the vas deferens, plays an important role in the development of the testis. Extirpation of the gland results in the feminization of the male: Testicular development and spermatogenesis cease and female secondary sexual characters appear. An ovary implanted into an intact or gonadectomized male is converted into a testis, but this conversion does not take place in the absence of the androgenic gland. Similarly, implanting the androgenic gland into an otherwise normal female masculinizes the female: Her ovary becomes a testis and male copulatory organs develop. In isopods, a separate androgenic gland is lacking, but its functions reside in the testis. On the basis of its metabolic capabilities, the hormone involved may be a steroid. Many of the phenomena associated with parasitic castration in Crustacea are probably a result of disturbances of the androgenic gland (29, 93).

Crustacean eggs contain a good deal of yolk, and vitellogenesis has been seen to involve two phases. During primary vitellogenesis, the yolk proteins are synthesized by the oocyte itself (9). More recently, secondary vitellogenesis has been recognized as a period during which the yolk proteins are synthesized in the fat body and released into the hemolymph to be taken up by the oocytes (34, 158).

Vitellogenin synthesis by the fat body is under the control of both stimulatory and inhibitory hormones. In amphipods, a vitellogenin stimulating ovarian hormone has been demonstrated by implanting ovaries into males lacking their androgenic glands; implanting the ovary initiated vitellogenin synthesis in the fat body. However, in isopods and amphipods, it is the Y organs which are essential for vitellogenin synthesis. This evidence suggests that ecdysteroids are involved in the control of vitellogenin synthesis, but their precise role, and the nature of any relationship between ecdysteroids and the stimulatory hormone from the ovary remains obscure. Extirpation of the eyestalks stimulates vitellogenesis and injection of extracts inhibits the process. These and other experiments have led to the conclusion that a gonad inhibiting hormone (GIH) is released from the X organ–sinus gland complex. Gonad inhibiting hormone appears to be different from MIH in terms of its behavior on high

performance liquid chromatography (HPLC). It is not known whether GIH acts directly on the ovary or via another tissue (34). Mimics of JH affect egg production in Crustacea, but there is no clear evidence that insect JH plays a role in crustacean reproduction (222).

Color Change

Many of the colors in the Crustacea are a result of pigments contained in chromatophores, specialized cells located primarily in the integument, but also occurring on some of the internal organs. The dispersion of the pigment granules in the chromatophores of Crustacea is under hormonal control, and the color of an individual can be controlled so as to conform to the background. The responses of the chromatophores have been widely investigated both *in situ* and *in vitro*, and an array of blood borne factors has been implicated in their control.

These blood-borne factors are of two types. There is a group of neuropeptides from the X organ–sinus gland complex, which act directly on the chromatophores, causing either dispersion or concentration of the pigment granules. Some of these chromatophorotropic hormones act rather generally on chromatophores. Thus, there is a pigment concentrating hormone (PCH), which is an octapeptide that was isolated originally as red cell pigment concentrating hormone, but which is now known to concentrate the pigments in several types of chromatophores. Similarly a pigment dispersing hormone (PDH) has been isolated as an octadecapeptide, which also controls the dispersion of pigments in the retinal cells of Crustacea. In other cases, electrical stimulation of eyestalks results in the release of peptides with activity on specific types of chromatophores (79, 125).

In addition, there are a group of compounds that cause pigment dispersion or concentration in chromatophores *in situ* by bringing about the release of the chromatophorotropic neuropeptides. While many of these are probably acting as neurotransmitters entirely within the nervous system, others, primarily amines, are known to be released into the hemolymph from the pericardial organs (44), and it is thus possible that the regulation of the release of the chromatophoric neuropeptides may itself be at least partially hormonal (79).

Other Hormones

Peptides from the X organ–sinus gland have been implicated in the control of osmoregulation and of metabolism in Crustacea. Classical endocrinological experiments involving extirpation and replacement have led to the conclusion that a hyperglycemic hormone exists in the X organ–sinus gland (29). Several peptides with hyperglycemic activity have been isolated from the X organ–sinus gland (162). Osmotic and ionic regulation also appear to be targets of peptides from the X organ–sinus gland (147).

Annelida

The brain and the ganglia of the ventral chain of annelid worms contain many neurosecretory cells, the staining properties of which have been correlated with various physiological and developmental events. A neurohemal area can be found on the ventral surface of the brain (98).

Growth

Growth in annelids results from the development of new segments from a zone of proliferation immediately anterior to the pygidium or terminal segment at the posterior end of the worm. Annelids have considerable powers of regeneration: Most species can regenerate a new tail, and many can replace a lost anterior portion. Regeneration consists of an acceleration of the normal growth processes, and the ability to regenerate has been used in explorations of the hormonal control of growth. In the polychaete *Nereis*, regeneration depends on the presence of a brain, and surgical studies involving removal and transplantation of brains has demonstrated that the influence is hormonal (38, 69, 70). Essentially similar results have been obtained in oligochaetes (32).

Reproduction

Reproduction in polychaete annelids is complicated by epitoky, a metamorphic process by which some segments of the worm become specialized, or in which new reproductive segments are produced. The degree of metamorphosis varies with the species: In syllid worms the posterior reproductive segments break off to lead an independent life, while in some nereids the transformation of the reproductive segments may be comparatively slight. These metamorphic changes are closely associated with gametogenesis, the later stages of which occur in the coelom. Only the mitotic divisions of the spermatogonia occur in the testis, with the meiotic divisions and spermiogenesis occurring in the coelom. Most of the vitellogenic growth of the oocytes occurs in the coelom.

In nereids, removal of the brain accelerates vitellogenesis, but has no effect on the proliferative stages, which occur in the ovary. Implanting a brain from an immature individual into a brainless worm reimposes the inhibition on vitellogenesis. These and similar experiments have led to the conclusion that the brain produces an inhibitory hormone (97). Similar conclusions have been reached concerning the control of spermatogenesis in nereids, using an organ culture technique

(70). The same inhibitory hormone appears to control the development of epitoky (69).

The nereid model does not apply universally in annelids. In syllid polychaetes it appears to be the proventriculus rather than the brain that is the source of the inhibitory hormone (253). In the lugworm *Arenicola marina*, a hormone from the brain stimulates the entry of oocytes into meiosis and the production of mature sperm (116). In oligochaetes, the brain appears to produce a hormone that accelerates gonadal development (112).

Mollusca

Endocrine Organs

The nervous system of molluscs is richly endowed with neurosecretory cells situated in the ring of ganglia that constitutes the central nervous system. These cells have been described in great detail in gastropods using classical histological staining procedures as well as electron microscopy and immunocytochemistry, but relatively few of these cells have proved to possess an endocrine function, and many appear to function as peptidergic neurons within the brain (17, 121). Neuropeptides have been demonstrated to have a role to play in the regulation of heart rate and in hydromineral regulation and kidney function, and are suspected of involvement in the control of energy metabolism, perhaps via a hyperglycemic effect (122).

Closely associated with the central nervous system, but of nonnervous origin, are the dorsal bodies (DB), which are located as compact or dispersed groups of cells on the dorsal part of the cerebral ganglia of gastropods. In cephalopods, the optic glands, nonnervous structures associated with the optic tract, are important in the control of reproduction.

Growth

Growth, or increase in body mass, may be under the control of neurohormones in many molluscs. Extirpation or cauterization of specific areas in the cerebral ganglia containing neurosecretory cells results in a diminished rate of growth, which can be restored by implanting an undamaged cerebral ganglion (89, 141). The growth of the shell is under neuroendocrine control in snails. Removal of the same neuroendocrine centers in the cerebral ganglion results in a diminished rate of growth of the shell, both in terms of the periostracum and of the incorporation of Ca^{2+} (61). The cells concerned, the so-called light green cells of the dorsal ganglion, are immunoreactive to antibodies raised against mammalian somatostatin, and somatostatin-like material has been found in the hemolymph in amounts that vary with the rate of growth of the snail. Since injected mammalian somatostatin will not stimulate growth of snails, it has been concluded that the growth hormone of snails is similar to, but different from the mammalian hormone (122).

Reproduction

Reproduction in molluscs is complicated by the appearance in many of the classes of hermaphroditism, usually accompanied by some degree of protandry. In some of the prosobranchs, such as *Crepidula fornicata*, sex reversal occurs. In this species, the male phase precedes the female in development, and the two phases are separated by a hermaphroditic phase. In this, and other protandrous prosobranchs, a factor originating in the cerebral ganglia is necessary for mitosis in the germ cells. In pulmonates, the ovotestis exhibits protandry: The spermatozoa are produced first, and then oocytes are formed. In transplantation experiments and *in vitro* incubations, the cerebral ganglion accelerates the maturation of the gonad, both by increasing mitosis and by accelerating the development of oocytes (122, 154, 235).

The further growth of the oocytes due to vitellogenesis is under the control of the DB. Extirpation of the DB in the snail *Lymnaea* and the slug *Agriolimax* slows vitellogenesis, and reimplantation of the DB restores vitellogenesis (90, 244). The DB themselves may be under neurosecretory or neurohormonal control from the lateral lobes of the cerebral ganglion. The DB also control the development and functioning of the female accessory sex organs, structures that contribute, for example, albumin to the eggs and a mucin to the egg masses (122).

There have been many suggestions of hormones secreted by the gonads: The most convincing evidence is provided by the effect of castration in a styllomatophoran. In *Euhadra peliomphala*, castration causes a pheromone producing gland between the optic tentacles to atrophy. The gland responds *in vitro* to testosterone (122).

The hormonal control of ovulation and oviposition is perhaps the best understood endocrine system in molluscs. In various opisthobranchs, an egg laying hormone (ELH) has been described. The hormone originates from a group of neurosecretry cells called "bag cells" in the abdominal ganglion. Injection of extracts of these cells results in egg laying, and in *in vitro* experiments, extracts stimulate the release of oocytes from pieces of the ovotestis (65). Egg laying hormone also affects behavior. The ELH has been isolated, and consists of two very closely related peptides. Because the bag cells are large, and because the hormone has been isolated and identified, they have become important model systems for studying the biochemistry and physiology of neurose-

cretion. The release of ELH is thought to depend on the action of identified peptides released from the atrial gland. These atrial gland peptides act on some center in the head ganglia, causing the activation of neurones that penetrate the abdominal ganglion (122). In basommatophorans, ELH appears to originate in the caudodorsal neurosecretory cells of the cerebral ganglia (123).

In cephalopods, endocrine control of reproduction is simpler and more easily comprehended. The sexes are separate, and the adults usually die after breeding once. The optic gland secretes one or more factors which are essential for the differentiation and functioning of both the male and female systems. In the absence of the optic gland, spermatogenesis and oogenesis are inhibited, spermatophore production ceases in the male, and vitellogenesis and development of the accessory sex organs is slowed in the female. Interestingly, removal of the optic gland does not interfere with the elaborate reproductive behavior of cephalopods (237).

Echinoderms

The radial nerves of starfish contain a peptide called "gonad stimulating substance" (GSS). Gonad stimulating substance, when injected into a mature starfish, induces immediate shedding of the gametes, spawning behavior, and meiosis in the oocytes. The peptide acts on the follicle cells of the ovary and the interstitial cells of the testis, bringing about the release of a second hormonal substance, termed maturation inducing substance (MIS). The MIS, which has been identified as 1-methyladenine, has a variety of effects. It disrupts the follicular layer, acts on the surface of the oocyte to cause germinal vesicle breakdown and the initiation of meiosis, and possibly acts on the central nervous system to induce brooding behavior and the contraction of muscles leading to expulsion of the gametes (127). The action of MIS on the oocyte is such that a third chemical, maturation promoting factor (MPF) appears in the cytoplasm of oocytes exposed to MIS. The MPF appears to be a compound that appears generally in cells, and which promotes the breakdown of the nuclear envelope and condensation of chromosomes.

Nematodes

Endocrine Organs

Neurosecretory cells of the peptidergic type have been identified in some nematodes on the basis of their anatomy as revealed in both the light and electron microscopes, although no release sites such as neurohemal structures have yet been described. These cells are typically associated with the central nervous system and are found in the ganglia associated with the circumpharyngeal nerve ring in the anterior end of the worm. Some of the primary sense cells associated with the lips appear to be neurosecretory. Neurosecretory cells are also seen associated with the ventral nerve cord (Fig. 16) (49, 50, 96, 183, 234).

Hormones and Development

While most nematodes are free-living, the parasitic nematodes have provided the best evidence for hormonal controls. During their development, parasitic forms move from one host to another or from a free-living environment to a parasitic one. Their development must be timed to coincide with the move to a new environ-

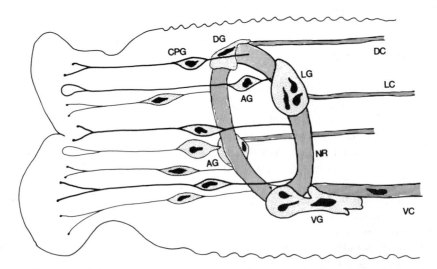

Figure 16. Diagram of the anterior nervous system of a nematode, illustrating the location of peptidergic neurosecretory cells. Neurons that stain with paraldehyde fuchsin and that exhibit the ultrastructural characteristics of neurosecretory cells can be found in the cephalic papillary ganglia, CPG, and the amphidial ganglia, AG, serving the sense organs on the lips of the nematode. They are also found in the dorsal ganglion, DG, ventral ganglion, VG, and lateral ganglia, LG, associated with the nerve ring, NR, which is the neuropil of the central nervous system. Neurosecretory cells have also been observed in the ventral nerve code, VC, although they have not been reported from the lateral, LC, or dorsal, DC, cords. [Redrawn after Davey (53).]

ment, and the characteristics of the new environment may require different physiological and biochemical strategies on the part of the nematode. Moreover, the switch from one environment to another is abrupt, suggesting that some element in the environment may act as a signal to alter the development and physiology of the worm. That signal might well act by effecting the release of one or more hormones.

The area near the nerve ring was implicated in the control of development in elegant studies on the sheep parasite *Haemonchus contortus*. The early stages of this nematode are free-living, and the product of the free-living phase is the third stage larva, which is the infective stage. However, the molt to the third stage is incomplete, so that the third stage larva remains ensheathed within the second stage cuticle. When the ensheathed larva is eaten by the sheep, ecdysis occurs, and the infective third stage larva resumes development within the abomasum of the sheep. Ecdysis is initiated by exposure of the worm to higher temperatures and increased CO_2: These stimuli result in the release into the space between the two cuticles of an ecdysial or exsheathing fluid. Although the worms are small, it was possible to irradiate an area near the nerve ring with a pencil of ultraviolet (UV) light: Under such circumstances, the worms fail to ecdyse. Ligaturing experiments supported these observations (184, 209). Ligaturing experiments (80), or laser surgery (195) on other species, indicate that an area in the anterior end of the worm is essential for normal ecdysis. These studies, however, were originally interpreted as interfering with the release of the ecdysial fluid from the excretory pore, which is situated in the same region as the nerve ring.

Studies on the cod worm, *Phocanema decipiens*, have provided more direct evidence of hormonal control. In this species, the third stage larva is dormant within the muscles of the cod. When an infected fish is eaten by a seal, the worm enters the digestive tract of the seal, development resumes, and molts to the fourth stage larva, and then the adult, occur in the seal. The molt to the fourth stage can be duplicated *in vitro*. Ecdysis to the fourth stage is accompanied by a cycle of changes in neurosecretory cells associated with the anterior ganglia, and the experimental prevention of these changes prevents ecdysis, but not cuticle formation (50, 55). Further experiments have demonstrated that the anterior end contains a factor that activates enzymes in the excretory cell of the worm, and brings about the secretion of these enzymes as the ecdysial fluid into the space between the old, third stage cuticle, and the new fourth stage cuticle. Some environmental signal, as yet unidentified, is considered to act via aminergic sense organs in the anterior of the worm (95). The release of noradrenalin from these neurones in turn brings about the release of the ecdysial hormone from peptidergic neurosecretory cells in the anterior end of the worm (53, 54, 96).

Other Hormones

There may be important endocrine centers in the posterior of nematodes, although there is no anatomical information suggesting an endocrine structure. Laser irradiation of the posterior end of *Panagrellus* leads to a cessation of development (195). In *Phocanema*, there is clear evidence for a hormone from the posterior end of the worm, which controls metabolic rate as the worm resumes development upon exposure to elevated temperatures (52).

Interestingly, ecdysteroids have been detected in nematodes: Ecdysone, 20-OH ecdysone, and 26-OH ecdysterone have all been detected (156). Since these have been found in adult forms, it is possible that, if they have an endocrine function, it will prove to be associated with reproduction.

Coelenterates

The nerve cells of *Hydra* contain a neuropeptide that stimulates specific growth. When the peptide is present in the medium in which fragments of *Hydra* are incubated, head regeneration is accelerated. The so-called head activator is also a stimulator of mitosis. The peptide has been isolated and sequenced: The identical peptide has been found from a wide variety of invertebrate and vertebrate sources, including the mammalian hypothalamus, although its function in other animals is not known (196a).

Conclusion

As in so many other subjects, the treatment of endocrinology has in comparative terms been unsystematic. Thus individual researchers have explored phenomena in individual organisms without a specific comparative perspective. As a result, the picture is discontinuous, with many of the pieces of the jigsaw missing. Nevertheless, enough is now known in one or two areas to make some tentative comparative statements possible.

Steroids appear as hormonal substances in most taxa, although there is wide variation in the structure of the molecules involved. If there is a common thread, it appears to lie in the fact that steroids appear as secretions of the gonads. Steroids are also secreted by other tissues, but in every taxon in which the matter has been investigated, the gonads have been shown to secrete steroids (vertebrates, insects, possibly nematodes, and crustaceans). As gonadal products, they appear to constitute a signal of re-productive readiness to the organism, orchestrating a suite of behavioral, morphological and physiological accompaniments to reproduction. Obviously, reproductive processes are among the most fundamental of the processes to be encountered in metazoans, and it is essential that the development, behavior, and physiology of the organism remain in step with gonadal development. A signal from the gonads to the rest of the organism, indicating reproductive readiness, must be among the most primitive of endocrine phenomena.

Also striking in the field of comparative endocrinology is the ubiquity of the neuroendocrine cell. Although it serves a wide variety of functions, there is no metazoan taxon from which the neuroendocrine cell is absent. Thus, the neuroendocrine cell must also be a very primitive structure. Indeed, the neuroendocrine cell might well have preceded in evolution the "normal" neuron. Thus, the origin of differentiated nervous tissue must have proceeded in a number of steps, of which the first would obviously be the development of a specialized receptor, monitoring changes in the external environment, such as light. In order for the detected changes to influence the organism, the receptor or primitive neuron would have to communicate in some way with the rest of the organism. Chemical communication is an obvious means of establishing the communication, and the primitive receptor neuron might be seen as secreting a chemical in response to appropriate stimuli. That scenario defines a minimal neuroendocrine structure, in which a receptor becomes also an independent effector, secreting a molecule that carries a message to all parts of the organism. Thus, the neurosecretory cells in the ganglia supplying the primary sense organs in nematodes appear to play an important role in the control of the complex of events associated with molting. According to this view, the neuroendocrine cell is the primitive neuron, and the normal neuron, in which the secretion of one neuron stimulates only a neuron which makes synaptic contact with it, is a more specialized evolutionary development.

Also ubiquitous in endocrine mechanisms are the various neurosecretory peptides. Generally speaking, many (but not all) of the hormonal peptides occur as neurosecretory products. The availability of the techniques of immunocytochemistry and of analysis of peptide structure has led to the realization that immunologically similar peptides are widely distributed in the animal kingdom (Chapter 1). Many of these peptides were originally discovered in mammals, but the discovery that peptides similar to, as an example, somatostatin, occur in lower invertebrates demonstrates that, at the very least, these compounds are, in evolutionary terms, very old, and that there has been very considerable conservatism in preserving their structure during evolution. Of course, the structural similarity does not necessarily imply functional similarity: Indeed, the work on molluscs implies that some of these peptides may not function as hormones, but as neurotransmitters (Chapter 1). The functional significance of the conservatism exhibited by the preservation of these structures escapes us because we lack complete information: We can only conclude that the significance exists. Given that most neuroendocrine hormones are peptides, and given that some neurosecretory peptides also function in more primitive organisms as neurotransmitters, the notion that the neuroendocrine cell is a primitive structure gains support from these observations.

References

1. Abu-Hakima, R. and K. G. Davey. *J. Exp. Biol.* 69:33–44, 1977. Juvenile hormone and volume changes of follicle cells.

2. Ahlquist, J. In *Psychoneuroimmunology* (R. E. Ader, ed.), pp. 355–403. Academic Press, Orlando, FL, 1981. Hormonal influences on immunologic phenomena.

3. Arai, K. et al. *BioEssays* 5:166–171, 1986. T-cell lymphokines.

4. Ashburner, M., C. Chihara, P. Meltzer, and G. Richards. *Cold Spring Harbor Symp Quant. Biol.* 38:655–662, 1974. Ecdysterone and puffing in polytene chromosomes.

5. Audhya, T. et al. *Science* 231:997–999, 1986. Structure of bursin.

6. Baird, D. T. and A. S. McNeilly. *J. Reprod. Fertil., Suppl* 30:119–133, 1981. Gonadotrophic control of follicular development and function.

7. Barker, J. F. and K. G. Davey. *Int. J. Reprod.* 3:291–296, 1981. Neuroendocrine regulation of male accessory gland in *Rhodnius*.

8. Barker, J. F. and K. G. Davey. *Insect Biochem.* 13:7–10, 1983. Neuropeptide regulates protein synthesis in male accessory gland.

9. Beams, H. W. and R. G. Kessel. *J. Cell Biol.* 18:621–649, 1963. Ultrastructure of yolk formation in crayfish oocytes.

10. Beckemyer, E. F. and A. O. Lea. *Science* 201:819–821, 1980. Ecdysterone on follicle formation in the ovary of the mosquito.

11. Bennet, A. et al. *Endocrinology (Baltimore)* 115:1577–1583, 1984. Glucocorticoid regulation of IGF receptors.

12. Bergot, B. J., G. C. Jamieson, M. L. Ratcliff, and D. A. Schooley. *Science* 210:336–338, 1980. JH in embryos.

13. Berridge, M. J. In *Endocrinology of Insects* (R. G. H. Downer and H. Laufer, eds.), pp. 615–624. Alan R. Liss, New York, 1983. Calcium as a second messenger.

14. Besedovsky, H. O. et al. In *Immunoregulation* (N. Fabris, E. Garaci, J. Hadden, and N. Mitchison, eds.), pp. 315–340. Plenum, New York, 1983. Neuroendocrine immunoregulation.

15. Bodenstein, D. *Wilhelm Roux' Arch. Entwicklungsmech. Org.* 137:636–660, 1938. Development in ligatured pupae.

16. Bodenstein, D. *J. Exp. Zool.* 123:413–423, 1953. Prothoracic gland and moulting in cockroaches.

17. Boer, H. H., E. Douma, and J. M. A. Koksma. *Z. Zellforsch. Mikrosk. Anat.* 87:435–450, 1968; *Symp. Zool. Soc. London* 22:237–256. Molluscan neurosecretory cells.

18. Bollenbacher, W. E. and M. F. Bowen. In *Endocrinology of Insects* (R. G. H. Downer and H. Laufer, eds.), pp. 89–99. Alan R. Liss, New York, 1983. Prothoracicotropic hormone.

19. Bollenbacher, W. E. and N. E. Granger. In *Comprehensive Insect Physiology Biochemistry and Pharmacology* (G. A. Kerkut and L. I. Gilbert, eds.), Vol. 7, pp. 109–151. Pergamon, New York, 1985. PTTH.

20. Bollenbacher, W. E., S. Smith, W. Goodman, and L. L. Gilbert. *Gen. Comp. Endocrinol.* 44:302–306, 1981. Hormone titres in metamorphosing *Manduca*.

21. Borst, D. and F. Engelmann. *J. Exp. Zool.* 189:413–419, 1974. Secretion of ecdysone by prothoracic glands of *Leucophaea*.

22. Bradshaw, R. A. *Annu. Rev. Biochem.* 47:191–216, 1978. Nerve growth factor.

23. Bradshaw, R. A. and G. M. Gill, eds. *Evolution of Hormone-Receptor Systems*. Alan R. Liss, New York, 1983.

24. Breslow, E. *Annu. Rev. Biochem.* 48:251–274, 1979. Neurophysins.

25. Burgess, A. W. *BioEssays* 5:15–18, 1986. Growth factors, receptors and cancer.

26. Butenandt, A. and P. Karlson. *Z. Naturforshc. B: Anorg. Chem., Org. Chem., Biochem., Biophys., Biol.* 93:389–391. Isolation of ecdysone.

27. Cahill, L. P. *J. Reprod. Fertil., Suppl.* 30:135–142, 1981. Folliculogenesis.

28. Canalis, E. *Endocrinology (Baltimore)* 118:74–81, 1986. Interleukin-I and collagen synthesis.

29. Carlise, D. B. and F. Knowles. *Endocrine Control in Crustaceans.* Cambridge Univ. Press, London, 1959.

30. Caspi, R. R. and R. R. Artalion. *Comp. Immunol.* 8:51–60, 1984. Lymphocyte growth factor in fish.

31. Cassier, P. *Int. Rev. Cytol.* 57:1–73, 1979. Corpora allata of insects.

32. Cazaux, M. and F. Andre. *C. R. Hebd. Seances Acad. Sci.* 274:1550–1553, 1972. Brain and growth in *Eisenia*.

33. Chang, E. S. and J. D. O'Conner. *Gen. Comp. Endocrinol.* 36:151–160, 1978. Ecdysone and molting in Crustacea.

34. Charniaux-Cotton, H. *Am. Zool.* 25:197–206, 1985. Control of vitellogenesis in crustacea.

35. Cheesman, K. L. In *Biochemistry of Mammalian Reproduction* (L. J. D. Zaneveld and R. T. Chatterton, eds.), pp. 401–454. Wiley, New York, 1982. Steroid hormones.

36. Chino, H., S. Sakurai, T. Outaki, N. Ikekawa, H. Miyazakai, M. Ishihashi, and H. Abuki. *Science* 183:529–530, 1974. Secretion of ecdysone by prothoracic glands of *Bombyx*.

37. Claret, J. *Ann. Endocrinol.* 27:311–320, 1966. Photoperiod action on brain of *Pieris*.

38. Clark, R. B. and S. M. Evans. *J. Embryol. Exp. Morphol.* 9:97–105, 1961. Brain and regeneration in *Nereis*.

39. Clemons, D. R. *Endocrinology (Baltimore)* 117:77–83, 1985. Somatomedin and PDGF.

40. Clever, U. *Science* 146:794–795, 1964. Ecdysone and chromosome puffing.

41. Cloutier, E. J., S. D. Beck, D. C. R. McLeod, and D. L. Silhacek. *Nature (London)* 195:1222–1224, 1962. Brain transplant and diapause.

42. Cohen, P. *Adv. Cyclic Nucleotide Res.* 14:345–359, 1981. Cyclic AMP as a second messenger.

43. Compton, M. M. and J. A. Cidlowski. *Endocrinology (Baltimore)* 118:38–45, 1985. *In vivo* effects of glucocorticoids on DNA.

44. Cooke, I. M. and R. E. Sullivan. In *The*

Biology of Crustacea (H. L. Atwood and D. C. Sandeman, eds.), Vol. 3, pp. 205–290. Academic Press, New York, 1982. Crustacean neurosecretion.

45. Cooper, E. L. *Am. Zool.* 25:649–664, 1985. Comparative immunology.

46. Cottrell, C. B. *J. Exp. Biol.* 39:449–458, 1962. Hormonal control of tanning in blowflies.

47. Coussens, L. et al. *Science* 130:1132–1139, 1985. Tyrosine kinase, EGF and oncogenes.

48. Daughaday, W. H. et al. *Gen. Comp. Endocrinol.* 59:316–325, 1985. IGF.

49. Davey, K. G. *Can. J. Zool.* 42:731–734, 1964. Neurosecretion in *Ascaris*.

50. Davey, K. G. *Am. Zool.* 6:243–249, 1966. Neurosecretion and molting in nematodes.

51. Davey, K. G. In *Juvenile Hormone Biochemistry* (G. E. Pratt and G. T. Brooks, eds.), pp. 233–240. Elsevier, Amsterdam, 1982. Regulation of follicle cell volume by JH.

52. Davey, K. G. *Gen. Comp. Endocrinol.* 64:30–35, 1986. Hormonal control of O_2 consumption in *Phocanema*.

53. Davey, K. G. In *Invertebrate Endocrinology* (R. G. H. Downer and H. Laufer, eds.), Vol. 2, pp. 63–86. Alan R. Liss, New York, 1987. Nematode hormones.

54. Davey, K. G. and S. L. Goh. *Can. J. Zool.* 62:2293–2296, 1984. Hormonal control of ecdysis.

55. Davey, K. G. and S. P. Kan. *Nature (London)* 214:737–738, 1967. Hormones and ecdysis in *Phocanema*.

56. Davey, K. G. and F. L. Kriger. *Gen. Comp. Endocrinol.* 58:452–457, 1985. Ovulation hormone in the hemolymph of *Rhodnius*.

57. Davey, K. G. and J. E. Kuster. *Can. J. Zool.* 59:761–764, 1980.

58. Dembinski, A. B. and L. R. Johnson. *Endocrinology (Baltimore)* 116:90–94, 1985. EGF and gastric mucosa.

59. Denlinger, D. L. In *Comprehensive Insect Physiology, Biochemistry and Pharmacology* (G. A. Kerkut and L. I. Gilbert, eds.),

Vol. 8, pp. 354–412. Pergamon, New York, 1985. Hormones and insect diapause.

60. Dodd, J. M. In *Fish Physiology*, (W. S. Hoar, D. J. Randall, and E. M. Donaldson, eds.) Vol. IXA, pp. 31–95. Academic Press, Orlando, 1983. Reproduction in cartilaginous fishes.

61. Dogterom, A. A. and A. Doderer. *Calcif. Tissue Int.* 33:505–508, 1981. Neurohormonal control of shell formation in snails.

62. Dorn, A. *Verh. Dtsch. Zool. Ges.* 67:85–89, 1975. Structure of embryonic corpus allatum in *Oncopeltus*.

63. Dorn, A. and F. Romer. *Cell Tissue Res.* 171:331–350, 1976. Prothoracic glands in *Oncopeltus*.

64. Downward, J. et al. *Nature (London)* 307:521–524, 1984. EGF and v *erb*-B oncogenes.

65. Dudek, F. E. and S. S. Tobe. *Gen. Comp. Endocrinol.* 36:618–627, 1978. Bag cells and ovulation in *Aplysia*.

66. Dumser, J. B. *Annu. Rev. Entomol.* 25:341–369, 1980. Regulation of insect spermatogenesis.

67. Dumser, J. B. and K. G. Davey. *Can. J. Zool.* 52:1682–1689, 1975. Hormonal effects on cell division in *Rhodnius* testis.

68. Durchon, M. *Arch. Zool. Exp. Gen.* 94:1–9, 1956. Brain and regeneration in nereids.

69. Durchon, M. *Bull. Soc. Zool. Fr.* 85:275–301, 1960. Polychaete endocrinology.

70. Durchon, M. and M. Porchet. *Gen. Comp. Endocrinol.* 16:555–565, 1971. Effect of brain hormone on nereid gametogenesis *in vitro*.

71. Ebendal, T. et al. *EMBO J.* 5:1483–1487, 1986. Expression of NGF gene.

72. Edery, M. et al. *Endocrinology (Baltimore)* 117:405–411, 1985. EGF levels in mouse mammary glands.

73. Eldridge, R. K. and P. A. Fields. *Endocrinology (Baltimore)* 117:2512–2519, 1985. Rabbit placental relaxin.

74. Engelmann, F. In *Endocrinology of Insects* (R. G. H. Downer and H. Laufer, eds.),

pp. 259–270. Alan R. Liss, New York, 1983. Juvenile hormone and vitellogenesis.

75. Fabris, N. et al., eds. *Immunoregulation*. Plenum, New York, 1983.

76. Farner, D. A., and K. Lederis, eds. *Neurosecretion*. Plenum, New York, 1981.

77. Farrar, W. L. In *Peptide Hormones as Mediators in Immunology and Oncology* (R. D. Hesch, ed.), pp. 135–148. Raven Press, New York, 1985. Lymphokine regulation and interleukins.

78. Fasolo, A. et al. *Boll. Zool.* 47(Suppl.):127–147, 1980. Hypothalamo-hypophysial regulation in tetrapods.

79. Fingerman, M. *Am. Zool.* 25:233–252, 1985. Crustacean chromaophores.

80. Fisher, J. M. *Aust. J. Biol. Sci.* 19:1073–1079, 1966. Control of moulting in *Paratylenchus*.

81. Fisher, J. W. *Proc. Soc. Exp. Biol. Med.* 173:289–305, 1983. Control of erythropoietin production.

82. Folkman, J., and M. Klagsbrun. *Science* 235:442–447, 1987. Angiogenic factors.

83. Fraenkel, G. and C. Hsaio. *Science* 138:27–29, 1962. Hormonal control of tanning in blowflies.

84. Friedel, T. and B. G. Loughton. In *Endocrinology of Insects* (R. G. Downer and H. Laufer, eds.), pp. 131–140. Alan R. Liss, New York, 1983. Insect neurophysins.

85. Froesch, E. R. et al. *Annu. Rev. Physiol.* 47:443–467, 1985. IGF.

86. Fukuda, S. *Proc. Imp. Acad. (Tokyo)* 16:417–420, 1940. Thoracic gland and moulting in *Bombyx*.

87. Fukuda, S. *J. Fac. Sci., Univ. Tokyo, Sect. 4* 6:477–532, 1944. Hormones and moulting in *Bombyx*.

88. Gallo-Payet, N. and J. S. Hugon. *Endocrinology (Baltimore)* 116:194–210, 1985. EGF in intestinal cells in vitro.

89. Geraerts, W. P. M. *Gen. Comp. Endocrinol.* 29:61–71, 1976. Control of body growth in snails.

90. Geraerts, W. P. M. and J. Joose. *Gen.*

Comp. Endocrinology 27:450–457, 1975. DB and control of vitellogenesis and accessory gland in *Lymnaea*.

91. Ghomeum, M. N. et al. *Dev. Comp. Immunol.* 10:35–44, 1986. Corticosteroid effect on the thymus.

92. Gilbertson, P. *Dev. Comp. Immunol.* 10:1–10, 1986. Lymphocyte heterogeneity.

93. Gilgan, M. W. and D. R. Idler. *Gen. Comp. Endocrinol.* 9:319–324, 1967. Testosterone production in lobster tissues.

94. Giminez-Gallego, G. et al. *Science* 230:1385–1388, 1985. Growth factor in brain.

95. Goh, S. L. and K. G. Davey. *Tissue Cell* 8:421–435, 1976. Aminergic cells in the nervous system of *Phocanema*.

96. Goh, S. L. and K. G. Davey. *Can. J. Zool.* 63:475–479, 1985. Noradrenalin and ecdysis in *Phocanema*.

97. Golding, D. W. *Gen. Comp. Endocrinol., Suppl.* 3:580–590, 1972. Hormones and reproduction in nereids.

98. Golding, D. W. *Biol. Rev. Cambridge Philos. Soc.* 49:161–224, 1974. Neurosecretion in non-arthropod invertebrates.

99. Gomez, E. D. et al. *Science* 179:813–814, 1973. JH and metamorphosis in barnacles.

100. Gorbman, A. *Can. J. Fish. Aquat. Sci.* 37:1680–1686, 1980. Evolution of the brain-pituitary relationship.

101. Gorbman, A. *Fish Physiol.* 9A:1–30, 1983. Reproduction in cyclostome fishes.

102. Gorbman, A. et al. *Comparative Endocrinology* Wiley, New York, 1983.

103. Gospodarowicz, A. et al. *Endocrinology (Baltimore)* 118:82–90, 1986. Growth factor in adrenal gland.

104. Grondel, J. L. and G. M. Harmsen. *Immunology* 52:477–482, 1984. GF produced by leucocytes.

105. Guillemin, R. et al. *Recent Prog. Horm. Res.* 40:233–299, 1984. Somatocrinin.

106. Gwadz, R. W. and A. Spielman. *J. Insect Physiol.* 19:1441–1448, 1973. JH and ovary development in mosquitoes.

107. Hachlow, V. *Wilhelm Roux' Arch. Entwicklungsmech. Org.* 125:26–49, 1931. Pupal development in Lepidoptera.

108. Hadley, M. E. *Endocrinology.* Prentice-Hall, New York, 1984.

109. Hagedorn, H. H. In *Comprehensive Insect Physiology, Biochemistry and Pharmacology* (G. A. Kerkut and L. I. Gilbert, eds.), Vol. 8, pp. 205–262. Pergamon, New York, 1985. Ecdysteroids and insect reproduction.

110. Hall, K. and V. R. Sara. *Vitam. Horm. (N.Y.)* 40:175–234, 1983. Growth and somatomedins.

111. Hall, N. A. and A. L. Goldstein. In *Immunoregulation* (N. Fabris, E. Garaci, J. Hadden, and N. Mitchison, eds.), pp. 141–163. Plenum, New York, 1981. Thymosin and the regulation of immunity.

112. Herlant-Meewis, H. *C. R. Hebd. Seances Acad. Sci.* 248:1405–1407, 1956. Hormones and reproduction in *Eisenia.*

113. Hertz, W. A. and E. S. Chang. *Int. J. Invertebr. Reprod. Dev.* 10:71–77, 1986. Effect of JH on lobster metamorphosis.

114. Hocks, P., G. Schulz, E. Watzke, and D. Karlson. *Naturwissenschaften* 54:44–45, 1967. Identification of ecdysterone.

115. Hoffman, J. A., M. Lagueux, C. Hetru, M. Charlet, and F. Goltzené. In *Progress in Ecdysone Research* (J. A. Hoffmann, ed.), pp. 431–465. Elsevier/North-Holland, Amsterdam, 1980. Ecdysteroids in eggs.

116. Howie, D. I. D. *Gen. Comp. Endocrinol.* 6:347–360, 1966. Brain hormone and maturation of gametes in *Arenicola.*

117. Huebner, E. In *Insect Endocrinology* (R. G. H. Downer and H. Laufer, eds.), pp. 314–329. Alan R. Liss, New York, 1983. Antigonadotropins in insects.

118. Ilenchuk, T. T. and K. G. Davey. *Can. J. Biochem. Cell Biol.* 63:102–106, 1985. JH binding and ATPase activity.

119. Ilenchuk, T. T. and K. G. Davey. *Insect Biochem.* 17:525–529, 1987. JH and development of follicle cells.

120. Johnsson, A. et al. *EMBO J.* 5:1535–1541, 1986. V-*sis* and PDGF.

121. Joose, J. In *Comparative Endocrinology: Developments and Directions* (C. L. Ralph, ed.), pp. 13–32. Alan R. Liss, New York, 1986. Neuropeptides as transmitters.

122. Joose, J. and W. P. M. Geraerts. In *The Mollusca* (A. S. M. Saleuddin and K. Wilbur, eds.), Vol. 4, pp. 318–406. Academic Press, New York, 1983. Molluscan hormones.

123. Joose, J. et al. *Prog. Brain Res.* 55:379–404, 1982. Neuropeptides in gastropods.

124. Jorgensen, C. B. et al. In *Role of Temperature in Environmental Endocrinology,* pp. 27–36. Springer-Verlag, Berlin, 1978. Environmental control of toad ovarian cycle.

125. Joseffson, L. *Am. Zool.* 23:507–516, 1983. Crustacean chromatophorotropins.

126. Kambysellis, M. P. and C. M. Williams. *Biol. Bull. (Woods Hole, Mass.)* 141:541–552, 1971. Action of ecdysone on spermatogenesis in *Samia.*

127. Kanatani, H. and Y. Nagahama. *Biomed. Res.* 1:273–291, 1980. Mediators of oocyte maturation.

128. Karlson, P., H. Hoffmeister, H. Hummel, P. Hocks, and G. Spiteller. *Chem. Ber.* 98:2394–2402, 1968. Structure of ecdysone.

129. King, D. S. In *Endocrinology of Insects* (R. G. H. Downer and H. Laufer, eds.), pp. 57–64. Alan R. Liss, New York, 1983. Biochemistry of JH.

130. King, D. S. and J. B. Siddal. *Nature (London)* 221:955–956, 1969. Conversion of ecdysone to 20-OH ecdysone by insects and crustacea.

131. King, D. S. et al. *Proc. Natl. Acad. Sci. U.S.A.* 71:793–796, 1974. Secretion of ecdysone by prothoracic glands of *Manduca.*

132. Kingan, T. G. *Life Sci.* 28:2585–2594, 1981. Purification of PTTH in *Manduca.*

133. Koolman, J. and K.-D. Spindler. In *Endocrinology of Insects* (R. G. H. Downer and H. Laufer, eds.), pp. 179–201. Alan

R. Liss, New York, 1983. Action of ecdysteroids.

134. Kopec, S. *Biol. Bull. (Woods Hole, Mass.)*
42:322–342, 1922. Necessity for brain in
molting in insects.

135. Kriger, F. L. and K. G. Davey. *Can. J.
Zool.* 62:1720–1723, 1984. Ovulation
hormone in *Rhodnius*.

136. Kurtz, A. et al. *Endocrinology (Baltimore)*
118:567–572, 1986. Erythroprotein production.

137. Lachaise, F. and J. A. Hoffmann. *C. R.
Hebd. Seances Acad. Sci.* 285:701–704,
1972. Ecdysone in crabs.

138. Laufer, H. et al. *Science* 235:202–205,
1987. Methyl farnesoate in crabs.

139. Lea, A. O. *Gen. Comp. Endocrinol.,
Suppl.* 3:602–608, 1972. EDNH in mosquitoes.

140. Lees, A. D. *The Physiology of Diapause in
Arthropods.* Cambridge Univ. Press,
London, 1955.

141. Le Gall, S. *Gen. Comp. Endocrinol.* 43:51–
62, 1981. Control of growth in *Crepidula*.

142. LeRoith, D. et al. *Endocrinology (Baltimore)* 117:2093–2097, 1985. Somatostatin in plants.

143. Licht, P. In *Hormones, Adaptation and Evolution* (S. Ishii, T. Hirano, and M.
Wada, eds.), pp. 167–174. Receptor
binding gonadotropins. Springer-Verlag, Berlin, 1980. Gonadotropin activity.

144. Licht, P. In *Marshall's Physiology of Reproduction* (G. E. Lamming, ed.), pp.
206–282. Churchill-Livingstone, London and New York, 1984. Reptiles.

145. Lofts, B. In *Hormones, Adaptation and Evolution* (S. Ishii, T. Hirano, and M.
Wada, eds.), pp. 175–184. Reproduction in male vertebrates. Springer-
Verlag, Berlin, 1980. Regulation of
reproduction in male vertebrates.

146. Londershausen, M. and K.-D. Spindler.
Am. Zool. 25:187–196, 1985. Binding of
ecdysone in crayfish.

147. Mantel, L. H. *Am. Zool.* 25:253–263,
1985. Hormonal regulation of osmotic
balance in crustacea.

148. Marrack, P. and J. Kappler. *Sci. Am.*

254:36–45, 1986. The T-cell and its receptor.

149. Marx, J. L. *Science* 221:1362–1364, 1983.
Chemical signals in the immune sys-
tem.

150. Marx, J. L. *Science* 223:806, 1984. Growth
factor and oncogenes.

151. Marx, J. L. *Science* 232:1093–1095, 1986.
Growth factor.

152. Mason, C. A. *Z. Zelforsch. Mikrosk. Anat.*
141:19–32, 1973. Retrocerebral complex
of locust.

153. Mattson, M. P. and E. Spaziani. *Biol.
Bull. (Woods Hole, Mass.)* 171:264–273,
1986. Action of ecdysteroid on sinus
gland in crabs.

154. McCrone, E. J. and P. G. Sokolove. *J.
Comp. Physiol.* 133:117–123, 1979. Brain
and gonad maturation in slugs.

155. McCruden, A. B. and W. H. Stimson.
Thymus 3:105–117, 1981. Androgen
binding cytosol receptors.

156. Mendis, A. H. W. et al. *Mol. Biochem.
Parasitol.* 9:209–226, 1983. Ecdysteroids
in nematodes.

157. Metcalf, D. *The Hemopoietic Colony Stimulating Factors.* Elsevier, Amsterdam,
1984.

158. Meusy, J. J. *Reprod. Nutr. Dev.* 20:1–21,
1980. Crustacean vitellogenin.

159. Mukku, V. K. and G. M. Stancet. *Endocrinology (Baltimore)* 117:149–154,
1985. EGF.

160. Naisse, J. *Arch. Anat. Microsc. Morphol.
Exp.* 54:417–426, 1965. Hormones and
sex determination in *Lampyris*.

161. Naisse, J. *J. Insect Physiol.* 15:877–882,
1969. Hormones and sex determination
in *Lampyris*.

162. Newcomb, R. et al. *Am. Zool.* 25:157–
171, 1985. Neuropeptides from crustacean sinus gland.

163. Normann, T. C. *Z. Zellforsch. Mikrosk.
Anat.* 67:461–501, 1965. Fine structure of
corpus cardiacum, *Calliphora*.

164. O'Connor, J. D. *Am. Zool.* 25:173–178,
1985. Titres of ecdysteroids during crustacean molt cycle.

165. O'Connor, J. D. In *Comprehensive Insect Physiology, Biochemistry and Pharmacology* (G. A. Kerkut and L. I. Gilbert, eds.), Vol. 8, pp. 85–98. Pergamon, New York, 1985. Molecular action of ecdysteroids.

166. O'Connor, J. D. and E. S. Chang. In *Metamorphosis: A Problem in Developmental Biology* (L. I. Gilbert and E. Frieden, eds.), 2nd ed., pp. 241–261. Plenum, New York, 1981. Ecdysteroid receptors in *Drosophila* cells.

167. Ong, J. et al. *Endocrinology (Baltimore)* 120:353–357, 1987. IGF action in muscles.

168. Orchard, I. et al. *J. Insect Physiol.* 29:387–391, 1983. Aminergic pathway for ecdysteroid action in ovulation.

169. Peter, R. E. and L. W. Crim. *Annu. Rev. Physiol.* 41:323–335, 1979. Reproductive endocrinology of fishes.

170. Peter, R. E. and J. N. Fryer. In *Fish Neurobiology*, Vol. 2. (R. E. Davis and R. G. Northcult, eds.), Univ. of Michigan Press, Ann Arbor, 1983. Endocrine functions of the hypothalamus of actinopterygians.

171. Pfeiffer, I. W. *Trans. Conn. Acad. Arts Sci.* 36:489–515, 1945. Role of Corpus allatum in metamorphosis in grasshopper.

172. Pflugfelder, O. *Z. Wiss. Zool.* 149:477–512, 1937. Corpus allatum and molting in the stick insect.

173. Piepho, H. *Naturwissenschaften* 27:301–302, 1939. Molting of implanted cuticle in the wax moth.

174. Pipa, R. L. *Cell Tissue Res.* 193:443–455, 1978. Central projections of neurons in retrocerebral complex.

175. Pipa, R. L. In *Endocrinology of Insects* (R. G. Downer and H. Laufer, eds.), pp. 39–53. Alan R. Liss, New York, 1983. Structure of neuroendocrine system.

176. Pipa, R. L. and F. J. Novak. *Cell Tissue Res.* 201:227–237, 1979. Nervi corporis allati of the cockroach.

177. Pratt, G. E. and K. G. Davey. *J. Exp. Biol.* 56:201–237, 1972. JH and oogenesis in *Rhodnius*.

178. Raabe, M. *Insect Neurohormones.* Plenum, New York, 1982.

179. Raison, R. L. and W. H. Hildemann. *Dev. Comp. Immunol.* 8:99–108, 1984. Immunoglobulin and leucocytes.

180. Rechler, M. M. *Annu. Rev. Physiol.* 47:425–442, 1985. IGF receptors.

181. Reynolds, S. E. In *Endocrinology of Insects* (R. G. H. Downer and H. Laufer, eds.), pp. 235–248. Alan R. Liss, New York, 1983. Hormones and cuhile.

182. Riddiford, L. M. In *Comprehensive Insect Physiology, Biochemistry and Pharmacology* (G. A. Kerkut and L. I. Gilbert, eds.), Vol. 8, pp. 37–84. Pergamon, New York, 1985. Action of ecdysone.

183. Rogers, W. P. *Parasitology* 50:329–348, 1968. Neurosecretion in *Haemonchus*.

184. Rogers, W. P. and R. I. Sommerville. *Parasitology* 50:329–348, 1960. Physiology of ecdysis in *Haemonchus*.

185. Roller, H. and K. H. Dahm. *Recent Prog. Horm. Res.* 24:651–680, 1968. Chemistry of juvenile hormone.

186. Romer, F. *Z. Zellforsch. Mikrosk. Anat.* 122:425–455, 1971. Innervation of prothoracic glands in *Tenebrio*.

187. Ross, R. et al. *Cell (Cambridge, Mass.)* 46:155–169, 1986. PDGF.

188. Rosse, W. E. and T. A. Waldmann. *Blood* 27:654, 1966. Factors controlling erythropoiesis.

189. Rothschild, I. *Recent Prog. Horm. Res.* 37:183–298, 1981. Regulation of mammalian corpus luteum.

190. Ruben, L. N. In *Immune Regulation: Evolutionary and Biological Significance* (L. N. Ruben and M. E. Gershwin, eds.), pp. 217–236. Dekker, New York, 1982. Evolution and immune regulation.

191. Ruben, L. N. *Dev. Comp. Immunol.* 8:247–256, 1984. Phylogeny of macrophage-lymphocyte immune regulation.

192. Ruegg, R. P., F. L. Kriger, K. G. Davey, and C. G. H. Steel. *Int. J. Invertebr. Reprod.* 3:357–361, 1981. Ecdysterone and release of ovulation hormone.

193. Ruegg, R. P., I. Orchard, and K. G. Davey. *J. Insect Physiol.* 28:243–248,

1982. Ecdysterone and electrical activity in the corpus cardiacum.

194. Saad A. H. et al. *Dev. Comp. Immunol.* 8:121–130, 1984. Effect of hydrocortisone on immune system.

195. Samoiloff, M. R. *Science* 180:976–977, 1973. Laser surgery on nematodes.

196. Saunders, D. S. *Insect Clocks.* Pergamon, New York, 1982.

196a. Schaller, H. C. and H. Bodenmuller. *Proc. Natl. Acad. Sci. U.S.A.* 78:7000–7004, 1981. Morphogenetic neuropeptide in *Hydra.*

197. Schally, A. V. et al. *Endocrinology (Baltimore)* 93:893–902, 1973. *In vitro* effects of sex steroids response.

198. Scharrer, B. *Z. Zellforsch. Mikrosk. Anat.* 60:761–796, 1963. Ultrastructure of the corpus cardiacum.

199. Scharrer, B. *J. Neuro-Visc. Relat., Suppl.* 9:1–20, 1969. Structure of neurosecretory cells.

200. Schauenstein, K. and Y. Hayari. *Dev. Comp. Immunol.* 7:767–768, 1983. Avian lymphokines.

201. Schneiderman, H. A. and L. I. Gilbert. *Biol. Bull. (Woods Hole, Mass.)* 115:530–535, 1958. JH activity in Crustacea.

202. Schoonhoven, L. M. *Science* 141:173–174, 1963. Electrical activity in the brain during diapause.

203. Schrader, W. T. et al. *Recent Prog. Horm. Res.* 37:583–633, 1981. Action of progesterone on chick oviduct.

204. Sedlak, B. J. *Gen. Comp. Endocrinol.* 44:207–218, 1981. Neurosecretory fibers in corpus allatum.

205. Sherwood, J. B. *Vitam. Horm. (N.Y.)* 41:161–212, 1984. Erythropoietin.

206. Sherwood, N. M. et al. *Gen. Comp. Endocrinol.* 61:313–322, 1986. Gonadotropin RH in amphibian brains.

207. Skinner, D. M. *Am. Zool.* 25:275–284, 1985. Control of Crustacean molt cycle.

208. Smith, D. S. *Insect Cells: Their Structure and Function.* Oliver & Boyd, Edinburgh, 1968.

209. Sommerville, R. I. *Exp. Parasitolol.* 6:18–

30, 1957. Control of ecdysis in *Haemonchus.*

210. Sower, S. A. et al. *Can. J. Zool.* 61:2653–2659, 1983. Ovulatory and steroidal responses in the lamprey.

211. Sower, S. A. et al. *Endocrinology (Baltimore)* 120:773–779, 1987. Lamprey gonadotropin-releasing hormone.

212. Soyez, D. and L. H. Kleinholz. *Gen. Comp. Endocrinol.* 31:233–242, 1977. Molt inhibiting hormone in crustacea.

213. Sporn, M. B., A. B. Roberts, L. M. Wakefield, and R. K. Assoian. *Science* 233:532–534, 1986. TGF structure.

214. Stay, B. et al. *Science* 207:898–900, 1980. Ecdysterone and JH synthesis.

215. Steel, C. G. H. et al. *J. Insect Physiol.* 28:519–525, 1982. Ecdysteroid titres in *Rhodnius.*

216. Steel, C. G. H. and K. G. Davey. In *Comprehensive Insect Physiology, Biochemistry and Pharmacology* (G. A. Kerkut and L. I. Gilbert, eds.), Vol. 8, pp. 1–36. Pergamon, New York, 1985. Hormonal integration in insects.

217. Stein, M. et al. In *Psychoneuroimmunology* (R. Ader, ed.), pp. 429–447. Academic Press, New York, 1981. Hypothalamic influences on immune response.

218. Stoschek, C. M. et al. *Endocrinology (Baltimore)* 116:528–535, 1985. EGF receptor synthesis.

219. Suttie, J. M. et al. *Endocrinol. (Baltimore)* 116:846–848, 1985. IGF and antler growth.

220. Takeda, N. *Gen. Comp. Endocrinol.* 34:123–141, 1978. Hormones and diapause in *Monema.*

221. Tam, J. P. et al. *Proc. Natl. Acad. Sci. U.S.A.* 83:8082–8086, 1986. TGF synthesis.

222. Templeton, N. S. and H. Laufer. *Int. J. Invertebr. Reprod.* 6:99–110, 1983. Effects of JH mimic on *Daphnia.*

223. Thoenen, H. and D. Edgar. *Science* 229:238–242, 1985. Neurotrophic factors.

224. Tobe, S. S. In *Insect Biology in the Future*

(M. Locke and D. S. Smith, eds.), pp. 345–367. Academic Press, New York, 1980. Regulation of CA in adult female insect.

225. Tobe, S. S. and B. Stay. *Gen. Comp. Endocrinol.* 31:138–147, 1977. CA activity and egg development in *Diploptera*.

226. Tobe, S. S., C. S. Chapman, and G. E. Pratt. *Nature (London)* 268:728–730, 1977. JH synthesis after nerve transection.

227. Truman, J. W. *Biol. Bull. (Woods Hole, Mass.)* 144:200–211, 1973. Eclosion hormone in giant silk moths.

228. Truman, J. W. In *Comprehensive Insect Physiology, Biochemistry and Pharmacology* (G. A. Kerkut and L. I. Gilbert, eds.), Vol. 8, pp. 413–440. Pergamon, New York, 1985. Hormonal control of insect ecdysis.

229. Tsuneki, K. *Acta Zool. (Stockholm)* 57:137–146, 1976. Effects of estradiol and testosterone in the hagfish.

230. Unnithan, G. C., H. A. Bern, and K. K. Nayar. *Acta Zool. (Stockholm)* 52:117–143, 1971. Neuroendocrine apparatus in *Oncopeltus*.

231. Vandenbunder, B. *Biochimie* 68:1149–1152, 1986. Oncogenes and cell growth.

232. van Tienhoven, A. *Reproductive Physiology of Vertebrates.* Cornell Univ. Press, Ithaca, New York, 1968.

233. Wang, F. F., C. K.-H. Kung, and E. Goldwasser. *Endocrinology (Baltimore)* 116:2286–2292, 1985. Human erythropoietin.

234. Ware, R. W., D. Clark, K. Crossland, and R. L. Russell. *J. Comp. Neurol.* 162:71–110, 1973. Ultrastructure of the CNS of *Caenorhabditis*.

235. Wattez, C. *Gen. Comp. Endocrinol.* 35:360–374, 1978. Brain and gonad maturation in gastropods.

236. Weisbart, M., W. Dickhoff, A. Gorbman, and D. Idler. *Gen. Comp. Endocrinol.* 41:506–519, 1980. Steroids in the sera of hagfish and lamprey.

237. Wells, M. J. and J. Wells. *Symp. Zool. Soc. London* 38:525–540, 1977. Optic glands and reproduction in the octopus.

238. Wigglesworth, V. B. *Q. J. Microsc. Sci.* 77:191–222, 1934. The head and moulting in *Rhodnius*.

239. Wigglesworth, V. B. *Q. J. Microsc. Sci.* 79:91–121, 1936. Corpus allatum, growth and reproduction in *Rhodnius*.

240. Wigglesworth, V. B. *Naturwissenschaften* 27:301, 1939. Molting of adults in *Rhodnius*.

241. Wigglesworth, V. B. *J. Exp. Biol.* 17:201–222, 1940. Brain and moulting and metamorphosis in *Rhodnius*.

242. Wigglesworth, V. B. *J. Exp. Biol.* 29:561–570, 1952. Prothoracic gland and moulting in *Rhodnius*.

243. Wigglesworth, V. B. In *Comprehensive Insect Physiology, Biochemistry and Pharmacology* (G. A. Kerkut and L. I. Gilbert, eds.), Vol. 7, pp. 1–24, 1985. History of insect endocrinology.

244. Wijdenes, J. and N. W. Runham. *Gen. Comp. Endocrinol.* 29:545–551, 1976. DB function in *Agriolimax*.

245. Willey, R. B. *J. Morphol.* 108:219–261, 1961. Morphology of stomodeal nervous system in Blattaria.

246. Willey, R. B. and G. B. Chapman. *J. Ultrastruct. Res.* 4:1–14, 1960. Ultrastructure of corpus cardiacum.

247. Williams, C. M. *Biol. Bull. (Woods Hole, Mass.)* 93:89–98, 1947. Brain and prothoracic gland in *cecropia*.

248. Williams, C. M. *Biol. Bull. (Woods Hole, Mass.)* 103:120–138, 1952. Brain and prothoracic gland in *cecropia*.

249. Williams, C. M. and P. L. Adkisson. *Biol. Bull. (Woods Hole, Mass.)* 127:511–525, 1965. Photoperiod and diapause in *Antherea*.

250. Williams, L. T. *Cancer Surv.* 5:233–241, 1986. *Sis* gene and PDGF.

251. Willis, J. H. *Am. Zool.* 21:763–773, 1981. Action of JH on chromatin.

252. Wingfield, J. In *Hormones, Adaption and Evolution* (S. Ishii, T. Hirano, and M. Wada, eds.), pp. 135–144. Springer-Verlag, Berlin, 1980. Sex steroid-binding proteins in vertebrate blood.

253. Wissocq, J. C. *C. R. Hebd. Seances Acad.*

Sci. 262:2605–2608, 1966. Proventriculus as a source of the hormone controlling reproduction in *Syllis.*

254. Woloski, B. M. et al. *Science* 230:1035–1037, 1985. Corticotropin releasing activity of monokines.

255. Wourms, J. P. *Am. Zool.* 17:379–410, 1981. Viviparity in fish.

256. Wourms, J. P. *Am. Zool.* 17:379–410, 1983. Reproduction in Chondrichthyan fishes.

257. Yu, J. Y. L. et al. *Gen. Comp. Endocrinol.* 43:492–502, 1981. Vitellogenesis and hormonal regulation in the hagfish.

258. Zapf, J. et al. *Endocrinol. Metab.* 13:3–30, 1984. IGF.

INDEX

755